THE CAMBRIDGE

DICTIONARY
OF SCIENTISTS

THE CAMBRIDGE

DICTIONARY
OF SCIENTISTS

DAVID MILLAR   IAN MILLAR

JOHN MILLAR   MARGARET MILLAR

CAMBRIDGE
UNIVERSITY PRESS

Published by the Press Syndicate of the University of Cambridge
The Pitt Building, Trumpington Street, Cambridge CB2 1RP
40 West 20th Street, New York, NY 10011–4211, USA
10 Stamford Road, Oakleigh, Melbourne 3166, Australia

Typeset by Technical Applications Group, Cambridge University Press
Printed in Great Britain at the University Press, Cambridge

A catalogue record for this book is available from the British Library

Library of Congress cataloguing in publication data
The Cambridge dictionary of scientists/David Millar . . . [et al.].
p.   cm.
Includes index.
ISBN 0-521-56185-X (hc). – ISBN 0-521-56718-1 (pb)
1. Scientists–Biography–Dictionaries. 2. Science–History.
I. Millar, David.
Q141.C128 1996
509.2'2–dc20 95-38471 CIP

ISBN 0 521 56185 X hardback
ISBN 0 521 56718 1 paperback

Contents

Panels

The authors share a common interest in science. Their individual scientific interests cover astronomy, biochemistry, chemistry, geology, geophysics, mathematics, physics and women in science, and between them they have authored over 90 research papers and five books on scientific topics. As a family they have collaborated on several writing projects, of which this is the latest.

David Millar has conducted research into the flow of polar ice sheets at the Scott Polar Research Institute, Cambridge, and in Antarctica. He has also written on a range of science and technology topics, and is editor of *Antarctica: the Next Decade*, a study of the politics of the southern continent. His professional career has largely been in the software industry, and he is now Vice-President of Sales and Marketing at Interactive Network Technologies Inc, Houston, TX, a company supplying software tools for the geoscience and financial markets.

Ian Millar is Professor Emeritus of Organic Chemistry and former Deputy Vice-Chancellor of Keele University, and previously conducted research at McGill University, Montreal, and the University of Cambridge, mainly on the chemistry of the reactive and toxic compounds of phosphorus and arsenic. His publications include *The Organic Chemistry of Nitrogen* and *Coordination Chemistry: Experimental Methods*.

John Millar graduated from Trinity College, Cambridge, and has a doctorate from Imperial College, London. He has worked for BP developing new geophysical methods for use in oil exploration and production. In 1994 he co-founded GroundFlow Ltd, which has developed electrokinetic surveying as a new technique for imaging and mapping subsurface water distribution.

Margaret Millar worked as a computor in X-ray crystallography at the Cavendish Laboratory, Cambridge, in the period when the work on the DNA double helix was being conducted there. She has worked in the Departments of Earth Science and Astronomy at Sheffield and Cambridge, and has a special interest in the women pioneers of science and medicine.

The central objective of *The Cambridge Dictionary of Scientists* is to survey the sciences through the lives of the men and women whose efforts have shaped modern science. Our focus is on chemical, physical, biological, earth and space science and also on the linked areas of mathematics, medicine and technology. The major outlines of all these disciplines had been drawn by early in the present century, but much of the work we describe has been done since then, with a large part of it developing at the interfaces between older subject areas; in this way newer areas such as computer science or molecular biology or astronautics have been created relatively recently.

The people profiled in this book are frequently associated, often by name, with scientific units, effects and laws or with chemical reactions, diseases and methods, and we have described these. Science before the First World War was dominated by the work of northern European males. It is often assumed that, apart from Marie Curie, there were no women scientists, or women interested in science, in the 18th, 19th and even early 20th centuries. The story of the struggle by women to gain access to libraries, lectures, scientific societies, education and careers in science perhaps needs to be told in more detail – but elsewhere. Here we have included an account of the work and lives of those women who cleared the path towards these objectives. The scientific work of some was modest, but their pioneering was important. The contributions of others were significant but, at a time when their exclusion from scientific societies and similar opportunities for publication was conventional, their work was only acceptable with that of their fathers, brothers and husbands.

As well as ideas, certain devices have been the key to advance in many areas of science: telescopes, microscopes and spectroscopes in various forms are notable examples. In some cases the discoverer is not known with any certainty and a sort of group awareness of an idea or device occurred rather than an isolated individual discovery. A related situation has increasingly been seen with large teams working in high-cost 'big science' projects whose success is collective rather than individual.

The foundation of this book lies with our *Concise Dictionary of Scientists* written for Chambers/Cambridge in 1989 and now out of print. The new book covers many more scientists (some 1300 from 38 countries) and older entries have been revised, updated and extended. In addition, 32 'panels' give summarizing accounts of major areas and some selected topics of current interest. In response to increasing interest in women's scientific contributions some 70 pioneer women in science have been added, with the story of their entry into the sciences described in panels.

The sources of information have been too many and varied to be listed usefully in much detail. Our interest in scientists' lives and personalities began partly with a study of autobiographical writing by scientists; there is more of this than might be expected, not only as books but as articles, Nobel Prize lectures, and interviews by us or recorded in print or on radio and TV. Living scientists have in many cases been able to check our accounts and we are grateful to them for doing so. Many biographies have been studied; they exist in great profusion for the dozen or so best-known scientists. *The Dictionary of Scientific Biography* (editor-in-chief, C C Gillispie, published by Charles Scribner's Sons, New York, 1970–80, 14 vols and two supplements) has been much consulted, as have the *Biographical Memoirs of Fellows of the Royal Society*, *Notes and Records of* the *Royal Society*, the *Nobel Lectures*, and also M B Ogilvie's *Women in Science* (MIT Press, 1986) and M Alic's *Hypatia's Heritage* (The Women's Press, 1986). In particular areas some individual studies have been of especial value; A and H Gernsheim's *History of Photography* (Thames & Hudson, 1969), R Benedict and J Modell's *Ruth Benedict – Patterns of a Life* (Chatto & Windus, 1984), M F and G W Rayner-Canham's *Harriet Brooks: Pioneer Nuclear Scientist* (McGill/Queens University Press, 1992) and *The Oxford Companion to Medicine*, edited by J Walton, P B Beeson and R B Scott (OUP, 1986).

As well as the more specialized scientific journals, we have found much helpful material in articles in the *Journal of Chemical Education*, *Chemistry in Britain*, *Nature*, *Endeavour*, *Scientific American*, *New Scientist*, *Science*, *Physics Today*, *Sky and Telescope*, *Isis*, *Chemia*, *British Journal for the History of Science*, *Journal of the British Astronomical Association*, *Acta Mathematica*, and *Annals of Botany*.

We are most grateful for the help and support given to us by many colleagues at the University of Keele, and by officers of the Cambridge University Press, especially Adrian du Plessis, Pauline Graham and Kevin Taylor; by Sukie Hunter; and by Dr John Creamer, consultant cardiologist at the North Staffordshire Hospital.

Symbols and Conventions

Scientists' biographical entries are given in alphabetical order of their surname. Where a name is an English version, the other-language form is also given. Names with prefixes such as von and de are listed under the prefix only where it is usual to do so in the person's native country (or adopted country in the case of scientists who emigrated early in life). Thus de Broglie (who was French) is listed under B, but de Moivre (who was British) is at D. Where there are multiple forenames, lesser-used ones are given in brackets. In the case of married women, their single name is also given; where they have combined their married and single surnames that form is shown (eg Maria Goeppert Mayer). However, if they have continued to use their unmarried surname in their work that form is given (eg Marie Stopes). Where a person's name has totally changed in adult life both the names are supplied (eg Hertha Ayrton, née (Phoebe) Sarah Marks). In some cases the headword covers a group (eg Bourbaki) or a family, connected either through the generations (eg the Monros), through marriage (eg the Coris) or as siblings (eg the Wright brothers). In such cases it is difficult, or not useful, to separate their work. Team work has become usual in the 20th-c, and where this is the case the work has been described fully in one entry only, with cross-reference to the principal co-workers, who did not necessarily play a smaller part. Small capitals are used for a scientist who has his/her own entry, on first mention in another entry.

People born in the British Isles have been classed as 'English' or 'Scottish' if born before 1707, otherwise 'British' (the need to use the term 'Welsh' did not arise). Individual decisions have been made for those born within the island of Ireland, whether officially 'British' or not. 'Russian' is used for those born before 1917, and 'Soviet' for those born after. In the case of early scientists, 'Greek' is used for those of Hellenic culture who wrote in Greek, even when they lived in Sicily, Asia Minor or Egypt. 'Arabic' is used for writers in Arabic in a similar situation. In giving nationalities we have taken note of where the scientist has lived and the nationalities s/he has adopted. For example, Einstein was born in Ulm, went to school in Munich, then disclaimed his German citizenship and went to college in Switzerland, where he worked until he was over 50 before moving to the USA and becoming an American citizen 10 years later; he is described as German–Swiss–US.

Pronunciations are given where they are difficult to predict from the spelling or where the reader might choose the wrong form. They appear with stressed syllables in bold type. No stress is given for unstressed languages such as French and Japanese. Dates for all scientists mentioned are given where available; dates of scientists before Christ (BC) are usually not precisely known, and the dates given are approximate. Where we describe a scientist as being educated in a city, we mean at the university or college there. Prizes and honours are not usually given, except for Nobel Prizes.

The International System of Units (SI) is used; and for chemical names the form most used by chemists, which is not always the IUPAC preferred name, is given. By the word 'billion' we mean 1000 million (10^9), not the older usage of 10^{12}. Laws are expressed in their modern form.

The following symbols have been used:

c	velocity of light in vacuum
e	unit of electrical charge
g	acceleration of free fall due to gravity (in vacuum)
h	Planck constant
K	thermodynamic temperature unit (kelvin)
kg	kilogram (SI unit of mass)
km	kilometre
m	metre (SI unit of length)
mi	mile
n	neutron
N_A	Avogadro constant
p	proton
p	pressure
s	second (SI unit of time)
t	tonne (megagram, ie 1000 kg)
STP	standard temperature and pressure: 298 K and 760 mmHg
T	temperature (on absolute scale)
V	volume
AU	astronomical unit of distance: the mean Earth–Sun distance
DNA	deoxyribonucleic acid
RNA	ribonucleic acid

Mathematical symbols

log	logarithm to base 10
π	pi: ratio of circumference to diameter of a circle
$=$	is equal to
$\neq$	is not equal to
$\approx$	is approximately equal to
$<$	is less than
$>$	is greater than
$\leq$	is less than or equal to
$\geq$	is greater than or equal to
ab	a multiplied by b
a/b	a divided by b
a^n	a raised to power n
i	$(-1)^{\frac{1}{2}}$
$\sum$	sum of the terms
e	base of Naperian logarithms, 2.71828 ... (Euler's number)
$\exp x$ or e^x	exponential of x

A

Abbe, Ernst [abuh] (1840–1905) German physicist and developer of optical instruments.

Abbe was professor of physics and observatory director at Jena. He worked on optical theory and with Carl Zeiss (1816–88), an instrument maker, and Otto Schott (1851–1935), a glass maker, was able to improve several devices. These include the Abbe condenser for converging light on microscope specimens; the achromatic lens, which is free from colour distortion (1886); and the Abbe refractometer. From 1888 he was the sole owner of the Zeiss company, whose optical instruments were of the highest standard. In 1893 he patented the now-familiar prismatic binocular.

Abegg, Richard [abeg] (1869–1910) German physical chemist.

Abegg's Rule (for which he is best remembered) states that each element has a positive valence and a negative valence, whose sum is 8. This idea reflects in primitive form the 'octet rule', ie the trend shown by most elements of the second and third (short) periods to attain an outer octet of electrons, but even as a mnemonic it applies only to elements of the fourth to seventh periodic groups.

Abel, Sir Frederick (Augustus) [aybel] (1827–1902) British chemist: expert on military explosives.

An early pupil of HOFMANN at the Royal College of Chemistry, he became chemist to the War Department in 1854. He showed that guncotton (obtained by nitrating cotton) could be made safe by removing traces of acid, which, if not removed, led to instability. In 1889 with DEWAR he invented 'cordite', a mixture of guncotton and nitroglycerin gelatinized with propanone and petroleum jelly, which became the standard British military propellant. It produces little smoke on firing, an important advantage on a battlefield.

Abel, John Jacob [aybel] (1857–1938) US biochemist: detected adrenalin, and crystallized insulin; isolated amino acids from blood.

An Ohio farmer's son, Abel studied very widely in Europe before returning to Johns Hopkins University equipped with a wide knowledge of chemistry, biology and medicine, as professor of pharmacology. He studied the adrenal hormone now known as adrenalin (epinephrine); and in 1926 first crystallized insulin and showed it was a protein and contained zinc. He was the first to isolate amino acids from blood, in 1914. He did this by passing blood from an artery through a Cellophane tube immersed in saline; the amino acids dialysed through the tube and the blood was returned to a vein of the animal. The proof that amino acids are present in blood is fundamental in animal biochemistry and the method used led the way towards dialysis in the treatment of kidney disease.

Abel, Neils Henrik [ahbel] (1802–29) Norwegian mathematician: pioneer of group theory; proved that no algebraic solution of the general fifth-degree equation exists.

Abel was the son of a Lutheran minister. In 1821 he went to Oslo to study at the university, but his father's death forced him to give this up in order to support the large family of which he was the eldest; he was extremely poor throughout his life. In 1825 he visited Germany and France and with Leopold Crelle (1780–1855) founded *Crelle's Journal* in which much of his work was published, since Abel could not persuade the French Académie des Sciences to do so. Having failed to find a university post in Germany, and with his health failing due to tuberculosis, he returned to Norway, where he died shortly afterwards aged 26. Two days later a letter from Crelle announced that the professorship of mathematics at Berlin, one of the most prestigious posts in the world, had been awarded to him.

Despite his tragically early death Abel largely founded the theory of groups, and in particular commutative groups, which were later known as Abelian groups. He also showed that the general fifth-degree equation is not solvable algebraically (ironically GAUSS threw this proof away unread when Abel sent it to him). He revolutionized the important area of elliptic integrals with his theory of elliptic and transcendental functions, and contributed to the theory of infinite series.

Adams, John Couch (1819–92) British astronomer: predicted existence of Neptune.

As the son of a tenant farmer, Adams had financial problems in entering Cambridge, but his career was successful and he remained there throughout his life.

By 1820 it had become apparent to astronomers that the motion of Uranus could not be explained by NEWTON's law of gravitation and the influence of the known planets alone, since a small but increasing perturbation in its orbit had been observed. While still an undergraduate, Adams proved that the deviation had to be due to the influence of an eighth, undiscovered, planet. He sent his prediction for its position to AIRY, the Astronomer Royal, who was sceptical of its value and ignored it. Only when LEVERRIER, in France, announced similar results 9 months later did Airy initiate a search by James Challis (1803–82) at the Cambridge Observatory, based on Adams's prediction. The planet, now named Neptune, was however found first by Johann Galle (1812–1910) in Berlin in 1846, using Leverrier's figures. A bitter controversy about the credit for the prediction soon developed. Adams's precedence was eventually recognized, despite his taking no part in the debate. He turned down the subsequent offers of a knighthood and the post of Astronomer Royal.

Adams, Walter Sydney (1876–1956) US astronomer: discovered first white dwarf star.

Adams was born in Syria, where his American parents were missionaries, but he returned with them when he was 9 and was educated in the USA and in Europe.

Adams's work was principally concerned with the spectroscopic study of stars. He showed how dwarf and giant stars could be distinguished by their spectra, and established the technique of spectroscopic parallax to deduce a star's distance. In 1915 he observed the spectrum of Sirius B, the faint companion of Sirius, and discovered it to be an exceptionally hot star. Since it is only 8 light years distant he realized that it must therefore be very small (otherwise it would be brighter), and hence of very high density. Sirius B proved to be a 'white dwarf' and the first of a new class of stellar objects; such stars are the final stage in the evolution of stars of similar mass to the Sun, which have collapsed to form extremely dense objects.

Adams also searched for the relativistic spectral shift expected from a heavy star's presumed intense gravitational field. This he succeeded in finding in 1924, thereby proving his hypothesis about the nature of Sirius B and strengthening the case for EINSTEIN's general relativity theory as well. Adams spent most of his working life at the Mount Wilson Observatory in southern California, and was its director from 1923 until 1946.

Addison, Thomas (1793–1860) British physician: a founder of endocrinology.

A graduate in medicine from Edinburgh and London, his early work included the first clear descriptions of appendicitis, lobar pneumonia and the action of poisons on the living body. In 1855 his small book *On the Constitutional and Local Effects of Disease of the Supra-renal Capsules* described two new diseases: one is 'pernicious' anaemia; the other, also an anaemia, is associated with bronzing of the skin and weakness, and is known as Addison's disease. He found that cases of the latter showed post-mortem changes in the suprarenal capsules (one on top of each kidney). Later, physiological studies by others showed that the suprarenal capsules are glands, now known as the adrenal glands, which produce a complex group of hormones. Addison's disease was the first to be correctly attributed to endocrine failure (ie disorder of the ductless glands of internal secretion).

Adrian, Edgar Douglas, Baron Adrian (1889–1977) British neurophysiologist: showed frequency code in nerve transmission.

Adrian began his research in physiology in Cambridge before the First World War, but in 1914 he speedily qualified in medicine and tried to get to France. In fact he was kept in England working on war injuries and his later work was a mixture of 'pure' research and applications to medical treatment.

In the 1920s he began his best-known work. Already, crude methods were available for detecting electrical activity in nerve fibres. Adrian used thermionic diode amplifiers to reliably record nerve impulses in a single nerve fibre, and to show that they do not change with the nature or strength of the stimulus, confirming work by his friend K Lewis (1881–1945) in 1905 on this 'all or none' law. He went on to show that a nerve transmits information to the brain on the intensity of a stimulus by frequency modulation, ie as the intensity rises the number of discharges per second (perhaps 10–50) in the nerve also rises − a fundamental discovery. He then worked on the brain, using the discovery by BERGER in 1924 that electrical 'brainwaves' can be detected.

From 1934 he studied these brainwave rhythms, which result from the discharge of thousands of neurones and which can be displayed as an electroencephalogram (EEG). Within a few years the method was widely used to diagnose epilepsy cases, and later to locate lesions, eg those due to tumours or injury.

Adrian was linked with Trinity College Cambridge for nearly 70 years and did much to advance neurophysiology. He was a very popular figure; as a student he was a skilful night roof-climber and an excellent fencer, and he sailed and rock-climbed until late in life. He helped to organize a famous hoax exhibition of modern pictures in 1913. He was never solemn, moved very quickly and claimed his own brainwaves were as rapid as a rabbit's; as a motorist his quick reflexes alarmed his passengers. When in a hurry he would use a bicycle in the long dark basement corridors of the Physiological Laboratory. He shared a Nobel Prize in 1932.

Agassiz, Jean Louis Rodolphe [agasee] (1807–73) Swiss–US naturalist and glaciologist: proposed former existence of an Ice Age.

Agassiz owed much of his scientific distinction to the chance of his birth in Switzerland. He studied medicine in Germany, but zoology was his keen interest. He studied under CUVIER in Paris and then returned home and worked with enthusiasm on fossil fishes, becoming the world expert on them (his book describes over 1700 ancient species of fish).

Holidaying in his native Alps in 1836 and 1837, he formed the novel idea that glaciers are not static, but move. He found a hut on a glacier which had moved a mile over 12 years; he then drove a straight line of stakes across a glacier, and found they moved within a year. Finding rocks which had been moved or scoured, apparently by glaciers, he concluded that in the past, much of Northern Europe had been ice-covered. He postulated an 'Ice Age' in which major ice sheets had formed, moved and were now absent in some areas – a form of catastrophism, in contrast to the extreme uniformitarianism of LYELL. We now know that a series of ice ages has occurred.

In 1846 Agassiz was invited to the USA to lecture, enjoyed it and stayed to work at Harvard. He found evidence of past glaciation in North America; it too had undergone an Ice Age. His studies on fossil animals could have been used to support DARWIN's ideas on evolution, but in fact Agassiz was America's main opponent to Darwin's view that species had evolved.

Agnesi, Maria Gaetana [anyayzee] (1718–99) Italian mathematician and scholar: remembered in the naming of the cubic curve 'the Witch of Agnesi'.

Born in Milan, Maria was one of the 24 children of a professor of mathematics at the University of Bologna. With his encouragement she spoke seven languages by the age of 11, and by the age of 14 she was solving problems in ballistics and geometry. Her interests covered logic, physics, mineralogy, chemistry, botany, zoology and ontology; her father arranged her public debates. From this time she suffered a recurring illness in which convulsions and headaches were symptoms. Her father agreed that she should in future lead a quiet life free from social obligations. Maria thereafter devoted herself to the study of new mathematical ideas. Her *Instituzioni analitiche ad uso della gioventù* (1748, Analytical Institutions) was published as a teaching manual. In 1750, she was appointed to the chair of mathematics and philosophy at Bologna. Maria Agnesi's work was one of promise rather than fulfilment: she made no original discoveries and her major work was written as a guide to students. The cubic curve named the Witch of Agnesi was formulated by FERMAT. A mistranslation caused the use of 'witch' for 'curve'.

Agricola, Georgius (*Lat*), Georg Bauer (*Ger*) [agrikola] (1494–1555) German mineralogist, geologist and metallurgist: described mining and metallurgical industries of 16th-c.

Born in Saxony, Agricola trained in medicine in Leipzig and in Italy. The link between medicine and minerals led to his interest in the latter, and his work as a physician in Saxony put him in ideal places to develop this interest and to extend it to mining and metal extraction by smelting, and related chemical processes. His book *De natura fossilium* (1546, On the Nature of Fossils) classifies minerals in perhaps the first comprehensive system. Later he wrote on the origin of rocks, mountains and volcanoes. His best-known book, *De re metallica* (1556, On the Subject of Metals) is a fine illustrated survey of the mining, smelting and chemical technology

of the time. An English edition (1912) was prepared by the American mining engineer H C Hoover (who became president of the USA, 1929–33) and his wife.

Airy, Sir George Biddell [ayree] (1801–92) British geophysicist and astronomer: proposed model of isostasy to explain gravitational anomalies.

Airy was successful early in life, his talent and energy leading to his appointment as Astronomer Royal in 1835, a post he held for 46 years. He much extended and improved the astronomical measurements made in Britain. Airy's researches were in the fields of both optics and geophysics. He experimented with cylindrical lenses to correct astigmatism (a condition he suffered from himself); and he studied the Airy discs in the diffraction pattern of a point source of light.

In geophysics he proposed that mountain ranges acted as blocks of differing thickness floating in hydrostatic equilibrium in a fluid mantle, rather like icebergs in the sea. He was thus able to explain gravitational anomalies that had been observed in the Himalayas, as due to the partial counteraction of the gravitational attraction of the topography above sea level with that of a deep 'root' extending into the mantle. His model of isostasy satisfactorily explains the gravity field observed over mountainous terrain in much of the world.

Airy was arrogant and unlucky in his failings, now almost better known than his successes. He failed to exploit ADAMS's prediction of a new planet, Neptune; he was against FARADAY's idea of 'lines of force' (a fruitful intuition, in fact); and although he expended great effort to ensure precise measurements of the transits of Venus, observed in 1874 and 1882, the results failed to give accurate measurements of the scale of the solar system because Venus's atmosphere makes the timing of its apparent contact with the Sun's disc uncertain. An ingenious inventor of laboratory devices, he was remarkably precise, to the extent of labelling empty boxes 'empty'.

Alembert, Jean Le Rond d' [dalãbair] (1717–83) French mathematician: discovered d'Alembert's principle in mechanics.

D'Alembert's forename comes from that of the church, St Jean le Rond, on whose steps he was found as a baby. He was probably the illegitimate son of a Parisian society hostess, Mme de Tenzin, and the chevalier Destouches; the latter paid for his education while he was brought up by a glazier and his wife. He studied law, and was called to the bar in 1738, but then flirted briefly with medicine before choosing to study mathematics and to live on his father's annuity.

Early research by d'Alembert clarified the concept of a limit in the calculus and introduced the idea of different orders of infinities. In 1741 he was admitted to the Académie des Sciences and 2 years later published his *Traité de dynamique* (Treatise on Dynamics), which includes d'Alembert's principle, that NEWTON's Third Law of Motion holds not only for fixed bodies but also for those free to move. A wide variety of new problems could now be treated, such as the derivation of the planar motion of a fluid. He developed the theory of partial differential equations and solved such systems as a vibrating string and the general wave equation (1747). He joined EULER, A I Clairault (1713–65), LAGRANGE and LAPLACE in applying calculus to celestial mechanics and determined the motion of three mutually gravitating bodies. This then allowed many of the celestial observations to be understood; for example, d'Alembert explained mathematically (1754) Newton's discovery of precession of the equinoxes, and also the perturbations in the orbits of the planets.

D'Alembert was then persuaded by his friend, Denis Diderot (1713–84), to participate in writing his encyclopedia, contributing on scientific topics. This project was denounced by the Church after one volume, and d'Alembert turned instead to publishing eight volumes of abstruse mathematical studies. Shortly before his death J H Lambert (1728–77) wished to name his 'newly discovered moon of Venus' after d'Alembert, but the latter was sufficiently acute to doubt (correctly) from calculations that it existed, and gently declined the offer.

Alfvén, Hannes Olof Gösta [alfvayn] (1908–95) Swedish theoretical physicist: pioneer of plasma physics.

Educated at Uppsala, Alfvén worked in Sweden until 1967, when he moved to California. Much of his work was on plasmas (gases containing positive and negative ions) and their behaviour in magnetic and electric fields. In 1942 he predicted magnetohydrodynamic waves in plasmas (Alfvén waves) which were later observed. His ideas have been applied to plasmas in stars and to experimental nuclear fusion reactors. He shared a Nobel Prize in 1970 for his pioneering theoretical work on magnetohydrodynamics.

Alhazen, Abu-Hassan ibn al Haytham (*Arabic*) [alhazen] (*c*.965–1038) Egyptian physicist: made major advances in optics.

Alhazen rejected the older idea that light was emitted by the eye, and took the view that light was emitted from self-luminous sources, was reflected and refracted and was perceived by the eye. His book *The Treasury of Optics* (first published in Latin in 1572) discusses lenses (including that of the eye), plane and curved mirrors, colours and the camera obscura (pinhole camera).

His career in Cairo was nearly disastrous. Born in Basra (now in Iraq), he saw in Cairo the annual flooding of the Nile and persuaded the caliph al-Hakim to sponsor an expedition to southern Egypt with the object of controlling the river and providing an irrigation scheme.

Alhazen's expedition showed him only the difficulties, and on his return he realized that the caliph would probably ensure an unpleasant death for him. To avoid this, he pretended to be mad, and maintained this successfully until the caliph died in 1021. Alhazen then considered studying religion, before turning fully to physics in middle age. His mathematical and experimental approach is the high point of Islamic physics, and his work in optics was not surpassed for 500 years.

Al-Khwarizmi [al-khwahrizmee] (*c*.800–*c*.850) Persian mathematician: introduced modern number notation.

Little is known of al-Khwarizmi's life; he was a member of the Baghdad Academy of Science and wrote on mathematics, astronomy and geography. His book *Algebra* introduced that name, although much of the book deals with calculations. However, he gives a general method (al-Khwarizmi's solution) for finding the two roots of a quadratic equation

$$ax^2 + bx + c = 0 \text{ (where } a \neq 0);$$

he showed that the roots are

$$x_1 = [-b + (b^2 - 4ac)^{\frac{1}{2}}]/2a$$
$$\text{and } x_2 = [-b - (b^2 - 4ac)^{\frac{1}{2}}]/2a$$

In his book *Calculation with the Hindu Numerals* he described the Hindu notation (misnamed 'Arabic' numerals) in which the digits depend on their position for their value and include zero. The term 'algorithm' (a rule of calculation) is said to be named after him. The notation (which came into Europe in a Latin translation after 1240) is of huge practical value and its adoption is one of the great steps in mathematics. The 10 symbols (1–9 and 0) had almost their present shape by the 14th-c, in surviving manuscripts.

Allen, James (Alfred) Van *see* **Van Allen**

Alpher, Ralph Asher (1921–) US physicist: (with Robert Herman) predicted microwave background radiation in space; and synthesis of elements in early universe.

A civilian physicist in the Second World War, Alpher afterwards worked in US universities and in industry. He is best known for his theoretical work concerning the origin and evolution of the universe. In 1948, Alpher, together with BETHE and GAMOW, suggested for the first time the possibility of explaining the abundances of the chemical elements as the result of thermonuclear processes in the early stages of a hot, evolving universe. This work became known as the 'alpha, beta, gamma' theory. As further developed in a number of collaborative papers with R Herman (1914–) over the years, and in another important paper with Herman and J W Follin Jr, this concept of cosmological element synthesis has become an integral part of the standard 'Big Bang' model of the universe, particularly as it explains the universal abundance of helium. The successful explanation of helium abundance is regarded as major evidence of the validity of the model. While this early work on forming the elements has been superseded by later detailed studies involving better nuclear reaction data, the ideas had a profound effect on later developments.

Again in 1948, Alpher and Herman suggested that if the universe began with a 'hot Big Bang', then the early universe was dominated by intense electromagnetic radiation, which would gradually have 'cooled' (or red-shifted) as the universe expanded, and today this radiation should be observed as having a spectral distribution characteristic of a black body at a temperature of about 5 K (based on then-current astronomical data). At that time radio astronomy was not thought capable of detecting such weak radiation. It was not until 1964 that PENZIAS and R W WILSON finally observed the background radiation. It was realized later that evidence for this radiation had been available in 1942 in the form of observed temperatures of certain interstellar molecules. The existence of this background radiation (current observed value 2.73 K), whose peak intensity is in the microwave region of the spectrum, is widely regarded as a major cosmological discovery and strong evidence for the validity of the 'Big Bang' model, to which Alpher, Gamow and Herman contributed the pioneering ideas.

Alter, David (1807–81) US physicist: contributed to spectral analysis.

A physician and inventor as well as a physicist, Alter was one of the earliest investigators of the spectrum. In 1854 he showed that each element had its own spectrum, conclusively proved a few years later by BUNSEN and KIRCHHOFF in their pioneer research on the FRAUNHOFER lines. He also forecast the use of the spectroscope in astronomy.

Altounyan, Roger (Ernest Collingwood) [altoonyan] (1922–87) British medical pioneer; introducer of the anti-asthma drug sodium cromoglycate.

Of Irish-Armenian and English parentage, he was also the grandson of W G Collingwood (1854–1932), the friend and biographer of John Ruskin (1819–1900). Roger Altounyan was born in Syria, but spent summer holidays sailing in the Lake District with his sisters. Here the family met Arthur Ransome (1884–1967) and became the model for his *Swallows and Amazons* children's books. Roger's asthma, however, was fact as well as fiction. He studied medicine, and practised it at the Armenian Hospital in Aleppo run by his father and grandfather. In 1956 he returned to England and joined a pharmaceutical company. He was concerned that asthma was not taken seriously and determined to find a cure. For the next 10 years he worked in his own time testing compounds on himself, inducing asthma attacks two or three times a week with a brew of guinea pig hair, to which he was allergic, almost certainly to the detriment

of his own health. Compound 670, sodium cromoglycate, has been much used to prevent attacks of allergic asthma and rhinitis. The Spinhaler device he invented to inhale the drug was based on aircraft propellers; he had been a pilot and flying instructor during the war.

Alvarez, Luis (Walter) [alvahrez] (1911–88) US physicist: developed the bubble-chamber technique in particle physics.

Alvarez was a student under COMPTON, and then joined LAWRENCE at the University of California at Berkeley in 1936. He remained there, becoming professor of physics in 1945.

Alvarez was an unusually prolific and diverse physicist. He discovered the phenomenon of orbital electron capture, whereby an atomic nucleus 'captures' an orbiting electron, resulting in a nuclide with a lower proton number. In 1939, together with BLOCH, he made the first measurement of the magnetic moment of a neutron. During the Second World War he worked on radar, developing such devices as microwave navigation beacons and radar landing approach systems for aircraft, and also worked on the American atomic bomb project. In 1947 he built the first proton linear accelerator, and later developed the bubble-chamber technique for detecting charged subatomic particles, which in turn led to a great increase in the number of known particles. For this he received the Nobel Prize for physics in 1968.

He was ingenious in the application of physics to a variety of problems. He used the X-ray component of natural cosmic radiation to show that Chephren's pyramid in Egypt had no undiscovered chambers within it; and he used physics applied to the Kennedy assassination evidence to confirm that only one killer was involved. With his son Walter (1940–), a geologist, he studied the problem of the catastrophe of 65 000 000 years ago which killed the dinosaurs and other fossil species; they concluded from tracer analysis that a probable cause was Earth's impact with an asteroid or comet, resulting in huge fires and/or screening of the Sun by dust. His interest in optical devices led him to found two companies; one to make variable focus spectacle lenses, devised by him to replace his bifocals; the other to make an optical stabilizer, which he invented to avoid shake in his cine camera and in binoculars. He was an engaging and popular personality.

Amici, Giovan Battista [ameechee] (1786–1868) Italian microscopist: improved the compound microscope.

Amici trained as an engineer and architect in Bologna; he became a teacher of mathematics but was soon invited to Florence to head the observatory and science museums there. His interest from his youth was in optical instruments, especially microscopes. At that time compound microscopes were inferior to simple types, partly because of aberrations and also because of the false idea that enlargement was the dominant target of design. Amici devised in 1818 a catadioptric (mirror) design that was free of chromatic aberration and used it to observe the circulation of protoplasm in *Chara* cells; at once he became distinguished as an optician and as a biologist. By 1837 he had a design with a resolving power of 0.001 mm and a numerical aperture of 0.4 that was able to magnify 6000 times. His objectives had up to six elements; he invented the technique of immersion microscopy, using oil.

He also much improved telescopes, but his main interest remained in biology, where he made the notable discovery of the fertilization of phanerogams, observing in 1821 the travel of the pollen tube through the pistil of the flower.

Amontons, Guillaume [amõtõ] (1663–1705) French physicist: discovered interdependence of temperature and pressure of gases.

In his teens Amontons became deaf, and his interest in mechanics seems then to have begun. He later improved the design of several instruments, notably the hygrometer, the barometer and the constant-volume air thermometer. In 1699 he discovered that equal changes in the temperature of a fixed volume of air resulted in equal variations in pressure, and in 1703 seemed near to suggesting that at a sufficiently low temperature the pressure would become zero. Unfortunately his results were ignored and it was almost a century later before CHARLES rediscovered the relationship. His work on the thermal expansion of mercury, however, contributed to the invention of the mercury thermometer by FAHRENHEIT.

Ampère, André Marie [āpair] (1775–1836) French physicist and mathematician: pioneer of electrodynamics.

Ampère was a very gifted child, combining a passion for reading with a photographic memory and linguistic and mathematical ability. He was largely self-taught. His life was disrupted by the French Revolution when, in 1793, his father, a Justice of the Peace, was guillotined along with 1500 fellow citizens in Lyon. For a year Ampère seems to have suffered a state of shock; he was aged 18. Ten years later, his adored young wife died following the birth of his son. His second marriage, undertaken on the advice of friends, was a disaster. His professional life ran more smoothly.

In 1802 Ampère was appointed to the first of a series of professorships, and in 1808 was appointed inspector-general of the university system by Napoleon, a post he retained until his death.

Ampère was a versatile scientist, interested in physics, philosophy, psychology and chemistry, and made discoveries in this last field that would have been important had he not been unfortunate in being pre-empted by others on several occasions. In 1820 he was stimulated by

OERSTED's discovery, that an electric current generates a magnetic field, to carry out pioneering work on electric current and electrodynamics. Within months he had made a number of important discoveries: he showed that two parallel wires carrying currents flowing in the same direction attracted one another while when the currents ran in opposite directions they were repelled; he invented the coiled wire solenoid; and he realized that the degree of deflection of Oersted's compass needle by a current could be used as a measure of the strength of the current, the basis of the galvanometer. Perhaps his most outstanding contribution, however, came in 1827, when he provided a mathematical formulation of electromagnetism, notably Ampère's Law, which relates the magnetic force between two wires to the product of the currents flowing in them and the inverse square of the distance between them. It may be generalized to describe the magnetic force generated at any point in space by a current flowing along a conductor. The SI unit of electric current, the ampere (sometimes abbreviated to amp) is named in his honour. The ampere is defined as that steady current which, when it is flowing in each of two infinitely long, straight, parallel conductors that have negligible areas of cross-section and are 1 metre apart in a vacuum, causes each conductor to exert a force of 2×10^{-7} N on each metre of the other.

Anaximander (of Miletus) [anaksimander] (611–547 BC) Ionian (Greek) natural philosopher: suggested Earth was a curved body in space.

A pupil of THALES, Anaximander's writings are now lost, but he is credited with a variety of novel ideas. He was the first Greek to use a sundial (long known in the Middle East), and with it found the dates of the two solstices (shortest and longest days) and of the equinoxes (the two annual occasions when day and night are equal). He speculated on the nature of the heavens and on the origin of the Earth and of man. Realizing that the Earth's surface was curved, he believed it to be cylindrical (with its axis east to west); and he was probably the first Greek to map the whole known world. He visualized the Earth as poised in space (a new idea).

Anderson, Carl David (1905–91) US physicist: discovered the positron and the muon.

Anderson, the only son of Swedish immigrants, was educated in Los Angeles and at the California Institute of Technology, where he remained for the rest of his career.

Anderson discovered the positron accidentally in 1932 (its existence had been predicted by DIRAC in 1928). As a result, Dirac's relativistic quantum mechanics and theory of the electron were rapidly accepted and it became clear that other antiparticles existed. Anderson shared the 1936 Nobel Prize for physics with V F HESS for this discovery.

Anderson discovered the positron while studying cosmic rays, which he did by photographing their tracks in a cloud chamber in order to find the energy spectrum of secondary electrons produced by the rays. A lead plate divided the chamber so that the direction of movement of the particles could be deduced (they are slowed or stopped by the lead). Also, a magnetic field was applied to deflect particles in different directions according to their charge and by an amount related to their mass. Many positive particles were seen which were not protons; they were too light and produced too little ionization. Anderson identified their mass as about that of an electron, concluding that these were positive electrons, or positrons. The discovery was confirmed by BLACKETT and OCCHIALINI the following year.

Anderson discovered another elementary particle within the same year, again by observing cosmic ray tracks. It had unit negative charge and was 130 times as heavy as an electron, and seemed a possible confirmation of YUKAWA's theory of a particle communicating the strong nuclear force (now called a pi-meson or pion). However a series of experiments by Anderson in 1935 revealed that it was not and the role of this mu-meson (or muon), as it is now called, remained unclear. The true pi-meson was first found by POWELL in 1947. Positrons are inherently stable, but as they are antiparticles of electrons the two annihilate each other. Mesons are intrinsically unstable and decay rapidly.

Anderson, Elizabeth, née Garrett (1836–1917) British physician; pioneered the acceptance of women into British medical schools.

Elizabeth Garrett was born in London, where her father had a pawnbroker's shop. He later built an expanding business malting grain at Snape in Suffolk. Educated by a governess at home, followed by boarding school in London, she settled to the duties of daughter at home, helping to run the large household. She joined the Society for Promoting the Employment of Women, whose aim was to improve the status of women through education and employment. ELIZABETH BLACKWELL, the first woman to graduate in medicine in America (1849) gave a lecture on 'Young Women Desirous of Studying Medicine', which impressed Elizabeth Garrett and her friend Emily Davies (1830–1921) who decided that Elizabeth should work to open the medical profession in England to women and that Emily Davies would pursue higher education for women (she founded Girton College, Cambridge). As she was the youngest, Elizabeth's sister Millicent (Fawcett) was to work for the vote for women.

After graduation in New York Elizabeth Blackwell was inscribed on the British Medical Register. The Medical Council decided in future to exclude all holders of foreign degrees. To practise medicine in Britain Elizabeth Garrett had to gain admission to a British medical school.

With the financial support of her father she became an unofficial medical student at the Middlesex Hospital in 1860. Although she had the approval of the Dean, students and staff objected and she had to leave. No medical school or university in Great Britain would admit a woman, so she applied to the Society of Apothecaries, which provided a minimum qualification in medicine. After taking counsel's opinion, the Society was unable to refuse her application because of the wording of its Charter, an opening closed soon after her success. She attended a course by T H HUXLEY on natural history and physiology and JOHN TYNDALL's course on physics at the Royal Institution, at their invitation. Only private tuition for the Apothecaries' medical course was open to her and she found tutors at the medical schools at St Andrews, Edinburgh and London. She passed the Apothecaries' Hall examination in 1865, becoming the first woman to complete a recognized course of medical training with legal qualifications in Britain.

As a woman Elizabeth Garrett was barred from any hospital appointment and was unacceptable as an assistant in general practice. She was dependent on an allowance from her father for some time. She became a consultant physician to women and children from her home in London. Although willing to attend male patients she feared that to do so might create a scandal. In 1865, just before a cholera epidemic reached London, she opened the St Mary's Dispensary for Women and Children in a poor area of London and became visiting medical officer to a children's hospital in East London.

The Sorbonne in Paris admitted women in 1868 and in 1870 Elizabeth Garrett became the first woman MD from that university. In 1871 she married James Skelton Anderson (died 1907) and combined family life with her work. The London School of Medicine for Women, which was initiated by SOPHIA JEX-BLAKE, opened in 1874 and Elizabeth Garrett Anderson served on the Executive Committee, taught in the school and worked for the students to be admitted to the University of London's examinations. In 1883 she was elected Dean, the same year that Mary Scharlieb and Edith Shove became the first women to gain medical degrees from London University. The School became a college of the University of London. From 1886 Elizabeth Garrett Anderson was concerned with the New Hospital for Women which served as the teaching hospital for the London School of Medicine for Women. In 1908 she was elected mayor of Aldeburgh, the first woman mayor in England. It was due to the efforts of Elizabeth Garrett Anderson that medical education and medical science was opened to women in Britain.

Anderson, Philip Warren (1923–   ) US physicist: discovered aspects of the electronic structure of magnetic and disordered systems.

Anderson studied at Harvard, doing doctoral research with VAN VLECK and spending 1943–45 involved in antenna engineering at the Naval Research Laboratory. Anderson's career was largely with Bell Telephone Laboratories, but he became professor of physics at Princeton in 1975, and he also held a visiting professorship at Cambridge, UK (1967–75). Under Van Vleck, Anderson worked on pressure broadening of spectroscopic lines. In 1958 he published a paper on electronic states in disordered media, showing that electrons would be confined to regions of limited extent (Anderson localization) rather than be able to move freely. In 1959 he calculated a model explaining 'superexchange', the way in which two magnetic atoms may interact via an intervening atom. In 1961 he published important work on the microscopic origin of magnetism in materials. The Anderson model is a quantum mechanical model that describes localized states and their possible transition to freely mobile states. This model has been used widely to study magnetic impurities, superconducting transition temperatures and related problems. Also, during his work on superconductivity and superfluidity, Anderson worked on the possible superfluid states of helium-3. For these investigations of electronic properties of materials, particularly magnetic and disordered ones, Anderson shared the 1977 Nobel Prize for physics.

Andrews, Roy Chapman (1884–1960) US naturalist and palaeontologist.

Andrews's career was mostly spent with the American Museum of Natural History, New York, and with its expeditions (especially to Asia) to collect specimens. His more dramatic finds included the fossil remains of the largest land mammal yet found, *Paraceratherium*, a relative of the rhino which stood 5.5 m high; and the first fossil dinosaur eggs. He had a special interest in whales and other cetaceans (aquatic mammals) and built up a fine collection of them; and he found evidence of very early human life in central Asia.

Andrews, Thomas (1813–85) British physical chemist: showed existence of critical temperature and pressure for fluids.

The son of a Belfast merchant, Andrews studied chemistry and medicine in Scotland. In Paris he studied chemistry under DUMAS and at Giessen he studied under LIEBIG. In Belfast he first practised medicine and later became professor of chemistry. He proved that 'ozone' is an allotrope of oxygen (ie a different form of the element; ozone was later shown to be O_3; ordinary oxygen is O_2). He was a fine experimenter and is best known for his work on the continuity of the liquid and gaseous states of matter (1869). Using carbon dioxide, he showed that above its 'critical temperature' (31°C) it cannot

be liquefied by pressure alone. This example suggested that at a suitably low temperature, any gas could be liquefied, as was later demonstrated by CAILLETET.

Anfinsen, Christian (Boehmer) (1916–95) US biochemist: made discoveries related to the shape and activity of enzymes.

Educated at Swarthmore and Harvard, Anfinsen afterwards worked at Harvard and from 1950 at the National Institutes of Health in Bethesda, MD. In 1960 MOORE and W H Stein (1911–80) found the sequence of the 124 amino acids which make up ribonuclease and it became the first enzyme for which the full sequence was known. However, it was clear that enzymes owe their special catalytic ability not only to the sequence of amino acid units but also to the specific shape adopted by the chainlike molecule. Anfinsen showed that, if this shape is disturbed, it can be restored merely by putting the molecule into the precise environment (of temperature, salt concentration, etc) favourable for it, when it spontaneously takes up the one shape (out of many possibilities) that restores its enzymic activity. He deduced that all the requirements for this precise three-dimensional assembly must be present in the chain sequence; and he showed that other proteins behaved similarly. He shared the Nobel Prize for chemistry with Moore and Stein in 1972.

Ångström, Anders (Jonas) (1814–74) Swedish spectroscopist: detected hydrogen in the Sun.

Ångström was educated at Uppsala and taught physics at the university there until his death. He was an early spectroscopist and deduced in 1855 that a hot gas emits light at the same wavelengths at which it absorbs light when cooler; this was proved to be so in 1859 by KIRCHHOFF. From 1861 he studied the Sun's spectrum, concluding that hydrogen must be present in the Sun, and mapping about 1000 of the lines seen earlier by FRAUNHOFER. A non-SI unit of length, the ångström (Å) is 10^{-10} m; it was used by him to record the wavelength of spectral lines.

Anning, Mary (1799–1847) British palaeontologist.

Mary Anning had the good fortune to be born in Lyme Regis in Dorset, a place of great geological interest and, when a year old, to survive a lightning strike. Her nurse sheltered with her beneath a tree during a thunderstorm with two others; only Mary survived.

Her father, a cabinetmaker, supplemented his income by selling local fossils to summer visitors. He died when Mary was 11 years old and she, apparently well trained by him, continued to help the family income by the same means. Her brother discovered the head of a marine reptile in the cliffs between Lyme Regis and Charnmouth in 1811 and Mary carefully excavated the complete remains, named *Ichthyosaurus* in 1817; she sold it to a collector for

£23. This was the beginning of a lifetime's fruitful fossil-hunting; her value to palaeontologists was in her local knowledge, her skill in recognition and the care she took to present her finds in an uninjured state. In 1823 she discovered the complete skeleton of an little-known saurian, named *Plesiosaurus* by William Conybeare (1787–1857) and described by him at a meeting of the Geological Society in London in 1824. Another major discovery of hers at Lyme was described as *Pterodactylus macronyx* by William Buckland (1784–1856) in 1829; this fossil of a strange flying reptile attracted much attention.

Mary Anning supplied fossils to palaeontologists, collectors and museums, as well as to the visitors to her shop in Lyme Regis and attracted the general public to fossil-collecting and to herself, by her successes. She became both knowledgeable and aware of the significance of her discoveries. She left no publications and her contribution to the 'golden age' of British geology has been largely neglected; her discoveries were described and collected by others. In her later years she was assisted by a small government grant awarded by the prime minister, Lord Melbourne, at the prompting of Buckland and the Geological Society of London. When she died her work was acknowledged by the president of the Society in his anniversary address. Later a stained-glass window to her memory was placed by the Fellows in the parish church at Lyme Regis.

Apollonius (of Perga) [apawlohneeuhs] (*c*.260–190 BC) Greek mathematician: wrote classic treatise on conic sections.

Apollonius was a student in Alexandria and later taught there, specialising in geometry. Of his books, one survives, *On Conic Sections*. It deals with the curves formed by intersecting a plane through a double circular cone (see diagram). These are the circle, ellipse, parabola and hyperbola (the last three were named by Apollonius). Much of the book on the properties of conics is original; it represents the high point of Greek geometry and, although at the time the work appeared to have no uses, KEPLER 1800 years later found that the planets moved in ellipses and the curves now have many applications in ballistics, rocketry and engineering.

Apollonius was also interested in astronomy and especially in the Moon, and proposed a theory of epicycles to describe the sometimes apparently retrograde motions of the outer planets.

Appert, Nicholas-Francois [apair] (*c*.1749–1841) French chef: devised an improved method of food preservation.

As a innkeeper's son, Appert was familiar with food preparation and he became a chef and confectioner. A government prize was offered for improved methods of preserving food, especially for military use, and from 1795 Appert experimented with sealing food into glass jars using waxed cork bungs. Using heat-sterilization of the vessel and contents he was successful and in 1810 he claimed the 12 000 franc prize. He used autoclaves (pressure cookers) to give a temperature a little above the boiling point of water. Tinplate cans came into use in England after 1810 and were sealed by soldering. Appert's work was highly praised, but he died in poverty. PASTEUR'S work, which rationalized Appert's success, came in the 1860s.

Appleton, Sir Edward (Victor) (1892–1965) British physicist: pioneer of ionospheric physics; discovered reflective layers within the ionosphere.

Appleton studied physics at Cambridge, but it was service in the First World War as a signals officer which led to his interest in radio. In 1924 he was appointed professor of experimental physics at King's College, London. In 1939 he was appointed secretary of the Department of Scientific and Industrial Research, and later became vice-chancellor of Edinburgh University.

In 1901 MARCONI had transmitted radio signals across the Atlantic, to the astonishment of many in the scientific community who believed that, since electromagnetic radiation travels in straight lines and the Earth's surface is curved, this was not possible. Shortly afterward, A E Kennelly (1861–1939) and HEAVISIDE proposed a reflecting layer of charged particles in the atmosphere as the explanation. In a classic experiment in 1925, Appleton became the first to demonstrate beyond doubt the existence of such a reflecting layer within the ionosphere. He transmitted signals between Bournemouth and Cambridge (a distance of 170 km); and by slowly varying the frequency and studying the received signal he showed that interference was occurring between the part of the signal that travelled in a straight line from transmitter to receiver (the direct, or ground, wave) and another part that was reflected by the ionosphere

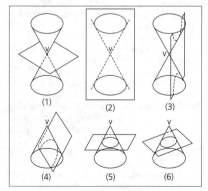

Conic sections – Cone (sometimes double) cut by a plane to give (1) a single point (2) a pair of straight lines (3) a hyperbola (4) a parabola (5) a circle (6) an ellipse

(the sky wave). Measurement of the interference caused by the different path lengths enabled him to measure the height of the reflecting layer, about 70 km. This was the first radio distance measurement. This layer is now known as the Heaviside layer or E layer. Further work revealed a second layer above the first, which is now called the Appleton layer or F layer. The E layer is more effective after dark, since the Sun's ultraviolet rays interact with the ionosphere, which is why distant radio stations are more readily picked up at night. For his achievements Appleton received the Nobel Prize for physics in 1947.

Arago, Dominique François Jean [aragoh] (1786–1853) French physicist.

Beginning his career as a secretary at the Bureau de Longitudes, Arago went with BIOT to Spain in 1806 to complete the geodetic measurements of an arc of the meridian. The return journey was eventful as the ship was wrecked and he was almost enslaved at Algiers. He made distinguished researches in many branches of physics, and in 1838 suggested a crucial experiment to decide between the particle and wave theories of light, by measuring its speed in air and in water. The experiment was tried by FOUCAULT in 1850 and pointed to the wave theory.

Arago was the first to discover that substances other than iron have magnetic properties. He also discovered the production of magnetism by electricity; a piece of iron, surrounded by a coil of wire, was briefly magnetized by passing a current from either a capacitor or a voltaic cell through the coil.

The device called Arago's disc consists of a horizontal copper disc with a central vertical spindle; the disc can be spun by a belt and pulley. Above it, and separately mounted, is a pivoted compass needle. When the disc is spun, the needle follows it; if the rotation of the plate is reversed, the needle slows, stops and also reverses. The effect is due to eddy currents in the disc, although Arago did not know this. The value of his experiments in electricity and magnetism is that they later inspired FARADAY to make his major discoveries.

Arago was a close friend of HUMBOLDT for half a century; the latter wrote of Arago as 'one gifted with the noblest of natures, equally distinguished for intellectual power and for moral excellence'.

Arber, Agnes, *née* Robertson (1879–1960) British botanist: her careful investigations of plant structure made a lasting contribution to botanical knowledge.

Agnes Robertson became an enthusiastic student of botany while attending the North London Collegiate School for Girls. The school was unusually good in its teaching of science and it was there that she learned about plant anatomy and classification. The school had a Science Club to which ETHEL SARGANT, who had also attended the school, gave talks on botany. After taking degree examinations at University College, London, and Newnham College, Cambridge, Agnes Robertson became a research assistant to Ethel Sargant at her private laboratory and worked on seedling structures. From 1903–08 she returned to London to take up research on gymnosperms.

After her marriage in 1909 she worked at the Balfour Laboratory in Cambridge until 1927 and thereafter in her own laboratory at home. For the next 50 years her researches were mainly concerned with the anatomy and morphology of monocotyledonous plants and her researches were gathered into book form. She published *Herbals, Their Origin and Evolution* (1912), *Water Plants: A Study of Aquatic Angiosperms* (1920), and *Monocotyledons* (1925). She also produced in the Cambridge Botanical Handbook series *The Gramineae: A Study of Cereal, Bamboo, and Grass* (1934). In 1946 she was elected to the fellowship of the Royal Society; she was the third woman to receive the honour. After the Second World War she turned more to philosophy and wrote *The Natural Philosophy of Plant Form* (1950), *The Mind and the Eye* (1954) and *The Manifold and the One* (1957).

Archer, Frederick (Scott) (1813–57) British inventor of the wet collodion photographic process.

Orphaned early in life, Archer was apprenticed to a London silversmith. This led him first to study coins and then to design them and to work as a portrait sculptor. To obtain likenesses for this he began in 1847 to use the primitive photographic methods of the time. He experimented to improve them and tried collodion (a solution of nitrocellulose in ether), on paper and then on glass, as part of the sensitive material. In 1851 he published his method: collodion containing iodide was flowed over a glass plate. This was followed by silver nitrate solution. The moist plate was quickly exposed in the camera and then developed, fixed and washed to give a glass negative from which positive paper prints could be made. The moist plate was much more sensitive to light than its predecessors and allowed exposures to be reduced to 2–20 s.

Archer was diffident, poor, generous and unworldly. TALBOT claimed (falsely) that the whole process was covered by his patents, but by 1854 his attempts to prevent the use of collodion by legal injunctions had failed and, as the daguerrotype patents had lapsed in 1853, the public could take up photography without restraint and did so enthusiastically. Despite the need to carry a tent and portable laboratory, the wet collodion process quickly supplanted all others and was widely used by amateurs and professionals until 1880 when the more convenient gelatine dry plate was introduced.

Archer died poor and unappreciated, and his family received niggardly provision from

Government and professional photographers who had profited by use of his unpatented methods. Only after many years was his contribution recognized.

Archimedes (of Syracuse) [ah(r)kimeedeez] (c.287–212 BC) Sicilian Greek mathematician and physicist: pioneer of statics and hydrostatics.

A member of a wealthy noble family, Archimedes studied in Alexandria but returned to Syracuse in Sicily, whose king Hieron II was a relative. Archimedes was the finest scientist and mathematician of the ancient world but little is firmly known of his life, although legends exist. He is known to have used experiments to test his theories, which he then expressed mathematically. He devised weapons against the Roman fleet when it attacked Syracuse in 215 BC; the Romans took the city in 212 BC and Archimedes was killed. Cicero found and restored his tomb in 75 BC.

In mathematics, Archimedes used geometrical methods to measure curves and the areas and volumes of solids (e.g. the volume of a sphere, $4\pi r^3/3$); he used a close approximation for π (he showed it to be between 223/71 and 220/70) and developed his results without the use of the calculus (which came nearly 2000 years later). He used a new notation to deal with very large numbers, described in his book *Sand-Reckoner*.

In applied mathematics, he created mechanics; his innovations ranged from the directly practical (eg the compound pulley and the Archimedian screw) to derivations of the theory of levers and centres of gravity, forming the basic ideas of statics. He founded hydrostatics, contributing ideas which included specific gravity and the Archimedes principle: this states that when a body is wholly or partly immersed in a fluid, it experiences a buoyant force (upthrust) which shows itself as an apparent loss of weight, equal to the weight of fluid displaced. (The fluid can be liquid or gas.)

GAUSS thought that Archimedes had only NEWTON as a mathematical equal.

Aristarchus (of Samos) [aristah(r)kuhs] (c.320–c.250 BC) Greek astronomer: proposed heliocentric cosmology; and made first estimate of astronomical distances.

Although little is known of the life of Aristarchus, he was perhaps the first to propose that the Earth moved around the Sun, in contrast to the accepted thinking of his day. He also attempted to estimate the relative distances of the Sun and the Moon, using the fact that when the Moon is exactly half light and half dark it forms a right angle with the Earth and the Sun. Although his result was wildly inaccurate, it was the first experimental attempt at measuring an astronomical distance. His work makes him the most original of the Greek astronomers and in the modern view the most successful. His heliocentric scheme was made precise by COPERNICUS in the 16th-c.

Aristotle [aristotl] (384–322 BC) Athenian (Greek) philosopher and naturalist: provided philosophical basis of science which proved dominant for 18 centuries.

Son of the court physician at Macedon, Aristotle was orphaned early and moved to Athens, where he became Plato's finest pupil. In 342 BC he returned to Macedon as tutor and then adviser to Philip II's son Alexander, who became Alexander the Great. Later he became a public teacher in Athens, using a garden he owned (the Lyceum). His collected lectures cover most of the knowledge of the time in science, and some other fields such as logic and ethics (but not mathematics), and include much of Aristotle's own work in zoology and anatomy. He was a first-class naturalist and marine biologist, whereas his record of older views in physics and cosmology contained many misguided, although defensible, ideas. Aristotle's books survived in the Arab world, and re-entered Christian Europe in Latin translation in the 12th and 13th-c. It was no fault of the writer that his books were accorded almost divine authority, and some of the erroneous ideas were not easily displaced (eg that bodies 'outside the sphere of the Moon' are perfect and unchanging). His status as a major figure in philosophy has never changed.

Armstrong, Edwin Howard (1890–1954) US radio engineer.

Many teenagers build radio receivers; Armstrong was unusual in also making a transmitter before he became a student of electrical engineering at Columbia. Then, during the First World War, he worked on the problem of locating aircraft by detecting the stray radio emission from their ignition systems; a side-result was his development of the superheterodyne circuit, which made radio tuning much easier and helped make radio popular. From 1934 he taught electrical engineering at Columbia and by 1939 he had devised a major advance in radio transmission: FM.

Previously, radio signals conveyed speech or music by changes in the amplitude of the carrier radio waves (amplitude modulation, AM). The snag of this is that electrical storms and appliances introduce random noise (static). Armstrong's method was to vary the carrier signal by changes in frequency (frequency modulation, FM) which is largely free from interference. This requires use of high frequencies that have only a limited range, but it has become the preferred mode for radio and TV use.

Armstrong, Neil (Alden) (1930–) US astronaut: the first man to walk on the Moon.

This most dramatic act of manned exploration occurred on 20 July 1969 and represented success in a curious international contest. In 1957 the USSR had placed an unmanned spacecraft (Sputnik) into orbit, and this blow to national self-esteem in the USA led to President

THE EXPLORATION OF SPACE

To transform dreams of space travel into reality and to explore space directly, rockets are needed: their power overcomes the force of gravity which pulls all objects towards the centre of the Earth.

Konstantin Edvardovich Tsiolkovsky (1857–1935) was the first to suggest multistage rockets. He also suggested the use of liquid hydrogen and liquid oxygen as propellants, burned in a combustion chamber. Independently, GODDARD successfully made rockets fuelled by gasoline (petrol) and liquid oxygen. In 1937, one of these reached a height of about 2.5 km/1.6 mi. Hermann Oberth (1894–1990) considered rocket propulsion and guidance systems theoretically. His work led to the ethanol/liquid-oxygen-fuelled German V2 rocket that was perfected by VON BRAUN during the Second World War.

After the war von Braun worked in the USA, developing rockets that could carry instruments into the uppermost part of the Earth's atmosphere. In 1954 he proposed using the Redstone rocket, similar in design to the V2, to put a satellite weighing 2 kg into orbit around the Earth. In the event, the first US satellite, the 14 kg Explorer 1, was launched by a Jupiter C rocket early in 1958. Using Geiger counters aboard it, VAN ALLEN discovered the radiation belts of energetic, charged particles that surround the Earth, and which now carry his name.

The first Earth-orbiting satellite was the 84 kg Soviet Sputnik 1. This was launched on 4 October 1957, using kerosene (paraffin oil)/liquid-oxygen fuel. The rocket engines were designed by a team led by Sergei Pavolvich Korolev (1907–66). His rockets also powered the first manned space flight on 29 April 1961 by Yuri Gagarin (1934–68), whose flight in the Vostok 1 capsule (4.7 tonnes) around the Earth lasted 108 minutes. Furthermore, Korolev's rockets launched the first space probe, Venera 3, to land on the planet Venus.

With the Soviet Luna programme and the US Ranger programme, television pictures of the Moon's surface were sent back to Earth before the spacecraft crashed into the Moon. Later, the soft landings of Surveyor spacecraft were a prelude to America's Apollo programme. It was during the Apollo 11 mission that NEIL ARMSTRONG became the first human being to set foot on the Moon, on 21 July 1969, speaking the now famous prepared words 'that's one small step for man; one giant leap for mankind'. The other Apollo 11 astronauts were Michael Collins (1930–) and Edwin E (Buzz) Aldrin (1930–).

For the Apollo programme, the gigantic 110 m-tall Saturn V rocket was developed by von Braun's team. It had to send 40 tonnes towards the Moon at a speed exceeding 11 km s^{-1}. It was needed to meet US President John F Kennedy's objective, stated in 1961, 'that this nation should commit itself to achieving the goal, before this decade is out, of landing a man on the Moon and returning him safely to Earth'. In successfully accomplishing this, scientists also investigated samples of lunar material, and studied the Moon's gravitational, magnetic, and seismic properties. The Apollo 15 mission, with David R Scott (1932–), Alfred M Worden (1932–) and James B Irwin (1930–), was the first to use a battery-powered lunar rover to travel up to 103 km/63 mi from the landing site. The Apollo missions, through the rock samples they provided, led to the general acceptance, from 1984, of a novel theory of the Moon's origin, as due to impact on the early Earth of a Mars-sized planetesimal, the resulting debris coalescing to form the Moon. This theory avoids some insurmountable problems linked with other, earlier, ideas on the Moon's origin.

Also during the 1960s, many Soviet cosmonauts and American astronauts orbited the Earth. The first American astronaut to do so was John H Glenn (1921–), later a senator. In February 1962 he travelled three times round the world in less than 5 hours in his Mercury Friendship 7 capsule, which was launched by an Atlas rocket. Confidence was building up as L Gordon Cooper (1927–) survived in space for more than 24 hours, with 22 orbits in the Mercury Faith 7 capsule during May 1963. Aboard Vostock 6, Valentina Tereshkova (1937–) was the first woman in space, in June 1963. In March 1965, Aleksei Leonov donned a spacesuit to emerge from Voskhod 2, a three-man spacecraft, and to walk in space for the first time. Edward H White (1930–67) became the first American to walk in space 3 months later, this time for 21 min. In 1965 and 1966, 10 two-man Gemini missions were launched by Titan 11 rockets, with Armstrong on Gemini 8, Michael Collins on Gemini 10, and Buzz Aldrin on Gemini 12. The Gemini programme demonstrated that humans could survive and operate in space for at least 2 weeks, and that one space vehicle could rendezvous and dock with another. Both astronauts and flight control crews on the ground gained valuable and necessary experience to prepare them for the Apollo programme.

Polar-orbiting satellites were used during the 1960s for research on both the Earth's atmosphere and the near-Earth space environment. Clouds were observed from the Tiros series of satellites, leading to improved weather forecasts. The Transit navigational satellite and Echo telecommunications satellite programmes commenced in 1960. Remote sensing of the Earth's surface for resource studies was successfully begun in July 1972 with Landsat 1.

From Earth-orbiting satellites in highly eccentric orbits, Norman F Ness (1933–) discovered and investigated the magnetopause, the boundary between the Earth's magnetic field and the interplanetary medium. This had been postulated by CHAPMAN in 1931. Konstantin Gringauz (1918–93) found the plasmapause, the outer boundary of the Earth's topside ionosphere, in 1963, and Lou A Frank (1938–) observed the ring of auroral light around the Earth's magnetic poles from the General Dynamics Explorer 1 satellite.

Realizing the 1945 idea of CLARKE, a Syncom 2 telecommunications satellite was launched into geostationary orbit in July 1963. At an altitude of 36 000 km/22 370 mi above the equator, the orbital period is one day, and so the satellite always remains above a certain geographical position. With at least three such satellites, point-to-point communications are possible anywhere in the world (except for the polar regions). The geostationary orbit is also invaluable for making continuous meteorological observations over most of the Earth.

From 1970s onwards, the USA explored the planets of the solar system with Mariner, Pioneer, Viking and Voyager spacecraft. Striking photographs were obtained and many scientific investigations pursued. The USSR concentrated on carrying out detailed studies of Mars and Venus. Detailed knowledge of both inner and outer planets was vastly increased by these programmes.

During 1973, several American scientists worked for weeks at a time in Skylab, an orbiting laboratory. First they corrected several mechanical problems with their equipment. Their subjects of study ranged from stellar astronomy and solar physics, to the behaviour of substances under almost weightless conditions, and the physiological effects of space on astronauts. From 1967, the then USSR conducted some 60 manned space flight missions, the longest lasting more than a year. In the Salyut or Cosmos spacecraft and, after 1986, the Mir space station, they accumulated a wealth of practical experience across many scientific and technological disciplines. Svetlana Savitskaia (1948–) worked outside the Salyut 7 space station for nearly 4 hours in July 1984, the first time that a woman performed extravehicular activity (EVA). The first British cosmonaut, Helen Sharman (1963–), worked aboard Mir for a week in 1991, and the second, Michael Foale (1957–) aboard the US Space Shuttle in 1992.

The US Space Shuttle flew for the first time in April 1981. It provides a recoverable and reusable launch vehicle rather than expendable rockets.

During launch its three main engines, fuelled by liquid hydrogen and liquid oxygen, are supplemented by two solid fuel (polybutadiene) rocket boosters. The Shuttle's cargo bay is vast, 18 m/59 ft long and 4.56 m/15 ft diameter. Once in space, the cargo of satellites can be launched either by the crew or by radio from Earth. Alternatively, the cargo bay can house a laboratory, such as the European Spacelab, in which experiments are performed during the mission, which typically lasts 10 days. Ulf Merbold (1941–) was the first European astronaut aboard Spacelab at the end of 1983. The Soviets used a rather similar space shuttle, called Buran ('Snowstorm'). This was launched for the first time in November 1988.

European, Chinese, Japanese and other national space programmes developed since the 1960s to complement the superpowers' space programmes. The Ariadne, Long March and Lambda or Mu rockets launched a wide variety of both civil and military satellites with scientific or technological payloads. For example, both Europe and Japan sent spacecraft (Giotto, and Suisei and Sakigake, respectively) as well as the Soviet Vega probes, to study HALLEY's comet when it neared the Sun in 1986. The European SPOT (*Satellite pour l'observation de la terre*) satellites have excellent spatial resolution and colour discrimination for geological and cartographic remote-sensing investigations of our planet. Besides downward-looking instruments for 'Mission to Planet Earth' to investigate such phenomena as the ozone hole, telescopes can view outwards to investigate space across the electromagnetic spectrum at wavelengths where the atmosphere absorbs radiation. New insights into the universe, and its origin, are obtained. Although initially plagued by an instrumental defect, the shuttle-launched Hubble space telescope, when repaired in orbit in 1994, gave superb observations.

For the future, Britain is proposing HOTOL, the Horizontal Take-Off and Landing, recoverable and air-breathing launcher; Germany is proposing a somewhat similar design – the Sänger concept – named after Eugen Sänger (1905–64), who studied it theoretically during the Second World War, and France proposed the Hermes, a mini-shuttle. The USA is forging ahead with its international space station, Freedom, and is considering the benefits of a manned mission to the Moon and/or to Mars, as well as smaller satellites dedicated to solving particular scientific problems. International collaboration should be the hallmark of future space exploration.

Prof M J Rycroft, College of Aeronautics, Cranfield

Kennedy in 1961 committing his country to a manned moon-landing within the decade.

The Manned Spacecraft Centre, at Houston, TX, worked for this result and success came with the Apollo 11 flight commanded by Armstrong. He had served as a naval aviator in the Korean War and later studied aeronautical engineering at Purdue University, IN, and the University of Southern California before joining the organization which became NASA (the National Aeronautics and Space Administration) in 1955. By 1969 he was well equipped to land a spacecraft on the Moon, under his manual control. The scientific results of the visit were limited and were exceeded by the results from unmanned space probes. Armstrong left NASA in 1971 and, after a period as professor of aeronautical engineering at Cincinnati, OH, became a businessman from 1980.

Arrhenius, Svante (August) [arayneeus] (1859–1927) Swedish physical chemist: proposed theory of ionic dissociation.

Arrhenius came from a family of farmers, and his father was an estate manager and surveyor. He attended Uppsala University and did very well in physical science, and then moved to Stockholm to work for a higher degree on aqueous solutions of electrolytes (acids, bases and salts); he concluded that such solutions conduct a current because the electrolyte exists in the form of charged atoms or groups of atoms (positive cations and negative anions), which move through the solution when a current is applied. He obtained good evidence for this during the 1880s but his theory was only slowly accepted, especially in Sweden. (Since then, further evidence has substantially confirmed his views, and has also shown that salts are largely ionic even in the solid state.) In 1903 he was awarded the Nobel Prize for chemistry. His work was surprisingly varied and included immunology, cosmic physics and the first recognition of the 'greenhouse effect' (heat gain by the atmosphere due to carbon dioxide). He also studied the effect of temperature on the rates of chemical reactions, and showed that

$$k = A \exp(-E/RT)$$

where k is the rate constant for the reaction, A is the frequency factor, E is the activation energy for the reaction, R is the gas constant and T the Kelvin temperature (this is the Arrhenius equation).

Aston, Francis William (1877–1945) British chemical physicist: invented mass spectrograph.

After graduating in chemistry in Birmingham, Aston worked for 3 years as a chemist in a nearby brewery. In his leisure at home he designed and made an improved vacuum pump and in 1903 he returned to physics as a career, working on discharge tubes in Birmingham and, from 1909, in Cambridge as J J THOMSON's assistant. They worked on the 'positive rays' which Thomson had found to be generated within one part of a vacuum tube through which an electric discharge is passed. Aston and Thomson believed that their experiments on positive rays from tubes containing neon gas showed it to contain atoms with masses of about 20 and 22 units. Proof of this, and extension of the work, was interrupted by the First World War.

Aston's war work at the Royal Aircraft Establishment linked him with a talented group of physicists, including LINDEMANN, TAYLOR, ADRIAN, G P THOMSON and H Glauert (1892–1934). Soon after the war he devised a mass spectrograph which was able to separate atoms of similar mass and measure these masses accurately (his third spectrograph to 1 in 10^5; 1 in 10^9 is now easily available on commercial machines). Aston showed clearly that over 50 elements consisted of atoms of similar but different relative atomic mass (eg for S; 32, 33 and 34) but the same atomic number (ie nuclear charge). The Aston rule is that the masses are approximately integers; the apparent deviations of relative atomic masses of the elements from integers results from the presence of isotopes.

Aston found that isotopic masses are not exactly integral (by about 1%) and he related the discrepancy (the 'packing fraction') to the force binding the nucleus together. Atomic energy generation from nuclear reactions, on Earth or in the stars, can be calculated from packing fractions.

The modern mass spectrograph has played a central part in nuclear physics and radiochemistry, and more recently in exact analysis in organic chemistry. Aston was a 'one device' investigator, but he chose a device whose value has been immense.

He was a shy man, a poor teacher, with a passion for sports and for sea travel. He won the Nobel Prize for chemistry in 1922.

Atkins, Anna, *née* Children (1799–1871) British botanist: the first to use photography to illustrate scientific studies.

Anna was the only child of J G Children, a Fellow of the Royal Society whose wife had died shortly after Anna's birth. Her father was a friend of the Herschel family and Anna knew JOHN HERSCHEL from childhood. She had a close relationship with her father and shared his scientific interests. No doubt this position helped her acceptance in the male scientific circle. She was a skilled illustrator and provided over 200 drawings for her father's translation of LAMARCK's book *The Genera of Shells*, published in 1823. Anna married J P Atkins in 1825; there were no children and she continued her scientific interests and collaboration with her father. Children chaired the Royal Society meeting at which TALBOT announced his 'calotype' photographic process, and father and daughter took up the process enthusiastically. In the same year she became an active member of the Botanical Society of London.

The calotype process was difficult (partly because Talbot's information was inadequate), but in 1842 Herschel described his 'cyanotype' process. To illustrate her large collection of algae Anna Atkins turned to photography. As she explains in her preface 'The difficulty of making accurate drawings of objects as minute as many of the Algae and Confervae, has induced me to avail myself of Sir John Herschel's beautiful process of Cyanotype, to obtain impressions of the plants themselves.' She made contact photograms of algae, totalling 389 pages of illustration and 14 of handwritten text, and making more than a dozen copies of the whole, which were sent to scientific friends and institutions. And so with her *Photographs of British Algae: Cyanotype Impressions* (3 vols, 1843–53) she became the first to apply photography to illustrate scientific studies, predating Talbot's *Pencil of Nature* (1844–46), although the latter includes photographs made with a camera. Atkins's work is both permanent and suited to its subject, with seaweeds shown as paler images on a rich blue background.

Auer, Carl, Freiherr (Baron) **von Welsbach** [ower] (1858–1929) Austrian chemist: invented the gas mantle.

Auer studied at Vienna Polytechnic and at Heidelberg, the latter under BUNSEN. In 1885 Auer succeeded in showing that the lanthanoid 'element' didymium was actually a mixture of two new elements, praseodymium and neodymium. He is perhaps better known, however, as the inventor of the gas mantle, a fabric net impregnated with thorium oxide and cerium oxide that glows incandescent when heated in a gas flame. Hitherto, gas lighting had relied on the luminescence of the flame itself. Unfortunately, the invention of electric lighting made his invention largely redundant within a few years. Following an unsuccessful attempt to improve EDISON's bulb with an osmium filament, he later found another use for cerium, as an alloy with iron as the 'flint' of cigarette lighters.

Auerbach, Charlotte [owerbakh] (1899–1994) German–British geneticist: discoverer of chemical mutagenesis.

Lotte Auerbach, born in Germany and the daughter and grand-daughter of scientists, herself studied science at four German universities and then taught in schools in Berlin until, in 1933, all Jewish teachers were dismissed. She escaped to Edinburgh, followed by her mother, worked for a PhD and obtained a lowly job at the Institute of Animal Genetics there, becoming a lecturer in 1947.

Discussions with the geneticist H J MÜLLER led her to study mutation in animal cells (a mutation is a change, spontaneous or induced, in a gene or a chromosome; such a change in the hereditary material leads to an abrupt alteration in the characteristics of an organism).

Müller had shown that X-rays produced mutations: Lotte Auerbach first showed that mustard gas ([$CH_2CH_2Cl]_2S$, used in the First World War) did so in the fruit fly, *Drosophila*. She became an authority on such chemical mutations, which have been of great value in research and in cancer treatment, and she directed the Medical Research Council Mutagenesis Research Unit 'for as long as she could conceal her age from her employers', not retiring until 1969. She was elected a Fellow of the Royal Society in 1957.

Auger, Pierre Victor [ohzhay] (1899–) French physicist: discovered the Auger effect.

Auger was educated at the École Normale Supérieure and subsequently became professor of physics at the University of Paris. After the Second World War he held a succession of posts in French and European science administration and was director general of the European Space and Research Organization at his retirement.

Auger is remembered for his discovery in 1925 of the Auger effect, in which an atom absorbs energy in the form of an X-ray photon, and loses it by emitting an electron. Auger spectroscopy uses the effect to yield information about the electronic structure of atoms, particularly if they form part of a crystal.

Avery, Oswald (Theodore) [ayveree] (1877–1955) US bacteriologist: showed that the genetic material of bacterial chromosomes is DNA.

Born in Canada, Avery went to New York when he was 10 and remained there for his working life; he qualified in medicine at Columbia in 1904, and from 1913 researched in bacteriology at the Rockefeller Institute Hospital. His special interest was pneumococci (the bacteria causing pneumonia). In 1928 he was intrigued by the claim of the British microbiologist F Griffith (1881–1941) that a non-virulent, 'rough' (ie unencapsulated) pneumococcus could be transformed into the virulent, smooth (capsulated) form in the mouse by the mere presence of some of the dead (heat-killed) smooth bacteria.

Avery found this so strange that he repeated the work, and also showed in 1944 that the substance that caused the transformation is deoxyribonucleic acid (DNA). Prudently, he did not go on to surmise that genes are simply DNA, which was surprising enough to be accepted only slowly after 1950 and formed the basic idea of molecular biology.

Avogadro, (Lorenzo Romano) Amedio (Carlo) [avohgadroh] (1776–1856) Italian physicist: proposed a method for finding molecular formulae of gases.

Trained in law like his forefathers and working as a lawyer for some time, after 1800 he turned to science and held professorships in physics for much of his life. His fame now rests on one brilliant and important idea. He considered GAY-LUSSAC's Law of combining volumes and with little evidence offered a daring explanation for it in 1811. His idea, Avogadro's Law, was that 'equal volumes of all gases, under the same conditions of temperature and pressure, contain the same number of smallest particles'. There is now ample evidence that he was right; in some cases (eg the noble gases) the smallest particles are atoms; for most other gases, they are combinations of atoms (molecules). The law gives a direct method of finding the molecular formula of a gas, and such a formula in turn gives the relative atomic masses of the elements present in it. Avogadro's Law shows that the simple gases hydrogen and oxygen are diatomic (H_2 and O_2) and that water is H_2O (and not HO as DALTON believed). However, the law was largely rejected or ignored for 50 years (although AMPÈRE accepted it) until CANNIZZARO in 1860 convinced a Chemical Congress at Karlsruhe of its value.

The SI base unit of amount of substance is the mole (which is related to Avogadro's Law). The mole is defined as containing as many elementary entities (usually atoms or molecules, and specified for each case) as there are atoms in 0.012 kg of carbon-12. Thus for a compound, 1 mole has a mass equal to its relative molecular mass in grams. The number of entities in a mole, the Avogadro constant, N_A, is 6.022×10^{23} mol^{-1}; and 1 mole of any ideal gas, at STP, (standard temperature and pressure) has a molar volume of 22.415 dm^3.

For example: since the relative atomic masses ('atomic weights') of carbon and oxygen are 12 and 16 respectively, a mole of carbon dioxide (CO_2) will weigh $12 + (2 \times 16) = 44$ g, and will have a volume at STP close to 22.4 dm^3.

Ayrton, Hertha, *née* (Phoebe) Sarah Marks (1854–1923) British electrical engineer; the first woman to present a paper to the Royal Society.

Sarah Marks (she later adopted the name Hertha) was the daughter of a Polish Jew who fled to England following persecution under the Tsarist regime and died when she was 7, leaving his widow with six sons and two daughters to care for. Mrs Marks was a strong-minded woman who believed that women needed a better, not worse, education than men because 'women have the harder battle to fight in the world'. Consequently, she took the offer of her sister Marion Hartog to raise and educate her elder daughter Sarah, then 9, at her school in London. The young Sarah was gritty, stubborn, undisciplined and disliked conformity, but she was educated by the talented family she had joined. She learned French from her uncle and mathematics and Latin from her cousin Numa, senior wrangler at Cambridge.

An introduction to Mrs Barbara Bodichon, who became a life-long friend, led to her entry to Girton College, Cambridge in 1876, and she sat the Tripos examination in 1880. At this time women students took the examination unofficially within their college; the names of the successful students were not published, nor were degrees awarded. She went to Finsbury Technical College, intending to follow a career of research and invention, having patented in 1884 an instrument for dividing a line into any number of equal parts.

In 1885 she married W E Ayrton (1847–1908), professor of physics at the college and Fellow of the Royal Society. During a visit to Chicago in 1893 Ayrton lost the only copy of 3 years' work on 'Variation of Potential Difference of the Electric Arc, with Current, Size of Carbons, and Distance Apart' (a servant used it to light a fire) and thereafter he lost any inclination to repeat the work. Hertha Ayrton, who had previously assisted him in the work, began the whole research afresh. She improved the technique, obtained consistent results expressed in curves and equations and published some of the results in the *Electrician* in 1895. She presented papers to the British Association and to the Institution of Electrical Engineers and became recognized as the authority on the electric arc.

She was elected a member of the Institution of Electrical Engineers in 1899, their first female member. In 1900 she spoke at the International Electrical Congress in Paris. In 1901 her paper 'The Mechanism of the Electric Arc' was read to the Royal Society by an associate of her husband, as women were not then permitted to do so. Her book *The Electric Arc* was published in 1902, a history of the electric arc from the time of HUMPHRY DAVY; it became the accepted textbook on the subject. She was proposed for the fellowship of the Royal Society in 1902, but was not accepted because she was a married woman.

In 1904 Ayrton read her own paper 'The Origin and Growth of Ripple Marks' before the Royal Society and became the first woman to do so. In 1906 she received the Hughes Medal for original research on the electric arc and on sand ripples.

From 1905–10 she worked for the War Office and the Admiralty on standardizing types and sizes of carbons for searchlights, both with her husband and, after his death, alone; her suggestions were adopted by the War Office and the Admiralty.

Baade, Wilhelm Heinrich Walter [bahduh] (1893–1960) German–US astronomer: classified stars into different population types; his work gave larger estimates for the size and the age of the universe.

Educated in Germany at Göttingen, Baade was on the staff of the University of Hamburg for 11 years before moving to the USA in 1931. He spent the Second World War at the Mount Wilson and Palomar Observatories studying the Andromeda galaxy (as a German immigrant he was excluded from military service). He used the 100 in telescope and had the advantage of the wartime blackout of Los Angeles, which cleared the night sky. He identified two fundamentally distinct classes of star in the galaxy – hot young blue stars in the spiral arms of the galaxy, which he called Population I stars, and older redder stars in the central region, which he called Population II (see HR diagram, p. 280). This distinction was to prove fundamental to theories of galactic evolution.

He showed that cepheid variable stars found in Andromeda, whose period/luminosity relationship had been discovered 30 years earlier by LEAVITT and quantified by SHAPLEY as a means of calculating their distance, could also be divided into the two categories. In 1952 he demonstrated that Leavitt and Shapley's period/luminosity relationship was only valid for Population I cepheids, and calculated a new relationship for Population II cepheids. HUBBLE, in the 1920s, had used the cepheid variable technique to calculate the distance of the Andromeda galaxy as 800 000 light years, from which he estimated the age of the universe to be 2000 million years. However, Hubble's estimate proved to have depended upon Population II cepheids, for which the original period/luminosity relationship was invalid; using his new relationship Baade showed that Andromeda was more than 2 million light years away and that the universe was therefore at least 5000 million years old. (This revised time scale came as a relief to geologists, who had estimated the age of the Earth as 3000–4000 million years or more.)

Baade also discovered two asteroids, Hidalgo and Icarus, which strangely are those with (respectively) orbits which take them farthest and nearest to the Sun of all known asteroids. He also worked on supernovae and the optical identification of radio sources.

Babbage, Charles (1791–1871) British mathematician and computer scientist: inventor of the programmable computer.

As the talented child of affluent parents, Babbage entered Cambridge in 1814 to study mathematics. He and his friend JOHN HERSCHEL put effort into spurring their teachers to achieve a better standard in mathematics teaching, translating continental textbooks for their use and advocating LEIBNIZ's calculus notation rather than NEWTON's. Babbage became professor of mathematics in Cambridge in 1828, and worked on the theory of functions and on algebra. However, he soon concerned himself with the poor quality of the mathematical tables then available, which were rich in errors. He was convinced that mechanical calculation could give error-free results; the subject became obsessive for him and was ultimately to change him from a sociable young man into an irascible elder who clashed even with the street musicians whose activities, he claimed, 'ruined a quarter of his work potential'.

After making a small-scale mechanical calculator in 1822, Babbage designed his 'Difference Engine No 1', which was to perform arithmetical operations using toothed wheels. Hand-powered, it was to work on decimal principles – the binary system is logically linked with electronics and was yet to come. Over 10 years later, in 1833, the project was abandoned with only 12 000 of the 25 000 parts made and the then large sum of £17 470 expended. AIRY, the Astronomer Royal, pronounced the project 'worthless' and the Government withdrew its support. A modest section of the engine with about 2000 components, made as a demonstration piece in 1832, works impeccably to this day and is the first known automatic calculator.

Babbage promptly began to design a more advanced 'Analytical Engine' and worked on this until his death. It was to have a punched card input, a 'store' and 'mill' (equivalent to the memory and processor in a modern computer) and would give a printed, punched or plotted output. Construction of its 50 000 geared wheels, to be mounted on 1000 vertical axles, never began in Babbage's time. In the late 1840s he also planned a simpler and more elegant calculator ('Difference Engine No 2') to work with numbers up to 31 digits, but could not get government support for either machine.

In 1985 the Science Museum in London began to build, in public view, Difference Engine No 2. Its 4000 bronze, cast iron and steel components were assembled to make the 3 ton machine with only modest changes in Babbage's design. It was completed in time for the Museum's exhibition commemorating his 200th birthday; its first full-scale calculation was to form the first

100 values in the table of powers of 7, and it has operated without error ever since then. It cost nearly £300 000. Its automatic printer is still to be built.

Babbage worked not only in mathematics and statistics, but also on cryptology, climatology, tree-rings as historic climatic records, and the theory of what we would now call mass production and operational research; but his main claim to fame is as the central ancestral figure in the history of computing. He was assisted in his work by Ada King, Countess of Lovelace (1815–52), the daughter of the poet Byron, who spent much time assisting him and publicizing his work; the best account of Babbage's views on the general theory of his 'engines' is due to her and the US Defense Department programming language ADA is named for her. The two of them also tried to devise a system for predicting the winners of horse races (she was a fearless horsewoman) and lost money in the process.

Babcock, Horace Welcome (1912–) US astronomer: made first measurements of stellar magnetic fields.

Horace Babcock was the son of Harold Delos Babcock (1882–1968), also an astronomer, in collaboration with whom his most profitable work was done. Both worked at the Mount Wilson Observatory, Horace as director from 1964–78. It had been known since 1896 that some spectral lines are 'split' in the presence of strong magnetic fields (the ZEEMAN effect), and in 1908 HALE had shown that light from sunspots is split in this way and that magnetic fields of up to 0.4 T (tesla) in strength must be present in sunspots. A generalized solar magnetic field could not, however, be detected at that time.

In 1948 the Babcocks developed equipment for measuring the Zeeman splitting of spectral lines far more precisely than had hitherto been possible. This allowed them to detect the Sun's magnetic field, which is about 10^{-4} T in strength. They discovered that the Sun's magnetic poles periodically flipped polarity, and went on to measure the magnetic fields of many other stars. Some of these were found to be 'magnetic variables', their field strength varying by several tesla over periods as short as a few days.

Backus, John (1924–) US computer scientist: developed first high-level computer language.

Born in Philadelphia and educated at Columbia, Backus was closely associated with IBM for much of his career.

The Second World War gave a great stimulus to the development of electronic computers, but until the early 1950s they still had to be programmed in a very basic fashion. Backus demonstrated the feasibility of high-level computer languages, in which a problem could be expressed in a readily understandable form, which was then converted into the basic instructions required by the computer via a 'compiler'. In 1954 he published the first version of FORTRAN (FORmula TRANslator) and by 1957 it was commercially available for use on IBM computers. High-level languages have greatly aided the use of computers in solving scientific problems, and FORTRAN itself remains the most widely used scientific programming language.

Bacon, Francis, Viscount St Albans (1561–1626) English statesman and natural philosopher: advocate of inductive method in science.

Son of a statesman and courtier, Bacon was trained in law to follow the same path; with much effort and little scruple, he succeeded and held office under James I, finally becoming Lord High Chancellor in 1618. Convicted of taking bribes, he was banished from Court and office in 1621.

His views of scientific method were influential and were expressed in a series of books and essays. He criticized ARISTOTLE and the deductive method and advocated 'induction', in which emphasis is on the exhaustive collection of scientific data (with careful choice and the exclusion of extraneous items) until general causes and conclusions emerge almost mechanically; Bacon was antagonistic to imaginative speculation. His ideas were certainly influential in science and probably even more in philosophy. His personality was unattractive and his writings abstruse, but his confidence that nature could be understood and even controlled was important, and as a critic and a prophet his role in the scientific development of the following centuries is significant. His own direct scientific work was limited; the best example is his conclusion on the nature of heat, which by argument and thought-experiments he decided was 'an expansive motion restrained, and striving to exert itself in the smaller particles'. Ahead of his time, he suggested in The Great Instauration (1620) that marine science be developed, including study of 'the Ebbs and Flows of the Sea ... its Saltness, its various Colours, its Depth; also of Rocks, Mountains and Vallies under the Sea and the like'.

Bacon, Roger (c.1214–92) English philosopher and alchemist: supporter of experimental method in science.

Probably a member of a wealthy family, Bacon studied at Oxford under Robert Grosseteste (c. 1175–1253) and in Paris, and joined the Franciscan Order as a monk about 1247. He was not himself an experimentalist nor a mathematician (although he did some work in optics), but he saw that these two approaches were needed for science to develop; and he foresaw a control of nature by man, as his namesake FRANCIS BACON was also to foresee 350 years later. He made imprecise predictions about mechanical transport – on land, above and below the surface of the sea, and in the air, circumnavigation of the globe and robots. He had a wide knowledge of the science of the time, together with alchemy, and was thought to have magic powers. He knew of gunpowder, but did not invent it. He saw theology as the supreme area of knowledge but his difficult personality led to conflict with his colleagues. Among 13th-c thinkers, his attitude to science is nearest to that of the present day.

Baekeland, Leo Hendrik [baykland] (1863–1944) Belgian–US industrial chemist: introduced Bakelite, the first widely used synthetic plastic.

Baekeland became an academic chemist in his native Ghent, but a honeymoon visit to the USA led him to settle there from 1889, working as an independent consultant. From 1893 he made 'Velox' photographic paper, but sold out to Kodak in 1899. A few years later he studied the already known reaction of phenol C_6H_5OH with methanal, H.CHO. Under suitable conditions the dark solid product is a thermosetting resin, rigid and insoluble. Baekeland manufactured it from 1909, and mixed with fillers as 'Bakelite', it has been much used for moulded electric fittings. It is now known to be a highly cross-linked three-dimensional polymer of high relative molecular mass, consisting largely of benzenoid rings linked by methylene ($-CH_2-$) groups at their 1-, 3- and 5- positions.

Baer, Karl Ernst von [bair] (1792–1876) Estonian embryologist: discoverer of the mammalian ovum.

Baer's wealthy family was of German descent, so it was natural for him to study in Germany after graduating in medicine at Dorpat in Estonia. He taught at Königsberg in Germany from 1817–34, when he moved to St Petersburg. His best-known discoveries, however, were made in Königsberg. There, in 1826, he studied the small follicles discovered in the mammalian ovary by R de Graaf (1641–73) in 1673, and named after him; they had often been assumed to be mammalian eggs. Baer showed that the Graafian follicle of a friend's bitch contained a microscopic yellow structure which was the egg (ovum). He identified structures within the embryo (the fertilized and developing egg), including the notochord, a gelatinous cord which develops into the backbone and skull in vertebrates, and he found the neural folds (which later form the central nervous system). In 1817 C H Pander (1794–1865) had noted three layers of cells in the vertebrate embryo, which were to be named by REMAK in 1845 ectoderm (outer skin) mesoderm (middle skin) and endoderm (inner skin). These 'germ layers' each develop into specialized organs later (eg the mesoderm forms muscles and bones); Baer emphasized that the embryos of various species are at first very similar and may not be distinguishable and that, as it develops, the embryo of a higher animal passes through stages which resemble stages in the development of lower animals. This idea was later to be fruitful in embryology and in evolution theory.

Baer led expeditions to Arctic Russia to collect plant and animal specimens, studied fishes and collected human skulls (in 1859 he suggested that human skulls might have a common ancestral type, but he never supported DARWIN's ideas). His fame rests on his position as a founder of modern embryology.

Baeyer, Adolf von [biyer] (1835–1917) German organic chemist: master of classical organic synthesis.

Baeyer's life spanned a period of rapid change in science and technology; from FARADAY's laws of electrolysis to X-ray crystallography and from the first rail services to regular air transport. His father was a Prussian soldier who became a general. The boy was a keen chemical experimenter, which prompted a poet visiting the family to write a verse on the dreadful smells he caused. When Baeyer was 12 he made his first new substance, the beautiful blue crystalline carbonate $CuNa_2(CO_3)_2.3H_2O$; and he celebrated his 13th birthday by buying a lump of the bronze-purple dye indigo.

After his military service in 1856 he went to study chemistry in Germany's best-known laboratory, that of BUNSEN in Heidelberg. However his interest soon focused on the organic side, which Bunsen had given up, and so he joined KEKULÉ as his first research student. His first independent work was done during 12 years spent teaching organic chemistry in a small Berlin technical college. He moved from there to Strasbourg and then to Munich, working there for 40 years.

Baeyer was a hugely talented organic chemist with an instinctive feel for structures and reactions. He was an experimenter who saw theory as a tool which was easily expendable after use: he wrote 'I have never planned my experiments to find out if I was right, but to see how the compounds behave'. His preference was for simple equipment, mainly test-tubes and glass rods; he was suspicious even of mechanical stirrers. He had no superior as an organic chemist in his Munich period and all the best men in the field worked with him.

His successes included the structure and synthesis of indigo. His work on the purine group began with studies on uric acid and included the synthesis of the useful barbiturate drugs (named, he said, after a lady friend named Barbara). Other work dealt with hydrobenzenes, terpenes and the sensitively explosive polyalkynes. It was in connection with the latter that he devised his strain theory to account for the relative stabilities of carbocyclic rings, which in modified form is still accepted. Absent-minded and genial, he was very popular with his students. He won the Nobel Prize in 1905.

Baily, Francis (1774–1844) British astronomer: discovered Baily's beads.

Baily was a stockbroker and amateur astronomer. He observed the phenomenon known as Baily's beads seen during total solar eclipses where, for a few seconds just before and after totality, brilliant beads of light are seen around the edge of the Moon. These are caused by the Moon's irregular surface allowing rays of sunlight to shine fleetingly down suitably aligned lunar valleys. Baily observed the effect during the solar eclipse of 1836.

Baird, John Logie (1888–1946) British electrical engineer: television pioneer.

Son of a Presbyterian minister, Baird was educated in Glasgow, almost completing a course in electrical engineering. His poor health made a career difficult and several ventures failed, including making and selling foods, boot-polish and soap. After a serious illness in 1922 he devoted himself to experimentation and developed a crude TV apparatus, able to transmit a picture and receive it over a range of a few feet. The first real demonstration was within two attic rooms in Soho in early 1926. In the following year he transmitted pictures by telephone line from London to Glasgow and in 1928 from London to New York. In 1929 his company gave the first BBC TV transmissions, soon achieving daily half-hour programmes with synchronized sound and vision. He used a mechanical scanning system, with 240 lines by 1936, but then the BBC opted to use the Marconi–EMI electronic scanning system, with 405 lines. Baird also pioneered colour, stereoscopic and big screen TV, and ultra-short-wave transmission. Television has no single inventor, but to Baird is due its first commercial success, although his methods have largely been replaced.

Balmer, Johann Jakob [balmer] (1825–98) Swiss mathematician: discovered relationship between hydrogen spectral lines.

Son of a farmer, Balmer studied in Germany and, from 1850, taught in a girls' school in Basle. Rather late in life he became interested in spectra and reported his first research when aged 60. The lines in the Sun's spectrum had earlier seemed to be randomly scattered, but KIRCHHOFF had shown that, if the spectrum of an individual element was considered, this was not so; for example, the spectrum of hydrogen consists of lines which converge with diminishing wavelength, λ. Balmer found in 1884 that one set of the hydrogen lines fitted the relation $\lambda = A\, m^2/(m^2-4)$ where m has integral values 3,4,5... for successive lines, and A is a constant. This is the Balmer series; originally empirical, it pointed to the need to find an explanation for the data, which led through RYDBERG's work to BOHR's theory and to quantum theory.

Baltimore, David (1938–) US molecular biologist: discovered reverse transcriptase enzyme.

Baltimore's interest in physiology was initiated by his mother (a psychologist) when he was a schoolboy. However, he studied chemistry at Swarthmore and later at the Massachusetts Institute of Technology and Rockefeller University; afterwards he moved into virology, and in 1972 he became professor of biology at MIT and later director of the Whitehead Institute at Cambridge, MA. In 1968 Baltimore showed how the polio virus replicates, with some detail on how its RNA core and protein coat are formed. In 1970 he announced his discovery of the enzyme 'reverse transcriptase', which can transcribe RNA into DNA and does so in some tumour viruses. This was a novel finding; the 'central dogma' of molecular biology, due to CRICK, is the scheme: DNA→RNA→protein, in which the first arrow is designated transcription, and the second translation. Before Baltimore's work it had been assumed that the converse of transcription did not occur. Baltimore shared a Nobel Prize in 1975 with H M Temin (1934–94) who had independently discovered the same enzyme.

Baltimore became president of Rockefeller University in 1990 and combined administration with fundraising and research. Then a report from the National Institutes of Health alleged that one of his co-authors in a paper published in *Cell* in 1986 had used falsified data. Baltimore first defended his colleague, but later changed his position on this and apologized to the 'whistle blower' who had first raised the question of research ethics in this matter and who had faced antagonism as a result.

Baltimore's position as president became difficult and he resigned in late 1991, continuing in full-time research in Rockefeller University. The US Secret Service carried out forensic work on the papers held to contain false data, but failed to convince all parties that fraud was proved, in face of the difficulties involved in a very complex case.

Banks, Sir Joseph (1743–1820) British naturalist and statesman of science.

Educated at Harrow, Eton and Oxford, Banks was wealthy and able to indulge his interest in science; he was a passionate and skilful botanist and this took him on several major expeditions at his own expense. The best known of these began in 1768; young Banks had learned that COOK was to sail to the south Pacific to observe the transit of Venus in 1769 and realized this would be a great opportunity to see entirely new plants and animals. He joined the expedition, which lasted 3 years, with his staff of eight, and returned with a large collection of new specimens to find himself a celebrity. The voyage was the first to be organized and equipped for biological work, even though the Government's secret plan was political – to secure a territorial advantage over the French. Banks brought back 1300 new plant species, as well as the idea that Botany Bay would form a suitable penal settlement.

He became president of the Royal Society in 1778 and held the post for 42 years, as the dominant personality in British science. His successes included the introduction of the tea plant in India (from China) and breadfruit in the Caribbean (after a frustrated first attempt in which HMS *Bounty*, carrying the breadfruit, was diverted by a mutiny). Banks did much to establish the Botanic Garden at Kew, which he planned as a major collecting centre and source of advice on all aspects of plants.

Banting, Sir Frederick (1891–1941) Canadian physiologist: co-discoverer of insulin.

Banting studied in Toronto for the church, but after a year changed to medicine and, after graduation in 1916, joined the Canadian Army Medical Corps, winning an MC for gallantry in action in 1918. After the war he set up a practice in London, Ontario, and also worked part-time in the physiology department of the University of Toronto.

Diabetes mellitus is a disease in which glucose appears copiously in the blood and urine, disturbing the metabolism. It is not curable and until Banting's work it was always fatal. It was known that the disease was linked to failure of the pancreas and probably to the cells in it known as the islets of Langerhans. In 1921 Banting devised a possible method for obtaining from these islets the unknown hormone that was suspected of controlling glucose levels and whose absence would cause the disease. J J R Macleod (1876–1935), professor of physiology at Toronto, was not impressed by the research plan or by Banting's skills, but eventually gave him the use of a university laboratory, experimental dogs and a recently qualified assistant, C H Best (1899–1978), to try the method while Macleod himself went on holiday. In 1922, after 8 months' work, they announced their success. Extracts of a hormone (insulin) were obtained, and with the help of a chemist, J B Collip (1892–1965), these extracts were purified sufficiently to inject and treat diabetic patients. The effect was dramatic, and since 1923 millions of diabetics have led manageable lives using insulin to control their glucose levels. Industrial production of insulin (from pig pancreas) began in 1923.

In 1923 a Nobel Prize was awarded to Banting and Macleod. Banting was furious at the omission of Best and shared his half-prize with him; Macleod shared his with Collip. Banting became a professor at Toronto. When the Second World War began he joined an army medical unit and researched on war gases, but was killed in an air crash in Newfoundland. In 1926 insulin was isolated in pure form, but it was a generation later before SANGER deduced its chemical structure and 1966 before it was made by synthesis; it is a protein molecule built of 51 amino acid units.

Insulins from different mammalian species differ in one or more amino acid units. For diabetics, insulin from pigs or oxen has long been used but from 1978 human insulin has been available by a genetic engineering process based on the bacterium *Escherichia coli*.

Bardeen, John [bah(r)deen] (1908–) US physicist: co-inventor of the transistor and contributor to the BCS theory of superconductivity.

Bardeen came from an academic family and studied electrical engineering at the University of Wisconsin. He worked as a geophysicist for 3 years at the Gulf Research Laboratories before obtaining a PhD in mathematical physics at Harvard under WIGNER in 1936. Following periods at the University of Minnesota and the Naval Ordnance Laboratory, Bardeen joined a new solid-state physics group at Bell Telephone Laboratories at the end of the Second World War. His major creative work then began, and continued after his move from Bell to a professorship at the University of Illinois in 1951. Bardeen together with BRATTAIN and SHOCKLEY received the Nobel Prize for physics in 1956, for the development of the point-contact transistor (1947). He won the Nobel Prize again in 1972, shared with COOPER and SCHRIEFFER, for the first satisfactory theory of superconductivity (1957), now called the BCS theory. Bardeen thereby became the first man to receive the Nobel prize for physics twice.

Superconductivity was discovered in 1911 by KAMERLINGH-ONNES. A metal brought into this state by low temperature (< 15 K) expels magnetic field and will maintain electric currents virtually indefinitely (it shows zero resistance). Work in 1950 had revealed that the critical temperature is inversely proportional to the atomic mass of the metal, and Bardeen inferred that the oscillations of the metal lattice must be interacting with the metal conduction electrons. Cooper (1956) at the University of Illinois showed that electrons can weakly attract one another by distorting the metal lattice around them, forming a bound pair of electrons (Cooper pair) at low temperature when thermal vibrations are much reduced. Bardeen, Cooper and Schrieffer then assumed that a co-operative state of many pairs formed and that these pairs carried the superconducting current. The members of a pair have a common momentum and the scattering of one electron by a lattice atom does not change the total momentum of the pair, so that the flow of electrons continues indefinitely.

The BCS theory not only greatly revived interest in superconductivity but showed how quantum theory can give rise to unusual phenomena even on a macroscopic scale.

Barkhausen, Heinrich Georg [bah(r)khowzen] (1881–1956) German physicist: developed early microwave components.

Barkhausen moved from his studies at Bremen and Göttingen to Dresden, where he became professor of electrical engineering. His early research established the theory of the amplifier valve (1911), and he went on to discover the Barkhausen effect (1919). This is the discontinuous way in which the magnetization of a piece of ferromagnetic material rises under an increasing applied field. It occurs because a ferromagnet is made up of many magnetic domains and these change direction or size in a sudden manner.

His work on ultra-high-frequency oscillators and early microwave components was done in 1920 with K Kurz, and was rapidly developed for military radar during the Second World War.

Barlow, Peter (1776–1862) British mathematician.

Self-educated, Barlow taught mathematics at the Royal Military College, Woolwich, from 1801. Of his books on mathematics, the best known is *Barlow's Tables* (1814) which gives the factors, squares, cubes, square and cube roots, reciprocals and hyperbolic logarithms of all integers from 1 to 10 000. Remarkably accurate, it was familiar to generations of students and was in print until the 1950s. He also worked on magnetism and devised a method for correcting ships' compasses for deviation due to iron in the ship's structure, by use of an iron plate suitably positioned. The 'Barlow lens' is a negative achromatic combination of flint and crown glass used to produce magnification of a photographic or telescopic image. Used in addition to a camera lens as a 'telescopic converter' it will give a magnification of 2× or 3× with acceptably little aberration. A similar use of a Barlow lens is between the objective and eyepiece of a telescope, where again it increases magnification by extending the focal length of the main lens.

Barnard, Edward Emerson (1857–1923) US astronomer: discovered Amalthea and Barnard's star.

Despite a background of poverty and poor schooling, Barnard became a professional astronomer with great skill as an observer; he discovered a variety of interesting celestial objects. By the time he was 30 he had found more than 10 comets, and in 1892 he discovered Amalthea, the first new satellite of Jupiter to be discovered for nearly three centuries. In 1916 he discovered the star with the largest proper motion, a red star 6 light years away which moves across the sky at 10.3" of arc per year and which is now known as Barnard's star. With M Wolf (1863–1932), he showed that 'dark nebulae' were clouds of dust and gas.

Barr, Murray Llewellyn (1908–) Canadian geneticist.

Working in 1949 with a research student in the medical school at London, Ontario, Barr found that a characteristic small mass of chromatin can be detected in the nuclei of the nerve cells of most female mammals, but it is absent in the males. So this 'Barr body' provides the marker in a simple test for the sex of an individual; previously sex could be detected at cell level only by examining chromosomes in dividing cells. It allows the sex of a fetus to be found long before birth, which is valuable if a parent carries a sex-linked genetic disorder. Barr also devised methods, using a smear of cells from a patient's mouth, to locate some chromosomal defects, such as certain types of hermaphroditism.

Bartholin, Erasmus [bah(r)tohlin] (1625–98) Danish mathematician: discovered double refraction of light.

Bartholin qualified in medicine in Leiden and Padua (his father and brother were both distinguished anatomists) and he taught medicine and mathematics at Copenhagen from 1656. His pupils included RÖMER and Prince George, who married the British Queen Anne. In 1669 he described in a book his study of the crystals of Iceland spar (a form of calcite, $CaCO_3$) including his discovery that it produces a double image of objects observed through it. He realized that the crystals split a light ray into two rays by what he called ordinary and extraordinary refraction. He gave no theory of this double refraction, which much puzzled other physicists; HUYGENS argued that the effect supported the wave theory of light, rather than NEWTON's idea that light consisted of particles. In the early 19th-c, work by E L Malus (1775–1812) and FRESNEL on polarized light made double refraction easier to understand.

Bartlett, Neil (1932–) British–American inorganic chemist; prepared first noble gas compounds.

Bartlett studied in Newcastle upon Tyne and later worked in Canada and the USA. Although it had been previously accepted that the noble gases were not chemically reactive (the valence theory of chemical bonding being in accord with this), Bartlett used platinum hexafluoride, PtF_6, a highly reactive compound, to prepare xenon hexafluoroplatinate $Xe^+[PtF_6]^-$ (he had shown in 1961 that blood-red PtF_6 combined with oxygen to give a red salt, $O_2^+PtF_6^-$). Since 1962 many other compounds, mainly of krypton and xenon, have been made using the noble gases.

Barton, Sir Derek (Harold Richard) (1918–) British organic chemist: distinguished for work on stereochemistry and organic natural products.

Educated at Imperial College, London, Barton returned there as professor for over 20 years and in 1985 became professor at Texas A & M University. In 1950, he deduced that some properties of organic molecules depend on their conformation: that is, the particular shape adopted by a molecule as a result of rotations about single carbon–carbon bonds. The study of these effects (conformational analysis) is applied mainly to six-membered carbon rings, where usually the conformers easily convert into each other; this interconversion is not easy if the rings are fused, eg as in steroids. Reactivity can be related to conformation in many such molecules. Barton studied many natural products, mainly phenols, steroids and antibiotics. He won a Nobel Prize (with O Hassell (1897–1981), who also studied six-membered carbon rings), in 1969.

Bary, (Heinrich) Anton de see **de Bary**

Basov, Nikolai (Gennediyevitch) [basof] (1922–) Soviet physicist: invented the maser and laser.

After service in the Red Army during the Second World War, Basov studied in Moscow and obtained his doctorate in 1956. Remaining there, he became head of his laboratory in 1962.

From 1952 onwards Basov developed the idea of amplifying electromagnetic radiation by using the relaxation of excited atoms or molecules to release further radiation. His colleague A M Prokhorov (1916–) had studied the precise microwave frequencies emitted by gases, and together they produced (1955) molecular beams of excited molecules that would amplify electromagnetic radiation when stimulated by incident radiation. Such a device is known as a maser (microwave amplification by stimulated emission of radiation). The 1964 Nobel Prize for physics went to Basov, Prokhorov and Townes (who did similar independent work in the USA) for the invention of the maser.

Rather than selecting excited molecules from a beam, Basov and Prokhorov found a way of using a second radiation source to 'pump' the gas into an excited state (the 'three-level' method). Basov then invented the laser (light amplification by stimulated emission of radiation; 1958) and even achieved the effect in semiconductor crystals. He afterwards worked on the theory of laser production in semiconductors, on pulsed lasers and on the interaction of light with matter.

Bassi, Laura (Maria Catarina) [basee] (1711–78) Italian physicist; the first female professor of physics at any university.

Laura Bassi was born in Bologna, the daughter of a lawyer, and was educated at home by the family physician, who was a professor at the university and a member of the Academy of the Institute for Sciences (Istituto delle Scienze). Bassi was instructed in mathematics, philosophy, anatomy, natural history and languages and news of her remarkable ability spread. She became an object of curiosity and was pressed to appear in public. In March 1732 she was elected to the Institute Academy and a month later engaged in a public debate with five scholars of the university and was awarded a degree from the University of Bologna. Bassi was given an official position at the university, but the Senate attempted to restrict her appearances as a lecturer to ceremonial public events and the

social circles of the city. She continued to study mathematics and gave private lessons at her home, while successfully petitioning for wider responsibilities and a higher salary to cover the cost of equipment for physical and electrical experiments. She was one of the first scholars to teach Newtonian natural philosophy in Italy. At the age of 65 she was appointed to the Chair of Experimental Physics at Bologna. She married and had eight children, five of whom survived childhood.

Bateson, William (1861–1926) British geneticist: a founder of genetics.

Bateson was described as 'a vague and aimless boy' at school and he surprised his teachers by getting first-class honours in science at Cambridge in 1883. He then spent 2 years in the USA. He returned to Cambridge, taught there and in 1910 became director of the new John Innes Institution. From the time of his US visit he was interested in variation and evolution, and by 1894 he had decided that species do not develop continuously by gradual change but evolve discontinuously in a series of 'jumps'. To support his view against opposition, he began breeding experiments, unaware of Mendel's work of 1866. When the latter was rediscovered in 1900, Bateson saw that it gave support for his 'discontinuity' theory and he translated and publicized Mendel's work and extended it to animals by his own studies on the inheritance of comb shape in fowls. He showed that Garrod's work on human inborn errors of metabolism had a Mendelian interpretation. He also found that some genes can interact; so that certain traits are not inherited independently, which is in conflict with Mendel's laws. This interaction results from 'linkage', that is genes being close together on the same chromosome, as Morgan and others showed. Bateson coined the word 'genetics' but he never accepted the ideas of natural selection or chromosomes.

Beadle, George Wells (1903–89) US geneticist: pioneer of biochemical genetics.

Born on a farm at Wahoo, NE, Beadle first planned to return there after graduation, but became an enthusiast for genetics and was persuaded to work for a doctorate at Cornell on maize genetics. In 1935 he worked with B Ephrussi in Paris on the genetics of eye-colour in the fruit fly *Drosophila*; as a result of this work, ingeniously transplanting eye buds in the larvae, they suspected that genes in some way controlled the production of the eye pigment. When he returned to the USA, to a job at Stanford, he met the microbiologist E L Tatum (1909–75), and in 1940 they decided to use the pink bread fungus *Neurospora crassa* for a study of biochemical genetics. It grew easily, reproduced quickly and has an adult stage which is haploid (only one set of chromosomes) so that all mutant genes show their phenotypic expression. (*Drosophila*, like other higher organisms, has two genes for every character, so dominant genes can mask recessives). Beadle and Tatum exposed *Neurospora* to X-rays to produce mutations and then examined the mutant strains to find their ability or inability to synthesize a nutrient needed for their own growth. They concluded that the function of a gene is to control production of a specific enzyme; they did not know that GARROD had reached the 'one gene-one-enzyme' idea 30 years earlier by studying human metabolic disease. The value of their work was in providing an experimental method that allowed biochemical genetics to develop. It did so speedily, and their central idea remains unchallenged. More precisely, we would now say that one functional unit of DNA controls the synthesis of one peptide chain. Beadle, Tatum and LEDERBERG shared a Nobel Prize in 1958.

Beaufort, Sir Francis [bohfert] (1774–1857) British hydrographer; inventor of Beaufort wind scale.

Born in Ireland, Beaufort joined the Royal Navy at an early age and saw active service for over 20 years. In 1806 he proposed the Beaufort wind scale, ranging from 0–12, and specifying the amount of sail that a ship should carry in each situation. Thus force 12 was a wind 'that no canvas can withstand'. He also devised a useful concise notation for meteorological conditions in general. The scale was officially adopted by the Admiralty some 30 years later. Beaufort became hydrographer to the Royal Navy and retired as a rear admiral.

Beaumont, William [bohmont] (1785–1853) US surgeon; made pioneer studies of human digestive physiology.

Beaumont was a farmer's son who became a village schoolmaster and later qualified in medicine. In the War of 1812 he became an army surgeon, rather minimally licensed to practise on the basis of his 2 years spent as an apprentice to a country doctor.

In 1822 at Fort Mackinac a young Canadian trapper was accidentally shot by a duck gun at close range, producing gross abdominal wounds and an opening into the stomach. Beaumont was nearby, saved his life, and tended him for 2 years. He was left with a permanent fistula (opening) into the stomach. Beaumont employed him and for 10 years was able to study digestion rather directly. Gastric juice could be obtained; and the lining of the stomach could be examined easily, and its movements, and the effects of different diets and emotions. Beaumont's 238 observations gave a firm basis to the physiology of gastric digestion (they also much exasperated the trapper). The work also suggested to BERNARD the value of artificial fistulas in experimental physiology, using animals. The trapper lived to be 82, greatly outliving his surgeon.

Beche, Sir Henry Thomas de la *see* **de la Beche**

Beckmann, Ernst Otto (1853–1923) German chemist: discovered a rearrangement reaction and a method for determining relative molecular mass in solution.

Beginning as an apprentice pharmacist, Beckmann turned to chemistry with success, being professor at three universities before being appointed first director of the Kaiser Wilhelm Institute for Chemistry at Berlin-Dahlem in 1912. His distinction began in 1886, when he discovered the Beckmann rearrangement – the reaction of ketoximes with acid reagents to give a substituted amide, often in high yield:

$$RR'C = NOH \rightarrow R'CONHR$$

The reaction has been used to prepare some amides, and also in studies on stereochemistry and on reaction mechanism. Beckmann's work led him to seek a general method for finding the relative molecular mass of a reaction product; he devised a method, using RAOULT's Law, by measuring the rise in boiling point of a solvent caused by dissolving in it a known amount of the substance whose molecular mass is required. To measure this small temperature rise he devised the Beckmann thermometer, which has a reservoir for adjusting its range and will measure accurately a small rise in temperature.

Becquerel, (Antoine) Henri [bekuhrel] (1852–1908) French physicist: discoverer of radioactivity.

Like his father and grandfather before him, Becquerel studied physics, and like them he was interested in fluorescence; he also succeeded to the posts they had held in Paris. Educated mainly at the École Polytechnique, he became professor of physics there in 1895. Partly by chance, he found in 1896 that a uranium salt placed on a wrapped photographic plate caused this to blacken. He soon found that this did not require light; that it was due to the uranium only; and that the radiation was not reflected like light. He found it was able to ionize air. Although similar to the X-rays discovered in 1895 by RÖNTGEN, it was not the same. His work was soon confirmed and was the starting point for all studies on radioactivity. He shared the Nobel Prize for physics in 1903 with the CURIES. His other studies, on magnetic effects and on light absorption by crystals, were valuable; but his work on radioactivity gave physics a new direction.

Radiotherapy (later used to treat cancer) began with his observation that radium carried in his pocket produced a burn. The SI unit of radioactivity is the becquerel (Bq), defined as an activity of one disintegration per second.

Beddoes, Thomas (1760–1808) British physician and chemist: mentor of Humphry Davy.

A man of wide talents, Beddoes studied classics, modern languages, science and medicine at Oxford and in 1788 was appointed reader in chemistry there. However, his sympathy with the French revolutionaries led to his resignation in 1792. He then turned to medicine, and linked this with his interest in the new gases ('airs') discovered in the previous few years, several by his friend PRIESTLEY. With the help of friends he set up his Medical Pneumatic Institution in Bristol to study the therapeutic uses of gases. In 1798 he appointed the 19-year-old DAVY to join him. A year later they observed the anaesthetic potential of 'nitrous oxide', N_2O (unhappily neglected for half a century). Beddoes then guided Davy in his early work on electrochemistry, as well as introducing him to influential friends in science and in literature. In 1801 Davy left for the Royal Institution and soon Beddoes also left for London and returned to medical practice. Beddoes's greatest discovery was Davy (as Davy's was FARADAY) but, although much overshadowed by his pupil, Beddoes's own talents, probably partly unused, were real. His Institution was perhaps the first specialized institute of a type now common.

Bednorz, (Johannes) Georg [bednaw(r)ts] (1950–) Swiss physicist: co-discoverer of a new class of superconductors.

Nobel prizes have usually been awarded many years after the work which led to them; but the Prize won by Bednorz and K A Müller (1927–) of the IBM Zürich Research Laboratory at Rüschlikon in 1987 followed quickly on their work on novel electrical superconductors. Superconductivity, the absence of resistance shown by some metals near 0 K, had been observed by KAMERLINGH-ONNES in 1911, and a theory for it was devised by BARDEEN and others (the BCS theory) in 1957. The effect was seen to be of immense value in electronic devices if materials could be found in which it occurs above, say, 77 K (the boiling point of liquid nitrogen, an easily obtainable temperature). In 1986, Bednorz and Müller showed that a mixed-phase oxide of lanthanum, barium and copper superconducted above 30 K, much above any previous temperature for this effect. A special meeting of the American Physical Society in New York in 1987 on superconductivity became known as 'the Woodstock of physics' and oxides of the type M-Ba-Cu-O (with M a rare earth metal, usually lanthanum or yttrium) were then announced which showed superconductivity up to 90 K.

Bednorz graduated at Münster in 1976 and worked for his doctorate under Müller at IBM Zürich, where he had joined the research staff in 1982.

Beebe, (Charles) William (1887–1962) US naturalist: pioneer of deep-sea exploration.

Graduating from Columbia (New York) in 1898, Beebe's first interest was in ornithology and he joined the staff of the New York Zoological Society. After service as a fighter pilot in the First World War, he returned in 1919 to direct the Society's Department of Tropical Research. Further work on birds was overtaken by his interest in deep-sea exploration. In his 'bathysphere' he reached a record depth of about 1000 m near Bermuda in 1934 and later went even lower. He found that light was absent below 600 m, and discovered previously unknown organisms at these depths. A A Piccard (1884–1962) later went even deeper.

Beer, Sir Gavin Rylands de see de Beer

Behring, Emil (Adolf) von [bayring] (1854–1917) German bacteriologist: co-discoverer of diphtheria antitoxin.

Behring studied at Berlin and after qualifying in medicine joined the Army Medical Corps. In 1889 he became assistant to KOCH, and from 1895 he was professor of hygiene at Marburg. It was already known that the bacteria causing tetanus produced a chemical toxin that was responsible for most of the illness of the patient; the toxin could be obtained from a culture. In 1890 Behring worked with KITASATO and showed that blood serum from an animal with tetanus could, if injected into other animals, give them a temporary resistance to the disease and so contained an antitoxin. Similar antitoxic immunity was found with diphtheria, then a major killer of children; this part of his work was done with EHRLICH. A diphtheria antitoxin to protect human patients was soon made (best from the blood serum of an infected horse) and found to be protective, and also to be useful for those already having the disease; it was possibly first used on an infected child on Christmas night, 1891, in Berlin.

Behring was awarded the first Nobel Prize in medicine or physiology for this work, in 1901. In 1913 he showed that a mixture of toxin and antitoxin gives more lasting immunity than the antitoxin alone, and later methods for preventing the disease used this method until it in turn gave way to the use of toxoid (which is toxin treated with formalin, introduced by G Ramon (1886–1963) in 1923). Since then, large-scale immunization of young children has given good control over the disease.

Much honoured, Behring ranks high in medical science; but he was always a lone researcher with few pupils, with much of his energy spent in disputes and in his unsuccessful search for a vaccine against tuberculosis.

Beilstein, Friedrich Konrad [bıylshtiyn] (1838–1906) German–Russian encyclopedist of organic chemistry.

A student of organic chemistry under several of the masters of the subject in Germany, Beilstein was lecturer at Göttingen and later professor at St Petersburg. His own experimental researches were modest. He is remembered for his *Handbook of Organic Chemistry* (1881), which formed a substantially complete catalogue of organic compounds. The compilation (in many volumes) has been continued by the German Chemical Society and is of great value to organic chemists.

Beilstein's test for halogen in an organic compound is quick and useful. An oxidized copper wire is coated with the compound and heated in a gas flame. If the flame is coloured blue-green, halogen is probably present. However, some nitrogen compounds give the colour, so the test is only decisive if negative.

Bell, Alexander Graham (1847–1922) British–American speech therapist; inventor of the telephone.

The son and grandson of speech therapists, Bell followed the same interest but he also studied sound waves and the mechanics of speech. He emigrated to Canada in 1870 and moved to the USA in 1871. From 1873 he was professor of vocal physiology at Boston and could experiment on his belief that if sound wave vibrations could be converted into a fluctuating electric current this could be passed along a wire and reconverted into sound waves by a receiver. Success produced the 'telephone', patented by him in 1876, and the start of the AT & T company. Soon EDISON much improved Bell's telephone transmitter. Bell made other improvements in telegraphy, improved Edison's gramophone, worked with LANGLEY and on Curtis's flying machines and founded the journal *Science*.

Bell, Sir Charles (1774–1842) British anatomist and surgeon: pioneer of neurophysiology.

Bell learned surgery from his elder brother John (a distinguished surgeon and anatomist) and at Edinburgh University. He moved to London in 1804 and became well known and liked as a surgeon and lecturer on surgery. He treated wounded from the battles of Corunna and Waterloo. From 1807 he showed that nerves are not single units, but consist of separate fibres within a common sheath; that a fibre conveys either sensory or motor stimuli, but not both (ie it transmits impulses in one direction only); and that a muscle must be supplied with both types of fibre. In this way Bell began modern neurophysiology. His work was as fundamental and as revolutionary as that of HARVEY on the circulation of the blood. Later he discovered the long thoracic nerve (Bell's nerve); and he showed that lesions of the seventh cranial nerve produce facial paralysis (Bell's palsy).

Bell Burnell, (Susan) Jocelyn, *née* Bell (1943–) British astronomer: discoverer of first pulsar.

It is probably no coincidence that Jocelyn Bell's father, a Belfast architect, designed the Armagh Planetarium. She decided, after studying physics at Glasgow, to work for a PhD in radioastronomy with HEWISH at Cambridge. He had built a large radiotelescope there, with 2048 fixed dipole antennas spread over $18\,000\,m^2$. Bell checked the recorders daily, examining the 30 m of chart paper, as part of a survey of radio-emitting quasars (remote quasistellar objects) looking in particular for scintillations due to the solar wind. In mid-1967 she saw a small unusual signal, and she found after a few weeks that it recurred. The radio pulsation had a precise period of over a second, and with some effort was shown not to be an equipment malfunction or to be man-made. The name pulsar was coined, and Bell found a second example late in 1967; over 500 are now known.

The explanation due to GOLD, and now accepted, is that pulsars are neutron stars, small but very massive stars, rapidly spinning. They form at the end of a star's life, before its final collapse to a black hole.

Bell Burnell went on to research in astronomy at London and Edinburgh, and a chair at the Open University.

Beneden, Edouard van [beneden] (1846–1910) Belgian embryologist and cytologist: discovered that the number of chromosomes per cell is constant for a particular species.

Van Beneden followed his father in taking charge of zoology teaching at Liège in 1870. His course of teaching was based largely on his own researches, which he did not publish, but one of his students published them after van Beneden's death. He showed in the 1880s that the number of chromosomes is constant in the cells of an animal body (except the sex cells) and the number is characteristic of the species (eg 46 in each human cell). He worked particularly with the chromosomes in the cell nuclei of an intestinal worm from horses; these chromosomes are conveniently large and few (four in the body cells, two in the sex cells). He found that the chromosome number is not doubled in the formation of the sex cells (the ova and spermatozoa) so that these have only half the usual number (a process called meiosis). When they unite, the normal number is restored, with results in accord with MENDEL's work in genetics. In fact, van Beneden misinterpreted some of his observations, which were clarified by the work of WEISMANN and DE VRIES.

Benedict, Ruth *née* Fulton (1887–1948) US anthropologist.

Ruth Fulton was born in New York City, the eldest daughter of a surgeon. She was 2 years old when her father died and her mother's grief was hysterical, and ritually repeated on every anniversary. Her childhood was spent with her maternal grandparents on their farm near Norwich, NY and later with her aunt.

Fulton graduated at Vassar in 1909, studying philosophy and English literature, moved to California to teach and in 1914 married Stanley Rossiter Benedict. The marriage was not a success; looking for occupation, Ruth Benedict chanced upon anthropology. In 1921 she went to Columbia to study for a doctorate under Franz Boas (1858–1942). From 1923, as lecturer in anthropology at Columbia, despite her partial deafness and acute shyness, she undertook fieldwork among several southwestern tribes of native Americans: the Zuñí, the Cochiti and the Pima.

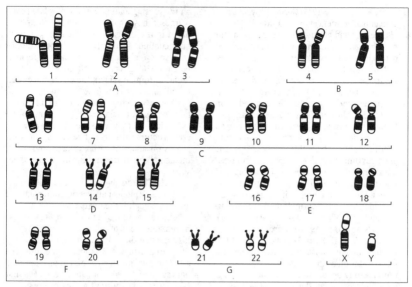

The 46 human chromosomes. The banding, characteristic of specific chromosomes, is developed by staining: magnified about 2500 times.

Her views were presented in *Patterns of Culture* (1934). Benedict saw cultures as 'personality writ large' and psychological normality as culturally defined, so that the misfit becomes one whose disposition is not contained by his culture. She made a plea for tolerance of all 'the co-existing and equally valid patterns of life which mankind has created for itself'. Sharply criticized by some, nevertheless *Patterns of Culture* was a most influential work, translated into 14 languages and frequently reprinted.

Benedict was made associate professor at Columbia in 1936 and served 3 years as head of the department. About this time she joined the protest against racism and intolerance, and published *Race: Science and Politics* (1940).

During the Second World War she moved to Washington as head of the Basic Analysis Section, Bureau of Overseas Intelligence, Office of War Information. With MARGARET MEAD and others, she applied anthropological methods to the study of complex societies and culture, working from documentary materials to make a number of national character studies. From her study of the Japanese she produced *The Chrysanthemum and the Sword* (1946), regarded as one of the best accounts of Japanese culture written by a westerner.

In 1947 Ruth Benedict became president of the American Anthropological Association and in 1948 was made a full professor at Columbia shortly before her death.

Bentham, George (1800–84) British plant taxonomist.

Son of a wealthy naval architect, Bentham became interested in botany at 17. He was trained in law and worked as secretary to his uncle, the philosopher Jeremy Bentham, from 1826–32. Thereafter, his studies in botany took up all his time; in 1854 he gave his herbarium and library to Kew Gardens and worked there for the rest of his life. His *Plant Genera* (3 vols, 1862–83) written with HOOKER has continued to be a standard work for British botanists; but he also wrote other floras, eg the seven-volume *Australian Flora*.

Berg, Paul (1926–) US molecular biologist: discovered first transfer RNA and pioneered recombinant DNA techniques.

Educated in the USA, Berg held chairs from 1970 at both Washington University (St Louis) and Stanford. In 1955 CRICK had suggested that the biosynthesis of proteins from amino acids, under the control of an RNA template, involved an intermediate 'adaptor' molecule. He thought it possible that a specific adaptor existed for each of the 20 amino acids. The next year Berg identified the first adaptor, now called a transfer RNA; it is a small RNA molecule which transfers a specific amino acid, methionine.

Later, Berg developed a method for introducing selected genes into 'foreign' bacteria, thereby causing the bacteria to produce the protein characteristic of the cells from which the genes had been taken. This technique of recombinant DNA technology ('genetic engineering') is of value because it can give a convenient bacterial synthesis of a desired protein such as insulin or interferon. However, it offers the potential danger that novel pathogens might be created, by accident or otherwise, and Berg was influential in warning of this problem. He shared a Nobel Prize in 1980.

Berger, Hans (1873–1941) German psychiatrist: pioneer of electroencephalography (EEG).

Berger studied physics for a year at Jena, but then changed to medicine and later specialized in psychiatry. He worked on the physical aspects of brain function (eg its blood circulation, and its temperature) to try and relate these to mental states; and in 1924 he recorded the electric currents he detected on the exposed brain of a dog. Then he found he could detect currents through the intact skull from his family as well as from patients with brain disorders, and from 1929 he published on this. He described the alpha rhythm (10 cycles per second, from certain areas of the brain at rest) and he recognized that the method could be useful in the diagnosis of diseases of the brain; since then this EEG method has become routinely used in neurological and psychiatric cases, especially since ADRIAN's work from 1934 onwards.

Bergeron, Tor Harold Percival (1891–1977) Swedish meteorologist: explained mechanism of precipitation from clouds.

After studying at Stockholm and Leipzig, Bergeron worked at the Bergen Geophysical Institute with BJERKNES. In 1947 he was appointed professor of meteorology at the University of Uppsala. His principal contribution to the subject was to suggest, in 1935, a mechanism for the precipitation of rain from clouds. He proposed that ice crystals present in the cloud grew by condensation of water vapour on to their surfaces and that at a certain size they fell, melted and produced rain. His ideas were soon borne out by the experimental studies and observations of W Findeisen and are now known as the Bergeron–Findeisen theory.

Bergius, Friedrich Karl Rudolf [bairgeeus] (1884–1949) German industrial chemist: devised conversion process from coal to oil.

Son of a chemical manufacturer, Bergius studied under NERNST and HABER. After 5 years in teaching he worked in the chemical industry from the start of the First World War to the end of the Second World War. His interest in high-pressure reactions of gases developed under Haber. Realizing that petroleum (crude mineral oil) differs from coal in the higher hydrogen content and lower relative molecular mass of the oil, Bergius developed a method (the Bergius process) for the conversion, by heating a mixture of coal dust and oil with hydrogen under pressure, with a catalyst. Hydrogen is taken up, and the product is distilled to give petrol (gasoline). The process was much used in Germany in the Second World War. He also developed industrial syntheses for phenol and ethane-1,2-diol. He shared a Nobel Prize in 1931.

Bergström, Sune Karl [bairgstroem] (1916–   ) Swedish biochemist.

Educated at the Royal Caroline Institute in Stockholm, Bergström returned there as professor of biochemistry in 1958. His interest focused on the prostaglandins, a group of related compounds whose biological effects were first noted in the 1930s. Their effects are complex, but a common feature is their ability to induce contraction of smooth muscle, and their high potency (10^{-9} g can be effective); originally found in human semen, they have since been found in many cells (one rich source is the Caribbean sea whip coral). Bergström first isolated two prostaglandins in pure form, in the 1950s. In 1962 they were shown to have a general structure pattern of a five-carbon ring with chains on adjacent carbon atoms, and much medicinal chemistry has been devoted to them since.

Bernard, Claude [bairnahr] (1813–78) French physiologist: pioneer of experimental medicine and physiological chemistry.

Bernard was the son of vineyard workers and he remained fond of country life; later he spent his time in either a Paris laboratory or, during the harvest, in the Beaujolais vineyards. His schooling was provided by his church, and at 19 he was apprenticed to an apothecary. His first talent was in writing for the theatre, but he was urged to qualify in a profession and chose medicine. He qualified for entry with some difficulty and emerged from his training in Paris as an average student. Then as assistant to MAGENDIE he found his talent in experimental medicine. He never practised as a physician, and an early problem for him was how to make a living. He solved this by marrying a successful Paris physician's daughter and living on the dowry until he succeeded to Magendie's job in 1852. His marriage was unhappy.

Bernard's discoveries were wide-ranging; many depended on his skill in vivisection, using mainly dogs and rabbits. In digestion he showed the presence of an enzyme in gastric juice; the nervous control of gastric secretion and its localization; the change of all carbohydrates into simple sugars before absorption; and the role of bile and pancreatic juice in the digestion of fats. He noted that the urine of herbivores is alkaline and that of carnivores acid, and he pursued the comparisons that these observations suggested. This led him to find that nutrition is complex and involves intermediate stages and synthesis as well as transport. He discovered glycogen, and sugar production by the liver. He studied the nervous system and discovered the vasomotor and vasoconstrictor nerves. Beginning with an attempt to prove LAVOISIER's simple ideas on animal combustion, Barnard showed that in fact the oxidation producing animal heat is indirect, and occurs in all tissues and not simply in the lungs. He studied the action of curare and other paralysing poisons and showed their use in experimental medicine. His approach to research was essentially modern; he combined experimental skill with theory and had a valuable talent for noting experimental results that were not in accord with existing ideas, which led to fruitful new concepts. Perhaps his greatest contribution to physiology was the idea that life is dependent on a constant internal environment (homeostasis); cells function best within a narrow range of osmotic pressure and temperature and bathed in a fairly constant concentration of chemical constituents such as sugars and metallic ions.

Bernoulli, Daniel [bernooyee] (1700–82) Swiss mathematician: pioneer of hydrodynamics and kinetic theory of gases.

This extraordinary family, in the century before and the century after this Daniel's birth, produced 11 substantial mathematicians in four generations. Most of them worked mainly in applied mathematics and analysis, had talents in some other areas, from astronomy to zoology, and quarrelled vigorously with their relatives.

Daniel studied medicine in Switzerland and Germany and qualified in 1724, and published some major work in mathematics in the same year. In 1725 he was appointed professor of mathematics in St Petersburg, but found conditions in Russia primitive and returned to Basle in 1733 as professor of anatomy and botany and, later, of physics. He worked on trigonometry, calculus and probability. His work on hydrodynamics used NEWTON's ideas on force

applied to fluids, and advanced both theory and a range of applications. One of his results (Bernoulli's principle) deals with fluid flow through pipes of changing diameter and shows that pressure in a narrow section (see diagram) is lower than in the wider part, contrary to expectation. A closely related effect leads to the uplift of an aircraft wing; since the distance from leading edge to rear edge is greater over the top of the wing than below it, the air velocity over the top must be higher and therefore its pressure is lower; the result is uplift. Again, when a golf ball is driven off, the loft of the club causes the ball to spin, and the resulting airflow gives it lift so that it has an asymmetric flight, rising in nearly a straight line. As the spin decreases, the lift diminishes and the ball moves into a path like that of a thrown ball.

Bernoulli also proposed a mental model for gases, showing that if gases consist of small atoms in ceaseless rapid motion colliding elastically with each other and the walls of their container, BOYLE's experimental law should result. This was both a very early application of the idea of atoms and the origin of the kinetic theory of gases.

Berthelot, Marcellin (Pierre Eugène) [bairtuhloh] (1827–1907) French chemist: pioneer in organic synthesis and in thermochemistry.

As the son of a Paris physician, Berthelot saw

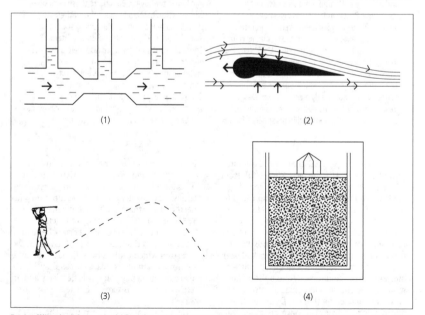

Bernoulli's principle. 1. – Liquid flow through a narrowed tube. 2. – Air flow supporting a wing: air has a longer path above the wing than below it, so it moves faster, and its pressure is lower, above the wing; hence there is an uplift acting on the wing. 3. – Trajectory of a golf shot: the dimpled, spinning ball counteracts gravity for much of its flight. 4. – Diagram from Bernoulli's *Hydrodynamics*, in which kinetic theory is used to show that *pV* is constant for a gas (Boyle's Law). In reality, the average distance between gas molecules is about 300 times the molecular diameter at STP (ie much more than in the diagram in relation to the apparent size of the 'atoms').

NAPOLEON AND SCIENCE

Napoleon Bonaparte (1769–1821) is unique among leading historical figures in many ways, and one of them is that he had an active interest in science. He expended time and effort in gaining personal knowledge of results and problems in the science and technology of the early 19th-c. Only after he became emperor (1804) did he become a mere patron of science, although even in this role he was vastly better informed than other national leaders who patronized science, such as Charles II in 17th-c England or Lenin in 20th-c Russia. In making scientific effort a part of national policy, he broke new ground.

Born in Corsica, in the years following the French Revolution Bonaparte was a young artillery officer; he had entered the École Militaire in 1785 and did exceptionally well in mathematics, which he was taught by MONGE's brother Louis; his examiner in finals was LAPLACE. His speedy military successes in Europe gave him the reputation that allowed him to propose and lead a military expedition to Egypt in 1798, and his political skills secured his election as First Consul in France by 1800. He easily obtained personal rule in this capacity, became emperor in 1804 and established his three brothers as kings in Holland, Germany and Italy. By 1810 his empire was vast, but his invasion of Russia in 1812 was disastrous and defeats in Germany soon followed. His abdication in 1814 and exile in Elba were the result; his return to Europe in 1815 and the 'Hundred Days' only led to his final defeat by Wellington at Waterloo, renewed exile in St Helena and his death there in 1821. During the 14 years of Bonaparte's dominance, France was nearly always at war. Despite this, it was a period of advance for some cultural and practical institutions within his sphere of influ-

ence, including most areas of science, in which France took a leading place, with a commanding position in chemistry and physics in the first decades of the 19th-c.

As a schoolboy and as a young artilleryman Bonaparte had a substantial interest in science. Then, as a very young general commanding the French army in Italy in 1796, he came to know the chemist BERTHOLLET, who had been sent there by the French Government (the *Directoire*) to secure spoils of war, both art treasures and scientific apparatus. The two became great friends, with a liking for one another that was to survive despite future difficulties. The next year a vacancy arose in the limited and exclusive Institut de France, in its First Class (ie the science division). Despite severe competition, the 28-year-old Napoleon was elected, and took a real part in its meetings thereafter.

His pleasure in membership was linked with his ambition that France should dominate the world not only militarily and politically but in scientific achievement also. At the same time as he joined the Institute, Bonaparte was appointed to command the army to invade England. In the event, this plan was abandoned and replaced by a scheme to invade Egypt. Its object was to extend French influence, embarrass England's link with India, secure spoil and civilize backward Egypt. To join him in the voyage Bonaparte chose Berthollet, who was also to recruit others: the mathematicians Monge and FOURIER, the zoologist Geoffroy Saint-Hilaire (1772–1844) and the inventor N-J Conté (1755–1805). Included also were a team of engineers, cartographers and interpreters. The invasion, in 1798, was a success and soon after the Battle of the Pyramids and the victorious entry into Cairo Bonaparte set up an Institute of Egypt, whose task was to educate and civilize north Africa and bring to it the benefits of French culture and,

the city life of the poor and the sick and was often unwell himself. His life was successful from school prizes to world-wide honours in old age, but his early impressions remained, and at 71 he wrote 'I have never trusted life completely'.

Originally a medical student, he turned to chemistry early. Previously, organic chemistry had been concerned with compounds derived from living nature and little synthesis had been attempted. From 1854 Berthelot used synthetic methods in a systematic way and built up large molecules from simple starting compounds. Thus he made methanol from methane, methanoic acid from carbon monoxide, ethanol from ethene, and fats (glycerides) from propane-1,2,3-triol and organic acids. He made ethyne from hydrogen passed through a carbon arc and benzene from ethyne. The former idea of a 'vital force' was banished; organic chem-

istry became simply the chemistry of carbon compounds and organic chemists had a new basis for their thinking and an emphasis on synthesis, increasingly making compounds (as Berthelot did) that do not occur in nature. In the 1860s he studied the velocity of reactions and, later, the heat they evolved (thermochemistry). He concluded that reactions are 'driven' in the direction which evolves heat. (In fact, the matter is not as simple as this, as GIBBS showed). He also worked on physiological chemistry and on explosives (he discovered the 'detonation wave').

He was scientific adviser during the siege of Paris by the Prussians in 1870 and later was a Senator, and Foreign Minister in 1895. He died a few hours after his wife, and the two had the unique state honour of a joint burial in the Panthéon.

especially, French science. The Institute's quarters in Cairo included laboratories for chemistry and for physics, a library and observatory. Much work of scientific value was done. However, French success in Africa was checked by the British Navy's dominance under Nelson in the Mediterranean, and Bonaparte soon returned to France, with Berthollet and Monge. The Rosetta stone, found by the French in 1799 and surrendered to the British in 1801, was to yield the long-awaited key to the written language of ancient Egypt.

By 1800 Bonaparte was combining his rule of France as First Consul with the presidency of the First Class of the Institute. His interest in science was well demonstrated by his treatment of the Italian physicist VOLTA; in 1801 Bonaparte attended three Institute meetings led by Volta on electricity, saw the potential importance of the subject, and awarded a substantial annual prize for new work in it, performed by a scientist of any nation. An early winner was DAVY, who was later given a passport to visit and work in France in 1813 (despite the Anglo-French war) and whose treatment there again confirmed that Bonaparte saw science and its practitioners as above mere national interests. Although he had a very direct interest in applications (he sat on committees assessing the value of new work on gunpowder, and on steam traction) he had a comparable interest in mathematics and in pure physics, spending time in 1808, for example, studying Chladni's work on acoustics and again making him a generous award. He ensured that Berthollet and some other leading scientists had such substantial and assured incomes that they could operate effective research laboratories. After 1807 Bonaparte set up the 'Continental System', excluding British trade from Europe, and Britain countered by imposing a blockade. This cut off the supply of cane sugar to France

and Bonaparte saw to it that work on sugar from beet was encouraged to make up the loss. A prize system encouraged advances in other areas of manufacture and technology.

Something is known of Bonaparte's reading in science. During the voyage to Egypt, he spent time in tutorials with Berthollet and others, and shared a tent with him during the campaign. Later he wrote that had he not became a general and national leader 'I would have thrown myself into the study of the exact sciences. I would have traced a path following the route of Galileo and Newton'. Returning from Egypt, his library contained 14 scientific books, along with some poetry, novels and political works (although he held art and literature in rather low regard).

Despite his notable success in encouraging science-based industry in France, and the fundamental research which underpins it, he was not able quickly to adjust the educational system to provide the body of scientifically trained men he desired, mainly because of the shortage of teachers at school level; eventually this was largely rectified, through the work of the École Normale and the École Polytechnique.

One period when Bonaparte had full leisure for reading was during his banishment to Elba, and again on St Helena, where he was able to study some of the scientific writing which he had himself commissioned, such as HAÜY's books on physics, along with natural history, astronomy and chemistry.

Although by modern standards Napoleon would be rated as a scientific amateur with major talents in other directions, his support for work in science and its funding was well ahead of its time, was unique among national leaders and certainly led to outstanding results for his country.

IM

Berthollet, Claude Louis, comte (Count) [bairtolay] (1748–1822) French chemist: worked on a range of inorganic problems.

Originally a physician, Berthollet moved to chemistry and was an early staff member of the École Polytechnique, but was not an effective teacher. He was a friend of Napoleon, and joined him in the attack on Egypt in 1798. In 1814 he helped depose Napoleon 'for the good of France', and was made a peer by Louis XVIII. In chemistry, he was an early supporter of LAVOISIER'S ideas; his research examined the nature of ammonia, the sulphides of hydrogen, hydrogen cyanide and cyanogen chloride, and the reactions of chlorine. He deduced that some acids did not contain oxygen (unlike Lavoisier's view). He discovered $KClO_3$, but his use of it in gunpowder destroyed a powder mill in 1788. His work on bleaching fabrics with chlorine, and on dyes and

steelmaking, was more successful.

He believed that chemical affinity resembled gravitation in being proportional to the masses of the reactants; he was wrong, but his work foreshadowed that of GULDBERG. Similarly, he had a courteous conflict with PROUST, attacking the latter's law of constant composition. For long, Proust's views seemed to have prevailed entirely, but since 1935 'berthollide' compounds of slightly variable composition have been proved to exist. Berthollet's chemical instincts were usually good and even when they were not the debate led to a valuable outcome.

Berzelius, Jöns Jacob, Baron [berzayleeus] (1779–1848) Swedish chemist; dominated chemical theory for much of his lifetime.

Orphaned, Berzelius was brought up by relatives. He was interested in natural history and medicine was his chosen career from his schooldays. After studying medicine he graduated at Uppsala in 1802. He had read and experimented in chemistry under J Afzelius and his interest focused upon the subject. The wars against France (1805–9 and 1812–14) gave him financial freedom, because the need for military surgeons led to an increase in pay for the medical faculty in Stockholm where Berzelius held the chair of medicine and pharmacy from 1807 (renamed chemistry and pharmacy in 1810). In 1808 he became a member of the Swedish Academy of Sciences. Berzelius married late in life; he was 56 and his bride 24; as a wedding gift the king of Sweden made him a baron.

Berzelius provided the first major systemization of 19th-c chemistry, including the first accurate table of relative atomic masses (for 28 elements in his list of 1828); the reintroduction and use of modern 'initial letter' symbols for elements; concepts including isomerism and catalysis, and the division of the subject into organic and inorganic branches; and, importantly, his theory of dualism, based on his work in electrochemistry. This theory proved first a spur and later an inhibitor to further development but can now be seen as a precursor to the later division of elements into the electropositive and electronegative classes. He was the discoverer of three new elements (selenium, cerium and thorium).

For many years Berzelius was a uniquely dominant figure in chemistry, with great influence through his research, his year-book on advances in chemistry and his many pupils.

Besicovitch, Abram Samoilovitch (1891–1970) Russian–British mathematician: contributed to mathematical analysis and periodic functions.

Besicovitch was educated in St Petersburg, studying under MARKOV, and became professor when 26 at Perm (renamed the Molotov University) and Leningrad. In 1924 he worked in Copenhagen, and then HARDY secured him a lectureship at Liverpool. After a year he obtained a similar post in Cambridge and in 1950 was elected to the Rouse Ball professorship.

The research in Copenhagen was done with Harald Bohr (1887–1951; younger brother of NIELS BOHR), and imbued Besicovitch with an interest in analysis and almost-periodic functions. He published a book on this area (1932) as well as work on real and complex analysis, and geometric measure theory.

Bessel, Friedrich Wilhelm (1784–1846) German astronomer and mathematician: made first measurement of a star's distance by parallax; detected that Sirius had a companion; introduced Bessel functions.

As a young trainee accountant in Bremen, Bessel prepared for travel by studying navigation and then astronomy. This in turn took him, aged 26, to be director of the new Königsberg Observatory.

Much of Bessel's work deals with the analysis of perturbations in planetary and stellar motions. For this purpose he developed the mathematical functions that now bear his name, publishing his results in 1824 in a paper on planetary perturbations. Bessel functions have subsequently proved to have wide application in other areas of physics. In 1838 he was the first to announce the measurement of a star's distance by measurement of its parallax. Stellar parallax is the displacement the nearer stars should show (relative to distant ones) over time, because they are viewed at varying angles as the Earth moves across its orbit. Such parallax had not been observed previously; COPERNICUS suggested that this was because all stars are so distant that the parallax is immeasurably small. Bessel was able to measure the parallax of the binary star 61 Cygni as 0.3" of arc and thus found its distance to be 10.3 light years (within 10% of the present value). Around the same time he observed a small wave-like motion of Sirius and suggested that it was the result of the gravitational influence of an unseen orbiting companion; the faint companion Sirius B was subsequently detected by Alvan Clark (1832–97), a telescope lens maker, in 1862. Bessel also succeeded in computing the mass of the planet Jupiter by analysing the orbits of its major satellites and showed its overall density to be only 1.35. He suggested that irregularities in the orbit of Uranus were caused by the presence of an unknown planet, but died a few months before the discovery of Neptune.

Bessemer, Sir Henry [besemer] (1813–98) British engineer and inventor: developed a process for the manufacture of cheap steel.

Bessemer's father was an English mechanical engineer at the Paris Mint who returned to England during the French Revolution. Bessemer first gained some knowledge of metallurgy at his father's type foundry. It was a time of rapid progress in industrial manufacture and Bessemer was interested in all new developments. Largely self-taught, he became a prolific inventor.

When he was 20 he produced a scheme for the prevention of forgeries of impressed stamps used on documents; these forgeries were then costing the Government some £100 000 per year. The Stamp Office adopted his suggestion, but did not reward him. It was an experience he did not forget.

Horrified at the price of hand-made German 'gold powder' purchased for his sister's painting, he devised a method of manufacturing the powder from brass. Unable to patent the process, for secrecy he designed a largely automatic plant, workable by himself and his three brothers-in-law. From this he made enough money to cover the expenses of his future inventions, which included improvements in sugar cane presses and a method for the manufacture of continuous sheet glass.

The Crimean War directed Bessemer's interest to the need for a new metal for guns. Cast iron (pig iron), which contains carbon and other impurities, is brittle, and the relatively pure wrought iron was then made from pig iron by a laborious and time-consuming method. Steel (iron with a small amount of carbon) was made in small quantities with heavy consumption of fuel and so was costly. He developed the Bessemer process for making cheap steel without the use of fuel, reducing to minutes a process which had taken days. Bessemer steel was suitable for structural use and was cheap enough to use for this, which greatly helped the industrial development then in progress; as well as being of value for railway systems and the machine-tool industry. The Bessemer process consisted of blowing air through molten crude iron, oxidizing carbon to blow-off gas and silicon and manganese to solid oxides. For this purpose he designed the Bessemer converter, a tiltable container for the molten metal, with holes for blowing air through its base. Some early users of his process were unable to reproduce his results, which led to legal disputes over royalty payments to him. The failures were due to ores containing phosphorus; Bessemer had by chance used iron ore free from phosphorus. This problem of phosphorus impurities in some ores was solved in 1878 by S G Thomas (1850–85) and P C Gilchrist (1851–1935). R F Mushet (1811–91) also improved the process, by the addition of an alloy of iron, manganese and carbon.

Bessemer set up his own steel works in Sheffield, using phosphorus-free ore. During his lifetime the Bessemer process was appreciated more abroad. Andrew Carnegie made his fortune by it in the USA, but the British steel manufacturers were loth to acknowledge its success. Bessemer was knighted in 1879.

Bethe, Hans Albrecht [baytuh] (1906–) German–US physicist: proposed mechanism for the production of stellar energy.

In 1939 Bethe proposed the first detailed theory for the formation of energy by stars through a series of nuclear reactions, having the net result that four hydrogen nuclei are converted into a helium nucleus and radiated fusion energy. The process (sometimes known as the 'carbon cycle' because of the key part played in it by carbon-12) gives good agreement with observation for some types of stars. Bethe also contributed, with ALPHER and GAMOW, to the alpha-beta-gamma theory of the origin of the chemical elements during the origin of the universe (see ALPHER for an account).

Bichat, (Marie-Francois) Xavier [beesha] (1771–1802) French pathologist: founder of animal histology.

Bichat followed his father in studying medicine. His studies were interrupted by a period in the army, and he returned to Paris at the height of the Terror in 1793. From 1797 he taught medicine and in 1801 worked in Paris's great hospi-

tal, the Hôtel-Dieu. He was struck by the fact that various organs consist of several components or 'tissues' and described 21 of them (such as connective, muscle and nerve tissue). He saw that when an organ is diseased, usually it was not the whole organ but only certain tissues which were affected. He distrusted microscopes and did not use one for most of his work; and cell theory was yet to come. Bichat's work, done with great intensity during the last years of his short life, had much influence in medical science. It formed a bridge between the 'organ pathology' of MORGAGNI and the later 'cell pathology' of VIRCHOW; and the study of tissues (histology) has been important ever since.

Biffen, Sir Rowland Harry (1874–1949) British geneticist and plant breeder.

Biffen graduated at Cambridge in 1896 and, after an expedition to South and Central America studying rubber production, he returned to Cambridge to teach agricultural botany. He was to dominate the subject there for a generation. In 1899 he began cereal trials intended to select improved types, then a chance process. A year later MENDEL's neglected work of the 1860s on plant genetics at last became known. Biffen quickly saw that plant-breeding could be rationalized and he guessed that physiological traits might be inherited, as well as the morphological traits studied by Mendel. In 1905 he showed this was the case for resistance by wheat to yellow rust, a fungal disease; it is inherited as a simple Mendelian recessive. Since then, improvement of crop plants by hybridization has been widespread. Biffen's own wheat variety 'Little Joss' was unsurpassed for 40 years.

Biot, Jean (Baptiste) [beeoh] (1774–1862) French physicist: pioneer of polarimetry.

A child during the French Revolution, Biot joined the artillery at 18 but soon left to study mathematics, and at 26 was teaching physics at the Collège de France. His research showed variety. With GAY-LUSSAC he made an early balloon ascent (1804) and made meteorological and magnetic observations up to 5 km; his nerve failed for a second attempt. He made a number of geodetic and astronomical expeditions, visiting Spain and Orkney.

His famous work is on optical activity. He showed, for the first time, that some crystals of quartz rotated the plane of polarized light while other crystals rotated it to the same extent but in the opposite direction. In 1815 he showed that some liquids (eg turpentine) will also rotate plane polarized light, and later he observed the same effect with some solids when dissolved in water (eg sugar, and tartaric acid). He realized that this ability of some substances in solution must mean that the effect is a molecular property ('optical activity'). He showed (Biot's Law) that the amount of rotation of the plane of polarization of light passing through an optically active medium is proportional to the length of its path, and to the concentration, if the medium is a solution of an active solute in an inactive solvent, and that the rotation is roughly inversely proportional to the square of the wavelength of the light.

Polarimetry (measurement of optical activity) was pioneered by Biot and after 1870 proved of great value in gaining information on molecular configuration (shape). Later still, the variation in optical rotation with the wavelength of the polarized light was also found to be useful in locating molecular shape.

Biot was a man of great talent, as was seen by his older friend LAPLACE; and in his old age Biot saw and appreciated the talent of his young friend PASTEUR.

Birkeland, Kristian Olaf Bernhard [beerkuh-lahnt] (1867–1917) Norwegian physicist: devised process for nitrogen fixation.

Birkeland studied physics in Paris, Geneva and Bonn before returning to his native Oslo to teach at the Christiania University. He studied the aurora borealis and in 1896 suggested (correctly) that it resulted from some charged solar radiation becoming trapped in the Earth's magnetic field near the North Pole. His theory was partly based on an experiment with a magnetized model of the Earth, which he placed in a beam of electrons in a cathode ray tube; he found luminous effects near the poles which resembled aurorae. However, he is best known as co-discoverer of the Birkeland–Eyde process. This was designed to meet the shortage of nitrate fertilizer, and used CAVENDISH's observation of 1784 that atmospheric nitrogen and oxygen combined in an electric spark to give nitrogen monoxide, NO. The process used an electric arc spread by a magnetic field and the NO was mixed with air and water to give nitric acid; it was used (with the benefit of cheap Norwegian hydro-electricity) from 1903–28.

Birkhoff, George (David) (1884–1944) US mathematician: proved the ergodic theorem of probability theory.

After taking his first degree at Harvard and a doctorate on boundary problems at Chicago, Birkhoff taught at Michigan and Princeton. He became an assistant professor at Harvard in 1912 and a full professor there at 35 in 1919, retiring in 1939.

An early interest in differential and difference equations allowed Birkhoff to apply matrix methods generally for the first time. He studied dynamics and POINCARÉ's celestial mechanics, and in 1913 obtained a now famous proof of Poincaré's 'last geometrical theorem' on the three-body problem.

Collaboration with VON NEUMANN gave rise to the 'weak form' of the ergodic theorem, which was shortly followed by Von Neumann's discovery of the 'strong form'. Ergodicity refers to whether a dynamical system will develop over time so as to return exactly to a previous configuration. In 1938 Birkhoff published several papers on electromagnetism and also argued that better alternatives to EINSTEIN's general theory of relativity were possible.

Overall Birkhoff is acknowledged as the greatest American mathematician of the early 20th-c; he excelled as a teacher and in developing celestial mechanics and the analysis of dynamical systems.

Bishop, (John) Michael (1936–) US virologist and oncologist.

A graduate in medicine from Harvard, Bishop became in the 1960s a virologist at the National Institutes of Health at Bethesda, MD, and from 1968 at the University of California. In the 1970s he worked with H E Varmus to test the idea that normal mammalian cells contain genes that, although dormant, can be activated to cause the uncontrolled proliferation characteristic of cancer. By 1976 they confirmed that healthy cells do indeed contain dormant viral genes similar to the cancer-inducing gene present in the ROUS sarcoma virus that causes cancer in chickens, which by faulty activation can lead to cancer. By 1989 Bishop, Varmus and others had found over 40 genes which can potentially induce cancer (oncogenes) and in that year the two shared the Nobel Prize.

Bjerknes, Vilhelm (Firman Koren) [byerknes] (1862–1951) Norwegian meteorologist: pioneer of dynamical meteorology.

Bjerknes's father was professor of mathematics at the Christiania University (now Oslo), and Bjerknes himself held professorships at Stockholm and Leipzig before founding the Bergen Geophysical Institute in 1917. Through his hydrodynamic models of the atmosphere and the oceans Bjerknes made important contributions to meteorology, and in 1904 showed how weather prediction could be achieved numerically using mathematical models. During the First World War Bjerknes established a network of weather stations throughout Norway, the results allowing him and his collaborators (who included his son, Jacob (1897–1975), and BERGE-RON) to develop their theory of polar fronts. They demonstrated that the atmosphere is composed of distinct air masses with different characteristics, the boundaries between such air masses being called 'fronts'. Their Bergen frontal theory, as it became known, explains how cyclones are generated over the Atlantic where warm and cold air masses meet.

Black, Sir James (Whyte) (1924–) British pharmacologist: designer of novel drugs.

A graduate in medicine from St Andrews, Black was a university lecturer there, in Malaya and in Glasgow before working in pharmacology first with ICI, and later with Smith, Kline & French and with Wellcome. At the time of his Nobel Prize in 1988 he had been professor of analytical pharmacology at King's College Hospital Medical School, London, from 1984.

He is best known for two major contributions

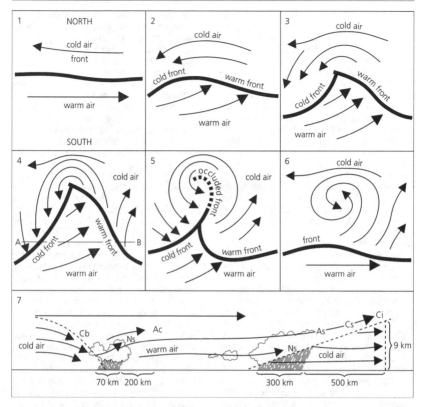

Depression – plan view of the six idealized stages in the development and final occlusion of a depression along a polar front in the Northern Hemisphere. Stage 4 shows a well developed depression system and stage 5 shows the occlusion. The cross-section (7) is taken along the line AB in stage 4. The cloud types are: Cb – cumulonimbus; As – altostratus; Ac – altocumulus; Cs – cirrostratus; Ns – nimbostratus; Ci – cirrus.

to medicine. His work on beta-blockers was based on the theory that heart muscle has specific beta-receptors that respond to hormonal control; Black reasoned that if these sites could be blocked, the effect of the hormones on the heart would be inhibited and its workload reduced; and he was able to find a very satisfactory antagonist, propranolol, in 1964. Since then such beta-blockers have been much used to control heart disease and hypertension. He went on to devise a comparatively rational approach to the control of stomach ulcers; he deduced in 1972 that a particular type of histamine receptor is located on the wall of the gut and stimulates acid secretion in the stomach, and then found a compound (cimetidine) that blocked the action of these H_2 receptors, so curbing stomach acidity and allowing healing. Black's successes much encouraged a rational approach in medicinal chemistry, as well as having provided drugs of value in two major areas.

Black, Joseph (1728–99) British physician, chemist and physicist: pioneer of modern chemical logic; discoverer of latent heat and specific heat.

Black was born in Bordeaux, where his Scots-Irish father was a wine merchant. Joseph was educated in Belfast, Glasgow and Edinburgh. Finally, he studied medicine. His work for his MD degree, expanded in a paper of 1756, is his major contribution to chemistry; it is a model of experiment and logic. In particular, he saw the importance of recording changes of weight and he recognized the importance of gases. He studied the cycle of changes we would now express in formulae as follows: (note that Black knew his compounds by the names given in brackets; formulae and atomic theory came much later, but he understood the key relationships between the compounds):

$CaCO_3$ (limestone) + heat
$\rightarrow CaO$ (quicklime) + CO_2 (fixed air)
$CaO + H_2O$ (water) $\rightarrow Ca(OH)_2$ (slaked lime)
$Ca(OH)_2 + CO_2 \rightarrow CaCO_3 + H_2O$

Black showed that fixed air (CO_2) is produced by respiration and fermentation, and by burning charcoal; that it behaves as an acid (eg in neutralizing an alkali); and he deduced its presence in small quantities in the atmosphere. He was a very popular lecturer at Glasgow and later in Edinburgh; one of his pupils was Benjamin Rush (1745–1813), who became the first professor of chemistry in America.

He taught LAVOISIER'S new views on chemistry when they appeared, but his own research moved to physics. About 1763 he showed that heat is necessary to produce a change of state from solid to liquid, or liquid to vapour, without a rise in temperature; e.g., ice at 0° requires heat to form water at 0°. He called this 'latent heat'. On this basis, he went on to distinguish clearly between heat and temperature; and he examined the different heat capacity of substances. Thus in physics, as in chemistry, he provided basic ideas essential for the subject to advance.

Blackett, Patrick (Maynard Stuart), Baron Blackett (1897–1974) British physicist: used an improved Wilson cloud chamber to make discoveries using cosmic rays.

Blackett, the son of a stockbroker, was educated at Osborne and Dartmouth Naval Colleges and saw action at sea in the battles of the Falkland Islands (1914) and Jutland (1916). He then studied physics at Cambridge and, continuing in research, made the first cloud chamber photographs (1924) of the transmutation of nitrogen into oxygen-17 by bombardment with alpha particles (helium nuclei). He was appointed to professorships in London (1933), Manchester (1937) and finally Imperial College, London (1953). During the Second World War Blackett pioneered the use of operational research to produce economies in military resources, including work on submarine warfare, and also invented a new bomb-sight for aircraft. After the war he was active in public affairs, opposing the growing role of nuclear weapons. He was awarded the Nobel Prize for physics in 1948 .

The prize winning work was the construction, with OCCHIALINI, of a cloud chamber that underwent vapour expansion and took a photograph when two aligned Geiger counters were triggered. Blackett used the apparatus to identify the first positron to be seen following their prediction by DIRAC; however while he sought further experimental confirmation C D ANDERSON'S discovery of the positron was published first. He was made a life peer in 1969.

Blackman, Frederick Frost (1866–1947) British plant physiologist: demonstrated that gas exchange occurs through leaf stomata.

Blackman was the eldest son in a family of 11 and followed his father in studying medicine. However, he never practised; he had been a keen botanist since his schooldays and in 1887 he moved to Cambridge as a science student, stayed to teach and never left.

In 1895 he showed experimentally that gas exchange between plant leaves and the air occurs through the stomata; it had been believed since 1832, but not proved, that these pores are the entry points for the exchange. In 1905 he put forward the principle of limiting factors: where a plant process depends on several independent factors, the overall rate is limited by the rate of the slowest factor. This idea was offered by LIEBIG in the 1840s, but Blackman demonstrated it clearly by work on the effect of temperature, light and carbon dioxide availability on photosynthesis in the aquatic willow moss.

Blackwell, Elizabeth (1821–1910) British–US physician: the first woman to graduate in medicine in the USA and the first woman on the British Medical Register.

Elizabeth Blackwell was born in Bristol and was taught by a tutor at home. When she was 11, the family emigrated to the USA and she went to school in New York. After the death of her father 6 years later, she became a teacher, but was not attracted to the work. She decided on a medical career, more as a challenge than as a vocation. After many fruitless applications she was accepted by Geneva College in New York. This happened because the application was thought to be a student joke by a rival college and was accepted in like spirit. Honourably, they kept to their commitment and she graduated in 1849. After further studies in Paris and London she returned to New York in 1851, but found she was prevented from practising. She gave lectures on hygiene, which brought useful social and professional contacts, and in 1853 opened a dispensary in a poor district of New York; from this the New York Infirmary for Women and Children emerged. With onset of the Civil War, plans for a medical school were shelved and Elizabeth Blackwell went to Europe to lecture on 'Medicine as a Profession for Ladies'. In 1859 she placed her name on the new British Medical Register, becoming the first women to do so. She returned to the USA where her medical institute was opened in 1868. In 1869 she returned to Britain to practise medicine and later retired to Scotland. Elizabeth Blackwell influenced many women, by her example and lectures, to battle for acceptance into medical schools in their own country.

Blagg, Mary Adela (1858–1944) British amateur astronomer; prepared the definitive uniform lunar nomenclature.

Mary Blagg was born in Cheadle, North Staffordshire, the eldest daughter of a solicitor, and lived there all her life. After education at home and boarding school in London she remained at home, occupying herself with voluntary social work. She had a natural aptitude for mathematics and taught herself using her brothers' school books; later she had no difficulty in manipulating harmonic analysis. When in middle age she attended a course of University Extension lectures given in Cheadle, she became seriously interested in astronomy. She wished to do original work and her tutor pointed out the great need for a uniform lunar nomenclature. At that time there were many discrepancies between the maps of the principal selenographers, the same name attached to different formations and different names given to the same formation. In 1905 a committee was appointed by the International Association of Academies to attack the problem. Mary Blagg was appointed by S A Saunder, a member of that committee, to collate the names given to all the lunar formations on existing maps of the moon and her resulting *Collated List* was published in 1913. In 1920 the International Astronomical Union was formed and Mary Blagg was appointed to its Lunar Commission. She continued to work on lunar nomenclature and was appointed with Dr Müller of Vienna to prepare a definitive list of names. *Named Lunar Formations* became the standard authority on lunar nomenclature (1935). The International Lunar Committee gave her name to a small lunar crater.

Bloch, Felix [blokh] (1905–83) Swiss–US physicist: invented nuclear magnetic resonance spectrometry.

Bloch was educated at Zürich and Leipzig, but following a short period of teaching in Germany moved to the USA in 1933. He spent the rest of his career at Stanford.

The theory of solid-state physics and of how electrons behave in solids was advanced by Bloch's research. The Bloch wavefunction describes an electron which is moving freely in a solid and the term Bloch wall describes the boundary between two magnetic domains in a ferromagnetic material.

In 1946 Bloch introduced the nuclear magnetic resonance (NMR) technique, also developed independently by PURCELL. Many types of atomic nucleus possess a magnetic moment and quantum mechanics indicated that the moment could only adopt one of a number of possible orientations with respect to an applied magnetic field. Each orientation requires a different energy and so transitions from one state to another can be accomplished if a photon of electromagnetic radiation (of radio frequencies) is absorbed. The magnetic moments of the proton and neutron were measured by this method and since then many complex molecules have been studied. The energy state of the nucleus gives information about its atomic neighbours in the molecule because of the effect of the surrounding electrons. Bloch shared the 1952 Nobel Prize for physics with Purcell, and the NMR method has since become a powerful analytical technique in chemistry.

Bode, Johann Elert [bohduh] (1747–1826) German astronomer: publicized numerological relationship between planetary distances.

Although Bode was director of the Berlin Observatory for almost 40 years and constructed a notable star atlas, his fame rests, strangely enough, on his popularization of a relationship discovered by someone else. In 1772 J D Titius (1729–96) pointed out that the members of the simple series 0,3,6,12,24,48,96, when added to 4 and divided by 10, give the mean radii of the planetary orbits in astronomical units, surprisingly accurately (even though only six planets were known at the time). (An astronomical unit (AU) is the mean distance of Earth from the Sun.) Through Bode's publicizing of the relationship it became named after him. It played a part in the discovery of Uranus, the asteroid belt (the fifth 'planet'), and Neptune (although its results are hopelessly inaccurate for Neptune and Pluto). It has never been proved whether Bode's Law has any real meaning, or is merely coincidental; if the latter, it is a remarkable coincidence. Bode was also responsible for the naming of the planet Uranus.

Bogoliubov, Nikolai Nikolaevich (1909–) Soviet mathematical physicist: contributed to quantum theory and the theory of superconductivity.

Bogoliubov worked at the Academy of Sciences in the Ukraine and at the Soviet Academy of Sciences. He contributed new mathematical techniques to physics, and the Bogoliubov transformation, by which variables are changed in quantum field theory, is named after him. A distribution function describing non-equilibrium processes is due to him, as are many developments in parallel to the BCS theory of superconductivity.

Bohr, Niels (Henrik David) [baw(r)] (1885–1962) Danish theoretical physicist: put forward the quantum theory of the electronic structure of atoms.

Bohr's family was distinguished, his father was professor of physiology at Copenhagen and his younger brother Harald a gifted mathematician. Niels and Harald were both footballers to a professional standard and Niels and his son Aage (born in 1922) both won the Nobel Prize for physics, in 1922 and 1975 respectively.

After Bohr had finished his doctorate at Copenhagen (1911) he spent 8 months in Cambridge with J J THOMSON, who was not attracted by Bohr's ideas on atomic structure, and so he moved to join RUTHERFORD at Manchester and spent 4 years there. Rutherford's model of the atom (1911) envisaged electrons as spread around the central positive nucleus, but according to classical physics this system would be unstable. Bohr countered this difficulty by suggesting that the electron's orbital angular momentum about the nucleus can only adopt multiples of a certain fixed value, ie it is quantized. Radiation is then only emitted or absorbed when an electron hops from one allowed orbit to another. On this basis Bohr calculated in 1913 what the emission and absorption spectra of atomic hydrogen should be, and found excellent agreement with the observed spectrum as described by RYDBERG and BALMER.

In 1916 Bohr returned to Copenhagen and 2 years later became the first director of its Institute of Theoretical Physics. This became the focal centre for theoretical physics for a generation, in which physicists throughout the world co-operated in developing quantum theory. In that first year Bohr established the 'correspondence principle': that a quantum description of microscopic physics must tend to the classical description for larger dimensions.

His 'complementarity principle' appeared in 1927: there is no sharp separation between atomic objects and their interaction with the instruments measuring their behaviour. This is in keeping with BROGLIE's belief in the equivalence of wave and particle descriptions of matter; HEISENBERG's uncertainty principle and BORN's use of probability waves to describe matter also fit naturally with this principle.

Rutherford's work had developed nuclear physics to the point by the 1930s where Bohr could apply quantum theory to the nucleus also. This was held to be of neutrons and protons coupled strongly together like molecules in a liquid drop (1936). The very variable response of nuclei to collisions with neutrons of different energies could then be explained in terms of the possible excited states of this 'liquid drop'. By 1939 Bohr and J A Wheeler (1911–95) had a good theory of nuclear fission and were able to predict that uranium-235 would be a more appropriate isotope for fission (and, as EINSTEIN pointed out, an atomic bomb) than uranium-238.

By the autumn of 1943 Bohr was in danger in occupied Denmark (his mother was Jewish) and he chose to escape to Sweden in a fishing boat. He was then flown to England in the bomb-bay of a Mosquito aircraft. Before he left Denmark he dissolved the heavy gold medal of his Nobel Prize in acid; the inconspicuous solution escaped detection in occupied Denmark and was later reduced to metal and the medal was recast from it. After his escape, he joined the atomic bomb programme. In 1944 he lobbied Roosevelt and Churchill on the danger inherent in atomic weapons and the need for agreements between the West and the USSR. This led to his organizing the first Atoms for Peace Conference in Geneva in 1955.

At his death in 1962 Bohr was widely acknowledged as the foremost theoretician of this century after Einstein. The Bohr model of the atom gave a good 'fit' with the observed spectra only for the simplest atoms (hydrogen and helium) and it was much modified later, but the concept was a milestone for physics and for chemistry. Similarly, the liquid drop model of the nucleus was to be much developed by others, and notably by Aage Bohr.

Unlike many physicists who have shaped their ideas alone, Bohr refined his ideas in discussions, which often became monologues. He was very popular with his fellow physicists, who produced a 5-yearly *Journal of Jocular Physics* in his honour to celebrate his birthdays.

Bois-Reymond, Emil du (1818–96) German physiologist: pioneer electrophysiologist.

Du Bois-Reymond's father was a Swiss teacher who moved to Berlin; he was an expert on linguistics and authoritarian enough to 'arouse his son's spirit of resistance'. The family spoke French and felt part of the French community in Berlin. Emil studied a range of subjects at Berlin for 2 years before he was fully attracted to medicine, which he studied under J P MÜLLER. He graduated in 1843 and was then already working on animal electricity (discovered by GALVANI) and especially on electric fishes. He introduced refined physical methods for measuring these effects and by 1849 had a sensitive multiplier for measuring nerve currents; he found an electric current in injured, intact and contracting muscles. He traced it correctly to individual fibres and found that their interior is negative with respect to the surface. He showed the existence of a resting current in nerve and suggested, correctly, that nerve impulses might be transmitted chemically. He was modest, but also confident, and his ideas aroused vigorous debate; his experimental methods dominated electrophysiology for a century.

Boksenberg, Alexander (1936–) British astrophysicist: inventor of an image photon counting system (IPCS) of great value in optical astronomy.

Boksenberg studied physics in London, and from the 1960s worked on image-detecting systems for use in space vehicles and ground-based telescopes. For a century, photographic plates or films have been used to accumulate light in faint telescopic images, but the method has disadvantages and electronic detectors (eg CCDs, charge-coupled devices) have been much used. Boksenberg's method uses a TV camera and image intensifier, whose amplified signals are processed by computer, to give a pixel picture of dots (each due to a photon) which can be both viewed and stored.

In this way, very distant and/or faint objects can be studied in the optical, UV and X-ray range with enhanced sensitivity and accuracy. Boksenberg and others have looked particularly at quasars, aiding understanding of stellar evolution. He became director of the Royal Greenwich Observatory in 1981; the use of CCDs, and especially his IPCS, has revitalized optical astronomy.

Boltwood, Bertram Borden (1870–1927) US radiochemist: developed understanding of uranium decay series.

Growing up fatherless, but in an academic family, Boltwood studied chemistry at Yale and then in Munich and Leipzig. From 1900 he operated a laboratory as a consultant on analytical and related problems. From 1904 he worked on radiochemistry, from 1906 at Yale, and became America's leading researcher on this. He did much to bring about understanding of the uranium decay series, to improve techniques in radiochemistry and to introduce Pb:U ratios as a method for dating rocks. A good friend of RUTHERFORD, he was a victim of depression and eventually killed himself.

Boltzmann, Ludwig Eduard [boltsmahn] (1844–1906) Austrian physicist: established classical statistical physics; and related kinetic theory to thermodynamics.

Boltzmann grew up in Wels and Linz, where his father was a tax officer. He obtained his doctorate at Vienna in 1866 and held professorships during his career at Graz, Vienna, Munich and Leipzig.

Theoretical physics in the 1860s was undergoing great changes following the establishment by CLAUSIUS and KELVIN of the Second Law of Thermodynamics, the kinetic theory of gases by Clausius and MAXWELL and the theory of electromagnetism by Maxwell. Boltzmann extended the kinetic theory, developing the law of equipartition of particle energy between degrees of freedom and also calculating how many particles have a given energy, the Maxwell–Boltzmann distribution.

Furthermore, Boltzmann used the mechanics and statistics of large numbers of particles to give definitions of heat and entropy (a measure of the disorder of a system). He showed that the entropy S of a system is related to the probability W (the number of 'microstates' or ways in which the system can be constructed) by $S = k \log W$ (Boltzmann's equation), where k is Boltzmann's constant $(k = 1.38 \times 10^{-23} \text{ J K}^{-1})$. Other contributions were a new derivation of STEFAN's law of black-body radiation; and his work on electromagnetism.

Throughout his life Boltzmann was prone to depression and this was intensified by attacks from the logical positivist philosophers in Vienna, who opposed atomistic theories of phenomena. However, he attracted students who became distinguished and had both many friends and honours. Depressed by lack of acceptance of his work, Boltzmann killed himself while on holiday on the Adriatic coast.

Bondi, Sir Hermann (1919–) Austrian–British mathematical physicist and astronomer: proponent of the steady-state theory for the origin of the universe.

Bondi was born in Vienna and had his schooling there, studied at Cambridge, where he held academic posts, and in 1954 was appointed professor of mathematics at King's College, London. From 1967–84 he was in the public service (European Space Agency, Defence, Energy,

Natural Environment Research Council). He was Master of Churchill College, Cambridge, 1983–90.

Bondi worked in many areas of theoretical physics and astronomy, especially the theory of gravitation (gravitational waves, etc). He is best known as one of the originators, with GOLD and HOYLE, of the steady-state theory of the universe, according to which the universe looks the same at all times. On this basis it is considered to have no beginning and no end, with matter being spontaneously created from empty space as the universe expands, in order to maintain an unchanging uniform density. Although the theory enjoyed support for a number of years, the discovery of the cosmic microwave background in 1964 by PENZIAS and WILSON gave conclusive support to the rival 'Big Bang' theory. However, in provoking new lines of discussion, the steady-state theory made an important contribution to modern cosmology.

Boole, George (1815–64) British mathematician: developed mathematical treatment of logic.

Largely self-taught, Boole was a schoolteacher for a number of years before being appointed professor of mathematics at Queen's College, Cork in 1849. His early work concerned the theory of algebraic forms, but it is for his pioneering of the subject of mathematical logic that he is best known. In 1847 he developed a form of algebra (Boolean algebra) that could be used to manipulate abstract logical functions and which for the first time bridged the hitherto separate disciplines of mathematics and formal logic. Boolean algebra was essential to the development of digital computers from the principles established by BABBAGE, and has important applications in other fields such as probability and statistics.

Bordet, Jules (Jean Baptiste Vincent) [baw(r)day] (1870–1961) Belgian immunologist: a founder of serology.

Bordet graduated in medicine in Brussels in 1892 and taught there from 1901. While working at the Pasteur Institute in Paris in 1898 he found that if blood serum is heated to 55°C its antibodies are not destroyed but its ability to destroy bacteria is lost. He deduced that some heat-sensitive component of serum is necessary, which EHRLICH called complement. In 1901 Bordet showed that this is used up when an antibody reacts with an antigen, a process called complement fixation and of importance in immunology. Ehrlich thought that each antigen had its own complement; Bordet thought there was only one. We now know that the immune system contains nine varieties of complement, each an enzyme system which is responsible for the destruction of a range of pathogens. For this and his other work on immunity, Bordet was awarded the Nobel Prize in 1919.

Born, Max (1882–1970) German physicist: invented matrix mechanics and put forward the statistical interpretation of the wavefunction.

Max Born was the son of a professor of anatomy at the University of Breslau and, following the death of his mother when he was 4, he was brought up by his maternal grandmother. He studied at Breslau, Heidelberg, Zürich and Cambridge, gaining his PhD at Göttingen (1907). He remained there as a teacher, becoming professor of physics in 1921 and establishing a centre of theoretical physics second only to the Niels Bohr Institute in Copenhagen. As a Jew he left Germany for Cambridge in 1933, becoming a professor in Edinburgh and finally returning in retirement to Göttingen in 1953. In 1954 he was awarded the Nobel Prize for physics for his fundamental contributions to quantum mechanics, together with BOTHE.

Initially Born's research interests were lattice dynamics and how atoms in solids hold together and vibrate. The Born–Haber cycle of reactions allows calculation of the lattice energy of ionic crystals. However, in 1923 the old quantum theory established by PLANCK, EINSTEIN, BOHR and SOMMERFELD remained inconsistent and unable to account for many observations. BROGLIE then made the startlingly apt suggestion that particles possess wave-like properties (1924) and Born, E P JORDAN, HEISENBERG and PAULI in collaboration rapidly developed a sequence of important ideas. With Jordan, Born constructed (1925) a method of handling quantum mechanics using matrices (matrix mechanics) and this was the first consistent version of the new quantum mechanics. DIRAC subsequently took this, and the equivalent wave mechanics due to SCHRÖDINGER, and blended them into a single theory (1926).

Born also put forward the probability interpretation of the wavefunction. In Schrödinger's wave mechanics a particle is represented by a wave packet, which unfortunately disperses in time. Born's solution was to state that the wave guides the particle in the sense that the square of the amplitude of the wavefunction is the probability of finding a particle at that point. Einstein opposed a move from deterministic to statistical physical laws and Born and Einstein discussed the issue from time to time over many years.

Born is buried in Göttingen, where his gravestone displays his fundamental equation of matrix mechanics:

$$pq - qp = h/2\pi i$$

where p is the momentum operator, q the position operator and h Planck's constant.

Borodin, Aleksander Porfiryevich [borawdeen] (1833–87) Russian chemist and musician.

Trained as a chemist in St Petersburg, Borodin later travelled in Europe and from 1864 held a professorship in the Russian Academy. His work was mainly in organic chemistry, where he devised methods for fluorinating organic compounds in the 1860s, and he worked on reactions of aldehydes. He showed that both polymerization and condensation of aldehydes occurs, and in this way made aldol and, from it, crotonaldehyde. His method for analysing urea was long used by biochemists. He is well known as a composer, notably of the heroic opera *Prince Igor* and many songs.

Bose, Satyendra Nath [bohs] (1894–1974) Indian physicist: discovered the quantum statistics of particles of integral spin.

An education at Presidency College in Calcutta led Bose to a lectureship at the Calcutta University College of Science, and another at the University of Dacca when it was formed in 1921. During his research career he made significant advances in statistical mechanics and quantum statistics, the description of all forces by a single field theory, X-ray diffraction and the interaction of electromagnetic waves with the ionosphere.

In 1924 Bose derived PLANCK'S black-body radiation law without the use of classical electrodynamics, which Planck had needed to use. Bose was able to obtain 2 years leave for research and travel, and in Europe he met BROGLIE, BORN and EINSTEIN. Einstein took up Bose's work and formed a general statistics of quantum systems from it (the Bose–Einstein statistics) which describes particles of integral spin, which may multiply occupy the same quantum state. Such particles are now known as bosons. An equivalent statistics for spin-$\frac{1}{2}$ particles which are limited to one particle per quantum state is called the Fermi–Dirac statistics, and the particles are called fermions.

Bothe, Walther [bohtuh] (1891–1957) German physicist.

A student of PLANCK at Berlin and a prisoner of war in the First World War, Bothe taught at three German universities before, in 1925, working with GEIGER to study the COMPTON effect, by using a pair of Geiger counters to examine, in an ingenious way, single photons and electrons from individual collisions. This work confirmed that radiation can, as predicted, behave like a stream of particles.

He used similar methods in 1929 to show that cosmic rays do not consist only of gamma rays, as believed previously, but also contain heavy particles. After becoming professor of physics at Giessen in 1930, he noted a new uncharged emission from beryllium bombarded with alpha particles; in 1932 CHADWICK showed this strange, penetrating emission to consist of neutrons.

Bothe led some work in Germany on nuclear energy during the Second World War and shared a Nobel Prize for physics in 1954.

Bouguer, Pierre [boogair] (1698–1758) French physicist and mathematician: pioneer of photometry.

A child prodigy, being appointed teacher of hydrography at Havre at the age of 15, Bouguer is chiefly remembered for laying the foundations of photometry. He invented the heliometer and later a photometer with which he compared the luminosities of the Sun and the Moon. He discovered that the intensity of a collimated beam of light in a medium of uniform transparency decreases exponentially with the length of its path through the medium, a result now known as Bouguer's Law but often incorrectly attributed to J H Lambert (1728–77).

In geophysics the correction required to adjust gravity measurements to sea level (approximately 0.1 mgal per metre of rock) is known as the Bouguer correction, following his work on gravity in the Andes in 1740.

Bourbaki, Nicholas [boorbakee] (c.1930–) French group of mathematicians.

This name was used as a pseudonym by an anonymous but eminent club of mainly French mathematicians. The membership of about 20 was not constant, but undoubtably included several men of great creative ability; retirement at age 50 was required. 'Bourbaki' published 33 parts of an encyclopedic survey of modern mathematics during 1939–67.

The treatment was formalized and abstract, required expert mathematical knowledge and was most influential in its earlier years. The attempt made by the group to persuade their readers that Bourbaki was a person failed; and the austerity of their work eventually reduced its popularity among French mathematicians, despite its elegant sophistication.

Boussingault, Jean-Baptiste Joseph [boosī-goh] (1802–87) French chemist: pioneer of experimental agricultural chemistry.

After a school career lacking distinction, Boussingault entered the École des Mines at Saint-Étienne and soon after graduation was employed to direct a mine in Venezuela. During his 10 years there he travelled and reported on the geology and geography of the area to the Institut de France, and on his return was appointed professor of chemistry at Lyon. His main work afterwards was in agricultural chemistry. He showed that legumes (peas, beans, etc) can secure nitrogen from the air (actually via root bacteria), whereas most plants, and all animals, cannot secure nitrogen from the air and must obtain it from their food. His work on the nutritional value of foods opened an area of study which later led to major discoveries on metabolism and on the vitamins. He found iodine in salt deposits claimed by South American natives to be curative for goitre, which led him to suggest the use of iodine in treatment, but this was not taken up for many years.

Boveri, Theodor Heinrich [boveri] (1862–1915) German cytologist: did basic work on relation of chromosomes to heredity.

Boveri began his university life as a student of history and philosophy at Munich, but soon changed to science and later taught zoology and anatomy at Munich and Würzburg. By 1884 it was known (largely from BENEDEN'S work) that in sexual reproduction the nuclei of spermatozoon and ovum provide equal numbers of chromosomes in the fusion which is the central feature of fertilization; that the chromosome number is constant for a given species; and that heredity is dependent on the nucleus. Boveri confirmed and extended Beneden's work on cell division using the roundworm *Ascaris*, and went on to study sea-urchin eggs. He was able to show that embryos that are deficient in chromosomes develop abnormally into the new individual and that normal development requires not only an appropriate number of chromosomes for the species but a particular selection of chromosomes. This implied that each chromosome carried in some way certain specific determiners for growth and development, and by 1910 it was fairly widely accepted that the chromosomes are the vehicles of heredity.

Bovet, Daniel [bohvay] (1907–92) Swiss–French–Italian pharmacologist: introduced antihistamines, and curare-type muscle relaxants for surgery.

After qualifying in Geneva, Bovet went to the Pasteur Institute in Paris, later moving to Rome. In Paris he was a member of a group that showed that the antibacterial drug Prontosil owed its effect to its conversion in the body to sulphanilamide, which is the parent of the sulphonamide group of drugs. Sulphanilamide was cheap and unpatented, and its derivatives have been widely used against streptococcal infections.

Later Bovet found compounds that antagonize the action of histamine; this opened the way to widespread use of such antihistamines for the relief of allergic symptoms and related conditions (such as the common cold). A visit to Brazil began his interest in the nerve poison curare; later he made simpler synthetic compounds which have a usefully short-acting curare-type activity. These have been much used as muscle relaxants in surgical operations since 1950. Bovet received a Nobel Prize in 1957.

Bowen, Ira Sprague (1898–1973) US astronomer: explained spectral lines seen in nebulae.

In the 1860s HUGGINS had observed spectral lines in nebulae that did not correspond to any known element. In 1928 Bowen was able to explain these as being due to doubly and triply ionized oxygen and nitrogen atoms, and not to a previously unknown element as had been thought. It is the transition from such highly excited atomic states to more stable forms that gives the characteristic red and green emission colours of nebulae.

Bowen, Norman Levi (1887–1956) Canadian geologist: used experimental petrology to reveal stages in formation of igneous rocks.

The son of English immigrants, Bowen entered Queen's University in Kingston as a student in 1903, and later worked there, at the Geophysical Laboratory at Washington, DC, and at Chicago. His main work was in experimental petrology. From 1915 he studied the crystallization of natural and synthetic minerals under controlled conditions of temperature and pressure; and he linked these results with field observations of igneous rocks. He was able to deduce a crystallization series in which differentiation occurs through the separation of crystals in stages from fused magma. The geochemistry of rock-forming silicates is complex, but Bowen's work allowed the pattern of their behaviour to be understood in general terms and was summarized in his book *The Evolution of Igneous Rocks* (1928).

Boyer, Herbert Wayne (1936–) US biochemist: developed recombinant DNA technique to synthesize proteins.

A graduate of Pittsburgh, Boyer became professor of biochemistry at the University of California at San Francisco in 1976. He showed in 1973 that a functional DNA can be constructed from two different gene sources, by splicing together segments of two different plasmids from the bacillus *Escherichia coli* (plasmids are deposits of extrachromosomal DNA found in some bacterial strains). The result of this recombinant RNA technique, known as a chimera, was then inserted into *E. coli* and was found to replicate and to show traits derived from both the original plasmids. By the late 1970s the method was in use by Boyer and others to give biological syntheses of costly proteins such as insulin and growth hormone.

Boyle, Robert (1627–91) Irish chemist: established the study of chemistry as a separate science and gave a definition of an element.

The youngest of the 14 children of the first Earl of Cork, Boyle was educated by a tutor at home (Lismore Castle) and at Eton. He showed an ability in languages before the age of 8, and in his interest in algebra he found a useful distraction during convalescence (he was to suffer ill-health throughout his life). His education was continued with a Grand Tour of France and Italy (1638–44), accompanied by his brother Francis and a tutor. In Italy he studied the work of the recently deceased GALILEO. On the death of his father, Boyle retired to live simply on his estate at Stalbridge in Dorset, where he took no part in the English Civil War then raging.

Boyle moved to Oxford in 1654. He worked on an improved air-pump (which HOOKE made for him), showing for the first time that Galileo was correct in his assertion that all objects fall at the same velocity in a vacuum. His most famous experiment was with trapped air compressed in the end of a closed shorter end of a U-shaped tube, by the addition of mercury to the open longer end of the tube, which showed that the volume of air halved if the pressure was dou-

bled. The work was published (1660) and became known as Boyle's Law (in Britain and the USA; credited to MARIOTTE in France): it states that for a fixed mass of gas at constant temperature, the pressure and volume are inversely proportional, ie pV = constant.

With the publication of *The Sceptical Chymist* (1661), Boyle prepared the way for a more modern view of chemistry, which put aside alchemical ideas and the Aristotelian doctrine of the four elements. He proposed the notion of elements as 'primitive and simple, or perfectly unmingled bodies' and that elements could be combined to make compounds and that compounds could be divided into their elements. Later LAVOISIER used this approach experimentally; but it was Boyle who changed chemical attitudes and prepared the way for PRIESTLEY and Lavoisier to create the Chemical Revolution. He also believed in the atomic theory and the importance of the shape of the atoms; his views here were taken from older writers. Boyle was a founder member of the Royal Society.

Boys, Sir Charles Vernon (1855–1944) British experimental physicist: ingenious inventor of sensitive instruments.

Educated at Cambridge and a Fellow of the Royal Society, Boys distinguished himself as a clever and original experimenter. In 1895 he designed a torsion balance, which was an improvement on previous models, and with this he determined the value of NEWTON's constant of gravitation, thus arriving at a value of 5.5270 for the mean density of the Earth. He invented the micro-radiometer, a combination of a thermocouple and a delicate suspended-coil galvanometer, and with it he was able to measure the heat radiation from the Moon and planets. He proved that the surface temperature of Jupiter is low. He used quartz fibres instead of silk for delicate suspension instruments and obtained them by shooting from a bow an arrow with the molten quartz attached. He also designed a calorimeter to measure the thermal power of coal gas, and a camera with a moving lens with which he obtained some remarkable photographs of lightning flashes.

Bradley, James (1693–1762) English astronomer: discovered stellar aberration; obtained first accurate measurement of the speed of light and direct proof of Earth's motion.

Bradley was HALLEY's successor as Astronomer Royal. While attempting to observe parallax in the position of γ Draconis (caused by the Earth's movement across the diameter of its orbit), Bradley found that the star did indeed appear to move, but that the greatest contrast was between September and March, not between December and June, as would be expected from parallax. He deduced that the movement (aberration) he saw was related to the ratio of the velocity of light to the velocity of the Earth about the Sun (the latter is about

30 km s^{-1}, and the ratio about 10 000:1). This discovery allowed him to estimate the speed of light to be 3.083×10^8 m s^{-1}, which is more accurate than RÖMER's value. It also gave the first direct evidence for the Earth's motion about the Sun. Bradley also discovered nutation, the wobble of the Earth's axis caused by the changing gravitational attraction of the Moon due to its slightly inclined orbit. It was not until BESSEL's work, a century later, that stellar parallax was observed.

Bragg, Sir (William) Lawrence (1890–1971) British physicist: founder with W H Bragg of X-ray crystallography.

Born in Adelaide, W L Bragg studied mathematics there and at Cambridge, and in 1910 moved his interest to physics. Like his father, he was attracted by VON LAUE's observation that X-rays could be diffracted by crystals. Bragg showed that the condition for diffraction by a crystal with lattice planes (layers of atoms) d apart, for X-rays of wavelength λ and angle of incidence θ, is that $n\lambda = 2d\sin\theta$ (Bragg's Law) where n is an integer. The atomic layers of a crystal acted as mirrors, reflecting X-rays, with interference resulting from reflections at different layers when the angle of incidence met the above condition. Using an X-ray goniometer made by the father (who had taken instruction in instrument-making in Adelaide) the pair were able to measure X-ray wavelengths and then to measure d, the interatomic distance, in crystals of diamond, copper, sulphur and salts such as KCl (which they found contained only ions and no molecules). Previously, crystallography had been concerned with the angles at the exterior of crystals; now X-ray crystallography could study their atomic interior.

At 25, Lawrence Bragg was the youngest Nobel prizewinner, sharing the prize in 1915 with his father. In 1919 he became professor at Manchester and in 1938 at Cambridge. He developed methods whereby X-ray diffraction by crystals (giving on a photographic plate or film a pattern of spots whose position and intensity could be measured) can be used to determine electron density within the crystal and therefore the position of the atoms. Modern metallurgy, crystallography and molecular biophysics owe much to his methods and to those of his co-workers in Cambridge. Like his father he became director of the Royal Institution (in 1954), did much to popularize science and was knighted.

Bragg, Sir William (Henry) (1862–1942) British physicist: discovered characteristic X-ray spectra; and developed (with his son) X-ray diffraction methods for determining crystal structures.

Bragg is unusual among noteworthy researchers in that his first significant research was done when he was over 40. However, he had as co-worker after 1912 his son Lawrence (see entry above) and their success brought a Nobel Prize in 1915; they are the only father–son pair to share one. William studied at Cambridge and did so well in mathematics that he was appointed professor in Adelaide in 1886.

In 1904 he gave a major lecture on the new subject of radioactivity and was spurred by this to research on the subject. In 1909 he took up his duties as professor at Leeds and there began work on X-rays, inspired by VON LAUE's recent work.

In 1913 Bragg found that, when X-rays are generated by the impact of high-energy electrons on a platinum target, the resulting continuous spectrum of X-rays contains some lines whose position is characteristic of the metal target. MOSELEY was shortly to use these X-ray spectra in a valuable way. Bragg, with his son Lawrence, went on to examine the wavelengths of X-rays by using crystals (see entry above); their method founded X-ray crystallography.

W H Bragg moved to University College London in 1915, worked on submarine detection in the First World War, and became director of the Royal Institution in 1923. (See panel overleaf.)

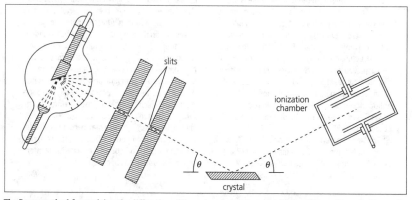

The Bragg method for studying the diffraction of X-rays by crystals. A narrow beam of X-rays from the X-ray tube (left) strikes the crystal; an ionization chamber is used to find the position of the diffracted beam.

THE THREE FORMS OF CARBON

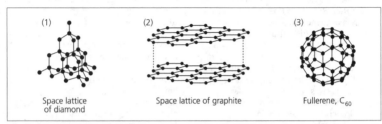

(1) Space lattice of diamond

(2) Space lattice of graphite

(3) Fullerene, C_{60}

Several of the chemical elements exist in more than one form, known as allotropic forms or simply allotropes. The allotropes of a given element consist of the same atoms, and the differences between them arise from differences in the arrangement of these atoms, the way they are bonded to one another.

Two forms of carbon occur in nature as mineral deposits and have long been known and used. One of them is diamond, the hardest, most brilliant and most prized of gemstones. In a diamond every atom (except the minority at the surface) is surrounded by four neighbours, equally spaced around it, as was first shown by W H and W L Bragg by X-ray diffraction methods. Graphite, the other long-known form of carbon, is very different: black, soft and used in pencils and as a lubricant. Its carbon atoms are arranged in net-like flat sheets, with each atom being a member of a six-atom ring; the sheets are fairly well apart and are only weakly bonded to one another. Both diamond and graphite have industrial uses, and both are made synthetically to supplement supplies from mineral deposits. Carbon black, soot and coke all have disorganized, graphite-like structures.

The third rather recently discovered form of carbon, the fullerenes, form a very different and highly surprising group. Probably the long delay in their discovery – compared with diamond and graphite -– stems from the fact that there is no natural path which concentrates them and exposes them for discovery. They exist as discrete molecules (whereas diamond and graphite are giant extended molecules, of indefinite size); their molecular structure is weirdly novel; and their discovery was achieved by research groups following rather disparate routes. The first suggestions that carbon atoms could form molecules in the form of these hollow closed cage-like structures were made in the 1960s and 1970s, but the first sample was obtained in 1990 .

The first good evidence for the existence of such compounds came in 1985, when R F Curl (1933–) and R E Smalley at Rice University in Texas, working with H W Kroto (1939–) of Sussex University, UK, laser-vaporized graphite in a jet of helium and showed that the carbon fragments were rich in C_{60} and C_{70} molecules. They proposed that these remarkably stable molecules were closed shells of carbon atoms, with C_{60} being a spheroidal structure with its carbon atoms linked to form 12 pentagons dispersed among 20 hexagons, so that it resembles a European football (an icosahedron). The larger C_{70} would have 25 hexagons, and be rugby-ball shaped. Because of the clear similarity to the geodesic buildings created by the architect Buckminster Fuller, the class of new compounds were named fullerenes (and informally buckyballs).

Earlier, in 1983, W Krätschner of the Max Planck Institute in Heidelberg, with D Huffman and L Lamb at Arizona University in Tucson, had studied the ultraviolet spectra obtained from an electric arc between graphite electrodes in a helium atmosphere, and they re-examined this in 1989, soon finding that C_{60} was formed in the arc and gathering evidence from its spectra confirming the 'buckyball' structure. By 1990 they showed that soot from their arc partly dissolved in benzene to give a magenta solution and evaporation of this gave orange crystals of C_{60}, now easily studied in a range of ways which left its structure in no doubt at all. And C_{70} can be made in the same way.

These strange substances, now easily available, offer a whole new range of interesting prospects to chemists and physicists. They are certainly present in sooty flames, and in interstellar dust. Larger molecules on the same pattern have been made and others can be expected, including the hyperfullerenes C_{240}, C_{540} and C_{960}. They offer novel possibilities both for chemistry and electronics. Compounds of C_{60} include an insulator, a conductor, a semiconductor, a superconductor, and a ferromagnet which strongly suggests an interesting future in layered microelectronic devices. The chemical properties which may become available from fullerene compounds likewise offer remarkable and exciting prospects, and a period of intensive investigation has begun.

IM

Brahe, Tycho, Tyge Brahe (*Dan*) [brah-hoe] (1546–1601) Danish astronomer: produced important star catalogue; the greatest pre-tele-scopic observer.

Brahe, son of a nobleman, was brought up by a childless uncle who effectively kidnapped him, gave him a good education and planned a political career for him. However, young Tycho at 14 saw the partial solar eclipse of 1560 and devoted his life to astronomy thereafter.

Brahe was without doubt the greatest astronomical observer of the pre-telescopic era. In 1572 he observed a nova (exploding star) in Cassiopeia, the first to be visible to the naked eye since 134 BC, and demonstrated that it was a 'fixed' star and outside the solar system. (It was brighter than Venus for more than a year.) This was cosmologically very important, as it had been believed since ARISTOTLE's time that the stars were eternal and immovable. His observations made his reputation. In 1577 the patronage of the king of Denmark, Frederick II, made possible his second great achievement. Frederick gave him the island of Hven as a gift for life, with funds to build the observatory of Uraniborg; Brahe furnished it with the best and largest instruments available, many of them designed by himself. He devoted the next 20 years to measuring the positions of 777 stars with unprecedented accuracy, thus providing an invaluable body of information for later astronomers, particularly KEPLER. He was probably the first to realize that multiple observations (such as he made) are much superior to single measurements in scientific work..

In 1596 Frederick's successor, Christian IV, forced Brahe to leave Hven. After 3 years of travelling he settled in Prague, sponsored by the mad emperor of the Holy Roman Empire, Rudolph II. He was given a castle near Prague as an observatory, and acquired the young Kepler as his assistant; the association was very fruitful, although stormy. Brahe died 2 years later, leaving Kepler to publish their star catalogue, the *Rudolphine Tables*, in 1627. Talented, energetic, eccentric and quarrelsome, Brahe lost most of his nose in a duel when at 19 he fought over a mathematical dispute; his false nose, made by himself from silver, can be seen in contemporary portraits. The nova of 1572 is known as 'Tycho's star' and the best-known of lunar craters is also named after him. The magnificent Uraniborg observatory was destroyed by fire in the Thirty Years War.

Brattain, Walter (Houser) (1902–87) US physicist: co-inventor of the transistor.

Born in China, Brattain grew up in the state of Washington on a cattle ranch, and gained his PhD in physics at Minnesota in 1929. In the same year he joined the talented team at Bell Telephone Laboratories, and soon began work on the surface properties of semiconductors; at first he used copper(I) oxide, but during the Second World War silicon became available, and this and germanium offered better prospects. Working with BARDEEN and SHOCKLEY, and using a mix of theory and experiment, the point-contact transistor was developed by 1947; it used a thin germanium crystal and both rectified and amplified current. For many purposes, the days of the vacuum tube or thermionic valve were numbered and the silicon micro-chip, smaller, cheaper and requiring less power, moved towards the dominant place it has held in electronics ever since. Brattain was very much a practical physicist, with a special interest in surfaces. When he left Bell in 1967, he went on to study the lipid surfaces of biological membranes at Whitman College, where he had once been a student. He shared a Nobel Prize with Bardeen and Shockley in 1956.

Braun, Karl Ferdinand [brown] (1850–1918) German experimental physicist: introduced crystal diodes and the cathode-ray oscilloscope.

Braun studied at Marburg and Berlin and taught at Tübingen and Strasbourg. In 1874 he found that some crystalline semiconductors (eg PbS) could be used as rectifiers to convert AC to DC. He used this from 1900 in his crystal diodes, which made possible the crystal radio receiver. He also modified a cathode-ray tube so that its electron beam was deflected by a changing voltage; the resulting cathode-ray oscilloscope has been much used in scientific work and is also the basic component of a TV receiver. He shared a Nobel Prize with MARCONI in 1909.

Braun, Wernher Magnus Maximilian von [brown] (1912–77) German–US rocket engineer: a pioneer of rockets and space travel.

The son of a baron and a former Government minister, von Braun was educated at the Zürich and Berlin Institutes of Technology. In 1932 he started working on rocket design for the German military, developing his first successful liquid-fuel rocket 2 years later. By 1938 he was technical head of the rocket research establishment at Peenemünde, where he was responsible for the V-2 supersonic ballistic missile used in the Second World War. At the end of the war he took his entire development team to surrender to the American army. He subsequently became a major figure in the American space programme, designing the Jupiter rocket that put America's first satellite, Explorer I, into orbit in 1958 and being influential in the Saturn rocket that put the first man on the Moon in 1969. Von Braun resigned from NASA in 1972, feeling that the American government was no longer strongly committed to space exploration.

It is probable that in 1945 he would have been charged as a war criminal but for his valuable expertise. There is evidence that he was active in the brutal treatment of slave labourers at the V-2 launch sites.

Breit, Gregory [briyt] (1899–1981) Russian–US

physicist: made contributions in quantum mechanics and nuclear physics.

Breit moved to America at the age of 16 and finished his doctorate at Johns Hopkins University at 22. After several years of travelling, with posts at Leiden, Harvard and Minnesota, he gained a position at the Carnegie Institute, Washington, DC. His later years were spent at New York, Wisconsin and Yale. With TUVE he measured in 1924 the height and density of the ionosphere by reflecting short bursts of radio waves from it, which was essentially the first use of radar imaging.

However, Breit's major research was in quantum mechanics, nuclear physics and the interaction of electrons and photons (quantum electrodynamics), often in collaboration with WIGNER. The formula giving the absorption cross-section of a nucleus as a function of energy of incoming particles is known as the Breit–Wigner formula.

Brenner, Sydney (1927–) South African–British molecular biologist: co-discoverer of triplet nature of genetic codons.

Brenner's parents had emigrated to South Africa from Russia and Lithuania. He qualified in medicine and medical biology at Johannesburg and then became a research student with HINSHELWOOD at Oxford, working on bacteriophage. In 1957 he joined the Medical Research Council's Molecular Biology Laboratory at Cambridge, becoming its director 1979–86.

In the 1950s Brenner did notable work in showing that the triplets (codons) of bases in DNA chains, each of which were believed to code for a specific amino acid destined for protein synthesis, do not form an 'overlapping' code. Thus in a sequence ... ATCGCATAG ... the codons could be ATC, GCA, TAG ... but not ATC, TCG, CGC.... By 1961 Brenner, CRICK and others had confirmed that codons are triplets (and not, for example, quadruplets) and that neither overlapping nor 'punctuation marks' appeared to exist in the code. In the same year Brenner and others also demonstrated that the ribosomes, which require an instructional code to carry out their task of protein synthesis, receive it in the form of a special type of RNA, messenger RNA (mRNA).

In the 1970s he began intensive studies of the nervous system of a type of nematode worm. Although less than 1 mm long, its nervous system is complex and roughly 100 genes contribute to the make-up of its nervous system, which has about 300 neurones and so is usefully intermediate between *Escherichia coli* (1 neurone) and man (10^{10} neurones). Most mutants of the worm show variations in the nervous system which can be informative, and Brenner's work has involved slicing a worm into up to 20 000 serial sections for electron microscopy in order to define the anatomy of the system and ultimately to relate molecular biology to its visible structure and development.

Brewster, Sir David (1781–1868) British physicist: discovered polarization by reflection.

Trained for the church, Brewster after graduation turned to physics. He produced much factual, non-speculative work on the reflection, absorption and polarization of light. When light is reflected from a non-metallic surface, partial polarization occurs. As the angle of incidence is increased towards a glancing angle the polarization increases, passes through a maximum (the Brewster, or polarization, angle) and then decreases. Brewster's law states that the tangent of the Brewster angle is equal to the refractive index of the reflecting substance (1815). The Brewster angle and the angle of refraction sum to a right angle.

Brewster improved the optics of lighthouses, invented the kaleidoscope and helped to found the British Association for the Advancement of Science.

Bridgman, Percy Williams (1882–1961) US experimental physicist: studied effects of very high pressure on materials.

A Harvard graduate, Bridgman stayed there in a variety of positions until retirement. Most of his research was concerned with very high pressures, for which he designed special equipment. He showed that most liquids become more viscous under high pressure and that some solid compounds (eg ice) and solid elements (eg phosphorus) then exist in novel forms. He achieved pressures of 10^{10} N m^{-2}.

The Bridgman effect is the absorption or evolution of heat when an electric current passes through an anisotropic crystal. He was awarded the Nobel Prize in 1946.

Briggs, Henry (1556–1630) English mathematician: introduced 'common logarithms' (ie to base 10).

Educated at Cambridge, Briggs became professor of mathematics at Oxford in 1620. Noted for his work on logarithms, which did more than anything to popularize their use, he suggested the decimal base instead of the Napierian or natural base, and undertook the tedious work of calculating and preparing the tables, which extended to the 14th place of decimals. He also introduced the method of long division which is in common use. His *Logarithmical Arithmetic* was published in 1624.

Bright, Richard (1789–1858) British physician: pioneer user of clinical biochemistry.

Bright studied and wrote on botany, zoology, geology and medicine, and travelled widely in Europe, from Iceland to Hungary. Working thereafter as a physician in London, he showed that disease can in some cases be linked with body chemistry and with post-mortem findings. He is best known for his recognition that kidney disease (nephritis) is linked with dropsy (accumulation of fluid in the body) and with the presence of albumin in the urine. The term Bright's disease was formerly used to cover the non-suppurative inflammatory renal (ie kidney) diseases that show these symptoms. Bright was also a skilful artist and travel writer.

Broensted, Johannes Nicolaus (1879–1947) Danish physical chemist.

Broensted qualified in chemical engineering in 1897 and then in chemistry in 1902 in Copenhagen. He taught there from 1905. He worked mainly in electrochemistry and reaction kinetics, applying thermodynamics to chemical problems. He is best known for a definition of acids and bases, due to him (and independently and concurrently in 1923 to T M Lowry (1874–1936) of Cambridge). This, the Broensted–Lowry definition, defines an acid as a substance with a tendency to lose a proton, and a base as a substance that tends to gain a proton.

Broglie, Louis-Victor Pierre Raymond, duc (Duke) **de , Prinz** (Prince) [broglee] (1892–1987) French physicist: discoverer of the wave nature of particles.

Louis de Broglie was a member of a Piedmontese family; in 1740 Louis XIV had conferred on the head of the family the hereditary title of *duc*, which Broglie inherited in 1960 on the death of his brother Maurice (who was also a physicist). The German title *Prinz* dated in the family from service to the Austrians during the Seven Years War (1756–63). Broglie studied history at the Sorbonne and acquired an interest in science by service at the Eiffel Tower radio station during the First World War. He then took a doctorate at the Sorbonne (1924) and taught there as the professor of theoretical physics at the newly founded Henri Poincaré Institute (1928–62).

Our ideas concerning quanta stem from PLANCK (1900), and modern ideas on the interaction of matter and energy had begun with EINSTEIN in 1905. Broglie's work began with a derivation of WIEN's electromagnetic radiation law, based on light quanta with frequency v, mass hv/c^2 and momentum hv/c (1922). It then occurred to him to go beyond the idea of waves acting as particles and to suggest that particles can behave as waves. A particle such as an electron should move at the group velocity of a number of matter waves, which have wavelength $\lambda = h/mv$. This revolutionary idea appeared in Broglie's doctoral thesis of 1924, which was published as a paper of over 100 pages in *Annales de Physique* in 1925. The waves were detected and agreement with λ found through the wave interference, using the atoms of a crystal lattice as a diffraction grating. This was done by DAVISSON and GERMER using slow electrons (59 eV) and by G P THOMSON using fast electrons in 1927. The wave-particle duality was used by SCHRÖDINGER in his formulation of quantum mechanics, and it also began the great debate as to whether there is determinacy in quantum mechanics. Broglie received the Nobel Prize for physics in 1929.

Brongniart, Alexandre [brónyahr] (1770–1847) French geologist and palaeontologist: pioneer of stratigraphic geology.

Brongniart spent some time as an army engineer before being appointed, in 1800, director of the famous porcelain factory at Sèvres, a post he was to hold all his life with considerable success. At the beginning of the 19th-c, working with CUVIER, he pioneered stratigraphic geology, being the first to use the fossils contained within a geological stratum to identify and date that layer. In 1811 they published a classic study of the geology of the Paris Basin, setting out the Tertiary rocks in order and classifying them according to the fossils they contained. Brongniart was also amongst the first to recognize strata containing alternately freshwater and seawater molluscs, and to interpret this as indicating periodic changes in sea level, an important discovery.

Brooks, Harriet (1876–1933) Canadian nuclear physicist.

Harriet Brooks went to McGill University in Montreal and gained a first class honours degree in 1898. She was invited to join RUTHERFORD's well-equipped physics research group at McGill, as his first graduate student, and gained a master's degree in 1901; the highest award at that time and the first awarded to a woman at McGill.

Radioactivity was then a very novel research area, the first radioactive effects having been observed by BECQUEREL in 1896. Brooks studied the 'radioactive substance' given off by thorium and, using a diffusion method, she identified the emanation as a radioactive gas of relative atomic weight in the 40–100 range. The method gave a low value for the gas, which is an isotope of radon, but it was shown that the gas had a significantly lower molecular weight than thorium and so could not be simply a gaseous form of thorium. This led Rutherford and SODDY to the realization that a transmutation of one element to another had occurred; this entirely novel idea was central to the whole development of nuclear physics and chemistry.

Later Brooks worked on a comparison of the beta-radiations from the elements thorium, uranium, radium and polonium, and showed it to consist of fast negative particles. She spent 1902–03 working with J J THOMSON at the Cavendish Laboratory in Cambridge and in a letter to Rutherford in 1903 refers to radioactivity decreasing to one-half of its value in about a minute, the first measurement of the half-life of the thorium emanation (radon-220).

Back at McGill Brooks observed the recoil of the radioactive polonium atom, although she attributed this phenomenon to volatility of the decay product from the polonium; Rutherford recounted Brooks's findings in his Bakerian Lecture in 1904. Later, this important recoil effect was rediscovered by HAHN and others.

In 1904 Harriet Brooks moved to Barnard College, New York City as tutor in physics, and 2 years later became engaged to be married to a physicist from Columbia. The dean of the college insisted that Brooks must resign, saying 'the good of the College and the dignity of the woman's place in the home demand that your marriage shall be a resignation.' This was common practice at the time and Brooks debated the decision, but the result of the dispute led to her resignation and a broken engagement. She went to the Curie Institute (1906–07) and worked with André Debierne (1874–1949) on the recoil of radioactive atoms using the radium decay series. In 1907 she chose to marry and abandoned her career.

Rutherford said of her that 'next to Mme CURIE she is the most pre-eminent woman physicist in the department of radioactivity'. If she had been allowed to combine research work and marriage her work might have been better appreciated. She died at the age of 56 of a 'blood disorder' and it is hard to avoid the suspicion that exposure to radiation was involved; in the 'golden age' of physics its hazards were not appreciated.

Broom, Robert (1866–1951) British–South African physician and palaeontologist: confirmed significance of *Australopithecus* as a hominid and proved his bipedality.

After graduating in medicine from Glasgow in 1889, Broom practised general medicine in Australia for some years before moving to South Africa in 1897. In 1934 he gave up medicine and was appointed palaeontologist at the Transvaal Museum, Pretoria.

Something of an eccentric (he buried dead Bushmen in his garden, exhuming them when decomposed), Broom did much to clarify the classification of the fossil reptiles of Africa. In 1936, at the age of 69, he turned his attention to hominid fossils and was almost immediately successful in finding at Sterkfontein a skull of *Australopithecus africanus*, a hominid first identified by DART in 1924. Two years later a small boy brought him the jaw of another early hominid, *Australopithecus robustus*, now believed to have lived about 1–2 million years ago. These two finds convinced a hitherto sceptical scientific community of the significance of Dart's earlier claim of *Australopithecus africanus* as an ancestor of man. In 1947, when over 80, Broom found a partial skeleton of *Australopithecus* that included the pelvis, giving the first conclusive evidence that he had walked upright.

Brown, Herbert Charles (1912–   ) US chemist: introduced organoboranes for organic synthesis.

Brown was born in London, but his family emigrated to Chicago in 1914. He obtained a university education with difficulty, but his talent secured a professorship at Purdue in 1947 which he held until retirement in 1978. His researches included studies on carbocations and on steric effects, and especially on boron compounds. He was co-discoverer of sodium borohydride ($NaBH_4$) and pioneered its use for the reduction of organic compounds; and he found a simple way of making diborane (B_2H_6) and discovered its addition to unsaturated organic molecules to form organoboranes. The latter are of great value in organic syntheses, the sequence of reactions being known as hydroboration. He was awarded a Nobel Prize in 1979.

Brown, Robert (1773–1858) British botanist: named cell nucleus, and advanced plant taxonomy.

While on service as an army medical officer, Brown met BANKS and as a result joined the Flinders expedition to Australia in 1801. This lasted 5 years; Brown as naturalist collected 4000 plant species and spent 5 years classifying them. In doing so he established the main differences between gymnosperms and angiosperms, and he also observed an essential part of living cells in which he named the nucleus (1831). In 1827 he noticed that a suspension of pollen grains in water showed, under the microscope, continuous erratic movement. He found this also with other small particles (for example, of dyes). He had no explanation for this Brownian movement, but much later it was recognized to be due to the molecular motion of the liquid; this was the first evidence for the existence of molecules based on direct observation rather than on deduction.

Bruce, Sir David (1855–1931) British microbiologist: investigated undulant fever and sleeping sickness.

Bruce belongs to a tradition of military medical men who worked on tropical diseases in an age of colonial concern. He studied medicine at Edinburgh and joined the Army Medical Service in 1883. The next year he was posted to Malta. There he studied undulant fever (now called brucellosis) and in 1886 he isolated the causal bacterium. Later he and his assistants showed that unpasteurized goat's milk carried the infection to the garrison there, and control followed. Later still it was found that the same organism caused contagious abortion in cattle and that it can be transmitted by a variety of animals.

In 1894 Bruce went to South Africa to study nagana, another disease of cattle, and soon showed it to be carried by the tsetse fly and to be due to a trypanosome, a protozoal parasite now named as *Trypanosoma brucei*. Soon Bruce and others showed that the human disease known as African sleeping sickness (trypanosomiasis) is due to the same organism, transmitted in the same way by the bite of the tsetse fly.

In 1912 Bruce was promoted to Surgeon-General and in the First World War was commandant of the Royal Army Medical College.

His research was always carried out with his wife, Mary Elizabeth, a skilled microscopist. She shared all his work, including 2 years in a primitive hut in the Zululand bush studying nagana and a period as theatre nurse during the siege of Ladysmith, with Bruce as the surgeon.

Brunel, Isambard Kingdom [broonel] (1806–59) British civil engineer; pioneer designer of large steamships.

Brunel revealed a talent for drawing and grasp of geometry by the age of 6. His father, Sir Marc Isambard Brunel (1769–1849), having fled his native France and the Revolution for America before settling in Britain, educated his son in England, Normandy and Paris. Brunel joined his father in his engineering projects and in 1825 helped him to construct the first tunnel under the Thames (designed for foot passengers but later used by the London Underground). Isambard Brunel (Kingdom was his mother's surname) was a short man with a commanding presence, an ability to lead and a capacity for hard work which contributed to his early death. He confessed to self-conceit.

In 1830 Brunel won the competition for a design for the Clifton Suspension Bridge, his first independent work. He was appointed engineer of the Great Western Railway in 1833. He surveyed the route, designed tunnels, bridges and the termini at Paddington and Temple Meads, Bristol. Then Brunel turned to the design of steamships to cross the Atlantic, the problem being carriage of sufficient fuel for the distance. Brunel realized that capacity for fuel increased with the cube of the ship's size; its power requirement increased with the square of the size, so a big enough vessel could succeed. He built the *Great Western* in oak in traditional manner, to make an extension to the Great Western Railway; it made the crossing to New York in 15 days in 1838. He designed and built the *Great Britain*, an iron-hulled, screw-driven vessel which was then the largest vessel afloat. He went on to design the *Great Eastern* to carry 4000 passengers around the world without refuelling. Immense, double-skinned with 10 boilers, the ship was beset with financial and other problems from the start; it was eventually used to lay the Atlantic cable of 1865.

The great liners which dominated intercontinental travel for a century stemmed from Brunel's confident approach to large-scale ship construction based on steel.

Bruno, Giordano (1548–1600) Italian philosopher: supporter of the Copernican (heliocentric) system.

Bruno entered but later left the Dominican Order, and spoke and wrote supporting radical views on religion, the infinity of space, the motion of the Earth and the Copernican system. He travelled widely in Europe, was arrested by the Inquisition (1592) and after a lengthy trial refused to recant. Details of the trial have been destroyed, but Bruno was burned at the stake in 1600. It is usually believed that this event influenced GALILEO in favour of recanting when he was similarly charged with heresy and supporting the heliocentric system in 1633.

In 1991 it was suggested that Bruno had an important position as a spy. In the early 1580s he was accredited as a chaplain in the French embassy in London, at a time when England saw itself in grave danger from Catholic conspiracy. Bruno secured information on Spanish and French schemes and reported, under the name 'Henry Fagot' to Elizabeth I's spymaster, Sir Francis Walsingham, and his espionage overturned the Throckmorton plot. The truth may never be shown with certainty, but it seems likely that Bruno has as significant a place in the political scene as in science.

Buchner, Eduard [bukhner] (1860–1917) German organic chemist: showed that fermentation does not require living cells.

Buchner's elder brother Hans (see below), first interested and guided him in science, succeeding so well that Eduard studied botany under NAEGELI and chemistry under BAEYER and became the latter's assistant. From 1893 he was professor at Kiel and, after several moves, at Würzburg from 1911 until he was killed in action in the First World War.

Until Buchner's work in 1897, it had been believed that fermentation required intact living yeast cells. Buchner tested this view by grinding yeast cells with sand and pressing from the mixture a cell-free extract. This extract when added to sugar solution, caused fermentation to ethanol and CO_2 much as would yeast cells. The vitalist view was defeated. Buchner named the active principle 'zymase'. We now call such biological catalysts 'enzymes' and recognize that they are proteins, highly specific in action and involved in nearly all biochemical changes. Buchner won the Nobel Prize for chemistry in 1907. His brother Hans Buchner (1850–1902) worked in bacteriology and showed that protein in blood serum was important in immunity.

Buffon, Georges-Louis Leclerc, comte (Count) **de** [büfõ] (1707–88) French naturalist and polymath: surveyed much of biology and had early ideas on evolution of species.

Buffon's mother was wealthy and, despite his father's desire that Buffon should study law, it is likely that he studied medicine and mathematics. A duel made him leave France in 1730 for 2 years, but on his return he became active in scientific and financial circles; he was highly energetic and both increased his fortune and contributed to most of the sciences of the time. His range was vast; he translated HALES and NEWTON into French, introduced calculus into probability theory, and worked on microscopy, tensile strength, cosmology and geology, and the origin of life. His ideas were non-theological, rational and ahead of their time, if not always correct. From 1739 he was in charge of the Jardin du Roi, the natural history museum and botanical garden of Paris, which he much improved and enlarged. His vast and beautifully illustrated *Natural History* (44 vols by 1804) attempted to provide a survey of the natural world and was much esteemed. In it he noted that animal species are not fixed but show variation, and he recognized vestigial features such as the pig's toes, a contribution to later theories of evolution, together with his view of 'common ancestors' for similar species. He devised some eccentric experiments: for example to check the legend that ARCHIMEDES fired the Roman fleet with mirrors and the Sun's rays when 'distant by a bowshot' he used 168 mirrors, and ignited timber at 50 m range.

Bullard, Sir Edward (Crisp) (1907–80) British geophysicist: made first measurement of geothermal heat flow through the oceanic crust, and proposed dynamo theory for the Earth's magnetic field.

Bullard served in naval research during the Second World War, afterwards working in Cambridge and North America before becoming director of the National Physical Laboratory, England. In 1964 he was appointed director of the Department of Geodesy and Geophysics at Cambridge. Bullard made the first successful measurements of geothermal heat flow through the oceanic crust, establishing that it is similar in magnitude to that of continental crust, and not lower as had been thought.

After the start of the Second World War in 1939 he devised a protection for ships against magnetic mines by 'degaussing' ships with an applied external current.

In the late 1940s and 1950s, independently of ELSASSER, he proposed the dynamo theory for the origin of the Earth's magnetic field, in which the field is generated by the motion of the Earth's liquid iron core undergoing convection. Providing that there is a small magnetic field to start with, the movement of the molten iron will set up electric currents which will in turn generate the observed magnetic field. In 1965 Bullard was also the first to use computer modelling techniques to study continental drift, finding an excellent fit between Africa and South America at the 500-fathom (close to 1000 m) contour: a valuable contribution to what became the theory of plate tectonics.

Bunsen, Robert Wilhelm (1811–99) German chemist: wide-ranging experimenter, and pioneer of chemical spectroscopy.

Bunsen's father was librarian and professor of linguistics in Göttingen, and Robert studied chemistry there before travelling and studying also in Paris, Berlin and Vienna. He became professor at Heidelberg in 1852 and remained there until retirement, 10 years before his death.

Bunsen was pre-eminently an experimentalist with little interest in theory. His first major research did much to support the radical theory, due largely to DUMAS and LIEBIG, which held that organic groups ('compound radicals') correspond, in part, to the simple atoms of inorganic compounds. He prepared a series of compounds all containing the cacodyl group $(CH_3)_2As-$; and did so despite their remarkably offensive character. They combine a repulsive and persistent odour with toxicity and flammability. The presence in all of them of the same cacodyl group effectively established the theory. During this work Bunsen lost the sight of an eye and nearly died of arsenic poisoning; he excluded organic chemistry from his laboratory thereafter.

With his fellow-professor KIRCHHOFF he discovered the use of spectroscopy in chemical analysis (1859) and within 2 years they had discovered the new elements caesium and rubidium with its aid. He devised the Bunsen cell, a zinc-carbon primary cell which he used to obtain metals (Cr and Mb) by electrodeposition from solution, and others (Mg, Al, Na, Ba, Ca, Li) by electrolysis of the fused chlorides. To find the relative atomic mass of these metals he measured their specific heat capacity (to apply DULONG's Law) and for this he designed an ice calorimeter. He was a master of gas analysis, and used it in many ways, eg his study of Icelandic volcanoes and the improvement of English blast-furnaces. He was a pioneer, working with ROSCOE, in photochemistry, and for this devised a photometer and an actinometer. His great interest in analysis led him to invent many laboratory devices, including the filter pump. The Bunsen gas burner was probably devised and sold by his technician, Peter Desdega, and based on one due to FARADAY.

Bunsen was a great teacher and his lecture courses were famous; his researches continued until he was 80. Like DALTON he admitted that he never found time to marry (although this may have been because he worked with odorous compounds). EMIL FISCHER's wife said of him, 'First, I would like to wash Bunsen and then I would like to kiss him because he is such a charming man.'

Burbidge, (Eleanor) Margaret, *née* Peachey (1922–) British astronomer: leading optical astronomer.

Burbidge studied physics in London and afterwards researched in astronomy at Chicago, the California Institute of Technology, and Cambridge, before becoming professor of astronomy at the University of California, San Diego, in 1964. In 1972, on leave, she became director of the Royal Greenwich Observatory, then at Herstmonceaux Castle in Sussex. But her high hopes for optical astronomy using the 2.5 m Isaac Newton telescope there were frustrated by administrative work as well as by poor seeing conditions, and she resigned after a year and returned to California.

From 1948 she worked with her astrophysicist husband Geoffrey Burbidge (1925–), and in 1957 with HOYLE and W A Fowler (1911–95) they published on the formation of atomic nuclei in stars. She had also worked on quasars and gave the first accurate values for the masses of galaxies, based on her own observations of their rotation.

Burkitt, Denis (Parsons) (1911–93) British epidemiologist: discovered Burkitt's lymphoma, a cancer caused by a virus.

Born and educated in Ulster, Burkitt entered Trinity College, Dublin to study engineering but changed to medicine and specialized in surgery. Working in Uganda in the 1950s, he discovered the type of cancer now known as Burkitt's lymphoma. This presents as swellings of the jaw, usually in children of 6–8 years. Burkitt toured Africa to examine its incidence and found it mainly in areas where malaria is endemic, but no microorganism linked with it could be detected initially, so a virus was clearly a possibility. However, attempts made in London to establish tissue cultures of the cancer cells were unsuccessful until 1964. Then the cells were grown in culture and electron microscopy showed them to be infected with the Epstein–Barr virus. It seems likely that the lymphoma is caused by a conjunction of factors, including the virus (which is very common, worldwide) and exposure to malaria.

Burkitt was a proponent of high-fibre diets, partly because some bowel diseases common in developed countries are rare in Africa, where such diets are usual.

Burnet, Sir (Frank) Macfarlane (1899–1985) Australian medical scientist: made studies of virus and the immune system.

A graduate in medicine from Melbourne, Burnet spent two year-long visits studying bacteriology in London, and the rest of his career in Melbourne. In the 1930s he worked on viruses, where his successes included studies on bacteriophages (viruses which attack bacteria) and a method for culturing some viruses in living chick embryos. This last work led him to the view that an animal's ability to produce antibody in response to an antigen is not inborn, but is developed during fetal life. (Evidence that this is correct was later found by MEDAWAR.) Burnet also worked on the mode of action and the epidemiology of the influenza virus, the cholera vibrio, polio and Q fever. His clonal selection theory (1951) offered a general scheme explaining how an immune system develops the ability to distinguish between 'self' and 'non-self' and initiated both controversy and further work by others. He shared a Nobel Prize with Medawar in 1960.

Bury, Charles [beree] (1890–1968) British physical chemist: little-known theorist on electronic structure of atoms.

BOHR is usually credited with the feat of giving the first clear account of the arrangement of electrons in atoms and its relation to chemical behaviour. In fact the first rough suggestion of electron 'shells' is due to J J THOMSON (1904), and LANGMUIR (1919) gave a more detailed shell model (partly incorrect), which he linked with chemical behaviour. In 1921 Bohr gave a better version, but it was very brief and was only a limited account (he gives electronic structures only for the noble gases). Within a month, a concise, clear and complete account was given by Bury (in the *Journal of the American Chemical Society*), who had written his paper before he saw Bohr's. All later accounts use the Bury scheme.

Bury was an Oxford graduate in chemistry who served 5 years in the First World War. His classic paper, written when he was 31, was his first, but he went on to study the relation of colour to structure in dyes and the properties of micelles. Again, his work on dyes appears to have preceded better-known work by others: Bury was a modest man.

Butenandt, Adolf Frederick Johann [bootuhnant] (1903–95) German organic chemist: developed chemistry of sex hormones.

A student at Marburg and Göttingen, Butenandt later held posts in Danzig, Berlin and Tübingen. His work was mainly in the field of sex hormones. In 1929 he isolated the first pure sex hormone, oestrone, from human pregnancy urine. He also isolated the male hormone androsterone from normal human male urine in 1931. These potent hormones are present in natural sources only in small amounts: eg 15 mg of androsterone was isolated from 15 000 litres of urine donated by Viennese policemen. Butenandt was awarded the Nobel Prize in 1939, but the German government forbade him to accept it. He secured progesterone, the mammalian pregnancy hormone (20 mg from the ovaries of 50 000 sows), in 1934; and later worked on an insect hormone (ecdysone) and other insect pheromones. He showed the relation between the above compounds, which are all members of the steroid group, and did much to establish the chemistry of this interesting and valuable group.

Buys Ballot, Christoph Hendrik Diederik [boyzbalot] (1817–90) Dutch meteorologist: described the direction of rotation of cyclones.

Buys Ballot was appointed professor of mathematics at the University of Utrecht in 1847, and in 1854 founded the Netherlands Meteorological Institute. In 1857 he showed that, in the northern hemisphere, winds circulate counterclockwise around low-pressure areas and clockwise around high pressure ones, a fact now known as Buys Ballot's Law; the situation is reversed in the southern hemisphere.

Cagniard de la Tour, Charles [kanyah(r)] (1777–1859) French physicist: discovered critical state of liquids.

Cagniard studied at the École Polytechnique, Paris. He is primarily remembered for his discovery in 1822 of the critical state of liquids. For certain liquids, when heated in a sealed tube, the meniscus disappears and liquid and vapour become indistinguishable at a critical temperature. Cagniard was also an inventor; his best-known device being the disk siren (in which a note is produced by blowing air through holes cut in a spinning disk).

Cailletet, Louis Paul [kiytay] (1832–1913) French physicist: pioneer of liquefaction of gases.

Cailletet studied in Paris and then returned to Chatillon-sur-Seine to manage his father's iron-works. His first interest was metallurgy, which led him to study blast furnace gas and to interest himself in gas properties. At that time, attempts to liquefy some gases (for example H_2, N_2 and O_2) had all failed and they were classed as 'permanent gases'. Cailletet learned of T ANDREWS's work on critical temperature, which suggested to him that more cooling was needed, as well as pressure, for success. He used the JOULE–THOMSON effect (the cooling which occurs when a gas expands through a nozzle), followed by pressure, and by 1878 had liquefied these 'permanent' gases. He was interested in flying (even in advance of the development of the aeroplane) and the sundry devices he invented included a high-altitude breathing apparatus and an aircraft altimeter.

Calvin, Melvin (1911–   ) US biochemist: elucidated biosynthetic paths in photosynthesis.

Calvin studied at Michigan, Minnesota and Manchester and then began teaching at the University of California at Berkeley in 1937. Except for war work on the atomic bomb, he remained there for the rest of his career. His interest in photosynthesis began in Manchester and developed from 1946, when new and sensitive analytical methods (notably the use of radioisotope labelling and chromatography) became available. Photosynthesis is the process whereby green plants absorb carbon dioxide from the air and convert it by complex stages into starch and into oxygen (which is discharged into the air, at the rate of about 10^{12} kg per year). This can be claimed as the most important of all biochemical processes, since animal life also depends on plant foods and on the oxygen-rich atmosphere which, over geological time, photosynthesis has provided.

Calvin allowed the single-celled green algae *Chlorella* to absorb radioactive CO_2 for seconds only, and then detected the early products of reaction. He identified a cycle of reactions (the reductive pentose phosphate or **Calvin cycle**) which form an important part of photosynthesis. He was awarded the Nobel Prize for chemistry in 1961.

Most recently his work involved the attempt to produce entirely synthetic sensitizers and catalysts which would permit the construction of a device to photochemically produce oxygen from water and reduce carbon dioxide to a useful chemical.

Camerarius, Rudolf Jakob [kamerayreeus] (1665–1721) German botanist: demonstrated sexual reproduction in plants.

Camerarius followed his father as professor of medicine at Tübingen, and he was also director of the botanic garden there. RAY and others had suggested that plants can reproduce sexually, but it was Camerarius who first showed by experiment that this is so. In 1694 he separated some dioecious plants (ie plants in which the male and female flowers are borne on separate plants) and showed that although the pistillate plants gave fruit, they did not produce seed in the absence of staminate flowers. He identified the stamens as the male plant organs and the carpels (consisting of the style, ovary and stigma) as the female apparatus of a flowering plant. He also described pollination.

Candolle, Augustin-Pyramus de [kādol] (1778–1841) Swiss botanist.

Candolle studied in Geneva and Paris, and from 1806–12 made a botanical survey of France as a government commission. His ideas on taxonomy, set out in his *Elementary Theory of Botany* (1813), developed from and replaced the schemes due to CUVIER and LINNAEUS and were used for 50 years. He believed that morphology, rather than physiology, should be the basis of taxonomy and that relationships between plants could be best seen by studying the symmetry of their sexual parts. From this he was led to the idea of homology. He also studied plant geography and the influence of soil type in his travels to Brazil and the Far East. He taught at Montpelier and, from 1816, at Geneva.

Cannizzaro, Stanislao [kaneedzahroh] (1826–1910) Italian chemist: resolved confusions on atomic and molecular mass.

Cannizzaro began his university life as a medical student, but attended a variety of courses and became attracted to chemistry, partly because he saw it as the basis of physiology. In 1847 he joined the rebel artillery in one of the frequent rebellions in his native Sicily, where his magistrate father was at the time chief of police. The rebellion failed, and Cannizzaro wisely continued his chemistry in Paris, with CHEVREUL. He returned to Italy 2 years later, teaching chemistry in three universities, all poorly equipped. In 1853 he discovered the Cannizzaro reaction, in which an aldehyde (aromatic, or having no alpha-hydrogen) is treated with a strong base to give an acid and an alcohol:

$$2\,RCHO + NaOH \rightarrow RCO_2Na + RCH_2OH$$

However, his main work was done in 1858, when he cleared the way to a single system of relative atomic and molecular mass. He did this by seeing the value of the theory due to AVOGADRO (then dead) and using it to deduce that common gaseous elements existed as molecules (H_2, N_2, O_2) rather than as single atoms. With this in mind, Avogadro's Law enables relative atomic and molecular mass to be deduced from the densities of gases and vapours. (Initially, as hydrogen gas was the lightest known, the hydrogen atom was assigned atomic mass = 1; we now use as a basis the common isotope of carbon = 12, which gives a very similar scale.) Cannizzaro's fervour as a speaker at a chemical conference in 1860, and a pamphlet he distributed there, convinced most chemists and removed basic ambiguities in chemical ideas during the 1860s. The half-century of confusion on atomic mass which had followed DALTON'S atomic theory had ended.

He became a Senator in 1871 and afterwards worked mainly on public health.

Cannon, Annie Jump (1863–1941) US astronomer: compiled Henry Draper catalogue of variable stars.

Originally interested in astronomy by her mother, Annie Cannon graduated from Wellesley College in 1884. For the next 10 years she lived at home but after the death of her mother she returned to study physics at Wellesley, and specialized in astronomy at Radcliffe College. In 1896 she was appointed by E C Pickering (1846–1919) to the staff at Harvard College Observatory where she began the study of variable stars and stellar spectra. These spectra were studied by Pickering's objective prism method, which made spectra visible even from faint (9th or 10th magnitude) stars. Colour film was not available and classification by eye was a skilled task.

Her first major publication (1901) was a catalogue of 1122 southern stars which built upon the early classification used by Williamina Fleming (1857–1911). Annie Cannon's system represented a sequence of continuous change from the very hot white and blue stars of types O and B, which showed many helium lines, through the less hot stars of types A, F, G and K to the very red stars of type M, which were cool enough that compounds of chemical elements, such as titanium and carbon oxides, could exist

in their atmospheres. However, it was not then known that it was a temperature sequence. In 1910 Annie Cannon's scheme was adopted as the official classification system at all observatories.

Her most important work was the *Henry Draper Catalogue of Stellar Spectra*, published by the Harvard Observatory between 1918 and 1924, which lists the spectral types and magnitudes of 225 300 stars, all those brighter than ninth magnitude, giving their positions and visual and photographic magnitude. In 1922 the classification system used by her in the catalogue was adopted by the International Astronomical Union as the official system for the classification for stellar spectra. She then began on an extension to fainter stars, in selected regions of the sky, down to about the 11th magnitude, and was occupied with this task until her death. She discovered 277 variable stars and five novae.

Annie Cannon received many academic honours; she received honorary degrees from four universities, including Oxford, where she was the first woman to be honoured with a doctor's degree (1925). (See panel overleaf.)

Cannon, Walter Bradford (1871–1945) US physiologist: introduced first radio-opaque agent.

Cannon was very much a Harvard man; he was an arts student there, then a medical student, and professor of physiology from 1906–42. RÖNTGEN discovered X-rays in 1896 and the next year Cannon, still a student, tried feeding a cat a meal containing a bismuth compound to give an X-ray 'shadow' of its alimentary tract. The method worked, and made the mechanics of digestion visible; and, with a barium compound in place of bismuth, it has been used in diagnostic radiography ever since. Cannon went on to work on the effect of shock and emotion on the nervous system and the transmission of nerve impulses. He developed BERNARD'S concept of the importance of a constant internal physiological environment (ie a narrow range of salt, sugar, oxygen and temperature in the living body) which he named homeostasis, and he studied the mechanism which achieves this essential equilibrium. Later he applied similar ideas to political and social organizations, but without the same success.

Cardan, Jerome, Girolamo Cardano (*Ital*), Hieronymus Cardanus (*Lat*) (1501–76) Italian mathematician and physician: gave general algebraic method for solving cubic equations.

Cardan was the illegitimate son of a Milanese lawyer, a situation which caused difficulty for him both practically and emotionally. He was taught mathematics by his father when young and educated at Pavia and Padua where he studied medicine. He was unable to enter the college of physicians because of his birth, but eventually gained recognition through his work and became professor of medicine at Pavia in 1544 and at Bologna in 1562. His work

THE ENTRY OF WOMEN INTO ASTRONOMY

The early route to professional astronomy for men in Britain was by a degree in mathematics or mathematical physics, then through paid work as an assistant at a major observatory. For women this path was not available, as positions in observatories were not open to them and the opening of women's colleges did not at first enable them to become professional astronomers. The opening of Queen's College (1848) and Bedford College (1849), London, provided tuition in mathematics and physics for women but unless they had access to instruments owned by their relatives, practical observation was not possible.

In the USA, Vassar, Smith, Mt Holyoke and other women's colleges had departments of astronomy and employed women to teach and run their observatories. Positions in computing, though poorly paid, allowed women in the USA to enter into astronomical work, while in Britain such work was done by young men.

For women interested in astronomy in Britain a scientific society was the main focus of activity. After 1838 women were permitted to attend the mathematics and physics section of the British Association for the Advancement of Science. The Royal Astronomical Society admitted women as honorary fellows (to attend lectures) from 1835 and as ordinary fellows (full members) in 1916.

In the 17th-c and 18th-c a number of women in Germany and France had made contributions to astronomy by assisting their families and friends in observations, calculations and catalogues of stars. NICOLE-REINE ETABLE DE LABRI-LEPAUTE in France had assisted A-Clairaut (1713–65) to determine the exact time of the return of HALLEY's comet.

CAROLINE HERSCHEL assisted and was trained by her brother. In 1835 she was one of the first two women elected to honorary Fellowship of the RAS; the other was MARY SOMERVILLE, who explained the latest astronomical discoveries in her books popularizing science and so made the subject more accessible to women. Unlike other sciences, astronomy caught the interest of the population with its giant telescopes, its new visible discoveries and royal connections. Caroline Herschel was granted a pension of £50 a year by King George III and the Gold Medal for Science by the King of Prussia.

MARIA MITCHELL in the USA learned mathematics and astronomy initially through her father and used his instruments. In 1847 she won the Gold Medal offered by the King of Denmark for the discovery of a new comet, which led to her further astronomical career. She became professor of astronomy and director of the observatory at Vassar College, New York.

MARY BLAGG became interested in the subject through University Extension lectures, worked on lunar nomenclature and was admitted as one of the first female Fellows of the RAS in 1916. Margaret Lindsay Murray (1849–1915) was interested in astronomy from childhood and built a small spectroscope. After she married WILLIAM HUGGINS she worked as his assistant for 30 years, and papers were published in their joint names. Mary Acworth Orr (1867–1949) joined the recently formed British Astronomical Association in 1895. She observed variable stars and the solar eclipses of 1896 and 1900. She married astronomer John Evershed and went with him to Kodaikanal Observatory in India to work on the distribution and motion of solar prominences.

In the USA, Williamina Fleming (1857–1911) was working as a maid in the household of Edward Pickering (1846–1919), director of the Harvard College Observatory, when she was offered part-time work at the Observatory in computing. She became a permanent member of the staff in 1881 and took charge of the classification of stars on the basis of their photographed spectra, developing a useful classification scheme, published as the *Draper Catalogue of Stellar Spectra* in 1890. This work was the result of a fund established by the widow of Henry Draper (a Harvard College physician and eminent amateur astronomer) as a memorial to her husband. ANTONIA MAURY, the

---

in medicine is now eclipsed by his distinction as one of the greatest algebraists of his century. He recognized negative and complex roots for equations, found the relations between the roots of an equation and the coefficients of its terms, and gave a general algebraic method for solving cubic equations (Cardan's solution). He has been accused of pilfering parts of this method from TARTAGLIA, but the accusation has been contested.

His contribution to chemical thought is more substantial than is often recognized. He wrote an encyclopedia of the sciences which discusses the major chemical theories of the time. He was credulous in many ways, but critical of alchemical claims. He recognized only three Aristotelian elements (earth, water and air), arguing ahead of his time that fire is not a substance but a form of motion; and he distinguished between electrical and magnetic attraction. His writing includes a variety of chemical recipes and his chemical and clinical interests are brought together in a text on toxicology.

Cardan's life was not easy: his childhood was marred by ill-health and harsh treatment while his talents emerged and were acknowledged late in life. His eldest son was convicted and beheaded for wife-murder and his second son was exiled at Cardan's instigation as 'a youth of evil habits'. Cardan describes himself in his

niece of Henry Draper, graduated from Vassar College in 1887 and worked at the Harvard Observatory classifying the bright northern stars according to their spectra. ANNIE JUMP CANNON graduated from Wellesley College in 1884 and joined the team of women, appointed by Pickering, working on variable stars and stellar spectra. She built upon the previous classification schemes to publish her own, and in 1922 her system was adopted by the International Astronomical Union as the official system for the classification of stellar spectra. As Agnes Mary Clerke (1842–1907), a writer on popular astronomy, foresaw in 1902, work in this area was to lead later to modern theories of stellar evolution.

Fiammetta Wilson (1864–1920) joined the British Astronomical Association and began observing planets, comets and meteors; she observed over 10 000 meteors between 1910 and 1920, was appointed joint acting head of the meteor section during the First World War and was admitted to the RAS in 1916. DOROTHEA KLUMPKE, an American who was educated in Europe, was the first woman to gain a PhD in maths at the Sorbonne. She joined the staff at the Paris Observatory, became the first woman to make astronomical observations from a balloon and remained in France throughout her working life.

Agnes Mary Clerke followed Mary Somerville into popular science writing, contributing to the *Dictionary of National Biography* from 1885–1901 and to the ninth edition of the *Encyclopaedia Britannica. A Popular History of Astronomy during the 19th Century* (1885) by her was valuable for its discussion of the introduction and application of the spectroscope, remained in print for 23 years, and had a German translation. She was elected an honorary Fellow of the RAS in 1903 and has a lunar crater named for her.

HENRIETTA LEAVITT graduated at Radcliffe and joined the Harvard Observatory in 1895. Studying stellar brightness by photographic methods, she deduced that Cepheid variable stars have a simple relationship between the period of a given star and its luminosity and that, based on this, their distances could be calculated. Previous to her discovery only the distance of relatively nearby stars (up to 100 light years away) could be found, by measurement of stellar parallax.

During the first 30 years of the 20th-c the number of women working as paid computors increased rapidly; many of them had no university education and they rarely made a career of their ill-paid work. Most women graduating in mathematics and physics in Britain went into teaching. Six women gained a PhD during this time. Dorothy Wrinch (1894–1976) researched at Oxford and Cambridge, became an X-ray crystallographer and emigrated to the USA. Bertha Swirles gained a PhD at Cambridge in 1929 and published on mathematical physics and theoretical astrophysics. She became director of studies in mathematics at Girton College, Cambridge and married the astronomer and geophysicist HAROLD JEFFREYS. CECILIA PAYNE-GAPOSCHKIN studied physics and astronomy at Cambridge, joined the BAA and was elected to the RAS. Her postgraduate research at Harvard College Observatory in the USA, on the elements present in stellar atmospheres, was described by Otto Struve (1897–1963) in 1962 as 'undoubtedly the most brilliant PhD thesis ever written in astronomy', presumably of those he had seen. After a brief return to Britain where she gave a paper to the 1925 meeting of the BAA , she returned to the USA and a distinguished career at Harvard. Openings for a career in astronomy in Britain were then still hard to find.

In the later 20th-c, several women have achieved distinction and careers in astronomy in Britain. Best known are JOCELYN BELL BURNELL, co-discoverer of the first pulsar, MARGARET BURBIDGE, who made major contributions to ideas on quasars, on the synthesis of the nuclei of atoms in the stars and on the masses of galaxies; and Heather Couper, prolific writer on astronomy.

MM

autobiography as 'timid of spirit, I am cold of heart, warm of brain and given to never-ending meditation. I ponder over ideas ...'. He was a man who made more enemies than friends.

Carlson, Chester (1906–68) US physicist: inventor of xerography.

Carlson worked for a printer before studying physics at the California Institute of Technology; he then worked for the Bell Telephone Company before taking a law degree and moving to the patent department of an electronics firm. During the Depression he decided that invention was a way to prosperity and in his spare time he searched for a cheap, dry method of copying documents. After three years he focused on a scheme using electrostatic attraction to cause powder to adhere to plain paper, and got his first copies in 1938. It took another 12 years and a team of co-workers to develop a commercial xerographic copier; he died a very wealthy man.

Carnot, (Nicolas Léonard) Sadi [kah(r)noh] (1796–1832) French theoretical physicist: a founder of thermodynamics through his theoretical study of an idealized heat engine.

Carnot's family was unusual. His father, Lasare Carnot (1753–1823), was the 'Organizer of Victory' for the Revolutionary Army in 1794 and became Napoleon's minister of war; unusually, he left politics for science in 1807 and did good work in pure and applied mathematics and in engineering. Sadi had one brother, Hippolyte, also a politician, whose son became president of France. Sadi was educated by his father and at the École Polytechnique, and served in the army as an engineer, leaving it as a captain in 1828. He was a cholera victim in the Paris epidemic of 1832.

His scientific work was highly original, and the single paper he published before his early death did much to create the new science of thermodynamics. His paper was *Reflections on the Motive Power of Fire* (1824) and it originated in Carnot's interest in steam engines, which had been developed by British engineers and, as the nationalistic Carnot realized, were generating an industrial revolution in the UK. However, their theory was non-existent and their efficiency very low. Carnot set out to deduce if the efficiency could be improved and whether steam was the best 'working substance'. His paper is a brilliant success, despite the fact that he used the caloric theory of heat, which presumed it to be a 'subtle fluid'. (This did not affect the main answers and, incidentally, Carnot's notes show that long before his death he was converted to modern heat theory.) He also used the correct idea that perpetual motion is impossible, a fact of experience.

In his paper, Carnot considers an idealized steam engine, frictionless, with its working substance passing from heat source to heat sink through a series of equilibrium states, so that it is truly reversible. The pressure–volume changes in it constitute a Carnot cycle. He was able to show that the efficiency of such an engine depends only on the temperature (T_1) of the heat source and the temperature (T_2) of the heat sink; that the maximum fraction of the heat energy convertible into work is $(T_1 - T_2)/T_2$; and that it does not depend at all on the working substance (Carnot's theorem). These ideas, which were eventually to mean so much for both engineers and theoreticians, were too abstract to attract much interest in 1824. In 1849 when W Thomson saw the paper he realized its importance, and he and Clausius made it widely known. The paper contains ideas linked with the laws of conservation of energy and the First Law of Thermodynamics, and led Thomson and Clausius towards the Second Law. Later still, Gibbs and others were to use thermodynamic ideas to forecast whether chemical reactions will occur.

Carothers, Wallace Hume [karuth/erz] (1896–1937) US industrial chemist: discovered fibre-forming polyamides (nylons).

The son of a teacher, Carothers graduated from a small college and later both studied and taught chemistry at three universities before moving in 1928 to the research department of the Du Pont Company at Wilmington, DE. His object was 'to synthesize compounds of high molecular mass and known constitution'; an early success was Neoprene, the first successful synthetic rubber, marketed from 1932. He then studied the linear polymers made by condensing a dibasic acid with a diamine. By heating adipic acid with hexamethylenediamine at 270° he obtained Nylon 6.6, which can be melt-spun into fibres whose strength is improved by cold-drawing:

$$HO_2C(CH_2)_4CO_2H + H_2N(CH_2)_6NH_2 \rightarrow$$
$$...CONH(CH_2)_6NHCO(CH_2)_4...$$

This polyamide has a relative molecular mass of $10–15 \times 10^3$, with useful properties as a textile fibre, and has had much commercial success. Carothers established useful principles in research on polymers. Despite his successes he suffered from depression, and soon after his marriage killed himself at the age of 41.

Carrel, Alexis (1873–1944) French-US surgeon: pioneer of vascular surgery and perfusion methods.

Carrel qualified in medicine at Lyons in 1900. He was a skilful surgeon, but lacked interest in routine surgery and in 1904 visited Canada, intending to become a cattle rancher. However, later in 1904 he moved to Chicago and in 1906 joined the Rockefeller Institute for Medical Research in New York. He remained there until retirement in 1939, except for an interlude as a French Army surgeon in the First World War (when he shared the introduction of the Carrel–Dakin solution (mainly NaOCl) for the antiseptic treatment of deep wounds). Even before the First World War he began to attack the problem of organ transplantation. One difficulty in this is the need to ensure a blood supply to the transplanted organ without failure due to thrombosis or stenosis. Carrel developed methods for suturing blood vessels with minimum damage and risk of infection or thrombosis. These techniques greatly advanced vascular surgery. He even suggested, in 1910, the coronary bypass procedure and carried it out on a cadaver. The method was not usable on living patients until half a century later. He won an unshared Nobel Prize in 1912.

He went on to study methods of keeping organs alive by perfusion (ie passage of blood or a blood substitute through the organ's blood vessels). With C Lindbergh (1902–74), the aviator, he produced a perfusion pump ('artificial heart') in 1935. Major advances (eg in dealing with rejection of donor tissues) were needed before transplants of organs such as the kidney could be achieved by others after the Second World War, but Carrel's methods were essential for that later success.

Carrington, Richard Christopher (1826–75) British astronomer: discovered differential rotation of Sun with latitude.

A wealthy amateur, Carrington made over 5000 observations of sunspots between 1853 and 1861, and showed that the Sun does not rotate as a solid body but that its rotational period varies from 25 days at the equator to 27.5 days at latitude 45°. He also discovered solar flares in 1859.

Carson, Rachel Louise (1907–64) US naturalist and science writer: warned of the dangers of modern synthetic pesticides.

Rachel Carson was born in Springdale, PA, and studied biology at Johns Hopkins University. After teaching at Maryland (1931–36) she worked as a marine biologist for the US Fish and Wildlife Service (1936–49). In *The Sea Around Us* (1951) she warned of the increasing danger of large-scale marine pollution. With *The Silent Spring* (1962), however, she created an awareness world-wide of the dangers of environmental pollution and roused public concern for the problems caused by modern synthetic pesticides and their effects on food chains. Her work was the starting point for the increasing ecological and conservationist attitudes emerging in the 1970s and 1980s. Although generally desirable these new 'green' attitudes can be unfortunate: eg in Sri Lanka there were 2.8 million cases of malaria in 1948 but by 1963 this had been cut to 17 cases by DDT spraying. As a direct result of Carson's *The Silent Spring* spraying was stopped in 1964 and by 1969 there were 2.5 million cases. Much study of DDT, introduced by P H MÜLLER, has shown it to be non-toxic to humans, although it can damage food chains for some birds and fishes.

Casimir, Hendrik (Brugt Gerhard) (1909–) Dutch physicist: originated the 'two-fluid' model of superconductivity.

Casimir was educated at the universities of Leiden, Copenhagen and Zürich. He held a variety of research positions until, in 1942, he began a career with Philips. He became director of the Philips Research Laboratories in 1946.

Casimir's papers cover aspects of theoretical physics, particularly low-temperature physics and superconductivity. W Meissner (1922–) had examined some properties of superconductors, such as the expulsion of a magnetic field below the superconducting transition temperature (the Meissner effect). Casimir and C Gorter proposed in 1934 that two sorts of electrons exist, normal and superconducting, and used this to explain the relation between thermal and magnetic properties in superconductors. When BARDEEN and others produced the BCS theory it was clear that the two categories represented unpaired electrons and paired electrons (called Cooper pairs).

Cassini, Giovanni Domenico [kaseenee] (1625–1712) Italian–French astronomer: greatly enhanced knowledge of the planets.

Born in Italy, Cassini became director of the Paris Observatory in 1669 and never returned to Italy. He added greatly to our knowledge of the planets of the solar system. It was he who worked out the rotational periods of Jupiter,

Mars and Venus and tabulated the movement of the Jovian satellites discovered by GALILEO (RÖMER subsequently used his results to calculate the speed of light). Between 1671 and 1674 he discovered four new satellites of Saturn (Iapetus, Rhea, Dione and Tethys) and in 1675 observed the gap in Saturn's ring system first noted 10 years before by William Balle and now known as the Cassini division. Most importantly, he was able to calculate the first reasonably accurate figure (only 7% low) for the Earth's distance from the Sun (the astronomical unit). To do this he observed Mars from Paris at the same time as Jean Richer (1630–96) observed it from Cayenne in French Guiana, 10 000 km away. (Jupiter's satellites provide a universal clock; when they are seen in the same positions at both sites, the time is the same.) The parallax gave the distance of Mars, and KEPLER's Third Law then gave the distances of the other planets. In later life he attempted to measure the shape of the Earth but concluded incorrectly that it was a prolate spheroid. Three generations of his descendants succeeded him as director of the Paris Observatory; all were highly conservative astronomers, resisting major new theories.

Cauchy, Augustin Louis, baron [kohshee] (1789–1857) French mathematician: founded complex analysis.

The Terror of 1793–4 drove the Cauchy family to their country retreat at Arcueil, and there Augustin was educated by his father. He also became badly malnourished, which affected his health for the rest of his life. In 1805 he entered the École Polytechnique, and after moving to the École des Ponts et Chaussées served as an engineer in Napoleon's army. In 1813 ill health caused his return to Paris; 3 years later he became a professor at the École Polytechnique. With the restoration of the Bourbons and departure of republicans such as MONGE and others, Cauchy was elected to the Académie des Sciences, which, in the same year (1816), awarded him its Grand Prix for his paper on wave modulation. His recognition and status increased (including a chair at the Collège de France), but all was lost with Charles X's abdication in 1830, following the July Revolution. Cauchy was extremely pious and, although sincere, was 'a bigoted Catholic' even according to ABEL. He refused to take a new oath of allegiance and went into exile.

A professorship at Turin followed, together with a tedious period as tutor to Charles X's son in Prague. Although Cauchy stuck to his principles, the Government fortunately turned a blind eye and in 1838 he returned to a professorship at the École Polytechnique, and at the Sorbonne in 1848. He died of a fever at the age of 68, after a highly creative lifetime in mathematics, and mathematical physics, which included seven books and over 700 papers. His tally of 16 named concepts and theorems compare with those of any other mathematician.

Cauchy played a large part in founding modern mathematics by his introduction of rigour into calculus and mathematical analysis. He published on convergence, limits and continuity and defined the integral as the limit of a sum. Together with GAUSS, Cauchy created the theory of real and complex functions, including complex analysis and contour integration. He recognized the theory of determinants and initiated group theory by studying substitution groups.

Cavendish, Henry (1731–1810) British chemist and physicist: studied chemistry of gases and of air, water, and nitric acid; made discoveries in heat and electricity and measured the density of the Earth.

As eldest son of Lord Charles Cavendish, Fellow of the Royal Society, and grandson of the 2nd Duke of Devonshire, Henry was wealthy and well-educated. His mother died when he was 2. He spent 4 years at Cambridge, took no degree and studied in Paris for a year before making his homes in London (for living, Gower Street; workshop and laboratory, Clapham; library, Dean Street, Soho). Thereafter he devoted his time and money to personal research in chemistry and physics. He had a most peculiar personality: although he enjoyed scientific friends and discussion, he otherwise avoided conversation to an extreme degree, especially with women. He was generous with money, but not to himself. He published only a part of his scientific work, although he was unperturbed by either jealousy or criticism. When he was 40 he inherited a large fortune, but he was not interested in it, although he did use part of it to form a library and apparatus collection. This was used by the public and by himself on the same terms, and characteristically was located well away from his house. He was described as 'the richest of the learned and the most learned of the rich' and as having 'uttered fewer words in the course of his life than any man who lived to fourscore years'.

In 1766 he described methods for handling and weighing gases. He studied 'fixed air' (CO_2), showing that it was produced by fermentation or from acid and marble, and he re-studied 'inflammable air' (H_2, which had been studied by BOYLE). He exploded mixtures of hydrogen and air with an electric spark, and found that no weight was lost and that the product 'seemed pure water' (1784). The volume ratio he found to be 2:1; and the synthesis of water in this way cast out the long-held idea that water was an element. These experiments also convinced Cavendish that heat was weightless. He examined air from different places, heights and climates, and showed it to be of nearly constant composition. He showed that nitric acid is formed by passing sparks through air (when N_2

and O_2 combine and the NO reacts with water). Cavendish, like PRIESTLEY, interpreted his results on the phlogiston theory and Cavendish thought hydrogen was phlogiston. Cavendish, unlike Priestley, realized that LAVOISIER's theory would also explain his results. He noticed that a small residue (1%) of air remained after long sparking; this was later found by RAMSAY and RAYLEIGH to be argon, a noble gas and a new element.

In physics, Cavendish used a method devised by MICHELL to determine the gravitational constant (G) in 1798. BOUGUER had earlier attempted to find the density of the Earth; Cavendish's value for G (from which the Earth's mass and density is easily calculated) gave a mean density of nearly 5.5 times that of water. (BOYS obtained a slightly more accurate value, by the same method, a century later.) Since most rock has a density in the range 3–4, a metal core for the Earth could be deduced. Most of Cavendish's work on heat and electricity was not published by him, but was revealed from his notes after 1879 by MAXWELL. He showed that Cavendish had distinguished between quantity and intensity of electricity and that he had measured the electrical conductivity of salt solutions. He had proved that the inverse square law (COULOMB's law) holds (within 2%) by showing that no charge exists inside a charged hollow spherical conductor, a result which is consistent only with that law. He worked on specific and latent heat (possibly knowing of JOSEPH BLACK's work) and believed heat to stem from 'internal motion of the particles of bodies'. After 60 years of research, he chose (characteristically) to die alone. In his long lifetime this eccentric recluse achieved most in chemistry: notably in showing that gases could be weighed, that air is a mixture and that water is a compound – all fundamental matters if chemistry was to advance. His work in physics was equally remarkable but was largely without influence because much was unpublished. The famous Cambridge physics laboratory named after him was funded by a talented mathematical kinsman, the 7th Duke of Devonshire, in 1871.

Cayley, Arthur (1821–95) British mathematician: developed n-dimensional geometry, and the theory of matrices and algebraic invariants.

Cayley, the son of an English merchant, spent his first 8 years in Russia, where his father was then working. He was educated at King's College School, London and Trinity College, Cambridge. Reluctant to be ordained, which was a necessary condition to remain a Fellow of Trinity College, he became a barrister. For 14 years he practised law, and only accepted the Sadlerian Chair of Pure Mathematics in Cambridge in 1863 when the requirement concerning religious orders was dropped.

He managed to publish over 300 papers while a barrister and by his death he had published over 900, covering all areas of pure mathematics, theoretical dynamics and astronomy. While they were both lawyers Cayley and his friend SYLVESTER established the theory of algebraic invariants.

Cayley also developed a theory of metrical geometry, linking together projective geometry and non-Euclidean geometry. Together with KLEIN he classified geometries as elliptic or hyperbolic depending on the curvature of space upon which the geometry was drawn (that is, whether a surface is saddle-like or dome-like).

The theory of matrices was an invention of Cayley's and allowed compact manipulation of the many components of a geometrical system. The movement of a vector (directed line) when the space in which it is embedded is distorted can be described by this theory.

Cayley was a prolific mathematician with a strong and dependable character, much in demand both as a lawyer and administrator.

Cayley, Sir George (1773–1857) British engineer: the founder of aerodynamics.

Cayley belongs to the group of gentleman amateurs, able to use their wealth and the peace of country life to advance a scholarly enthusiasm. His school at York, and a clergyman tutor who trained him as a mechanic and in mathematics, gave him some skills needed for his later work, and his teenage enthusiasm for models became a life-long interest in flying. He knew of the first balloon flights, made in France by the MONTGOLFIER brothers when he was 10 years old. Cayley improved a helicopter-type toy (then known as a 'Chinese top') and later made one which rose to 30 m, and he devised model gliders powered by twisted rubber. By 1799 he realized, ahead of all others, the basic problems of heavier-than-air flight and the relation of the forces of lift, drag and thrust.

By 1804 he had made a whirling-arm device for testing purposes, and he saw the advantages of a fixed-wing aircraft design; other experimenters, for the rest of the century, focused on flapping-wing designs which, despite the example of birds, were to prove a dead end. His experiments with gliders were extensive and led him to a man-carrying design able to fly short distances. However, although he foresaw, correctly, that propulsion using a light engine driving a screw propeller was the way to success, no such light engine was then available.

His publications, which form the basis of aerodynamics, appeared from 1809 to the 1840s. He also found time to serve as an MP, enjoy his 10 children and invent the caterpillar tractor, automatic signals for railways, the self-righting lifeboat and the tension wheel, which he designed for aircraft undercarriages and which is now most familiar in bicycles.

It was not until 1903 that the WRIGHT brothers, using Cayley's ideas and adding their own talents, were able to achieve effective powered heavier-than-air flight; by then, a satisfactory light power unit, driven by petrol, could be made. As Wilbur Wright wrote in 1909, 'about 100 years ago an Englishman, Sir George Cayley, carried the science of flying to a point which it never reached before and which it scarcely reached again during the last century'.

Celsius, Anders [selseeus] (1701–44) Swedish astronomer: devised Celsius scale of temperature.

Celsius devised a thermometric scale in 1742, taking the boiling point of water as $0°$ and the melting point as $100°$. Five years later, colleagues at Uppsala observatory inverted the scale, to its present ('centigrade') form.

In thermodynamics, temperatures are measured on the absolute or Kelvin scale. However, the Celsius scale is often used for other purposes, and is now defined by the relation (temperature in °C) = (temperature in K) – 273.15

Chadwick, Sir James (1891–1974) British physicist: discoverer of the neutron.

Chadwick graduated in physics in Manchester in 1911 and stayed there to do research under RUTHERFORD. He won an award in 1913 to allow him to work with GEIGER in Berlin, and when the First World War began in 1914 he was interned. Although held in poor conditions in a racecourse stable for 4 years, he was able to do some useful research as a result of help from NERNST and others.

In 1919 he rejoined Rutherford, who had moved to Cambridge, and for 16 years was to be his principal researcher. Chadwick's research with him was mainly with alpha particles (helium nuclei $^4_2$He); from the way these were scattered by heavier nuclei he could work out the positive charge of the scattering nucleus and show it to be the same as the atomic number. They also used alpha particles to bombard light elements and induce artificial disintegration. Then, in 1932, he was able to reinterpret an experiment reported by the JOLIOT-CURIES, which he saw as evidence for the existence of the neutron (charge 0, mass 1), which Rutherford had foreseen in 1920. Chadwick quickly did his own experiments to confirm his deduction; the neutron allowed a massive advance in knowledge of atomic nuclei and was one of a series of major discoveries in atomic physics made in the 'marvellous year' of 1932, largely in Rutherford's laboratory.

In 1935 Chadwick won the Nobel Prize for his discovery of the neutron, but soon afterwards friction with Rutherford arose because Chadwick wanted to build a cyclotron and Rutherford opposed this. Chadwick went to Liverpool as professor and soon had his cyclotron (the first in the UK) and made the department there a leading centre for atomic physics. When the Second World War came, he was the natural leader of the UK's effort to secure an atomic bomb before the enemy succeeded in this. Clearly the work had to be done in the USA, as the UK was exposed to German bombing. Chadwick made a masterly job of first propelling the work there into effectiveness, and then ensuring that collaboration proceeded smoothly.

Back in the UK after the war, advising government on nuclear matters and increasingly doubtful of the wisdom of its policies, he had an unsatisfying last phase in his career as Master of his old Cambridge college.

Chain, Sir Ernst Boris [chayn] (1906–79) German–British biochemist: member of the team which isolated and introduced penicillin for therapeutic use.

Having studied physiology and chemistry in his native Berlin, Chain left Germany in 1933 and worked in London and Cambridge. He joined FLOREY's staff in Oxford in 1935 and from 1938 worked with him and HEATLEY on the production, isolation and testing of the mould product penicillin, which by 1941 was shown to be a dramatically valuable antibacterial. He shared a Nobel Prize in 1945, moved to Rome in 1948 and returned to Imperial College, London in 1961. His work on penicillin led him to discover penicillinase, an enzyme which destroys penicillin; he later worked on variants of penicillin which were resistant to such destruction. Chain was a talented linguist and musician, with forceful but unpopular views on the organization of science.

Chamberlain, Owen (1920–) US physicist: discovered the antiproton.

Chamberlain was educated at Dartmouth College and the University of Chicago, and was appointed professor of physics at the University of California at Berkeley 1958–89.

Like many physicists at the time, Chamberlain worked on the Manhattan atomic bomb project during the Second World War, studying spontaneous fission of heavy elements. After the war he conducted experiments with the bevatron particle accelerator at Berkeley and in 1955, together with SEGRÈ and others, discovered the antiproton, a new elementary particle with the same mass as the proton, but of opposite charge. Antiparticles had been predicted theoretically by DIRAC in 1926. Chamberlain and Segrè shared the 1959 Nobel Prize for physics for their discovery.

Chandler, Seth Carlo (1846–1913) US geophysicist: discovered variation in location of the geographic poles (Chandler wobble).

By occupation both a scientist and an actuary, Chandler became interested in the possible free nutation (oscillation) of the Earth's axis of rotation. By re-analysing repeated measurements of the latitudes of different observatories he discovered an annual variation in latitude (due to the motion of air masses) and also another variation with a period of roughly 14 months. Despite initially hostile reaction from the scientific establishment his conclusions were soon

fully borne out. The cause of the secondary variation was subsequently explained and it has since become known as the Chandler wobble – the apparent motion of the Earth's axis of rotation across the Earth's surface (detectable as a variation of latitude with time), with a period of approximately 14 months. It is caused by the precession (or free nutation) of the Earth's axis of symmetry about its axis of rotation. For a rigid planet the period would be exactly one year; the observed slightly longer period, and its broad spectral peak (428 ± 17 days), is due to elastic yielding of the Earth's interior.

Chandrasekhar, Subrahmanyan [chandrah-sayker] (1910–95) Indian–US astrophysicist: developed theory of white dwarf stars.

Chandrasekhar studied in India and then in Cambridge before moving to the USA in 1936. Chandrasekhar's interest has been the final stages of stellar evolution. He showed that when a star has exhausted its nuclear fuel, an inward gravitational collapse occurs, which will normally be eventually halted by the outward pressure of the star's highly compressed and ionized gas. At this stage the star will have shrunk to become an extremely dense white dwarf, which has the peculiar property that the greater its mass, the smaller its radius. This means that massive stars will be unable to evolve into white dwarves; this limiting stellar mass is called the Chandrasekar limit and is about 1.4 solar masses. It has been shown that all known white dwarves conform with this limit. Like his uncle, C V RAMAN, Chandrasekhar won a Nobel Prize, in 1983.

Chapman, Sydney (1888–1970) British applied mathematician: developed the kinetic theory of gases and worked on gaseous thermal diffusion, geomagnetism, tidal theory and the atmosphere.

Chapman studied engineering at Manchester and mathematics at Cambridge, graduating in 1910. During his career he held professorships in Manchester, London and Oxford, and from 1954 worked at the High Altitude Observatory, Boulder, CO, and the Geophysical Institute in Alaska.

Chapman's interests were broad. He made a notable contribution to the kinetic theory of gases, taking the theory beyond the earlier work of MAXWELL and BOLTZMANN, to the Chapman–Enskog theory of gases.

Thermal diffusion refers to heat transfer between two parts of a solid, liquid or gas that are at different temperatures, in the absence of convection. He applied his theory to a variety of problems, notably in the upper atmosphere. (Later, isotopes for atomic fission were separated by use of gaseous thermal diffusion.) On geomagnetism, his other main interest, he investigated why the Earth's magnetic field varies with periods equal to the lunar day (27.3 days) and its submultiples; he showed this was due to a tidal movement in the Earth's atmosphere due to the Moon. The Chapman–Ferraro theory of magnetic storms predated modern plasma theory. He also studied the formation of ozone in the atmosphere and the ionizing effect of solar ultraviolet light on the ionosphere (the Chapman layer being named for him). In his later years he developed, with S I Akasofu (1930–), the modern theory of geomagnetic storms, the ring current and the aurora.

Charcot, Jean-Martin [sha(r)koh] (1825–93) French neurologist: related many neurological disorders to physical causes.

Charcot was born and studied medicine in Paris, and spent his career at its ancient and famous hospital, the Salpêtrière. Appointed there in 1862, he found it full of long-stay patients with diseases of the nervous system about which little was known. By careful clinical observation and later autopsy he was able to relate many of their conditions with specific lesions; for example the paralysis of polio with the destruction of motor cells in the spinal cord; the paralysis and lesions of cerebral haemorrhage; and a type of arthritis with neurosyphilis. He was the major figure in the Paris Medical School for many years and his many pupils included Sigmund Freud (1856–1939), who developed Charcot's special interest in hysteria. After his death his only son, Jean, gave up medicine and became the leading French polar explorer.

Chargaff, Erwin [chah(r)gaf] (1905–) Czech–US biochemist: discovered base-pairing rules in DNA.

Chargaff studied at Vienna, Yale, Berlin and Paris, and worked in the USA from 1935 at Columbia University, New York. His best-known work is on nucleic acids. By 1950 he had shown that a single organism contains many different kinds of RNA but that its DNA is of essentially one kind, characteristic of the species and even of the organism. The nucleic acids contain nitrogenous bases of four types: adenine, thymine, guanine and cytosine. Chargaff showed that the quantities of the bases are not equal, as some had thought, but that, if we represent the number of the respective bases in a DNA by A, T, G and C respectively, then (very nearly) A = T, and C = G. These Chargaff rules were of great value as a clue to the double helix structure for DNA put forward by CRICK and WATSON in 1953, in which the two helical nucleic acid strands are linked by bonds between complimentary bases, adenine linking with thymine and cytosine linking with guanine, by hydrogen bonds. Chargaff's relations with Watson and Crick were marked by mutual antipathy.

Charles, Jacques Alexandre César [shah(r)l] (1746–1823) French physicist: established temperature–volume relationship for gases.

Originally a clerk in the civil service, an interest in ballooning and the physics of gases together with a flare for public lecturing brought Charles fame, and ultimately a professorship of physics in Paris.

In 1783, together with his brother Robert, Charles made the first manned ascent in a hydrogen balloon, a feat which brought him considerable public acclaim. On a later flight he was to reach an altitude of 3000 m. His interest in gases subsequently led him to formulate Charles's Law in 1787, that the volume of a given amount of gas at constant pressure increases at a constant rate with rise in temperature. Further experimental work by GAY-LUSSAC and DALTON confirmed the relationship, which holds best at low pressures and high temperatures (ie it applies to ideal gases). Incidentally, AMONTONS had also discovered the relationship almost a century before but, failing to publicize the fact, did not receive the credit for it.

Charney, Jule Gregory (1917–81) US meteorologist: pioneer of numerical techniques in dynamic meteorology.

Charney's work was principally concerned with dynamic meteorology. In 1947 he analysed the problem of the formation of mid-latitude depressions, in particular the dynamics of long waves in a baroclinic westerly current. He went on to work on numerical methods of weather prediction with VON NEUMANN, developing a system of quasi-geostrophic prediction equations and the concept of the 'equivalent barotropic level'. Charney also tackled problems concerned with the flow of the Gulf Stream, the formation of hurricanes and the large-scale vertical propagation of energy in the atmosphere.

Charnley, John (1911–82) British orthopaedic surgeon who devised a satisfactory replacement hip joint.

Charnley's parents were a pharmacist and a nurse, so it is unsurprising that he studied medicine. He specialized in orthopaedic surgery, qualified young and operated throughout his life in the Manchester area where he had always lived.

His career was dominated by one problem, the treatment of the painful and disabling condition of osteoarthritis of the hip joint, common in the elderly. Until his success in the 1960s surgical reconstruction of this joint was, in his words, 'no great credit to orthopaedic surgery'. In the 1950s he set up a workshop in the attic of his home and a biomedical testing laboratory at the Wrightington Hospital. His systematic studies on possible replacement joint materials and their friction and lubrication led him, by 1963, to settle on a replacement joint consisting of a socket made of the then novel plastic HMWP (high molecular weight polyethylene) in which moved a rather small polished steel head which replaced the diseased head of the patient's femur. Both were fixed into the bone with acrylic cement. Charnley's procedure, widely used by him and others, transformed many lives.

Chatelet-Lomont, Gabrielle-Emilie, marquise (Marchioness) **du**, *née* le Tonnelier de Breteuil [shatuhlay lohmõ] (1706–49) French writer on physics and mathematics.

The youngest child of the chief of protocol at the court of Louis XIV, Emilie de Breteuil was born into an aristocratic society that expected its women to be beautiful, intelligent and witty. As she was considered too tall (175 cm/5 ft 9 in), her father believed she would remain single and, unusually for the period, provided her with the best tutors. Her marriage (1725) survived a succession of lovers, lawsuits and separations; her husband owned a number of large estates, had a passion for war and was frequently absent. After the birth of her third child Emilie du Chatelet began serious studies in Newtonian physics with P L de Maupertuis (1698–1759) and A-C Clairaut (1713–1765). She became the mistress of Voltaire and provided him with a safe retreat at the du Châtelet estate at Cirey-sur-Blaise, where they established a laboratory. Cirey became the French centre of Newtonian science, with du Châtelet providing Voltaire with the mathematical expertise he lacked. In *Institutions de physique* (1740), written as a textbook for her son, she tried to reconcile Newtonian and Leibnizian views. In 1744 she began a translation of NEWTON's *Principia mathematica*, which remains a standard version. At the age of 42 she became pregnant again and, fearing that she would not survive the birth of her child, she worked long hours, taking little sleep; she died of puerperal fever. She had deposited the manuscript with the librarian of the Bibliothèque du Roi in Paris; it was published in 1759. Emilie du Châtelet's work, which made Newtonian and Leibnizian ideas available in France, was followed by strong development of celestial mechanics there.

Chatelier, Henri Louis Le *see* **Le Chatelier**

Cherenkov, Pavel Alekseyevich [cherengkof] (1904–90) Russian physicist: discoverer of the Cherenkov effect.

A graduate of Voronezh State University, Cherenkov worked at the Lebedev Institute of Physics from 1930. In 1934 he first saw the blue light emitted from water exposed to radioactivity from radium, which had been observed by many earlier workers who had assumed it to be fluorescence. Cherenkov soon found that this could not be the explanation because the glow is shown by other liquids; and he found it was caused by fast electrons (beta rays) from the radium and that it was polarized. By 1937, working with I M Frank (1908–   ) and I E Tamm (1895–1971), he was able to explain the effect. They showed that in general the effect arises when a charged particle traverses a medium (liquid or solid) when moving at a speed greater than the speed of light in that medium, and they were able to predict its direction and polar-

ization. The effect is dramatically visible in the blue glow in a uranium reactor core containing heavy water; and it is used in a method for detecting high-energy charged particles. A counter of this type, using a photomultiplier, can detect single particles. The effect has some analogy with the shock wave and sonic boom produced when an aircraft exceeds the speed of sound in air. Cherenkov, Frank and Tamm shared a Nobel Prize in 1958.

Chevreul, Michel Eugène [shevroei] (1786–1889) French organic chemist: investigated fats and natural dyes.

A surgeon's son, Chevreul learned chemistry as assistant to VAUQUELIN, and by 1824 became director of dyeing at the famed Gobelins tapestry factory. His best-known work is on animal fats, which he showed by 1823 could be separated into pure individual substances that, with acid or alkali, break down to give glycerol and a fatty acid. (The fatty acids were later shown to be long-chain monocarboxylic acids.) Chevreul showed that soap-making (saponification) of animal fats by alkali could be understood and improved chemically, and that soaps are sodium salts of fatty acids. In 1825 Chevreul with GUY-LUSSAC patented a method of making candles using 'stearin' (crude stearic acid) in place of tallow, which was odorous, less luminous and unreliable; when developed, the improvement was of substantial importance. Chevreul showed that the urine of diabetic patients contained grape-sugar (ie glucose). He worked on organic analysis and the chemistry of drying oils (used in paints), on waxes and natural dyes; on theories of colour; on the use of divining rods and (after he was 90) on the psychological effects of ageing. As a child of 7, he had watched the guillotine in action; after his centenary, he watched the construction of the Eiffel Tower. He never retired.

Clarke, Arthur Charles (1917–   ) British inventor of the communication satellite and science fiction writer.

Very few individuals have single-handedly devised a technical advance of world-wide importance; to combine this in one career as a leading science fiction writer is not only exceptional but unique. A radar instructor in the Second World War, Clarke published a seminal article in 1945 in the popular non-academic technical journal *Wireless World*, outlining a full scheme for a novel concept, the communication satellite. He deduced that the satellite must be in a geostationary orbit (ie centred over a fixed earthly location) at a precise distance, which he calculated. The first such satellite was in use in 1964, and by the late 1980s over 400 satellites in Clarke orbits had been launched, and were transmitting 4000 million telephone calls annually and linking TV transmissions in 100 countries. After a science degree in London,

Clarke wrote the non-fictional *The Exploration of Space* (1951); but thereafter his novels made him best known as a writer of science fiction. As co-writer of the film *2001: A Space Odyssey* he had a major success in the genre in 1968. From 1956 he lived in Sri Lanka.

Clausius, Rudolf [klowzeeus] (1822–88) German theoretical physicist: a founder of thermodynamics, especially linked with its Second Law.

Clausius's father was a Prussian pastor and proprietor of a small school which the boy attended. Later he went to the University of Berlin to study history, but changed to science; his teachers included OHM and DEDEKIND. He was short of money, which delayed his graduation, but his ambition was to teach university physics and he did so at Zürich, Würzburg and Bonn. In the Franco-Prussian War of 1870 he and his students set up an ambulance service and he was badly wounded.

By the 1850s a major problem had arisen in heat theory: CARNOT's results were accepted, but while he believed correctly that, when a heat engine produces work, a quantity of heat 'descends' from a higher to a lower temperature, he also believed that it passed through the engine intact. The First Law of Thermodynamics, largely due to JOULE, visualizes some heat as being lost in a heat engine and converted into work. This apparent conflict was solved by Clausius, who showed in 1850 that these results could both be understood if it is also assumed that 'heat does not spontaneously pass from a colder to a hotter body' (the Second Law of Thermodynamics). The next year W THOMSON arrived at the same law, differently expressed, and there are now several other equivalent formulations of the same principle. Clausius developed this concept, of the tendency of energy to dissipate, and in 1865 used the term entropy (S) for a measure of the amount of heat lost or gained by a body, divided by its absolute temperature. One statement of the Second Law is that 'the entropy of any isolated system can only increase or remain constant'. Entropy was later seen (eg by BOLTZMANN) as a measure of a system's disorder. The Second Law generated much controversy, but Clausius, MAXWELL and Thomson led a vigorous and successful defence, although we would not now fully accept Clausius's crisp summaries of 'the energy of the universe is constant' (First Law) and 'the entropy of the universe tends to a maximum' (Second Law), thereby predicting a 'heat-death' for the universe.

Clausius also did valuable work on the kinetic theory of gases, where he first used the ideas of 'mean free path' and 'effective molecular radius' which later proved so useful. In the field of electrolysis, Clausius was the first to suggest (in 1851) that a salt exists as ions in solution before a current is applied. In each area he attacked, he showed outstanding intuition, and

his work led to major developments by others; but Clausius was strangely little interested in these developments.

Cockcroft, Sir John Douglas (1897–1967) British physicist: pioneered the transmutation of atomic nuclei by accelerated particles.

Cockcroft had completed only his first year at Manchester University when the First World War broke out and he joined the Royal Field Artillery as a signaller. Remarkably he survived unscathed through 3 years and most of the later battles. Afterwards, he joined the Metropolitan Vickers Electrical Company and took his degree at Cambridge in mathematics (1924). He then became part of RUTHERFORD's research team at the Cavendish and in 1932 made his reputation by a brilliant experiment with E T S Walton (1903–95), for which they received the 1951 Nobel Prize for physics.

Cockcroft was methodical in his work, genial and decisive, and no waster of words. He soon became mainly interested in research management, and in 1940 was a member of the Tizard Mission to the USA to negotiate wartime technological exchange. He then became head of the Air Defence Research and Development Establishment (1941–4). He was also Jacksonian Professor at Cambridge (1939–46). He became founding director of the Atomic Energy Research Establishment at Harwell (1946) and led the establishment of the Rutherford High-Energy Laboratory at Harwell (1959). In 1959 he became founding Master of Churchill College, Cambridge. Receiving many honours, Cockcroft became a leading statesman of science, combining research and administrative skills.

The experiment conducted by Cockcroft and Walton was triggered by GAMOW mentioning (1928) to Cockcroft that bombarding particles may enter a nucleus by quantum mechanical 'tunnelling'. This could occur at much lower incident energies than those required to overcome COULOMB repulsion between the two. Using skilfully built voltage-doublers, protons were accelerated to 0.8 MeV and directed at a lithium target. Alpha particles (helium nuclei) were found to be released; the first artificially induced nuclear reaction (transmutation) was occurring; and was later shown to be:

$$^7_3\text{Li} + ^1_1\text{H} \rightarrow ^4_2\text{He} + ^4_2\text{He} (+ 17.2 \text{ MeV})$$

(In his experiments on transmutation, Rutherford had used, as projectiles, particles from a natural radioactive source.) With the publication of this exciting result the nuclear era began, and cyclotrons and linear accelerators were built to study nuclear physics. Rutherford, on seeing proof of the alpha particle generation, called it 'the most beautiful sight in the world'. When, later, Cockcroft suggested the generation of power by nuclear fission, Rutherford said the idea was 'all moonshine'.

Cockerell, Sir Christopher Sydney (1910-   )
British engineer: inventor of the hovercraft.

A Cambridge graduate in engineering, Cockerell's early career was in radio, with the Marconi Company from 1935 and working there mainly on radar in the Second World War. Leaving them in 1950 for a new career in commercial boat building and hiring, he turned to the long-studied problem of reducing drag on boat hulls. Both theory and his early experiments pointed to an air-cushion as a possible answer, an approach first considered a century earlier but never made effective. Cockerell had some success with models by 1955, and later the flexible skirt was devised, which retained a cushion of air well enough to give the first satisfactory hovercraft. A prototype (the SR-N1) built by the Saunder–Roe Company was completed in 1959; it weighed 7 tonnes and achieved manned crossings of the English Channel at speeds up to 95 kph/60 mph. Although hovercraft afterwards had some commercial success, their use has been more limited than was initially expected.

Cohn, Ferdinand Julius (1828–98) German botanist and bacteriologist.

Cohn was a precocious child and, despite the difficulties caused by German antisemitic rules, he was awarded a doctorate at Berlin for his work in botany when he was 19. He returned to his home city of Breslau (now in Poland) and became professor of botany there in 1872. A keen microscopist, he came to the important conclusion that the protoplasms (cell contents) of plant and animal cells are essentially similar. He was the first to devise a systematic classification for bacteria and did much to define the conditions necessary for bacterial growth.

Cohnheim, Julius [kohnhiym] (1839–84) German pathologist: a pioneer of experimental pathology.

A graduate in medicine from Berlin, Cohnheim became an assistant to Virchow and was probably his most famous pupil. He attracted many students himself, as a teacher of pathology at Kiel, Breslaw and finally Leipzig. His early work was in histology: soon after graduating he devised the freezing technique for sectioning fresh tissue and later a method of staining sections with a solution of gold. From 1867 he published a masterly series of studies on inflammation; he showed by experiments with frogs how blood vessels responded in its early stages and proved that the leucocytes (white cells) pass through the walls of capillaries at the site of inflammation and later degenerate to become pus corpuscles. Mechnikov and others were later to confirm and extend these studies.

Despite evidence, tuberculosis (then a major cause of death in Europe) was not easily accepted as infectious. Cohnheim provided new and convincing evidence by injecting tuberculosis matter into the chamber of a rabbit's eye and then observing the tuberculous process through its cornea. He also studied heart disease, examining obstruction of the coronary artery and deducing correctly that the resulting lack of oxygen led to myocardial damage (infarction). This work was reviewed and the condition named as 'coronary thrombosis' by J D Herrick (1861–1954) in 1912.

Colombo, Matteo Realdo (c.1516–59) Italian anatomist: a discoverer of the lesser circulation of the blood.

Son of an apothecary, Colombo was a student of anatomy, medicine and surgery under Vesalius and succeeded him at Padua and Pisa. In his book *On Anatomy* (1559) he gives more modern descriptions (without illustrations) than earlier anatomists. He describes the lens at the front of the eye (not in the middle, as earlier anatomists had believed) and the pleura and peritoneum. He describes clearly the lesser circulation through the lungs, and in a vivisection on a dog he cut the pulmonary vein and showed that it contained blood and not air; and its bright red colour made him believe that the lungs had made it 'spiritous' (ie oxygenated) by air. He did not understand the general circulation, as Harvey did later.

Columbus, Christopher, Cristoforo Colombo (*Ital*), Cristóbal Colón (*Span*) (1451–1506) Italian explorer: first nameable discoverer of the New World.

The eldest of the five children of a weaver, Columbus probably first entered his father's trade, but before 1470 he went to sea and for some years voyaged and traded for various employers based in Genoa, his birthplace. His work took him to England in 1477, and probably to West Africa in 1482, and about this time he began to seek financial support for a major Atlantic expedition. Classical writers (including Aristotle, Ptolemy and Pliny) had accepted that the Earth was spherical, and so it followed

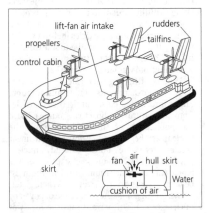

Hovercraft – inset shows method of operation

that China and Japan (known through the Polo family's descriptions) could be reached by sailing west. In accepting this idea, Columbus made two major errors. Firstly, he believed the Asian landmass to extend more to the east than is actually the case. Also he estimated the Earth's radius at only three-quarters of its true value. As a result, he believed Japan to be located in roughly the position where the West Indies are placed. Aside from these miscalculations, Columbus was a very competent navigator and he and one of his brothers had a business as chartmakers.

For some years Columbus failed to obtain support for a transatlantic expedition but in March 1492 the catholic monarchs of Spain, Isabella and Ferdinand, approved his voyage and awarded him the title of Admiral of the Ocean Sea and the governorship of any new land he might discover. Two wealthy ship's outfitters prepared his flagship *Santa Maria* at their own expense and in August 1492 he sailed in her from Palos near Huelva in southern Spain, with the *Pinta* and *Niña* also under his command, a total of about 100 men and a letter from the Spanish sovereigns to the 'grand khan of China'.

The fleet went south to the Canary Islands and then due west, making landfall in the Bahamas in October 1492 after a 5-week voyage from the Canaries. Columbus's difficulties had included maintaining the confidence of his crews and solving his navigational problems; these arose in part from the deviation of the magnetic compass from true north, which he may have been the first to observe.

From the Bahamas Columbus sailed to Cuba, which he thought was Japan, and believed he could soon reach China: then west to Haiti, where he began a settlement and traded with the native population. The *Santa Maria* was lost, a party was left in Haiti to study its inhabitants and their produce and Columbus sailed for Spain in January 1493 still convinced he had been in Asia. He was back in Palos in March to an enthusiastic welcome, bringing from the West Indies new plants and animals, a little gold and six natives.

Returning to Haiti in September with a much enhanced expedition, he found his settlement destroyed and the men killed, but he sailed on to discover Jamaica before returning to Spain in 1495, leaving his brother Bartolomeo in charge of a restored colony in Haiti. He found that his prestige in Spain had fallen, largely because the commercial profits on the first voyage was less than extravagant hopes had foreseen, so that his third expedition, in 1498, was on a reduced scale in men and ships. Nevertheless, it led to his discovery of Trinidad and, notably, of the South American mainland, the coast of Venezuela. In 1500 a newly appointed royal govenor visited him, was critical of affairs in the colony and sent Columbus back to Spain in irons. Tensions had arisen, in part because the 'gentleman adventurers' who had accompanied Columbus not only traded but, unlike him, took gold and girls by force. They resented the fact that Columbus and his brothers were not Spanish, and they accorded no rights to the natives because the latter were not Christians. Indiscipline had reached the point where some Spaniards had been hanged, and the new governor viewed the whole situation as highly unsatisfactory.

Fortunately the Spanish sovereigns repudiated Columbus's disgrace, restored him to favour and supported his fourth and last great voyage in 1502, during which he explored the southern coast of the Gulf of Mexico in search of a passage to Asia, which he still believed to be nearby. Much hardship and difficulty arose and this objective was inevitably not attained; but the coast of Central America was extensively explored, until hostile natives and disease forced Columbus to take refuge in Jamaica, with his ships in a poor state and their crews mutinous. In 1504 he returned to Spain dispirited and ill, still ignorant of the real nature of his discoveries. He died in 1506 at Valladolid and his remains, after several removals and confusions, were interred at Seville in 1902 in a mausoleum, honoured for something he had not meant to do and never knew he had done.

An authentic portrait of him probably does not exist, but he is known to have been tall and red-haired. In personality he was eccentric, impetuous, highly religious and driven by social ambition and the pursuit of gold. His explorations fall within a period of great discoveries: in 1487 Diaz had rounded the Cape of Good Hope; in the 1490s Columbus explored the West Indies, Central America and parts of South America; in 1498 da Gama reached India; and by 1521 Magellan had crossed the Pacific and circumnavigated the Earth. In only 35 years all the previously unknown oceans were crossed, and the existence of the continents was proved except for Australia and Antarctica, with Portugese seamen, inspired by Prince Henry 'the Navigator' taking leading parts.

Columbus's credentials as a scientist are modest, but he remains without a peer as a mariner and as discoverer and explorer of new islands and, unknowingly, of a New World which he was the first to link with the Old.

Compton, Arthur Holly (1892–1962) US physicist: discovered the Compton effect concerning the wavelength of scattered photons.

Compton was the son of a Presbyterian minister who was also a professor of philosophy, and inherited a deep religious faith from him. He obtained his doctorate at Princeton, and spent two years with Westinghouse Corporation. On travelling to Britain he spent a year doing research under RUTHERFORD at Cambridge before returning to America as head of the Physics Department at Washington University, St Louis, MO (1920). A professorship at Chicago followed in 1923. In 1945 he returned to Washington as chancellor.

In 1923 Compton observed that X-rays scattered by passing through paraffin wax had their wavelength increased by this scattering. Compton and DEBYE explained this in detail, stating that photons (electromagnetic waves) behave as particles as well as waves; they lose energy E and momentum on making elastic collisions and as

$$E = hc/\lambda$$

where c is the speed of light, their wavelength λ increases. Here h is PLANCK's constant. Compton found tracks in photographs taken in a Wilson cloud chamber, showing electrons recoiling from collisions with the invisible (because uncharged) photons of an X-ray beam. This work established EINSTEIN's belief that photons had energy and momentum, and also BROGLIE's assertion (1925) that in quantum mechanics objects display both wave and particle properties. Compton and C T R WILSON received the 1927 Nobel Prize for physics for their work, which is now part of the foundation of the new quantum theory (as opposed to BOHR's old quantum theory).

Compton developed an ionization chamber for detecting cosmic rays and in the 1930s used a world-wide survey to demonstrate that cosmic rays are deflected by the Earth's magnetic field and some are therefore charged particles (and not radiation). Variation of ray intensity with time of day, year and the Sun's rotation also indicated that the cosmic rays probably originate outside our Galaxy (1938).

In 1941 Compton was asked to take part in feasibility studies and the development of plutonium production for the atomic bomb. His religious faith made him question what was happening, but he felt that only such a weapon would quickly end the massive slaughter of the war. He became director of a major part of the Manhattan Project at Chicago and built the first reactor with FERMI (1942), publishing an account in his book *Atomic Quest* (1958).

Cook, Sir James (1728–79) British explorer: founder of modern hydrography and cartography; explored the Pacific and showed that scurvy was preventable on a long voyage.

The son of an agricultural labourer, Cook joined the Royal Navy in 1755 and was given his own command 2 years later. He is remembered for his voyages of discovery, which transformed knowledge of the Pacific and set the pattern for the great scientific expeditions of the 19th-c. After much hydrographic work of the highest quality, Cook was charged with taking the *Endeavour* to Tahiti in 1768 with observers (including BANKS) for the transit of Venus, on behalf of the Royal Society. At that time observations of transits of inner planets across the face of the Sun were one of the principal means of estimating the Earth–Sun distance. Cook went on to chart the east coast of Australia and the coast of New Zealand, showing it to consist of two main islands, and his voyage set an upper limit to the size of any possible southern continent. Both for this voyage and his second expedition, the Admiralty's secret orders to Cook required him to explore the South Pacific where they had 'reason to imagine that a continent, or land of great extent, may be found', and 'to take possession of it in the King's name'. These expeditions had both scientific and political objectives. Cook's second expedition in 1772–5 further delineated the possible extent of Antarctica and also demonstrated that fresh fruit and vegetables were all that were needed to prevent scurvy, a major problem on long sea voyages at the time. (See panel overleaf.)

In 1776 he was made a Fellow of the Royal Society. His last expedition, begun in 1776, was intended to discover a northern route between the Atlantic and the Pacific but ended in his tragic death when he was attacked by natives in Hawaii.

Improved sextants and other instruments, and especially Cook's talent and energy, ensured that more survey work and scientific research was done by him than by any previous expeditions. Modern maps of the Pacific with its coasts and islands owe much to him, and he set new standards of cartography and hydrography.

Cooper, Leon Neil (1930–) US physicist: contributed to BCS theory of superconductivity.

Leon Cooper was educated at Columbia University, obtaining his doctorate in 1954. He collaborated with BARDEEN and SCHRIEFFER at Illinois on the BCS theory of superconductivity.

Soon after his doctoral work in quantum field theory, Cooper made a theoretical prediction of the existence of bound pairs of electrons at low temperature. Although two electrons repel each other, they may behave differently in a solid with a sea of electrons with an embedded lattice of positive ions. One electron distorts the lattice, pulling it in about it, and the other electron is attracted to the locally higher concentration of positive ions. This effect can be imagined from the similarity to two cannonballs on a mattress rolling together into the same depression. Thus at low temperature, when thermal vibrations do not disturb this process, bound pairs (called Cooper pairs) of electrons form. The BCS theory then accounts for superconductivity as being due to the fact that these pairs can move through a lattice with zero scattering by impurities because the pair is much larger than any impurity atom. For this work Bardeen, Cooper and Schrieffer shared the 1972 Nobel Prize for physics.

Copernicus, Nicolaus, Mikolaj Kopernik (Pol) [kopernikuhs] (1473–1543) Polish astronomer: proposed heliocentric cosmology.

THE EXPLORATION OF AUSTRALIA

Australia was first colonized from Asia at least 40 000 and probably 100 000 years ago, but it remained unknown to Europeans until the east coast was charted by Portuguese sailors during the 1520s. Abel Janszoon Tasman (1603–c.1659) made two voyages to Australia, in 1642 and 1644. He was the first person to discover Tasmania (which he named after the Governor of Batavia, Antony Van Diemen) and New Zealand. However, he failed to discover whether Australia was an island or not, or what the relationship was between the pieces of land he had discovered.

The major work of charting Australia and New Zealand was carried out by Cook. On his first voyage (1768–70), he charted the coast of New Zealand and the entire east coast of Australia. He landed on Possession I off the tip of Cape York on 22 August 1770, and claimed the east coast for Britain under the name of New South Wales. Banks sailed as a botanist with Cook, and was later largely responsible for the establishment of the Botany Bay penal colony.

The first fleet landed in Botany Bay during Cook's third voyage, on 26 January 1788, commanded by Arthur Phillip, first Governor of New South Wales. Expeditions were sent to fill in the gaps in the charts of the coastline. George Bass (? –1812) circumnavigated Tasmania in the late 1790s, while his associate Matthew Flinders (1774–1814) later charted the coastline of South Australia and, during 1801–3, was the first person to circumnavigate the continent.

During the governorship of Lachlan Macquarie, the interior of New South Wales was explored by William Charles Wentworth (1793–1872) and other explorers were quick to follow into the depths of the continent. During 1824–5 Hamilton Hume (1797–1873) discovered the Murray River, and in 1827 Allan Cunningham (1791–1839) carried out a good deal of exploration of the remoter parts of New South Wales in the search for botanical specimens. Charles Sturt (1795–1869), together with Hume, discovered the Darling and lower Murray rivers between 1829–30, but failed to find the inland sea which he believed to exist west of the Darling. Thomas Livingstone Mitchell (1792–1855) discovered the rich grazing lands of Victoria, which he called Australia Felix, and settlers moved into them to raise ever-increasing numbers of sheep.

Settlements had already been established in Western Australia, and much energy was expended on trying to find a practical stock route between South Australia and Western Australia. Finally, in 1840–1, Edward John Eyre (1815–1901) established that there was no such route, but discovered Lake Eyre and Lake Torrens in the process. Later in the decade, Ludwig Leichhardt (1813–c.1848) explored the coastal region of N Queensland and the Northern Territory. His methods were somewhat haphazard, and he disappeared in 1848 while on an expedition into the interior of Queensland. No trace of his party has ever been found.

The fate of Leichhardt illustrated that Australia was not a safe place for inexperienced and ill-prepared explorers. The most disastrous expedition was that, in 1861, of Robert O'Hara Burke (1820–61) and William John Wills (1834–61) who managed to cross Australia S-N with relative ease, because it was an unusually wet season; but having left most of their supplies at base camp, they died of starvation on the way back. The sole survivor of the expedition, John King (1838–72), was the only one to accept help offered by local Aborigines.

The fate of Burke and Wills encouraged other explorers of the centre of Australia to be much more cautious. Over the course of several years (1858–60), John McDouall Stuart (1815–66) established a practical route S-N, supplied throughout with waterholes and passing by Alice Springs. Further exploration of the centre was carried out by Peter Egerton Warburton (1813–89) and John Forrest (1847–1918, later premier of Western Australia), and especially by William Ernest Powell Giles (1835–97) who crossed the centre of Australia twice (1875–6), travelling E-W and back again.

Sukie Hunter

Copernicus was the nephew of a prince bishop. Having studied mathematics, law and medicine in Poland and Italy, Copernicus was for most of his life a canon at Frauenburg Cathedral, his duties being largely administrative. Working mainly from the astronomical literature rather than from his own observations, he showed that a cosmology in which Earth and the planets rotate about the Sun offered a simpler explanation of planetary motions than the geocentric model of Ptolemy, which had been universally accepted for well over 1000 years. He circulated his preliminary ideas privately in a short manuscript in 1514 and continued to develop the theory over the next 30 years. Among his suggestions was the idea that the fixed stars were much further away than had previously been thought and that their apparent motion at night (and the Sun's motion by day) was due to Earth's daily rotation about its axis, but he retained the conventional idea that the planets moved in perfectly circular orbits. His ideas were first fully described in his book *De revolutionibus orbium coelestium* (The Revolution of the Heavenly Spheres) which, although complete by 1530, was not published until

1543. Copernicus himself may only have seen the published book on the day he died.

Copernicus's ideas were immediately criticized by other astronomers, notably BRAHE, who argued that if the Earth was moving then the fixed stars ought to show an apparent movement by parallax also. Copernicus's answer to this, that the stars were too far away for parallax to be apparent, was rejected on the grounds that it was inconsistent with the accepted size of the universe. The idea of a moving Earth was also hard to accept. The Church later officially banned De revolutionibus in 1616 and did not remove it from its Index of Forbidden Books until 1835.

Copernicus's view that the Sun was the centre of the solar system gained credence from GALILEO's work on Jupiter's moons in 1609; but the parallax of a fixed star was not measured until 1838 by BESSEL. However, the idea of a heliocentric (Sun-centred) system, with a moving Earth, had been accepted as a reality and not a mere mathematical device long before that; and Copernicus's circular orbits for planets had been replaced by KEPLER's elliptical orbits by 1609. The 'Scientific Revolution' is often dated from Copernicus's work, reaching its climax with NEWTON about 150 years later. In the same year (1543) that Copernicus's De revolutionibus appeared, VESALIUS's book On The Structure of the Human Body was published; men's views of nature were changing fast. (See panel overleaf.)

Cori, Carl Ferdinand [koree] (1896–1984) Czech–US biochemist.

Cori graduated in medicine in Prague in 1920 and in the same year married his classmate Gerty Radnitz (1896–1957). They formed a close team until her death (their research collaboration had begun as students), moving to the USA in 1922 and sharing a Nobel Prize in 1947, the only other husband and wife pairs to do so being the CURIES in 1903 and the JOLIOT-CURIES in 1935. Gerty Cori became the first woman medical graduate to receive a Nobel Prize.

Their best-known joint research concerned the conversion of glucose to glycogen in the animal body and the reverse breakdown. BERNARD had shown in 1850 that glycogen forms an energy reserve held in the liver and muscles, which is converted to the simpler sugar, glucose, when needed. The Coris discovered the precise steps involved in this essential biochemical process and revealed the part played by sugar phosphates for the first time.

Coriolis, Gaspard Gustave de [koriohlis] (1792–1843) French physicist: discovered the Coriolis inertial force.

Coriolis was educated at the École Polytechnique in Paris, where he became assistant professor of analysis and mathematics and eventually director of studies.

Coriolis was responsible for defining kinetic energy as $\frac{1}{2}mv^2$ and introducing 'work' as a

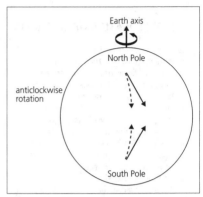

Deflections (dotted lines) from expected paths (solid lines) resulting from the Coriolis force. They are observed in the flight of missiles and artillery shells.

technical term of precise meaning in mechanics. In 1835 Coriolis discovered the Coriolis force, an inertial force which acts on rotating surfaces at right-angles to their direction of motion, causing the elements of the surface to follow a curved, rather than straight, line of motion. Such effects are particularly important in oceanography and meteorology (eg the Ekman effect), and account for the movement of ocean currents near the Equator.

Cormack, Allan Macleod (1924–) South African–US physicist: pioneer of X-ray tomography.

Cormack studied at the University of Cape Town and then worked on the medical applications of radioisotopes in Johannesburg before moving to the USA in 1956.

In 1963, independently of HOUNSFIELD, Cormack developed the mathematical principles for the X-ray imaging of 'soft' biological tissue, and demonstrated its viability experimentally. Hitherto, X-rays had only been used for obtaining 'silhouettes', primarily of bone structure. Cormack showed that by effectively combining many X-ray images, taken in different directions through the human body, it was possible to build up a picture of a slice through the soft tissue. This technique, known as computer-assisted tomography (CAT) or computed tomography (CT), is the basis for the modern body scanners which have become an invaluable medical tool. Cormack and Hounsfield shared the Nobel Prize for physiology or medicine in 1979.

Coulomb, Charles (Augustin de) [koolôb] (1736–1806) French physicist: discovered inverse square law of electric and magnetic attraction.

Coulomb trained as a military engineer and served in Martinique for 9 years. He eventually returned to France as an engineering consultant but resigned from the Army altogether in 1791

THE HISTORY OF ASTRONOMY

The rising and setting of the Sun, Moon and stars must have attracted attention early in human history, but astronomy can be said to have begun around 3000 BC when the Mesopotamians, Egyptians, and Chinese grouped the stars into constellations. By 2500 BC, stone monuments such as Stonehenge and the pyramids attest to our concern with astronomical matters. Astronomy provided the calendar for ancient civilizations. The heliacal rising of the star Sirius and the passage of the Sun through the equinoxes defined the year of 365 days. The time between similar lunar phases led to the month. The interval between new, first quarter, full, and last quarter Moon became the week. And the fact that there are about 12 lunar months in a year triggered the division of night and day into 12 intervals – the hours. HIPPARCHUS of Rhodes accurately measured the length of the seasons, discovered the precession of the Earth's spin axis and produced the first catalogue of 1080 star positions, before 100 BC.

The measurement of astronomical size and distance started with ERATOSTHENES of Cyrene, who obtained excellent values for the radius of the spherical Earth, and ARISTARCHUS of Samos who used eclipse timing to measure the Earth–Moon size and distance. However, Aristarchus obtained an Earth–Moon distance that was a mere 5% of its true value. ARISTOTLE introduced the first practical proofs that the Earth was spherical, but he also dogmatically insisted that the sphere was the perfect solid shape, and the sphericity of heavenly bodies and orbits convinced him that the heavens were a region of perfection and as such were unchangeable. This distinction between a perfect heaven and a corruptible Earth survived for nearly 2000 years.

Mercury, Venus, Mars, Jupiter and Saturn were the five known planets, and the explanation of their complicated paths amongst the fixed stars provided a challenging intellectual exercise. Eudoxus of Cnidus (*c.*408–355 BC) not only had a spherical spinning Earth at the centre of the universe, but he also introduced a mathematical scheme to explain the apparent motion of the Sun, Moon, and planets. Each was given a series of transparent crystal spheres. The axis of each sphere was embedded in the next sphere out. By a judicious choice of 26 spheres with different axis directions and spin periods, the planetary motion could be modelled. Claudius Ptolemaeus (PTOLEMY) of Alexandria revised the system using eccentric and epicyclic circles that combined two perfect spherical motions of differing periods and amplitudes. This complicated, workable, and intellectually satisfying system was in vogue for nearly 1400 years.

Little progress was made between AD 200 and 1500. COPERNICUS, a Polish monk, suggested that the Sun-at-centre (heliocentric) model for the Solar System would be simpler than the Earth-centred (geocentric) scheme. His book *De revolutionibus orbium coelestium* (1543, On the Revolutions of the Celestial Spheres) set out methods for calculating the size of the system and predicting the motion of the planets. The Copernican model was at the centre of one of the most violent intellectual controversies the world had known, and it took a century or two for the truth to be accepted. It became clear that verification required the accurate measurement of stellar and planetary positions, and this task fell to the Danish astronomer TYCHO BRAHE. He improved the accuracy by a factor of nearly 20. He also collected together a vast number of planetary positional observations, and these prepared the way for the work of KEPLER, who discovered that planets had elliptical orbits and that the Sun was at a focus of these ellipses. He also derived a relationship between the orbital size and period. These three laws formed the connecting link between the geometrical makeshifts of Copernicus and the ancients and the gravitational discoveries of NEWTON.

GALILEO GALILEI made a series of spectacular discoveries by observing the sky through the newly discovered telescope. His 1609–10 observations of the four major satellites of Jupiter and the phases of Venus firmly established the Copernican doctrine. His measurements of the height of lunar mountains and the rotation of the Sun, as exhibited by the movements of the sunspots, broke the bonds of Aristotelian perfection. Telescopes quickly improved. HORROCKS observed a transit of Venus across the solar disc; Johannes Hevelius (1611–87) constructed lunar charts; HUYGENS discovered the true nature of Saturn's rings; and CASSINI discovered a gap in these rings and four Saturnian satellites.

In 1687 Newton published his *Principia*, and at a stroke unified celestial and terrestrial science. His gravitational force, proportional to the product of the masses divided by the square of the distances, not only described the fall of an apple but also the orbit of the Moon and planets, and Kepler's three laws. Newton used gravitation theory to measure the mass of the Sun. He also calculated the orbit of a comet. HALLEY continued this cometary work; he not only showed that comets were periodic but also accurately predicted the 1759 return of 'his' comet, thus confirming Newton's breakthrough. Halley mapped the southern sky and discovered that stars moved. Over many centuries the closer stars moved against a background of the more distant ones. The first modern star catalogue was produced by FLAMSTEED, the first Astronomer Royal.

By his discovery of binary stars, WILLIAM HERSCHEL showed that Newton's laws extended throughout the observable universe. Herschel also discovered the planet Uranus (in 1781) and estimated the shape of the Milky Way. Astrophysics dawned when FRAUNHOFER catalogued the dark lines of the solar spectrum. The scale of the universe started to be established when BESSEL measured (in 1838) the distance to the star 61 Cygni. And Newtonian gravitation had a crowning distinction when ADAMS and LEVERRIER used gravitational perturbation analysis to predict the existence and position of Neptune, which was discovered in 1846.

The foundations of astronomical spectroscopy were laid in 1859 when KIRCHHOFF correctly interpreted the process responsible for the Fraunhofer lines in the solar spectrum. Stellar spectra were first classified by Pietro Angelo Secchi (1818–78), and this was greatly developed by Edward Pickering (1846–1919) at Harvard with his group of women assistants, notably Williamina Fleming (1857–1911), ANNIE JUMP CANNON and ANTONIA MAURY. Variations in the stellar spectra were caused by changes in surface temperature. HUGGINS not only showed that the Earth, stars and planets were made from the same atoms, but he also used Doppler shifts to measure the movement of astronomical objects. LOCKYER inaugurated solar physics by measuring the spectra of solar prominences at the time of total eclipse. The work of MAUNDER on the variability of solar activity and HALE on solar magnetism continued this advance.

In 1866 SCHIAPARELLI demonstrated the connection between the comets and meteor showers, and also noticed (in 1877) 'canals' on Mars. This sparked a great increase in planetary observational work under the guidance of people such as LOWELL. A new planet, Pluto, was discovered in 1930 by TOMBAUGH.

During the late 19th-c photography played a large role in the advancement of astronomy. Unlike the eye, a film is able to 'accumulate' light. The first photograph of a star was taken in 1850. Solar prominences were photographed in 1870, a stellar spectrum in 1872, the nebula M 42 in 1880, and the Milky Way in 1889. In 1891 and 1892 the photographic discovery of an asteroid and a comet was achieved. The source of solar energy and the age of the Earth were also topics of great interest in the late 19th-c. EINSTEIN's theory of mass-energy commensurability pointed to the energy source of the Sun and other stars, but astronomers had to wait until 1939 for the detailed solution put forward by BETHE.

In 1904 Jacobus Kapteyn (1851–1922) initiated important research into the motion of stars in the Milky Way, and the size of the Galaxy was defined by the work of SHAPLEY on the distribution of star clusters. This relied on the discovery of the period–luminosity relationship in Cepheid variable stars by HENRIETTA LEAVITT.

Around 1914 HERTZSPRUNG and RUSSELL modified and extended the classification of stellar spectra. Knowing stellar distances, they were able to introduce what was to become known as the Hertzsprung–Russell diagram. This not only stressed that there are giant and dwarf stars, but also acted as the basis for studies of the finite lifetime of stars and their evolution. Great strides in the study of stellar interiors were made by EDDINGTON. In 1916 he discovered that stellar energy transport was by radiation as opposed to convection, and that stellar equilibrium depended on radiation pressure as well as gas pressure and gravitation. His model of the interior of a typical star indicated that the central temperature was tens of millions of degrees. In 1924 he found that the luminosity of a star depended almost exclusively on its mass. In the mid-1920s CECILIA PAYNE-GAPOSCHKIN showed that the major constituents of stars are hydrogen and helium. HOYLE and William Fowler (1911–95) put forward a theory showing how nuclear fusion synthesized other heavier elements in the centre of stars.

The ever-increasing size of astronomical telescopes enabled detailed spectroscopy to be applied to fainter and fainter objects. HUBBLE not only discovered (in 1924) that extragalactic nebulae were galaxies like our own but, together with SLIPHER, measured their radial velocities and in 1929 discovered that the universe was expanding.

Cosmology, the study of the origin of the universe, became a popular subject, with the 'Big Bang' theory, first proposed by LEMAÎTRE, vying with the 'Continuous Creation' theory put forward by BONDI, GOLD and Hoyle. The discovery in 1964 of the 3.5 K microwave background by PENZIAS and R W WILSON strongly favoured the 'Big Bang'.

Radio astronomy started in 1937, when REBER built a 9.4 m diameter radio antenna. In 1942 HEY (1909–   ) discovered radio waves from the Sun, and in 1944 VAN DER HULST predicted that interstellar hydrogen would emit 21-cm radio waves. Unusual radio sources such as Cassiopeia A and Cygnus A were found, and the detection of the 21-cm radiation in 1951 by PURCELL soon enabled OORT to map the Galaxy using radio waves.

The launch of the space race in the late 1950s saw telescopes and telescopic devices lifted above Earth's blanketing atmosphere. The first X-ray source was discovered in 1962, and in 1969 two men walked on the Moon. Infrared- and ultraviolet-detecting satellites have followed, and space probes have to date flown past all the major planets, revealing a host of new details. Unusual stellar and galactic objects continue to be found. White dwarfs, neutron stars, black holes, pulsars and quasars are typical examples. Today, astronomers are still considering such diverse topics as the large scale structure of the universe, the origin of galaxies and the origin of the solar system. There is evidence that planetary systems other than our own exist, but messages from extraterrestrial life forms have yet to be detected.

Carole Stott

and moved from engineering to physics. During the French Revolution he was obliged to leave Paris, but returned under Napoleon and became an inspector-general of Public Instruction.

Not surprisingly in view of his military service, much of Coulomb's early work was concerned with engineering problems in statics and mechanics. He showed that friction is proportional to normal pressure (Coulomb's Law of Friction) and introduced the concept of the thrust line. However, he is primarily remembered for his work on electrical and magnetic attraction and repulsion. From 1784 onwards he conducted a series of very delicate experiments, using a torsion balance he had invented himself and capable of detecting forces equivalent to 10^{-5} g. He discovered that the force between two charged poles is inversely proportional to the square of the distance between them and directly proportional to the product of their magnitude (Coulomb's Law of Force). This was a major result, paralleling NEWTON's law of gravitational attraction. The SI unit of electric charge, the coulomb (C), is named in his honour. It is the charge crossing any section of a conductor in which a steady current of 1 ampere flows for 1 second.

Coulson, Charles Alfred (1910–74) British mathematician: a founder of modern theoretical chemistry.

Coulson was unusual in holding professorships in theoretical physics (King's College, London 1947–52), applied mathematics (Oxford 1952–72) and theoretical chemistry (Oxford 1972–4). He also played a major role in creating the third of these subject areas and wrote useful books on *Waves* and on *Electricity*. He also published on meteorology, biology and theology.

Coulson developed the application of quantum mechanics to the bonds between atoms in molecules. These bonds originate in the interaction between the outer electrons of the bonded atoms. He showed how to calculate those molecular bond-lengths and energies of interest to chemists. The method he used is called molecular orbital (MO) theory (1933). He also showed how bonds intermediate between single and double bonds could arise (1937). This then allowed him and H C Longuet-Higgins (1923–) to explain the delocalized (ie multicentre) bonding in such aromatic molecules as benzene. In 1952 he wrote his classic textbook *Valence*, which proved valuable in the development of the subject. Later he studied bonding in molecules of biochemical importance.

Coulson influenced his generation not only as a theoretical chemist, where his methods have proved of great value, but also as a leading Methodist and writer on science and Christianity. He was chairman of the charity Oxfam from 1965–71.

Couper, Archibald Scott [kooper] (1831–92) British organic chemist: pioneer of structural organic chemistry and victim of misfortune.

After leaving school Couper studied a variety of subjects; classics at Glasgow and philosophy at Edinburgh were separated by visits to Germany, where he learned German speedily. As the son of a wealthy manufacturer, he seems to have studied whatever interested him; he concentrated on chemistry somewhere between 1854 and 1856. By 1858 he had spent 2 years in Paris, researching on benzene compounds, and early in that year he completed a paper 'On a New Chemical Theory' and asked WURTZ to present it at the French Academy. However, Wurtz delayed and KEKULÉ published his theory of organic structure shortly before Couper's paper appeared. Couper's views were similar, but much more clearly expressed. He argued that carbon had a valence of 2 or 4; and that its atoms could self-link to form chains. He showed chemical structures with broken lines to connect bonded atoms and he saw these structures as representing chemical reality; in these respects his ideas were ahead of Kekulé's. However, the latter had priority of publication and forcefully pressed his superiority. Couper quarrelled with Wurtz, returned to Edinburgh and soon his depression led to illness. He never recovered, although he lived in mental frailty for another 33 years, ignored as a chemist.

Credit for the idea of a ring structure for benzene is rightly given to Kekulé (1865), but it is hardly known that the first ring structure for any compound was proposed 7 years earlier by Couper, for a heterocyclic reaction product from salicylic acid and PCl_5.

Modern structural formulae were first widely used by A C Brown (1838–1922) of Edinburgh from 1861, who included double and triple bonds between carbon atoms.

Cousteau, Jacques (Yves) [koostoh] (1910–) French oceanographer: pioneer of underwater exploration.

Cousteau was in the French navy when the Second World War broke out, having graduated from the École Navale at Brest. Following distinguished service in the Resistance (during which time he designed and tested his first aqualung) he was awarded the Légion d'Honneur and the Croix de Guerre. After the war he became head of the Underwater Research Group of the French navy and afterwards made many notable advances in the technology and techniques of underwater investigation, constructing a diving saucer capable of descending to 200 m for long periods and working with Auguste Piccard (1884–1962) on the design of the first bathyscaphes. He set a world record for free diving in 1947, but is probably best known for pioneering underwater cinematography and for his studies of marine life.

Cray, Seymour R (1925–) US computer engineer: leading designer of supercomputers.

After graduating in 1950 from Minnesota, Cray worked on UNIVAC I, the first commercially available electronic computer, and went on to design large-scale computers for Control Data Corporation, which he had helped to found. In 1972 he left them to found Cray Research Inc., and to design the Cray-1 (1976), price $8 million, the world's fastest computer, which could perform 240 million calculations per second. Later models were even faster; the Cray-2 (1985), supercooled by liquid nitrogen, achieved 1200 million calculations per second. Such supercomputers are invaluable for complex scientific and some governmental work.

His circuit designs have very short electrical connections between internal components to increase their speed. After founding the Cray Computer Corporation in Colorado Springs he continued his dominance in the supercomputer field with his work on the gallium-arsenide-based Cray 3. His passion for supercomputer design has been legendary.

Crick, Francis (Harry Compton) (1916–) British molecular biologist: co-discoverer with WATSON of double-helix structure of DNA.

The outstanding advance in the life sciences in this century has been the creation of a new branch of science: molecular biology. In this, Crick has been a central figure and its key concept, that the self-replicating genetic material DNA has the form of a double helix with complementary strands, is due to him and J D Watson.

Crick graduated in physics in London but his first research was interrupted by war service, working on naval mines. After the war he was attracted to Cambridge and to biology and by 1949 was with the Cambridge Medical Research Council Unit, then housed in the Cavendish physics laboratory. His field of expertise was the use of X-ray crystal diffraction methods (originally devised by the BRAGGS) to examine the structure of biopolymers. The overall head of the Cavendish Laboratory was then Sir Lawrence Bragg. In the 1950s and under his patronage, the team led by PERUTZ and including J C Kendrew (1917–), Watson, H E HUXLEY, Crick and later BRENNER were to have as dramatic an effect on molecular biology as RUTHERFORD'S team had on particle physics in the 1930s, and in the same building.

In 1951 Watson joined the group. He was 23, a zoologist with experience of bacterial viruses and an enthusiasm for genetics. He and Crick quickly became friends; they shared an optimistic enthusiasm that it should be possible to understand the nature of genes in molecular terms, and in under 2 years they were to succeed. Important background material was available for them. There was good evidence from AVERY's work that the DNA of genes formed the key genetic material. A R Todd (1907–) had shown that DNA consists of chains of sugar residues (deoxyribose) linked by phosphate groups and carrying base molecules (mainly of four types) attached to the sugar rings. CHARGAFF had shown that the number of these bases had a curious ratio relation. Helical structures had been met with; PAULING had shown, as had Crick, that the protein keratin consists of chains of protein arranged in helical form; Pauling, like Crick, was an enthusiast for making molecular models as an aid to deducing possible structures.

Crick had devised a general theory that would show whether a given X-ray pattern was due to a helical structure; and his friendship with M H F Wilkins (1916–) at London gave him limited access to the X-ray pictures made there by Wilkins's colleague, ROSALIND FRANKLIN. With all this in mind, Crick and Watson built their models and in 1953 focused on a model in the form of a double helix, with two DNA chains. It could accommodate all known features of DNA, with acceptable interatomic angles and distances, and would accord with Franklin's observed X-ray diffraction pattern. The sequence of atoms in them give the DNA chains a direction, and the pair of chains forming the double helix run in opposed directions. Also, the helices are right-handed. The model had its sugar and phosphate chains on its outside and the bases (linked in pairs, A with T, C with G) on the inside (see diagram). The model explains how DNA replicates, by the uncoiling of its double-helical strands, with these strands then acting as templates. It also suggested how genetic information could be encoded, by the sequence of bases along the chains. Crick proposed as a 'central dogma' the scheme DNA → RNA → protein, with the first arrow representing transcription and the second representing translation. (The conversion DNA → DNA, shown in the diagram, is known as replication.)

Crick and Watson had found the broad answer to the question 'how do genes replicate and carry information?' and in the succeeding years most work on molecular biology has been directed to confirming, refining and extending these ideas. Crick himself has done much in this area, for example in work with Brenner demonstrating that the code is read in triplets of bases (codons) each defining one specific amino acid used to make a protein, and in showing that adjacent codons do not overlap. He also studied the structure of small viruses and collagen, and the mechanism by which transcription and translation occurs; and he has offered novel ideas on the origin of life on Earth and on the nature of consciousness. He worked mainly in Cambridge until 1977 when he moved to the Salk Institute in

San Diego, CA. He shared the Nobel Prize for physiology or medicine in 1962 with Watson and Wilkins.

Crile, George Washington (1864–1943) US surgeon: advanced knowledge of surgical shock and methods for reducing this.

A graduate of Ohio Northern University, Crile studied in Europe and afterwards practised in Cleveland, OH. He was an early user of blood transfusion in surgery, with his own method of direct linkage for this. He was particularly interested in surgical shock, in which bodily functions (cardiac output, blood pressure, temperature, respiration) are depressed with possibly fatal results. His animal experiments and studies on patients led him to emphasize shock prevention; he advocated greater care in anaesthesia, the use of an epidural anaesthetic where possible, the maintenance of blood volume (eg by transfusion) and the monitoring of blood pressure during surgery. These techniques improved control in general surgery and became routine; their effectiveness is not reduced by Crile's incorrect belief that surgical shock originates in the nervous system.

Cronin, James Watson (1931–   ) US particle physicist: demonstrated the non-conservation of parity and charge conjugation in particle reactions.

Cronin was educated at the University of Chicago, later working at Brookhaven National Laboratory and Princeton University before returning to Chicago as professor of physics in 1971.

LEE and YANG had shown in 1956 that parity was not conserved in weak interactions between subatomic particles. In 1964 Cronin, together with V L Fitch (1923–   ), J Christensen and R Turlay, made a study of neutral kaons and discovered the surprising fact that a combination of parity and charge conjugation was not conserved either. This was an important result since it was known that a combination of parity, charge conjugation and time is conserved, implying that the decay of kaons is not symmetrical with respect to time reversal.

Crookes, Sir William (1832–1919) British chemist and physicist: discovered thallium; studied 'cathode rays'; predicted need for new nitrogenous fertilizers.

The eldest of 16 children of a London tailor, little is known of his childhood. Crookes was a student in the Royal College of Chemistry from 1848, and became HOFMANN'S assistant. After two modest teaching jobs he inherited some money, returned to London and set up a personal chemical research laboratory. He was also editor and proprietor of the influential *Chemical News* from 1859–1906.

In 1861 he examined the spectrum of crude selenium and found a new bright green line. From this clue he was able to isolate a new element, thallium; he studied its rather strange

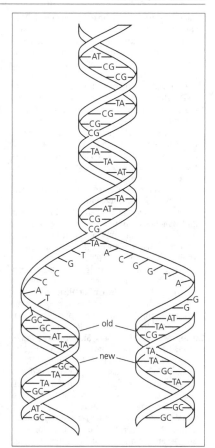

DNA replication, following Watson and Crick. The two strands of the double helix separate, and a daughter strand is laid down alongside each with a constitution determined by the base sequence of its parent strand.

chemistry and measured its relative atomic mass. The accurate weighings for this (done in a vacuum) led to his invention of the Crookes radiometer, in which four light vanes, each with one face blackened, are pivoted in a glass container with gas at a low pressure. In light, the vanes rotate; the device helped to confirm the kinetic theory of gases. He also studied electrical discharges in vacuum tubes, already studied by J Plücker (1801–68) and J W Hittorf (1824–1914). Crookes found that the 'cathode rays' travelled in straight lines, could cast shadows, heat obstacles and be deflected by a magnet; he concluded they were negatively charged particles but this found little support until J J THOMSON'S studies (20 years later) firmly identified them as electrons. Crookes also invented the spinthariscope (1903; Greek for 'spark-viewer') to detect the alpha-particles (helium nuclei) emitted by radioactive elements. This consists of a screen coated with ZnS and viewed by a lens: each impacting particle causes a visible light flash. Crookes also studied a variety of problems in technical chemistry (sugar from beet; textile dyeing; electrical lighting; antiseptics; sanitation; diamond formation) and especially the need to produce fertilizer from atmospheric nitrogen if soil fertility was to be maintained (1898). He had much scientific imagination, and he also experimented in spiritualism, suggesting that telepathy resulted from wave communication between brains. His long, active life covered a most interesting period in science.

Cross, Charles Frederick (1855–1935) British chemist: co-discoverer of viscose process.

Cross studied chemistry at London, Zürich and Manchester. With E J Bevan (1856–1921) as partner, he worked on the chemistry of wood (which consists largely of cellulose and lignin). Their viscose process (1892) involves extracting cellulose from wood pulp (or other cheap source) by treating it with aqueous sodium hydroxide followed by carbon disulphide. The cellulose solution can then be squirted through holes into dilute acid to regenerate the cellulose as fibre (rayon) or as film (Cellophane).

Curie, Marie, *née* Manya Sklodowska (1867–1934) Polish–French physicist: discovered the radioelements polonium and radium.

Manya Sklodowska grew up in Russian-dominated Poland; her family were intensely patriotic and took part in activities furthering the Polish language and culture. Manya's father was a teacher of mathematics and physics and her mother the principal of a school for girls. She developed an interest in science, but her parents were poor and there was no provision for higher scientific education for women in Poland. She and her sister Bronya, however, were determined to gain their education. Manya took a post as a governess and helped Bronya go to Paris to study medicine, after which she in turn was to help Manya.

In 1891 Manya went to Paris to study physics. By nature she was a perfectionist, tenacious and independent. She graduated in physics in 1893 from the Sorbonne, coming first in the order of merit. The following year, with a scholarship from Poland, she studied mathematics and graduated in second place. During this year she met PIERRE CURIE, then 35 and working on piezoelectricity at the School of Industrial Physics and Chemistry, and her plans to return to teach in Poland changed; they married in July 1895.

In 1896 BECQUEREL had discovered radioactivity in a uranium salt. Marie Curie (as she was now known), looking for a research topic for a doctoral thesis, decided to study the 'new phenomena' discovered by Becquerel. Working in her husband's laboratory, she showed that radioactivity is an atomic property of uranium and discovered that thorium emitted rays similar to uranium. In 1897 she gave birth to their daughter Irène (who also became a Nobel Prize winner in physics). When she examined the natural ores Marie discovered that the radioactivity of pitchblende and chalcolite was more intense than their uranium or thorium content implied, and correctly concluded that they must contain new radioactive elements. To find the new elements she began to separate the components of pitchblende to determine where the radioactivity lay, by a laborious process of fractional crystallization. Pierre Curie left his own research to join his wife in the work. No precautions against radioactivity were taken, as the harmful effects were not then known. Her notebooks were subsequently discovered to be highly radioactive and are still too dangerous to handle.

In July 1898 they announced the discovery of the existence of an element they named polonium, in honour of Marie's native country, and in December the even more radioactive radium. In order to isolate pure radium they obtained waste ore rich in uranium from mines in Bohemia and, working in an old shed, they purified and repurified the ore, work mostly undertaken by Marie. By 1902 they had obtained one tenth of a gram of radium chloride from several tonnes of ore. It was intensely radioactive, ionizing the surrounding air, decomposing water, evolving heat and glowing in the dark.

In 1903 Marie Curie presented her doctoral thesis (and became the first woman to be awarded such a degree in France). In 1903 she was awarded the Nobel Prize for physics jointly with Pierre Curie and Henri Becquerel for their work on radioactivity. The following year their second daughter Eve was born and the Curies appear to have begun to suffer from radiation sickness. Pierre Curie was named in 1904 as the new professor of physics at the Sorbonne and Marie was appointed 'chief of work' in the laboratory that was to be built for him; it was opened in 1915. In 1906 Pierre was killed in a street accident and the professorship was offered to Marie; she became the first woman professor at the Sorbonne. She continued to work on radium and attempted to isolate polonium, but most of her time was spent in supervising the research of others and raising funds, along with caring for her two daughters.

In 1910 Marie was proposed for the decoration of the Légion d'Honneur, but refused it, as her husband had refused a previous offer of the honour. At the same time she was a candidate for election to the Académie des Sciences in Paris, (she would have been the first woman member), but was not elected.

In 1911 she was awarded a second Nobel Prize for chemistry for her discovery of polonium and radium. The original unit of measurement of the activity of a radioactive substance was named the curie (Ci); it is now defined as a decay rate of exactly 3.7×10^{10} disintegrations per second. Characteristically, she insisted on defining the unit herself. In 1914 she organized X-ray services for military hospitals; radiography had hardly begun and there was as yet no provision for it. She died at the age of 67 from leukaemia; her exposure to radioactivity is suggestive in this.

Marie Curie was no theoretician, but she was a remarkably skilful radiochemist and her discoveries did much to focus research on the new and major field of radioactivity; she was the first woman scientist of international distinction.

Curie, Pierre (1859–1906) French physicist: discovered piezoelectric effect; pioneer in study of radioactivity.

The son of a physician, Pierre Curie was educated at the Sorbonne where he became an assistant teacher in 1878. He was appointed laboratory chief at the School of Industrial Physics and Chemistry in 1882 and in 1904 was appointed to a new chair of physics at the Sorbonne. He and his brother Jacques (1855–1941) first observed the phenomenon they named piezoelectricity; this occurs when certain crystals (eg quartz) are mechanically deformed; they develop opposite charges on opposite faces and, conversely, when an electric charge is applied to a crystal a deformation is produced. If a rapidly changing electric potential is applied, the faces of the crystal vibrate rapidly. This effect can be used to produce beams of ultrasound. Crystals with piezoelectric properties are used in microphones, pickups, pressure gauges and quartz oscillators for timepieces. Jacques and Pierre Curie used the effect to construct an electrometer to measure small electric charges; this was later used by his wife in her investigation into radioactivity. For his doctorate (1895), Pierre Curie studied the effect of heat on ferromagnetism and showed that at a certain temperature, specific to a substance, it will lose its ferromagnetic properties and become paramagnetic; this is now known as the Curie point (eg 1043 K for iron) He had already shown that magnetic susceptibility for diamagnetic materials is generally independent of temperature, but for paramagnetic materials the susceptibility is inversely proportional to absolute temperature (Curie's Law).

He married Manya Sklodowska and thereafter followed her into research on radioactivity. Together with BECQUEREL they were awarded the Nobel Prize for physics in 1903 for this work. Pierre Curie showed that 1 g of radium gave out about 500 J h^{-1}, the first indication of the energy available within the atom and the dangers of radioactivity.

Curtius, Theodor [kurtyus](1857–1928) German organic chemist.

A pupil of KOLBE, he later held professorships at Kiel, Bonn and Heidelberg. He first made hydrazine N_2H_4 (1887) and hydrogen azide HN_3 (1890); and he studied organic azides and aliphatic diazo compounds. All these compounds are toxic or unstable (or both) but have proved of great value in organic synthesis. Hydrazine also has industrial uses, and methylhydrazines are used as rocket fuels (eg in the Apollo probes), with liquid oxygen as oxidant.

Cushing, Harvey Williams (1869–1939) US physiologist and neurosurgeon: pioneered investigations of the physiology of the brain.

A physician's son, Cushing studied medicine at Yale and Harvard, finally specializing in neurosurgery. He experimented on the effects of raised intracranial pressure in animals; his improved methods for diagnosis, localization and surgical removal of intracranial tumours stemmed from this work. For a long time his personal surgical skill in this field was unsurpassed. Measurement of blood pressure in his patients began in 1906 and knowledge of hypertension and its effects begins with his work. From 1908 he also studied the function and pathology of the pituitary gland at the base of the brain, again working first with dogs. He showed that acromegaly is linked with one type of pituitary overactivity in the growing animal and dwarfism with its underactivity. Cushing's syndrome, which is associated with chronic wasting and other symptoms, he showed to be linked with a type of pituitary tumour; it is now known that other disorders that increase the production of corticosteroid hormones by the adrenal glands also lead to this syndrome.

Cuvier, Georges (Léopold Chrétien Frédéric Dagobert), baron [küvyay] (1769–1832) French zoologist and anatomist: pioneer of comparative anatomy and vertebrate palaeontology.

Son of a Swiss soldier, Cuvier was educated in Stuttgart. He was a brilliant student and from early childhood had been fascinated by natural history. From Stuttgart he went as tutor to a family in Normandy and from 1785 taught in Paris, at the Museum of National History, then the largest scientific establishment in the world.

He did much to establish the modern classification of animals, extending that of LINNAEUS by adding another broader level, the phylum. Thus he divided the invertebrates into three phyla. His work on molluscs and fish was particularly notable. In 1811, working with BRONGNIART on the Tertiary rocks of the Paris Basin, he became the first to classify fossil mammals and reptiles, thus founding vertebrate palaeontology.

Before this, he had developed comparative anatomy and the technique of showing, from a few bones, a probable reconstruction of the entire animal of an extinct species. His emphasis was always on the facts and he derided general theories. In long conflicts with LAMARCK and E Geoffroy St-Hilaire (1772–1844) (both precursors of DARWIN) he attacked theories of evolution: he believed in catastrophes, with the Biblical flood as the most recent. After each catastrophe, life was created anew. Cuvier became the world's most eminent biologist in his lifetime, with an authority akin to that of BERZELIUS in chemistry.

D

Daguerre, Louis Jacques Mandé [dagair] (1787–1851) French inventor of the daguerrotype.

No other invention in the 19th-c produced as much popular excitement as photography, for which the first practical process was devised by Daguerre. He was trained as a scene painter and stage designer at the Paris opera and in 1822 he devised the Diorama: this was an entertainment based on large (12 × 20 m) semi-transparent painted linen screens, which were hung and ingeniously lit to create illusions of depth and movement. This strange precursor of the cinema was a great popular success and 'Diorama Theatres' opened in Paris, London and other capital cities. Daguerre used a camera obscura (a box with a lens at one end and a small screen at the other end) to assist him in making his design sketches and he soon began experiments to mechanize or capture the image and so avoid the laborious tracing.

In 1826 he met Niépce, who was working on similar lines, and in 1830 they formally became partners. In 1831 after extensive experiments based on the known sensitivity of some silver compounds to light, he began to use a silvered copper plate iodized by exposure to iodine vapour as the light-sensitive surface. This was Niépce's idea; and then in 1835 Daguerre made by chance a momentous discovery. A plate which had been exposed in a camera obscura without any visible result was left in a cupboard; a few days later it was found to bear a visible picture. Amazed, he soon found that this result was due to a latent image being 'developed' by mercury vapour from a broken thermometer. This trick of 'development' allowed a picture to be made after a photographic exposure time as short as 20 minutes.

Further, by 1837 he found that the result could be 'fixed', ie rendered permanent, by washing with common salt in water. Confidently, Daguerre sought publicity, and used his experience as an entrepreneur to seek public and governmental support. He named the scheme 'the daguerrotype process' (although the profits were shared equally with Niépce) and succeeded, through the efforts of Arago, in attracting government praise and finance. Grandly, the French government offered his invention freely to the world on 19 August 1839, although Daguerre had patented his scheme in London 5 days before this. Public interest was intense, even though the invention of the new art was widely held to be blasphemous.

'From today painting is dead' said the artist Delaroche. It was claimed that by 1839 'all Paris was seized with daguerrotypomania'. Cartoons showed engravers hanging themselves. The invention was claimed to be 'a mirror with a memory' and 'the first to conquer the world with lightning rapidity', despite its defects. These were many: the equipment was bulky and exposures took many minutes in strong sunlight. The resulting daguerrotype was difficult to see; it was laterally reversed; and the plate could not be replicated. But despite being a photographic dead-end in a technical sense, it initiated a technique of visual recording using silver which through the efforts of Archer, Talbot and others has progressed ever since. Daguerre's instruction manual appeared in 32 editions in eight languages in 1839 and he retired in 1840 in honour and some glory to a modest estate at Bry-sur-Marne.

Dale, Sir Henry Hallett (1875–1968) British physiologist and pharmacologist: worked on histamine and on acetylcholine.

Educated in medicine in Cambridge, London and Frankfurt, Dale joined the Wellcome Laboratories in 1904 and at once began (at the suggestion of Sir H Wellcome (1853–1936)) to study the physiological action of ergot (a potent extract from a fungal infection of rye) on test animals. This work led, through fortunate and shrewd observations and the skill of his co-worker G Barger (1878–1939), to the two research themes especially linked with their names. These are, firstly, the work on histamine, a compound released by injured cells or in reaction to foreign protein, and secondly the work on the neurotransmitter acetylcholine. Both these areas have been fruitful for extended investigations, leading to fuller understanding of allergy and anaphylactic shock, and the nature of chemical transmission of nerve impulses. Dale directed the Medical Research Council from 1928–42; in 1936 he shared a Nobel Prize. For many years he was a dominant spokesman for science in the UK, especially in the medical and allied sciences.

d'Alembert, Jean Le Rond *see* **Alembert**

Dalton, John (1766–1844) British meteorologist and chemical theorist: proposed an atomic theory linked to quantitative chemistry.

Dalton was the son of a weaver and a Quaker and grew up in an isolated village in Cumbria. He left his village when he was 15 for Kendal in central Cumbria and thereafter made his living as a teacher. In 1793 he moved to Manchester and taught science, and from 1799 he worked as a private tutor, giving short courses to groups of students for a modest fee. Throughout his life, from 1781, he kept daily meteorological records.

This interest in weather and the atmosphere led to his work on gaseous mixtures generally.

In 1794 he wrote an excellent paper on colour vision (he was colour blind); in 1799, returning to his interest in the weather, he showed that springs arise from stored rainfall. This concern with rain and the water content of the atmosphere, which appears as the origin of all his work on gases and on atomic theory, arose through his life in the wet Lake District and the influence of a childhood teacher. He made his reputation in science in 1801 with his Law of Partial Pressures. This states that the pressure of a gas mixture is the sum of the pressures that each gas would exert if it were present alone and occupied the same volume as the whole mixture. He also found the law of thermal expansion of gases, now known as CHARLES'S Law although Dalton published it first. In 1803, at the end of a paper on gas solubility, he noted rather casually his first table of relative atomic masses. The interest this aroused led him to develop his theory further, in lectures and in his book *A New System of Chemical Philosophy* (1808). Briefly, his atomic theory proposed that every element consists of very small particles called atoms, which are indivisible and indestructible spheres. The atoms of one element were presumed to be identical in all respects, including mass, but to differ from atoms of other elements in their mass. Chemical compounds are formed by the union of atoms of different elements, in simple ratios (ie elements A and B would form a compound AB; and possibly A_2B, AB_2, A_2B_3). This is known as the Law of Simple Multiple Proportions.

The theory was able to interpret the laws of chemical combination and the conservation of mass; it gave a new basis for all quantitative chemistry. Each aspect of Dalton's theory has since been amended or refined, but its overall picture remains as the central basis of modern chemistry and physics.

Dalton assumed that when only one compound of two elements exists (for example water was the only compound of hydrogen and oxygen then known) it had the simplest formula; ie HO for water. On this basis, relative atomic masses ('atomic weights') could easily be found; the early lists were on the scale $H = 1$, but now a scale on which the common isotope of carbon $= 12$ is used. After AVOGADRO's work, corrections were needed (eg water is H_2O, not HO) and discussion by chemists on these changes, and improved analyses, greatly occupied 19th-c chemists. The unit of relative atomic mass is named the dalton for him. The dalton is equal to one-12th of the mass of a neutral carbon-12 atom.

Dalton himself remains a strangely dull personality. His main work came after he was 30. He was a gruff lecturer, a poor experimenter and his writing seems old-fashioned. Apart from the bril-

liant insight of his atomic theory, his other work seems pedestrian. He was independent, modest and attributed his success to 'perseverance'.

Manchester was strongly aware of Dalton's fame. His lying in state in their Town Hall was attended by 40 000 people, his funeral was a major public event and memorials (sculptural and financial) were made, marking the surprising regard in an industrial city at that time for a scientific theorist.

He had instructed that his eyes be studied to find the cause of his colour blindness. This was done 150 years after his death, and DNA from them was found to lack the genes giving the green-sensitive pigment present in the normal human eye.

Dam, (Carl Peter) Henrik, (1895–1976) Danish biochemist, discoverer of vitamin K.

Dam studied in Copenhagen and later in Freiburg and Zürich before being appointed Assistant Professor of Biochemistry in Copenhagen in 1928. Soon after this he was studying the effect of a restricted diet on chicks, and found that within 2–3 weeks they got haemorrhages under the skin and their blood showed delayed coagulation. They were not cured by feeding the then-known vitamins, but by 1934 he found that hempseed, tomatoes, green leaves and hog liver contained something that prevented the bleeding and by 1935 it was recognized as a new vitamin, 'K'. Later DOISY showed it to consist of two closely related compounds, K_1 and K_2, and Dam and others showed that its deficiency leads to lack of prothrombin, which is needed in the series of changes involved in blood coagulation. Dam and Doisy shared the Nobel Prize for physiology or medicine in 1943.

Daniell, John Frederic (1790–1845) British meteorologist and chemist: devised Daniell cell.

Although his early research was in meteorology (he devised a dew-point hygrometer and theorized on the atmosphere and trade winds), Daniell is best known for his work on primary cells. The earliest types quickly lost power. The Daniell cell (1836) uses amalgamated zinc as negative electrode and copper as positive electrode, and gives a nearly constant emf of ≈ 1.08 V. It proved a great asset in telegraphy and in the study of electrolysis. In 1831 Daniell became the first professor of chemistry at King's College, London.

Dart, Raymond Arthur (1893–1988) Australian anatomist and palaeoanthropologist: discovered *Australopithecus africanus*.

After qualifying as a physician from the University of Sydney in 1917, Dart served in France before being appointed professor of anatomy at the newly formed University of the Witwatersrand, Johannesburg, in 1922. The work there he found to be a most depressing experience, until in 1924 one of his students showed him a fossil baboon skull that had been found in a lime quarry at Taung, Botswana. Dart arranged with the quarry managers for any other similar items to be preserved and sent to him, and soon afterwards the skull of a hitherto unknown hominid, named by Dart *Australopithecus africanus* ('southern ape of Africa'), was discovered. Dart's claim that *Australopithecus* was the 'missing link' between man and the apes was rejected by authorities of the day, however, until BROOM found further hominid remains in the Transvaal in 1936. It is now thought that *Australopithecus* lived about 1.2–2.5 million years ago but it is still a matter of debate whether modern man is directly descended from him or if he only represents an unsuccessful evolutionary branch from a much earlier common ancestor.

Darwin, Charles (Robert) (1809–82) British naturalist: developed a general theory of evolution and natural selection of species.

Young Darwin must have been a disappointment to his talented family. His 7 years at Shrewsbury School in his home town led to no career choice and his 2 years at Edinburgh as a medical student he found 'intolerably dull'. His father, a successful physician, tried again and sent him to Cambridge to study for the church but, although he made some good friends, his 3 years were 'sadly wasted there' and his main interests were still insect-collecting and bird-shooting. Then, when he was 22, he learned that Captain Robert FitzRoy (1805–65) had been commissioned by the Admiralty to take the naval survey ship *HMS Beagle* on a scientific expedition to circumnavigate the southern hemisphere and was looking for an unpaid volunteer naturalist to join him. Darwin was attracted; his father was against it, but his uncle Josiah Wedgwood (1769–1843) approved and, after some doubts, so did FitzRoy. Darwin's voyage on the *Beagle* began in 1831, and was to last 5 years and to stir a revolution in biology.

At that time, biologists in general believed either that species in natural conditions had continued without change since their original creation, or else (like LAMARCK) they thought that a characteristic acquired in life could simply be inherited by the offspring. Darwin's experience on his voyage made him doubtful of both theories. For example, he studied the life of the Galápagos Islands, off the western coast of South America. These 10 rocky islands are typically about 80 km apart, with a similar climate, and are separated by deep and fast sea. They are free from gales, and their geology suggests they were never united and are geologically quite young, The few plants and animals resemble those in South America, but are different. Remarkably, each island has to a large extent its own set of plants and animals; there are tortoises, finches, thrushes and many plants which correspond in several islands but are detectably different, so that as the vice-gover-

nor Lawson told Darwin, speaking of tortoises, 'he could with certainty tell from which island any one was brought'.

Darwin published the *Journal* of his voyage in 1839, and from then on he gathered his notes on species and read extensively. He read T R Malthus's (1766–1834) ideas of 1798 on human populations and their survival in the contest for food, and Darwin concluded that all plant and animal species undergo variation with time and that some variations tend to be preserved and others destroyed as a result of the inexorable contest for survival among all living things. Darwin's collection of material on this subject was made while he lived as a country gentleman in Kent, with his wife Emma Wedgwood (his first cousin) and their 10 children. He discussed his views with his two close friends, the geologist LYELL and the botanist HOOKER, but he was in no hurry to publish them.

Then in 1858 he had a shock; WALLACE, then in Malaya, sent him an essay offering the same essential idea and inviting his opinion. As a result, he and Wallace published at the same time in 1858 by agreement. The next year Darwin's book *The Origin of Species by Means of Natural Selection* appeared, giving his ideas in detail; it created excitement among biologists and widespread discussion. Many churchmen were shocked by it, since Darwin's theory of evolution entailed no special need for divine intervention and the theory implied also that man had evolved like other organisms and was not a product of a Biblical creation.

Darwin was diffident (and manipulative), and the forceful arguments for his ideas were pressed by his friends, especially T H HUXLEY. Interestingly, Darwin had no understanding of mutation, or of heredity in the modern sense and, although MENDEL'S work on heredity appeared in 1865, it was neglected then and effectively rediscovered only in 1900. The modern development of much of biology, anthropology and palaeontology is based on the idea of evolution of species, while discussion still continues on aspects of the subject such as whether the rate of evolutionary change is broadly uniform or includes periods of both sluggish and rapid change.

Darwin was a very careful observer and his theorizing showed both independence of mind and a desire (combined with caution) to reach general theories in biology. His famous work is in his best-selling books, *The Voyage of the Beagle* and *The Origin of Species* (their short titles), but he also wrote on the evolution of man, on emotion in men and animals and on climbing and insectivorous plants; he worked hard despite recurrent illnesses. He had ideas on the origin of life and in a letter of 1871 to Hooker wrote that 'if (and oh what a big if) we could conceive in some warm little pond, with all sorts of ammonia and phosphoric salts, light, heat, electricity, etc,

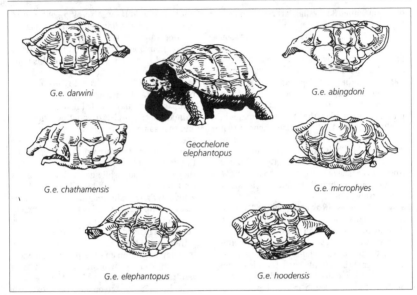

G.e. darwini

G.e. abingdoni

Geochelone
elephantopus

G.e. chathamensis

G.e. microphyes

G.e. elephantopus

G.e. hoodensis

Galápagos giant tortoise, showing shell shape of subspecies

present, that a protein compound was chemically formed' but he recognized that such speculation was then premature. In fact, ideas a century later were broadly in accord with his. Darwin also contributed to geology, but his valuable work on coral atolls and on land elevation has been overshadowed by his massive contribution to biology. He was never honoured by the Crown, a rather remarkable omission; he was the most influential biologist in history. He was buried in Westminster Abbey, near NEWTON.

Other members of Darwin's family contributed to work on evolution. His grandfather Erasmus Darwin (1731–1802) was a physician, biologist, engineer and poet, and the presiding genius of the Lunar Society; he had ideas on evolution that were ahead of their time. Erasmus's second wife was the grandmother of GALTON, who examined the statistics of inherited talent, in his own and other families (see family tree, p. 123).

Dausset, Jean [dohsay] (1916–) French immunologist: made important investigations into blood transfusion reactions.

From the time during the Second World War when he served in a blood transfusion unit, Dausset was mainly interested in transfusion reactions. This led him in the early 1950s to discover that the belief that blood of group O can be used for all patients is false. If the donor has recently been given antidiphtheria or antitetanus vaccine, the resulting antibodies can produce shock reactions when the blood is transfused. Continuing his study of transfusion responses, Dausset found that patients who had many blood transfusions were prone to produce antibodies against the white cells. The antigen (human lymphocyte antigen, HLA) is, he suggested, related to the mouse H-2 system. The work led to 'tissue typing' by simple tests and proved of great value in reducing rejection risks in organ transplant surgery. Dausset shared a Nobel Prize in 1980.

Davis, William Morris (1850–1934) US physical geographer: pioneer of geomorphology.

After a period as a meteorologist in Argentina, Davis worked with the North Pacific Survey before being appointed to a post at Harvard in 1877. He pioneered the study of landforms, conducting a classic study of the drainage system of the Pennsylvania and New Jersey area in 1889 in which he illustrated his idea of erosion cycles. He proposed that the erosive action of rivers causes first the cutting of steep V-shaped valleys, which mature into broader valleys and lead eventually to the formation of a rolling lowland landscape that he termed a 'peneplain'.

Davisson, Clinton Joseph (1881–1958) US physicist: discovered experimentally the diffraction of electrons by crystals.

After graduating from the University of Chicago and taking his PhD at Princeton, Davisson worked at the Carnegie Institute of Technology (1911–17). He then joined the Bell Telephone Laboratory (then the Western Electric Co Laboratory) for wartime employment after being refused enlistment in 1917, and subsequently stayed until his retirement in 1945.

The Davisson and GERMER experiment, which confirmed BROGLIE'S hypothesis that particles could behave like waves (and thus fundamentally altered modern physics), was initially accidental and in part due to a patent suit. Western Electric were protecting their patent for DE FOREST'S three-element vacuum tube (with an oxide-coated filament) against LANGMUIR'S similar tube with a tungsten filament developed by him at General Electric Co. In order to help settle the suit (which dragged on for a decade), Davisson and Germer measured electron emission from oxide-coated platinum under ion bombardment. The purpose was to establish that the electron emission did not depend upon positive ion bombardment due to oxygen traces in the tube, and therefore that Langmuir's tube did not fundamentally differ from that already under patent. This they did, and the Supreme Court eventually ruled in Western Electric's favour. In the meantime Davisson and C H Kunsman investigated electron emission under electron bombardment as an easy extension to the work, and found a small number of primary electrons with the full energy of the incident beam deflected back alongside the many low energy secondary electrons. In 1925 an accidental explosion of a liquid-air bottle heavily oxidized a nickel surface which Davisson was investigating, and after heating to clean it (which also recrystallized it from polycrystalline into a few large crystals) it displayed a maximum scattering at a particular angle.

On visiting Oxford in 1926 and hearing of Broglie's recent work, postulating wave behaviour for an electron, Davisson realized that he had seen diffraction maxima in the electron wave pattern. In 1927 Davisson, with Germer, obtained conclusive evidence that electron beams were diffracted on reflection by nickel crystals and had the wavelength predicted by Broglie. For this he shared the 1937 Nobel prize for physics with G P THOMSON, who had observed similar electron diffraction with high-energy electrons passing through metal foil.

Davy, Sir Humphry (1778–1829) British chemist: discoverer of sodium and potassium, exploiter of electrochemistry and propagandist for science.

Son of a Cornish woodcarver and small farmer, Davy became an apprentice pharmacist. However, in 1798 he was employed by BEDDOES to work in his Medical Pneumatic Institution in Bristol with the task of developing the medical uses of some newly discovered gases. Davy made N_2O ('nitrous oxide' or 'laughing gas') in quantity, studied it fully and, through this work and some useful friendships, was appointed as chemist by RUMFORD in the new Royal Institution in London, in 1801. He quickly became famous as a lecturer. His ideas on the uses of science appealed to the serious-minded, and demonstrations (especially of the inhibi-

tion-releasing effects of N_2O) attracted others. Davy made the Royal Institution a social and financial success and thereby acquired the equipment (especially a large voltaic cell) to develop his interest in electrochemistry. In 1807 he made the reactive metals potassium and sodium by electrolysis; and soon he secured other new and reactive metals. These exciting discoveries were followed by experiments that showed that chlorine was probably an element (and not a compound); and further work related it to iodine, newly found by B Courtois (1777–1838), and to fluorine.

In 1812 he was knighted, and 3 days later married Jane Apreece, a wealthy Scottish widow. He was now established as Britain's leading scientist and he embarked on the first of many European tours. In 1813 he hired FARADAY as an assistant (and also tried to use him as a valet on his travels). In 1815 he was asked to devise a safe lamp for use in gassy coalmines. This was the sort of problem that showed his talent well. In 6 months he had made the first thorough study of flame combustion and devised his safety lamp, which made mining of deep coal seams possible even where firedamp (CH_4) was present.

Davy's reputation outstrips his chemical achievements, substantial though they were. He had great energy and talent, especially in attacking limited but important chemical problems. He was also snobbish, excitable and ungenerous to other scientists, unskilled in quantitative work and uneven in his knowledge or interest in theories (he doubted DALTON'S new atomic theory). His early death left 'brilliant fragments' (said BERZELIUS), much interest in electrochemistry and perhaps his finest 'discovery', Faraday. An important achievement was that he had sold science to the industrialists, especially through his success with the miner's safety lamp.

He had an intense interest in angling; his younger brother John (also a chemist) says he was 'a little mad about it'. He was an enthusiastic poet and had friends with real literary talent, including Coleridge, Southey and Wordsworth, who thought better of his poetry than do modern critics.

de Bary, (Heinrich) Anton [duh baree] (1831–88) German botanist: a founder of mycology.

De Bary left medicine to teach botany in three German universities before settling in Strasbourg in 1872. His main work was in mycology, where he showed that fungi are the cause of rust and smut diseases of plants (and not a result, as others had thought). He went on to show that lichens consist of a fungus and an alga in intimate partnership, forming a remarkably hardy union with mutual benefits. De Bary named this symbiosis; the term now is used to cover three kinds of specialized association between individuals of different species, including parasitism (one organism gains, the other loses) as well as commensalism (one organism gains, the other neither loses nor gains) and mutualism (a mutually beneficial association, as in the lichens).

De Bary's excellent descriptions and classifications of fungi, algae, 'moulds and yeasts', established them as plants that happen to be small and did much to create modern mycology.

de Beer, Sir Gavin Rylands [duh beer] (1899–1972) British zoologist: refuted germlayer theory in embryology, and theory of phylogenetic recapitulation.

De Beer served in both world wars, in Normandy in 1944 with the Grenadier Guards as a Lieutenant-Colonel; in the interval he graduated from Oxford and he afterwards taught there. After the Second World War he became professor of embryology in London and from 1950 director of the British Museum (Natural History) until he retired in 1960. In 1926 he much injured the germ-layer theory in embryology by showing that some bone cells develop from the outer (ectodermal) layer of the embryo (the theory had them form from the mesoderm). In 1940 he also refuted E Haeckel's (1834–1919) theory of phylogenetic recapitulation; according to this theory an organism in its embryonic stage repeats the adult stages of the organism's evolutionary ancestors. De Beer showed that in fact the situation is rather the converse; adult animals retain some juvenile features of their ancestors (paedomorphism). His many other researches included studies of the earliest known bird, *Archaeopteryx*, and led him to propose a pattern of 'piecemeal' evolution to explain its possession of both reptilian and avian features (eg teeth and wings); he worked on the origin of the Etruscans from blood group data and on Hannibal's route over the Alps.

de Broglie, Louis Victor Pierre Raymond, duc de, *see* **Broglie**

de Buffon, Georges-Louis Leclerc, comte (Count) *see* **Buffon**

Debye, Peter (Joseph William) [duhbiy] (1884–1966) Dutch–US chemical physicist; developed ideas on dipole moments, and on solutions of electrolytes.

Debye was educated in the Netherlands and in Germany, and then held posts in theoretical physics in several European countries in rapid succession. Despite these frequent moves, he produced in 1911–16 a theory of the change in specific heat capacity with temperature, a method for X-ray diffraction analysis using powdered crystals (with P Sherrer) and the idea of permanent molecular electric dipole moments. He showed how these moments can be measured and how they can be used to find the shape of simple molecules; eg the molecule of water, H–O–H, is not linear but bent. He was also able to show that the benzene ring is flat.

The unit of electric dipole moment, the debye (D), is the electronic charge $(e) \times 10^{-10}$ m. His work with HÜCKEL led in 1923 to the Debye–Hückel theory of electrolytes, which deals with the behaviour of strong solutions of electrolytes by taking account of the mutual interaction of the charged ions (previous theories had dealt only with very dilute solutions). In 1934 he moved to Berlin and in 1940 to the USA, where he was professor of chemistry at Cornell until 1950. His work on light scattering in solutions, on polymers and on magnetism is also important. He was awarded the Nobel Prize for chemistry in 1936, and in 1939 had the strange experience of seeing a bust of himself unveiled in his native city of Maastricht.

de Candolle, Augustin-Pyramus *see* **Candolle**

du Chatelet-Lomont, Gabrielle-Emilie, marquise (Marchioness) *see* **Chatelet-Lomont**

de Coriolis, Gaspard Gustave *see* **Coriolis**

de Coulomb, Charles (Augustin) *see* **Coulomb**

Dedekind, Julius Wilhelm Richard [dayduhkint] (1831–1916) German mathematician: made far-reaching contributions to number theory.

Dedekind studied at Brunswick and then formed a close association with RIEMANN, DIRICHLET and GAUSS at Göttingen, with each influencing the others. Dedekind learned about the method of least squares from Gauss, the theory of numbers, potential theory and partial differential equations from Dirichlet. After a short time he moved briefly to Zürich, and then returned to spend the rest of his long life as a professor at the Technical High School, Brunswick. He lived long enough for much of his influential work (eg on irrational numbers) to become familiar to a generation of students in his later years, and he became a legend. Some 12 years too soon, Teubner's Calendar for Mathematicians recorded him as having died on 4 September 1899. Much amused, Dedekind wrote to the editor: 'According to my own memorandum I passed this day in perfect health and enjoyed a very stimulating conversation ... with my ... friend Georg Cantor of Halle'.

Dedekind was one of the first to recognize the value of the work of Cantor (1845–1918) on infinite qualities. Dedekind himself made major steps towards modern standards of rigour and was ahead of his time in his approach to number theory.

In 1858 he produced an arithmetic definition of continuity and clarified the concept of an irrational number (that is, roughly, a number that cannot be represented as a fraction). In the first of three books he used Dedekind cuts (the categorization of irrational numbers by fractions) to rigorously examine the real number system.

Then in his second major work (1888) he established a logical foundation for arithmetic and described axioms that exactly represent the logical concept of whole numbers (these are now, incorrectly, called PEANO axioms). Finally, Dedekind described in 1897–1900 the factorization of real numbers using modern algebra.

de Duve, Christian (René) *see* **Duve**

de Fermat, Pierre *see* **Fermat**

De Forest, Lee (1873–1961) US physicist: inventor of the thermionic triode 'valve' (electron tube) and pioneer of radio.

De Forest studied at Yale University, writing his doctoral thesis on radio waves (probably the first thesis on the subject of radio in America). He went to work for the Western Electric Company and in 1907 developed and patented the thermionic triode. This device, essentially a diode with an additional electrode between cathode and anode, could be used to amplify weak electrical signals and was crucial to the development of radio communication, radar, television and computers. The triode remained an essential component in all kinds of equipment for 50 years before being largely superseded by the transistor. De Forest also worked on a film soundtrack system and a medical diathermy machine, the former being a commercial failure at the time but later widely adopted.

de la Beche, Sir Henry Thomas [beech] (1796–1855) British geologist: conducted first systematic geological survey of the British Isles.

After the early death of his father, de la Beche lived with his mother for a time in the fossil-rich area of Lyme Regis in Dorset. He entered military training school at the age of 14 and later travelled widely, having given up an army career and devoted himself to geology. He mapped and wrote the first descriptions of the Jurassic and Cretaceous rocks of the Devon and Dorset area, the geology of the Pembrokeshire coast and that of Jamaica. During the late 1820s he began his most significant work, the first systematic geological survey of Britain. Working at first as an amateur, his efforts led to the establishment of the Geological Survey of Great Britain in 1835, with himself as its first director.

de Laplace, Pierre Simon, marquis *see* **Laplace**

de la Tour, Charles Cagniard *see* **Cagniard de la Tour**

Delbrück, Max [delbrük] (1906–81) German–US biophysicist: pioneer of molecular biology.

Delbrück is unusual in 20th-c science for practising both physics and biology and for the fact that his place, although substantial as a discoverer, is largely that of an inspirer of others in the creation of molecular biology.

He began in physics, with a PhD from Göttingen in 1930, and spent 2 years on atomic physics with BOHR, then 5 years as assistant to LISE MEITNER at the Kaiser Wilhelm Institute in Berlin; and from 1937 he was at the California Institute of Technology, where he moved into biology. MORGAN and the *'Drosophila'* geneticists

had gone to Pasadena from New York in 1928, taking with them the conviction that genetic problems should be solvable by chemistry and physics. Delbrück agreed, with the proviso that new concepts in these sciences would be needed, and his ideas were developed in the physicist SCHRÖDINGER's influential book *What is Life?* in 1945. Delbrück decided to work on viruses as the simplest life form. He did much to create bacterial and bacteriophage genetics, and in 1946 he showed that viruses can exchange (recombine) genetic material, the first evidence of recombination in primitive organisms. His firm belief in an 'informational basis' in molecular biology bore fruit in other hands, but with much help from his forceful catalytic ideas. He shared a Nobel Prize in 1969, which led to another of his famous parties.

Democritus (of Abdera) [demokrituhs] (c.470–c.400 BC) Greek philosopher: pioneer of atomic theory.

Almost nothing is firmly known of Democritus's life, and his ideas have survived through the writings of others, either supporting or attacking him. His idea of atoms seems to have begun with his teacher Leucippus (5th-c BC), but Democritus much extended the theory. He proposed that the universe contains only a vacuum and atoms, and that these atoms are invisibly small and hard, eternal and are in ceaseless motion. On this adaptable, materialist view he explained taste, smell, sound, fire and death. He supposed that in their form and behaviour lay the natural, godless cause of all things and all events. Plato and ARISTOTLE were not in favour of these ideas, which never formed a part of the mainstream of Greek philosophy, but they were adopted by the Greek philosopher Epicurus about 300 BC and well recorded in a long poem (*On the Nature of Things*) by the Roman Lucretius (c.99–55 BC). In the 17th-c BOYLE and NEWTON were aware of these ideas; it is doubtful if they contributed at all directly to modern atomic theory, which began with DALTON about 1800.

de Moivre, Abraham [duh mwahvruh] (1667–1754) French–British mathematician: founded analytical trigonometry and stated de Moivre's theorem.

De Moivre had the misfortune to be a Huguenot (Protestant) at the time that Roman Catholic France revoked the Edict of Nantes and began to persecute them (1685). He was imprisoned in Paris for a year and moved to England on his release. Friendship with NEWTON and HALLEY aided his election to the Royal Society (1697). However, de Moivre remained poor, working as a tutor or consultant to gambling or insurance syndicates, and never obtained a university post. He died blind and disillusioned, with his work unrecognized.

His book *The Doctrine of Chances* (1718) is a masterpiece, and sets out the binomial probability or Gaussian distribution, the concept of statistical independence and the use of analytical techniques in probability. Deriving an expansion for $n! = n(n-1)(n-2)...3.2.1$, de Moivre summed terms of the binomial form. He established many of the elements of actuarial calculations. Above all he discovered the trigonometric relation

$$(\cos\theta + i\sin\theta)^n = \cos n\theta + i\sin n\theta$$

called de Moivre's theorem (1722), which is a powerful step in developing complex number theory.

de Montgolfier, Joseph(-Michel) and (Jacques-)Étienne *see* Montgolfier

de Réaumur, René-Antoine Ferchault *see* Réaumur

Descartes, René [daykah(r)t] (1596–1650) French philosopher and mathematician: creator of analytical geometry.

Descartes has a dominant position in shaping modern philosophy, but this is not our concern here. With enough modest inherited wealth to live as he chose, he spent his life in travel, on his work in philosophy, mathematics, physics and physiology and as a soldier serving in Holland, Bohemia and Hungary. In 1621 he left the army and in 1629 settled in Holland for some 20 years, before being persuaded to become tutor to Queen Christina of Sweden, a headstrong and athletic 19-year-old. From childhood Descartes had risen late and claimed to do his best thinking in a warm bed; the Queen's insistence on tutorials in philosophy at 5 am in a freezing library either hastened or produced the lung disease which killed him within 5 months of arrival.

Although Descartes theorized extensively in physics and physiology, his lasting influence outside philosophy is in mathematics, where he created analytical or coordinate geometry, also named (after him) Cartesian geometry. This translates geometrical problems into algebraic form, so that algebraic methods can be applied to their solution; conversely he applied (for the first time) geometry to algebra. His methods made a massive change in mathematical thought and remain familiar today, as in the equation of the straight line, $y = mx + c$ and the equations of familiar curves such as the conic sections. The thinker J S Mill (1806–73) claimed that Cartesian geometry 'constitutes the greatest single step ever made in the progress of the exact sciences'.

de Sitter, Willem *see* Sitter

Desmarest, Nicolas [daymaray] (1725–1815) French geologist: demonstrated the igneous origin of basalts.

Desmarest was by profession a trade and industry inspector for the department of commerce. In the 1760s he became interested in the large basalt deposits of central France (discovered by GUETTARD a decade before) and succeeded in tracing their origins to ancient volcanic activity in the Auvergne region. In 1768 he produced a detailed study of the geology and eruption history of the volcanoes responsible. His work was important because it demonstrated that basalts were igneous in origin, which led to the abandonment of the widely held belief that all rocks were sedimentary (the Neptunist theory of A G WERNER).

Deville, Henri Étienne Sainte-Claire (1818–81) French chemist: developed methods for making light metals in quantity; studied high temperature reactions.

Deville is one of the few major 19th-c scientists to be born in the West Indies, where his family had been leading citizens for two centuries. With his older brother Charles he was educated in Paris. He chose medicine but was soon attracted to chemistry; in the 1840s he worked on essential oils and obtained methylbenzene and methyl benzoate from balsam of Tolu, but his chemical fame began in 1849 when he made the crystalline and highly reactive dinitrogen pentoxide by treating warm silver nitrate with chlorine. From 1851 he held a post at the École Normale Supérieure, with PASTEUR as a colleague and close friend from 1857. The main work of the institution was to train senior school teachers, which Deville did for 30 years while maintaining a major research output. He was a masterly lecturer and experimentalist, uninvolved in disputes over theory. Realizing that sodium metal in quantity would be of great use, he developed a large-scale method for making it by reduction of sodium carbonate with carbon. One target was to use sodium to make aluminium by reduction of $AlCl_3$. In 1855 Deville was summoned to show aluminium, then rare, to the Emperor; the latter was attracted by the idea of fitting his troops with helmets of the new metal, and a government grant to set up a pilot plant was arranged. Success followed; bars of aluminium were shown at the 1855 Exposition, and it quickly ceased to be mainly used for jewellery. Soon Deville made pure magnesium in quantity, and titanium, and crystalline boron and silicon, all by reduction of chlorides with sodium metal. He worked on the platinum metals and in 1872 was given the task of making the platinum–iridium (90–10) alloy for the standard kilogram and metre.

His interest in high temperature chemistry had begun in the 1850s and his oxy-hydrogen blowpipe method led to technical welding methods as well as studies of minerals and high-melting metals. From 1857 he also studied vapour densities, using porcelain bulbs and the vapour of boiling metal (Hg,Cd or Zn) to give a constant high temperature. He found that relative molecular mass could change with temperature; thus aluminium chloride is mainly Al_2Cl_6 at 500°C, but $AlCl_3$ at 1000°C. Chemical changes due to heat which reverse on cooling were described as dissociations by Deville, who first fully examined such changes. He found that at high temperatures H_2O, CO_2, CO, HCl and SO_2 all dissociated.

de Vries, Hugo *see* **Vries**

Dewar, Sir James [dyooer] (1842–1923) British chemist and physicist: pioneer of low temperature studies and the chemistry of metal carbonyls.

Son of a wine-merchant, Dewar became attracted to chemistry at Edinburgh University and spent a summer in Ghent in KEKULÉ's laboratory. In 1869 he went to teach chemistry at the Royal Veterinary College in Edinburgh. Although he was never interested in teaching students, he was a popular society lecturer on a wide range of scientific topics, and in 1875 became Jacksonian Professor at Cambridge. He was probably elected because of his work in physiological chemistry and was perhaps expected to do more of it; in fact he found the laboratory so poor that he was glad to be appointed also in 1877 to the Professorship of Chemistry at the Royal Institution, London. Thereafter he lived and worked in London, visiting Cambridge only briefly and infrequently in order to abuse his staff there for idleness and to visit G D Liveing (1827–1924), professor of chemistry; they collaborated in spectroscopic research for 25 years. Dewar had a strangely wide range of research interests, which he maintained with his own hands or with assistants, never having students or founding a 'school'. In the early 1870s he invented the Dewar flask (domestically a Thermos flask), a double-walled glass flask with the inner walls reflective and the space between them evacuated; heat is only slowly passed to or from the contents. From 1877 he worked on the liquefaction of gases, using the flasks for storage. He used CAILLETET's method for making oxygen, on a scale which allowed him to study low temperatures; and by 1898 he made liquid hydrogen in quantity and the solid in 1899, at a temperature below 14 K. At this temperature all known substances become solid, except helium. He tried to liquefy helium, discovered on Earth in 1895, but did not succeed. With F A ABEL he invented cordite. He worked on specific heat capacities, electrical effects of very low temperatures, metal carbonyls, diffusion, high vacua, coal tar bases, dissociation of molecules at high temperatures, emission and absorption spectra, soap films (he made them over 1 m in diameter) and the Sun's temperature. A small, brusque Scot, he was unsurpassed in diversity and productivity as an experimentalist.

Dicke, Robert Henry (1916–) US physicist: predicted existence of cosmic microwave background.

Dicke studied at Princeton University and at Rochester, and in 1957 was appointed professor of physics at Princeton, where he became Albert Einstein Professor of Science in 1975.

In 1964, Dicke made the prediction that, assuming the universe had been created by a cataclysmic explosion (the 'Big Bang'), there ought to be a remnant radiation, observable in the microwave region of the spectrum. At almost the same time, PENZIAS and R W WILSON did in fact observe this cosmic microwave background, although they were unaware of Dicke's theoretical work at the time. Together, their work established the 'Big Bang' model of the origin of the universe as far more plausible than the rival steady-state theory. Interestingly, and unknown to Dicke at the time, GAMOW and others had made a similar prediction in 1948.

Dicke was also interested in gravitation and established that the gravitational mass and inertial mass of bodies are equivalent to an accuracy of at least one part in 10^{11}, an important result for general relativity. However, in 1961 the Brans–Dicke theory suggested that the gravitational constant (G) was not in fact constant, but varied slowly with time (by about 10^{-11} per year). Unfortunately the experimental observations required to verify this hypothesis are not yet sufficiently precise to prove it or disprove it.

Diels, Otto (Paul Hermann) [deels] (1876–1954) German organic chemist: co-discoverer of Diels–Alder reaction.

A member of an academically talented family, Diels studied chemistry in Berlin with E FISCHER. Most of his life was spent in the University of Kiel. In 1906 he discovered a new oxide of carbon (the monoxide and dioxide were long known); he made tricarbon dioxide (C_3O_2) by dehydrating malonic acid with P_2O_5:
$$CH_2(CO_2H)_2 - 2H_2O \rightarrow O=C=C=C=O$$
He showed in 1927 that cholesterol is dehydrogenated by heating with selenium and that the hydrocarbon products include one of melting point 127°C, now known as Diels's hydrocarbon. The method can be applied generally to steroids, and many yield Diels's hydrocarbon; in this way and by interconversions this biologically important steroid group was shown to all have the same carbon skeleton. By 1934 Diels's hydrocarbon had been synthesized by others and its structure was the critical clue that allowed other steroid structures to be assigned, all containing the four-ring skeleton of Diels's hydrocarbon: this skeleton became the defining feature of steroids.

In 1928, with his assistant K Alder (1902–58), Diels discovered the valuable general synthesis now known as the Diels–Alder reaction. In this, a conjugated diene reacts by 1,4-addition with one of a large group of unsaturated compounds (dienophiles) to give, usually, a six-membered ring compound as the product. The method has proved of great value in the synthesis of complex organic compounds. Diels and Alder shared a Nobel Prize in 1950.

Diesel, Rudolph (Christian Karl) (1858–1913) German engineer: devised the compression-ignition internal combustion engine.

TELESCOPES

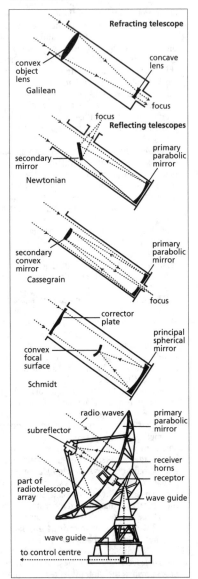

Refracting telescope

convex object lens

Galilean

concave lens

focus

focus

Reflecting telescopes

secondary mirror

Newtonian

primary parabolic mirror

secondary convex mirror

Cassegrain

primary parabolic mirror

focus

corrector plate

convex focal surface

Schmidt

principal spherical mirror

radio waves

subreflector

part of radiotelescope array

primary parabolic mirror

receiver horns

receptor

wave guide

wave guide

to control centre

It is not clear who made the first telescope. It may well have been LEONARD DIGGES, who probably made a telescope about 1550, with a convex lens whose image was reflected to the side by a concave mirror for viewing. A claim has long been made for Hans Lippershay (*c.*1570–*c.*1619), a Dutch spectacle maker who sought a patent in

Section through five types of telescope

1608 for a design we would now class as Galilean. What is certain is that such telescopes were on sale in Paris, Milan and London in 1609 and that GALILEO quickly heard of the device, deduced its construction (see diagram), made several for himself and within a year had seen mountains on the Moon, sunspots, the phases of Venus and the four larger satellites of Jupiter, all for the first time. His best telescope had a magnification of about 30 ×. In the next century many more refracting telescopes (ie using lenses only) were made. In KEPLER's design, a positive lens forms the eyepiece; the image is inverted; but for terrestrial use a second lens in the eyepiece gives an erect image. These early telescopes were awkward to use; in particular, as the red and blue ends of the spectrum are not brought to the same focus by a simple lens, the image is seen to have colour fringes.

NEWTON became interested in telescopes in the 1660s, but soon decided (incorrectly) that colour fringes were unavoidable with lenses, and in 1668, when he was 26, he made for himself a small reflecting telescope which solved this problem (see diagram).

The value of a telescope in astronomy depends largely on the area of its main lens or mirror, which gathers light and so allows the eye (or a camera) to observe small, dim, or distant objects by concentrating their light. F W HERSCHEL cast, ground and polished mirrors for his telescopes (which were of Newtonian or related type) up to 48 in (122 cm) across, with a focal length of up to 40 ft (12 m), and used them to discover a new planet (Uranus), to examine nebulae (clouds of luminous gas or dust) and to show that the Milky Way is dense with stars. The mirrors were made of an alloy of copper and tin, and soon tarnished. In the 1840s the Earl of Rosse made similar mirrors 72 in (183 cm) across, but the unsuitable climate of Birr Castle and Parsonstown, his estate in Eire, gave few clear nights to use his massive instruments.

Refracting telescopes were restored to favour after DOLLOND began making achromatic lenses made of two components: one of crown glass and one of flint glass, a combination that largely solved the colour problem. However, a large lens requires two good surfaces and fairly flawless glass between them, and the largest in use is a 40 in (1 m), with focal length 62 ft (19 m), in the USA.

In this century reflectors have proved dominant: usually of Pyrex glass, silvered or aluminized, driven electrically to follow the apparent motion of the stars, and linked to a spectroscope, a camera or an electronic detector (a CCD, charge-coupled device) rather than a human eye. Their massive mountings provide a substantial engineering problem: in the first half of this century high locations in California were favoured, and 60 in (1.5 m), 100 in (2.5 m) and 200 in (5 m) instruments are used in the Mount Wilson and Mount Palomar Observatories there. But atmospheric pollution and light pollution from cities have proved a problem, and recent large instruments have been placed at high points in Hawaii, Las Palmas and Chile to reduce this. One of the largest (but of inferior optical quality) is the 6 m (236 in) reflector installed in 1969 on Mt Pastukhov in the northern Caucasus.

Telescopes can also be designed to detect radiation outside the visible range, and the radio, ultraviolet, infrared and X-ray regions are all much studied. The Second World War gave a great impetus to astronomy, wartime radar and rocket missiles giving a basis for major advances. Radio astronomy stemmed from work by JANSKY and REBER, who showed that the stars included radio sources, studied after the war by HEWISH, OORT, LOVELL and others. An infrared astronomical satellite (IRAS) was launched in 1983, with a 24 in (61 cm) reflecting telescope for infrared rays; it survived for 10 months and located many infrared sources. Telescopes for use in the ultraviolet spectrum had been used in satellites in the 1970s. By 1986 the space probe Giotto passed close to HALLEY's comet and transmitted TV pictures of its nucleus to Earth (see chronology); in 1990

satellites were launched carrying a telescope to detect X-ray sources among stellar objects (ROSAT), and a 95 in (2.4 m) reflecting telescope (named after HUBBLE) to observe deep space. The Hubble Space Telescope, after repairs in space had corrected some defects, proved to be excellent, and entirely free from unwanted effects due to the Earth's atmosphere; the data are transmitted by radio. The period 1950–2050 should be a golden age for astronomy, with new information becoming available at an unprecedented rate. At last, after 380 years, astronomers have escaped from the great restriction on their observational work, the Earth's atmosphere.

At the time of writing, the world's largest optical telescope is the Keck I, on the summit of Mauna Kea (4200 m/13 780 ft), Hawaii. It weighs 297 t, and the mirror system is 10 m in diameter and consists of 36 hexagonal segments fitted together. Its huge size and good location enable it to record very distant and faint objects, allowing better estimates to be made of the size and age of the universe than any made previously. When its nearby sibling Keck II is completed in 1996 the paired system, used as an optical interferometer, will give a tenfold improvement in the already high resolution: and the key role of optical astronomy will be reaffirmed.

IM

Born in Paris of German parents, Rudolph and his family left for London when the Franco-Prussian war of 1870 began but he soon moved to Germany to continue his education, eventually studying at the Munich Polytechnic.

From 1880–90 he worked on refrigeration plant, but his interest was in engines. He realized that on thermodynamic principles an internal combustion engine should desirably operate with a large temperature range, which implies a high pressure. His patent of 1893 and his engines produced in the late 1890s use a four-stroke cycle like the OTTO engine (induction, compression, combustion, exhaust) but in the Diesel engine a higher-boiling petroleum fraction is used. In the induction stroke, air alone is drawn into the cylinder. On the compression stroke this air is compressed by up to 25:1 (unlike the 10:1 compression of the petrol:air mixture in a petrol engine) and this raises its temperature to near 600°C. Then an injector admits a fine spray of fuel into the heated air; it ignites spontaneously, and the combustion stroke provides power. An exhaust stroke to remove the burned gas completes the cycle. Diesel prospered, but in 1913 he vanished from the Antwerp–Harwich mail steamer and

was presumed drowned; his body was never found.

Diesel engines are more efficient than petrol engines and were used in the First World War in submarines, and later in ships and rail locomotives. The smaller units for buses, tractors, trucks and small electrical generators were developed in the 1930s, with important design contributions by two British engineers, C B Dicksee (1888–1981) and H R Ricardo (1885–1975); by the 1980s larger commercial vehicles were normally powered by high-speed compression-ignition units.

Digges, Leonard (c.1520–59) and Thomas Digges (c.1546–1595) English mathematicians, surveyors and probable inventors of the telescope.

Following his studies at Oxford, Leonard Digges published his book *Prognostication* which sold well in various editions from 1553. It contained astronomical, astrological and meteorological tables and advice, and was followed in 1556 by a book on practical surveying. However, in early 1554 he took part, with others from his county of Kent, in a nationwide revolt against a Catholic royal marriage. The revolt was easily suppressed and its leaders beheaded; Digges was also condemned but later reprieved.

When he died aged 39, he left his 13-year old son Thomas in the care of the mathematician John Dee (1527–1608) and the boy grew up with the same interests as his father. In 1571, when he was 25, Thomas published *Pantometria*, a book on surveying originally drafted by his father and 'lately finished' by Thomas himself. It contains a description of a telescope using a lens and a mirror; this is amplified in a report prepared for Elizabeth I's government in 1585, which credits the device to Leonard Digges and his son and which clearly has in mind its military uses. It is likely that Thomas Digges used it to observe the sky and was impressed by the great number of stars it made visible.

However, the effective beginning of observational astronomy came with GALILEO's work in 1610, when he used a refracting telescope made by himself and first saw mountains on the moon, the phases of Venus, Jupiter's satellites and sunspots. Improved designs of reflecting telescopes due to GREGORY and to NEWTON set these reflectors on the path of dominance for astronomical use which they later achieved (see panel 'Telescopes').

Dines, William Henry (1855–1927) British meteorologist: devised methods for study of the upper atmosphere.

A meteorologist's son, Dines studied at Cambridge and afterwards worked for the Meteorological Office. In 1901 he invented his pressure-tube anemometer, the first device to measure both the direction and the velocity of wind. He was an early user of kites and balloons to study the upper atmosphere, devising an ingenious meteorograph, weighing only 60 g, to measure and record upper-air soundings. From about 1907 his device gave data on pressure, temperature and humidity at heights up to and including the stratosphere, and became a standard instrument for this purpose. His work on these data showed new correlations between measurable properties of the upper air and gave improved understanding of cyclones and anticyclones, showing for example that their circulation is not greatly affected by thermal processes near the Earth's surface.

Diophantus (of Alexandria) [diyohfantuhs] (lived *c.*250) Greek mathematician: discoverer of the Diophantine equations.

Although an outstanding mathematician of his time, very little is now known about Diophantus's life. His work is preserved in the six surviving chapters of his *Arithmetica* (a further seven chapters have been lost), which was probably the earliest systematic treatise on algebra. Diophantus was primarily interested in number theory and the solution of equations and did much to advance algebra by his use of symbols for quantities, mathematical operations and relationships; previously such quantities had been described in words. He is perhaps best remembered as the discoverer of the Diophantine equations, indeterminate equations with rational coefficients for which a rational solution is required.

Dirac, P(aul) A(drien) M(aurice) (1902–84) British theoretical physicist: major contributor to quantum mechanics; predicted existence of the positron and other antiparticles.

Dirac, the son of a Swiss father and English mother, studied electrical engineering at Bristol and mathematics at Cambridge. After teaching in America and visiting Japan and Siberia, Dirac was appointed in 1932 to the Lucasian professorship in mathematics at Cambridge, where he remained until his retirement in 1969. He was then a visiting lecturer at four US universities before becoming professor of physics at Florida State University in 1971.

A uniquely gifted theoretician, Dirac contributed creatively to the rapid development of quantum mechanics. In 1926, just after BORN and JORDAN, he developed a general theoretical structure for quantum mechanics. In 1928 he produced the relativistic form of the theory, describing the properties of the electron and correcting the failure of SCHRÖDINGER's theory to explain electron spin, discovered by UHLENBECK and GOUDSMIT in 1925.

From the relativistic theory he proposed in 1930 that the theoretically possible negative energy solutions for the electron exist as states but these states are filled with particles of negative energy so that other electrons cannot enter them. He predicted that a sufficiently energetic photon could create an electron–positron pair apparently from nowhere by knocking an electron out of one of these negative energy states. The positively charged hole left is the antiparticle to an electron, called a positron. Also, an electron meeting a positron can give mutual annihilation, releasing energy as a photon (light). All these predictions were observed experimentally by C D ANDERSON in 1932. Dirac's argument applies to all particles, not just electrons, so that all particles possess corresponding antiparticles.

In 1930 Dirac published *The Principles of Quantum Mechanics*, which is a classic work which confirmed his stature as a 20th-c NEWTON in the minds of many physicists. The Nobel Prize for physics for 1933 was shared by Dirac and Schrödinger.

Dirichlet, Peter Gustav Lejeune [deereeshlay] (1805–59) German mathematician: contributed to analysis, partial differential equations in physics and number theory.

Dirichlet studied at Göttingen under GAUSS and JACOBI and also spent time in Paris, where he gained an interest in FOURIER series from their originator. He moved to a post at Breslau but at 23 became a professor at Berlin, remaining for 27 years. He was shy and modest but an excellent teacher; he was a close friend of Jacobi and spent 18 months in Italy with him when

Jacobi was driven there by ill-health. On Gauss's death in 1855 Dirichlet accepted his prestigious chair at Göttingen but died of a heart attack only 3 years later.

Dirichlet carried on Gauss's great work on number theory, publishing on Diophantine equations of the form $x^5 + y^5 = kz^5$, and developing a general algebraic number theory. Dirichlet's theorem (1837) states that any arithmetic series a, $a + b$, $a + 2b$, $a + 3b$, where a and b have no common divisors other than 1, must include an infinite series of primes. His book *Lectures on Number Theory* (1863) is a work of similar stature to Gauss's earlier *Disquisitiones* and founded modern algebraic number theory.

Dirichlet also made advances in applied mathematics. In 1829 he stated the conditions sufficient for a Fourier series to converge (those conditions necessary for it to converge are still undiscovered). He worked on multiple integrals and the boundary-value problem (or Dirichlet problem), which is the effect of the conditions at the boundary on the solution of a heat flow or electrostatic equation.

It is not only Dirichlet's many specific contributions that give him greatness, but also his approach to formulating and analysing problems for which he founded modern techniques.

Döbereiner, Johann Wolfgang [doeberiyner] (1780–1849) German chemist: introduced Law of Triads and studied catalysis.

A coachman's son, Döbereiner was largely self-educated, but secured a teaching post at Jena, possibly through aristocratic influence. He held the teaching post through his lifetime; one of his pupils was the philosopher Goethe. He improved organic analysis and made the first estimates of the abundance of elements in the Earth's crust. He used an earlier observation by DAVY (that platinum caused organic vapours to react with air) to devise Döbereiner's lamp, a toy or demonstration device in which a jet of hydrogen was ignited by contact with platinum sponge. His main claim to fame is his observation of 'trias' (later, triads) of elements. These are groups such as Cl, Br, I; or Ca, Sr, Ba; or S, Se, Te; in which the atomic mass of the middle element is close to the mean of the first and last elements in its group and its physical and chemical properties likewise appear average. By 1829 this was developed as the Law of Triads. It then attracted little attention, but can now be seen as a step towards the periodic classification of the elements.

Dobzhansky, Theodosius [dobzhanskee] (1900–75) Russian–US geneticist: linked DARWIN's ideas of evolution with MENDEL's laws of genetics and so was a creator of evolutionary genetics.

He studied zoology in Kiev and later taught there and in St Petersburg before, in 1927, joining MORGAN's research group working on *Drosophila* genetics in Columbia University, New York

City. Morgan, a major figure in modern genetics, moved to California in 1928, and Dobzhansky joined him the following year and spent his later career there and in New York.

Dobzhansky's book *Genetics and the Origin of Species* (1937) brought together the ideas of field naturalists and geneticists, both experimental and mathematical, to create a single argument on the process of evolution. His work with *Drosophila* had made him very aware that natural mutations can lead to large or small apparent changes, that these are acted upon by natural selection and that genes may interact and so give rise to variation which appears continuous rather than stepwise. He showed that natural selection could be seen in action, for example in the spread of scale insects resistant to control by cyanide gas, used as an insecticide in Californian citrus groves.

Dobzhansky developed good arguments for his view that a new species cannot arise from a single mutation and that, for a new species to form, it must for a period be isolated to protect it from disruption. The isolation could be geographical or due to differences in habitat or in the breeding season.

His work on human evolution led him to define races as 'Mendelian populations which differ in gene frequencies', as outlined in his *Genetics of the Evolutionary Process* (1970).

Doisy, Edward Adelbert [doyzee] (1893–1986) US biochemist: isolated vitamin K.

Educated at Illinois and Harvard, Doisy spent most of his life at St Louis University Medical School. In 1923 he devised a bioassay for the female sex hormone and secured potent extracts, but it was BUTENANDT who first isolated oestrone. Soon Doisy moved to the study of vitamin K, discovered but not isolated by DAM in 1934, deficiency of which leads to the blood failing to coagulate. Doisy was able to isolate a factor (K for koagulation) from alfalfa grass, and a related but different K factor from putrefied fish meal. These two potent antihaemorrhagic vitamins, K_1 and K_2, were shown by Doisy to be derivatives of 1,4-naphthoquinone; such compounds are valuable in therapy, for example to reduce bleeding in patients with an obstructed bile duct. Doisy shared a Nobel Prize with Dam in 1943.

Doll, Sir Richard (1912–) British epidemiologist: showed relationship of smoking with lung cancer.

Trained in medicine in London, Doll served in the RAMC (Royal Army Medical Corps) throughout the Second World War and afterwards worked with the Medical Research Council (1946–69) and in Oxford as Regius Professor of Medicine (1969–79). His first job in epidemiology in 1944 was directed, unsuccessfully, to finding if peptic ulcers were linked with long working hours. Then, with R Peto (1943–), he established in the 1950s that cigarette smoking is causally linked with lung cancer, a result in accord with the fact that tobacco smoke, like tar and soot, contains the carcinogen benzo[a]pyrene. A massive study was begun of the health of 35 000 doctors, continued for over 40 years.

Then he studied the relation between exposure to low levels of high energy radiation and cancer; the world wide nuclear test ban treaty owed something to this work. In the 1960s he examined the side effects of the contraceptive pill; the oestrogen level in these was later changed to reduce the slight risk of thrombosis his work had revealed. His later work examined the effects of radiation in inducing cancers in servicemen who attended the early nuclear weapon tests, in populations living near nuclear power plants and in households living under overhead power lines or above radon-emitting rock. His work showed that risks were very low in these cases, except for the last, in which underfloor ventilation should be installed.

Dolland, John (1706–61) British optician: introduced achromatic lenses for telescopes and microscopes.

Dolland was for many years a silk-weaver; he was the son of a French Huguenot refugee. In 1752, however, he joined his son Peter in his business as an optician. They attacked the problem of chromatic aberration, ie the colour fringes in the images produced by a simple lens, which NEWTON had considered inherent in lenses. In fact C M Hall (1703–71), a London lawyer, had designed a compound lens of crown and flint glass which was largely achromatic (colour-free) and had telescopes made using them, from 1733. The Dollands almost certainly knew of this; however they did much experimental work, secured improved glass and, after John Dolland's patent of 1758, produced good quality achromats. This led to nearly colour-free refracting telescopes being made; although the reflecting type ultimately became dominant in astronomy, as a mirror is completely achromatic, requires only one flaw-free surface and can be supported from the back. In microscopy Dolland-type lenses were of great value, as there is no easy alternative to a refracting system for obtaining optical magnification in the microscope.

Domagk, Gerhard [dohmak] (1895–1964) German biochemist: discoverer of sulphonamide antibacterial drugs.

EHRLICH'S success in treating some protozoal diseases by chemotherapy had led to high hopes of similar success in the treatment of bacterial diseases. Diseases due to protozoa are common in the tropics; in temperate regions, diseases due to the smaller bacteria are major problems. However, by 1930 hopes had faded; trial compounds usually failed to be effective in the presence of blood or pus. This was the position when Domagk began work on the problem. He had

qualified in medicine at Kiel in 1921 and in 1927 began to direct research at the bacteriology laboratory of I G Farbenindustrie at Wuppertal, while retaining a position at the University of Münster. His scheme was to test a series of new carpet dyes made by I G Farben, trying them as drugs against streptococcal infections in mice, and in 1932 he found that the dye Prontosil Red was highly effective. Human trials soon followed and included a dramatic cure of Domagk's daughter, who had a serious sepsis following a needle prick. In 1936 a French group including BOVET found that Prontosil is converted in the body into the rather simple compound sulphanilamide, which is the effective agent. It had been known since 1908, was cheap and unpatentable and does not discolour the patient. Treatment of bacterial infections (for example pneumonia and streptococcal infections) was vastly improved by the use of sulphanilamide and related 'sulpha' drugs such as M&B 693. After 1945 penicillin and other antibiotics became dominant, but sulpha drugs remain valuable. Domagk was awarded a Nobel Prize in 1939, but was not able to accept the medal until 1947 as his country was at war. The Nobel rules did not allow him to have the prize money after such a delay.

Doppler, Christian Johann (1803–53) Austrian physicist: discovered the Doppler effect.

Doppler was educated at the Vienna Polytechnic and, despite his ability, for some time could only gain rather junior posts in tutoring or schoolteaching. At 32 he decided to emigrate to America, but on the point of departure was offered a senior teaching post in a school in Prague. After 6 years he became professor of mathematics at the State Technical Academy there, and in 1850 professor of experimental physics at Vienna.

His claim to fame rests on a single important discovery, the Doppler effect (1842). This proposed that the frequency of waves from a source moving towards an observer will be increased above that from a stationary source; and waves from a source moving away from an observer will be decreased in frequency. In 1845 a test was made at Utrecht in which an open railway carriage carrying a group of trumpeters was taken at speed past a group of musicians with perfect pitch. It was one of the extraordinary occasions that made the 19th-c approach to physics entertaining and, while unsubtle, it demonstrated the correctness of Doppler's idea.

Doppler recognized that the effect applies not only to sound but also to light, and FIZEAU (1848) pointed out that the spectral lines of stars should be shifted towards the red end of the spectrum according to the speed at which they are receding from us (the Doppler shift). HUGGINS observed this for the star Sirius (1868) and HUBBLE later used the red shift to infer the speed of recession of other galaxies from us: the 'expanding universe' of cosmology.

The Doppler effect has also been used to measure the speed of the Sun's rotation and Saturn's rings, and the rotation of double stars. It forms the basis of police radar speed traps for vehicles; and 'Doppler satellites', emitting a fixed radio frequency and whose position is known, are used by ships and aircraft to locate their position and by mapmakers and surveyors to give precise locations using the global positioning system (GPS). Doppler reflection is also used in echocardiography. The frequency of a beam of transmitted ultrasound is compared with the frequency of the beam reflected from the moving blood cells in the blood vessel under examination. This allows the velocity of the blood flow (around $1 \mathrm{~m~s}^{-1}$) to be measured and is of great value in locating valve and other heart defects, especially in children.

Douglass, Andrew Ellicott (1867–1962) US astronomer and dendrochronologist: devised a tree-ring dating technique (dendrochronology).

Douglass worked at the Lowell Observatory in Arizona. His interest in the Sun led to an interest in climate. While trying to construct a historical record of sunspot activity, Douglass recalled that climatic conditions affected the width of the annual growth rings of trees, and that these distinctive patterns could often be recognized and seen to overlap in timber buildings. (Fortunately in the dry climate of Arizona ancient wood is well preserved.) He developed this idea into an important dating technique and succeeded in constructing a continuous dendrochronological time scale for timber building materials back to the first century (later workers, using the Californian bristlecone pine, extended this to about 5000 BC). Although the technique of radiocarbon dating has to a large extent superseded it, dendrochronology has proved vital in calibrating the radiocarbon time scale.

Draper, John William (1811–82) British–US chemical physicist: a pioneer of scientific photography.

Draper's life and his scientific interests were both oddly disperse. His father was an itinerant Methodist preacher whose possession of a telescope attracted the boy to science. He began premedical studies in London in 1829 but emigrated to Virginia in 1832. Helped by his sister Dorothy's earnings as a teacher, he qualified in medicine by 1836 and then taught chemistry in New York. When DAGUERRE'S process for fixing photographs was published in 1839, Draper took it up and in 1840 he made what is probably the oldest surviving photographic portrait; it shows his sister Dorothy (exposure, 65 s). In the same year his photograph of the Moon began astronomical photography and in 1850 he made the first microphotographs, to illustrate his book on physiology. In 1841 he proposed the principle that only absorbed radiation can produce chemical change (Draper's Law; this principle was also known to T C J D Grotthus (1785–1822) in 1817). He made early photographs in the infrared and ultraviolet regions; and he showed that all solids become incandescent at the same temperature and, when heated sufficiently, give a continuous spectrum. His later work was on the history of ideas.

Dubois, Marie Eugène Francois Thomas [dübwah] (1858–1940) Dutch anatomist and palaeoanthropologist: discovered Java Man.

After graduating in medicine from the University of Amsterdam in 1884, Dubois was appointed lecturer in anatomy, but resigned in 1887 after some disagreements with his professor. His great interest in the 'missing link' between apes and man prompted him to join the Dutch East Indian Army as a surgeon, this being a convenient way of getting to Java, where he believed that the remains of such a hominid might be found (on the grounds that it is the only place in which the orang-utan and gibbon are found). In 1891, having obtained support from the army in the form of a gang of convict labour, he eventually succeeded in finding the skullcap, femur and two teeth of Java Man (*Homo erectus*), a hominid who lived approximately 0.5–1.5 million years ago. Dubois's belief that Java Man represented the missing link was at first widely ridiculed but was ultimately accepted after the announcement in 1926 of the discovery of Peking Man (also *Homo erectus*). Irritated by the lack of support for his theory, Dubois refused to allow study of his specimens until 1923, by which time he had convinced himself that they were merely the bones of a giant gibbon.

du Bois-Reymond, Emil *see* **Bois-Reymond**

Dufay, Charles [düfay] (1698–1739) French chemist: discovered positive and negative charges of static electricity.

Dufay came from an influential family, which secured an army career for him; he left as a captain to become a chemist at the Académie des Sciences when he was 25. He had no training in science, but he began to study electricity in 1733. He showed that there are two kinds of electricity (and only two) generated by friction; he called them vitreous and resinous because they were obtained by rubbing glass (or rock crystal, hair or wool) or resin (or amber, silk or paper) respectively; they are the positive and negative charges of today. Dufay showed that like types repelled and unlike kinds attracted one another. The 'two-fluid' theory of electricity was linked with these results in opposition to FRANKLIN's later one-fluid theory. Dufay's experiments included suspending a boy by silk cords, electrifying him by friction and drawing sparks from him. GRAY in London had done similar experiments, and had also distinguished conductors from insulators (eg silk from metal wire).

Dulong, Pierre Louis [dülõ] (1785–1838) French chemist: co-discoverer of law of constant atomic heat.

Originally a physician, Dulong moved to chemistry as assistant to BERTHOLLET. In 1811 he discovered NCl_3, which cost him an eye and two fingers. He was an early supporter of the hydrogen theory of acids. From 1815 he worked with A T Petit (1791–1820) on thermometry; and in 1819 they published Dulong and Petit's Law. This stated that the specific heat capacity of a solid element, when multiplied by its atomic weight, gives a constant which they called the atomic heat. The law is only approximately observed and is best followed at or above room temperature (eg C, B and Si only have specific heat capacities in accord with it at high temperature). It had some use, however, for easily giving rough atomic weights for new metals, at a time when this was valuable. In modern parlance we can express the law as:

relative atomic mass (ie atomic weight) $\times$ specific heat capacity $\approx 25\,\mathrm{J\,K^{-1}\,mol^{-1}} = 3R$ where R is the gas constant.

Dumas, Jean Baptiste André [dümah] (1800–84) French organic chemist; classified organic compounds into types.

Originally an apprentice apothecary, Dumas improved his knowledge of chemistry in Geneva and also attracted the notice of some eminent scientists, with the result that he was encouraged to go to Paris. There he got a post as assistant at the École Polytechnique, and by 1835 a senior post there. He initially worked on atomic weights but his main distinction is that he was a leader in the group of mainly French chemists who partly rejected the authoritative views of BERZELIUS and offered new views on the relations between organic compounds, setting the stage for the major advances made later by KEKULÉ. Dumas's work in this began with his study of the choking fumes from candles used in the Tuileries. He found that these had been bleached with chlorine and from this clue examined the reaction of chlorine with other organic compounds. In some cases he showed that the reaction had replaced hydrogen by chlorine on an atom-for-atom basis and yet gave a product of essentially the same type (eg acetic acid, CH_3CO_2H, gives a series of three chlorine-substituted acids CH_2ClCO_2H, $CHCl_2CO_2H$, CCl_3CO_2H, which are not greatly unlike their parent in their chemistry). This was in direct conflict with Berzelius's dualism theory, which did not allow for atoms of opposite electrical type replacing one another in this manner. Dumas pressed his theory of substitution and his theory that organic compounds exist as 'types' (eg the alcohols) and argued that a type may contain a series of compounds whose formulae differ by a constant unit (eg CH_2). Somewhat similar views were developed by others (notably LAURENT, GERHARDT and WURTZ in France, LIEBIG and HOFMANN in Germany and WILLIAMSON in England). During the period of debate many new and useful organic compounds were made and theory was advanced, apparently with rejection of Berzelius's views. However, after 1930, it was seen that the Berzelius approach to organic reactions (in a much modified form) had an important part in understanding why organic reactions occur, while his opponents had also been right in their criticisms. Dumas, ambitious and energetic, followed a pattern more familiar in France than elsewhere by moving from science to politics, holding various ministerial posts after 1848.

Dutrochet, Henri [dütrohshay] (1776–1847) French plant physiologist: discovered some basic features of plant physiology.

Born into a wealthy family, Dutrochet's early life was blighted by a club foot, ultimately fully corrected by a local healer (also the hangman) after medical men had failed. After the Revolution he became an army medical officer but had to retire after catching typhoid in the Peninsular War.

After his resignation from the army in 1809 he seems to have spent his time researching in animal and especially plant physiology. He held the view that life processes are explicable in chemical and physical terms and that cellular respiration is essentially similar in plants and animals. In 1832 he found the small openings (stomata) on the surface of leaves, later found to be the entry points for gas exchange in plants. INGENHOUSZ had shown that plants absorb carbon dioxide and emit oxygen, and Dutrochet found that only those parts of plants containing the green pigment chlorophyll can do this. He was the first to study successfully the production of heat during plant growth. Although osmosis had been observed previously, it was he who first studied it fully and proposed that it was the cause of sap movement in plants.

Duve, Christian (René) de [düv] (1917–) Belgian biochemist: discovered lysosomes.

Born in England and educated in medicine in Louvain, de Duve worked in Sweden and the USA before returning to Louvain in 1947 and later holding a dual post also at Rockefeller University, New York.

From 1949, de Duve obtained ingenious experimental evidence that some at least of a cell's digestive enzymes must be enclosed in small organelles within the cell. By 1955 these were positively identified with the aid of electron microscopy and named lysosomes. These serve both to isolate the enzymes from attack on their own animal or plant cells and to concentrate their attack when the lysosome fuses with a food vacuole. After digesting the macromolecules present in the food, the resulting small molecules of sugar or amino acid pass through the lysosome wall into the cell. Another function of lysosomes is to destroy worn-out cell organelles, or even cells. Some hereditary metabolic diseases (eg cystinosis) are due to absence of a lysosomal enzyme. De Duve shared a Nobel Prize in 1974.

Du Vigneaud, Vincent [doo veenyoh] (1901–78) US biochemist: researcher on sulphur-containing vitamins and hormones.

Originally a student of chemistry at Illinois, Du Vigneaud's postgraduate work in the USA and in the UK became increasingly biochemical; from 1938 he was head of biochemistry in Cornell Medical School and his research became 'a trail of sulphur research'. This began with studies on the hormone insulin in the 1920s. In the 1930s he worked on the sulphur-containing amino acid methionine and showed that its function is particularly to transfer methyl (-CH_3) groups in biochemical reactions. In 1941 he isolated vitamin H from liver and showed that it was identical with the growth factor biotin, which had been isolated in 1936 by F Kögl (1897–1960) (1 mg from 250 kg of dried duck egg yolk). Du Vigneaud deduced the complete (and rather complex) structure of biotin in 1942. He next studied two pituitary hormones, oxytocin and vasopressin, the first of which induces labour and milk flow. Both structures were determined, and in 1953 he synthesized oxytocin – the first synthesis of an active polypeptide hormone (it contains eight amino acids). For this work in particular he was awarded a Nobel Prize for chemistry in 1955.

Dyson, Freeman John (1923–) British–US theoretical physicist: unified the independent versions of quantum electrodynamics.

Dyson, the son of a distinguished English musician, graduated from Cambridge and spent the Second World War at the headquarters of Bomber Command. In 1947 he did research at Cornell and joined the staff at Princeton in 1953.

Shortly after the war several people began to apply quantum mechanics to systems in which particles (particularly electrons) interact with electromagnetic radiation (photons). In 1946 Willis Lamb (1913–) observed a shift (the Lamb shift) in the lowest energy levels of the hydrogen atom, away from the previously predicted levels. SCHWINGER, TOMONAGA and FEYNMAN rapidly developed independent theories correctly describing how electrons behave when interacting with photons, and accounted for the Lamb shift. Dyson then showed how the formulations related to each other and produced a single general theory of quantum electrodynamics ('QED', dealing with the interactions of subatomic particles with photons).

Subsequently Dyson was involved in many areas of physics, in cosmology and even speculations on space travel.

Eddington, Sir Arthur (Stanley) (1882–1944) British astrophysicist: pioneered the study of stellar structure; and discovered mass–luminosity relationship.

Eddington was the son of the head of a school in Cumbria where, a century earlier, DALTON had taught. He was an outstanding student at Manchester and then at Cambridge, where he later became Director of the Observatory. The internal structure of stars is an area of study pioneered by Eddington. In 1926 he demonstrated that, in order to remain in equilibrium, the inward gravitational pressure of a star must balance the outward radiation and gas pressure. He realized that there was consequently an upper limit on the mass of a star (of about 50 solar masses), because above this the balance between gravitation and radiation pressure could not be achieved. (Some stars, verging on instability, pulsate; these are the Cepheid variables). He discovered the mass–luminosity relationship, which shows that the more massive a star the greater its luminosity and which allows the mass of a star to be determined from its intrinsic brightness. Eddington provided some of the most powerful evidence for the theory of relativity by observing that light from stars near to the Sun's rim during the total solar eclipse of 1919 was slightly deflected by the Sun's gravitational field in accordance with EINSTEIN's predictions.

Edelman, Gerald (Maurice) (1929–) [aydlman] US biochemist: pioneer in study of molecular structure of antibodies.

Edelman originally planned a career as a concert violinist, but came to realize that he lacked the extroversion needed for success as a performer. He had also been attracted to science and, believing rather ingenuously that medical school was a suitable start (his father was a physician in New York), he entered the University of Pennsylvania. After qualifying he worked as a US Army physician in Paris, where his interest in proteins, physical chemistry and immunology began, and he returned to New York to work at Rockefeller University in this field.

During his doctorate studies at Rockefeller University Edelman investigated the immunoglobulins and after he joined the staff there he continued his interest in these compounds. They are formed on the surface of B-lymphocytes and when released into the body fluids are known as antibodies. They form a class of closely related proteins, each specific in its ability to bind with a particular antigen; the system forms a major part of the vertebrate animal's defence against infection. Edelman found that human immunoglobulin, a large protein molecule, is a combination of two kinds of protein chains ('light' and 'heavy') linked by sulphur bridges. He went on to study the sequence of amino acids in the chains of the immunoglobulin IgG and by 1969 had achieved this; the 1330 amino acids form a Y-shaped structure, in which the amino acids in the tips are very variable but the main part of the structure is constant. This result could be linked with R R PORTER'S biochemical and immunological study of IgG to give a more detailed picture of this molecule, which is likely to be typical of antibodies (see Porter's entry for diagram). His later work was on neural networks and the computerized simulation of brain function. Edelman and Porter shared the Nobel Prize for 1972.

Since then Edelman has worked on embryonic development and has identified the adhesive molecules that have a central place in morphogenesis. Always attracted by large problems in rather unexplored fields, he has since attacked the difficult problem of the nature and origin of consciousness. He visualized brain development as a process in which random neurone connections are progressively refined by a process akin to Darwinian selection, to give a brain in which effective activities are retained and useless ones discarded.

Edison, Thomas (Alva) (1847–1931) US physicist and prolific inventor.

Edison received virtually no formal education, having been expelled from school as retarded, and was educated by his mother. During the American Civil War he worked as a telegraph operator, during which time he invented and patented an electric vote recorder. Some 3 years later, in 1869, he invented the paper tape 'ticker', used for communicating stock exchange prices across the country, sold it for $30 000 and opened an industrial research laboratory. He was thereafter to apply himself full-time to inventing, filing a total of 1069 patents before his death. His more notable inventions include the carbon granule microphone, to improve A G BELL's telephone, the phonograph (a device for recording sound on a drum covered in tin foil, invented in 1877) and the electric light bulb. The latter required an extraordinary amount of trial and error testing, using over 6000 substances until he found a carbonized bamboo fibre that remained lit for over 1000 hours in a vacuum. This led in turn to improved electricity generators (he increased their efficiency from 40% to over 90%), power cables, the electricity meter and the revolutionizing of domestic lighting and public electricity supply. During his work on light bulbs he also discovered the Edison effect, that electricity flows from a heated filament to a nearby electrode but not in the reverse direction, which was later to form the basis of the thermionic diode. Edison's impact on 20th-c life was immense and his reputation as a prolific inventive genius remains unrivalled.

Edwards, Robert (Geoffrey) (1925–   ) British physiologist: pioneer of human IVF.

After qualifying in medicine at Edinburgh and research on mammalian reproduction in the UK and USA, Edwards became professor of human reproduction at Cambridge (1985–89), with special interests in fertility and infertility and the process of conception. Attempts to fertilize mammalian eggs outside the body (in vitro fertilization, IVF) were made in 1878; but it was not until 1978 that the first IVF baby, Louise Brown, was born. Success was due to work by Edwards in collaboration, from 1968, with the gynaecologist P C Steptoe (1913–88) whose expertise with the laparoscope allowed eggs to be removed from the womb. Edwards systematically studied the factors needed to preserve and ripen the immature eggs, which were then fertilized with sperm; the resulting embryo was matured in 'a magic culture fluid' before being implanted in the uterus. Much work was needed to define this optimum culture medium and conditions for success, and to show that the artificially fertilized embryos would not result in abnormal offspring. The IVF method has proved very successful in dealing with some types of infertility.

Ehrlich, Paul [ayrlikh] (1854–1915) German medical scientist: pioneer of chemotherapy, haematology and immunology.

Ehrlich was born in eastern Germany, the son of an eccentric Jewish innkeeper and his talented wife. Undistinguished at school (where he hated examinations) he did well enough to enter university to study medicine, and qualified at Leipzig in 1878. With difficulty, partly because he was Jewish, he got a hospital post in Berlin. He spent his entire career there, except for a year in Egypt using its dry air as a cure for his tuberculosis; at the time it was probably the best treatment.

While Ehrlich was a student the aniline dyes had recently been discovered and his mother's cousin Carl Weigert (1845–1904) had used them for microscopic staining. Ehrlich worked with him and was impressed by the way in which some dyes would stain selectively. The study of this linked his interest in chemistry with his medical work and was to form the basis of all his later research. He found how to stain and classify white blood cells, discovered the mast cells later found to be important in allergy and worked with Behring and Kitasato on anti-

toxins. His work on antibodies largely began modern immunology. It led him to think that, although the search for vaccines against malaria and syphilis had failed, it might be possible to attack the parasites causing these diseases in another way, since they could be selectively stained. He also knew that when an animal died from lead poisoning the lead was found concentrated in certain tissues. He hoped that he could find a synthetic chemical which would bind on to and injure the parasites. He was encouraged by his discovery that the dye Trypan Red was fairly effective against trypanosomes (the pathogens causing trypanosomiasis) in mice, although he also discovered that drug resistance soon developed. Both discoveries were important.

From 1905 he and his assistants began trials using compounds with molecules not unlike dyes but containing arsenic, as a part of his programme to find a 'magic bullet' that could locate and destroy the invading pathogenic cells. Their organoarsenical compound No. 606 (which had failed against trypanosomes) was eventually found by them in 1909 to be effective against *Treponema pallidum*, the bacterium that causes syphilis, and it was soon used in patients as 'Salvarsan'. The principles used by Ehrlich came to guide this new approach (chemotherapy) to disease, in which a compound is sought that will seek out and destroy the disease organisms with only minor damage to the patient. In the event, it was over 20 years later before Domagk achieved the next major success.

Ehrlich inspired loyalty in some of his co-workers and high exasperation in others. He was dictatorial and impatient and appeared to live largely on cigars and mineral water. He shared a Nobel Prize in 1908 for his work on immunity, which is only a part of his contribution to medical science.

Eijkman, Christiaan [aykman] (1858–1930) Dutch physician: discovered cure for beriberi.

Eijkman served as an army medical officer in the Dutch East Indies in the early 1880s and was sent back there in 1886 to study beriberi, then an epidemic disease in south Asia. This paralysing and often fatal disease is in reality a deficiency disease, whose rise was linked with increased use of polished (white) rice as the major diet in some 'closed' communities. When Eijkman began his work in Java, it was assumed to be an infection, but he noticed that some laboratory birds showed similar symptoms to beriberi victims, and they had been fed on leftover rice from a military hospital kitchen. Then a change occurred and the birds recovered. Eijkman discovered that a new cook refused to give 'military rice' to civilian birds and had changed to less refined rice. Eijkman went on to show that the disease could be cured by adding rice husks to the diet and could be caused by feeding polished rice. He did not interpret his results correctly (he thought the bran contained a substance which protected against a poison) but his work was a valuable step towards the full recognition of vitamin deficiency diseases by F G HOPKINS after 1900. Isolation and synthesis of vitamin B_1 (thiamin) was achieved by R R Williams (1886–1965) in the 1930s; deficiency of it in the diet causes beriberi. Eijkman and Hopkins shared a Nobel Prize in 1929.

Einstein, Albert [iynshtiyn] (1879–1955) German–Swiss–US theoretical physicist: conceived the theory of relativity.

Einstein's father was an electrical engineer whose business difficulties caused the family to move rather frequently; Einstein was born while they were in Ulm. Despite a delay due to his poor mathematics he entered the Swiss Federal Institute of Technology in Zürich at the age of 17, and on graduating became a Swiss citizen and sought a post in a university, or even in a school. However, he had great difficulty in finding any job and settled for serving in the Swiss Patent Office in Bern. It worked out well; he was a good patent examiner and the job gave him enough leisure for his research. In 1903 he married a fellow physicist, Mileva Maric; their illegitimate daughter, born in 1902, was adopted; two sons followed. This marriage ended in divorce in 1919 and he then married his cousin Elsa, who had two daughters by a previous marriage. It was while at the Patent Office that he produced the three papers published in 1905, each of which represented an enormous achievement, covering Brownian motion, the photoelectric effect and special relativity.

Einstein's first university post was secured in 1909, when he obtained a junior professorship at the University of Zürich, and a full professorship at Prague (1910) and Zürich (1912) followed. In 1913 he was made Director of the Institute of Physics at the Kaiser Wilhelm Institute in Berlin. The general theory of relativity was completed during the First World War and following its publication (1915) Einstein was awarded the 1921 Nobel Prize for physics for his work of 1905.

He began to undertake many lecture-tours abroad and was in California when Hitler came to power in 1933. He never returned to Germany, resigning his position and taking up a post at the Institute of Advanced Study, Princeton. Einstein put much effort into trying to unify gravitational, electromagnetic and nuclear forces into one set of field equations, but without success. He had some involvement in politics, in that he helped initiate the Allied efforts to make an atomic bomb (the Manhattan project) by warning Roosevelt, the American president, of the possibility that Germany would do so, in a letter in 1939. In 1952 Einstein was offered, and sensibly declined, the presidency of Israel. He was also active in promoting

nuclear disarmament after the Second World War. He led a simple life, with sailing and music as his main relaxations.

The first of his papers of 1905 considered the random movement of small suspended particles (Brownian motion, discovered in 1828). The bombardment by surrounding molecules will make a tiny particle in a fluid dart around in an erratic movement, and Einstein's calculations provided the most direct evidence for the existence of molecules when confirmed experimentally by PERRIN (1908).

The next paper by Einstein tackled the photoelectric effect by considering the nature of electromagnetic radiation, usually thought of as waves obeying MAXWELL'S equations. Einstein assumed that light energy could only be transferred in packets, the quanta used by PLANCK to derive the black body radiation spectrum. Einstein then was able to explain fully the observations of LENARD (1902), in which the energy of electrons ejected from a metallic surface depended on the wavelength of light falling on it but not on the intensity. The result became a foundation for quantum theory and clothed Planck's quanta with a physical interpretation.

Finally, Einstein set out the special theory of relativity (restricted to bodies moving with uniform velocity with respect to one another) in his third paper. Maxwell's electromagnetic wave theory of light indicated that the velocity of a light wave did not depend on the speed of the source or observer and so contradicted classical mechanics. LORENTZ, FITZGERALD and POINCARÉ had found a transformation of Maxwell's equations for a region in uniform motion which left the speed of light unchanged and not altered by the relative velocity of the space and observer (the Lorentz transformation).

Einstein correctly proposed that the speed of light is the same in all frames of reference moving relative to one another and, unknown to him, this had been established by the MICHELSON–MORLEY experiment (1881, 1887). He put forward the principle of relativity, that all physical laws are the same in all frames of reference in uniform motion with respect to one another. When applied it naturally gives rise to the Lorentz–FitzGerald transformation, with classical mechanics obeying this rather than simple addition of velocity between moving frames (the Galilean transformation). A further consequence derived by him was that if the energy of a body changes by an amount E then its mass must change by E/c^2 where c is the velocity of light.

From 1907 Einstein sought to extend relativity theory to frames of reference which are being accelerated with respect to one another. His guiding principle (the principle of equivalence) stated that gravitational acceleration and that due to motion viewed in an accelerat-

THE HISTORY OF NUCLEAR AND PARTICLE PHYSICS

One of the continuing themes of science is to discover and understand the constituents of matter. Nuclear physics is one aspect of this endeavour and deals with the nucleus at the centre of atoms. It grew naturally from attempts at the end of the 19th-c to understand the structure and composition of atoms and has had profound effects on the development of science as a whole and on society in general.

Although the existence of a nucleus in atoms was established only in 1911, the phenomenon of radioactivity, which is essentially a nuclear process, had been discovered earlier by BECQUEREL in Paris in 1896. He had been investigating the mysterious X-rays discovered by RÖNTGEN in Germany in 1895. Becquerel left some uranium salts for several days on top of a wrapped photographic plate intended for his X-ray experiments, which had been held up by damp conditions. On developing the plate, he found that it had been fogged in the region where the uranium had been resting on it. He concluded that there was some kind of radiation emanating from the uranium which could penetrate the paper around the photographic plate. This phenomenon came to be called radioactivity, and its discovery caused a great stir in the scientific community. One prominent investigator of radioactivity was MARIE CURIE, a Polish scientist working in Paris with her French husband, PIERRE CURIE. In 1898, as the main subject of her doctoral thesis, and after painstaking work in finding ways of chemically isolating other radioactive substances in uranium ore by detecting their radioactivity (using electrometers devised by Pierre Curie), she discovered two very radioactive new elements to which she gave the names polonium and radium.

The discovery of radium and its means of isolation made available a far more powerful source of radiation for experiments than had been available before. Several of the early workers with radioactive materials, including Becquerel and the Curies, noticed that it could cause red patches and burns to appear on the skin. It is likely that Pierre Curie was suffering from radiation sickness during his last years before his untimely death in a road accident in Paris in 1906, and that radiation damage was a factor in Marie Curie's death from aplastic anaemia in 1933. Its use as a luminous material on watches led to cases of cancer among the workers who painted the dials. Nevertheless, radium was found to be effective in treating some cancers, and its controlled use in this way has continued.

The systematic study of the nature of radioactivity began in 1898, when RUTHERFORD showed that there were at least two types of radiation emitted by radioactive elements: alpha rays, which carried positive electric charge and were not very penetrating; and beta rays, which carried negative electric charge and were more penetrating. In 1900 Paul Villard (1860–1934) found a third component, called gamma rays, which carried no electric charge and were not easily stopped or detected. More work, principally by Rutherford, showed that alpha rays were in fact helium ions, that beta rays were electrons, and that radioactive emissions caused the transmutation of one element into another. (Gamma rays are ultra-high-energy X-rays.)

The fact that atoms contained very light negatively charged particles called electrons had been shown by J J THOMSON in 1897 at the Cavendish Laboratory in Cambridge. In 1907 Thomson put forward what he called the 'plum pudding' model of the atom, in which he postulated that the electrons were dispersed in a uniform distribution of positive charge in an atom, like plums in a pudding. This model was refuted by a famous series of experiments carried out in 1911 at Manchester University by GEIGER and Ernest Marsden (1889–1970), working under the supervision of Rutherford. In these experiments a beam of alpha particles was directed at thin foils of gold or platinum, and the number of alpha particles scattered through various angles was recorded. It was found that a few alpha particles were scattered through large angles. Rutherford realized that the very strong electrical forces needed to do this could only be produced if, in contradiction to Thomson's model, all the positive charge of the atom was concentrated in a tiny 'nucleus' at the centre of the atom. Thus, the nuclear model of the atom was born. If an atom is imagined as a large concert hall, the nucleus would be the size of a pea at its centre. Most of the volume of an atom is empty space, with a number of electrons moving around a tiny but heavy central nucleus. This suggests that electrons orbit the central nucleus, just as planets orbit the Sun.

The next major step was taken by Danish physicist NIELS BOHR. He realized that the basic force holding electrons in an atom was the electric attraction exerted by the positively charged nucleus, but that a radically new approach was needed to understand what determined the size of electron orbits and what stopped them from collapsing, given that the electrons should continuously radiate energy away.

The key step which he took was to postulate that angular momentum was quantized, ie only able to take on integer multiples of some constant value. The idea of quantization had been introduced, but only for the energy of electromagnetic radiation, in 1900 by PLANCK. In 1913, Bohr produced a convincing theory of the hydrogen atom, and explained the origin of its spectral lines as due to what we today call 'quantum jumps', in which the energy of an atom suddenly changes, resulting in the emission of a quantum of energy now called a photon. The 'Bohr atom' was a great step forward in both physics and chemistry, and led to the new subject of quantum mechanics, established between 1920 and 1930, which eventually gave a rather different but far more satisfactory theoretical understanding of atoms. The key people in the development of quantum mechanics were SCHRÖDINGER and HEISENBERG.

In the process of beta radioactivity electrons emerge from the nucleus, so it was natural to assume that nuclei contain electrons whose negative charge cancelled out some of the positive charge of the protons (the name given to the nucleus of hydrogen). However, in 1932 CHADWICK discovered a neutral particle, the neutron, of about the same mass as a proton. From then on, it was realized that nuclei were built up from neutrons and protons, and that they did not contain electrons. The problem of where the electrons came from in beta radioactivity was solved in 1934 by FERMI, working in Rome. His theory used quantum mechanics, and also involved the neutrino, a new neutral particle postulated by PAULI in 1929. Fermi proposed that an electron and a neutrino are created in a nucleus at the instant of beta decay by a new force called weak interaction.

Very soon after the discovery of the nucleus, it was realized that there must be a new strong attractive force operating in the nucleus to overcome the electrical repulsive force between the protons. This new force would have to have a very short range, as its effects were not observed outside the nucleus. The first theoretical understanding of how a strong short-range force could be produced was provided in 1934 by YUKAWA. His meson exchange theory of nuclear forces provided a spring-board for the development of theoretical nuclear and particle physics. The new particle, named the pi meson (pion), predicted by Yukawa, was not discovered until 1947 by POWELL.

Nuclear fission, the process by which a nucleus splits into two, was discovered in Berlin in 1938 by HAHN and Fritz Strassman (1902–80). They had found traces of light elements such as barium in the products of the bombardment of uranium by neutrons. The interpretation of this as the splitting of the uranium nucleus into two roughly equal parts was mainly due to LISE MEITNER and her nephew, OTTO FRISCH. Meitner had been a colleague of Hahn in Berlin, but had fled to Sweden to escape the Nazis. Many physicists soon confirmed these important results, and within a few months in

1939 it was established that a large amount of energy was released in fission and that a number of free neutrons were also produced. This meant that it was likely that a chain reaction was possible in which the neutrons produced in the fission of one uranium nucleus could cause the fission of other uranium nuclei. One of the physicists involved was Fermi, by then working in the USA. He and others realized that it might be possible both to construct a power plant and also a very powerful explosive (a million times more powerful than TNT) based on nuclear fission. These developments occurred at the start of the Second World War, and a major secret programme (the Manhattan project, directed by OPPENHEIMER) was started in the USA to produce what became known as the atom bomb. As part of this project, Fermi constructed the world's first nuclear reactor in Chicago in 1942 as a source of plutonium for bombs and also to provide experimental information on fission chain reactions. The project led to the testing of the first atom bomb in 1945, and the dropping of a uranium bomb on Hiroshima and a plutonium bomb on Nagasaki in August 1945, so ending the Second World War.

After the war, the production of even more powerful nuclear bombs based on nuclear fusion reactions was pursued. In nuclear fusion, light nuclei such as hydrogen and tritium fuse together to produce heavier nuclei, while in fission heavy nuclei such as uranium and plutonium split into two. Both these processes release large amounts of energy. One chief scientist working on the so-called hydrogen bomb based on fusion was EDWARD TELLER. The purely scientific study of nuclear physics also continued after the war, with the main emphasis on nuclear structure. In 1948, MARIA GOEPPERT MAYER put forward the shell model of nuclei, which explains many features of nuclei in terms very similar in principle to the analogous atomic shell model: this formed the basis for most subsequent work.

Particle physics may be regarded as the study of the most fundamental constituents of matter and the forces which act between them. As a subject in its own right, it diverged from nuclear physics in the early 1930s. Until then, the study of the atomic nucleus was regarded as the most fundamental part of physics, and in fact the term particle physics came into use only in the 1960s. As noted above, in 1932 Chadwick discovered the neutron, a particle similar to the proton but without electric charge. This similarity prompted Heisenberg to describe the proton and neutron as different states of a particle called the nucleon, with spin up and spin down. So began an investigation of a new layer of nature, more fundamental even than the realm of neutrons and protons. It was also the start of a mathematical approach based on symmetries, which proved to be very fruitful in the 1960s.

By the early 1930s the existence of the neutrino came to be taken seriously, and the idea of anti-particles was introduced by DIRAC as a necessary consequence of his relativistic theory of quantum mechanics. In particular, Dirac predicted in 1931 that an anti-electron should exist. This particle was discovered by CARL ANDERSON at the California Institute of Technology in 1932, although he was not aware of Dirac's prediction at the time. Soon this positive electron was identified with Dirac's prediction by BLACKETT and was named the positron. Thus, fairly quickly in the early 1930s, the number of supposedly elementary particles increased from two (the electron and proton) to at least four.

In 1938, studies of cosmic radiation (which is a flux of high-energy particles from outer space) revealed the tracks of a particle whose mass was intermediate between that of the electron and the proton. This was subsequently called the mu-meson (muon). It is now known to be one of the class of fundamental particles called leptons. At first it was thought to be the particle, predicted by Yukawa, whose exchange causes nuclear forces, but the fact that it appeared not to undergo nuclear interactions argued against that interpretation. It was not until 1947 that Yukawa's predicted particle, the pi-meson, was discovered in cosmic rays by Powell. Between 1947–53 several other new particles were discovered in cosmic rays. The discovery techniques used cloud chambers and photographic emulsions, and were mostly carried out on mountain tops in the Alps, where the flux of cosmic rays is higher. Prominent in this work were Clifford Butler (1922–) and George Rochester (1908–) who, in 1947, discovered the so-called strange particles, later known as K and Λ.

Two important developments in experimental methods were made in the 1950s and early 1960s which changed dramatically the scale of experiments and the rate of progress. These were the invention of a particle detector called the bubble chamber by GLASER (and its large scale implementation by LUIS ALVAREZ), and the development of large-particle accelerators, notably the proton synchrotron. Most of this work was carried out in the USA, but from 1959 the European Organization for Nuclear Research (CERN) became a major contributor. As a result of these developments in experimental methods, many new particles (collectively called hadrons) were discovered during the 1960s and 1970s. It became hard to believe that they

were all fundamental and several physicists searched for theoretical explanations of them in terms of more elementary constituents. A break-through came with the development of the quark model by GELL-MANN and, independently, by George Zweig (1937–), which explained the large number of hadrons in terms of combinations of just a few elementary particles called quarks, which had fractional electric charge.

A new large electron linear accelerator (SLAC) at Stanford, CA, enabled experiments to be done that were analogous to Rutherford's scattering of nuclei. These experiments scattered electrons from protons inelastically, and showed that the proton could indeed be regarded as a composite of much smaller entities as envisaged in the quark model. The key experiments were carried out by a team led by Richard Taylor (1929–), Henry Kendall (1926–) and Jerome Friedman (1930–). The interpretation of these experimental results was largely the work of FEYNMAN. His other important contributions to theoretical particle physics included the renormalization technique for quantum electrodynamics and the theory of weak interactions.

A milestone in particle physics was the unification of the theories of electromagnetic and weak interactions in the early 1970s, principally by WEINBERG, SALAM and GLASHOW. A consequence of their work was the prediction of new, very heavy particles called the W and Z bosons. These were discovered in a large scale experiment at CERN in 1983 involving colliding beams of protons and anti-protons. The team which made the discovery was led by Carlo Rubbia (1934–), later the director-general of CERN.

Since then, most advances in particle physics have come about through the work of large experimental teams, and a biographical account is no longer appropriate. The present situation is that there is now a 'standard model' of particle physics expressed in terms of six quarks, six leptons (which are the constituents of all matter) and a small number of so-called gauge bosons (which are responsible for fundamental forces). This model has been subjected to (and has survived) precision tests by a series of large experiments carried out principally on the electron-positron colliding beam machine, LEP, at CERN. The scope of particle physics has much increased since the early 1980s and it is now indispensable for an understanding of astrophysics and cosmology, and particularly of the 'Big Bang' model of the early stages of the universe.

Dr Gareth Jones, Imperial College, London

ing frame are completely equivalent. From this he predicted that light rays should be bent by gravitational attraction. In 1911 he reached a specific prediction: that starlight just grazing the Sun should be deflected by 1.7" of arc. During a total eclipse of the Sun in 1919 EDDINGTON measured this in observations made at Principé in West Africa, finding 1.61" of arc. This dramatic confirmation immediately made Einstein famous world-wide and made it clear that he had moved the foundation of physics.

In 1915 he had published the general theory of relativity in complete form, using Riemannian geometry and other mathematical ideas due to H Minkowski (1864–1909) (in 1907), RIEMANN (1854) and C Ricci (1853–1925) (in 1887). Mass was taken to distort the 'flatness' of space-time and so to give rise to bodies in space moving along curved paths about one another. While the resulting 'gravitational' attraction is very close to that predicted by NEWTON's law, there are small corrections. Einstein and M Grossmann (1878–1936) estimated that the ellipse traced out by Mercury around the Sun should rotate by 43" arc per century more than that given by Newtonian theory. The observed value is indeed 43" of arc larger and Einstein reported: 'I was beside myself with ecstasy for days'.

General relativity produced many other startling predictions, such as that light passing from one part of a gravitational field to another would be shifted in wavelength (the Einstein redshift). This was observed astronomically in 1925 and terrestrially, with a 23 m tower on Earth using the MOSSBAUER effect, by R Pound and G Rebka in 1959. Gamma rays moving from the bottom to top of the tower were found to have a longer wavelength.

Cosmological models of the universe were also completely changed by general relativity and FRIEDMANN (1922) put forward a model that represented an expanding universe obeying Einstein's equations.

During the 1920s and 1930s Einstein engaged in debate over quantum theory, rejecting BORN's introduction of probability ('God may be subtle, but He is not malicious'). He also sought to find a unified theory of electromagnetic and gravitational fields, without success. By 1921 he had been prepared to say 'Discovery in the grand manner is for young people... and hence for me a thing of the past'.

Einthoven, Willem [aynthohven] (1860–1927) Dutch physiologist: introduced clinical electrocardiography.

Einthoven's father was a physician in Java, where the family lived until he was 10, afterwards settling in Utrecht. He studied medicine there and was appointed professor of physiology at Leiden in 1886. The next year A D Waller (1856–1922) in England showed that a current was generated by the heart, but his recording device was cumbersome and insensitive. Einthoven was interested in physics, and he devised a sensitive string galvanometer. It used a fine wire stretched between the poles of a magnet. When a current passed through the wire it was deflected and an optical system magnified this for recording. Einthoven made electrocardiograms (ECGs) from the chest wall and from contacts on the arms and legs, and described his results from 1903. Soon afterwards, cardiologists gave full accounts of coronary artery disease and Einthoven and others (especially Sir T Lewis (1881–1945) in London) related the ECG tracings to clinical data for this and other heart diseases. This became an important diagnostic method, and Einthoven won the Nobel Prize for 1924.

Ekman, Vagn Walfrid (1874–1954) Swedish oceanographer: explained the variation in direction of ocean currents with depth.

After graduation, Ekman worked at the International Laboratory for Oceanographic Research in Oslo for several years before returning to Sweden in 1908. He was appointed professor of mathematical physics at Lund in 1910. In the 1890s the Norwegian Arctic explorer NANSEN had noted that the path of drifting sea ice did not follow the prevailing wind direction, but deviated about 45° to the right. In 1905 Ekman was able to explain this as an effect of the Coriolis force caused by the Earth's rotation. He went on to describe the general motion of near-surface water as the result of the interaction between surface wind force, the Coriolis force and frictional effects between different water layers. Ekman flow thus accounts for situations in which near-surface water moves in the opposite direction to that at the surface, with the net water transport at right angles to the wind direction. The resulting variation of water velocity with depth is known as the Ekman spiral. An analogous situation exists in atmospheric flow.

Elion, Gertrude (Belle) (1918–) US pharmacological chemist.

Elion studied biochemistry at Hunter College, New York and after graduating in 1937 worked in industry and as a high-school science teacher for 7 years. At the same time she was a part-time research student at New York University, obtaining her master's degree in 1941. She began to work for a doctorate, but could not continue because full-time study was required for this.

She joined the Burroughs Wellcome laboratory in 1944 and the next year began working with HITCHINGS. At that time the usual path to new drugs was to synthesize variations on natural plant drugs and then to look for useful therapeutic effects when they were given to test animals. Elion and Hitchings used a different approach. They looked for differences between the biochemistry of normal human cells and the cells of bacteria and other infective agents, or of cancer cells, and then used the differences to deduce chemical structures that would damage the infective or cancerous cells only. This rational programme of drug design gave important successes for the team led by Hitchings and Elion for over 20 years, including drugs for treatment of leukaemia, malaria, gout and autoimmune disorders. They shared with JAMES BLACK the Nobel Prize for physiology or medicine in 1988. After 1967 Elion led the Wellcome group, and in 1974 she announced another major therapeutic and commercial success, the antiviral drug acyclovir – evidence of her position in research, recognized by the award of the US National Medal of Science in 1991.

Elsasser, Walter Maurice (1904–   ) German–US theoretical physicist: developed theory of Earth's magnetic field.

Elsasser was born and educated in Germany; he left that country in 1933 and spent 3 years in Paris, where he worked on the theory of atomic nuclei. In 1936 he settled in the USA and began to specialize in geophysics. During the 1940s he developed the dynamo model of the Earth's magnetic field, which attributes the field to the action of electric currents flowing in the Earth's fluid metallic outer core. These currents are amplified through mechanical motions in the same way that currents are maintained in power station generators. The analysis of past magnetic fields, frozen in rocks, has since turned out a very powerful tool for the study of geological processes.

Eméleus, Harry Julius (1903–93) British inorganic chemist: revitalizer of experimental inorganic chemistry.

A student at London, Karlsruhe and Princeton, Eméleus was professor of inorganic chemistry at Cambridge from 1945–70. He worked on a wide variety of topics and his experimental work did much to dispel the pre-1945 view that rather little of interest remained to be done in inorganic chemistry. His early work was on phosphorescent flames and on photochemistry. In the 1940s he made novel silicon compounds and after 1950 many new halogen compounds, especially trifluoromethyl (CF_3-) derivatives of metals and non-metals.

Emiliani, Cesare [emeelyahnee] (1922–   ) Italian–US geologist: demonstrated the cyclic nature of ice ages and established the climatic history of the Quaternary period.

Emiliani emigrated to the USA in 1948, graduating from the University of Chicago in 1950, where he remained until moving to the University of Miami in 1956. Following the suggestion of UREY that the isotopic ratio of oxygen ($^{18}O/^{16}O$) in sea water depends upon the prevailing temperature (due to isotopic fractionation), Emiliani pioneered a technique for determining the past temperature of the oceans by measuring the $^{18}O/^{16}O$ ratio in the carbonate remains of microorganisms in ocean sediments. By selecting for study only pelagic species (ie those that

live near the ocean surface) he was able to establish, in 1955, that there had been seven glacial cycles during the Quaternary period, almost double the number of ice ages that were formerly thought to have occurred. Oxygen isotope methods are now an established technique in palaeoclimatic studies.

Empedocles (of Acragas) [empedokleez] (c.490–c.430 BC) Greek philosopher: proposed early view on nature of matter.

Active in politics, poetry, medicine and mysticism, Empedocles is credited with the suggestion that all substances are derived from four 'roots' or elemental principles: fire, air, water and earth. These are joined or separated by two forces, attraction and repulsion (or love and strife). This view, especially as developed later by ARISTOTLE, was influential for 2000 years, until BOYLE's work. Empedocles is said to have ended his life by jumping into the volcanic crater on Mount Etna, possibly in an attempt to prove his divinity.

Enders, John Franklin (1897–1985) US virologist: developed improved method for culturing viruses.

Enders had several early career changes. He left Yale in 1917 to become a flying instructor in the First World War; began a career as an estate agent and left it to study languages at Harvard, and then changed to microbiology, thereafter staying at the Harvard Medical School through a long career. Before his work, few laboratory cultures of viruses were available and these were inconvenient (eg cultures in a living chick embryo). Enders argued that living cells should be adequate, without the whole animal, if bacterial growth was prevented by adding penicillin. In 1948 together with F C Robbins (1916–   ) and T H Weller (1915–   ) he cultured the mumps virus using a homogenate of chick embryo cells and ox serum with added penicillin. The next year a similar method was used for the polio virus and in the 1950s for the measles virus. For measles they were able to develop a vaccine by 1951 that came into widespread use in 1963. The trio shared a Nobel Prize in 1954 and their methods of culturing viruses allowed virology to advance with successes such as the Salk and the SABIN polio vaccines.

Eratosthenes (of Cyrene) [eratostheneez] (c.270–c.190 BC) Greek astronomer and polymath: gave first accurate measurement of the Earth's circumference.

Eratosthenes was educated in Athens and became chief librarian of the Alexandrian museum. He devised an ingeniously simple way of measuring the circumference of the Earth. Eratosthenes knew that on a certain day the Sun at its highest point (midday), at Cyrene (now Aswan), was exactly overhead (it was known to shine down a deep well). He determined that on the same day at Alexandria, when the Sun was at its highest point, it was at an angle corresponding to 1/50th of a circle south of its zenith. Knowing the distance between the two places he therefore calculated that the Earth's circumference was 50 times that length. His result was probably fairly accurate, perhaps within 50 miles of the correct value.

Among his other discoveries Eratosthenes suggested a method of separating primes from composite numbers (known as the sieve of Eratosthenes); he obtained an improved value for the obliquity of the ecliptic (the tilt of the Earth's axis), and he produced the first map of the world based on meridians of longitude and parallels of latitude. In later life he became blind and, no longer able to read, committed suicide.

Erxleben, Dorothea Christiana Leporin [erkslaybn(1715–62) German physician: the first woman to gain a full medical degree in Germany.

Dorothea Erxleben's father was a doctor in the small town of Quedlinburg, in Germany. He grieved at the waste of talented women being confined to household duties and taught his daughter alongside his son, teaching them Latin, basic science and medicine, preparing them both for a medical career. Dorothea petitioned King Frederick II for consent to accompany her brother to Halle to study for a medical degree in 1740; this was granted. However, the prospect of a woman studying medicine caused outrage; it was pointed out that, as women were forbidden by law to hold public office, they could not practise medicine and they did not need a medical degree.

War with Austria broke out and Dorothea's brother left for military service. Alarmed at the prospect of going to university on her own, she married a widower with five children and continued to study. Her father died 6 years later leaving debts and her husband became ill; financial responsibility for the family fell to her. Her attempts to practise medicine once more fell foul of the licensed doctors of Quedlinburg, who demanded that she must sit an examination. The rector of the University of Halle decided that on the matter of women's entry to university - 'one designates the sex to which the degree most often applies, but by affirming the one sex the other is not excluded'; he also ruled that the profession of medicine was not the same as holding public office, so that Dorothea Erxleben was free to take her examination. She sat her final examinations without delay and was granted her degree on 12 June 1754.

Esaki, Leo (1925–) Japanese physicist: discovered the tunnel (Esaki) diode.

While working for his doctorate on semiconductors at the University of Tokyo (1959), Esaki was also leading a small research group at the Sony Corporation. He chose, in 1957, to investi-gate conduction by quantum mechanical 'tunnelling' of electrons through the potential energy barrier of a germanium p-n diode. Such conduction is in the reverse direction to the normal electron drift and, using narrow junctions (only 100 Å wide) with heavy impurity doping of the p-n junction, Esaki observed the effect. He realized that with narrower junctions the effect would become so strong that the total current would actually fall with increasing bias (negative resistance) and succeeded in making such devices (tunnel or Esaki diodes) in 1960. These devices have very fast speeds of operation, small size, low noise and low power consumption; they have widespread electronic applications in computers and microwave devices. In 1960 Esaki joined IBM's Thomas J Watson Research Centre and in 1973 was awarded the Nobel Prize with JOSEPHSON and I Giaever (1929–) for work on tunnelling effects.

Eskola, Pentti Elias (1883–1964) Finnish geologist.

Eskola graduated in chemistry in Helsinki in 1906 but then turned to petrology, especially the mineral facies of rocks. The term facies refers to the appearance of a rock and the total of its characteristics.

Eskola was professor of geology at Helsinki from 1928–53. He concluded in 1914 that in metamorphic rocks the mineral composition when equilibrium is reached at a particular temperature and pressure is controlled only by the chemical composition. Later work has confirmed this view. For example, a zeolite results when the appropriate chemical components are equilibrated in a region where temperature is 100–220°C and pressure 150–400 kbar.

Euclid [yooklid] (lived c.300 BC) Greek mathematician: recorded, collated and extended mathematics of the ancient world.

Euclid offers strange contrasts: although his work dominated mathematics for over 2000 years, almost nothing is known of his life and personality. One alleged remark survives, his reply to Ptolemy Soter, King of Egypt, who hoped for an easy course of tuition: 'in geometry there is no straight path for kings'. Working in Alexandria, then a new city but a centre of learning, Euclid brought together previous work in mathematics and his own results and recorded the whole in a systematic way in 13 'books' (chapters), entitled *Elements of Geometry*. Others are lost.

The system attempted to be fully rigorous in proving each theorem on the basis of its predecessors, back to a set of self-evident axioms. It does not entirely succeed, but it was a noble attempt, and even the study of its deficiencies proved profitable for mathematicians. His work was translated into Arabic, then into Latin and from there into all European languages. Its style became a model for mathematicians and even for other fields of study. Six of the chapters deal with plane geometry, four with the theory of numbers (including a proof that the number of primes is infinite) and three with solid geometry, including the five Platonic solids (the tetrahedron, octahedron, cube, icosahedron and dodecahedron – Euclid finally notes that no other regular polyhedrons are possible).

Only in the 19th-c was it realized that other kinds of geometry exist. This arose from the fact that, while most of the Euclidean postulates are indeed self-evident (eg 'the whole is greater than the part'), the fifth postulate ('axiom XI') is certainly not so. It states that 'if a point lies outside a straight line, then one (and only one) straight line can be drawn in their plane which passes through the point and which never meets the line'. Then in the 19th-c it was accepted that this certainly cannot be deduced from the other axioms, and LOBACHEVSKY and others explored geometries in which this 'parallel axiom' is false. In the 20th-c, EINSTEIN found that his relativity theory required that the space of the universe be considered as a non-Euclidean one; it needed the type of geometry devised by RIEMANN. For all everyday purposes, Euclidean space serves us well and the practical differences are too small to be significant.

Euclid's achievement was immense. He was less talented than ARCHIMEDES but for long-lived authority and influence he has no peer. Within the limits of his time (with its inadequate concepts of infinity, little algebra and no convenient arithmetic) his attempt at an unflawed, logical treatment of geometry is remarkable.

Euler, Leonhard [oyler] (1707–83) Swiss mathematician: the most prolific mathematician in history.

Euler was the son of a Calvinist pastor who gave him much of his early education, including mathematics. Later he studied at the University of Basle, where he became close friends with members of the Bernoulli family, and DANIEL BERNOULLI in particular. Because he was still rather young (he graduated at 16), Euler could not obtain a post at the university. However, Daniel persuaded Euler to join him at Catherine I's Academy of Science at St Petersburg in 1727. The Empress died the day Euler arrived in Russia and the future of the academy became uncertain. After an unhappy period working in the Naval College and medical section of the Academy he became professor of physics in 1730. When Bernoulli returned to Switzerland in 1733 Euler succeeded him as professor of mathematics.

The repressive reign of a boy Tsar led Euler, now married, to retreat into reclusive mathematical work, and this solitariness increased during the reign of Anna Ivanovna (1730–40) which was one of the bloodiest in Russian history. During this time Euler lost the sight of his

right eye, perhaps due to looking at the Sun accidentally during his astronomical studies. Although conditions eased in Russia after Anna's death, Euler departed to join Frederick the Great's Berlin Academy of Science in 1741. Despite great authority in mathematics Euler frequently engaged ineptly in philosophical discussions and Frederick sought a replacement. In 1766 Euler took up Catherine the Great's offer of the directorship of the St Petersburg Academy, accompanied by his family and servants (18 people in all). He became totally blind soon after his arrival, but due to his remarkable ability to calculate in his head his productivity did not diminish and he successfully carried out his work for another 15 years. He remained in Russia for the rest of his life.

Euler was the most prolific mathematician in history and contributed to all areas of pure and applied mathematics. In analysis he lacked GAUSS's or CAUCHY's rigour but he had a gift for deducing important results by intuition or by new ways of calculating quantities. He systematized much of analysis, cast calculus and trigonometry in its modern form and showed the important role of e (Euler's number, 2.718 28...). Euler developed the use of series solutions, paying due regard to convergence; he solved linear differential equations and developed partial differential calculus. He applied these analytical tools to great effect in problems in mechanics and celestial mechanics and introduced the principle of virtual work. The formidable three-body problem of the Earth, Sun and Moon system was solved approximately by him (1753, 1772), leading to an award of £300 by the British Government for the resulting improvement in navigational tables. In the course of this he developed much of classical perturbation theory.

He worked on number theory, fluid flow, geometry and acoustics. A large number of theorems are named after this extraordinarily creative and productive man. One of the best-known is Euler's rule, which shows that for a polyhedron with v vertices, f faces and e edges, then $v + f - e = 2$.

He was active in mathematics to the moment of his death, on a day spent partly in calculating the laws of ascent of the recently invented hot-air balloons.

Euler, Ulf Svante von [oyler] (1905–83) Swedish physiologist.

Son of a physiologist who won a Nobel Prize in chemistry, von Euler was a student and later a professor at the Royal Caroline Institute in Stockholm. In 1903 T R Elliott (1877–1961) of Cambridge made the novel suggestion, based on experiments, that nerve transmission is at least partly chemical. For a time this idea was largely ignored but it led to later successes by DALE and by O Loewi (1873–1961) and in 1946 von Euler isolated a neurotransmitter of the sympathetic nervous system and showed it to be noradrenalin, and not adrenalin as had been believed. Already, in 1935, he had initiated work in another area by showing that human semen contained a potent chemical, which lowered blood pressure and contracted muscle, which he named prostaglandin. (BERGSTROM later isolated two prostaglandins; more are now known and they form an important biochemical group). Von Euler shared a Nobel Prize in 1970.

Ewing, William Maurice (1906-74) US marine geologist: made first measurements of the thickness of the oceanic crust and discovered the global extent of mid-ocean ridges.

Ewing joined the Lamont–Doherty Geological Observatory, New York, in 1944 and was instrumental in making it one of the world's leading geophysical research institutes. Using marine seismic techniques he discovered that the oceanic crust is much thinner (5-8 km thick) than the continental crust (c.40 km thick). He also demonstrated the global extent of mid-ocean ridges and in 1957 discovered the presence of a deep central rift in them. His studies of the ocean sediment showed that its thickness increases with distance from the mid-ocean ridge, which added support for the sea-floor spreading hypothesis proposed by H H HESS in 1962.

Eyring, Henry (1901-81) US physical chemist: developed the theory of reaction rates.

Trained as a mining engineer, Eyring changed to chemistry for his PhD and worked thereafter on chemical kinetics and the theory of liquids; his career was spent in Princeton and Utah.

In a chemical reaction, some chemical bonds are broken and new bonds are formed. Eyring developed methods based on quantum mechanics for calculating the energies involved from which the rate of the chemical reaction can be calculated (in selected cases) and also the effect of temperature on the rate.

Fabre, Jean Henri [fabruh] (1823–1915) French entomologist.

Always poor, Fabre spent his working life as a science teacher and was aged 50 before he could spend all his time as a field entomologist. In 1878 he bought a small plot of land in Serignan, Provence, to make an open-air laboratory. There he observed and wrote about the insect world in a way which revitalized interest in it by others and which made him the best known of all entomologists. His early research was on parasitic wasps but his close studies of a variety of groups led him to write a 10-volume survey of insects, which remains a classic.

Fabricius, David [fabreesyus] (1564–1617) German astronomer: discovered first variable star.

A clergyman and an amateur astronomer, Fabricius discovered that the brightness of the star o Ceti regularly varied from magnitude 9 to magnitude 3 over a period of about 11 months. This was the first variable star to be found, causing him to name it Mira (the marvellous). In fact, Mira's change in luminosity is the result of a true change in surface temperature, rather than the eclipsing effect of a binary companion (eg Algol).

Fabrizio, Girolamo (*Ital*), Fabricius ab Aquapendente (*Lat*) [fabritsioh] (*c.*1533–1619) Italian anatomist: pioneer of scientific embryology.

Fabrizio is often named in Latin, coupled with the Tuscan village of Aquapendente where he was born. He was a student in Padua and later taught there for 50 years. He first studied the classics and then medicine; his teacher of anatomy and surgery was FALLOPIUS, whom he succeeded as professor in 1565. He researched and wrote on the larynx, the eye, muscular action and respiration. He supervised the building of the anatomy theatre in Padua, which was the first of its kind and which still exists. It was there that he demonstrated the valves in the veins to his students, including HARVEY, who became interested in the problem of blood circulation; Fabrizio did not understand the function of the valves, which were to be a key in Harvey's work.

Fabrizio's most original research was in embryology. In 1600 he wrote a comparative study of the late fetus in various animals and in 1604 he described the formation of the chick in the hen's egg from the sixth day. His well-illustrated descriptions mark the beginning of embryology as a new branch of biology.

After he officially retired in 1613 he continued as an active researcher until his death, aged about 86.

Fahrenheit, Gabriel Daniel [fahrenhiyt] (1686–1736) German physicist: developed the mercury thermometer and the Fahrenheit temperature scale.

Fahrenheit worked as a glassblower in Holland, specializing in the construction of meteorological instruments. He succeeded in improving the reliability and accuracy of the alcohol thermometers of the day and in 1714 constructed the first successful mercury thermometer, following AMONTONS's work on the thermal expansion of the metal. Using these instruments he discovered that different liquids each have their own characteristic boiling point, which varies with atmospheric pressure. He also discovered the phenomenon of supercooling of water, whereby water may be chilled a few degrees below its freezing point without solidification.

He is best remembered, however, for devising the Fahrenheit scale of temperature, which used as its reference points the melting temperature of a mixture of ice and salt (the lowest temperature he could obtain), and the temperature of the human body. This range was subdivided into 96 equal parts, with the freezing point of water falling at 32°F and the boiling point at 212°F.

Fajans, Kasimir [fahyans] (1887–1975) Polish–US physical chemist: devised rules for chemical bonding.

Born in Warsaw, Fajans studied in Germany and in England, and worked in Munich from 1917–35, when he emigrated to Chicago. His early research was in radiochemistry, where he had ideas on isotopes, and the displacement law, simultaneously with others. Although he worked in several areas of physical chemistry, he is best known for Fajans's rules on bonding between atoms. The first rule is that as highly charged ions are difficult or impossible to form, so covalent bonds are more likely to result as the number of electrons to be removed or donated increases; the second rule is that ionic bonding is favoured by large cations and small anions. Both rules follow from simple electrostatic energy requirements; and they lead to the 'diagonal similarities' shown by elements in adjacent periodic groups (eg Li and Mg; Be and Al; B and Si).

Fallopius (*Lat*), Gabriello Falloppio (*Ital*) (1523–62) Italian anatomist.

Fallopius first studied to become a priest, but changed to medicine and was taught anatomy by VESALIUS. From 1551 he taught in Padua, and 10 years later his textbook extended and corrected Vesalius's work. His discoveries included structures in the human ear and skull and in the female genitalia. He first described the tubes from the ovary to the uterus; he did not know their function. It was almost 300 years later that the ovum was discovered; ova are formed in the ovary and pass down these Fallopian tubes to the uterus; if they are fertilized on their way, the embryo develops in the uterus. Fallopius is claimed to have invented condoms. Made of linen, they were intended to prevent syphilis rather than conception.

Faraday, Michael (1791–1867) British chemist and physicist: discovered benzene and the laws of electrolysis; invented an electric motor, dynamo and transformer; creator of classical field theory.

Faraday had an intuitive grasp of the way physical nature may work, combined with a genius for experiment and great energy. EINSTEIN had the view that physical science has two couples of equal magnitude; GALILEO and NEWTON, and Faraday and MAXWELL: an interesting equation.

Faraday's talents ripened late (he was at his best in his 40s; many scientists have their major ideas behind them at 30), but he began his education late. His father was an ailing blacksmith and the boy became a bookseller's errand boy at 13. He learned bookbinding, read some of the books and was captivated by an article on electricity in an encyclopedia he had to rebind and by JANE MARCET's *Conversations on Chemistry*. These books were to shape his life, and he soon joined a club of young men who met weekly to learn elementary science. He was given tickets to attend DAVY's last course of lectures at the Royal Institution and he wrote elegant notes of these and bound them. These notes he sent to Davy, and applied for a job with him. Davy firmly recommended him to stay with bookbinding, but he had injured an eye (making NCl_3) and took Faraday as a temporary helper. After a few weeks he gave him a permanent job as assistant; Faraday was later to become his co-worker, then his successor at the Royal Institution and in time his superior as a scientist. Faraday learned quickly and he was lucky, because Davy decided to make a grand European tour, taking Faraday with him as helper and valet. The young man was to meet most of the leading scientists during a one-and-a-half year tour, made despite the Anglo-French war by special permission; it was a strange education, but it gave him an awareness of most of the physical and chemical science of the time and he became a skilful chemical analyst. The main omission was mathematics, a shortcoming which he never repaired. His first solo research, made when he was 29, was the synthesis of the first known chlorocarbons (C_2Cl_6 and C_2Cl_4) and until 1830 he was mainly a chemist. In 1825 he discovered benzene, which was later to be so important in both theoretical and technical chemistry. He worked on alloy steels and

he liquefied chlorine and a range of other gases by pressure and cooling. He was established at the Royal Institution, became an excellent lecturer and never left until retirement; he could have become rich from consultant work, but he belonged to a fervent religious group and he declined both wealth and public honours.

From about 1830 he increasingly studied electricity. An early venture was the study of electrolysis, and in 1832 and 1833 he reported the fundamental laws of electrolysis: (1) the mass of a substance produced by a cathode or anode reaction in electrolysis is directly proportional to the quantity of electricity passed through the cell, and (2) the masses of different substances produced by the same quantity of electricity are proportional to the equivalent masses of the substances (by equivalent mass is meant the relative atomic mass divided by the valence). Faraday had an excellent set of new words devised for him by WHEWELL for work in this area: electrolysis, electrolyte, electrode, anode, cathode, ion. It follows from the laws of electrolysis that an important quantity of electricity is that which will liberate one mole of singly charged ions. This amount, the Faraday constant, F, is defined by $F = N_A e$ where N_A is the Avogadro constant and e is the charge on an electron. F can be measured accurately (eg by electrolysis of a silver solution) and has the value 9.648×10^4 C mol^{-1}. Also named for Faraday is the unit of capacitance, the farad (F). It is the capacitance of a capacitor (condenser) having a charge of one coulomb (C) when the potential difference across the plates is one volt. This is a large unit and the more practical unit is the microfarad, equal to 10^{-6} F.

Faraday's work on electricity in the 1830s largely developed the subject. OERSTED had shown that a current could produce a magnetic field; Faraday argued that a magnetic field should produce a current. He found this to be so, provided that 'a conductor cut the lines of magnetic force'. He had discovered electromagnetic induction (independently discovered by J HENRY) and to do it he used his idea of lines and fields of force producing a strain in materials, an idea which was to be highly productive. With it he was able to devise primitive motors, a transformer and a dynamo; he cast off the old idea of electricity as a fluid (or two fluids) and moved to solve some basic problems. For example, he showed that current from an electrostatic machine, a voltaic cell and a dynamo is the same, and devised methods to measure its quantity. He examined capacitors and the properties of dielectrics, and he discovered diamagnetism. In the early 1840s he was unwell for 5 years with 'ill health connected with my head'. It may have been mercury poisoning.

Back at work in 1845, he worked on his idea that the forces of electricity, magnetism, light and gravity are connected and was able to show that polarized light is affected by a magnetic field. He failed to get a similar result with an electric field (KERR succeeded in 1875) and the general theme of the 'unity of natural forces' has been pursued to the present day. In 1846 WHEATSTONE was due to speak at the Royal Institution, but at the last moment panicked and Faraday had to improvise a lecture. He included his 'Thoughts on Ray Vibrations', which Maxwell claimed were the basis of the electromagnetic theory of light that Maxwell, with new data and more mathematical skill, devised 18 years later.

Faraday had a very strange mind, but it well fitted the needs of physics at the time. His personality offers curious contrasts; he had much personal charm, but no social life after 1830. He had great influence on later physicists, but no students, and worked with his own hands helped only by a long-suffering ex-soldier, Sergeant Anderson. He had highly abstract ideas in science, but he was a most effective popularizer; his Christmas lectures for young people, begun in 1826, are still continued and are now televised. In quality and in quantity, he remains the supreme experimentalist in physics.

Farman, Joseph C (1930–) British atmospheric scientist: found the Antarctic ozone hole.

After graduating from the University of Cambridge in mathematics and natural sciences, Farman worked briefly in the aerospace industry before embarking on a career with the British Antarctic Survey in 1956.

In 1984 Farman and his colleagues discovered an 'ozone hole' in the stratosphere above the Antarctic. This finding, made using ground-based instruments, was confirmed by American satellite observations and a review of past satellite data revealed that winter ozone levels had been declining for the past 10 years and by the mid-80s were down to about 50% of the 1957 level. This had not been noticed earlier because the computers that processed the satellite measurements had been programmed to ignore such low values as 'impossible', although ozone loss was foreseen by P Crutzen, M Molina and S Rowland in the 1970s, who won a Nobel Prize for this in 1995.

The stratospheric ozone layer plays a vital role in protecting the Earth's life from the more harmful effects of the Sun's ultraviolet rays. Increased levels of UV light have been shown to cause skin cancers and eye cataracts, to kill phytoplankton and therefore disrupt the marine food chain, and to decrease crop yields. Long-term depletion of the ozone layer could have very serious effects for life on Earth.

For the first few years after Farman's discovery it was hoped that the depletion of the ozone layer, which undergoes natural seasonal variations, was a natural and transient phenomenon. However, the depletion has both continued and increased in severity, and has also been observed in the ozone layer of the more heavily populated northern hemisphere, which is losing its ozone at a rate of about 5% per decade.

It is now widely accepted that the depletion is due to the effects of man-made chemicals, notably chlorofluorocarbons (CFCs), released into the atmosphere over the past 20–30 years. CFCs have been widely used in refrigeration and insulation materials, but also have the unforeseen property of catalysing the breakdown of ozone (O_3). Although governments are now taking measures to reduce the release of CFCs into the atmosphere, it is estimated that even if CFC production were completely halted immediately, it would take another 70 years before this had a significant effect on restoring the ozone layer.

Farman's discovery has been of the utmost importance in highlighting the fact that man's polluting activities are now of such a scale as to jeopardize the whole future of life on Earth. Within 3 years of his discovery the Montreal Protocol took the first steps to limit worldwide CFC production. Farman has remained extremely active in lobbying governments and other bodies on the consequences of ozone depletion and in campaigning for increased restrictions on the manufacture of CFCs.

Fehling, Hermann Christian von [fayling] (1812–85) German organic chemist.

A pupil of LIEBIG, Fehling taught in Stuttgart. He is best known for the test reagent Fehling's solution, which contains a deep blue copper(II) complex in aqueous alkaline solution. If the organic test sample on boiling with this removes the blue colour and reduces the copper to brick-red copper(I) oxide, it is likely to be an organic aldehyde or reducing sugar. However, formates, lactates, haloforms and some esters and phenols also give a positive test.

Fermat, Pierre de [fairmah] (1601–65) French mathematician: 'the prince of amateurs'.

As a senior Government law officer (a job he did not do very well) it is remarkable that Fermat found time to maintain his skills as a linguist, amateur poet and, most notably, as an amateur mathematician. After 1652, when he nearly died of plague, he did give most of his time to mathematics, but he still did not publish his work in the usual sense, and his results are known through his letters to friends, notes in book margins and challenges to other mathematicians to find proofs for theorems he had devised.

His successes included work on probability, in which he corresponded with PASCAL and reached agreement with him on some of its basic ideas; on analytical geometry, where again he achieved parallel results with another talented researcher, DESCARTES, and went further in extending the method from two dimensions to three; and on the maxima and minima of curves and tangents to them, where his work

THE HISTORY OF MATHEMATICS

The earliest mathematical writer whose name we know was the Egyptian scribe Ahmes, who in *c.*1650 BC copied an earlier text on handling fractions and solving arithmetical problems. But for at least 1000 years before that, scribes in the great river civilizations of Egypt and Mesopotamia were developing ways of representing numbers and solving problems that are recognizably precursors of today's mathematical activity.

A significant change was introduced by Greek-speaking people around the E Mediterranean during 500–200 BC: the development of the notion of proving results as a fundamental characteristic of mathematical activity. A research tradition in geometry grew up, whose basic results were codified by EUCLID in *Elements of Geometry* (*c.*300 BC), and culminated in the work of ARCHIMEDES and APOLLONIUS. Later, the idea that the cosmos is intrinsically mathematical, an influential idea found in the work of Plato, was retrospectively attributed to the semi-mythical figure of PYTHAGORAS. The Greek mathematical tradition lasted for several further centuries, notably in Alexandria, and ranged from the astronomical and geographical work, exemplified by PTOLEMY, to the arithmetical investigations of DIOPHANTUS, who raised the solving of number-problems to a new height. It was in Alexandria, too, that the first well-attested woman mathematician, HYPATIA, lived and where she was murdered in 415, probably by a mob of zealous early Christians.

Meanwhile, mathematical activity had been vigorously pursued in the civilizations of China and India. It is not easy to reconstruct Chinese mathematics before the Emperor Shih Huang Ti's great book-burning of 212 BC. Nonetheless, during the following centuries scholars such as Lui Hui (*c.*260) worked to reconstruct and comment on earlier mathematical works, besides developing both geometry and a characteristically Chinese arithmetical-algebraic computational style, which they brought to bear on such areas as computing bounds for π and solving determinate and indeterminate problems.

India, too, has a long mathematical tradition, initially in a religious context of astronomy and altar-construction, first recorded by Baudhayana *c.*800–600 BC. The later mathematician-astronomer Aryabhata (late 5th-c) and Brahmagupta (early 7th-c) wrote important works involving arithmetic, algebra, and trigonometry, whose influence spread to the West in succeeding centuries.

Baghdad in the 9th-c was an important centre for the pivotal Islamic contributions to mathematics. There AL-KHWARIZMI wrote a number of books, drawing together Babylonian, Greek and Indian influence. His *Arithmetic* introduced the Indian decimal-place-value numerals, and his *Algebra* has given the subject its name (his own name still being remembered in the word algorithm). By the late 12th-c much mathematical knowledge had been developed and held in the Islamic culture around the southern shores of the Mediterranean, and it was beginning to percolate into Christian Europe through trading posts in such places as Sicily and Spain. In particular, the Islamic world was using for its numerals a decimal-place-value system. An Italian with trading links to Sicily, FIBONACCI, noticed that the Arabs were using much more efficient numerals – which one could calculate with as well as record number values – and wrote his book *Liber abaci* (1202, Book of the Abacus) to publicize this.

This period is one of rich mathematical activity in many parts of the world. China at this time was home to mathematicians of the calibre of Yang Hui, who had explored the binomial pattern several centuries before 'Pascal's triangle' became known in the West; and Chu Shih Chieh, who took the Chinese arithmetical-algebraic computational style to new heights. In India, Bhaskara (12th-c) wrote valuable works on arithmetic, algebra and trigonometry; and Madhava (*c.*1340–1425) headed a research tradition in Kerala whose work in infinite series and trigonometrical functions, anticipating later European work in mathematical analysis, is only now beginning to be discovered. In Iran, Omar Khayyam (*c.*1048–*c.*1122) worked to develop arithmetic, algebra and geometry, as well as astronomy and philosophy.

During the late Renaissance, European mathematics had begun to absorb ancient Greek mathematical works, made available by the efforts of such scholars as Regiomontanus (1436–76), Maurolico (1494–1575), and Commandino (1509–75). An important development was the solution of cubic and quartic equations in Italy, notably in the *Ars magna* (1545, Great Art) of CARDANO, and the further development of algebraic analysis by VIÈTE.

During the 17th-c, Europe saw not only a spectacular flourishing of mathematical creativity – with such mathematicians as NAPIER, DESCARTES, FERMAT, HUYGENS, NEWTON and LEIBNITZ – but also the growth of institutions for promoting scientific activity and journals to communicate and broadcast the results. During the next century, mathematization of many diverse fields of human interest became fashionable in the wake of the enormous success of Newton's *Principia* (1687). Mathematicians of the calibre of the BERNOULLI family, EULER and LAGRANGE consolidated the methods of calculus, applied them to mechanics and developed new mathematical areas and approaches – notably, an increasing movement from geometry to algebra as the natural language of mathematics.

One consequence of the French Revolution was the promotion of mathematics in education. About this time text-books were increasingly used, along with tests and examinations, to create mathematics syllabuses and new educational practices. Mathematical activity in France (MONGE, LAPLACE, CAUCHY) and subsequently in Germany (GAUSS, JACOBI, DIRICHLET, RIEMANN) was strongly developed and professionalized, in both research and teaching directions. The works of Niels Hendrik and GALOIS, two remarkable talents who died young, proved influential on later generations, as did the non-Euclidean geometry of Janos Bolyai (1802–60) and LOBACHEVSKY. The growing importance of the numerical data in society led to the development of statistical thinking, notably in the work of Adolphe Quételet (1796–1874), with applications in both the physical sciences (eg MAXWELL) and biological and human sciences (eg PEARSON). The work of BABBAGE on computing engines also attests to the recognized need for accurate mathematical tables and efficient handling of numbers.

The foundations of mathematics received growing attention through the 19th-c, from the need to teach and explain the theorems of analysis. Georg Cantor (1845–1918) explored the infinite and founded the theories of sets, while DEDEKIND, with his definition of real numbers, helped consolidate the

process of arithmetizing analysis, as did the work of WEIERSTRASS. One of Weierstrass's pupils who benefited from the slowly opening higher educational opportunities for women was SONYA KOVALEVSKY.

During the 20th-c much new mathematics has developed, partly through exploring structures common to a range of theories, thus counteracting the tendency to split into more and more distinct specialized areas. Topology has, under the considerable influence of POINCARÉ, reached new heights of geometrical generality and unifying power, while algebra too has become even more general in its exploration of structural depth, as in the work of EMMY NOETHER. HILBERT set the agenda for much 20th-c research through his prescient outline, at the International Mathematical Congress in 1900, of the major problems for mathematics. In this century, too, mathematics has become applied more diversely than ever before, from the traditional applications in physical science to new applications in economics, biology and the organization of systems. The development of electronic computers, particularly associated with TURING and VON NEUMANN, grew out of mathematical, logical and number-handling activity and has affected mathematics in a variety of ways. Recent explorations of mathematics which use the computer as a research tool can be seen as restoring mathematics to its roots as an experimental science.

Dr John Fauvel, Open University

was seen by NEWTON as a starting point for the calculus. In optics he devised Fermat's principle and used it to deduce the laws of reflection and refraction and to show that light passes more slowly through a dense medium. He worked on

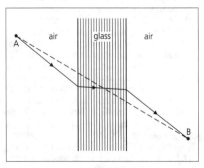

Fermat's principle – the dotted line shows the shortest path between A and B. A light beam follows the solid line, consistent with the laws of refraction, because the velocity of light in the glass is less than in air. The solid line is the path of least time.

the theory of equations and especially on the theory of numbers. Here he was highly inventive and some of his results are well known but, as he usually did not give proofs, they teased other mathematicians in seeking proofs for a long time, with much advantage to the subject. Proofs were eventually found, but not in every case. Fermat's last theorem, noted in one of his library books, states that the equation $x^n + y^n = z^n$ where n is an integer greater than 2, can have no solutions for x, y and z, and records 'I have discovered a truly marvellous demonstration which this margin is too narrow to contain'. Since he wrote this, in about 1637, generations of mathematicians have attempted to find a proof. By 1960 it had been shown to be true for all values of n less than 125 000; and it is likely that work by A J Wiles of Princeton, announced in Cambridge, UK in 1993 and 1994, will provide a full proof.

Fermi, Enrico [fairmee](1901–54) Italian–US nuclear physicist: built first atomic reactor.

Enrico Fermi was the greatest Italian scientist of modern times and was highly creative both as a theoretical and experimental physicist. The son of a railway official, he showed ability from an early age and earned his PhD at Pisa, researching on X-rays. Fermi then worked with BORN at Göttingen and with P Ehrenfest (1880–1933) at Leiden, before returning to a professorship at Rome in 1927. He had already published over 30 papers, including some on quantum statistics (Fermi–Dirac statistics) followed by work on spin-$\frac{1}{2}$ particles (now called fermions) such as the electron.

Fermi worked hard to build up Italian physics, but the circle of talent around him was dispersed by the growth of Fascism, and Fermi and his wife, who was Jewish, left for Columbia University, New York in 1938. While in Rome, Fermi worked on the RAMAN effect, hyperfine structure, cosmic rays and virtual quanta. In 1933 he produced the theory of radioactive beta-decay, whereby a neutron emits an electron (beta-particle) and an anti-neutrino and becomes a proton. The following year he showed that, rather as the JOLIOT-CURIES had used helium nuclei (alpha-particles) to induce nuclear transmutations, neutrons were even more effective. This led to his rapid discovery of over 40 new radioactive isotopes. He then, by chance, discovered that paraffin wax could be used to slow down the neutrons and make them more effective, by a factor of hundreds, in causing transmutations of nuclei (they remain close to the target nucleus longer and are thus more likely to be absorbed). For all this work he received the 1938 Nobel Prize for physics.

Fermi had misinterpreted the transmutation of uranium with neutrons, but the ideas of FRISCH and MEITNER in 1938 corrected this, and proposed that nuclear fission with production of additional neutrons was occurring. HAHN and F Strassmann (1902–80) in Berlin also obtained these results, and both parties realized that vast amounts of energy could be released in such a chain reaction. Fermi, SZILARD and EINSTEIN moved quickly to warn Roosevelt and urge him to develop a nuclear weapon before Germany did so. The Manhattan project was set up (at a final cost of $2 billion) and Fermi's group at Chicago obtained the first controlled self-sustaining nuclear reaction (in a graphite-moderated reactor or 'pile' at Stagg Field stadium) on 2 December 1942. A historic telephone call was made by COMPTON to the managing committee at Harvard stating that 'the Italian navigator has just landed in the New World'.

Fermi continued to work on the project and attended the first test explosion of the fission bomb (A-bomb) in the New Mexico desert. While approving of its use against Japan, he, like OPPENHEIMER, opposed the development of the fusion bomb (H-bomb). He defended Oppenheimer, the director of Los Alamos Laboratory, where the work was done, against charges of disloyalty and of being a security risk.

Fermi took a professorship at Chicago after the war, but died young of cancer. He was much liked as an inspiring teacher and warm and vivacious character, enjoying sports and displaying clarity as a lecturer and research leader. Element number 100 was named fermium after him.

Fernel, Jean François (c.1497–1558) French physician: made systematic survey of physiology and pathology.

Fernel did not begin to study medicine until he was 27, after he had studied philosophy and the classics. He was an innkeeper's son and his life was spent in Paris. He was very successful in medicine and became personal physician to Henry II of France. His main contribution to medical science was his textbook (1554), which was a standard work for over a century, with about 30 editions, reprintings and translations. Its first part deals systematically with physiology; the second part with pathology, giving an account of human organs in a diseased state – a method which, like the word 'pathology', was new. The final part dealt with treatment. Fernel was an excellent observer who rejected the use of astrology and his influence, through his book, was extensive and useful.

Fessenden, Reginald Aubrey (1866–1932) Canadian–US physicist: devised amplitude modulation for radio transmission.

Born and educated in Canada, Fessenden's first job was in Bermuda but in 1886 he joined EDISON's laboratory in New Jersey. Then he moved in 1890 to Edison's great rival, Westinghouse, for 2 years and then to academic life at Purdue and later at Pittsburgh; in 1900 he moved to the US Weather Bureau and then back to industry.

Before 1900 he began to work on radio, which MARCONI and others were using to transmit messages by sending signals mimicking MORSE code. Fessenden sent out a continuous signal or 'carrier wave' at a steady high frequency and varied the amplitude of the waves to correspond to the sound waves of voices or music. At the receiver, these amplitude modulations could be reconverted to reproduce the sound. By Christmas Eve 1906 he was able to broadcast what was probably the first sound programme in the USA. He also devised the 'heterodyne effect' to improve amplification, and established two-way radio communication between the USA and Scotland. He was second only to Edison in the number and variety of his patents (over 500) and like him was involved in many lawsuits. Modern radio is largely based on the work of Marconi, Fessenden, DE FOREST (who devised the triode) and E H ARMSTRONG, who devised frequency-modulated (FM) transmission.

Feynman, Richard Phillips [fiynman] (1918–88) US theoretical physicist: developer of mathematical theory of particle physics.

Feynman's father, a New York maker of uniforms, developed the boy's interest in scientific ideas and logical observation. The young man graduated from Massachusetts Institute of Technology and Princeton, worked on the atomic bomb (the Manhattan project) and later joined the staff of the California Institute of Technology.

During the late 1940s Feynman developed new techniques for considering electromagnetic interactions within quantum theory, contributing methods on field theory which have been used widely. He showed that the interaction between electrons (or between positrons, the positively charged antiparticle to an electron, see DIRAC) could be considered by regarding them as exchanging virtual photons (electromagnetic radiation). This electron–electron scattering can be described quantitatively as a sum of terms, with each term coming from a matrix element describing a topologically distinct way in which a photon can be exchanged. Each term can be written as a Feynman diagram consisting of lines, called Feynman propagators, which describe the exchange of particles. This work contributed greatly to a new theory of quantum electrodynamics (QED), which deals with nuclear particle interactions and which is in excellent agreement with experiment. Feynman received the 1965 Nobel Prize for physics for fundamental work on QED, together with SCHWINGER and TOMONAGA.

Feynman was a colourful character in modern physics whose originality and showmanship made him a highly regarded lecturer (Feynman's *Lectures in Physics* are a delight to students). He enthused over any kind of puzzle and enjoyed the company of a wide variety of people; he was renowned as a storyteller and practical joker. His recreations he listed as 'Mayan hieroglyphics, opening safes, playing bongo drums, drawing, biology experiments and computer science'. One of his few antagonisms was pomposity. When authority tried to close a topless restaurant in Pasadena, he went to court to defend it, and claimed to use it frequently to work on physics.

Fibonacci, Leonardo [fibonahchee] (*c.*1170–*c.*1250) Italian mathematician: introduced the Arabian numeral system to Europe.

Fibonacci's father was an Italian consul in Algeria, where Fibonacci himself was educated from the age of 12 by an Arabian mathematician. No doubt because of this he learned of the Arabian (originally Hindu) system of numerals, and through his *Book of the Abacus* (1202) he demonstrated how they could be used to simplify calculations. This resulted in the widespread adoption of the system in Europe.

Fibonacci was primarily interested in the determination of roots of equations, his main work, in 1225, dealt with second order Diophantine equations (indeterminate equations with rational coefficients for which a rational solution is required) and was unsurpassed for 400 years. He also discovered the Fibonacci sequence (1, 1, 2, 3, 5, 8 ...), a series in which each successive number is the sum of the preceding two. This series has many curious properties, its appearance in leaf growth patterns being one biological example.

Fischer, Emil (Hermann) (1852–1919) German organic chemist: the unsurpassed master of natural product chemistry.

Fischer was born near Cologne, to a grocer who acquired a wool-spinning mill and a brewery, hoping his son would follow him in business or, failing that, become a chemist. The father, a Rhinelander, passed to his son his cheery temperament and an appreciation of wine. Fischer did so well at school that he passed the leaving examination too young to enter the university, and so he joined his uncle's timber trade. To the sorrow of his relatives, he then set up a private laboratory to work in during the day and spent the evenings playing the piano and drinking. In the family judgment 'he was too stupid for a business-man and therefore he must become a student'. This he did and read physics and botany, and a little chemistry under KEKULÉ, in 1871 in Bonn. In the following year he moved to Strasbourg to study under BAEYER, and in 1875 he went to Munich following Baeyer. He had already discovered phenylhydrazine, which was to become so useful to him 10 years later. It also gave him chronic eczema.

He had become a single-minded and successful organic researcher. However, since he 'could not give up smoking and did more wine-drinking than was good for him' he had to recuperate every year. Nevertheless, his researches went well – on purines, sugars, dyes and indoles. But the dreadful odour of skatole so adhered to him and his students that they encountered difficulties in hotels when travelling. In 1885 he moved to Würzburg as professor, and in 1892 succeeded HOFMANN as professor in Berlin. The chair carried much work in administration and he complained bitterly about the loss of time and energy during his 12 years in Berlin.

His work on natural products was superb. As well as bringing order to carbohydrate chemistry, partly by use of phenylhydrazine, and synthesizing a range of sugars, including glucose, his studies on glycosides, tannins and depsides are outstanding: especially those on the peptides and proteins, begun in 1899. These compounds are fundamental to biochemistry. It was he who clearly grasped their essential nature as linear polypeptides derived from amino acids; he laid down general principles for their synthesis and made an octadecapeptide (having 15 glycine and 3(-)-leucine residues, relative molecular mass 1213) in 1907. He had much personal charm and wrote with great clarity and brevity. For his contributions to natural product chemistry, he received the second Nobel Prize awarded in chemistry, in 1902.

Fischer, Hans (1881–1945) German organic chemist: synthesized porphyrins.

Fischer graduated in Marburg in chemistry and in Munich in medicine, where he was professor of organic chemistry from 1921. In the last stage of the Second World War his laboratory was destroyed by bombing and Fischer killed himself. His work from 1921 was almost entirely on the porphyrin group of compounds, which contain four pyrrole rings linked together. His first major success with these difficult but important compounds was with haemin, the red non-protein part of haemoglobin, which carries oxygen from the lungs to the tissues. In it the four pyrrole rings are linked together to form a larger (macrocyclic) ring, with an iron atom at the centre. Fischer found the detailed structure, and synthesized it in 1929. He went on to study the chlorophylls, the green pigments of plants which are key compounds in photosynthesis. He showed that they are porphyrins related in structure to haemin, with magnesium in place of iron. For this porphyrin work he was awarded the Nobel Prize for 1930. He went on to find the structure of bilirubin (the pigment of bile, related to haemin) and synthesized it in 1944.

Fisher, Sir Ronald Aylmer (1890–1962) British statistician and geneticist: pioneer of modern statistical methods.

Fisher's father was a successful auctioneer and the boy was one of eight children; he also had eight children, 'a personal expression of his genetic and evolutionary convictions', although he and his wife later separated. He was small, forceful, eloquent and eccentric.

Fisher graduated in Cambridge in 1912 in mathematics and physics, and spent from 1913–19 in a variety of jobs (his poor eyesight excluded him from service in the First World War). Then he was appointed as the only statistician at the Rothamsted Experimental Station, with 66 years of data on agricultural field trials to examine. He was there for 14 years before moving to London, and to a Cambridge chair of genetics in 1943. When he was 69 he joined the Commonwealth Scientific and Industrial Research Organisation (CSIRO) staff in Australia. Before he went to Rothamsted he worked on the statistics of human inheritance, showing that MENDEL's laws must lead to the correlations observed. He went on to show that Mendel's work on genetics and DARWIN's on natural selection are in good accord, rather than in conflict as some had believed. His work on the Rothamsted field data led to major advances on the design of experiments and on the best use of small samples of data. He unravelled the genetics of the Rhesus blood factor. A smoker himself, he argued to the end that smoking should not be causally related to disease. In nearly all other matters his views and methods have been adopted and extended, and

used in all the many areas where statistical analysis is possible.

FitzGerald, George Francis (1851–1901) Irish physicist: suggested the Lorentz–FitzGerald contraction to explain the failure of the Michelson–Morley experiment to detect the 'ether'.

FitzGerald was educated at Trinity College, Dublin, remaining there as a professor for the rest of his life. Although he did not publish much original work himself, he was influential in 19th-c physics through his informal suggestions and discussions with others, tending to pass on ideas to experimenters rather than publish them himself. He is chiefly remembered for his hypothesis that, in order to explain the failure of the MICHELSON–MORLEY experiment to detect the 'ether', bodies moving through an electromagnetic field contracted slightly in their direction of motion, in proportion to their velocity. This accounted for light appearing to move at the same speed in all directions. The same idea was also developed by LORENTZ, becoming known as the Lorentz–FitzGerald contraction, and was an important stepping stone towards EINSTEIN's theory of relativity. FitzGerald also proposed that the tails of comets are made up of small rock particles, and that solar radiation pressure is responsible for the fact that their tails always point away from the Sun.

Fizeau, Armand Hippolyte Louis [feezoh] (1819–96) French physicist: determined the velocity of light experimentally.

Born into a wealthy family, Fizeau turned from studying medicine to research in optics. With FOUCAULT he improved the early photographic process introduced by DAGUERRE and obtained the first detailed pictures of the Sun (1845).

Fizeau made the first accurate determination of the speed of light (1849). In an experiment using an 8 km light path between the hilltops of Suresnes and Montmartre, he sent a light-beam through a rotating toothed wheel and reflected it from the far hilltop so that it returned through the wheel. At a certain speed of the wheel the return signal was blocked by a tooth of the wheel, enabling the speed of light to be calculated.

The following year, Fizeau, working with L F C Brequet (1804–83), showed that light travels more slowly in water than in air: Foucault simultaneously made the same discovery. This was strong evidence against NEWTON's particle theory of light and for HUYGENS's and FRESNEL's wave theory. In 1851 Fizeau measured the effect on light when it is passed through a moving medium (water); the result was in excellent agreement with Fresnel's prediction. Fizeau was the first to apply DOPPLER's results on moving wave sources and observers to light as well as sound. Other experiments by Fizeau included using the wavelength of light for measurement of length by interferometry and using interference to find the apparent diameter of stars.

Many honours were given to Fizeau for his work, including membership of the French Académie des Sciences (1860) and foreign membership of the Royal Society of London (1875).

Flamsteed, John (1646–1719) English astronomer: constructed first comprehensive telescopic star catalogue.

Flamsteed's poor health, which was to hinder his work, led to frequent absence from school and he was largely self-educated until he entered Cambridge in 1670. Some youthful publications impressed Lord Brouncker (1620–84; first president of the Royal Society) with his knowledge of navigational astronomy.

Charles II appointed Flamsteed as the first Astronomer Royal in 1675, charging him with the construction of accurate lunar and stellar tables, needed to enable seafarers to determine longitude at sea, a major problem in the 17th-c. The Royal Greenwich Observatory was created for the purpose and the task was to occupy Flamsteed for the rest of his life. His desire not to publish anything until his work was complete led to bitter disputes with other scientists; he irritated NEWTON in particular (then president of the Royal Society) and this led in 1712 to the virtual seizure of his papers by the Royal Society. Flamsteed did eventually finish the work, which contained the positions of nearly 3000 stars to an accuracy of 10" of arc, but it was not published until 6 years after his death.

Fleming, Sir Alexander (1881–1955) British bacteriologist: discoverer of penicillin.

An Ayrshire farmer's son, Fleming spent 4 years as a clerk in a London shipping office before a small legacy allowed him to study medicine at St Mary's. His later career was spent there, except for service in the Royal Army Medical Corps in the First World War. In that war he saw many fatal cases of wound infection and the experience motivated his later interest in a non-toxic antibacterial; his first result in the search was lysozyme, an enzyme present in nasal mucus, tears and saliva. It pointed to the possibility of success, but could not be got in concentrated form and was inactive against some common pathogens.

Then in 1928 he made the observations which were eventually to make him famous and which are often claimed as the major discovery in medical science in this century. Fleming had left a culture dish of staphylococci uncovered, and by accident it became contaminated with an airborne mould. He noticed that the bacteria were killed in areas surrounding the mould, which he identified as *Penicillium notatum*. He cultured the mould in broth and confirmed that a chemical from it (which he named penicillin) was bactericidal and did not injure white blood cells (a pointer to its lack of toxicity). He saw it as a possibly useful local antiseptic. However, the chemical methods of the time were inadequate to allow concentrated penicillin to be obtained; it is easily destroyed and present only in traces in a culture broth. That success came through work in the Second World War by a team led by FLOREY, and Fleming had no real part in the later work and was cool in his attitude to it.

The widespread use of penicillin from the 1940s onwards made a vast change in the treatment of many infections; it also led to a successful search for other antibiotics, and it made Fleming a near-legendary figure, partly because of a wartime need for national heroes. The legend portrayed him as lucky and diffident; both were exaggerations, since his discovery was a part of his systematic work and good observation and he fully enjoyed his retrospective fame.

Fleming, Sir John Ambrose (1849–1945) British physicist and electrical engineer: inventor of the thermionic diode.

Fleming had a mixed education, studying at times at University College and at the Royal College of Chemistry, London, and at Cambridge under MAXWELL. After intermittent periods of teaching and study he was appointed professor of electrical technology at University College, London.

Although he worked on a number of electrical engineering problems, including radio telegraphy and telephony, Fleming's outstanding contribution was the invention of the thermionic valve, in 1900. This was based on an effect noticed by EDISON (to whose company Fleming had been a consultant) and consisted of a vacuum tube containing a cathode heated to incandescence by an electric current, and an anode. When the anode was maintained at a positive potential with respect to the cathode, an electric current could flow from cathode to anode but not in the opposite direction. This electric 'valve', or diode, was to form an essential component in electronic devices such as radios, television sets and computers for half a century, until eventually superseded by the

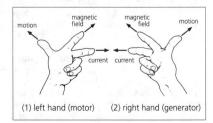

(1) left hand (motor)   (2) right hand (generator)

Fleming's rules – Left hand (1) A current-carrying wire in a magnetic field undergoes motion in the direction indicated. Right hand (2) A moving wire in a magnetic field produces a current flowing in the direction indicated.

cheaper and more robust transistor. The right-hand rule, a useful mnemonic, is due to Fleming.

Flemming, Walther (1843–1905) German cytologist: pioneer of cytology and discoverer of mitosis.

Flemming studied medicine in five German universities and later became professor of anatomy at Kiel. Using the new aniline dyes as microscopic stains, and the improved microscopes of the 1860s, he found that scattered fragments in an animal cell nucleus became strongly coloured; he named this substance chromatin. In cell division, the chromatin granules coalesce to form larger threads (named chromosomes in 1888 by WALDEYER-HARTZ). He went on to show that simple nuclear division as described by REMAK is not the rule, and the more common type of cell division he named 'mitosis'. In this, the chromosomes split lengthwise and the identical halves move to opposite sides of the cell, entangled in the fine threads of the starlike aster (in animal cells only). The cell then divides, giving two daughter cells with as much chromatin as the original. Flemming gave a fine account of the process in 1882 and it has been intensively studied ever since. He did not know of MENDEL's work and so he could not relate his work to genetic studies; that realization did not come for 20 years.

Fletcher, Alice (Cunningham) (1838–1923) US ethnologist and pioneer in the field of American Indian music.

Alice Fletcher was born in Cuba where her father, Thomas, a lawyer from New York City, was staying in an attempt to recover from tuberculosis; he died the following year. She was educated at private girls' schools in New York City and toured Europe before beginning work as a governess. She was attracted to various reform movements and in 1873 helped to found the Association for the Advancement of Women. In the course of this work she became a successful and skilful popular lecturer. When researching for a lecture on 'Ancient America' at the Peabody Museum she was encouraged by the curator F W Putnam to work in anthropology and ethnology. She studied with Putnam at the museum. He impressed on her the need to base anthropology on empirical data and interested her in the early inhabitants of the American continents.

In 1879 she met the Omaha Indian Susette La Flesche who, with her husband, was touring the country speaking for Indian rights. Alice Fletcher's reforming interests were now channelled into the granting of lands to Indians. Francis La Flesche, Susette's half-brother, became her interpreter and assistant when she went to live among the Omaha Indians, primarily to study the life of Indian women. Perhaps at his suggestion, and with his help, she began a study of Indian ceremonies and became a pioneer in the study of American Indian music. In the process she made a complete record of the ritual and music of a Plains Indian religious ceremony, the Hako ceremony of the Pawnee Indians, and was the first white observer permitted to do so.

Florey, Howard Walter, Baron Florey of Adelaide (1898–1968) Australian pathologist: central figure in introduction of penicillin as useful antibiotic.

Born and educated in Adelaide, Florey studied physiology in Oxford and pathology in Cambridge; in 1931 he became professor of pathology in Sheffield and in 1934 at Oxford. In the early 1930s he began to study lysozyme, an antibacterial enzyme present in mucus discovered by FLEMING in 1922. In 1935 CHAIN joined the department and in 1936 HEATLEY; both were chemists and so the group had the skills to begin a study of the antibacterials formed in small amounts by certain moulds.

Fleming in 1928 had discovered a good candidate, penicillin, which he had been unable to isolate because of its instability, and this came early on the group's programme. By 1941 they had isolated enough penicillin to try on nine human patients and the results were good. The war was at a critical stage and large-scale production was begun in the USA out of range of enemy bombers, as a result of Florey's discussions with US drug companies. By 1944 enough was available to treat casualties in the Normandy battles, as well as severe civilian infections, with impressive results. Fleming, Florey and Chain shared a Nobel Prize in 1945 for their work on penicillin.

Florey went to work on other antibiotics, but he was always involved in other areas of experimental pathology, especially on the lymphatic and vascular systems. Although he lived in the UK from 1922 and in Oxford from 1935 he retained his Australian outlook and accent, and he did much to found the Australian National University at Canberra.

Flourens, (Marie Jean) Pierre [flooräs] (1794–1867) French anatomist and physiologist: early experimental physiologist.

Flourens qualified in medicine in 1813 at Montpellier and then went to Paris, where CUVIER befriended him, secured teaching and research posts for him and ensured that after his own death his appointments would pass to Flourens.

From 1820 Flourens began to work on the central nervous system, using pigeons and later dogs, whose sacrifice yielded fundamental information. He found that vision depends on the integrity of the celebral cortex, and that removal of part of it produces blindness on the opposite side. Removal of the cerebellum causes loss of coordination of movement; he also found that loss of the semi-circular canals of the ear causes loss of balance while respiration is controlled by a centre in the medulla oblongata. He did not attempt to remove or to stimulate small centres of the cerebellum and it was HITZIG in 1870 who established cerebral localization. Flourens attacked the pseudo-science of phrenology and largely demolished it. In 1847 he showed that trichloromethane (CHCl$_3$) is an effective anaesthetic for small animals, and later in the year the British obstetrician J Y Simpson (1811–70) first used it for human patients in childbirth. In his old age Flourens mounted a forceful attack on DARWIN's theory of evolution, describing his ideas as 'childish and out of date'.

Fock, Vladimir Alexandrovich (1898–1974) Russian theoretical physicist: advanced the quantum mechanics of many-electron systems.

Fock received his training at St Petersburg (1922) and then moved to the Institute of Physics. He was appointed professor there in 1961.

Simultaneously with D R Hartree (1897–1958), Fock first developed a means of solving quantum mechanical problems for atoms in which more than a single electron is present (ie atoms other than hydrogen). The possible energy levels of an electron in a hydrogen atom had been solved by SCHRÖDINGER in 1926, and the Hartree–Fock approximation for other atoms appeared in 1932.

Other important research by Fock includes work on general relativity.

Forbes, Edward (1815–54) British naturalist: showed that marine life existed at great depths.

Forbes studied medicine at Edinburgh but soon became more interested in natural history. He became curator and later palaeontologist to the Geological Society and subsequently professor of natural history at Edinburgh and at the Royal School of Mines.

Travelling widely in Europe and in the region bordering the Eastern Mediterranean (he joined a naval expedition as naturalist in 1841), Forbes collected much fauna and flora, particularly molluscs (which he classified systematically), studying their migration habits and the inter-relationships between animals. He divided British plants into five groups and proposed that Britain had once been joined by land to the continent, whence the plants had migrated in three distinct periods. He also discounted the contemporary belief that marine life existed only near the sea surface by dredging a starfish from a depth of 400 m in the Mediterranean. His *Natural History of European Seas* (1859) was the first general study of oceanography.

Forest, Lee De see **De Forest**

Forssman, Werner (Theodor Otto) (1904–79) German physician: introduced cardiac catheterization.

Born in Berlin, Forssman attended school and studied medicine there before going in 1929 to a hospital at nearby Eberswalde to continue clinical studies in surgery. He became interested in the problem of delivering a drug rapidly and safely to the heart in an emergency; at that time, the best but unsafe technique was to inject directly through the chest wall. Forssman tried passing a catheter (a narrow flexible tube) into a vein near the elbow, using corpses in the hospital mortuary for his first trials. Then he tried the method on himself, passing the tube into his own antecubital vein for about 65 cm before walking to the X-ray department, where a radiographer held a mirror to the X-ray screen while Forssman fed the catheter into the right atrium of his heart. The method had clear potential, but it was severely criticized in Germany as dangerous and Forssman had little support. It was used to a limited extent in Lisbon and in Prague, but was undervalued and full use began only in New York after 1940. Catheters with pressure gauges, or a device to collect samples of blood gases, enabled further study of heart action in health and disease. Passing a radio-opaque dye into the catheter while taking X-ray pictures on cine film allows the blood vessels and chambers of the heart to be studied (angiography). This method of contrast radiography is a routine procedure for examination of the heart valves before and after surgical repair, and of the coronary arteries as a preface to bypass surgery in cases of obstruction of these arteries.

Forssman served as an army surgeon in the Second World War, became a prisoner of war and afterwards specialized in urology. He shared the Nobel Prize for physiology or medicine in 1956.

Foucault, (Jean Bernard) Léon [fookoh] (1819–68) French physicist: measured the speed of light; and demonstrated rotation of the Earth.

Foucault was the son of an impoverished bookseller; he began to study medicine but was revolted by the sight of blood. He was attracted to physics by DAGUERRE's discovery of photography, and then became editor of the *Journal de Débats* (1845) and a physicist at the Paris Observatory (1855). A gifted experimentalist with great originality and instinct, he died of paralysis aged 48, having been elected to the Académie des Sciences (1865) and the Royal Society of London (1864).

Foucault collaborated with FIZEAU on the toothed-wheel experiment which first measured the speed of light terrestrially. In 1850 he took over ARAGO's experimental equipment and first measured the speed of light in water, showing that it was slower than in air. This was important evidence in favour of the wave theory of light and contrary to the prediction of NEWTON's particle theory of light. Foucault then constructed a rotating mirror experiment for measuring the speed of light and used it to obtain a more accurate value (1862).

The simplicity and imaginativeness of Foucault's measurement in 1850 of the rotation of the Earth by a swinging pendulum (Foucault's pendulum) make it an outstanding achievement. The effect occurs because the Earth rotates, leaving the plane of the swinging pendulum fixed with respect to the stars. In 1852 Foucault demonstrated this with a 67 m pendulum with a 28 kg ball hung from the dome of the Panthéon in Paris. The experiment was carried out there with the help of Napoleon III, before an admiring crowd who watched a needle attached to the ball inscribe a mark in sand; the mark moved as the Earth rotated about the plane of the pendulum's swing. This was the first direct (ie non-deductive) demonstration of the Earth's rotation.

Other valuable work by Foucault included invention of the gyroscope in 1852, and improvements to reflecting telescopes. He discovered the yellow (sodium D) lines in emission spectra corresponding to the dark lines seen in absorption spectra by FRAUNHOFER; the value of this, however, was only realized later by KIRCHHOFF.

Fourier, (Jean Baptiste) Joseph, Baron [fooryay] (1768–1830) French mathematician: discovered Fourier series and the Fourier Integral Theorem.

Fourier, the son of a tailor, was orphaned at age 8. He had a mixed education, at military school, an abbey and later (after a narrow escape from the guillotine during the French Revolution in 1794) at the École Normale. He joined the staff of the École Normale, newly-formed to train senior teachers, and the École Polytechnique in Paris. When Napoleon invaded Egypt in 1798 Fourier accompanied him, but it is unlikely that he became governor of Lower Egypt as is often stated. He became the prefect of the *département* of Grenoble for 14 years but resigned during Napoleon's Hundred Days campaign. Fourier died in 1830 of a disease contracted while in Egypt.

Fourier established linear partial differential equations as a powerful tool in mathematical physics, particularly in boundary value problems. For example, to find the conduction of heat through a body of a given shape when its boundaries are at particular temperatures, the heat diffusion equation can be solved as a sum of simpler trigonometric components (Fourier series). This way of solving the linear differential equations that often occur in physics has led to much use of the method on many new problems to the present day. Importantly, any arbitrary repeating function may be represented by a Fourier series; for instance, a complex musical waveform can always be represented as the sum of many individual frequencies. At a different level, the understanding of Fourier series and integrals has contributed greatly to the development of pure analysis, particularly of functional analysis. On the problem of heat conduction through a uniform solid, Fourier's Law states that the heat flux is given by the product of the thermal conductivity and the temperature gradient.

Fracastoro, Girolamo [frakastawroh] (*c.*1478–1553) Italian logician and physician: proposed early theory of germ origin of disease.

Having studied a variety of subjects at Padua, Fracastoro became lecturer in logic there in 1501. After moves due to war and plague, he settled in Verona from 1516, practising as a physician until 1534. Thereafter he spent his retirement in research. His major medical book, *On Contagion and Contagious Diseases* (1546), gave the first logical explanation of the long-known facts that some diseases can be passed from person to person or passed by infected articles. Fracastoro had in 1530 described and named syphilis (previously 'the French disease'), which was epidemic in Europe from about 1500. He proposed that infection is due to minute self-multiplying bodies which can infect by direct contact or indirectly through infected articles, or which can be passed at a distance. His ideas were not widely adopted, many preferring the notion of miasmata, exhalations from earth or air which caused disease. Only much later did PASTEUR and others show the essential correctness of Fracastoro's proposals.

Franck, James (1882–1964) German–US physicist: gave first experimental demonstration of the quantized nature of molecular electronic transitions.

Franck studied law at Heidelberg but left after a year in order to study physics at Berlin instead. After service in the First World War (he was awarded the Iron Cross), he became professor of experimental physics at Göttingen. In 1933, being a Jew, he felt obliged to leave Germany, settling eventually in the USA, where he became professor of physical chemistry at the University of Chicago. During the Second World War he worked on the American atomic bomb project, proposing in the Franck Report that the bomb be demonstrated to the Japanese on uninhabited territory before being used on a city.

Franck's major work concerned the quantized nature of energy absorption by molecules. In 1914, together with Gustav Hertz (1887–1975), he demonstrated that gaseous mercury atoms, when bombarded with electrons, absorb energy in discrete units (or quanta). For mercury atoms this unit of energy is 4.9 eV. Following the absorption, which leaves the mercury atom in an energetically excited state, the atom returns to its original (or ground) state by emitting a photon of light. This experiment constituted the first experimental proof of BOHR'S ideas on energy levels in atoms, and Franck and Hertz were awarded the 1925 Nobel Prize for physics for their work.

Later, in conjunction with E Condon (1902–74), Franck also studied the energy requirements for vibration and rotation of diatomic molecules, showing that these were also quantized and that dissociation energies (the energy required to break the chemical bond between the two atoms), could be extrapolated from them. The Franck–Condon principle states that the most probable electronic transitions are those in which the vibrational quantum number is preserved, since electronic transitions take place on a much shorter time-scale than vibrational ones.

Frankland, Sir Edward (1825–99) British organic chemist: originator of the theory of valence.

Frankland was apprenticed to a druggist in Lancaster for 6 years in an ill-advised attempt to enter the medical profession. Guidance from a local doctor led him to study chemistry under Lyon Playfair (1819–98) at the Royal School of Mines in London (1845). He later studied with BUNSEN at Marburg and LIEBIG at Giessen. At 28 Frankland became the founding professor of chemistry at the new Owens College (1851–57) that became the University of Manchester. He moved to St Bartholomew's Hospital, London (1857); the Royal Institution (1863) and the Royal School of Mines (1865).

He prepared and examined the first recognized organometallics, the zinc dialkyls. In his view, their reaction with water gave the free alkyl radicals (in fact, the corresponding dimeric alkanes). Vastly more significant was his later recognition, after thinking about a range of compounds, of numerical, integral limitations in atomic combining power: the theory of valence – this being the number of chemical bonds that a given atom or group can make with other atoms or groups in forming a compound. He used the word 'bond' and modern graphic formulae (Frankland's notation) and through these ideas did much to prepare foundations for modern structural chemistry. He also made major contributions to applied chemistry, notably in the areas of water and sewage purification – paramount requirements for good public health.

Franklin, Benjamin (1706–90) US statesman: classic experimenter and theorist on static electricity.

Franklin had an unusually wide range of careers: he was successful as a printer, publisher, journalist, politician, diplomat and physicist. Trained as a printer and working in New England and for nearly 2 years in London (UK), Franklin found he also had talents as a journalist; when he was 27 he published *Poor Richard's Almanac*. In this, he 'filled all the little spaces that occurred between the remarkable days in the calendar with proverbial sentences' which he concocted. Most are trite platitudes of the 'honesty is the best policy' kind, but some show the sly irony of his journalism (of which the best-known sample is probably his advice to

young men to take older mistresses, 'because they are so *grateful*'). The almanac made him both famous and prosperous and, by franchises in printing shops and other businesses in which he provided a third of the capital and took a third of the profit, he made himself wealthy.

When he was nearly 40 he became interested in electricity, which at that time provided only amusing tricks, but at least the dry air of Philadelphia made these more reproducible than in damper climates. Franklin's experiments and ideas turned electrical tricks into a science, made him the best-known scientist of his day and, perhaps for the first time, showed that what we would now call pure research could have important practical uses.

Franklin proposed that electrical effects resulted from the transfer or movement of an electrical 'fluid' made of particles of electricity (we would now call them electrons) that can permeate materials (even metals) and which repel each other but are attracted by the particles of ordinary matter. A charged body on this theory is one that has either lost or gained electrical fluid, and is in a state he called positive or negative (or plus or minus). Linked with this 'one fluid' idea was the principle or law of conservation of charge: the charge lost by one body must be gained by others so that plus and minus charges appear, or neutralize one another, in equal amounts and simultaneously. Franklin's logic had its defects, but it was a major advance; and he continued to experiment, and worked on insulation and grounding. He examined the glow that surrounds electrified bodies in the dark, and it may have been this which caused a friend to show that a grounded metallic point could quietly 'draw off' the charge from a nearby charged object. This led Franklin to his idea that it should be possible to prove whether clouds are electrified (as others had suggested) and also to propose that 'would not these pointed rods probably draw the electrical fire silently out of a cloud before it came nigh enough to strike, and thereby secure us from that most sudden and terrible mischief?' He planned in 1750 an experiment using a metal rod passing into a sentry box mounted on the steeple of a new church; but delay in building it led him to try using a kite instead, and he found that the wet string did indeed conduct electricity from the thundercloud and charge a large capacitor. The electrical nature of such storms was proved, lightning conductors became widely used, Franklin became famous and others trying such experiments were killed.

He was in London for most of the years 1757–75 representing the American colonies and trying to prevent the rising conflict. When, despite his efforts, war began, he was active in support of the revolution and was one of the five men who drafted the Declaration of Independence in 1776. In the same year he went as ambassador to France and, largely through his fame as a scientist and his popularity, secured an alliance in 1778. Afterwards he continued to be active in American politics; despite being an Anglophile, a womanizer and a tippler, he was increasingly seen as the virtuous, homespun, true American sage. His work in physics amused him but was always directed to practical use; he devised the Franklin stove with an efficient underfloor air-supply for heating, invented and used bifocal spectacles, used a flume for testing ships by using models, and his work on the Gulf Stream was a pioneering study in oceanography. By having ship's captains record its temperature at various depths and its velocity, he mapped the Gulf Stream and studied its effects on weather.

In the 1780s he was present at the first ascents by hydrogen balloons made by CHARLES at Versailles and enthused over the possibility of studying the atmosphere and of aerial travel; he foresaw aerial warfare. His book, *Experiments and Observations on Electricity made at Philadelphia in America* (1751) not only founded a new science, but had the incidental result of interesting PRIESTLEY in science, with momentous results for chemistry.

Franklin, Rosalind (Elsie) (1920–58) British physical chemist and X-ray crystallographer.

The daughter of prosperous Jewish parents, Franklin studied chemistry and physics at Cambridge and afterwards, as war work, took a job in coal research, studying its pore size and properties as a solid colloid and being awarded a PhD for this in 1945. Then, during a happy period of 4 years in Paris at the Sorbonne, she moved from physical chemistry to X-ray diffraction, using this method to study colloidal carbon.

Armed with this new skill, she moved in 1951 to join the biophysics group at King's College, London, to extend the X-ray diffraction work there. Soon she began to use this method on DNA, which was known to have a central place in the mechanism of heredity. In the department M H F Wilkins (1916–) was already working with DNA, using X-ray and other methods; unhappily, relations between the two were never good. In part, this was because their differing responsibilities in the DNA work at King's were unclear, a problem for which Sir John Randall, director of the unit, must bear the blame. DNA is a difficult substance to work on; a sticky, colloidal nucleic acid, its precise properties depend upon its origin and history. However, Franklin used her past experience with awkward materials to design an X-ray camera suitable for low-angle reflections, and she used specimens of DNA which were drawn into thin fibres under carefully controlled conditions, notably of hydration. In this way she defined two forms of DNA and obtained X-ray diffraction photographs that showed the DNA to have an ordered, crystalline structure. They

were the best X-ray diagrams then obtained for DNA, and she deduced from them that the long, chain-like DNA molecules might be arranged in a helical form, with the phosphate groups on the outside.

Knowledge of her results passed by several routes (some of which have been considered of questionable propriety) to WATSON and CRICK in Cambridge, who were working on the structure of DNA also, mainly by use of model-building methods rather than the experimentation and formalized calculation technique used by Franklin. Her interpretation of her excellent pictures was at times for and at other times against a helical DNA; but after Watson and Crick had formed their inspired view of DNA, her results could be seen to give valuable support to their deductions. Their idea of a double helix structure for DNA also gave pointers to its mode of action in the transfer of genetic information. The famous issue of *Nature* in April 1953 which proposed their double helix was accompanied by her paper, with R Gosling, supporting their views.

Unhappy at King's, she moved in 1953 to Birkbeck College, London, again to work with X-rays on biological macromolecules, this time viruses; initially TMV (tobacco mosaic virus, a much-studied subject) and, from 1954, working with Aaron Klug (1926–), who was to win the Nobel Prize for chemistry in 1982. Again she secured X-ray photographs superior to any obtained previously and used them to show that the TMV virus is not solid, as had been thought, but a hollow tubular structure. Despite being a cancer victim from 1956, she began work on the polio virus.

Her death aged 37 excluded her from consideration for the Nobel Prize for physiology and medicine in 1962, which was shared by Crick, Watson and Wilkins.

Frasch, Herman [frash] (1851–1914) German–US industrial chemist: devised process for extraction of sulphur from underground deposits.

Frasch was the son of a prosperous pharmacist in Württemberg and began his training in pharmacy when he emigrated to the USA in 1868 at age 17. Industry was expanding after the Civil War and he soon interested himself in the new petroleum industry. One of its problems was that some wells yielded a 'sour' oil (nicknamed skunk oil) containing sulphur and organic sulphides. Frasch found a method for removing these by reaction with metal oxides. His interest in sulphur changed direction in 1891, when he began work on the problem of obtaining sulphur from deposits overlain by a limestone caprock and by quicksand. His method (the Frasch process) was to sink a trio of concentric pipes into the deposit; superheated water was pumped down one to melt the sulphur (m.p. 119°C) and air down another, to bring up a froth of molten sulphur through the remaining pipe. The process gave an abundant supply of 99.5% pure sulphur and broke the Sicilian supply monopoly after about 1900.

Fraunhofer, Josef von [frownhohfer] (1787–1826) German physicist and optician: described atomic absorption bands in the solar spectrum.

Fraunhofer was apprenticed as a mirror-maker and lens-polisher in Munich, rising to become a director of his company in 1811. His miserable time as an apprentice was relieved when he was buried under the collapsed workshop and the Elector celebrated his survival with a gift of money which gave him independence. His interest in the theory of optics, and his scientific discoveries, led him eventually to become director of the Physics Museum of the Bavarian Academy of Sciences in 1823, but he died of tuberculosis 3 years later.

Fraunhofer's main interest, and the motivation behind his experiments, was in producing a good quality achromatic lens. During the course of his investigations into the refractive properties of different glasses he used a prism and slit to provide a monochromatic source of light. In doing so, he noticed that the Sun's spectrum was crossed by many dark lines, and he proceeded carefully to measure the wavelengths of almost 600 of them. Later, he used a diffraction grating to prove that the lines were not due to the glass of the prism but were inherent in the Sun's light. These Fraunhofer lines were subsequently shown by KIRCHHOFF to be due to atomic absorption in the Sun's outer atmosphere, and tell us a great deal about its chemistry. Similar lines are observed for other stars and have been equally informative. Fraunhofer's work was also important in establishing the spectroscope as a serious instrument, rather than merely a scientific curiosity.

Fredholm, Erik Ivar (1866–1927) Swedish mathematician: founded the theory of integral equations.

Fredholm studied at Uppsala and subsequently worked as an actuary until gaining his doctorate 10 years later. He then became a lecturer in mathematical physics at Stockholm (1898) and moved from studying partial differential equations to their inverse, integral equations. In the next 5 years he rapidly established the field that was later extended by HILBERT in his work on eigenfunctions and infinite-dimensional spaces. In 1906 Fredholm became a professor at Stockholm.

Fredholm built upon earlier research by G Hill (1838–1914) and V Volterra (1860–1940). He studied two integral equations (denoted of the first and second kind) named after him: and later found a complete algebraic analogue to his theory of integral equations in linear matrix equations.

Fresnel, Augustin Jean [fraynel] (1788–1827) French physicist: established and developed the wave theory of light.

When the French Revolution arrived Fresnel's father took his family to a small estate near Caen. After showing his practical skills Fresnel entered the École Polytechnique and, despite ill-health, gained distinction. He qualified as an engineer at the École des Ponts et Chaussées, but was removed from his post during 1815 for supporting the Royalists. He spent the Hundred Days of Napoleon's return at leisure in Normandy, and began his main work on the wave theory of light.

He performed some new experiments on interference and polarization effects. HUYGENS and YOUNG had suggested that light consisted of longitudinal waves; in order to explain polarization effects Fresnel replaced this with a theory of light as transverse waves. He steadily established his theory as able to account for light's observed behaviour.

Fresnel also applied his skills to the development of more effective lighthouses. The old optical system for them consisted of metal reflectors and he introduced stepped lenses (Fresnel lenses). This work still forms the basis of modern lighthouse design. Fresnel was made a member of the Académie des Sciences (1823) and of the Royal Society of London (1827).

Friedel, Charles [freedel] (1832–99) French mineralogist and organic chemist: co-discoverer of Friedel–Crafts reaction.

Friedel was born in Strasbourg and as a student there was taught by PASTEUR. Later he studied under WURTZ in Paris, before becoming curator at the École des Mines and professor of mineralogy there in 1876. His early work was on the synthesis of minerals, but then he moved to organic chemistry and succeeded Wurtz at the Sorbonne in 1884.

His famous work was done in 1877 with an American, J M Crafts (1839–1917). The Friedel–Crafts reaction is of great value in organic synthesis. An aromatic hydrocarbon is heated with an alkyl halide and a Lewis acid (typically $AlCl_3$), and alkylation of the hydrocarbon is the result, eg

$$C_6H_6 + C_2H_5Cl \rightarrow \quad C_2H_5C_6H_5 \quad + HCl$$
$$\text{(ethylbenzene)}$$

In modified forms, such reactions are used in the petrochemical industry. Also useful is the acylation reaction, in which the hydrocarbon reacts with an acyl chloride and $AlCl_3$, eg

$$C_6H_6 + CH_3COCl \rightarrow \quad CH_3COC_6H_5 \quad + HCl$$
$$\text{(methylphenyl ketone)}$$

Friedman, Herbert [freedman] (1916–) US astronomer: pioneered X-ray astronomy.

From the 1940s onwards Friedman and his colleagues used rockets to launch X-ray devices above the Earth's absorbing atmosphere to observe the Sun, studying solar X-ray and ultraviolet activity through a complete 11-year solar cycle, and producing the first X-ray and ultraviolet photographs of the Sun in 1960. In 1962 the first non-solar X-ray source was discovered by a team led by Rossi, and 2 years later Friedman made the first attempt to identify accurately such a source with an optical object, making use of the occultation of Tau X-1 by the Moon. He was able to show that this X-ray source coincided with the Crab nebula, a supernova remnant.

Friedmann, Aleksandr Alexandrovich [freedman] (1888–1925) Russian cosmologist: developed a mathematical model of the expanding universe.

Friedmann applied EINSTEIN's field equations to cosmology and showed that general relativity did permit solutions in which the universe is expanding. He did this work in 1917, during the Siege of Petrograd. At the time Einstein preferred static solutions; he and DE SITTER had to introduce arbitrary terms into their solutions. The Friedmann solution gave model universes that are more physically reasonable, and laid the foundations after his death for the 'Big Bang' theory of modern cosmology. Friedmann also made significant contributions to fluid mechanics.

Frisch, Karl von (1886–1982) Austrian ethologist: observer of the dancing bees.

Frisch studied zoology at Munich and Trieste and later taught zoology at four universities, spending the longest period in Munich. For over 40 years he made a close study of the behaviour of the honey bee. His results showed that bees can use polarized light to navigate back to the hive; that they cannot distinguish between certain shapes; and that they can see some colours, including ultraviolet (invisible to man) but not red. He also concluded that a forager bee can inform other workers of the direction and distance of a food source by means of a coded dance, a circular or figure-of-eight movement performed at the hive. He believed that information on food was also communicated by scent.

In the 1960s, A M Wenner claimed that sound as well as scent is used; bees emit a range of sounds audible to other bees. Later still, the problem was further examined by J L Gould, who used ingenious ways to test if false information could be transmitted by bees. By the 1980s it seemed that dance, scent and sound are all used in bee communication, at least by for-

agers reporting on a food source. Frisch is regarded as a key figure in developing ethology, by combining field observation with experiment. He shared a Nobel Prize in 1973 with LORENZ and TINBERGEN.

Frisch, Otto Robert (1904–79) Austrian–British physicist: early investigator of nuclear fission of uranium.

Frisch studied physics in his home city, Vienna, and then took a job at the national physical laboratory in Berlin. In 1930 he went to Hamburg to work with STERN, but he was sacked in 1933 as a result of the Nazi anti-Jewish laws and worked for a year in London and then in Copenhagen (with BOHR) until the Second World War. At Christmas 1938 he visited his aunt, LISE MEITNER, then a refugee in Stockholm and a former co-worker with HAHN in Berlin. She had a letter from Hahn reporting that uranium nuclei bombarded with neutrons gave barium. Frisch and Meitner walked in the snow and talked about it 'and gradually the idea took shape that this was no chipping or cracking of the nucleus but rather a process to be explained by Bohr's idea that the nucleus was like a liquid drop; such a drop might elongate and divide itself'. This division would give lighter elements such as barium, and more neutrons, so a chain reaction should occur. Frisch worked out that this should be easiest for heavy nuclei such as uranium; and Meitner calculated that it would release much energy (about 200 MeV). Back in Copenhagen, Frisch confirmed the energy of the fragments in experiments in an ionization chamber. He named the effect 'nuclear fission'. Working in Birmingham from 1939, he and PEIERLS confirmed Bohr's view that a chain reaction should occur more readily with the rare isotope uranium-235, rather than the common uranium-238. They also calculated that the chain reaction would proceed with huge explosive force even with a few kilograms of uranium. If an atomic bomb based on this was made in Germany it would clearly decide the war, and they wrote to the British scientific adviser on this; their letter probably spurred government to action, and soon Frisch was working at Los Alamos on the A-bomb project, which reached success in 1945.

Gabor, Dennis (1900–79) Hungarian–British physicist: invented holography.

The son of a businessman, Gabor studied electrical engineering in Budapest and Berlin. He worked as a research engineer with the firm of Siemens and Halske but in 1933 he had to flee from the Nazis and spent the rest of his life in Britain. He was initially with the British Thomson–Houston Co at Rugby and from 1948 at Imperial College, London.

In 1947–8 Gabor conceived the idea of using the phase (or position in the wave's cycle) as well as the intensity of received waves to build up a fuller image of the object in an electron microscope. In this way he hoped to extract better electron images so that atoms in a solid might be resolved, but soon he developed the method for use in an optical microscope also. The phase of the electron or light waves was obtained by mixing them with coherent waves directly from the wave source, and the waves then form a standing wave that is larger or smaller according to whether the two are in phase or out of phase. This interference pattern is recorded on a photographic plate. The waves from different parts of the object travel a varying number of wavelengths to the plate, and so the interference pattern and phase of the waves give information on the three-dimensional shape of the object.

When the plate (hologram, from the Greek *holos*, whole) is placed in a beam of coherent waves, a three-dimensional image of the object is seen and as the observer moves a different perspective appears. (A coherent wave is one in which the wave train consists of waves exactly in phase, not several waves with different phases and intensities.) For light Gabor achieved this crudely, using a pinhole in a screen in front of a mercury lamp. In 1960 lasers were invented and these powerful coherent sources allowed high quality holograms to be made by E Leith and J Upatnieks (1961). Gabor was awarded a Nobel Prize in 1971.

Gajdusek, Daniel Carleton (1923–) US virologist: pioneer in study of slow virus infections.

Educated in physics at Rochester and in medicine at Harvard, Gajdusek later worked with PAULING at the California Institute of Technology, and in Iran and Papua New Guinea before returning to the USA. It was in New Guinea in the 1950s that he studied the Fore people, who frequently died from a disease they called *kuru*. He found that it could be passed to other primates (eg chimpanzees) but that it took 12 months or more to develop after infection. Since then other diseases have been shown, or suspected, to be due to slow and persistent virus infections (one example is the herpes 'cold sore', others include scrapie in sheep, 'mad cow disease' (BSE) in cattle and Creutzfeldt–Jakob dementia in humans). Recent evidence shows that some of these are conveyed by peculiar proteins (prions), which seem to be the smallest of all infective agents. Kuru was the first of the group to be observed and identified in humans and was doubtless transmitted among the Fore by their cannibal rituals in which the brains of the dead were eaten by their relatives. Gajdusek shared a Nobel Prize in 1976.

Galen [gaylen] (129–199) Roman physician, anatomist and physiologist: his ideas on human anatomy and physiology were taught for 15 centuries.

Born in Pergamon (now in Western Turkey) Galen began studying medicine early; at 21 he went to Smyrna to study anatomy and later to Asia Minor to study drugs; later still he visited Alexandria where he examined a human skeleton. In his time human dissection was no longer carried out (although Galen may have done some) and his practical anatomy and physiology was based in part on his work on animals, including the Rhesus monkey. His lifetime coincided with a high point in the success of the Roman Empire and the army had its medical service; but science did not flourish in Rome and Galen's interest in medical science was unusual. He had a large practice in Rome and was physician to four successive emperors. His extensive writing was partly based on the ideas of HIPPOCRATES and ARISTOTLE, and he added his own results and theories. His descriptions of the anatomy of the muscular system are excellent and his studies on the physiology of the spinal cord and the effects of injury at various levels were a major advance. In his mind, every organ and all its parts have been formed for a purpose, and he theorizes at length on this. However, most of his physiological theories were erroneous and, like others of his time, he had no knowledge of the circulation of the blood. He was fully aware of the existing medical experience and theory, he was highly industrious and his authority was very long-lived.

Galileo (Galilei) (1564–1642) Italian astronomer and physicist: discovered Jupiter's moons and laws governing falling bodies.

Usually known by his first name, Galileo was born in Pisa; he was the son of a musician and became a medical student, but his interest moved to mathematics and physics. He became the ill-paid professor of mathematics at Pisa when he was 25, moved to Padua in 1591 and later to Florence. Galileo never married but when he was 35 Marina Gamba (a Venetian girl) came to live with him, and they had two daughters and a son. When he moved to Florence in 1610 he left Marina behind and she married soon after.

Galileo's fame rests partly on the discoveries he made with the telescope, an instrument which he did not invent but was certainly the first to exploit successfully. His design used a convex object glass and a concave eyepiece and gave an erect image. In 1610 he observed for the first time mountains on the Moon, four satellites around Jupiter and numerous stars too faint to be seen with the naked eye. These observations he described in his book *Sidereal Messenger* (1610), which made him famous. He also discovered the phases of Venus, the composite structure of Saturn (although he was unable to resolve the rings as such: it looked to him like a triple planet) and sunspots. His discovery of heavenly bodies that were so demonstrably not circling Earth, together with his open public support for the COPERNICAN heliocentric cosmology, was to bring him into conflict with the Church. He wrote his *Dialogue on the Two Chief World Systems, Ptolemaic and Copernican* in 1632. He tried in the book to make his support for the Copernican view diplomatic, and he seems to have believed that the Church authorities would be sympathetic, but he misjudged their resistance to such novel ideas. The next year he was before the Inquisition, and was shown the torture chamber and forced to recant. He was sentenced to house arrest for life at the age of 69.

Among his notable non-astronomical findings were the isochronism (constant time of swing, if swings are small) of a pendulum, which he timed with his pulse when he was a medical student. (He designed a clock with its escapement controlled by a pendulum and his son constructed it after his death). He also found that the speed at which bodies fall is independent of their weight. The latter was the result of experiments rolling balls down inclined planes, and not by dropping weights from the leaning tower of Pisa as was once widely believed. His work on mechanics is in his *Discourses Concerning Two New Sciences* (1638), which completes the claim to regard him as 'the father of mathematical physics'. (The two 'new sciences' which he described are now known as 'strength of materials' and 'dynamics'.) He died in the year in which NEWTON was born. His work sets the modern style; observation, experiment and the full use of mathematics as the preferred way to handle results.

The gal, named after him, is a unit of acceleration, $10^{-2}\,\text{m s}^{-2}$. The milligal is used in geophysics as a measure of change in the regional acceleration due to gravity (g).

Galileo was an able musician, artist and writer – a true man of the Renaissance. His massive contribution to physics makes him one of the small group of the greatest scientists of all time and his startling discoveries, his forceful personality and his conflict with the church help to make him the most romantic figure in science.

Galois, Évariste [galwah] (1811–32) French mathematician: founded modern group theory.

Galois was born with a revolutionary spirit, politically and mathematically, and at an early age discovered he had a talent for original work in mathematics. He was taught by his mother until the age of 12, and at school found interest only in exploring books by creative mathematicians such as A-M Legendre (1752–1833), LAGRANGE and later ABEL. Twice (in 1827 and 1829) Galois took the entrance examinations for the École Polytechnique (which was already one of the foremost colleges for science and mathematics) but failed on each occasion. At 17 he submitted a paper to the French Académie des Sciences via CAUCHY, but this was lost. Following his father's suicide Galois entered the École Normale Supérieure to train as a teacher. During 1830 he wrote three papers breaking new ground in the theory of algebraic equations and submitted them to the Academy; they too were lost.

In the political turmoil following the 1830 revolution and Charles X's abdication Galois chided the staff and students of the École for their lack of backbone and he was expelled. A paper on the general solution of equations (now called Galois theory) was sent via POISSON to the Academy, but was described as 'incomprehensible'. In 1831 he was arrested twice, for a speech against the king and for wearing an illegal uniform and carrying arms, and received 6 months' imprisonment. Released on parole, Galois was soon challenged to a duel by political opponents. He spent the night feverishly sketching out in a letter as many of his mathematical discoveries as he could, occasionally breaking off to scribble in the margin 'I have not time'. At dawn he received a pistol shot through the stomach, and having been left where he fell was found by a passing peasant. Following his death from peritonitis 8 days later he was buried in the common ditch of South Cemetery, aged 20.

The letter and some unpublished papers were discovered by LIOUVILLE 14 years later, and are regarded as having founded (together with Abel's work) modern group theory. It outlines his work on elliptic integrals and sets out a theory of the solutions (roots) of equations by considering the properties of permutations of the roots. If the roots obey the same relations after permutation they form what is now called a Galois group, and this gives information on the solvability of the equations.

Galton, Sir Francis (1822–1911) British geogra-

THE DARWIN/WEDGWOOD/GALTON RELATIONSHIPS

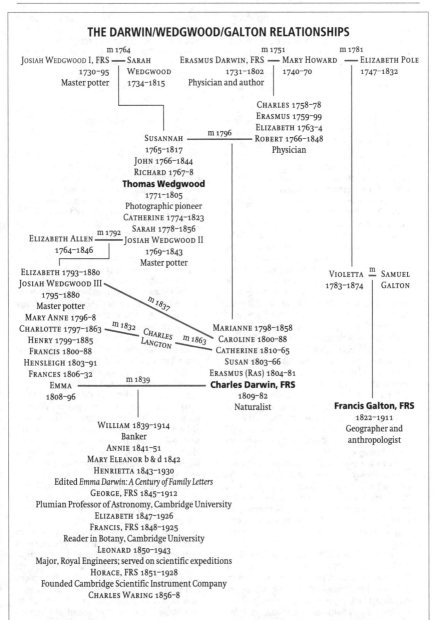

m 1764
JOSIAH WEDGWOOD I, FRS — SARAH
1730–95　　　　WEDGWOOD
Master potter　　1734–1815

m 1751
ERASMUS DARWIN, FRS — MARY HOWARD
1731–1802　　　1740–70
Physician and author

m 1781
— ELIZABETH POLE
1747–1832

CHARLES 1758–78
ERASMUS 1759–99
ELIZABETH 1763–4
SUSANNAH — m 1796 — ROBERT 1766–1848
1765–1817　　　　　　Physician
JOHN 1766–1844
RICHARD 1767–8
Thomas Wedgwood
1771–1805
Photographic pioneer
CATHERINE 1774–1823
SARAH 1778–1856

m 1792
ELIZABETH ALLEN — JOSIAH WEDGWOOD II
1764–1846　　　1769–1843
　　　　　　　　Master potter

ELIZABETH 1793–1880
JOSIAH WEDGWOOD III
1795–1880
Master potter
MARY ANNE 1796–8
CHARLOTTE 1797–1863 — m 1832 — CHARLES
HENRY 1799–1885　　　　　LANGTON
FRANCIS 1800–88
HENSLEIGH 1803–91
FRANCES 1806–32
EMMA — m 1839
1808–96

m 1837

m 1863

MARIANNE 1798–1858
CAROLINE 1800–88
CATHERINE 1810–65
SUSAN 1803–66
ERASMUS (RAS) 1804–81
Charles Darwin, FRS
1809–82
Naturalist

VIOLETTA — m — SAMUEL
1783–1874　　　GALTON

WILLIAM 1839–1914
Banker
ANNIE 1841–51
MARY ELEANOR b & d 1842
HENRIETTA 1843–1930
Edited *Emma Darwin: A Century of Family Letters*
GEORGE, FRS 1845–1912
Plumian Professor of Astronomy, Cambridge University
ELIZABETH 1847–1926
FRANCIS, FRS 1848–1925
Reader in Botany, Cambridge University
LEONARD 1850–1943
Major, Royal Engineers; served on scientific expeditions
HORACE, FRS 1851–1928
Founded Cambridge Scientific Instrument Company
CHARLES WARING 1856–8

Francis Galton, FRS
1822–1911
Geographer and
anthropologist

By studying mental ability within families, FRANCIS GALTON concluded that intelligence is predominantly due to inheritance rather than environment, a view he presented in his book *Hereditary Genius* (1869). His own very inter-related family provided the basis for his observations. More recent work, such as studies of identical twins brought up in contrasting environments, has led to a different view: most psychologists now see heredity and environment as comparable contributors in the shaping of individual abilities.

Three of Galton's family have entries in this book (shown in heavy type) and others have arguable claims for inclusion.

MM

pher and anthropologist: invented the statistical measure of correlation.

Galton was born near Birmingham. His family included prosperous manufacturers and bankers as well as scientists and, from an early age, he developed a life-long passion for scientific investigation. In 1844 he graduated from Cambridge and, in that same year, his father's death left him with the independence of a financial fortune. Galton wanted to undertake scientific geographical exploration and a cousin (Douglas Galton) introduced him to the Royal Geographical Society in London. With the Society's advice, Galton financed and led a 2-year expedition to an uncharted region of Africa. On his return to England in 1852, his geographical work brought him recognition in scientific circles. He was made a Fellow of the Royal Society in 1856 and took up the life of a London-based scientist-at-large. He never held, or sought, paid employment but was an active officer in the Royal Geographical Society, the Royal Society, the Anthropological Institute and the British Association for the Advancement of Science. These societies gave him contact with leading scientists and also the opportunity to report his own investigations.

Galton's investigations were many and varied but he consistently stressed the value of quantitative evidence. His motto was: 'whenever you can, measure and count'. For example, in order to construct large-scale weather maps, he sent a questionnaire to several weather stations around Europe asking for specified measurements on specified dates. When he received and mapped these data, in 1863, he discovered and named the now familiar 'anticyclone'. In 1875 he published, in The Times, the first newspaper weather map.

Throughout his life, Galton energetically pursued a variety of investigations. However, in 1859, a book appeared which stimulated him to concentrate more and more on the measurement of human individual differences. This book was The Origin of Species by DARWIN, who was another of Galton's cousins (see panel on p. 123).

People obviously differ greatly in their physical and mental characteristics, and the question that intrigued Galton was: to what extent do these characteristics depend on heredity or on environmental conditions? He pursued this question by various investigations, eg selectively breeding plants and animals and collecting the medical histories of human twins, who are genetically identical. In his human investigations, he faced the challenge that there was, at that time, no reliable body of measurements across generations. For example, how do the heights of parents relate to the adult heights of their children? He assembled large amounts of intergenerational data about height and other characteristics. Then he faced the further prob-

lem that there was, at that time, no mathematical way of expressing compactly the extent to which, say, the heights of offspring vary as a function of the heights of parents. By working over his accumulated measurements Galton solved this problem and, in 1888, he presented to the Royal Society his technique for calculating the correlation coefficient.

Galton's technique of 1888 was basically sound, but crude by modern standards. It was much improved by later workers. It provided a powerful new tool which nowadays is widely used, for example in medical science. Galton was a pioneer in several areas: he invented the term 'eugenics' to describe the science of production of superior offspring and was largely responsible for the introduction of fingerprinting as a means of identifying individuals in criminal investigations. But his most enduring contribution is perhaps his invention of the correlation coefficient.

Galvani, Luigi [galvahnee] (1737–98) Italian anatomist: discoverer of 'animal electricity'.

Galvani taught anatomy at Bologna, where he had graduated. His best-known work concerns his study of the effects of electricity on frogs. Among his systematic studies, a chance observation was important; this was that dead frogs being dried by fixing by brass skewers to an iron fence showed convulsions. He then showed that convulsions followed if a frog was part of a circuit involving metals. He believed that electricity of a new kind (animal electricity, or galvanism) was produced in the animal, but in 1800 VOLTA devised the voltaic pile and resolved the problem: the current originated in the metals, not the frog. Galvani's name lives on in the word galvanized (meaning stimulated as if by electricity; also used for the zinc-coating of steel) and in the galvanometer used from 1820 to detect electric current.

Gamow, George [gamov] (1904–68) Russian–US physicist: explained helium abundance in universe; suggested DNA code of protein synthesis.

Born in Odessa and a student in Petrograd (now St Petersburg), Gamow worked in Copenhagen with BOHR and in Cambridge with RUTHERFORD, felt oppressed on his return to Russia and worked in the USA from 1934. He made important advances in both cosmology and molecular biology. In 1948, together with ALPHER and BETHE, he suggested a means by which the abundances of chemical elements observed in the universe (helium in particular) might be explained (see Alpher). Gamow also showed, in 1956, that the heavier elements could only have been formed in the hot interiors of stars. He had showed in the 1930s that our Sun is not cooling down but is slowly heating up and in the 1940s he was a major expounder of the 'Big Bang' theory of the origin of the universe. A large and enthusiastic person, and a keen joker, it was his idea to include Bethe (legitimately) in the classic paper on the 'Big Bang' in 1948.

In molecular biology Gamow made a major contribution to the problem of how the order of the four different kinds of nucleic acid bases in DNA chains could govern the synthesis of proteins from amino acids. He realized (after some false starts) that short sequences of the bases could form a 'code' capable of carrying information for the synthesis of proteins; and that, since there are 20 amino acids making up proteins, the code must consist of blocks of three nucleic acid bases in order to have a sufficient vocabulary of instructions. Some details were wrong, but this central idea was known by 1960 to be correct.

Garrod, Sir Archibald Edward (1857–1936) British physician: discovered nature of congenital metabolic disorders.

While studying four human disorders (alcaptonuria, albinism, cystinuria and pentosuria) Garrod discovered that in each case a chemical substance derived from the diet was not being completely metabolized by the body, with the result that a product that is normally only an intermediate was excreted in the urine. He deduced that this metabolic failure was due to the absence of an enzyme (in 1958 this was proved to be correct). The family histories of patients also showed that the disorders were not due to infection or some random malfunction but were inherited on a Mendelian recessive pattern. Garrod's results thus showed that Mendelian genetics applied to man and were the first to suggest a connection between an altered gene (mutation) and a block in a metabolic pathway. This major concept, the biochemical basis of genetics, was surprisingly ignored (as MENDEL's original work on genetics had been) for 30 years.

Garrod's daughter Dorothy, an archaeologist, was the first woman to hold a Cambridge professorship.

Gasser, Herbert Spencer (1888–1963) US physiologist.

Gasser studied physiology in Wisconsin and in Europe; from 1935 to 1953 he directed the Rockefeller Institute for Medical Research in New York City, but in 1921–31 he was professor of pharmacology at Washington University (St Louis) and worked on nerve conduction with his former teacher Joseph Erlanger (1874–1965). DU BOIS-REYMOND had shown by 1850 that a nerve impulse was an electrical wave of negativity passing along the nerve, whose average speed was first measured by HELMHOLTZ, and ADRIAN later found that nerve cells discharge rapid series of such impulses. Gasser and Erlanger, using the then newly perfected low-voltage cathode ray oscillograph, found in the 1920s that nerve fibres differed in their conduction velocities, which fell in three main groups. The thickest mammalian fibres (such as those activating the muscles) conduct at 5–100 m s^{-1}, while pain is felt through thin slowly conducting fibres (below 2 m s^{-1}). Many other properties of nerves vary with conduction speed, and Gasser and Erlanger's methods did much to generate new work in electrophysiology; they shared the Nobel Prize for physiology or medicine in 1944.

Gauss, Karl Friedrich [gows] (1777–1855) German mathematician: one of the greatest of all mathematicians.

Gauss was of the stature of ARCHIMEDES and NEWTON and in range of interests he exceeded both. He contributed to all areas of mathematics and to number theory (higher arithmetic) in particular. His father was a gardener and merchant's assistant; the boy showed early talent, teaching himself to count and read, correcting an error in his father's arithmetic at age 3, and deducing the sum of an arithmetic series (a, a + b, a + 2b...) at the age of 10. Throughout his life he had an extraordinary ability to do mental calculations. His mother encouraged him to choose a profession rather than a trade, and fortunately friends of his schoolteacher presented him to the Duke of Brunswick when he was 14; the Duke thereafter paid for his education and later for a research grant. Gauss was grateful, and was deeply upset when the Duke was mortally wounded fighting Napoleon at Jena in 1806. Gauss attended the Collegium Carolinum in Brunswick and the University of Göttingen (1795–98). He devised much mathematical theory between the ages of 14 and 17; at 22 he was making substantial and frequent mathematical discoveries, usually without publishing them. After the Duke's death he became director of the Observatory at Göttingen, and was able to do research with little teaching, as he preferred.

Up to the age of 20 Gauss had a keen interest in languages and nearly became a philologist; thereafter foreign literature and reading about politics were his hobbies (in both he had conservative tastes). When at 28 he was financially comfortable he married Johanne Osthof; unbelievably happy, Gauss wrote to his friend W Bolyai (1775–1856), 'Life stands before me like an eternal spring with new and brilliant colours.' Johanne died after the birth of their third child in 1809, leaving her young husband desolate and, although he married again and had three more children, his life was never the same and he turned towards reclusive mathematical research. This was done for his own curiosity and not published unless complete and perfect (his motto was 'Few, but ripe') and he often remained silent when others announced results that he had found decades before. The degree to which he anticipated a century of mathematics has become clear only since his death, although he won fame for his work in mathematical astronomy in his lifetime. Of the many items named after him, the Gaussian error curve is perhaps best known.

During his years at the Collegium Carolinum, Gauss discovered the method of least squares for obtaining the equation for the best curve through a group of points and the law of quadratic reciprocity. While studying at Göttingen he prepared his book *Disquisitiones arithmeticae* (Researches in Arithmetic), published in 1801, which developed number theory in a rigorous and unified manner; it is a book which, as Gauss put it, 'has passed into history' and virtually founded modern number theory as an independent discipline. Gauss gave the first genuine proof of the fundamental theorem of algebra: that every algebraic equation with complex coefficients has at least one root that is a complex number. He also proved that every natural number can be represented as the product of prime numbers in just one way (the fundamental theorem of arithmetic). The *Disquisitiones* discusses the binomial congruences $x^n \equiv A \pmod{p}$ for integer n, A and p prime; x is an unknown integer. The algebraic analogue of this problem is $x^n = A$. The final section of the book discusses $x^n = 1$ and weaves together arithmetic, algebra and geometry into a perfect pattern; the result is a work of art.

Gauss kept a notebook of his discoveries, which includes such entries as

EYPHKA! num $= \Delta + \Delta + \Delta$

which means that any number can be written as a sum of three triangular numbers (ie $\Delta = \frac{1}{2}n(n+1)$ for n integral). Other entries such as 'Vicimus GEGAN' or 'REV. GALEN' inscribed in a rectangle have never been understood but may well describe important mathematical results, possibly still unknown.

The notebook and Gauss's papers show that he anticipated non-Euclidean geometry as a boy, 30 years before J Bolyai (1802–60, son of Wolfgang) and LOBACHEVSKY; that he found CAUCHY'S fundamental theorem of complex analysis 14 years earlier; that he discovered quaternions before HAMILTON and anticipated A-M Legendre (1752–1833), ABEL and JACOBI in much of their important work. If he had published, Gauss would have set mathematics half a century further along its line of progress.

From 1801–20 Gauss advanced mathematical astronomy by determining the orbits of small planets such as Ceres (1801) from their observed positions; after it was first found and then lost by PIAZZI, it was rediscovered a year later in the position predicted by Gauss.

During 1820–30 the problems of geodesy, terrestrial mapping, the theories of surfaces and conformal mapping of one domain to another aroused his interest. Later, up to about 1840, he made discoveries in mathematical physics, electromagnetism, gravitation between ellipsoids and optics. He believed that physical units

should be assembled from a few absolute units (mainly length, mass and time); an idea basic to the SI system. Gauss was a skilled experimentalist and invented the heliotrope, for trigonometric determination of the Earth's shape, and, with W E Weber (1804–91), the electromagnetic telegraph (1833). From 1841 until his death Gauss worked on topology and the geometry associated with functions of a complex variable. He transformed virtually all areas of mathematics.

Gay-Lussac, Joseph Louis [gay lüsak] (1778–1850) French chemist: established law of combining volumes of gases; discovered a variety of new chemical compounds, including cyanogen, and developed volumetric analysis.

An adventurous child and a brilliant student, Gay-Lussac grew up during and after the French Revolution. LAVOISIER had done much to create modern chemistry in the 1780s, but he had been guillotined in 1794. From then on the time was ripe for the subject to develop.

Gay-Lussac studied engineering before becoming interested in physics and chemistry. He became well known through his hot-air balloon ascents in 1804. These were intended to find if magnetism persisted at height, and if the composition of air changed. The first ascent was with BIOT; in the second (alone) he rose to 7 km (23 000 feet) the highest then achieved. (The composition of air, and magnetism, appeared to be unchanged.) The next year he made a tour of Europe, visiting scientists and scientific centres, and had the luck of observing a major eruption of Vesuvius.

In 1808 he published the law of combining volumes; this states that the volumes of gases that react with one another, or are produced in a chemical reaction, are in the ratios of small integers. This law clearly gave support to DALTON'S atomic theory, which had so recently appeared, although Dalton failed to grasp this, or even to accept Gay-Lussac's experimental results that led to the law. Earlier, Gay-Lussac had found that all gases expand equally with rise of temperature, a result discovered by CHARLES but not published by him. These two laws regarding gases formed the basis for AVOGADRO'S Law of 1811.

By 1808 Gay-Lussac had an established reputation as a scientist and Paris was then the world's centre for science. It proved an eventful year for him; he married Joséphine Rogeot, then a 17-year-old shop assistant, whom he had seen reading a chemistry book between serving customers. He also began to work with his friend L J Thenard (1777–1857), a collaboration which was very fruitful. With Thenard, in 1808, Gay-Lussac made sodium and potassium in quantity (by reduction of the hydroxides with hot iron), discovered the amides and oxides of these metals and isolated the element boron (9 days ahead of DAVY). In 1809 they made the dangerously reactive fluorides HF and BF_3. Gay-Lussac was temporarily blinded by a potassium explosion which demolished his laboratory, but he never lost his enthusiasm for experimentation. In 1814 he published his research on the new element iodine, discovered by B Courtois (1777–1838), which was a model study. In 1815 he first made cyanogen (C_2N_2), and showed it to resemble the halogens and to be the parent of a series of compounds, the cyanides. He developed volumetric analysis as an accurate method and devised new industrial methods in chemistry.

He usually worked with his own hands, which at that time some thought inappropriate for such an eminent scientist. Davy described him in 1813 as 'lively, ingenious, and profound, with ... great facility of manipulation... the head of the living chemists of France'. He remains one of chemistry's immortals, like Lavoisier, BERZELIUS, Davy and his own pupil LIEBIG.

Geber (*Lat*), Jabir ibn Hayyan (*Arabic*) [jayber] (*c*.721–*c*.815) Arabic alchemist.

Son of a druggist, and orphaned young, Geber became the best-known of Arab alchemists. One result of his fame is that later writers used his name (perhaps to provide authority, or as a mark of respect). He was the resident physician and alchemist in the court of the Caliph Haroun al-Rashid (of *Arabian Nights* fame). His writings give detailed (but mystical) accounts of the principles whereby base metals could possibly be transmuted into gold; and he was familiar with a range of chemical substances and methods, including distillation, sublimation and crystallization.

Geiger, Hans (Wilhelm) [giyger] (1882–1945) German physicist: invented a counter for charged nuclear particles.

Having studied electrical discharges through gases for his doctorate, Geiger moved from Erlangen in Germany to Manchester, where soon he began work under RUTHERFORD. Together they devised a counter for alpha-particles (1908), which consisted of a wire at high electric potential passing down the centre of a gas-filled tube. The charged alpha-particles cause the gas to ionize, and the gas briefly conducts a pulse of current which can be measured. They showed that alpha-particles have two units of charge, and Rutherford later established that they are helium nuclei. In 1909 Geiger and E Marsden (1889–1970) demonstrated that gold atoms in a gold leaf occasionally deflect alpha-particles through very large angles, and even directly back from the leaf. This observation led directly to Rutherford's nuclear theory of the atom as like a small solar system rather than a solid sphere (1913). In 1910 Geiger and Rutherford found that two alpha-particles are emitted when uranium disintegrates. Work by Geiger and J M Nuttall (1890–1958) showed that there is a linear relation between the logarithm of the

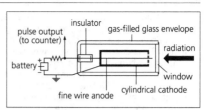

Geiger counter tube

range of alpha-radiation and the radioactive time constant of the emitting atoms (the Geiger–Nuttall rule). Geiger took part in the identification of actinium-A (1910) and thorium-A (1911). Both are isotopes of element number 84, polonium.

Geiger served in the German artillery during the First World War. Following this he was head of the National Physicotechnical Institution in Berlin and in 1925 used his counter to confirm the COMPTON effect by observing the scattered radiation and the recoil electron. Geiger became a professor at Kiel later that year and in 1928, together with W Müller, produced the modern form of the Geiger–Müller counter. In this, a metal tube acts as the negative cathode and contains argon at low pressure and a central wire anode. A window of thin mica or metal admits charged particles or ionizing radiation, which ionize the gas. The current pulse is amplified to operate a counter and produce an audible click. From 1936 he worked on cosmic rays, artificial radioactivity and nuclear fission. Geiger was ill during the Second World War, and died soon after losing his home and possessions in the Allied advance into Germany.

Gell-Mann, Murray (1929–) US theoretical physicist: applied group theory to understanding of elementary particles.

Gell-Mann was educated at Yale University and Massachusetts Institute of Technology, gaining his PhD at 22. Work with FERMI followed and he then moved to California Institute of Technology, where he became professor of theoretical physics in 1967.

At 24, Gell-Mann made a major contribution to the theory of elementary particles by introducing the concept of 'strangeness', a new quantum number which must be conserved in any so-called 'strong' nuclear interaction event. Using strangeness Gell-Mann and NE'EMAN (independently) neatly classified elementary particles into multiplets of 1, 8, 10 or 27 members. The members of the multiplets are then related by symmetry operations, specifically unitary symmetry of dimensions 3, or SU(3). The omega-minus particle was predicted by this theory and was observed in 1964, to considerable acclaim. Their book on this work was entitled *The Eightfold Way*, a pun on the Buddhist eightfold route to nirvana (loosely, heaven).

Gell-Mann and G Zweig (1937–) introduced in 1964 the concept of quarks, which have one-third or two-thirds integral charge and baryon number. From these the other nuclear particles (hadrons) can be made. The name is an invented word, associated with a line in Joyce's *Finnegan's Wake*: 'Three quarks for Muster Mark!' Six types of quark are now recognized. Five were detected indirectly after 1964, but the sixth ('top') quark eluded detection until 1995.

Another major contribution was Gell-Mann's introduction (with FEYNMAN) of currents for understanding the weak interaction. For all this work he was awarded the Nobel Prize for physics in 1969.

Gerhardt, Charles Frédéric [gairah(r)t] (1816–56) French chemist: classified organic compounds according to type.

Gerhardt studied chemistry in Germany, but after quarrelling with his father he became a soldier. He was 'bought out' by an unknown friend and returned to chemistry with LIEBIG, and later with DUMAS in Paris, where he met LAURENT. Together they did much to advance organic chemical ideas. The 'theory of types' reached a high point in their hands. This was a formal system of classifying organic compounds by referring them all to one (or more) of four types (hydrogen, hydrogen chloride, water and ammonia) by formal replacement of hydrogen by organic radicals. Examples based on the water type would be

H C_2H_5 C_2H_5
O, water O, ethanol O, diethyl ether
H H C_2H_5

Combined with the idea of homologous series (compounds differing by CH_2 units), this gave a general system of classification with some predictive power. Type formulae (such as those above) were not thought to represent structures but were formal representations of relationships and reactions. The theory rejected BERZELIUS's idea of 'dualism' (opposed charges within two parts of a molecule); and this, combined with Gerhardt's antiauthoritarian and quarrelsome personality, ensured controversy (which was fruitful in leading to new results).

Germain, Sophie [zhairmī] (1776–1831) France's greatest female mathematician.

Born into a liberal, educated, merchant family, Sophie Germain did not share their interest in money and politics. She retreated to her father's library and taught herself mathematics, to the dismay of her parents, who did their best to thwart her. There followed a battle of wills, which she won. She spent the years of the Terror (1793–4) teaching herself differential calculus. Unable to attend the newly opened École Polytechnique (1795) because of her sex, she obtained notes for many of the courses, including analysis given by LAGRANGE. Using a pseudonym she submitted work to Lagrange and started a correspondence with GAUSS, discussing FERMAT's 'last theorem' (his conjecture that $x^n + y^n = z^n$ has no positive integral solutions if n is an integer greater than 2).

In 1808 she wrote to Gauss describing what was to be her most important work in number theory. She proved that Fermat's conjecture is true if x, y and z are prime to one another and to n, if n is any prime less than 100. However, Gauss had become professor of astronomy at the University of Göttingen and did not respond. Germain's theorem remained largely unknown, although A-M Legendre (1752–1833) mentions it in a paper of 1823. Germain's theorem was the most important result related to Fermat's last theorem from 1738 to the work of E E Kummer (1810–93) in 1840.

Sophie Germain was a talented mathematician without sufficient training to fulfil her potential. She was regarded as a phenomenon rather than a serious mathematician in her time. In 1809 Napoleon urged the First Class of the Institut de France to establish a *prix extraordinaire* for anyone who could devise a theory that explained E F F Chladni's (1756–1827) experiments (the vibration patterns of elastic plates). There were no outright winners, though Germain was awarded the prize in recognition of her competence. After 1820 Germain began to be accepted by Parisian scientific society. She worked with Legendre and, through friendship with FOURIER, attended the sessions of the Académie des Sciences, the first woman to do so in her own right. She never received a degree; in 1830 Gauss failed to persuade the University of Göttingen to award her an honorary doctorate.

Germer, Lester Halbert [germer] (1896–1971) US physicist: demonstrated experimentally the wave-like nature of electrons.

After starting his career at Western Electric Co. and Bell Telephone Laboratories, Germer moved to Cornell University. His career was spent studying thermionics, erosion of metals and contact physics. Germer, with DAVISSON, carried out one of the crucial experiments in physics in 1927. This demonstrated that particles, in their case electrons, also display wave-like properties.

Giauque, William Francis [jeeohk] (1895–1982) US physical chemist: pioneer of low temperature techniques.

Giauque hoped to become an electrical engineer when he left school at Niagara Falls, but he failed to find a job in a power plant and for 2 years worked in a laboratory at a chemical plant, which moved his interest to chemistry. This became his main study at the University of California at Berkeley, where he subsequently remained for all his professional life.

In 1925 he proposed a method known as adiabatic demagnetization for achieving temperatures below 1 K, which had hitherto been unattainable. The method consists of placing a sample of a paramagnetic substance, at as low temperature as possible, in a very strong magnetic field; this causes the elementary magnetic ions in the substance to become aligned. When the magnetic field is switched off the elementary magnetic ions tend to increase their entropy by becoming randomly aligned, but since this requires energy the temperature of the sample will drop. Despite considerable practical difficulties, Giauque himself achieved a temperature of 0.1 K by this technique in 1933, and soon afterwards temperatures of a few thousands of a kelvin had been reached. The method remains the basis for reaching very low temperatures today. Giauque was awarded the Nobel Prize for chemistry in 1949 for his discovery. He was also the first to discover, in 1929, that atmospheric oxygen contains the isotopes oxygen-17 and oxygen-18.

Gibbs, Josiah Willard (1839–1903) US physical chemist: founder of chemical thermodynamics.

Before 1850, the Americas had produced few physical scientists of renown, with only FRANKLIN and RUMFORD in the pre-revolutionary USA; but the next 30 years saw the work of J HENRY, H A Rowland (1848–1901) and Gibbs, who was perhaps the most original of all of them and the only theorist. From youth he maintained the family tradition of skill in classical languages but also won prizes in mathematics, and in 1863 he gained the first Yale PhD in engineering, the second PhD awarded in the USA.

The next 3 years he spent at Yale as a tutor (2 years in Latin, and one in physics) before spending 2 years in France and Germany, with the two survivors of his four sisters, attending lectures by leading chemists, mathematicians and physicists. In 1871 he was appointed professor of mathematical physics at Yale. He held the job until his death, despite being unsalaried for the first 9 years on the curious grounds that he was not in need of money. He was not a good teacher and few understood his work. He never married, and lived with his sisters in New Haven, close to the college. His ideas, which founded chemical thermodynamics and statistical mechanics, were expressed in elegantly austere mathematical form in lesser-known journals, so few chemists understood them; some of his results were later rediscovered by PLANCK and EINSTEIN (among others), to their disappointment, and even POINCARÉ found reading his papers 'difficult'.

His ideas were of permanent use and were also still giving new insights a century after their appearance. In the 1870s he derived the Gibbs phase rule, which deals elegantly with heterogeneous equilibria, and devised the concept now known as the Gibbs function, which enables prediction of the feasibility and direction of a hypothetical chemical change in advance of direct trial. His later work covered chemical potential (an idea invented by him),

surface adsorption and the deduction of thermodynamic laws from statistical mechanics. It could be said that his fellow American Rumford began to solve the problem of heat and Gibbs completed the solution. In the 1890s his work was translated into French and German, and recognition and public honours followed. He remains probably the greatest theoretical scientist born in the USA.

Gilbert, Walter (1932–) US molecular biologist: isolated the first gene repressor.

Gilbert made a remarkable transition: from a basis of physics and mathematics at Harvard and Cambridge and a post at Harvard as a theoretical physicist, he changed in 1960 to biochemistry and molecular biology and became professor of biophysics at Harvard in 1964 and of molecular biology in 1968.

MONOD and F Jacob (1920–) had proposed in 1961 that gene action is controlled by a 'repressor substance' whose function is to 'turn off' the gene when it is not needed. In 1966 Gilbert and B Muller-Hill devised and successfully used an ingenious method for isolating one of these hypothetical substances, which are present only in traces in cells. They purified their sample of the *lac*-repressor (ie the repressor that represses the action of the gene that forms an enzyme that acts on lactose) and showed it to be a protein.

Later he worked on the problem of finding the sequence of bases in DNA and devised an elegant method, broadly similar to SANGER'S but suitable for either single-or double-stranded DNA; the two methods are complementary and each is best suited to particular cases. Gilbert shared a Nobel Prize for chemistry in 1980.

Gilbert, William (1544–1603) English physician and physicist: pioneer of the study of magnetism and the Earth's magnetic field.

Gilbert was a physician by profession, being royal physician to both Elizabeth I and James I, but is remembered for his extensive work on magnetism. He discovered how to make magnets by stroking pieces of iron with naturally magnetic lodestones, or by hammering iron while it is aligned in the Earth's magnetic field, and also found that the effect was lost on heating. From his investigations of magnetic dip he concluded that the Earth acted as a giant bar magnet, and he introduced the term 'magnetic pole'. His book *De magnete* (Magnets, 1600) is a classic of experimental science and was widely read throughout Europe. It is often considered to be the first great scientific work written in England.

Glaser, Donald Arthur [glayzer] (1926–) US physicist: invented the bubble chamber for observing elementary particles.

Graduating in 1946 from the Case Institute of Technology in his home town of Cleveland, OH, Glaser then did cosmic ray research at California Institute of Technology for his doctorate

(1950). For 10 years he worked at the University of Michigan and from 1959 at the University of California at Berkeley. In 1964 he turned to molecular biology.

By the early 1950s the WILSON cloud chamber was failing to detect the fastest high-energy particles available. Glaser realized that particles passing through a superheated liquid will leave tracks of small gas bubbles nucleated along the trajectory. In 1952 he produced a prototype bubble chamber a few centimetres across, filled with diethyl ether. The tracks were observed and recorded with a high-speed camera. Bubble-chambers up to several meters across and filled with liquid hydrogen were developed by ALVAREZ and used in many of the major discoveries of the 1960s and 1970s in particle physics. Glaser received the 1960 Nobel Prize for physics. He had calculated which liquids would be suitable for use in a bubble chamber, but as he 'wanted to be sure not to omit simple experimental possibilities' he also tried beer, ginger beer and soda water. None worked; water is unsuitable because it has a high surface tension and a high critical pressure.

Glashow, Sheldon Lee [glashow] (1932–   ) US physicist: produced a unified theory (QCD) of electromagnetism and the weak nuclear interaction.

After Cornell and Harvard Glashow spent a few years in postdoctoral work at the Bohr Institute, at the European Organization for Nuclear Research (CERN) in Geneva and in the USA, and in 1967 returned to Harvard as professor of physics.

Glashow produced one of the earliest models explaining the electromagnetic and weak nuclear forces. The WEINBERG-SALAM theory then developed this further and was a coherent theory for particles called leptons (electrons and neutrinos). Glashow extended their theory to other particles such as baryons and mesons by introducing a particle property called 'charm'. He used GELL-MANN's theory that particles were made up of smaller particles called quarks, and postulated that a fourth 'charmed' quark was necessary, giving a group of particles described by SU4 (Unitary Symmetry of Dimension 4). The dramatic discovery of the J (or psi) particle that confirmed this approach was made by TING and RICHTER in 1974. They found other predicted particles during the next 2 years. Since then the quark theory has been extended to include a 'coloured' quark, and the theory (which is known as quantum chromodynamics, QCD) is now discussed on the basis of Glashow's approach. He shared a Nobel Prize in 1979.

Glauber, Johann Rudolph [glowber] (1604–68) German chemist: understood formation of salts from bases and acids.

Glauber developed an interest in chemistry after having apparently been cured of typhus

through drinking mineral waters. He was one of the first to have clear ideas about the formation of salts from bases by the action of acids, and prepared a wide range of chemicals. He is credited with the discovery that sulphuric acid and common salt react to form hydrochloric acid and sodium sulphate (Glauber's salt). As Glauber was something of a charlatan, the latter was soon on sale as a cure for a wide range of ailments (and was used as a laxative). He wrote several useful treatises on industrial chemistry and noticed the peculiar precipitates known as 'chemical gardens'.

Goddard, Robert Hutchings (1882–1945) US physicist: pioneered the liquid-fuel rocket.

Goddard was born and educated in Worcester, MA, attending the Polytechnic Institute and Clark University. He held the position of professor of physics at Clark University for most of his life.

Goddard was interested in the practical aspects of space travel from an early age and in 1919 published a classic paper outlining many of the basic ideas of modern rocketry. Unlike some other pioneers of the space age, he was not content merely to test his ideas on paper and in 1926 built and tested a rocket propelled by gasoline and liquid oxygen. In 1929 he established a research station in New Mexico, backed by the Guggenheim Foundation, and soon sent up instrumented rockets, the forerunners of those used for atmospheric research, and developed a gyroscopic guidance system. By 1935 his rockets had broken the sound barrier and demonstrated that they could function in the near-vacuum of space. His pioneering work was not, however, publicly acknowledged by the US Government until 15 years after his death, when it awarded his widow $1 million for its numerous infringements of his 214 patents, incurred during its space and defence programmes.

Gödel, Kurt [goedl] (1906–78) Austrian–US mathematician: showed that mathematics could not be totally complete and totally consistent.

Growing up, Gödel was frequently ill and had a life-long concern with his health. He studied mathematics at Vienna, saw much of the development of the positivist school of philosophy and was apparently unconvinced. In 1930 he received his PhD, proving in his thesis that first-order logic is complete – so that in first-order logic every statement is provable or disprovable within the system. He then investigated the larger logical system put forward by B Russell (1872–1970) and A N Whitehead (1861–1947) in their *Principia mathematica*, and his resulting paper of 1931 may well be the most significant event in 20th-c mathematics.

The paper was titled 'On Formally Undecidable Propositions of *Principia mathematica* and Related Systems' and showed that arithmetic was incomplete. In any consistent formal system able to describe simple arithmetic there are propositions that can be neither proved nor disproved on the basis of the system. Gödel also showed that the consistency of a mathematical system such as arithmetic cannot necessarily be proved within that system. Thus a larger system may have to be used to prove consistency, and its consistency assumed; all pretty unsatisfactory. The programme for developing mathematical logic suggested by HILBERT, F L G Frege (1848–1925) and Russell was therefore untenable, and it is now clear that there is no set of logical statements from which all mathematics can be derived.

Between 1938 and 1940 Gödel showed that restricted set theory cannot be used to disprove the axioms of choice or the continuum hypothesis; this was extended in 1963 when P J Cohen showed that they are independent of set theory. Gödel also contributed to general relativity theory and cosmology, and was a close friend of EINSTEIN at Princeton. Gödel had married and emigrated to Princeton in 1938 when the Nazis took Austria, and was a professor there from 1953–76. He was an unassuming man with a variety of interests.

Goeppert Mayer, Maria, *née* Goeppert [goepert mayer] (1906–72) German–US mathematical physicist; discovered and explained the 'magic numbers' of nucleons in some atomic nuclei.

Maria Goeppert succeeded in making major contributions to science despite many obstacles. She first studied mathematics at Göttingen, perhaps because HILBERT was a family friend, but in 1927, attracted by BORN's lectures, she switched to physics. She worked on electronic spectra and in 1930 married an American chemist, Joseph Mayer (1904–). They went to Johns Hopkins where she had a lowly job with 'one of the only two people there who would work with a woman' – her husband. Her work on molecular spectra by quantum methods gained the respect of UREY and FERMI, who invited the couple to Columbia, but they could find a paid job only for 'Joe'. However, Maria continued to research, especially on the newly discovered elements heavier than uranium; so when the Second World War began and the atom bomb project developed, she was much in demand and soon led a team of 15 people. It was she who calculated the properties of UF_6, the basis of a separation method for the 'fission isotope,' uranium-235.

After the war both Mayers went to Chicago and it was there in 1948 that Maria found the pattern of 'magic numbers': atomic nuclei with 2, 8, 20, 28, 50, 82 or 126 neutrons or protons are particularly stable. (The seven include helium, oxygen, calcium and tin.) At first Mayer offered no theory to account for this, although she saw the analogy with electron shell structures. However, by 1950 and aided by a clue from her friend Fermi, Mayer worked out a complete shell model for atomic nuclei, in which spin orbit coupling predicted precisely the 'magic number' stable nuclei actually observed. In this model, the magic numbers describe nuclei in which certain key nucleon shells are complete. For this work she shared a Nobel Prize in 1963 with J H D Jensen (1907–73) of Heidelberg, who had arrived independently at rather similar conclusions on nuclear shell structure.

Gold, Thomas (1920–) Austrian–US astronomer: proponent of steady-state theory; contributor to the theory of pulsars.

An Austrian emigré, Gold studied at Cambridge, UK, subsequently working there and at the Royal Greenwich Observatory before moving to the USA in 1956, where he later became director of the Centre for Radiophysics and Space Research at Cornell University.

Gold, a leading proponent of the steady-state theory for the origin of the universe, published in 1948 the 'perfect cosmological principle' with BONDI and HOYLE, according to which the universe looks the same from every direction and at all times in its history. It is considered to have no beginning and no end, with matter being spontaneously created from empty space as the universe expands, in order to maintain a uniform density. Although the theory enjoyed support for a number of years, the discovery of the cosmic microwave background in 1964 by PENZIAS and R W WILSON gave conclusive support to the rival 'Big Bang' theory.

Pulsars, discovered by BELL (BURNELL) and HEWISH in 1968, were another area of interest to Gold, who was quick to propose an explanation for their strange radio signatures – he suggested that they were rapidly rotating neutron stars, sweeping out a beam of radio energy like a lighthouse. His hypothesis was verified when the gradual slowing down of their rate of spin, a phenomenon that he had predicted, was detected.

More recently, Gold was involved with a Swedish project to drill deep into the Earth's mantle to find commercial amounts of methane. He believed that significant amounts of hydrogen and helium remained within the Earth's interior from the time of the planet's formation and rejected the conventional organic theories of hydrocarbon formation, believing that oil fields are formed by the outward migration of this primordial gas.

These views were supported by the drill findings in 1991, under the Siljan ring, a meteorite crater north-west of Stockholm. Oil and gas were found there at 2800 m depth, in a crystalline granite rock. This location and depth could never have held organic remains whose decomposition would yield oil, and the find provides initial evidence in support of Gold's view that oil can have an abiogenic origin.

Goldbach, Christian (1690–1764) Russian mathematician: originator of 'Goldbach's conjecture'.

Goldbach studied mathematics and medicine and then travelled widely in Europe from 1710, meeting the leading scientists of the time and so preparing a basis for his later career as secretary of the Imperial Academy of Science at St Petersburg. He was erudite in mathematics, science, philology, archaeology and languages, but this wide range of interests, combined with his work as adviser on the education of the tsar's children and his activity as a privy councillor and courtier, inhibited sustained work in mathematics.

He is best known for his conjecture of 1742, noted in a letter to EULER, that every even number can be expressed as a sum of two primes (including 1 as a prime, if needed). Despite its apparent simplicity, this conjecture in number theory has so far defied all attempts to find a proof.

Goldbach also proposed that every odd number can be expressed as a sum of three primes (excluding 1 as a prime); this also is unproved.

Goldschmidt, Hans [goltshmit] (1861–1923) German chemist.

Goldschmidt invented the alumino-thermic or thermite process named after him, which consists of the reduction of metallic oxides using finely divided aluminium powder fired by magnesium ribbon. A similar mixture was much used in magnesium-cased incendiary bombs in the Second World War.

Goldschmidt, Victor Moritz [goltshmit] (1888–1947) Swiss-Norwegian chemist: pioneer of geochemistry and crystal chemistry.

Born in Zürich, Goldschmidt graduated from the University of Christiania (now Oslo) in 1911, becoming director of the Mineralogical Institute there when he was 26. He had already had great success in applying physical chemistry to mineralogy. In 1929 he moved to Göttingen, but returned to Norway in 1935 when the Nazis came to power. As a Jew he was fortunate that, although imprisoned when Germany occupied Norway in 1940, he was temporarily released; he escaped to Sweden in a haycart and moved to England in 1943.

Goldschmidt is regarded as the founder of modern geochemistry. Using X-ray techniques he established the crystal structures of over 200 compounds and 75 elements and made the first tables of ionic radii. In 1929, on the basis of these results, he postulated a fundamental law relating chemical composition to crystal structure: that the structure of a crystal is determined by the ratio of the numbers of ions, the ratio of their sizes and their polarization properties; this is often known as Goldschmidt's Law. It enabled Goldschmidt to predict in which minerals various elements could or could not be found. He also showed that the Earth's crust is made up largely of oxy-anions (90% by volume) with silicon and the common metals filling the remaining space.

Goldstein, Joseph Leonard [gohldstiyn] (1940–) US medical scientist: co-discoverer with M S Brown of the origin of one type of heart disease.

Goldstein graduated MD at the University of Texas in Dallas in 1966 and worked thereafter at the Massachusetts General Hospital in Boston and then at the National Institutes of Health, Bethesda, MD. At Boston he had become a close friend of Michael Stuart Brown (1941–), who also became an MD in 1966, and then worked in Boston and Bethesda. From 1972 both worked in the university medical school at Dallas and they jointly planned a study of familial hypercholesterolaemia (FH). Victims of this develop heart disease as children or below age 35, and Goldstein and Brown developed a full understanding of its nature within a year.

Cholesterol in the blood is always largely combined with a protein to form small droplets of low-density lipoprotein (LDL), which is normally absorbed by receptors on the surface of cells and so removed from the blood and employed in necessary biochemical processes within them. However, in victims of FH a genetic defect has caused the cells to lack these receptors and as a result the cholesterol in the blood remains at too high a level, and it initiates coronary artery disease; this path is responsible for about 5% of heart attacks in people under age 60.

Goldstein and Brown's work led to a new approach to the interactions between blood and cells, and to their own further studies on the fate of LDL in cells and on new methods for managing FH patients. Their work forms a striking example of the value of close collaboration between complementary personalities and of techniques linking biochemistry, genetics and clinical medicine. In 1985 they shared the Nobel Prize for physiology or medicine.

Golgi, Camillo [goljee] (1843–1926) Italian histologist: classified nerve cells and discovered synapses.

Golgi followed his father in pursuing a medical career, and studied at Padua and Pavia; he was later a physician in Pavia for 7 years and from 1875 taught there. He was interested in the use of organic dyes for histological staining (much used by KOCH, EHRLICH and others) and in 1873 he made, through a spillage accident, his own major discovery: the use of silver compounds for staining. Using this method with nerve tissue he was able to see new details under the microscope, allowing him to classify nerve cells and to follow individual nerve cells (which appeared black under the microscope when treated with silver). His method showed that their fibres did not join but were separated by small gaps (synapses). In the 1880s he studied the asexual cycle of the malaria parasite (a protozoon) in the red blood cells and related its stages to the observed stages of the various forms of malaria. In 1898 he described a peculiar formation in the cytoplasm of many types of cell (the Golgi body), which has been much studied since, especially by electron microscopy; it appears to be a secretory apparatus, producing glycoproteins and other essential cell materials. Golgi shared a Nobel Prize in 1906.

Goodricke, John (1764–86) British astronomer: explained nature of variable stars.

After careful observation of the variable star Algol, Goodricke suggested that its rapidly varying brightness was caused by a dark body orbiting a brighter one and partially eclipsing it. This was the first plausible explanation to be made for the nature of variable stars. He received the Copley Medal for his work, the Royal Society's highest honour, and was made a Fellow. Goodricke was a deaf mute. He died when he was only 21.

Goudsmit, Samuel Abraham [gowdsmit] (1902–78) Dutch–US physicist; first suggested that electrons possess spin.

After attending university at Amsterdam and Leiden, Goudsmit obtained his PhD in 1927 and emigrated to the USA, holding a post at Brookhaven National Laboratory from 1948–70.

At the age of 23, with fellow-student UHLENBECK, he developed the idea that electrons possess intrinsic quantized angular momentum (known as spin), with an associated magnetic moment, and used this to explain many features of atomic spectra. Spin later emerged as a natural consequence of relativistic quantum mechanics in DIRAC's theory of the electron (1928) and was found to be a property of most elementary particles, including the proton and neutron.

After first working on radar during the Second World War, Goudsmit was appointed head of the Alsos mission in 1944. This was to accompany and even to precede front-line Allied troops, seeking indications of the development of a German atomic bomb. He found that there was little danger of the Germans possessing such a weapon before the war ended. He was awarded the Medal of Freedom for this work, and later published an account in his book *Alsos* (1947).

Gould, Stephen Jay (1941–) US palaeontologist: developer of novel theories of evolution; prolific and skilful popularizer of evolutionary biology.

Educated at Antioch College, OH, and Columbia University, Gould taught and researched at Harvard from 1967. His early research on land snails was a precursor to his work, from 1972, in developing his theory of punctuated equilibria. This modifies DARWIN's theory of evolution by proposing that new species are created by evolutionary changes which occur in rapid bursts over periods as short as a few thousand years, separated by periods of stability in which there is little further change. This contrasts with Darwin's classical theory, in which species develop slowly over millions of years at fairly constant rates.

Gould's many books and essays on palaeontology and biological evolution have been remarkably effective in presenting these subjects with great clarity, accuracy and attractiveness to non-specialist audiences, as well as to biologists. (See panel overleaf.)

Graham, Thomas (1805–69) British physical chemist; studied passage of gases, and dissolved substances in solution, through porous barriers.

Son of a Glasgow manufacturer, Graham studied science in Glasgow and Edinburgh, despite his forceful father's desire that he should enter the church. He held professorships in Glasgow and London, and became a founder of physical chemistry, and first president of the Chemical Society of London (the first national chemical society). One part of his work deals with the mixing of gases separated by a porous barrier (diffusion) or allowed to mix by passing through a small hole (effusion). Graham's Law (1833) states that the rate of diffusion (or effusion) of a gas is inversely proportional to the square root of its density.

The density of a gas is directly proportional to its relative molecular mass, M. So if the rate of diffusion of one gas is k_A and its density d_A, and that of a second gas k_B and d_B, it follows that:

$$k_A/k_B = d_B^{\frac{1}{2}}/d_A^{\frac{1}{2}} = M_B^{\frac{1}{2}}M_A^{\frac{1}{2}}$$

Diffusion methods were used in 1868 to show that ozone must have the formula O_3; and more recently were used to separate gaseous isotopes.

Graham studied phosphorus and its oxyacids (which led to the recognition of polybasic acids, in which more than one hydrogen atom can be replaced by a metal); and he examined the behaviour of hydrogen gas with metals of the iron group. He found that H_2 will pass readily through hot palladium metal and that large volumes of the gas can be held by the cold metal.

His work on dialysis began the effective study of colloid chemistry. He found that easily crystallizable compounds when dissolved would readily pass through membranes such as parchment, whereas compounds of a kind which at that time had never been crystallized (in fact, of high relative molecular mass, such as proteins) would not dialyse in this way. This gives a method (today using polymers such as Cellophane rather than parchment) for separating large, colloidal molecules from similar compounds; this is useful in biochemistry, and in renal dialysis, where the blood of a patient with kidney failure is purified in this way.

Gram, Hans Christian Joachim (1853–1938) Danish physician and microbiologist.

Gram's life and work was based in

MNEMONICS

These are devices for aiding memory, in cases where a string of words (or symbols) needs to be recallable in a certain sequence, but is otherwise difficult to remember. They are usually constructed so that the components of the mnemonic are mutually suggestive; they may be rhymed or alliterative, and the imagery is often bizarre. Some are easily recalled nonsense-words. Mnemonics are of interest to psychologists in relation to the nature of memory, whose mechanism is still largely mysterious. Some examples follow.

In astronomy, **My Very Efficient Memory Just Sums Up Nine Planets** clues the sequence of the planets of the solar system, from the Sun outwards (Mercury, Venus, Earth, Mars, Jupiter, Saturn, Uranus, Neptune, Pluto).

In organic chemistry, the nonsense-word **omsgap** helps with the series of dicarboxylic acids of increasing chain length (oxalic, malonic, succinic, glutaric, adipic and pimelic acids).

Medical students are helped to recall the most common victims of gall bladder disease: **female, fat, forty, fertile and flatulent**, and have similar jingles (some obscene!) for a range of other conditions.

In geology, the student needs to be able to recall the order of rocks, in terms of age. This geological time scale is shown below. The best-known mnemonic for the sequence of geological periods (starting with the Cambrian) is: **Camels Often Sit Down CARefully. PERhaps Their Joints CREak.** And within the two most recent periods, the order

of epochs (starting with the oldest), **Previous Early Oiling May Prove Positively Helpful**.

Stephen Jay Gould in his *Wonderful Life* (1989) tells how he set an annual competition for his students to improve on this, the best attempt being the verse

Cheap Meat performs passably,
Quenching the celibate's jejune thirst,
Portraiture, presented massably,
Drowning sorrow, oneness cursed.

This yields the geological time scale, from the most recent to the oldest, listing all the eras first, and then the periods (with a few omissions). *Cheap Meat* refers to a pornographic film of the time.

Equally politically incorrect is a mnemonic in astrophysics where a key sequence is that of the spectral class (or colour index) of stars of progressively lower temperature, as in the horizontal scale of the familiar Hertzsprung–Russell ('H–R') diagram. The sequence is OBAFGKMRNS, and the mnemonic **Oh Be A Fine Girl, Kiss Me Right Now Sweetheart**. The last three classes (RNS) are now seen as subdivisions of the M class.

The H–R diagram effectively plots the brightness of stars (strictly, their absolute magnitude) as vertical axis against their temperature. It is intimately related, in a variety of ways, to the paths of stellar evolution.

In physics, the colours of the visible spectrum are recalled either by the nonsense-word VIBGYOR, or by **Richard of York gains battles in vain** (red, orange, yellow, green, blue, indigo, violet).

IM

Copenhagen; but during a visit to Berlin in 1884, he devised his famous microbiological staining method. He showed that bacteria can be divided into two classes; some (eg pneumococci) will retain aniline–gentian violet after treatment with iodine solution by his method; others (the Gram-negative group) do not.

Gray, Stephen (*c*.1666–1736) English physicist: distinguished between electrical conductors and insulators.

Gray was a dyer's son who began to follow the same trade but after meeting FLAMSTEED (the Astronomer Royal) was attracted to astronomy and obtained a job as an observer at Cambridge for a year. Back in London, he experimented with electrical devices, and in 1729 (when he was over 60) he made a major discovery. He had electrified a glass tube by friction and found by chance that the electricity was conducted along a stick or thread mounted in a cork inserted in one end of the tube. Led by this, he found that a string resting on silk threads would conduct for over 100 m, but if the string was supported by brass wires the transmission failed. Based on

this, he distinguished conductors (such as the common metals) from insulators (such as silk and other dry organic materials). In the 18th-c, experiments with electricity were difficult; primitive electrostatic machines, changes in humidity and induction effects easily led to confusion. Gray also observed that two spheres of the same size, one solid and one hollow, had the same capacity for storing electric charge.

Green, George (1793–1841) British mathematician: established potential theory in mathematical physics.

Green left school early to work in the family corn mill and bakery and studied mathematics on his own. When his father died the mill was sold and he became financially independent. At age 40 he began to study at Cambridge; he graduated in 1837 and received a fellowship, but became ill and died soon afterwards.

In 1828 he published a paper in a local journal of which only a few copies were issued. It was discovered by W THOMSON after Green's death and shown to leading physicists including MAXWELL; both realized its great value. In it Green

Eon	Era	Period	Epoch	Million years before present	Geological events	Sea life	Land life
Phanerozoic	Cenozoic	Quaternary	Holocene		Glaciers recede. Sea level rises. Climate becomes more equable.	As now.	Forests flourish again. Humans acquire agriculture and technology.
				0.01			
			Pleistocene		Widespread glaciers melt periodically, causing seas to rise and fall.	As now.	Many plant forms perish. Small mammals abundant. Primitive humans established.
				2.0			
		Tertiary	Pliocene		Continents and oceans adopting their present form. Present climatic distribution established. Ice caps develop.	Giant sharks extinct. Many fish varieties.	Some plants and mammals die out. Primates flourish.
				5.1			
			Miocene		Seas recede further. European and Asian land masses join. Heavy rain causes massive erosion. Red Sea opens.	Bony fish common. Giant sharks.	Grasses widespread. Grazing mammals become common.
				24.6			
			Oligocene		Seas recede. Extensive movements of Earth's crust produce new mountains (eg Alpine–Himalayan chain).	Crabs, mussels, and snails evolve.	Forests diminish. Grasses Pachyderms, canines, and felines develop.
				38.0			
			Eocene		Mountain formation continues. Glaciers common in high mountain ranges. Greenland separates. Australia separates.	Whales adapt to sea.	Large tropical jungles. Primitive forms of modern mammals established.
				54.9			
			Palaeocene		Widespread subsidence of land. Seas advance again. Considerable volcanic activity. Europe emerges.	Many reptiles become extinct.	Flowering plants widespread. First primates. Giant reptiles extinct.
				65			
	Mesozoic	Cretaceous	Late		Swamps widespread. Massive alluvial deposition. Continuing limestone formation. S America separates from Africa. India, Africa and Antarctica separate.	Turtles, rays, and now-common fish appear.	Flowering plants established. Dinosaurs become extinct.
			Early	97.5			
				144			
		Jurassic	Malm	163	Seas advance. Much river formation. High mountains eroded. Limestone formation. N America separates from Africa. Central Atlantic begins to open.	Reptiles dominant.	Early flowers. Dinosaurs dominant. Mammals still primitive. First birds.
			Dogger	188			
			Lias				
				213			
		Triassic	Late	231	Desert conditions widespread. Hot climate slowly becomes warm and wet. Break up of Pangea into supercontinents Gondwana (S) and Laurasia (N).	Ichthyosaurs, flying flying fish, and crustaceans appear.	Ferns and conifers thrive. First mammals, dinosaurs, and flies.
			Middle	243			
			Early				
				248			
	Palaeozoic	Permian	Late		Some sea areas cut off to form lakes. Earth movements form mountains. Glaciation in southern hemisphere.	Some shelled fish become extinct.	Deciduous plants. Reptiles dominant. Many insect varieties.
			Early	258			
				286			
		Carbon-iferous	Pennsylvanian	320	Sea-beds rise to form new land areas. Enormous swamps. Partly-rotted vegetation forms coal.	Amphibians and sharks abundant.	Extensive evergreen forests. Reptiles breed on land. Some insects develop wings.
			Mississippian				
				360			
		Devonian	Late	374	Collision of continents causing mountain formation (Appalachians, Caledonides, and Urals). Sea deeper but narrower. Climatic zones forming. Iapetus ocean closed.	Fish abundant. Primitive sharks. First amphibians.	Leafy plants. Some invertebrates adapt to land. First insects.
			Middle	387			
			Early				
				408			
		Silurian	Pridoli	414	New mountain ranges form. Sea level varies periodically. Extensive shallow sea over the Sahara.	Large vertebrates.	First leafless land plants.
			Ludlow	421			
			Wenlock	428			
			Llandovery				
				438			
		Ordovician	Ashgill	448	Shore lines still quite variable. Increasing sedimentation. Europe and N America moving together.	First vertebrates. Coral reefs develop.	None.
			Caradoc	458			
			Llandeilo	468			
			Llanvirn	478			
			Arenig	488			
			Tremadoc				
				505			
		Cambrian	Merioneth	525	Much volcanic activity, and long periods of marine sedimentation.	Shelled invertebrates. Trilobites.	None.
			St David's	540			
			Caerfai				
				590			
Proterozoic	Precambrian	Vendian			Shallow seas advance and retreat over land areas. Atmosphere uniformly warm.	Seaweed. Algae and invertebrates.	None.
				650			
		Riphean	Late	900	Intense deformation and metamorphism.	Earliest marine life and fossils.	None.
			Middle	1300			
			Early				
				1600			
		Early Proterozoic			Shallow shelf seas. Formation of carbonate sediments and 'red beds'.	First appearance of stromatolites.	None.
				2500			
Arch-aean		Archaean (Azoic)			Banded iron formations. Formation of the Earth's crust and oceans.	None.	None.
				4600			

uses the term 'potential' and developed this mathematical approach to electromagnetism. He also included his famous theorem, which gives a way of solving partial differential equations by reducing a volume integral to a surface integral over the boundary (Green's theorem).

Green's theorem has found recent use in pure mathematics, in particle physics, in various branches of engineering and in soil science. Appreciated at last, the 200th anniversary of his birth was celebrated, and included the unveiling of a plaque in Westminster Abbey among the memorials to the greatest of his fellow scientists.

Green, Michael Boris (1946–) British theoretical physicist.

Educated in Cambridge and afterwards working in Princeton, Cambridge, Oxford and London, Green became professor of theoretical physics in Cambridge in 1993. He is best known for his work on the superstring theory, a novel approach to the nature of nuclear particles and the forces of nature.

As background to Green's work, it should be noted that interactions within and between atoms depend upon four field forces. The electromagnetic force can be attractive or repulsive. The strong nuclear force is even stronger at very short distances (10^{-15} m) within an atomic nucleus. The weak nuclear force, only 10^{-4} as powerful, is also very important. Gravitational force is very much weaker (10^{-40}) and is only attractive: it is of course of great importance on the cosmic scale, as between planets or stars where, although distances are large, the bodies are uncharged and the masses are large.

From the 1970s grand unified theories ('GUTs') attempted to embrace the first three forces, with partial success: work by GLASHOW, SALEM and WEINBERG has linked electromagnetic and weak nuclear forces in one theory. The strong and weak nuclear forces have resisted unification.

The simplest and most elegant of the GUTs requires the proton to decay with a lifetime of 10^{30} years, and experiments to detect this have failed to do so. Gravity, also, which is not quantized (unlike the others) has resisted incorporation, so a 'theory of everything' ('TOE') has proved elusive.

The nearest to success has been superstring theory, devised in the 1980s with Green as the main proponent. This treats the interaction of subnuclear particles not in terms of points, but of one-dimensional curves (superstrings) having mass, and a length only 10^{-35} m (ie 10^{-20} the proton diameter), in 10-dimensional space-time (nine in space, plus time). The theory deals with all four field forces (ie it includes gravity), involves sophisticated mathematics (including symmetry properties) and has implications for cosmology as well as particle physics. It is consistent with special relativity and quantum theory. Despite its mathematical

attractions, wider acceptance by physicists must depend on experimentally testable predictions evolving from it.

Gregory, James (1638–75) Scottish mathematician: contributed to discovery of calculus.

A graduate of Aberdeen, Gregory went on to study at Padua, and became the first professor of mathematics at St Andrews when he was 30. After 6 years he moved to Edinburgh, but died a year later. While in Italy he published a book in which he discussed convergent and divergent series (terms he used for the first time), the distinction between algebraic and transcendental functions and circular, elliptic and hyperbolic functions. He found series expressions for the trigonometric functions and gave a proof of the fundamental theorem of calculus (in 1667), although he did not note its significance. In letters of 1670 he used the binomial series and NEWTON's interpolation formula (both independently of Newton) and the series named after B Taylor (1685–1731).

He also contributed to astronomy; when he was 25 he suggested that transits of Venus (or Mercury) could be used to find the distance of the Sun from the Earth, and the method was later used. He proposed in the same book that telescopes could be made using mirrors in place of lenses, avoiding the optical aberration inevitably introduced by a lens. About 5 years later Newton made such a reflecting telescope, and large telescopes have usually been reflectors ever since.

Grignard, (Francois Auguste) Victor [green-yah(r)] (1871–1935) French organic chemist: discovered use of organomagnesium compounds in synthesis.

Grignard, a sailmaker's son, studied at Lyon to become a teacher of mathematics. Later he moved to chemistry and to research in organic chemistry under P A Barbier (1848–1922), who gave him a research project 'as one throws a bone to a dog'. He found that magnesium would combine with reactive organic halogen compounds in dry diethyl ether solution to give an organomagnesium compound. Such compounds could be used (without isolating them) in reactions with a variety of carbonyl and other compounds to give organic alcohols, organometallics and other useful products. These Grignard reactions became the most useful of organic synthetic methods, were greatly used in research, and led also to increased interest in other organometallic compounds.

In the First World War Grignard (who had done military service in 1892 and become a corporal) began by guarding a railway bridge but was soon transferred to chemical warfare; he worked on the detection of mustard gas and the making of phosgene ($COCl_2$). In 1919 he succeeded Barbier at Lyon, and remained there, largely working on extensions of his major discovery. He shared a Nobel Prize in 1912.

Grove, Sir William Robert (1811–96) British lawyer and physicist; devised first fuel cell.

It is rare for anyone to practice law and science together, as Grove did. He became a barrister, but poor health was thought to point to a less active life and he turned to electrochemistry. He invented the Grove Zn-Pt cell, which was popular and survived in a form modified by BUNSEN. Then, to improve his income, Grove returned to legal work. He defended Palmer, the 'Rugeley poisoner', in a famous murder trial in 1856 and became a judge in 1871, meanwhile continuing his scientific interests. He devised in 1842 what he called a 'gas battery'; in fact the first fuel cell, not to be confused with the Grove cell described above. The fuel cell had two platinum strips, both half-immersed in dilute H_2SO_4; one strip was half in H_2 gas, the other in O_2 gas. When a wire connected the ends, a current flowed. Other pairs of gases (eg H_2 and Cl_2) also gave a current. Grove realized that the current came from a chemical reaction: it was not until the 1950s that F T Bacon (1904–92; a descendant of FRANCIS BACON) devised a practical, useful fuel cell.

In 1845 Grove made the first electric filament lamp and in 1846 showed that steam is dissociated (to H_2 and O_2) on hot platinum. He also studied lighting in mines, and discharge tubes, and offered early ideas on energy conservation.

In 1891, at the jubilee of the Chemical Society, of which he was a founder, he said, '...for my part, I must say that science to me generally ceases to be interesting as it becomes useful'.

Guericke, Otto von [gayrikuh] (1602–86) German engineer and physicist: inventor of the air pump; created and investigated properties of a 'vacuum'.

After a wide-ranging education, Guericke became one of the four *Burgermeisters* of Magdeburg in 1646, in acknowledgement of his service to the town as an engineer and diplomat during its siege in the Thirty Years War. His interest in the possibility of a vacuum (which ARISTOTLE had denied) led him to modify and improve a water pump so that it would remove most of the air from a container. He showed that in the resulting 'vacuum' a bell was muffled and a flame was extinguished. Most dramatically, at Regensburg in 1654 he showed that when two large metal hemispheres were placed together and the air within was pumped out, they could not be separated by two teams of eight horses.

He also built the first recorded electrostatic machine, consisting of a globe of crude sulphur which could be rotated by a crank and which was electrified by friction.

Guettard, Jean Etienne [getah(r)] (1715–86) French geologist: proposed the igneous origin of basalts.

Guettard was the keeper of the natural history collection of the duc d'Orleans. In 1751 he proposed, while travelling in the Auvergne region of France, that a number of the peaks in the area were former volcanoes, on the basis of the basalt deposits he found nearby. Although he later withdrew this hypothesis, his suggestion led DESMAREST to investigate and map the area in detail. His findings were, in turn, to lead to the abandonment of the Neptunist theory of A G WERNER, which had stated that all volcanic activity was recent and all rocks were sedimentary in origin. His geological map of France was constructed with the help of his young friend LAVOISIER and probably began the latter's interest in science.

Guldberg, Cato Maximilian [gulberg] (1836–1902) Norwegian physical chemist: deduced law of mass action for chemical reactions.

Guldberg spent his career in Oslo, where he was professor of applied mathematics from 1869. He and P Waage (1833–1900), his friend and brother-in-law, and professor of chemistry, were interested in BERTHELOT's work on the rates of chemical reactions, and this led them in 1864 to deduce the law of mass action. This states that, for a homogeneous system, the rate of a chemical reaction is proportional to the active masses of the reactants. The molecular concentration of a substance in solution or in the gas phase is usually taken as a measure of the active mass. The theory did not become known until the 1870s, largely because it was published in Norwegian, as also happened with his other work in physical chemistry.

Gutenberg, Beno [gootnberg] (1889–1960) German–US geophysicist: demonstrated that the Earth's outer core is liquid.

Gutenberg studied at Darmstadt and at Göttingen before being appointed professor of geophysics at Freiberg in 1926. In 1930 he moved to America to join the California

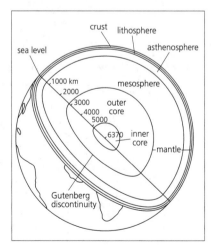

The layers of the Earth, showing the Gutenberg discontinuity

Institute of Technology, becoming director of its seismological laboratory in 1947.

By 1913 it was known that, on the opposite side of the Earth to an earthquake, there is a shadow zone in which compressional P (push–pull, primary) waves arrive later than expected and with reduced amplitude, and in which shear S (shake, secondary) waves are absent; OLDHAM had previously interpreted the delayed arrival of P waves as evidence for the existence of a core. Gutenberg realized that the absence of shear waves meant in addition that the core was liquid. He showed that the boundary between the solid mantle and the liquid core occurs at a depth of 2900 km; this interface is known as the Gutenberg discontinuity. It is now believed that there is an additional solid inner core, at a depth of about 5150 km (see diagram).

Guth, Alan (Harvey) [gooth] (1947–) US astrophysicist; made a major contribution to our understanding of the origin of the universe.

A student at MIT, Guth worked at Princeton, Columbia, Cornell and Stanford before taking posts in physics at MIT from 1980 and concurrently at the Harvard–Smithsonian Centre for Astrophysics from 1984.

For some time the 'Big Bang' theory of the creation of the universe, proposed by ALPHER, BETHE and GAMOW in 1948, has been generally accepted as the best explanation. A key observation about the universe, the existence of a background microwave radiation, can be explained in no other way. There were, however, questions that were not adequately accounted for by the basic theory. The standard 'Big Bang' model required a very hot early universe; why? Why also is the present universe so uniform on a large scale when it might be expected to be more irregular? Above all, it requires extraordinarily precise values of some physical properties (to within one part in 100 000 million million) to give the rate of expansion that we see today without either a much greater rate of expansion or a gravitational collapse.

It was to try and resolve these problems that, in 1980, Guth proposed a refinement to the 'Big Bang' theory. He proposed that the universe underwent an exponential expansion (a 10^{30} increase in radius) in the first microsecond of its existence – the so-called 'inflation hypothesis' of cosmology – as a result of a kind of 'supercooling', which in turn led to an antigravitational effect that enhanced the 'Big Bang' expansion. With subsequent refinements by HAWKING, and the Russian scientist Andrei Linde and others, the inflationary hypothesis accounts very well indeed for the problems found with the 'basic' 'Big Bang' model.

The proof of the inflation hypothesis was found in 1992 by a team led by George F Smoot (1945–) of the University of California at Berkeley. Using the Cosmic Background Explorer (COBE) satellite they observed very faint ripples in the background microwave radiation, which forms the 'afterglow' of the 'Big Bang' when the universe originated about 15 billion years ago. The ripples are related to the slight 'clumpings' of thinly dispersed matter, which later grow to form galaxies. This observation, which was subsequently confirmed by balloon-borne observations, is considered conclusive evidence for the 'Big Bang' theory with inflation, and was described by Hawking as 'the discovery of the century, if not of all time'.

Haber, Fritz [hahber] (1868–1934) German physical chemist: devised nitrogen 'fixation' process.

Haber's father was a dye manufacturer and so he studied organic chemistry to prepare him for the family firm. However, physical chemistry interested him more, and he worked on flames and on electrochemistry. By 1911 he was well known and was made director of the new Kaiser Wilhelm Institute for Physical Chemistry at Berlin–Dahlem. From about 1900 he worked on the problem of ammonia synthesis. CROOKES had shown that, if the world continued to rely on Chile nitrate deposits to provide nitrogenous fertilizer for agriculture, famine was inevitable. Haber solved the problem by 1908, showing that nitrogen from air could be used to make ammonia; the reaction $N_2 + 3H_2 \Leftrightarrow 2NH_3$ could be used at c. 400°C under pressure with a modified iron catalyst. With C Bosch (1874–1940) to develop the process to an industrial scale, production was established by 1913; the Haber–Bosch process made about 10^8 tonnes of ammonia annually by the 1980s. About 80% of this is used to make fertilizers. In the First World War it also solved the problem of making explosives for Germany, since nitric acid (essential for their production) can be made by oxidizing ammonia. Haber was also in scientific control of Germany's chemical warfare and devised gas masks and other defence against the Allies' gas warfare. A Nobel Prize was awarded to him in 1918 for the ammonia synthesis.

In 1933 he resigned his post and emigrated in protest against anti-Semitism, but he did not resettle well and worked only briefly in Cambridge. He died while on his way to a post in Israel.

Hadamard, Jacques [adamah(r)] (1865–1963) French mathematician: developed theory of functionals.

Hadamard's parents recognized his mathematical ability and he attended the École

SCIENCE AND THE FIRST WORLD WAR (1914–18)

This conflict embraced more science than any previous war. All pre-atomic explosives are based on nitrates, which at that time were largely derived from natural deposits of caliche ($NaNO_3$) from Chile, which would have been denied to Germany and the other powers at war with Britain. However, HABER and BOSCH designed a plant which began to produce synthetic ammonia in Ludwigshafen in 1913, from air and water. The ammonia could then be oxidized to nitric acid and hence nitrate supply was assured.

Chemical warfare began with the use of chlorine (Cl_2) by Germany in 1915; later in 1915 phosgene ($COCl_2$) was used as well and by 1917 mustard gas $S(CH_2CH_2Cl)_2$ was the shell-filling favoured by both sides. Haber in Germany and W J Pope (1870–1939) in England were key figures.

Advances in aircraft design and theory owe much to LINDEMANN and H Glauert (1892–1934). Improved petrol engines made the tank effective from 1917. Sound-ranging to locate gun positions was developed by W H BRAGG and sonar to detect submarines by RUTHERFORD and by LANGEVIN. Radio communications were used in war for the first time, as was blood transfusion against wound shock, immunization against typhoid, paratyphoid and enteric fever, and tetanus antitoxin after wounding; MARIE CURIE led motorized X-ray units to set bones and locate fragments in wounds.

For the first time, war was dominated by trench warfare, rendered near-static by the combination of artillery, machine guns and barbed wire. Repeated attempts to introduce mobility led to the tank, but it was air operations that proved decisive in the land battles of 1918.

The blockade of Britain by German submarines, and shortage of agricultural manpower everywhere in Europe, led to poor levels of nutrition, which in turn exposed the population to infection by tuberculosis and notably to the world pandemic of influenza in 1918–19.

Much work on nutrition before and during the war focused on the calorific value of diet and its content in terms of fat, protein and carbohydrate. In 1920 a British ministry revealed that a large proportion of young men called up in the war were unfit, but it was after 1930 before malnutrition was recognized as a major factor in this; one result was improved nutrition during the Second World War (1939–45).

Another curious delay resulted from the '30-year rule' that kept the medical reports on mustard gas victims secret from 1916 until 1946. It was then revealed that one effect of the gas was to reduce the white cell count of the blood, and this knowledge led to the use of 'nitrogen mustard' as a valuable treatment for leukaemia, a malignant disease characterized by progressive overproduction of a type of white cell.

IM

Normale Supérieure in Paris. His doctoral thesis was on function theory; he taught at the Lycée Buffon and then at Bordeaux. At 44 Hadamard became professor of mathematics at the Collège de France in Paris, and later at the École Polytechnique and École Centrale. In 1941, aged 76, he left occupied France for the USA and then joined the team in London using operational research for the RAF. Returning to France after the war, he retired to his interests in music, ferns and fungi.

Hadamard produced new insights in most areas of mathematics and influenced the development of the subject in many directions. He published over 300 papers containing novel and highly creative work. In the mid-1890s he studied analytic functions, that is those arising from a power series that converges. He proved the Cauchy test for convergence of a power series. In 1896 he proved the prime number theorem (first put forward by GAUSS and RIEMANN) that the number of prime numbers less than x tends to $x/\log_e x$ as x becomes large. This is the most important result so far discovered in number theory; it was independently proved by C J Poussin in the same year.

Hadamard investigated geodesics (or shortest paths) on surfaces of negative curvature (1888) and stimulated work in probability theory and ergodic theory. He then considered functions $f(c)$ that depend on the path c, and defined a 'functional' y as $y = f(c)$. The definitions of continuity, derivative and differential become generalizations of those for an ordinary function $y = f(x)$ where x is just a variable. A new branch of mathematics, functional analysis, with relevance to physics and particularly quantum field theory grew out of this.

Hadamard also analysed functions of a complex variable and defined a singularity as a point at which the function is no longer regular. A set of singular points may still allow the function to be continuous – and such regions are called 'lacunary space', the subject of much modern mathematics. Finally he initiated the concept of a 'well-posed problem' as one in which a solution exists that is unique for the given data but depends continuously on those data. A typical example is the solution of a differential equation written as a convergent power series. This has proved to be a powerful and fruitful concept and since then the neighbourhood and continuity of function spaces have been studied. Hadamard published books on the psychology of the mathematical mind (on POINCARÉ in particular) and was an inspiring lecturer who influenced several generations of mathematicians.

Hadfield, Sir Robert (Abbot) (1858–1940) British metallurgist: discovered several new steel alloys.

After a local schooling and training as a chemist, Hadfield started work in his father's small steel foundry in Sheffield. Due to his father's ill-health he took over the firm when he was 24 and inherited it 6 years later. Hadfield continued the research he had begun in the early 1880s into steel alloys, publishing 150 scientific papers.

The BESSEMER process of steel-making relied on the use of phosphorus-free iron ore; R F Mushet (1811–91) had added *Spiegeleisen* (Fe-C-Mn), but this resulted in a metal which, although hard, was too brittle. Hadfield found that, by adding large amounts of manganese (12–14%) and subsequent heating (to 1000°C) and quenching in water, he could produce a steel alloy both hard and strong and suitable for metal-working and railway points. His firm took out the patent on this in 1883. Continuing his work on steel alloys he produced silicon steels and showed that they have valuable magnetic properties and were suitable for use in transformers. His firm also produced armour-piercing and heat-resisting steels. He was knighted in 1908 and created a baronet in 1917.

Hadley, George (1685–1768) English meteorologist: explained the nature of trade winds.

Initially trained as a barrister, Hadley became more interested in science, and took over responsibility for the Royal Society's meteorological observations. HALLEY had proposed in 1686 that the trade winds were due to hot equatorial air rising, and pulling in colder air from the tropics, but was unable to account for their direction. In 1735 Hadley suggested that the reason that these winds blew from the northeast in the northern hemisphere, and from the south-east in the south, was the Earth's rotation from west to east. This form of circulation is now known as a Hadley cell.

Hadley, John (1682–1744) English optical instrument maker (brother of George, above).

Nothing is known of Hadley's early life, but by the 1720s he was making fine reflecting telescopes, being the first to develop NEWTON's

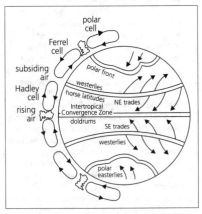

The three-cell general circulation model of the atmosphere

design of 1668. After Hadley's work such instruments dominated optical astronomy.

In 1731 he produced his reflecting quadrant (strictly, an octant), the precursor of the modern nautical sextant. With it, the altitude of a star (or the Sun) above the horizon could be found within 1' of arc. The octant was based on a mirror design first proposed by Newton but not published until 1742 (by HALLEY). From the 1780s good chronometers were available (due to HARRISON) which, used in conjunction with a sextant and astronomical tables, allowed a ship's position to be found satisfactorily. Two centuries later, radio- and satellite-based techniques again changed methods for ships and aircraft to define their position.

Hahn, Otto (1879-1968) German radiochemist.

Like KEKULÉ before him, Hahn originally was intended for a career in architecture, but this was overruled by his interest in chemistry, which he studied at Marburg. In 1904 he spent 6 months in London, which he much enjoyed and which introduced him, under RAMSAY's guidance, to the then novel field of radiochemistry. Its attractions led him to spend the next year with RUTHERFORD at Montreal; by this time he had real radiochemical expertise, and had characterized several new radioisotopes.

Soon after his appointment in Berlin in 1906 he was joined by LISE MEITNER. As a woman she was denied access to the all-male laboratories and there was no separate radiochemical laboratory. Both problems were solved by conversion of a basement woodworking shop, in which Meitner as a physicist and Hahn as a chemist began a collaboration that was to last for 30 years. In 1912 the new Kaiser Wilhelm Institute opened at Berlin-Dahlem, Hahn was appointed head of a radiochemical section and Meitner joined him. Taking advantage of the absence of significant radioactive contamination in the new laboratory, Hahn began study of the very weak beta-emitters potassium and rubidium. Later he showed that this radioactive breakdown of rubidium (to an isotope of strontium) gives a method for dating some mineral deposits, as also does the conversion of potassium to argon. During the First World War both Hahn and Meitner were in war-related work (Hahn on gas warfare under HABER; Meitner as a nurse), but some research could be continued during their leaves and they discovered the new radioelement protactinium.

From 1934 Hahn became interested in FERMI's discovery that slow neutrons could be captured by some atomic nuclei to give new, heavier, elements. Hahn, with Meitner and F Strassmann (1902-80), tried this with uranium, then the element with the heaviest known nucleus. However, by 1938 Meitner, being Jewish, was unsafe in Germany and with Hahn's help escaped, settling in Sweden. Hahn sent to her his results of the work with uranium and it

was she and her nephew OTTO FRISCH who published the novel idea that nuclear fission had occurred and showed that Hahn's explanation for these experiments was inadequate.

During the Second World War Hahn continued radiochemical work outside Germany's nuclear weapon programme; after the war he was active in the cause of nuclear disarmament. Awarded the Nobel Prize for chemistry in 1944, his name is commemorated in the Hahn-Meitner Institute for Nuclear Research in Berlin, in another institute at Mainz, in Germany's first nuclear ship the *Otto Hahn* and in the name (hahnium, Ha) officially given to the element of atomic number 108 by the International Union of Pure and Applied Chemistry in 1994.

Haldane, John (Burdon Sanderson) (1892-1964) British physiologist and geneticist: showed that enzyme reactions obey laws of thermodynamics.

Haldane is one of the most eccentric figures in modern science. If his life has a theme, it is of bringing talents in one field of work to the solution of problems in quite a different area. He was self-confident, unpredictable and difficult to work with. His family was wealthy and talented, and his father was Britain's leading physiologist.

The latter (John Scott Haldane, 1860-1936) was led to study death in coalmine disasters and from this to discover how poisoning by carbon monoxide arose; and then to discover the part played by carbon dioxide in controlling breathing. His work on deep sea diving and mountain ascents added to his work on respiration, and with his Oxford pupils he laid the basis of modern respiratory physiology.

His son J B S Haldane went to Eton, where he began to conflict with authority; service in the First World War made him an atheist. He went to Oxford to study mathematics and biology, but graduated in classics and philosophy. From 1910 his interest moved to genetics as a result of studying his sister's 300 guinea pigs. He began to teach physiology; he knew a good deal about respiration through helping his father and he had also worked on defence against poison gas in the war, but otherwise he had 'about 6 weeks start on my future pupils'. He researched on respiration and the effect on it of CO_2 in the blood. For this he used himself as an experimental animal, changing his blood acidity by consuming $NaHCO_3$, and by drinking solutions of NH_4Cl to get hydrochloric acid into his blood. Later he turned to biochemistry, applying his mathematical ability to calculate the rates of enzyme reactions and giving the first proof that they obey the laws of thermodynamics. Then he turned to genetics, and the mathematics of natural selection and of genetic disease and mutation in man. In 1938 he began work on deaths in submarine disasters and regularly risked his life in experiments on underwater escape.

He wrote on popular science; including over 300 articles in the *Daily Worker* (he was a Communist as well as the nephew of a viscount). In 1957 he emigrated to India, claiming that this was a protest against the Suez affair but probably because of the opportunity to work on genetics there; as usual, he quarrelled with his colleagues. Dying of cancer, he wrote comic poems about the disease, which inspired praise and offence in their readers in about equal numbers.

Hale, George Ellery (1868–1938) US astronomer: discovered that sunspots are associated with strong magnetic fields.

Son of a wealthy engineer, Hale had an early interest in astronomy and studied physics at Massachusetts Institute of Technology. Whilst still an undergraduate he invented the spectroheliograph, an instrument capable of photographing the Sun at precise wavelengths and now a basic tool of solar astronomy. In 1908 he discovered that some lines in the spectra of sunspots are split, and correctly interpreted this as being due to the presence of strong magnetic fields (the Zeeman effect). Together with W S ADAMS he discovered in 1919 that the polarity of the magnetic fields of sunspots reverses on a 23-year cycle.

Hale realized that larger telescopes were essential in order for astronomy to advance and put much of his energy and organizational ability into providing them. In 1892 he persuaded Charles Yerkes, a Chicago businessman, to fund a 40 in (1 m) refracting telescope for the University of Chicago, still the largest refractor ever built. Hale followed this by arranging for the Carnegie Institute to fund a 60 in (1.52 m) reflector for the Mount Wilson Observatory, and for a 100 in (2.54 m) reflector financed by John D Hooker that was to remain the largest in the world for 30 years. His greatest achievement, however, was to persuade the Rockefeller Foundation to provide the money for a telescope that would be the ultimate in size for Earth-based observations – the 200 in (5.08 m) Mount Palomar reflector. Construction began in 1930 and was to take 20 years. It remains perhaps the most famous telescope in the world, although by 1975 it was second in size to the 236 in (6 m) telescope at the Soviet Special Astrophysical Observatory in the Caucasus. Future giant telescopes will have mirrors built of close-fitting segments or will be formed by optical linking of an array of several separate mirrors.

Hales, Stephen (1677–1761) English chemist and physiologist: developed gas-handling methods; classic experimenter on plant physiology and on blood pressure.

Hales studied theology at Cambridge and in 1709 became perpetual curate of Teddington. There he stayed, refusing preferment, so that he could maintain his work in chemistry and biology. He seems to have been much influenced by

NEWTON's work, and his own is marked by careful measurement and the early use of physics in biology.

His book *Vegetable Staticks* (1727) describes 124 experiments on gases, which he made in several ways and collected using a pneumatic trough. This device was a major advance in gas manipulation. Oddly, he assumed all the gases he made were air; it was PRIESTLEY, using and improving Hales's methods, who examined their different properties. In the same book, Hales describes his experiments showing that plants take in a part of the air (actually CO_2) and that this is used in their nutrition. He measured growth rates and showed that light is needed, and that water loss (by transpiration) is through the leaves and causes an upward flow of sap, whose pressure he measured. Later he examined blood pressure, inserting and tying a vertical tube 11 feet long into an artery of a horse to measure the height to which the blood rose, and calculating also the output from the heart and the flow-rate in arteries, veins and capillaries. He showed that capillaries are liable to constriction and dilation, which later was seen to be of great significance.

His inventiveness was wide-ranging; he worked on the preservation of foods, water purification, the ventilation of buildings and ships, and the best way to support pie crusts. His theme was usually the application of physics to problems in biology.

Hall, Asaph (1829–1907) US astronomer: discovered the moons of Mars.

Hall discovered the two Martian satellites in 1877, naming them Phobos and Deimos (after the sons of Mars, meaning 'fear' and 'terror'). By a curious coincidence Jonathan Swift in *Gulliver's Travels* 150 years earlier had suggested that Mars had two satellites, whose size and orbital period accurately matched those of Phobos and Deimos. Hall was also the first to measure accurately the period of rotation of Saturn.

Haller, Albrecht von (1708–77) Swiss anatomist and physiologist: pioneer of neurology.

A child prodigy, Haller is claimed to have written a Greek dictionary at age 10. Haller's working life was divided into a period spent founding the medical school at Göttingen, and his last 24 years spent back in his native Bern. A man of wide-ranging talents, Haller was the first to offer views on the nervous system of a modern kind. He recognized the tendency of muscle-fibres (which had been discovered by LEEUWENHOEK) to contract when stimulated, or when the attached nerve is stimulated, and he named this 'irritability'. He showed that only the nerves can transmit sensation and that they are gathered into the brain. His work in neurology was extended by C BELL and by MAGENDIE. He wrote the first textbook of physiology and worked in the circulation, respiration and digestion; always with an emphasis on experiment. He was also a poet, a bibliographer, a botanist and a writer on politics.

Halley, Edmond [halee, hawlee] (1656–1742) English astronomer and physicist: made numerous contributions to astronomy and geophysics.

Son of a wealthy businessman, Halley was an experienced observer as a schoolboy before he entered Oxford in 1673. He became a remarkable and prolific scientist, who made important discoveries in many fields. He was also highly likable and even good-looking. He made his name as an astronomer by travelling at the age of 20 to St Helena, where he remained for 2 years to produce the first accurate catalogue of stars in the southern sky (also the first telescopically determined star survey), which was published in 1679. His interest in comets was kindled by the great comet of 1682, which prompted him to compute the orbits of 24 known comets; noting that the orbits of comets seen in 1456, 1531, 1607 and 1682 were very similar, he deduced that they were the same body and predicted its return in 1758. (It is now known by his name.) It was the first correct prediction of its kind and demonstrated conclusively that comets were celestial bodies and not a meteorological phenomenon, as had sometimes been believed.

He was a keen explorer and commanded small naval ships (not very competently) in expeditions, including one in the Southern Ocean in which he failed to discover Antarctica and nearly lost his ship.

His other astronomical discoveries were numerous: in 1695 he proposed the secular acceleration of the Moon; in 1718 he observed the proper motion of the stars after observing Sirius, Procyon and Arcturus; he was the first to suggest that nebulae were clouds of interstellar gas in which formation processes were occurring. He succeeded FLAMSTEED as Astronomer Royal in 1720 at the age of 63 and commenced a programme of observation of the 19-year lunar cycle, a task that he completed successfully and which confirmed the secular acceleration of the Moon. Halley's celebrated friendship with NEWTON enabled him to persuade Newton to publish his *Principia* through the Royal Society, of which Halley was clerk and editor. When the Society was unable to finance the book Halley paid for its printing himself.

If his interests within astronomy were broad, so were his achievements in other branches of science. In 1686 he published the first map of the winds on the Earth's surface and formulated a relationship between height and air pressure; between 1687 and 1694 he studied the evaporation and salinity of lakes and drew conclusions about the age of the Earth; between 1698 and 1702 he conducted surveys of terrestrial magnetism and of the tides and coasts of the English Channel; in 1715 he correctly proposed that the salt in the sea came from riverborne land deposits. He realized that the aurora borealis was magnetic in origin, constructed the first mortality tables, improved understanding of the optics of rainbows and estimated the size of the atom. He devised, and personally used, the first diving bell. If Halley was fortunate in his talents, wealth and personality, he certainly made good use of his assets.

Hamilton, Sir William Rowan (1805–65) Irish mathematician: invented quaternions and a new theory of dynamics.

Hamilton was born in Dublin. His father, a solicitor, sent him to Trim to be raised by his aunt and an eccentric clergyman-linguist uncle when he was 3. He showed an early and astonishing ability at languages (he had mastered 13 by age 13) and later also wrote rather bad poetry and corresponded with Wordsworth, Coleridge and Southey. At 10 he developed an interest in the mathematical classics, including those by NEWTON and LAPLACE, and later went to Trinity College, Dublin. There he did original research on caustics (patterns produced by reflected light).

At Trinity College Hamilton was one of the very few people to have obtained the highest grade (*optime*) in two subjects: Greek and mathematical physics. His work on caustics led to his discovery of the law of least action: for a light path the action is a simple function of its length and the light travels along a line minimizing this. Such least action principles dependent on a function of the path taken can powerfully express many laws of physics previously given in more clumsy differential equation form.

At age 22, and before he graduated, Hamilton was appointed professor of astronomy at Dublin (1827) and made Astronomer Royal of Ireland, so that he would be free to do research. He was not a good practical astronomer, despite engaging three of his many sisters to live at the Dunsink observatory to help him.

The field of complex numbers interested him, and he invented quaternions. From 1833 he had considered $a + ib$ as an ordered pair (a,b) and considered how rotations in a plane were described by the algebra of such couples. The quaternion refers to a triple and describes rotations in three dimensions. This was an algebra in which (for the first time) the commutative principle that $ij = ji$ broke down. For a quaternion $a + bi + cj + dk$, $i^2 = j^2 = k^2 = ijk = -1$ and $ij = -ji$. While important for the way that the concepts of algebra were generalized, the subject never had major uses in physics, which was better served by vector and tensor analysis.

In his later years Hamilton became a recluse, working and drinking excessively. His name remains familiar in the Hamiltonian operators of quantum mechanics.

Hanson, (Emmeline) Jean (1919–73) British biophysicist: co-deviser of sliding-filament theory of muscle contraction.

As a zoologist who became a biophysicist, Jean Hanson first became interested in how muscle contracts during the 1940s when she was a research student in London, studying the blood vessels of annelids. Thereafter she spent her career with the Medical Research Council Biophysics Unit there, except for a fruitful period at MIT in the 1950s. Her skill in classical microscopy led her to phase contrast microscopy and then to electron microscopy in its early days, and she applied all three methods to the study of muscle. With H E HUXLEY her results led to the sliding-filament theory of muscle contraction, first applied by them to striated muscle. This is made up of fibres, which are built up of myosin filaments and more slender actin filaments, and the theory was essentially that contraction is due to an interdigitated sliding motion of these which shortens the fibre. In the late 1950s Jean Hanson went on to show that in the smooth muscle of invertebrates a similar mechanism operated. She was elected to fellowship of the Royal Society in 1967.

Harden, Sir Arthur (1865–1940) British biochemist.

Educated at Manchester and Erlangen, Harden worked at the Lister Institute in London throughout his career. He is best known for his work on the alcoholic fermentation of sugars, which BUCHNER had shown could be brought about by a cell-free extract of yeast, thought to contain an enzyme, 'zymase'. Harden showed that zymase is actually a mixture of enzymes, each a protein that catalyses one step in the multi-step conversion of sugar to ethanol; and that non-protein co-enzymes are also present in zymase and are essential for the process. He found that sugar phosphates are essential intermediates in fermentation and that conversion of carbohydrate to lactic acid in muscle is intimately related to fermentation. These are both key matters in the development of biochemistry. Harden shared a Nobel Prize in 1929.

Hardy, Godfrey Harold (1877–1947) British mathematician: developed new work in analysis and number theory.

Hardy was the son of an art teacher; he was a precocious child, whose tricks included factorizing hymn-numbers during sermons. His early mathematical ability won him a scholarship to Winchester School and then another to Trinity College, Cambridge, where he was elected a Fellow. In 1919 he became Savilian Professor of Geometry at Oxford, but returned to Cambridge in 1931 as professor of pure mathematics.

Hardy's early research was on particularly difficult integrals and he also produced a new proof of the prime number theorem: that the number of primes not exceeding x approaches $x/\log_e x$ when x approaches infinity. He began to collaborate with his close friend J E Littlewood (1885–1977) on research into the partitioning of numbers, on the GOLDBACH conjecture (that every even number is the sum of two prime numbers, still unproved) and later the RIEMANN zeta-function. Together they wrote nearly 100 papers during 35 years.

In 1908 Hardy and W Weinberg (1862–1937) discovered independently a law fundamental to population genetics. It describes the genetic equilibrium of a large random-mating population and shows that the ratio of dominant to recessive genes does not vary down the generations. This was strong support for DARWIN's theory of evolution by natural selection. It was Hardy's only venture into applied mathematics.

Hardy was an excellent teacher and introduced a modern rigorous approach to analysis. He encouraged the young Indian genius Srinivasa Ramanujan (1887–1920), bringing him to Cambridge to do research. Hardy was a staunch anti-Christian, a firm friend of Bertrand Russell and a passionate and talented cricketer and 'real tennis' player.

Harrison, John (1693–1776) English navigator and horologist.

A Yorkshireman, Harrison was a carpenter-turned-clockmaker who invented the marine chronometer and so revolutionized marine navigation and exploration.

With the colonial and naval ambitions of the European nations in the late 17th and early 18th-c, marine navigation was an issue of key importance. In 1707, 2000 lives were lost when an English fleet unexpectedly struck rocks off the Scilly Isles, over 100 miles off course. Such was the need for reliable marine navigation that in 1714 the British government put up a prize of £20 000 for a practical way of measuring longitude at sea to within 30 miles after a voyage of 6 weeks, corresponding to keeping time to within 3 seconds a day.

In those days position at sea was found by measuring the angles of certain stars at an accurately known time. Since the rotation of the Earth means that an error of just 1 minute in time leads to an error of 15 nautical miles in longitude, accurate and reliable measurement of time was the limiting factor.

Harrison set about winning the prize and, after making several remarkable wooden clocks that were the most accurate timekeepers of their day, devised in 1737 a 'sea clock' which used a pair of dumb-bells linked by springs in place of a swinging pendulum. H1, as it is known, was tested by the Admiralty and proved a great success, correctly predicting landfall when the ship's master thought he still had 60 miles to run. Something of a perfectionist, Harrison then proceeded to improve his design, making a number of innovations along the way, including the bimetallic strip to compensate for temperature variations, and 24 years later produced his masterpiece, H4.

During sea trials lasting a year, H4 was found

to be a mere 39 seconds out on its return to Britain, well within the error limits required for the prize. Unfortunately for Harrison, now almost 70, the government then proceeded to set him a series of further tasks before it would agree to pay him the prize money. After Harrison had spent another 10 years on these tasks, King George III heard of his plight and intervened, spending 10 weeks personally testing H5 at his private observatory together with Harrison and the Astronomer Royal. They found H5 to be accurate to within a third of a second a day and the following year, in 1773, Harrison was eventually awarded his money, less expenses. Shortly before Harrison's death in 1776, COOK proved the real value of his work by taking his chronometer on his second voyage to the Pacific, where he used it to accurately map Australia and New Zealand.

Hartmann, Johannes Franz (1865–1936) German astronomer: discovered interstellar gas.

In 1904 Hartmann discovered the first strong evidence for interstellar matter. While observing the spectrum of δ Orionis, a binary star, he noticed that the calcium lines were not DOPPLER-shifted like the other lines in its spectrum (as would be expected for an orbiting pair of stars). This must mean that the calcium lines come from other gaseous matter between δ Orionis and Earth.

Harvey, William (1578–1637) English physician: founded modern physiology by discovering circulation of the blood.

Harvey was the eldest of seven sons in the close family of a yeoman farmer. After Cambridge he went to the greatest medical school of the time, at Padua, and studied there in 1600 under FABRIZIO, who discovered but did not understand the valves in the veins. Back in London from 1602, Harvey was soon successful and was physician to James I (and later to Charles I), but his main interest was in research. By 1615 he had a clear idea of the circulation, but he continued to experiment on this. He did not publish his results until 1628 in the poorly printed, slim book *Exercitatio anatomica de motu cordis et sanguinis in animalibus* (On the Motions of the Heart and Blood in Animals), usually known as *De motu cordis*. It is one of the great scientific classics. By dissection and experiment Harvey had shown that the valves in the heart, arteries and veins are one-way; that in systole the heart contracts as a muscular pump, expelling blood; that the right ventricle supplies the lungs and the left ventricle the rest of the arterial system; that blood flows through the veins towards the heart; these facts, and the quantity of blood pumped, led to his conclusion that 'therefore the blood must circulate'. This idea refuted the earlier views of GALEN and others, and Harvey was ridiculed, but his work was accepted within his lifetime. He was not able to show how blood passed from the arterial to the venous system, as there are no connections visible to the eye. He supposed correctly that the connections must be too small to see; MALPIGHI observed them with a microscope soon after Harvey's death. Modern animal physiology begins with Harvey's work, which was as fundamental as NEWTON's work on the solar system.

Harvey was an enthusiastic, cautious and skilful experimenter. Another area of his work was embryology; his book *On the Generation of Animals* (1651) describes his work on this, which was soon superseded by microscopic studies. His work on animal locomotion was not found until 1959.

In appearance, Harvey was short and round-faced with dark hair. He wore a dagger and was said to be 'quick-tempered'. (See panel overleaf.)

Haüy, René Just [hahwee] (1743–1822) French mineralogist: founder of crystallography.

Although Haüy's father was an impoverished clothworker, the boy's interest in church music secured an education for himself through the help of the church. He became a priest, and professor of mineralogy in Paris, in 1802. STENO had shown in 1670 that the angle between corresponding faces in the crystals of one substance is constant (irrespective of crystal size or habit), but he had not studied crystal cleavage. HOOKE and also HUYGENS proposed that a crystal must be built of identical particles piled regularly 'like shot' or, as NEWTON phrased it, 'in rank and file'. Haüy developed these ideas; he accidentally broke a calcite crystal and noted

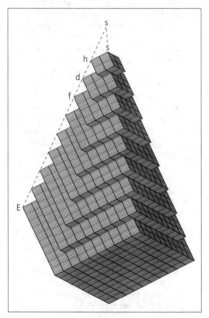

Haüy's drawing of a crystal built up of rhomboidal units

THE HISTORY OF MEDICINE

The learned Western medical tradition traces its origin to the works of a mysterious Greek physician, HIPPOCRATES. Some 60 writings survive, most dating from the period 430–330 BC, which have since antiquity been attributed to Hippocrates, but it is impossible to say with certainty which, if any, he actually wrote. The Oath, originally serving the interests of a Pythagorean medical sect, has been adopted by some modern medical schools as embodying the ethical ideals at which a physician should aim. The *Aphorisms* opens with the famous statement, 'Life is short, the art is long, opportunity fleeting, experience deceptive, judgment difficult'. The Hippocratic texts are very disparate and even contradictory, but they all present a 'naturalistic' view of diseases as having natural causes and being amenable to natural treatments, rather than being caused by gods or spirits and needing religious or magical treatments.

The elements of the 'humoral' theory can be found in the Hippocratic writings. The four humours are blood, yellow bile, black bile and phlegm. Their balance supposedly produced a state of health and their imbalance a state of disease, and they could be adjusted by diet, bleeding and other interventions in order to maintain health or restore it when lost. The Hippocratic doctrines, and particularly the humoral theory, were interpreted and developed in an idiosyncratic fashion by GALEN, a Greek physician of the 2nd-C AD who lived and worked in Rome, and it is his medical system that came to dominate Western medicine for well over 1000 years. No physician has been more important historically. To the humoral theory of Hippocrates, Galen added a stress on anatomizing and a rationalistic approach to healing and pharmacy. Galenic medicine is aimed at the welfare of the individual, not of people *en masse*, and is of little use in the treatment of epidemic disease. Galen's writings survive in great quantity, and became the basis of what was taught in the medical schools of Alexandria and Byzantium. His works were later systematized and developed by the Arab physicians Ibn Sina (Avicenna) (980–1037) and Ibn Rushd (Averroës) (1126–98), and the Jewish physician Moses Maimonides (1135–1204). Galen's works were also adopted in the medieval Western universities as the basis of the medical curriculum.

The vernacular and anti-authoritarian medicine of PARACELSUS, which was based on a mystical reading of nature and disease and on chemical cures, proved very popular in the 16th-C, and was developed later by HELMONT. But in general, Renaissance medicine was concerned with attempts to recover the full Greek texts of Galen and Hippocrates, and the practices of these ancients were held up as models to copy. In this tradition, FRACASTORO analysed the new venereal disease in Galenic terms, and named it after a character from Greek classical myth, Syphilis. Similarly, in writing his great illustrated book on the fabric of the human body (1542), VESELIUS was trying to put Galen's anatomical project into practice, and in doing so found many errors in Galen. Meanwhile, FABRICIUS sought to put ARISTOTLE's anatomical project back into practice, and in doing so discovered the 'valves' in the veins. Fabricius's pupil HARVEY took this Aristotelian investigation further, and discovered the circulation of blood in animals (1628), perhaps the most important physiological discovery ever made. DESCARTES, also trying to reinstate an ancient understanding of the world, that of the Greek atomists, adopted Harvey's discovery into his 'mechanistic' account of physiology (in his *Discourse on Method*, 1637, and *On Man*, 1651), in which everything is explained in terms simply of matter in motion, an approach of enormous influence.

These two traditions sparked off much investigative work in the second half of the 17th-C, including that of Christopher Wren (1632–1723), and LOWER on the blood and the heart, MALPIGHI on the microscopic structure of the lungs, and Giovanni Alfonzo Borelli (1608–79) on the mechanism of the muscles. In contrast, SYDENHAM investigated epidemics among Londoners, interpreting Hippocratic medicine as demanding the physician to watch the course of disease and cure in nature and not to be concerned about the inner working of the body. The mechanistic–vitalistic synthesis of all these traditions by Hermann Boerhaave (1668–1738) dominated medical thinking throughout the 18th-C and was developed by Gerard van Swieten (1700–72) and by HALLER. Medical practitioners extended their interests to normal childbirth during the 17th-C. The investigation into normal and abnormal birth by François Mauriceau (1637–1709) and Hendrick van Deventer (1651–1724), and the invention of obstetric forceps by the Chamberlen family in London, were of great importance. William Smellie (1697–1763) and William Hunter (1718–83) took the work on childbirth further during the 18th-C.

Paris played an important role in the great transformation of medicine at the end of the 18th-C. The field of 'clinical' medicine was defined, based on the Enlightenment desire to base all knowledge on sense-experience alone, following the work of the Abbé de Condillac (1715–80). Here, in the course of the French Revolution, the great public hospitals were used in a systematic attempt to correlate signs and symptoms during life with post-mortem dissections, thus enabling the physician to know what internal changes were happening during disease. Among the leaders of this development were Philippe Pinel (1745–1826) and Jean Nicholas Corvisart (1755–1821). Many new disease syndromes were identified, and investigation of lesions was taken to the level of tissue examination by BICHAT. This 'clinical' revolution was epitomized by the invention of the stethoscope ('chest-examiner') by LAËNNEC, which has come to be the emblem of the physician.

During the 19th-c, nursing was transformed from a religious to a secular vocation, and it was especially the exploits of Florence Nightingale (1820–1910) in the Crimean War that raised public consciousness about the desirability of trained nurses for patient care in hospital. Women became engaged in a parallel struggle to become doctors, and among the pioneers were ELIZABETH BLACKWELL in the USA and ELIZABETH GARRETT ANDERSON and SOPHIA JEX-BLAKE in Britain.

Laboratory medicine was built on techniques to cultivate and control micro-organisms, and was developed from the mid 19th-c, led by PASTEUR and KOCH, who worked in rivalry to develop techniques to identify, culture, and control disease-causing bacteria and other pathogenic micro-organisms. In the last three decades of the century these two investigators and their followers identified the causal agents of many infectious diseases, such as malaria (Alphonse Laveran, 1845–1922), tuberculosis, cholera (both Koch), and plague (Alexandre Yersin, 1863–1943, and KITASATO). Pasteur's school was particularly concerned to develop preventive vaccines and Pasteur's own vaccine against rabies in 1885 was an achievement of great importance. Serums were developed that destroy the toxins produced by certain micro-organisms and produce immunity, and BEHRING received the first Nobel Prize for physiology or medicine in 1901 for developing serum therapy against diphtheria. EHRLICH and MECHNIKOV later received Nobel Prizes for their work on immunity and LISTER applied the insights of Pasteur to produce aseptic conditions that would permit safe surgery.

Laboratory medicine was spurred on by the desire of industrialized Western countries to develop colonies and make them safe. Before the end of the century the causal agents and modes of transmission of many tropical diseases had been discovered, including the transmission of yellow fever by mosquitoes, discovered by REED and his collaborators, and sleeping sickness by the tsetse fly, discovered jointly by Aldo Castellani (1877–1971), BRUCE, and David Nabarro (1874–1958). RONALD ROSS received the second Nobel Prize for physiology or medicine for identifying the life-cycle of the causative agent of malaria. After the triumph of germ-theory, a number of earlier workers acquired posthumous fame, being credited with having been early bacteriologists, such as Fracastoro from the 16th-c and SEMMELWEIS who, in the early 19th-c, urged the washing of the doctor's hands to prevent the spread of childbed fever in labour wards. Laboratory medicine came to dominate the public health movement which, under the leadership of people such as Edwin Chadwick (1801–90) and John Simon (1816–1904) in Britain and Max von Pettenkofer (1818–1901) in Bavaria, had been seeking to improve the sanitary conditions of the towns.

During the 20th-c an increasing degree of specialization in medical research and the advent of the high-tech hospital have meant less opportunity for individuals to play a decisive role in medical advance. But the work of ALEXANDER FLEMING in discovering the antibiotic properties of the penicillin mould, and the efforts of FLOREY and CHAIN in developing it for medical use, was a further great turning point in modern medicine. All three shared the Nobel Prize in 1945. The more dramatic nature of laboratory medicine and surgery has tended to overshadow developments in clinical and social medicine, pharmacy, and epidemiology – though all contribute to the high state of medicine today. However, although life expectancy has greatly increased in the last two centuries, it is a matter of debate quite how much of this can be attributed to improvements in the scientific understanding of physiology and the greater availability of medical care. Better nutrition, clean water, and effective sewers have probably been just as important.

Andrew Cunningham, Wellcome Unit for the History of Medicine, Cambridge

that the pieces were all rhombodedral, which implied a common underlying structure. He showed in 1784 that the faces of a calcite crystal might be formed by stacking cleavage rhombs regularly, if the rhombs are assumed to be so small that the face appears smooth (see diagram). Similar principles would lead to other crystal shapes built from appropriate structural units. In developed form, this is still the modern view.

As a priest, Haüy was in some danger in the French Revolution, but friends protected him and Napoleon (the first world leader with a scientific or engineering training) appointed him to a post and directed him to write a textbook of physics for general use.

Hawking, Stephen (William) (1942–) British theoretical physicist: advanced understanding of space-time and space-time singularities.

Hawking graduated from Oxford in physics and, after a doctorate at Cambridge on relativity theory, remained there to become a Fellow of the Royal Society (1974) and Lucasian Professor of Mathematics (1979). He developed a highly disabling and progressive neuromotor disease while a student, limiting movement and speech. His mathematical work is carried out mentally and communicated when in a developed form. His life and work is an extraordinary conquest of severe physical disability.

Hawking began research on general relativity, recognizing that EINSTEIN's theory takes no account of the quantum mechanical nature of physics and is not adequately able to describe gravitational singularities such as 'black holes' or the 'Big Bang'. In *The Large Scale Structure of Space-Time* (with G F R Ellis, 1973) he showed that a space-time singularity must have occurred at the beginning of the universe and space-time itself, and this was the 'Big Bang' (a point of indefinitely high density and space-time curvature). The universe has been expanding from this point ever since.

Hawking greatly advanced our knowledge of black holes – these are singularities in space-time caused by sufficient mass to curve space-time enough to prevent the escape of light waves (photons). The boundary within which light cannot escape is called the event horizon and is given by the SCHWARZSCHILD radius. Hawking established that the event horizon can only increase or remain constant with time, so that if two black holes merge the new surface area is greater than the sum of that of the components. He showed that black hole mechanics have parallels with thermodynamic laws (in which entropy must increase with time). He also showed that black holes result not only from the collapse of stars but also from the collapse of other highly compressed regions of space.

During 1970–74 Hawking and his associates proved J Wheeler's (1911–) conjecture (known as the 'no-hair theorem') that only mass, angular momentum and electric charge is conserved once matter enters a black hole.

In 1974 Hawking deduced the extraordinary result that black holes can emit thermal radiation. For example if a particle–antiparticle pair are created close to an event horizon, and only one falls inside, then the black hole has effectively emitted thermal radiation. A finite temperature can therefore be associated with a black hole, and the analogy between black hole mechanics and thermodynamics is real.

More recently Hawking has sought to produce a consistent quantum mechanical theory of gravity, which would also link it with the other three basic types of force (weak nuclear, strong nuclear, and electromagnetic interaction). His non-technical book *A Brief History of Time* (1988) was an outstanding publishing success. He was admitted as a Companion of Honour by the Queen in 1989.

Hayashi, Chusiro (1920–) Japanese astrophysicist.

Educated at Kyoto and Tokyo, Hayashi returned to Kyoto in 1954 and taught there throughout his career. In 1948 the 'Big Bang' theory offered an explanation for the relative abundance of the chemical elements in the universe, using in part the idea due to GAMOW in 1946 that thermonuclear reactions would occur in the very hot and dense first stages of the 'Big Bang'. In these conditions they assumed that neutrons could combine with protons. In 1950 Hayashi modified 'Big Bang' theory by calculating that within the first 2 s the temperature would be above 10^{10} K, which is above the threshold for the formation of electron–positron pairs from photons. Working through the consequences of these ideas, he and others showed that they would lead to a fixed hydrogen:helium ratio in stars, with only small amounts of heavier elements. This was the first of the many variants on the 'Big Bang' idea, whose common feature is that the universe began as an explosive event from some primordial state, with space itself expanding along with the matter in it. The concept is broadly accepted by cosmologists, but whether it will ultimately be followed by contraction and repetition (ie an oscillating universe) or by limitless expansion, or by a violent final contraction (the 'Big Crunch') remains an open question.

Heatley, Norman (George) (1911–) British biochemist: made possible the bulk production of the first antibiotic, penicillin.

Heatley was born in Suffolk, the son of a veterinary surgeon. After graduation at Cambridge he continued there, taking a doctorate in biochemistry. His first job, in 1936, was in the School of Pathology at Oxford, where his subsequent career was largely spent. He was working there with FLOREY when, from 1939, the latter showed that the antibiotic penicillin had important potential clinical value. In the early work on it in Oxford it was secured only in small quantity, barely sufficient for even limited testing. However, Heatley devised the cylinder-plate assay method and a solvent-transfer extraction process, which made penicillin available on a substantial scale. Without this work it would have remained a laboratory curiosity; in fact, large-scale production in the UK and the USA began from 1942, with dramatic results in the treatment of infections. From that time, penicillin and later antibiotics dominated such treatment.

Heaviside, Oliver (1850–1925) British physicist: developed theoretical basis of cable telegraphy.

Lacking a university education, Heaviside worked initially as a telegraph operator until deafness forced him to stop. Unmarried, he lived with his parents, never obtained an academic position (although he received several honours) and eventually died in poverty.

Working alone, Heaviside developed much of the mathematics behind the theory of telegraphy and electric circuits, formulating the now familiar concepts of impedance, self-inductance and conductance and using complex numbers in the analysis of alternating current networks many years before others did so. He showed how audio signals could be transmitted along cables without distortion and proposed a method of using a single telephone line to carry several conversations simultaneously (multiplexing). Following MARCONI's success in transmitting radio signals across the Atlantic, he suggested, independently of A E Kennelly (1861–1939), that there had to be a reflecting layer in the upper atmosphere, otherwise the curvature of the Earth would prohibit the signals from being received. The existence of the Heaviside layer was demonstrated experimentally over 20 years later by APPLETON. Also known as the E-region or middle layer of the ionosphere, it usually lies between 90 km and 150 km above the Earth.

Although most of Heaviside's earlier work was ignored, leading him to become embittered and a recluse, his valuable contributions were later acknowledged and he was elected a Fellow of the Royal Society in 1891. The last, unpublished, volume of his *Electromagnetic Theory* was torn up by burglars a few days after his death, but is known to have described a unified field theory combining electromagnetism and gravitation.

Heezen, Bruce Charles (1924–1977) US oceanographer: demonstrated existence of turbidity currents.

Educated at Iowa State University and Columbia University, New York, Heezen worked at the Lamont–Doherty Geological Observatory at Columbia from 1948 until his death. In 1952 he used records of the times of the breakage of underwater communications cables off the Grand Banks, Newfoundland, during the 1929 Grand Banks earthquake to demonstrate that a sediment 'slump', or turbidity current, travelling at up to 85 km per hour had taken place. Such turbidity currents transporting large amounts of sediment down the continental slope had previously been proposed as the cause of submarine canyons, but not observed. In 1957 Heezen and EWING demonstrated the existence of a central rift in mid-ocean ridges.

Heisenberg, Werner Karl [hiyznberg] (1901–76) German physicist: developed quantum mechanics and discovered the uncertainty principle.

Heisenberg, son of a professor of Greek at the University of Munich, was educated at Munich and Göttingen. He worked with BORN in Göttingen, and BOHR in Copenhagen. In 1927 he returned to a professorship in Germany at Leipzig.

Heisenberg was a major creative figure among those who revolutionized physics by quantum mechanics. At 24 he formulated a non-relativistic form of the theory of quantum mechanics, producing the matrix mechanics version, and received the 1932 Nobel Prize for physics for this work. An equivalent theory called wave mechanics was produced independently by SCHRÖDINGER in 1925 and they were shown to be equivalent by VON NEUMANN.

Heisenberg broke away from the visual concept of the atom and avoided problems such as the apparent wave-particle duality by considering only observable quantities of the atom as 'real'. He separated in the theory the system of interest and operations on that system to produce an observable quantity. Respectively these were expressed as a matrix and a mathematical operation on the matrix to give a value. He used the theory to predict successfully the observed frequencies and intensities of atomic and molecular spectral lines. He concluded that two forms of molecular hydrogen, called ortho- and para-hydrogen, exist with their nuclear spins aligned in the former and opposed in the latter.

In 1927 Heisenberg discovered a further aspect of quantum mechanics, the principle of uncertainty; that it is impossible to determine exactly both the position and momentum of a particle simultaneously. The uncertainty in position Δx and in momentum Δp obey $\Delta x \Delta p \geq h/4\pi$ where h is the PLANCK constant. This relation removed absolute determinacy, or cause and effect, from physics for the first time and replaced it with a statistical probability. This deeply troubled EINSTEIN and some others, but is now generally accepted. LAPLACE's claim that the future of the universe could in principle be deduced from the position and velocity of all particles if given at one instant was rejected. For example: to try to locate the accurate position of an electron, radiation of short wavelength (such as gamma rays) might be bounced off it. However, such energetic rays will radically alter the electron's momentum on collision, so that certainty in its position is attempted at the expense of that in momentum. In 1932, after CHADWICK had discovered the neutron, Heisenberg proposed that a nucleus of protons and neutrons was a more satisfactory model than one of protons and electrons, as had been assumed. The components of the nucleus should be held together by quantum mechanical exchange forces, which was later confirmed by YUKAWA's theory of the strong nuclear interaction by which pi-mesons were exchanged. Later Heisenberg put forward a unified field theory of elementary particles (1966) which received little general support.

During the Nazi period, Heisenberg chose to remain and preserve the German scientific tradition, though he was not a Nazi supporter. He was attacked by the Nazis for refusing to reject in any way Einstein's physics. As a consequence he lost the chance of the professorship at Munich in 1935 as SOMMERFELD's successor. During the war Heisenberg was called to lead the atomic energy and weapons programme, becoming director of the Kaiser Wilhelm Institute in Berlin in 1941. After the war he helped establish the Max Planck Institute at Göttingen and moved with it to Munich in 1955 as its director.

Heisenberg's wartime role is controversial. He claimed to have had no intention of allowing an atomic bomb to reach Hitler's hands, and stated that in such a key role he could have diverted the programme if it ever neared success. He claimed to have revealed this to BOHR in 1941, but Bohr said that he failed to understand Heisenberg's guarded comments. The weapon was not produced by Germany probably because of a higher priority for planes and flying bombs.

Helmholtz, Hermann (Ludwig Ferdinand) von (1821–94) German physicist and physiologist: a discoverer of the law of conservation of energy; achieved major results in theories of electricity and magnetism, and on the physiology of vision and hearing.

Helmholtz must be the most versatile scientist of his century: he did first-class work in physics and physiology, both theoretical and experimental, and he was no mean mathematician. He has been claimed to be the last scholar whose work ranged over the sciences, philosophy and the arts. He believed that his diversity of interests was helpful to him in giving novel viewpoints in his researches.

As a child he was 'delicate' and often ill, but his parents did their best to amuse him. At school he found he did badly at memory work and rote-learning, but he enjoyed the logic of geometry and was delighted by physics. However, his father knew of no way of studying physics except as a medical student, for which he could get a university grant, provided he followed it by some service as an army surgeon. He must have done well as a medical student at Berlin, because after a short time as an army surgeon he became professor of physiology at Königsberg and later at Bonn and Heidelberg, and of physics at Berlin.

He was 26 and working in medicine when he published his pamphlet *On the Conservation of Force* in 1847; his ideas in it on the law of conservation of energy were much more precise than the ideas on the law given by J R Mayer (1814–78), and more wide-ranging than those of JOULE; Helmholtz gave examples of the law in mechanics, heat, electricity and chemistry, with numerical values.

By 1850 he had moved to physiological optics and colour vision. An early success for him in this area was his invention of the ophthalmoscope for viewing the human retina. (BABBAGE had invented a similar device 3 years earlier, but his medical friends failed to use it.) Helmholtz was more successful, and his device not only revolutionized the study of diseases of the eye but also was of value to physicians generally in giving the only direct view of the circulatory system. His study of the sense organs was continued in work on the ear and the mechanism of hearing, where he argued that the cochlea resonates for different frequencies and analyses complex sounds; and he developed a theory on the nature of harmony and musical sound (he was a skilful musician).

Earlier he had worked on the speed of nerve impulses and showed that this was of the order of a tenth of the speed of sound. He was a masterly experimenter, but in later life he gave up physiology for physics and became more interested in theoretical work, including MAXWELL's on electromagnetic radiation. He encouraged his pupil and friend HERTZ to work in this area, with important results: Hertz discovered radio waves in 1888. Other pupils included BOLTZMANN and MICHELSON; he had a great many pupils, his fame as a physicist compared with that of his friend KELVIN in England, and in Germany he was said to be 'the most illustrious man next to Bismarck and the old Emperor'.

Helmont, Jan Baptista van (c.1579–1644) Flemish alchemist, chemist and physiologist: made early studies of conservation of matter.

A member of a noble and wealthy family, Helmont studied the classics, theology and medicine before turning 'for 7 years to chemistry and the relief of the poor'. He believed in alchemy, but his own work represents a transition to chemistry proper. He used a chemical balance and understood clearly the law of indestructibility of matter (eg that metals, dissolved in acid, can be recovered). He knew a fair range of inorganic salts and the acids H_2SO_4 and HNO_3. He believed that all matter was based on two elements or principles, air and water. In an experiment on this, he grew a willow tree for 5 years, when its weight increased from 5 to 169 lb (2.27 to 76.66 kg); the earth it had grown in had hardly lost weight, and he had given the tree only rainwater. So in his view the tree (and presumably all vegetation) was made of water. He was half-correct (willow is about 50% water); and he failed to realize that the plant had taken in CO_2 from the air. He studied gases (he was the first to use the word 'gas', based on the Greek *chaos*) but he had no method of collecting them; and better distinctions between different gases had to wait for PRIESTLEY's work. He had rather confused ideas on animal digestion but, in directing thought to animal chemistry, his views were valuable.

Henderson, Thomas (1798–1844) British astronomer: first measured stellar parallax.

Henderson was a legal clerk and amateur astronomer who became director of the Cape of Good Hope Observatory in 1831. He was one of the first to detect and measure the parallax of a star, α Centauri, in 1832. (α Centauri is actually three stars and, at 4 light years distance, is still the closest star system to us). Unfortunately, his hesitation in publishing his discovery before he had thoroughly checked his result meant that BESSEL, who in 1838 made similar measurements on 61 Cygni, received most of the credit for the first determination of a stellar distance.

Henry, Joseph (1797–1878) US physicist: pioneer of electromagnetism.

Strangely, no American after FRANKLIN did much for the study of electricity for 75 years, when Henry did a great deal. In many ways Henry is the traditional American of folklore; tall, handsome and healthy, he was still a strenuous researcher at 80. Growing up in Albany in New York State with a widowed mother, he was not fond of schoolwork and at 15 was apprenticed to a watchmaker, but the business soon failed. For a year he wrote plays and acted in them, and then by chance read a book on science which reshaped his life. He attended the Albany Academy, did well and spent a period as a road engineer before taking a job as teacher of mathematics at the Academy, researching on electricity in his spare time. His research gave him enough reputation to secure a post at the College of New Jersey (which became Princeton) in 1832, teaching a full range of sciences.

In 1825 STURGEON devised an electromagnet with a varnished soft iron core wrapped by separate strands of uninsulated wire. Henry in 1829 much improved this by using many turns of thin, insulated, wire. (He supervised the making of one at Yale in 1831 that would lift a tonne.) Also in 1831 he made the first reciprocating electric motor, as 'a philosophic toy'. Like his magnets it was powered by batteries, his only source of current. He had wire in plenty, from an unknown source, and used miles of it. In 1830 he discovered electromagnetic induction, 'the conversion of magnetism into electricity'. FARADAY also discovered it independently soon after, and published first. However, Henry, in 1832, was the first to discover and to publish on self-induction, and the unit is named after him. A coil has a self-inductance of 1 henry (H) if the back emf in it is 1 volt when the current through it is changing at 1 ampere per second. In 1835 he introduced the relay, which made long-distance electric telegraphy practical, an important step in North America.

When he was 49, Henry became first director of the Smithsonian Institution. This had a curious history. James Smithson was an unrecognized bastard son of the Duke of Northumberland. Resentful of his position, he was determined that 'my name shall live in the memory of man when the Northumberlands ... are extinct and forgotten' and he therefore left a large fortune to go to the USA (with which he

had no links of any kind) to found 'an Establishment for the increase and diffusion of knowledge'. Henry shaped it well; it became 'the incubator of American science' and he was the model administrator. A strict Calvinist, he resisted patents or wealth for himself and refused for 32 years to increase his salary of $3500.

Henry, William (1774–1836) British chemist: discovered law of gas solubility.

Henry was the third and most successful son of Thomas Henry, whose profitable ventures in chemistry had included the early use of chlorine for bleaching textiles, the preparation and use of bleaching powder (Cl_2 absorbed in lime) as a useful alternative to the gas, and the making of 'calcined magnesia' (ie $Mg(OH)_2$) for medicinal use. Young William was injured by a falling beam at the age of 10; the injuries gave him ill health and pain throughout his life, and he finally killed himself. He qualified in medicine in 1807 but his research was mainly in chemistry. He is now best known for Henry's Law, which states that the mass of a gas dissolved by a given volume of a solvent, at a constant temperature, is directly proportional to the pressure of gas with which the solvent is in equilibrium. The law holds well only for slightly soluble gases at low pressures.

Henry was a close friend of DALTON but, despite superior skill and range as an experimenter, lacked his friend's boldness as a theorist and never committed himself to the atomic theory whose birth he had assisted.

Hermite, Charles [airmeet] (1822–1901) French mathematician: developed theory of hyperelliptic functions and solved the general quintic equation.

Hermite had an inability to pass examinations, and congenital lameness, but he greatly influenced his generation of mathematicians. Having entered and then been dismissed from the École Polytechnique, Hermite finally, but only just, graduated at 25; by then he was clearly an innovative mathematician. He gained a teaching post at the Collège de France, was elected to the Paris Académie des Sciences (1856) and immediately caught smallpox. Recovering, he eventually received a professorship at the École Normale (1869) and the Sorbonne (1870).

Hermite's creative work included extending ABEL's theorem on elliptic functions to hyperelliptic functions and using elliptic functions to give a solution of the general equation of the fifth degree, the quintic (1878). He proved that the number e is transcendental (that is, not a solution of any algebraic equation with rational coefficients) and used the techniques of analysis in number theory. He set out the theory of Hermite polynomials (1873), which are polynomial functions now much used in quantum mechanics, and of Hermitian forms, which are a complex generalization of quadratic forms.

Hero (of Alexandria) (lived c.AD 62) Greek physicist: invented a steam-powered engine.

Apart from writings ascribed to him, nothing is known of Hero. Some have argued that he had little scientific knowledge and was simply a recorder of ingenious devices, but recent study shows that he grasped all the mathematics of his time. His books *Pneumatics* and *Mechanics* make it clear that he was a teacher of physics, and these and other books survey the knowledge and devices known in his day, including pumps, siphons, a coin-operated machine and surveying instruments. He devised Hero's engine, a primitive reaction turbine in which steam emitted by two nozzles facing in opposite directions caused rotation.

Herschel, Caroline (Lucretia) [hershl] (1750–1848) German–British astronomical observer; discovered eight new comets.

Caroline Herschel's introduction to astronomy was entirely due to her devoted affection for her elder brother WILLIAM HERSCHEL but she became the most famous woman astronomer of her time. In 1835 she was one of the first two women elected to honorary fellowship of the Royal Astronomical Society.

Caroline, born in Hanover, was brought up to be the household servant, with minimal education; her mother believed that it was her daughter's duty to look after her brothers. Her father included Caroline in the musical instruction he gave his sons but warned her 'against all thoughts of marrying, saying I was neither hansom (*sic*) nor rich, it was not likely that anyone would make me an offer'. After her father's death Caroline continued to practise her singing and longed for independence. William Herschel proposed to his mother and elder brother that she should join him in Bath and take singing lessons, but this was only agreed to in 1772 after he had paid for a servant to replace Caroline.

Caroline's brief career as an oratorio singer faded as she became indispensable to William in his increasing interest in astronomy. Lessons in mathematics and astronomy were given at the breakfast table and Caroline became a valuable assistant. She was involved in every aspect of his work: in pounding and sieving horse manure to make material for moulds and in the grinding and polishing of the metal mirrors for telescopes. They observed as a team: she assisted in the recording, prepared catalogues and assisted in writing papers for publication.

In 1783 she began to 'sweep for comets' with a small refracting telescope and discovered three new nebulae. Her opportunities for independent observation occurred only when William was away and not needing her assistance. However, between 1786 and 1797 she discovered eight comets and gained a reputation as an astronomer in her own right. As a result of her increasing fame, she was awarded £50 a year by George III in recognition of her work as William's assistant, the first official female assistant to the Court Astronomer. William found FLAMSTEED's star catalogue very difficult to use; Caroline's revision of this, *Index to Flamsteed's Observations of the Fixed Stars*, was published by the Royal Society in 1798.

On William's death in 1822 Caroline returned to Hanover. She took a keen interest in her nephew JOHN HERSCHEL's work and for his use she compiled a new catalogue of nebulae arranged in zones, from William's work, but it was not published. In 1828 she was awarded the Gold Medal of the Royal Astronomical Society and in 1846 the Gold Medal for Science by the King of Prussia.

Caroline Herschel's contributions to astronomy were made for the love of her brother, but she brought perseverance, a sharp eye and notable accuracy to the work.

Herschel, Sir John (Frederick William) [hershl] (1792–1871) British astronomer and physicist: surveyor of the southern sky.

Although his father WILLIAM HERSCHEL, the 'gauger of the heavens', did notable scientific work, John did more and ranged outside his father's astronomical interests. For an only child his famous father was rather overwhelming, but his close friendship with his aunt CAROLINE HERSCHEL ended only with her death at 98.

He graduated from Cambridge in mathematics in 1813 with the highest distinction, then began to study law, but physics attracted him. His home experiments with polarized light gave some valuable results; he also deduced that polarized light should be rotated by an electric field, as was later confirmed experimentally by FARADAY. For a time, following two failed love affairs, he was uncertain of what career to follow. From 1816 he assisted his father in the study of double stars and nebulae, and thereafter he fixed on a scientific career, which was surprisingly wide-ranging. Astronomy continued to attract him and in the 1830s he decided to extend their survey of the sky to the Southern Hemisphere. He arrived at the Cape in 1834, accompanied by his young wife, two large telescopes, a mechanic and a children's nurse, and in 4 years of energetic work he mapped most of the southern sky.

He worked in meteorology and geophysics and planned SIR J ROSS's geomagnetic survey of the Antarctic. He was an expert chemist, and major contributions to early work on photography are due to him: he devised the cyanotype process. This used paper impregnated with an iron compound (ammonium ferricitrate), which on exposure to light formed Prussian blue. The process was simple and cheap and gave a permanent print, but the response to light was too slow for camera exposures and prints were made by placing the subject in contact with the sensitized paper and exposing to strong light (the process became much used in the 20th-c to make the 'blueprint' copies of engineer's drawings and was also used at Mafeking during the Boer War, when Baden-Powell directed that banknotes and postage stamps be reproduced during the siege).

For conventional silver-based photography he introduced 'hypo' as a fixing agent and was a pioneer in astronomical photography. In 1839 he made the first photograph on a glass plate (previously sensitized paper had been used), prepared the first coloured photographs of the Sun's spectrum and introduced the words 'negative' and 'positive' into 'photography' (also his word), and the word 'snapshot'. In 1850 he became Master of the Mint, a move which he was persuaded to make only by forceful appeals to his patriotism; he did the job well but without enjoyment for 5 years. At his death he was 'mourned by the whole nation, as a great scientist and one of the last of the universalists'.

Herschel, Sir (Frederick) William [hershl] (1738–1822) German–British astronomer: discovered Uranus, the Sun's intrinsic motion through space and the true nature of the Milky Way.

Herschel followed his father in becoming a musician in the Hanoverian Guards, entering as an oboist at 14. At 19 he came to England, working as a freelance musician before appointment as an organist in Bath. He was a keen amateur telescope maker and observer. His sister CAROLINE HERSCHEL joined him in Bath in 1772.

Herschel's first important discovery was the planet Uranus in 1781. This achievement, helped by his desire to name it after George III, resulted in his appointment the following year as Court Astronomer. This enabled him to finance the construction of a reflecting telescope 20 feet in length and with an aperture of 20 inches, with which he was to make many further discoveries. In 1787 he found two satellites of Uranus, Titania and Oberon, and soon afterwards two of Saturn, Mimas and Enceladus. In 1783 he discovered the intrinsic motion of the Sun through space by careful analysis of the proper motions of seven bright stars, showing them to converge towards a point. He had a special interest in double stars, cataloguing 800 of them and discovering in 1793 that many were in relative orbital motion. In 1820 he published a catalogue of over 5000 nebulae (a task that his son, John, was to continue in the southern hemisphere). He was the first to recognize the true nature of the Milky Way as a galaxy by counting the number of stars visible in different directions, finding that the greatest number lie in the galactic plane, and the least toward the celestial poles. Investigating the effect of parts of the Sun's spectrum on a thermometer, he discovered infrared radiation, outside the visible range.

Hershey, Alfred (Day) (1908–   ) US biologist: demonstrated information-carrying capability of bacteriophage DNA.

A graduate of Michigan State College, Hershey taught at Washington University, St Louis, until 1950 and then worked at the Carnegie Institution of Washington. His best-known work was done in the early 1950s with Martha Chase, when they proved that DNA is the genetic material of bacteriophage (the virus that infects bacteria). They used phage in which the DNA core had been labelled with radioactive phosphorus and the protein coat of the phage with radioactive sulphur. The work showed that when phage attacks a bacterial cell it injects the DNA into it, leaving the protein coat on the outside; but the injected DNA causes production of new phage, complete with protein. The DNA must carry the information leading to the formation of the entire phage particle. AVERY had been cautious on the status of DNA as an information-carrier; Hershey proved it. Hershey was a phage expert already, having shown that spontaneous mutations occurred in it in 1945.

DELBRÜCK's early impression of Hershey was that he preferred whisky to tea, liked living on his boat for months at a time and was very independent. Delbrück, S E Luria (1912–91) and Hershey were the nucleus of the 'phage group' who contributed so much to molecular biology. The trio shared a Nobel Prize in 1969.

Hertz, Heinrich Rudolph (1857–94) German physicist: discovered radio waves.

Hertz studied at the universities of Munich and Berlin, at the latter under HELMHOLTZ, whom he served as an assistant. In 1885 he was appointed professor of physics at Karlsruhe Technical College and later held a professorship at the University of Bonn.

Influenced by Helmholtz and by MAXWELL's electromagnetic theory, he was the first to demonstrate the existence of radio waves, generated by an electric spark. In 1888 he showed that electromagnetic waves were emitted by the spark and could be detected by a tuned electric circuit up to 20 m away. Further experiments demonstrated that the waves, which had a wavelength of about a foot, could be reflected (from the laboratory walls), refracted (through a huge prism of pitch), polarized (by a wire mesh) and diffracted (by a screen with a hole in it) in the same way as light, and travelled at the same speed. This was an important verification of Maxwell's ideas. Hertz also discovered in 1887 that an electric spark occurs more readily when the electrodes are irradiated with ultraviolet light (the Hertz effect), a consequence of the photoelectric effect. Hertz died at 36 (from blood poisoning) and did not live long enough to see MARCONI turn radio transmission into a means of worldwide communication. The SI unit of frequency, the hertz (Hz, one cycle per second), is named in his honour.

Hertzsprung, Ejnar [hertsprung] (1873-1967) Danish astronomer: discovered stellar spectral type/luminosity relationship.

Trained as a chemical engineer, Hertzsprung did research in photochemistry before appointment as an astronomer at the Potsdam Observatory in 1909. His work on photography led to his success in classifying stars. Hertzsprung was the first to realize that there was a relationship between the spectral colour of stars (as defined and classified by ANTONIA MAURY) and their luminosity; for most stars the more blue the colour, the brighter the star. He also found that a small proportion of stars did not fit this pattern, being far brighter than might be expected for their colour. These two groups are now called main sequence stars (the numerous faint dwarfs) and red giants (fewer, more luminous) respectively. His results were published in 1905 and 1907 in obscure journals. Independently, in 1913, RUSSELL came to the same conclusions, the usual representation of their results being known as the Hertzsprung-Russell diagram (see p. 280). This discovery had a great effect on ideas about stellar evolution.

His second important achievement was to utilize the period-luminosity relationship of Cepheid variable stars, discovered by HENRIETTA LEAVITT in 1912, as a means of calculating stellar distances. In 1913 Hertzsprung was able to determine the distance of some nearby Cepheids from their proper motion (the only method available for measuring stellar distance up to that time), and using Leavitt's results he began to calibrate the Cepheid variable technique. These first measurements of distances outside our Galaxy were developed by SHAPLEY. Hertzsprung was an active researcher until he was over 90.

Herzberg, Gerhard [hertsberg] (1904-) German-Canadian physical chemist: devised methods of analysing electronic spectra to detect new molecules and to find molecular dimensions.

Born and educated in Germany, Herzberg taught at Darmstadt from 1930 until 1935 when he emigrated to Canada. For 20 years until 1969 he was head of the physics division for the National Research Council in Ottawa, which he made into a centre of international renown in spectroscopy. He developed and used spectroscopic methods for a variety of purposes, including the measurement of energy levels in simple atoms and molecules for use in testing theories of their structure and for the detection of unusual molecules and radicals, some of which could be detected by flash photolysis in the laboratory and others by astrophysical methods (for example CH and CH$^+$ in interstellar space and the flexible C_3 in comets). For his work on the electronic structure and geometry of molecules, particularly free radicals, he was awarded the Nobel Prize for chemistry in 1971.

Hess, Germain Henri (1802-50) Swiss-Russian chemist; pioneer of thermochemistry.

When Hess was 3 years old his father, a Swiss artist, became tutor to a rich Moscow family and the boy moved from his birthplace (Geneva) to Russia. He was there for the rest of his life, taking a medical degree at Tartu in 1825 and then visiting BERZELIUS in Stockholm. The visit was only for a month, but its influence was permanent. From 1830 he studied the heat evolved in chemical reactions, as a route to the understanding of 'chemical affinity'. Rather little had been done in thermochemistry since the work by LAVOISIER and LAPLACE. Hess's Law (the law of constant heat summation) of 1840 states that the heat change accompanying a chemical reaction depends only on the final and initial states of the system and is independent of all intermediate states. The law enables the heat of reaction to be calculated in a case where direct measurement is impractical. Hess's law follows from the law of conservation of energy, but the latter was not clearly understood in 1840.

Hess researched in other areas, and did much for the development of chemistry in Russia, where he taught in St Petersburg.

Hess, Harry Hammond (1906-69) US geologist and geophysicist: proposed sea-floor spreading hypothesis.

Hess spent most of his academic life at Princeton University, moving there in 1934. During the Second World War he distinguished himself in the US Navy by conducting echo-sounding work in the Pacific, during which he discovered a large number of strikingly flat-topped seamounts, which he interpreted as sunken islands, naming them guyots (after Arnold Guyot (1807-84), an earlier Princeton geologist).

Following the war there was a great increase in knowledge about the sea bed, and it became apparent that parts of the ocean floor were anomalously young. In 1962, following the discovery of the global extent of the mid-ocean ridges and their central rift valleys by EWING, Hess proposed his sea-floor spreading hypothesis to account for these facts. He suggested that material was continuously rising from the Earth's mantle to create the mid-ocean ridges, which then spread out horizontally to form new oceanic crust; the further from the mid-ocean ridge, therefore, the older the crust would be. He envisaged that this process would continue as far as the continental margin, where the oceanic crust would sink beneath the lighter continental crust into a subduction zone, the whole process thus forming a kind of giant conveyor belt (see diagram opposite). Palaeomagnetic and oceanographic work, notably by MATTHEWS and VINE, confirmed the hypothesis. Later, as chairman of the Space Science Board of the National Academy of Sciences, Hess also had an influential effect on the American space programme.

Hess, Victor Francis (1883–1964) Austrian–US physicist: discovered cosmic rays.

Son of a forester, Hess was educated at Graz, receiving his doctorate in 1906. He worked on radioactivity at Vienna until 1920, and afterwards at Graz, New Jersey and Innsbruck. In 1931 he set up a cosmic ray observatory on the Hafelekar mountain. When the Nazis occupied Austria in 1938 Hess was dismissed, as his wife was Jewish, and he became professor of physics at Fordham University, New York City.

In 1910 T Wulf measured the background radioactivity of the atmosphere at the top of the 300 m Eiffel Tower with a simple electroscope and showed that it was greater than at ground level, indicating that it came from an extraterrestrial source; but the results were not conclusive. A Gockel in 1912 also used the rate of discharge of a gold-leaf electroscope to measure the radioactive ionization of the air, this time from a balloon; again the results were inconclusive. However in 1911–12 Hess made 10 balloon flights and showed that the ionization is four times greater at 5000 m than at ground level. Night ascents and an ascent during an eclipse of the Sun in 1912 showed that the radiation could not be from the Sun. MILLIKAN in 1925 named these high-energy particles 'cosmic rays'. Their study led to C D ANDERSON's discoveries of the positron and muon and POWELL's discovery of the pi-meson. Hess shared the 1936 Nobel Prize for physics with Anderson.

Hess, Walter Rudolf (1881–1973) Swiss neurophysiologist: showed that localized areas in the brain control specific functions.

Hess studied medicine at five universities in Switzerland and Germany and became a specialist ophthalmologist, but gave up this career to work in physiology and from 1917–51 headed physiology in the university of Zürich. The precision surgery he had learned as an ophthalmologist was to prove useful; in the 1920s he began his study of the autonomic nervous system, which controls involuntary functions such as breathing, blood pressure, temperature and digestion.

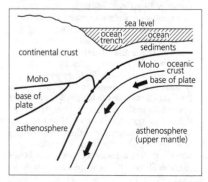

Subduction zone

It was already known roughly which parts of the brain are involved in this control; but Hess made this knowledge much more precise. He used cats into which, under anaesthetic, a fine insulated wire with a bare end was inserted so that the end was located at a defined point in the midbrain. When the animal was again conscious, a very small current was passed into the wire. Hess found that by this stimulation of small groups of cells in the midbrain he could induce a variety of reactions, including sleep, rage, evacuation and changes in blood pressure and respiration. Similarly, in the hypothalamus he located centres that appeared to control other parts of the sympathetic and parasympathetic components of the autonomic system. This influential work led to detailed mapping of the brain and began to relate physiology to psychiatry. He shared a Nobel Prize in 1949.

Hevesy, György [heveshee] (1885–1966) Hungarian–Swedish radiochemist: introduced use of radioactive 'tracers' in analysis.

Hevesy was a highly mobile chemist; he worked in at least nine research centres in seven countries. His visit to work with RUTHERFORD in Manchester (1911–13) established his interest in radiochemistry. While there he found that ordinary lead and radioactive 'radium-D' are chemically inseparable; later it was realized that radium-D is an isotope of lead, with relative atomic mass 210, that happens to be radioactive. Consequently, very small amounts of lead can be 'traced' by mixing into the lead some radium-D and then taking advantage of the fact that minute levels of radioactive material are easily located by using a counter, or by photography. In this way Hevesy and F A Paneth (1887–1958), in 1913, were able to find the solubility in water of lead sulphide and lead chromate; both are insufficiently soluble for traditional methods to measure their solubility accurately. In 1934 he used a stable but trackable isotope (deuterium) in heavy water, D_2O, to measure the water exchange between goldfish and their surroundings. In 1934 also he used radiophosphorus to locate phosphate absorption in human tissue. The technique has since been much used and suitable 'marker' isotopes for use as tracers are now widely available. In 1935 he devised a variant of this, activation analysis.

In 1922 BOHR predicted the existence of a new element and suggested that Hevesy should look for it in zirconium ore. Working with D Coster (1889–1950), who had experience of MOSELEY's X-ray method, Hevesy found the new element (atomic number 72) and it was named hafnium (Hf). He was awarded a Nobel Prize in 1943 for his work on tracers.

Hewish, Antony (1924–) British radio astronomer: identified first pulsar.

Hewish studied physics at Cambridge and worked with RYLE on radio telescopes. He became particularly interested in the scintillation of quasars, the radio equivalent of twinkling stars, and used this to examine the solar wind and clouds in interplanetary space.

In 1967 he completed a radio telescope of unusual design for further work on scintillation. Together with his student JOCELYN BELL (BURNELL), he discovered remarkably regular pulsed signals coming from a tiny star within our Galaxy; they had found the first pulsar. Many other pulsars have since been found and are believed to be rapidly rotating neutron stars, typically only 10 miles in diameter, which emit beamed radiation like a lighthouse. Hewish was awarded the Nobel Prize for physics for this discovery in 1974.

Hey, James Stanley (1909–   ) British physicist: pioneer radio astronomer.

Hey studied physics at Manchester, graduating in 1930, and obtained his master's degree in X-ray crystallography the next year. The Second World War began in 1939, and in 1942 Hey joined the Army Operational Research Group (AORG) after a 6-week course at the Army Radio School. His task was to work on radar anti-jamming methods; for a year German jamming of allied radar had been a problem and the escape of two German warships (*Scharnhorst* and *Gneisenau*) through the English Channel, aided by enemy radar jamming from the French Coast, had highlighted the problem. In February 1942 Hey had reports of severe noise jamming of anti-aircraft radars in the 4–8 m range. Realizing that the direction of maximum interference seemed to follow the Sun, he checked with the Royal Observatory and found that a very active sunspot was traversing the solar disc. He concluded that a sunspot region, which was believed to emit streams of energetic ions and electrons in magnetic fields of around 100 G (gauss), could emit metre-wave radiation. In 1942, G C Southworth in the USA also linked the Sun with radio noise, this time in the centimetre-wave region.

Later, in 1945, Hey used radar to track the paths of V2 rockets approaching London at about 100 miles high. A problem here arose from spasmodic transient radar echoes at heights of about 60 miles, arriving at a rate of five to 10 per hour. When the V2 attacks ceased, the echoes did not; Hey concluded that meteor trails were responsible and that radar could be used to track meteor streams, and could of course do so by day as well as by night.

He went on to locate in 1946 a radio source identifiable with Cygnus A, a powerful discrete stellar radio source. JANSKY, in 1933, had shown that a radio source exists in our Galaxy (the Milky Way) and REBER, using his homemade equipment, had made the first contour maps of cosmic radio noise distribution in 1944 and had shown that the Sun was a radio emitter, unaware of Hey's results of 1942, which could

not be published until after the war. Jansky and Reber moved on to other work after their initial discoveries and Hey became Head of the AORG in 1949. From 1950 radio astronomy expanded enormously, in the hands of LOVELL, HEWISH, RYLE and others.

Heyrovsky, Jaroslav [hiyrofskee] (1890–1967) Czech physical chemist: inventor of polarography.

Heyrovsky studied physical science at Prague and in 1910 came to London as a research student in physical chemistry. It was then that he began work on polarography, but this was interrupted by the First World War and the method was perfected by him in Prague in the 1920s. It is an electrochemical method of analysis, applicable to ions or molecules that can be electrolytically oxidized or reduced in solution using mercury electrodes. By plotting the voltage versus current curve (a polarogram) as the voltage between the electrodes is increased, different species are revealed as steps in the curve. The method is able to analyse several substances in one solution and is capable of high sensitivity. Heyrovsky won a Nobel Prize in 1959.

Higgs, Peter (Ware) (1929–   ) British cosmologist and particle physicist: devised the 'Higgs field' theory.

Educated at Bristol, London and Edinburgh, Higgs taught and researched in Edinburgh from 1960 and became professor of theoretical physics there in 1980.

The concept of mass is complex. In classical physics, the mass of a body is a measure of its inertia, its reluctance to undergo a change of velocity. In this sense, mass can be defined operationally as the ratio of the magnitudes of the force F and the acceleration a it produces in a body of mass m; so $m = F/a$. This definition, due to EULER, was later used by NEWTON, who recognized also that this 'inertial' mass is apparently universally proportional both to the active gravitational mass of the body (the measure of the gravitational field it produces) and to its passive gravitational mass (the measure of the gravitational pull exerted on it by other bodies). EINSTEIN's work embraced mass in his theory of relativity, which assumed the identity of inertial and gravitational mass and also showed that a mass m is related to the energy E involved in its destruction or generation by $E = mc^2$.

The elementary particles from which matter is made have masses over a large range, from the light electron to the heavier W and Z particles and the top quark. It is not obvious why particles, and therefore matter, should possess mass at all. In this context, a theory due to Higgs provides a possible explanation. The theory proposes that all space is permeated by a field (the Higgs field) which has some similarity to an electromagnetic field. A particle moving through this field, and interacting with it, will appear to have mass; and the greater the inter-action the larger the mass.

In general a field has a particle associated with it; the electromagnetic field is associated with the photon (which has no mass). Analogously, the Higgs field may be linked with a particle (the Higgs boson, believed to be heavy) or possibly with more than one particle. Proof or disproof of the existence of such a particle and other aspects of the theory would be of the highest value in theoretical physics. It would illuminate the nature of mass and gravitational attraction and perhaps link the latter with the other known physical forces (electromagnetic force, and the strong and weak nuclear forces).

Hilbert, David (1862–1943) German mathematician: originated the concept of Hilbert space.

Hilbert was educated at the universities of Königsberg and Heidelberg, spending short periods also in Paris and Leipzig. After 6 years as a *Privatdozent* (unsalaried lecturer) at Königsberg he became a professor there in 1892. In 1895 he was given the prestigious chair in mathematics at Göttingen, which he retained until 1930. He was a talented, lucid teacher and the university became a major focus of mathematical research. Hilbert contributed to analysis, topology, geometry, philosophy and mathematical physics and became recognized as one of the greatest mathematicians in history.

His earliest research was on algebraic invariants, and he both created a general theory and completed it by solving the central problems. This work led to a new and fruitful approach to algebraic number theory which was the subject of his masterly book *Der Zahlbericht* (trans The Theory of Algebraic Number Fields, 1897). He gathered and reorganized number theory and included many new and fundamental results; this became the basis for the later development of classical field theory.

Abandoning number theory while many problems remained, Hilbert wrote another classic, *Grundlagen der Geometrie* (Foundations of Geometry), in 1899. It contains fewer innovations but describes the geometry of the 19th-c, using algebra to build a system of abstract but rigorous axiomatic principles. Later, Hilbert developed work on logic and consistency proofs from this. Most important of all, he developed within topology (using his theory of invariants) the concept of an infinite-dimensional space where distance is preserved by making the sum of squares of co-ordinates a convergent series. This is now called Hilbert space, and is much used in pure mathematics and in classical and quantum field theory. His ideas on operators in Hilbert space prepared the way for WEYL, SCHRÖDINGER, HEISENBERG and DIRAC.

Hilbert's work also gave rise to the 'Hilbert programme' of building mathematics axiomatically and using algebraic models rather than

intuition. While a productive controversy arose, greatly influencing mathematical philosophy and logic, this formalistic approach was later displaced by GÖDEL's work. Hilbert's views on proof theory were later developed by G Gentzen. In 1900 Hilbert proposed 23 unsolved problems to the International Congress of Mathematicians in Paris. The mathematics created in the solution of many of these problems has shown Hilbert's profound insight into the subject.

Hinshelwood, Sir Cyril (Norman) (1897–1967) British physical chemist: applied kinetic studies to a variety of problems.

Hinshelwood's career, except for war service from 1916 working on explosives in an ordnance factory, was spent almost entirely in Oxford. His early research, developed from his war work, was on the explosion of solids, but he soon turned his interest to explosive gas reactions. In the 1920s he made a close study of the reaction of hydrogen with oxygen, which was a model for such research and led to a shared Nobel prize in 1956. He also studied the rates and catalytic effects in other gas reactions and reactions in the liquid phase. His later work applied the ideas of chemical kinetics to the growth of bacterial cells. In 1950 he made the suggestion, little noticed at the time, that in the synthesis of protein in living cells it is nucleic acid that guides the order in which amino acids are linked to form protein. The suggestion was correct.

Hinshelwood was an expert linguist and classical scholar, and was simultaneously president of both the Royal Society and the Classical Association – the only man, to date, to hold both offices. His own paintings were given a London exhibition a year after his death and he was an expert collector of Chinese ceramics.

Hipparchus (of Rhodes) [hipah(r)kuhs] (*c*.170–*c*.125 BC) Greek astronomer and geographer: discovered the precession of the equinoxes, constructed the first star catalogue and invented trigonometry.

Stimulated by the observation of a new star in 134 BC, Hipparchus constructed a catalogue of about 850 stars and was the first to assign a scale of 'magnitudes' to indicate their apparent luminosity, the brightest being first magnitude and the faintest visible to the naked eye being sixth magnitude. His scale, much refined, is still used. Comparison with earlier records of star positions led him to the realization that the equinoxes grew progressively earlier in relation to the sidereal year. (The equinoxes are the twice-yearly times when day and night are of equal length; they are the points when the ecliptic, the Sun's path, crosses the celestial equator.) He evaluated the amount of precession as 45″ of arc per year and determined the length of the sidereal and tropical years, the latter accurate to within 6 minutes. Hipparchus suggested improved methods of determining latitude and longitude on the Earth's surface, following the work of ERATOSTHENES. He constructed a table of chords, a precursor of the sine, and is therefore credited with the invention of trigonometry. All his major writing is lost, but his work was preserved and developed by PTOLEMY.

Hippocrates (of Cos) [hipokrateez] (c.460–370 BC) Greek physician: traditional founder of clinical medicine.

Little is known of Hippocrates's life with any certainty, except that he taught at Cos, travelled widely and had exceptional fame in his lifetime. The many writings under his name must include work by others, since over 100 years separate the earliest and the latest items in the 'collection'. The best of them represent a stage where medicine was emerging from a magical and religious basis and was seeking to become rational and scientific in its approach to diagnosis, prognosis and treatment. Success in this attempt was limited, but for nearly 2000 years no better work was done; like ARISTOTLE's work, Hippocratic ideas were to dominate their field and become sanctified by time. Many diseases listed in the Hippocratic Collection were ascribed to imbalance of the four 'humours' of the body and treatment was largely restricted to rest, diet and exercise rather than drugs. His case histories are admirably concise, and many of his descriptions and comments are still valid.

His, Wilhelm (1831–1904) Swiss anatomist and physiologist: introduced the microtome.

His qualified in medicine in 1855, and taught anatomy at Basle and later at Leipzig. His great practical innovation was the microtome for cutting very thin serial sections for microscopy (1866). He used it especially in his study of embryos; he gave the first accurate description of the human embryo. His son, also Wilhelm (1863–1934), first described the specialized bundles of fibres in the heart, 'the bundles of His', which are part of its electrical conducting mechanism.

Hitchings, George Herbert (1905–) US pharmacologist: deviser of new drugs.

Educated at Washington and Harvard, Hitchings worked in universities for 9 years before he joined the Wellcome company in 1942 and spent his main career there. He was notably successful in devising new drugs; in 1942 he began a programme of pharmacological study of the long-known group of purine compounds, which had first been made from uric acid and whose chemistry had been intensively examined by BAEYER and E FISCHER. Hitchings argued that the place of the purines in cell metabolism could lead to their use in the control of disease, and his approach was well rewarded. In 1951 he and GERTRUDE ELION made and tested 6-mercaptopurine (6MP) and found it to inhibit DNA synthesis and therefore cell division; it proved useful in the treatment of some types of cancer, especially leukaemia. Further work on it showed in 1959 that it inhibited production of antibodies in the rabbit; and a related compound was used from 1960 to control rejection in kidney transplantation, where the normal body processes would treat the transplant as a foreign protein and form antibodies against it. The work on 6MP also led Hitchings and Elion to the discovery of allopurinol, which blocks uric acid production in the body and which therefore forms an effective treatment for gout (which is due to uric acid deposition in the joints). Other drugs introduced by the group include the antimalarial pyrimethamine and the antibacterial trimethoprim, acyclovir, the first drug to be effective against viruses, and zidovudine, used against AIDS.

Hitchings shared a Nobel Prize in 1988 with Gertrude Elion, his principal associate from the 1940s onwards, and with JAMES BLACK.

Hitzig, Eduard [hitsik] (1838–1907) German psychiatrist and physiologist.

Although FLOURENS had shown in 1824 that removal of parts of the brain in animals led to loss of functions such as sight, he did not experiment on the effect of electrical stimulation of the brain and it was not seen as a source of muscular action. Hitzig, working as a psychiatrist in Zürich, reported in 1870 that electrical stimulation of points in the cerebral cortex of a dog produced specific movements in the opposite side of its body. With G T Fritsch (1838–97) he identified five such centres in the region now called the motor area, and such studies to relate movement to a map of the brain were continued by him and others into the 20th-c. Hitzig was less successful in his attempts to identify the site of abstract intelligence in the frontal lobes of the brain.

Hodgkin, Sir Alan Lloyd (1914–) British neurophysiologist: major contributor to understanding of nerve impulses.

As a student of biology and chemistry in Cambridge, Hodgkin became interested in the basis of nervous conduction, and he found by accident that it was easy to obtain single nerve fibres from a shore crab and that these could be used in experiments despite their small size (diameter about 35 µm). In the USA in 1938 he was impressed by the possibility of using larger nerve fibres from the squid. Some squids (the genus *Loligo*) are half a metre long and highly active, and have giant nerve fibres up to 1 mm in diameter.

It had long been known that a nerve impulse is electrical and that a major nerve fibre (axon) acts as a cable, but detailed knowledge was much advanced by the work of Hodgkin and his colleagues, especially A F Huxley (1917–). Their study of the squid axon began in 1939, was interrupted by their war service, and continued after 1945. They were able to insert a fine microelectrode into an axon and place a second electrode on the outer surface of its surrounding membrane. Even in a resting state there is a potential difference between the electrodes: the negative 'inside' has a resting potential compared with the positive exterior surface. When an impulse passes, this is reversed by the action potential for about a millisecond; the nerve impulse is a wave of depolarization passing along the axon. Hodgkin developed a detailed theory of the origin of this membrane potential, relating it to the presence of sodium and potassium ions and their distribution across the membrane. This knowledge of the biophysics of nervous conduction is basic to further understanding of the nervous system, and Hodgkin and Huxley shared a Nobel Prize in 1963 with J C Eccles (1903–   ), who also worked on nerve transmission. Hodgkin became president of the Royal Society in 1970, and Master of Trinity College, Cambridge, 1978–84.

Hodgkin, Dorothy, *née* Crowfoot (1910–94) British X-ray crystallographer: applied X-ray crystal analysis to complex biochemical molecules.

Dorothy Crowfoot was born in Cairo, where her father worked in the Education Service. Soon he moved to the Sudan, to become director of both Education and Antiquities; she visited her parents in Khartoum and always retained an interest in the region and in archaeology. From an early age her interest in chemistry (especially in crystals and identifying minerals) competed with archaeology and after school in England she went to Oxford and studied both subjects. She was advised to specialize in X-ray crystallography, did so and then went to Cambridge to work with J D Bernal (1901–71): after 2 years there she returned to an Oxford post in 1934 and remained there. She married the historian Thomas Hodgkin in 1937.

Dorothy Hodgkin developed the X-ray diffraction method of finding the exact structure of a molecule (originally devised by the Braggs) and applied it to complex organic molecules. Among her most striking successes were the antibiotic penicillin, whose structure she deduced in 1956 (before it had been deduced by purely chemical methods), and vitamin B_{12}, lack of which leads to pernicious anaemia. This vitamin has over 90 atoms in a complex structure, and her analysis in 1956 (after 8 years work) was a high point for X-ray methods. Until then, computing aid for X-ray crystallographers was primitive and Hodgkin used remarkable chemical intuition combined with massive computation. When modern computers became available, she was able to complete a study of insulin (with over 800 atoms) that she had begun in the 1930s, and in 1972 described its detailed structure. She won the Nobel Prize for chemistry in 1964.

Hoff, Jacobus Henrikus van 't (1852–1911) Dutch physical chemist: a founder of stereochemistry.

At the age of 17 van 't Hoff informed his mother and father, a physician, that he wished to become a chemist; their reaction was very unfavourable. Despite this, he entered Delft Polytechnic, and received his diploma in 2 years rather than the usual 3. He went on to study chemistry in Leiden, Bonn and Paris. He also became intensely interested in the philosophical ideas of Comte and Taine, in Byron's poetry and in the biographies of scientists.

Back in the Netherlands, aged 22 and ready to begin his doctoral work, he published a paper which founded stereochemistry. It had been known since Biot's work that many organic compounds are optically active (ie rotate the plane of polarized light). Pasteur had been able to relate this property, for crystalline solids, to

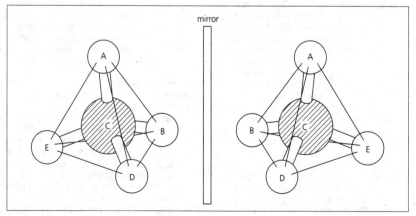

Four different atoms or groups (A, B, D, E) attached to a central carbon atom can be arranged in two different ways, which are non-superposable mirror images. The tetrahedra are imaginary: the double lines represent bonds with the carbon atom.

the dissymmetry of the crystals; but interest in the organic compounds was in their optical activity in solution. Van 't Hoff took up an idea of KEKULÉ's (1867) that the four groups usually linked to a carbon atom can be expected to be equally distributed in the space around it (a 'tetrahedral' distribution; see diagram). Van't Hoff saw that, if the four groups are all different from each other, they can be arranged about the carbon atom in two ways; and these two variants of a molecule are non-superimposable mirror images of each other (stereoisomers). He proposed that one form would rotate polarized light to the left and the other form to the right. On this basis a general theory of molecular shapes could be developed and, despite some initial doubts, his ideas of stereoisomerism were soon shown to be both correct and fruitful. The same ideas were offered independently by J A Le Bel (1847–1930) soon after, but he did not develop them. They had known one another slightly, in WURTZ's laboratory in Paris.

At 23 van 't Hoff tried for a job as a schoolteacher, but was turned down because he appeared to be a 'daydreamer'. He got a junior post in a veterinary college in 1876 but 2 years later took a professorship in Amsterdam until 1896, when he moved to Berlin. In the 1880s and later, his work on physical chemistry was as valuable as his stereochemistry. He studied reaction rates, mass action, transition points, the phase rule and especially the chemistry of dilute solutions and the application of thermodynamic theory to chemistry. He was awarded the first Nobel Prize in chemistry, in 1901.

Hofmann, August Wilhelm von (1818–92) German organic chemist: major discoverer of new organic compounds of nitrogen.

Hofmann began his studies in Giessen as a law student, but attendance at some of LIEBIG's chemistry lectures changed his interests and Liebig welcomed this, perhaps because Hofmann's father, an architect, was overseeing the building of the new chemical laboratory. Young Hofmann's first research was on coal tar aniline ($C_6H_5NH_2$) and began the interest in organic amines which was to prove so important. He won prizes, became engaged to Liebig's wife's niece and came to London as the first head of the new Royal College of Chemistry in Oxford Street in 1845. He stayed for 20 years, and he and his students created organic chemistry in England. One of these students, PERKIN, made the first synthetic dye produced on any scale (mauve, from aniline) and founded the British organic chemical industry. Other students, and Hofmann himself, made a variety of new organic dyes. From them, in turn, medicinal chemicals were developed.

In 1850 Hofmann showed that ammonia can be progressively alkylated by a reactive alkyl halide to give a mixture of amines. Thus ethyl iodide with ammonia in a sealed container (he

was a large-scale user of champagne bottles from Windsor) gives the ethylamines, which he represented as follows, basing them on the 'ammonia type' in a way which advanced the 'Type Theory':

H	C_2H_5	C_2H_5	C_2H_5
HN	HN	C_2H_5N	C_2H_5N
H	H	H	C_2H_5

He made similar compounds from phosphine (PH_3) in 1855. He moved to a professorship in chemistry in Berlin in 1865. Hofmann was not a theorist, but he had excellent instincts as an experimentalist and this, combined with his use of the theory available to him, led to his high output of new results. The Hofmann rearrangement (1881) gives a primary amine as the organic product (via an isosyanate) when an amide is heated with bromine (or chlorine) and alkali:

$$RCONH_2 \rightarrow [RNCO] \rightarrow RNH_2$$

Also named after him is the Hofmann exhaustive methylation reaction, which allows a complex nitrogen-containing organic compound to be degraded to simpler, identifiable, products; its early use was to determine structures but later it became a subject for the study of reaction mechanism. Hofmann produced hundreds of research papers; he had many assistants and a large number of friends; he was married four times and had 11 children.

Hofmeister, Wilhelm (Friedrich Benedict) [hohfmiyster] (1824–77) German botanist.

Hofmeister followed his father into his prosperous music and bookselling business in Leipzig and in his interest in plants, and by age 27 became well known as a botanist through his work on mosses and ferns (cryptogams). He went on to show that this group is related to the higher seed-bearing plants (phanerogams), and that the gymnosperms (conifers) lie between the cryptogams and the angiosperms (flowering plants); this prepared the way for a unified view of the plant kingdom. This work was linked with his discovery of the alternation of generations between sporophyte and gametophyte in lower plants. He was appointed professor at Heidelberg in 1863 and at Tübingen in 1872.

Hofstadter, Robert [hofstater] (1915–90) US physicist: used electron scattering by nuclei to give details of nuclear structure.

Graduating from New York and Princeton, Hofstadter afterwards worked at the Norden Laboratory Corporation (1943–6), Princeton and Stanford (1950) with a full professorship at 39; he was director of the Stanford high-energy physics laboratory from 1967–74.

In 1948 Hofstadter invented an improved scintillation counter using sodium iodide, activated by thallium. At Stanford he used linearly accelerated electrons scattered by nuclei to study nuclear structure. The charge density in the nucleus was revealed to be constant, but falling sharply at the nuclear surface, with a radial distribution related to the nuclear mass. Neutrons and protons were shown to have size and shape (ie were not 'points') and could be regarded as made up of charged shells of mesons, with the total charge cancelling out in a neutron. Hofstadter was led to predict the rho-meson and omega-meson, both of which were later observed experimentally. He shared a Nobel Prize in 1961.

Hollerith, Herman [holerith] (1860–1929) US computer scientist: introduced the modern punched card for data processing.

A graduate of the Columbia University School of Mines in New York City, Hollerith did some teaching at MIT and worked on air brakes and for the US Patent Office before joining one of his former teachers to assist with the processing of the US Census of 1880. By 1887 he had developed his machine-readable cards and a 'census machine' that could handle up to 80 cards per minute, enabling the 1890 census to be processed in 3 years. Punched cards had been used by JACQUARD before 1800 to mechanically control looms, but Hollerith introduced electromechanical handling and used the cards for computation: the Hollerith code relates alphanumeric characters to the positions of holes in the punched card. After the 1890 census, Hollerith adapted his device for commercial use and set up freight statistics systems for two railroads, founding the Tabulating Machine Company in 1896 to make and sell his equipment; by later mergers this became the International Business Machines Corporation (IBM). Hollerith's ideas were initially more used in Europe than in the USA, but from the 1930s punched card methods became widespread.

Holmes, Arthur (1890–1965) British geologist and geophysicist: devised modern geological time-scales.

Holmes pioneered the use of radioactive decay methods for dating rocks, whereby careful analysis of the proportions of elements formed by radioactive decay, combined with a knowledge of the rates of decay of their parent elements, yields an absolute age. He was the first to use the technique, in 1913, to systematically date fossils whose stratigraphic (ie relative) ages were established, and was thus able to put absolute dates to the geological time scale for the first time. A modern scale is on p. 135.

In 1928 he suggested that convection currents within the Earth's mantle, driven by radiogenic heat, might provide the driving mechanism for the theory of continental drift, which had been advanced by WEGENER some years earlier. He also proposed that new oceanic rocks were forming throughout the ocean basins, although predominantly at ocean ridges. Little attention was given to his ideas until the 1950s, when palaeomagnetic studies established continental drift as a fact.

He wrote several influential textbooks on geology, in particular *The Principles of Geology* (1944).

Hooke, Robert (1635–1703) English physicist: ingenious inventor of devices and of ideas then developed by others.

Born in the Isle of Wight, Hooke was intended for the church and went to Oxford as a chorister. However, his poor health was thought to make him unsuited to the church and he turned to science, becoming assistant to BOYLE in Oxford and making an improved air pump for him. From childhood on, Hooke was an ingenious and expert mechanic. In 1660 he moved to London, and was one of the founders of the Royal Society in 1662. He was made curator; one of his tasks was to demonstrate 'three or four considerable experiments' for each weekly meeting. He later added other posts (one of these, as a Surveyor of London after the great fire, made him rich), but the Royal Society work helped to shape Hooke's life as a prolific experimenter whose ideas were usually fully explored by others. As he was combative, this led to many disputes on priority, notably with NEWTON.

In the 1660s he found Hooke's Law: this is now often given in the form that, provided the elastic limit is not exceeded, the deformation of a material is proportional to the force applied to it. He did not publish this until 1676 (as a Latin anagram) and in intelligible form not until 1678. Also in the 1660s, he realized that a spiral spring can be used to control the balance-wheel of a timepiece, but HUYGENS made the first working model in 1674. He was fascinated by microscopy and in his book *Micrographia* (1665) Hooke describes the use of the compound microscope, which he had devised. He used the word 'cell' to describe the angular spaces he saw in a thin section of cork, and since then the word has come to be used for the membrane-bounded units of plant and animal life. The book includes also the idea that light might consist of waves; but further work on this was mainly by Huygens. Also in the book is Hooke's theory of combustion, which is good enough to make it very likely that he would have discovered oxygen if he had continued with chemistry. In the 1660s Hooke had ideas on gravity, as did many others, and he even suggested (in 1679) that its force obeys an inverse square law. These ideas may have been useful to Newton; what is certain is that Newton's toil and his mathematical genius succeeded in developing the idea brilliantly and that Newton forcefully resisted Hooke's claims of priority.

Hooke had no rival as a deviser of instruments; the microscope, telescope and barometer were all much improved by him and his other inventions include a revolving drum recorder for pressure and temperature, and a universal joint. His contribution to science is unusual; he did much, but his devices and ideas

SCIENTIFIC SOCIETIES

Scientific societies have played a major part in science. The exchange of ideas within and between societies encourages experiment and the testing of theory; an isolated researcher can too easily become complacent. Some societies were also engaged from their beginning in the organization and funding of science and in the selection of research projects, such as those directed to improving ship design and navigation at sea in the 17th-c.

Usually classed as the first scientific society, the **Accademia dei Lincei** (Academy of the Lynxes) was founded in 1609 by the 18-year-old Count Federico Cesi in Rome; so devoted to scientific study that they swore to remain unmarried, the 'lynxes' were named for their claimed clear-sightedness. Their membership rose to 32 and included GALILEO, but the society failed to survive Cesi's death in 1630.

The **Royal Society** of London for the Improvement of Natural Knowledge was officially founded in 1660, but its key figures had met in London, and later in Oxford, from about 1645. It was encouraged (but never funded) by the newly-restored King Charles II and its membership included BOYLE, HOOKE, LOWER, NEWTON and WALLIS. Independent of the state, it elected its own Fellows, and soon included many gentleman-amateurs, such as Samuel Pepys the diarist. Its only resources were the Fellows' shilling-a-week subscriptions, often in arrears. The Society appointed a Curator (Hooke), who had to provide a demonstration at each meeting; they acquired some apparatus (such as Boyle's first air-pump and Newton's reflecting telescope), bent their minds to anything new or strange (excluding religion and politics), such as blood transfusion or the habits of fish; exchanged research results with societies abroad and published their own in their *Philosophical Transactions*. To this day the Royal Society retains its seniority in British science, providing the highest-level advice to government, and electing to its fellowship only the most distinguished professional and a few amateur scientists (under 1000 Fellows in all, including about 3% women) in Britain, and a handful working abroad.

The Parisian **Académie Royale des Sciences**, founded by King Louis XIV in 1666 (after 1816 the Académie des Sciences) was funded by the state for state purposes. Its very select and active membership included LAVOISIER. The talented young artillery officer Napoleon Bonaparte, elected in 1797, was an active member and later its patron until his power ended in 1814. Academies on the Paris model followed in Berlin (1700), St Petersburg (1725) and Stockholm (1739).

The **American Philosophical Society**, founded in Philadelphia in 1743, also covered science and was modelled on the Royal Society rather than the European academies, encouraging the amateur enthusiast. It stemmed from BENJAMIN FRANKLIN's Junto, a secret literary and scientific club active from 1727. (It was not the oldest in the USA: that was probably the Boston Philosophical

---

were largely developed by others. He certainly did more than anyone else to change the Royal Society from a club of virtuosi to a professional body. The frequent comment that he was much disliked seems to have arisen because he quarrelled with Newton. He certainly had many friends and his large library attests to his wide interests.

Hooker, Sir Joseph Dalton (1817–1911) British botanist: plant taxonomist, phytogeographer and explorer.

Educated at Glasgow in medicine, Hooker became assistant surgeon and naturalist on Ross's Antarctic expedition of 1839–43 on board HMSS *Erebus* and *Terror*. His books *Flora Antarctica*, *Flora Novae-Zelandiae* and *Flora Tasmaniae* were a result. He travelled widely in India, Palestine and the United States, where he spent 3 years collecting plants. There followed his *Himalayan Journals* (1854), *Rhododendrons of the Sikkim Himalaya* (1849) and his seven-volume *Flora of British India* (1872–97). He undertook further expeditions to Syria and Palestine, and to the Atlas Mountains in Morocco. He became director of Kew Gardens, succeeding his father,

Sir William J Hooker (1785–1865), who had created the gardens.

His friendship with DARWIN led to his being instrumental, with LYELL, in presenting the joint communication of Darwin and WALLACE on the origin of species to the Linnean Society and in persuading Darwin to publish *The Origin of Species*. He joined with BENTHAM in producing the magisterial *Genera plantarum* (7 vols, 1862–83), giving their important system of classification; and was an authority on Antarctic flora. He became president of the Royal Society in 1873.

Hopkins, Sir Frederick Gowland (1861–1947) British biochemist: made first general scientific study of vitamins.

Hopkins believed firmly that chemical reactions in living cells, although complex, are understandable in normal chemical and physical terms. This faith, and his amiable forcefulness, made him 'the father of British biochemistry'. His beginnings were not indicative of his future fame. His widowed mother chose a career for him at 17, in an insurance office. Later he worked as an analyst, especially on forensic cases. When he was 27 he inherited

Society, founded by Increase Mather in 1683, but which soon expired.) An early project of the American Philosophical Society was the first accurate measurement of the Earth–Sun distance, by observing the transit of Venus in 1769.

The American Civil War made evident a need for a **National Academy of Sciences** and this was duly created in 1863. A later conflict, the 'Cold War' of the 1950s, generated the National Aeronautic and Space Agency (NASA) in 1957, which, although not a scientific society, has been linked with increasingly dramatic explorations of the solar system and whose scientific effort in support of government intentions recalls in scale the Manhattan Project of the Second World War.

Local scientific societies existed from the 18th-c in Europe and the USA. One such was the **Lunar Society**, which met in the English Midlands in the 1780s and was so called because its monthly meetings were held on the night of the full moon so it could light members home. These included PRIESTLEY; Josiah Wedgwood (1730–95) the industrialist potter; WATT; Matthew Boulton (1728– 1809), his partner and first manufacturer of steam engines; Erasmus Darwin (1731–1802), poet, engineer, medical man and grandfather of CHARLES DARWIN; and William Murdock (1754–1839) the inventor of gas lighting. Local societies such as the Liverpool Astronomical Society (1882), tapped the enthusiasm and expertise of amateurs to good effect, as did the comparable local scientific societies of the USA.

Specialized societies followed the model of the Linnean Society (1788) for botany and the Geological Society of London (1807). The Astronomical and Chemical Societies began in London in the 19th-c, with very many other specialized societies being formed in the USA. Through their publications and meetings they nurture expertise and publicize new discoveries for information and discussion.

The lack of encouragement of women in scientific societies is noteworthy. Before the First World War the poor quality of the education in science available to women excluded all but a very few from senior positions and if they achieved original work in science despite the difficulties, it tended to be eclipsed through their position as someone's wife, sister, daughter or assistant and subsumed into a male's publication. After the Second World War the proportion of women graduating in science rose steadily, and membership of scientific societies became more open to them, but a gender bias has remained. At senior levels in university, industry and the learned societies the proportion of women remains low; they have made up about 3% of the Royal Society fellowship for over 25 years. More encouragingly, the US National Academy of Sciences has 4.7% women, and for the last 5 years 10.7% of the newly elected members have been female. In 1991 the embryologist ANNE MCLAREN became Foreign Secretary of the Royal Society, the first woman to hold an office in its 330-year history.

IM

---

some money and entered the medical school at Guy's Hospital and after qualifying worked with GARROD, founder of biochemical genetics. At 37 he went to Cambridge, but his teaching load was so great that his health broke down in 1910. He recovered fully, and his college (Trinity) then gave him a research post. Recovered in health, and almost 50, he began the work for which he is famous, on 'accessory food factors' (vitamins), and in 1914 became the first professor of biochemistry in Cambridge. His classic studies on nutrition showed that young rats failed to grow on a diet of pure protein, carbohydrate, fat, salts and water; but the addition of small amounts of milk (2–3 ml per rat per day) caused them to thrive. The 'vitamin hypothesis' followed, as did much work on vitamins in Cambridge and elsewhere. Hopkins worked also on the biochemistry of muscle, on enzymes, on -SH groups and on glutathione. He shared a Nobel Prize with EIJKMANN in 1929.

Hopkins, Harold Horace (1918–94) British optical physicist.

Hopkins was educated at Leicester and London and was professor of applied optics at Reading from 1967–84. After N S Kapary's early proposals in 1955, work by many groups greatly advanced fibre-optics, with contributions by A C S Van Heel and by Hopkins of high value. Optical fibres of plastic or glass (pure silicon dioxide is suitable) are used; they guide light by trapping it through total internal reflection. Their diameter is of the order 0.01 mm and the light-carrying core has an outer cladding to confine the light. A bundle of such fibres is used in flexible endoscopes, which can be inserted through natural or surgical apertures to give direct viewing within the body, of great value in medical diagnosis and in 'keyhole' surgery.

Optical fibres also find application in communication systems; undersea fibre-optic cables for telephone services cross the Atlantic and Pacific oceans. Light signals, like electrical signals, can be transmitted and processed in either analogue or digital form. Optical fibres have advantages over electrical wiring in their smaller size and lower weight and in their freedom from electrical interference. A variety of information-carrying uses will certainly continue to develop.

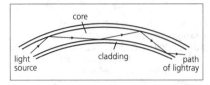

Section through an optical fibre; the cladding has a lower refractive index than the core

As well as his major work on flexible endoscopes, Hopkins did much to develop the variable zoom lens now widely used by photographers and in television cameras.

Hopper, Grace (Murray) (1906–92) US computer programming pioneer.

Grace Hopper graduated from Vassar College in mathematics and physics, received a PhD from Yale and taught at Vassar before joining the US services in the Second World War. In 1943 she joined the Naval Reserve and remained a reservist for the rest of her career. Having worked for the wartime navy using large-scale pre-electronic calculators, she joined the Eckert-Mauchly Computer Corp. in 1949: thereafter she completed the first compiler (A-O, the software for the Univac computer) in 1952, developed Flow-Matic (a language suitable for business data processing) in 1958 and provided a major input to the development of COBOL, a high-level computer language. She was recalled to the Navy in 1967 to help standardize its computer languages.

She had some unusual distinctions; she was the first computer scientist to be named Man of the Year (in 1969), and she was the oldest officer on active US naval duty when she retired in 1986, having been promoted to rear admiral in 1985.

Hoppe-Seyler, Ernst Felix [hopuh ziyler] (1825–95) German biochemist: first to isolate nucleic acid.

Orphaned early, Hoppe-Seyler was brought up by his brother-in-law and followed him by studying medicine. He only practised briefly, however, and after some research training with VIRCHOW he worked in Tübingen and later Strasbourg, mainly on the use of chemical and physical methods in physiology. He isolated haemoglobin, the red pigment of blood, and studied its reaction with carbon monoxide; he was also the first to obtain pure lecithin, and he studied oxidation in animal tissues and enzyme action. His Swiss pupil J F Miescher (1844–95) first isolated a nucleic acid, in 1869, and later Hoppe-Seyler extended this work. He founded in 1877 a journal for physiological chemistry, editing it in a characteristically autocratic way, and was a major figure in establishing classical biochemistry as a separate branch of science.

Horrocks, Jeremiah (1618–41) English astronomer.

Educated at Cambridge University but self-taught in astronomy, Horrocks became a curate at Hoole, in Lancashire; a brilliant amateur astronomer, he achieved a great deal in his short life. He is particularly remembered as the first observer (1639) of a transit of Venus across the face of the Sun, which he had predicted using KEPLER's Rudolphine tables. From the transit the value of planetary distances can be calculated by Kepler's laws; hence its importance. In the 2 years before his death, Horrocks achieved a remarkable range of accurate measurements, especially in the solar system.

Hounsfield, Sir Godfrey Newbold (1919–) British physicist: developed first X-ray tomographic body scanner.

Hounsfield received no formal university education but studied at the City and Guilds College and at the Faraday House College for Electrical Engineering in London. In 1951 he joined Electrical and Musical Industries (EMI), becoming head of the medical research division.

Independently of CORMACK, Hounsfield developed the technique of X-ray computer-assisted tomography (CAT), whereby high-resolution images of the soft body tissues (which are normally almost transparent to X-rays) are built up by computer from many measurements of the absorption of X-ray beams in different directions through the body. In the early 1970s he developed the first commercial CAT body scanner at EMI; such machines are now an invaluable medical tool. Hounsfield has continued to lead research into medical imaging, pursuing in particular the use of magnetic resonance imaging (MRI) techniques. In 1979 Hounsfield shared the Nobel Prize for physiology or medicine with Cormack.

Hoyle, Sir Fred (1915–) British cosmologist and astrophysicist: jointly proposed the steady-state theory of the universe.

Even as a small boy in a Yorkshire village, Hoyle was 'at war with the system' and became a long-term truant. Later he had problems in entering Cambridge. In 1973 he resigned his chair after disputes with the university authorities.

Together with GOLD and BONDI, Hoyle proposed the steady-state theory for the origin of the universe in the 1950s. This theory assumes that the universe is not only homogeneous and isotropic in space, but also unchanging with time. The known expansion of the universe is explained by the continuous and spontaneous creation of matter to maintain the mean mass density at a constant value. Newer evidence has left that theory with few supporters. Hoyle also suggested (in 1957, with W A Fowler (1911–95) and Geoffrey and MARGARET BURBIDGE) how elements heavier than helium and hydrogen might have been created by nuclear synthesis in the interior of stars, eventually being ejected into space and incorporated into new stars formed from clouds of interstellar matter. All this work has led to fruitful advances, directly or through their effect on others: and the 'B^{2}FH paper' is now accepted as a major contribution to cosmology.

More recently Hoyle, with Geoffrey Burbidge and J V Narlikar, developed a quasi-steady-state theory, in which the creation of matter is not continuous but intermittent. These creation events, the theory suggests, are linked with strong gravitational fields and can occur on various scales, with our part of the universe being created about 15 billion years ago. These ideas have not found wide support.

He is also a believer in an extraterrestrial origin for life, suggesting that biological molecules such as amino acids are synthesized in space on dust particles. His view that infective agents such as viruses can arrive from space has found little support. He has, however, been notably successful as a theorist and as a writer both of popular science and of science fiction.

Hubble, Edwin Powell (1889–1953) US astronomer and cosmologist: discovered expansion of universe and measured its size and age.

Hubble was trained in Chicago and Oxford in law; he was also a distinguished athlete and boxer. After a short career in law he turned to astronomy, working for most of his life at Mount Wilson Observatory. Using the 100 in telescope at Mount Wilson, Hubble was able in 1923 to resolve the nebulous outer part of the Andromeda galaxy into individual stars, obtaining a distance of 900 000 light years for several Cepheid variable stars he found there. Together with further work this proved for the first time that what were then thought of as 'spiral nebulae' were in fact spiral galaxies, and lay well beyond our own Galaxy. In 1929 he was able to measure the recessional velocities of 18 galaxies, and discovered that these velocities increased in proportion to their distance from Earth. This relationship ($v = Hd$) is now known as Hubble's Law, the constant of proportionality being Hubble's constant (H). This work gave the first direct evidence supporting the idea of an expanding universe, a concept that had been proposed a few years earlier by the cosmologists FRIEDMANN and LEMAÎTRE and is now fundamental to our understanding of the universe. Hubble's observations meant that two fundamental quantities of the universe could be calculated for the first time: its 'knowable' size, or the distance at which the recession velocity reaches the speed of light, which is about 18 billion light years; and the age of the universe, which Hubble himself estimated as 2 billion years, although modern values range between 12 and 20 billion years. Hubble also introduced a widely used system of classification for the shape of galaxies.

Hubel, David Hunter (1926–) Canadian–US neurophysiologist: investigator of the basis of visual perception.

Hubel qualified in medicine at McGill Univer-

sity, Montreal, and from 1959 worked at Harvard. With T N Wiesel (1924–) he did much to aid understanding of the mechanism of visual perception at the cortical level, using microelectrodes and modern electronics to detect the activity of individual neurones, especially in area 17 of the visual cortex. The cells of this striate cortex lie in several layers arranged in columns, which run through the thickness of the cortex (a few millimetres). Hubel and Weisel found that stimulation of cells on the retina by light causes excitation of particular cells in the striate cortex. The cell activation in the cortex by visual stimulation is very specific; some cells respond to spots of light, others to a line whose tilt is critical, so that a change of 10° in its angle greatly alters the response. Still others respond only to specific directions of movement or to specific colours. The visual cortex has become the best-known part of the brain through studies of this kind. Hubel and Wiesel established in addition that many of the connections responsible for these specific response patterns are present already at birth but may be modified or even destroyed if the young animal is visually deprived. These results have had an important influence on treatment of congenital cataracts and strabismus (squint). Hubel and Wiesel shared a Nobel Prize in 1981.

Hückel, Erich (1896–1980) German physicist and theoretical chemist: developed molecular orbital theory of bonding in organic molecules.

Hückel's study of physics at Göttingen was interrupted by the First World War, when he spent 2 years on aerodynamics before returning to finish his course. He worked as assistant first to the mathematician HILBERT and then to BORN, but he did not like the work and he also wished to travel. He joined his former teacher DEBYE at Zürich and began working on what is now known as the Debye–Hückel theory of electrolyte solutions. This assumes that strong electrolytes are fully dissociated into ions in solution, and calculates properties (such as electrical conductivity) for dilute solutions on this basis. Then, after a major illness, he worked on colloid chemistry, first with his father-in-law ZSIGMONDY and later with F G Donnan (1879–1956) in London, and then moved to Copenhagen to work on quantum theory with BOHR. The latter suggested that Hückel should try to calculate properties of the CC double bond by wave mechanics. Hückel classified the electrons making up such bonds as σ (sigma) and π (pi) types on a symmetry basis and was able to make useful calculations of bond properties. Hückel molecular orbital (HMO) theory has been widely applied to organic molecules. One result is Hückel's rule, which proposes that aromatic stability will be shown by planar monocyclic molecules in which all the cyclic atoms are part of the π-system only if the number of such π-electrons is $4n + 2$, where n is an integer.

The rule has provoked much fruitful study of molecules predicted by the rule to show 'aromaticity'. Hückel's work on such compounds probably began with a suggestion from his elder brother Walter, an organic chemist. From 1937 Erich was professor of theoretical physics at Marburg.

Huggins, Sir William (1824–1910) British astronomer and astrophysicist: pioneered stellar spectroscopy and discovered stellar red shifts.

Huggins was a wealthy amateur who used his private observatory in South London to study a full range of celestial objects. Like LOCKYER, he was attracted by spectrum analysis and its possible use in astronomy; aided by his wife Margaret (1848–1915), Huggins pioneered the study of the spectra of stars, finding them to contain elements already known on Earth and in the Sun. He went on to investigate nebulae, making the important discovery that they were composed of luminous gas; he later showed that a comet contained hydrocarbon molecules. In 1868 he made perhaps his most profound discovery, observing that the spectrum of Sirius is shifted towards the red end of the spectrum. He correctly interpreted this as being due to the DOPPLER effect, obtaining a recessional velocity of about 40 km s^{-1} (25 miles per second), and proceeded to measure the red shifts of many other stars. With the advent of the gelatine dry plate he pioneered the technique of spectroscopic photography, from 1875. (The particular advantage of photography over the eye is that a faint image can be 'accumulated'.)

Hulst, Hendrik Christofell van de (1918–) Dutch astronomer: predicted interstellar 21 cm hydrogen emission.

Educated in Utrecht and in the USA, van de Hulst became director of the Leiden Observatory. In 1944 he suggested that interstellar hydrogen might be detectable at radio wavelengths, because of the 21 cm radiation emitted when the orbiting electron of a hydrogen atom flips between its two possible spin states. Because of the war it was not until 1951 that such emissions were first detected, by PURCELL and H I Ewen (1922–). The technique has since proved invaluable in detecting neutral hydrogen in both our own and other galaxies, as well as in interstellar space. Since the 21 cm wavelength is not absorbed by interstellar dust, it has also enabled a much better picture of the centre of our Galaxy to be built up where optical methods have failed.

Humboldt, (Friedrich Wilhelm Heinrich) Alexander, Freiherr (Baron) **von** (1769–1859) German explorer: pioneer of geophysics and meteorology.

Humboldt had wide scientific interests and a passion for travel, and was wealthy enough to indulge both his enthusiasms. His father, a Prussian soldier, wished him to enter politics, but the boy preferred to study engineering; while doing so he was attracted to botany and moved to Göttingen to study science. He seems to have been happiest with geology, and spent 2 years at a school of mining before working as a mining engineer, when he devised and tested safety lamps and rescue apparatus. Then in 1796 he inherited enough money to travel, but the Napoleonic Wars frustrated him until 1799, when he began an epic exploration of central and south America. He covered over 6000 often dangerous miles with his friend, the botanist A Bonpland (1773–1858), before returning to Europe with a large collection of scientific specimens and observations after 5 years of absence. Analysis of his results, along with some diplomatic missions, kept him busy for the next 20 years. His wide-ranging interests were largely in geophysics, meteorology and geography. He studied the Pacific coastal currents, and was the first to propose a Panama canal. He introduced isobars and isotherms on weather maps, made a general study of global temperature and pressure and eventually organized a world-wide scheme for collecting magnetic and weather observations. He studied American volcanoes and showed that they followed geological faults, and deduced that volcanic action had been important in geological history and that many rocks are of igneous origin. He set a world record by climbing the Chimorazo volcano (5876 m) and was the first to link mountain sickness with lack of oxygen, to study the fall of mean temperature with rising altitude and to relate geographical conditions to its animal life and vegetation. In 1804 he discovered that the Earth's magnetic field decreases from the poles to the equator. His writing has been said 'to combine the large and vague ideas, typical of the 18th-c thought, with the exact and positive science of the 19th'.

Hume-Rothery, William (1899–1968) British metallurgist.

Soon after beginning a military career Hume-Rothery had meningitis and was left totally deaf so he entered Oxford to study chemistry, graduated well, spent 3 years researching in metallurgy in London and then returned to Oxford, which was his base thereafter. He brought together a range of ideas and techniques for the understanding of alloys; his first researches were on intermetallic compounds and he went on to study alloy phases, compositions and crystal structures, using modern electronic theory, V M GOLDSCHMIDT's findings on the importance of atomic size, and both microscopy and X-ray methods of examination. His books set out the empirical rules governing alloy formation and behaviour very fully and did much to advance the subject and to convert it from an art to a science.

Hutton, James (1726–97) British geologist: proposed the uniformitarian principle in geology.

Hutton had a disorganized start to his career. After leaving school he was apprenticed in a lawyer's office, but left for the continent to train as a doctor, and qualified as an MD at Leiden. However, it was a profession that he failed to take up, turning instead to farming; after studying agriculture in England and abroad, he returned to his native Scotland and a family farm near Edinburgh. After 14 years the success of a business extracting NH_4Cl from soot gave him an independent income and in 1768 he returned to Edinburgh to pursue science.

Hutton is widely regarded as the founder of geology as a modern science. He rejected the scriptural time scales hitherto accepted, which dated the Earth as only a few thousand years old, and argued that it was immeasurably ancient, with 'no vestige of a beginning, no prospect of an end'. He considered the erosive action of rivers to be a major agent in creating continental topography and believed that sediments washed into the sea by rivers accumulated and were metamorphosed via geothermal heat to form new rocks, which would eventually be uplifted and form new land masses. Such ideas ran contrary to previous ideas of a 'catastrophic' origin of the continents, at a fixed point in time corresponding to the biblical Creation. Hutton's concept of a cyclic process of denudation, transport, sedimentation, lithification, uplift and renewed denudation is often referred to as the Plutonic theory, due to the crucial part played in it by terrestrial heat; or as uniformitarianism, since it assumes that geological processes act in a continuous manner over a long time.

Huxley, Hugh (Esmor) (1924–) British physiologist: developed sliding filament theory of muscle contraction.

Huxley studied physics at Cambridge, worked on radar in the Second World War and afterwards was attracted to biophysics, working at the Massachusetts Institute of Technology and in London and in Cambridge from 1961. From the 1950s he was especially associated with the sliding filament model of muscle contraction and with the development of methods in X-ray diffraction and in electron microscopy designed for this work but also applicable in other studies.

Skeletal muscle is a very abundant animal tissue: it makes up some 40% of the human body mass. Its main purpose is to convert chemical energy into mechanical work, under neural control. Skeletal muscle tissue is a parallel array of myofibres, each consisting of some hundreds of myofibrils, which form long cylinders. The myofibril is divided into sarcomeres arranged end to end and, since the myofibrils are arranged with the sarcomeres in register, this gives to skeletal muscle its striated appearance. (The fibres of smooth muscle are not arranged in sarcomeres.) Within each sarco-

mere are the filaments which form the contractile apparatus; it can shorten by some 10%. These filaments are of two kinds: the thicker myosin filaments interdigitate with the slender actin filaments. With JEAN HANSON, Huxley developed the theory that these thick and thin protein filaments slide past each other in muscle contraction. The process has a complex system of regulation through changes in calcium ion concentration, and uses ATP as the energy source. When the muscle relaxes the crossbridges which project from the thick filaments and which have drawn the structure together are detached, and the whole structure regains its original length.

Huxley, Thomas Henry (1825–95) British biologist: forceful supporter of theory of evolution.

Despite having a schoolmaster father, young Huxley had only 2 years of regular schooling and was mainly self-taught. He was attracted to medicine, and attended a post-mortem at 14, but he may have contracted an infection there which recurred throughout his life. He became an apprentice to a medical brother-in-law, did well, studied medicine and surgery in London and joined the Royal Navy.

Although his duties on HMS *Rattlesnake* on a 4-year voyage around Australia were as surgeon and he had only a microscope and makeshift net as equipment for natural history, he did useful new work on plankton and after a discouraging interval this established him on the scientific scene in London. However, these interests exasperated the Admiralty and he became a self-employed writer on science in 1850 and a lecturer on natural history from 1854 at the School of Mines. This gave him an income to marry his Australian girlfriend of 8 years before; they eventually had seven children. Their son Leonard was the father of Julian (biologist), Aldous (writer) and Andrew Fielding Huxley (physiologist). For 30 years, while waiting for a job in physiology, Huxley worked in zoology and palaeontology.

These interests led him to his best-known place in science, that of advocate for his friend DARWIN's ideas on evolution; in famous debates and essays on this Huxley showed his forceful expertise. However, in a debate with Bishop Wilberforce at the British Association meeting in Oxford in 1860, his reply to the Bishop's query on whether Huxley's ancestry was from an ape on his grandfather's or his grandmother's side is variously reported and it is unclear who was the victor. He was a lucid and elegant writer and a charming man. His careful study of the primates established man as one of them and made evolution a matter of public debate in terms of science rather than emotion. Aside from all this, Huxley did much excellent work of his own in zoology and palaeontology and in shaping biological education. One of his students in the 1880s was H G Wells, who admired him enormously and whose early novels were much influenced by Huxley's teaching.

Huygens, Christiaan [hoykhenz] (1629–95) Dutch physicist and astronomer: proposed wave theory of light; discovered Saturn's rings; introduced the pendulum clock; worked on the theory of dynamics and the compound pendulum.

Well educated as a member of a wealthy family in The Hague, Huygens studied law before turning to science and mathematics. He was, after NEWTON, the most influential physical scientist of the late 17th-c. In 1655, using an improved home-made telescope, he was the first to describe correctly Saturn's ring system, also discovering Titan, its largest moon. He announced the discovery and observation of Saturn's rings in the form of a cypher. The following year he obtained the first solution to the problem of the dynamics of colliding elastic bodies. GALILEO had discovered the constancy of a simple pendulum's period; and Huygens showed that for small swings $T = 2\pi(l/g)^{\frac{1}{2}}$ where T = period, l = length, g = acceleration due to gravity. He designed a pendulum clock and later invented the more accurate compound pendulum (which moves in a cycloidal arc). Physics could not have moved on without accurate time measurement.

Huygens's greatest achievement, however, was his wave theory of light, first expounded in 1678. He described light as a vibration spreading through an all-pervading 'ether' consisting of microscopic particles, and he considered every point on the wave-front to be the source of a series of secondary spherical wavelets, the envelope of which defined the wave-front at the next instant (known as Huygens's construction). He was thus able to give a simple explanation for the laws of reflection and refraction of light, and for the double refraction of some minerals. He correctly predicted that light travelled slower in denser media. Newton preferred a particle theory of light. The present view that each concept can be appropriate, depending on the experimental situation, came only in the 20th-c.

Huygens found a value for the distance of a star (Sirius) by assuming that it had the same actual brightness as the Sun, and making a hole in an opaque plate so small that the Sun's light seen through it matched Sirius. Simple calculation then gave a distance of 27 664 AU, about one-18th of the correct value and the best then obtained. He also convinced himself that the planets were populated and wrote in detail about shipbuilding and other engineering on Jupiter and Saturn.

Hypatia of Alexandria [hiypaysha] (c.370–415) The earliest known female scientist.

Hypatia's reputation as a mathematician and philosopher has survived through the accounts of her given by three sources: the 5th-c historian of Constantinople, Socrates Scholasticus, and excerpts from earlier Greek writers collected in a lexicon-encyclopedia of the 11th-c, together with the writings of the 9th-c theologian Photius. None of her own writings have survived and little is reliably known of her life; the ancient accounts are often ambiguous and not in agreement, except on the dramatic manner of her death; an event which no doubt has assisted in keeping her reputation alive.

Hypatia's father, Theon, was a mathematician and astronomer attached to the museum at Alexandria; her education probably took place there and included mathematics and astronomy and a training in the Neoplatonic School. She is reputed to have written books on mathematics that included a commentary on the *Conics* of APOLLONIUS OF PERGA and a commentary on DIOPHANTUS. She lectured on astronomy and mathematics and the philosophies of Plato and ARISTOTLE. She taught in Alexandria and among her students was Synesius, later bishop of Ptolemais, who wrote of Hypatia's mechanical and technological skills in assisting him to invent a hydrometer and a silver astrolabe.

During Hypatia's lifetime the Roman Empire was converting to Christianity and Alexandria was in a state of dangerous confusion and of conflicting ideas. Although there seems to be no agreement among the ancient writers as to the reasons for her murder, there does seem to be consensus that Hypatia was set upon by a mob and murdered. One reason suggested for this violence is that, as a neoplatonist, she was regarded as dangerous by the more fanatical Christians. Another reason given is that she was a close friend of Orestes, the Roman Prefect of Egypt, also a former student of hers, that he relied heavily on her judgement and that she was caught in a political power struggle.

Ingen-Housz, Jan [eenggenhows] (1730–99) Dutch plant physiologist: early student of photosynthesis.

Ingen-Housz studied physics, chemistry and medicine and researched in all three; his early career was guided by a British army surgeon who met the family when encamped near their home in Breda. He travelled widely in Europe, as a popular and expert user of the pre-Jenner inoculation method (a risky affair, using live virus) against smallpox. He spent his last 20 years in London, where he published his work on gas exchange in plants. He showed that the green parts of plants absorb carbon dioxide and give off oxygen only in the light; in darkness they release carbon dioxide. This process, photosynthesis, is perhaps the most fundamental reaction of living systems, since it is the source of much plant substance, and animal life depends on the life of plants. Ingen-Housz made a number of curious inventions and discoveries. They include: a device for giving oxygen to a patient with chest disease; a pistol which used an explosive mixture of air and diethylether vapour and which was fired electrically; a hydrogen-fuelled lighter to replace the tinderbox; and thin glass microscope cover plates.

Ingold, Sir Christopher (Kelk) (1893–1970) British physical organic chemist: developed electronic theory of organic chemistry.

A student of chemistry at Southampton and London, Ingold spent 2 years in industry before returning to Imperial College London as a lecturer. In 1924 he became professor at Leeds and in 1930 at London, where he stayed. At Leeds he developed ideas on the electronics of organic reactions, somewhat parallel to those of ROBINSON, and much controversy followed. Ingold thereafter worked on reaction rates and the details of reaction mechanism, usually with E D Hughes (1906–63). Much physical organic research has followed on this line and the terminology, at least, has followed Ingold's preference rather than Robinson's.

Ipatieff, Vladimir Nikolayevich [eepatyef] (1867–1952) Russian–US chemist: pioneer of catalytic and isomerization reactions of hydrocarbons.

Ipatieff trained for a military career in Tsarist Russia, as was usual for those, like him, from an aristocratic family. However, he became interested in chemistry through the influence of an uncle, and initially was self-taught, with the help of MENDELAYEV's book *The Fundamentals of Chemistry*. He attended a military school, became an officer of the Imperial Russian Army in 1887 and entered the Mikhail Artillery Academy (1889–92), studying chemistry and mathematics. After graduation, he became an instructor and eventually professor of chemistry at the Academy. He studied in Munich under BAEYER (1896) and there synthesized isoprene, so beginning his interest in hydrocarbons. This led to further work at the Academy on high-pressure catalytic and hydrogenation reactions. He developed the Ipatieff 'bomb' for this work.

During the First World War Ipatieff co-ordinated Russia's chemical industries; he was a Lieutenant-General by 1916. The Revolution of 1917 interrupted Ipatieff's work (it was within his brother's house that the Tsar and his family were murdered), and he was fortunate to survive when officers of the Imperial Army were at risk. His skills were needed by the Bolsheviks and after a difficult period he worked for Soviet Russia, helping to rebuild its chemical industry.

By 1929 he began to worry about his own safety in Soviet Russia and in 1930 left Russia for America. At the age of 64 he remade his life and career, learned English, was appointed professor of chemistry at Northwestern University (Illinois) and acted as consultant to the Universal Oil Products Company of Chicago, who established the Ipatieff High Pressure Laboratory at Northwestern University, which he directed. In the USA he continued his work on hydrocarbons; he studied their formation, hydrogenation and dehydrogenation, cyclization and isomerization, with the emphasis on high-pressure catalysed reactions. Such processes are of great importance in the petroleum industry.

When aged 40, Ipatieff had the unusual experience of meeting his formerly unknown half-brother Lev Chugaeff (1873–1922), also an organic chemist, and they remained good friends.

Isaacs, Alick (1921–67) British virologist: discoverer of interferon.

A graduate of Glasgow, Isaacs studied at Sheffield and Melbourne before returning to London and the Virology Division of the National Institute for Medical Research in 1950. He was much concerned with the way viruses apparently interact with each other and in 1957 he reported on the substance interferon, which he found to be released from cells in response to viral infection and which inhibits the replication of viruses. Interferon is now known to consist of a group of related proteins, able to block the action of viral m-RNA. Fuller understanding of this action will clearly be of value in virology and interferons have been used in clinical trials as anticancer agents; the difficulty of obtaining them has delayed assessment of their value in therapy.

Issigonis, Alec (1906–88) British engineer: the most successful car designer of his time.

Issigonis's father had a marine engineering factory in Smyrna (then in Turkey) and engines were familiar to the boy from childhood; but in 1919 Smyrna was ceded to Greece, only to be invaded by the Turks in 1922. As a result the family fled, first to Malta and then to Britain. Later in the same year Alec was living in London with his widowed mother, as a rather poorly-educated refugee studying engineering at Battersea Polytechnic. From the mid-1920s he was working in the motor car industry, then dominated (as it had been from 1900) by designs with a petrol engine placed in-line at the front and driving the rear wheels. By the mid-1950s Issigonis had much experience in car design based on these principles, but his most innovative ideas had been frustrated, mainly by the Second World War. Then in 1956 the Suez crisis led to petrol shortage, and economical but unattractive 'bubble cars' appeared. Issigonis was given the task by the British Motor Corporation of designing a new and economical small car to compete with them; his response was highly original.

His answer (the Austin Design Office project 15) was to design and produce within 2 years a small car with its small wheels close to the four corners, and with the novel feature of a front transverse engine and transmission driving the front wheels. The suspension was an all-independent rubber system designed by A Moulton. The body shape of this 'Mini Minor' was functional and distinctive, and surprisingly roomy for its overall size; and its road-holding, stability and freedom from pitching were outstanding. From its launch in 1959 the Mini dominated the small car market for two decades and even after 25 years its market share was still growing. Its general layout has since become standard for all except large-size motor cars. Issigonis led design teams responsible for other successful cars, but none was as novel and influential as the Mini. He became a Fellow of the Royal Society in 1967 and was knighted in 1969.

Jacobi, Karl Gustav Jacob [yahkohbee] (1804–51) German mathematician: contributed to the theory of elliptic functions, analysis, number theory, geometry and mechanics.

Jacobi was the son of a Jewish banker, and showed wide-ranging talent from childhood. He became a lecturer at Königsberg in 1826 and in 1832 he became a professor there. He encouraged students to do original work before they had read all the previous work on a topic. As he said to one student: 'Your father would never have married, and you wouldn't be here now, if he had insisted on knowing all the girls in the world before marrying one'. In 1848 he made a brief but disastrous foray into politics and lost for a time the royal pension on which he, his wife and seven children lived; in 1851 he died of smallpox.

Jacobi, together with ABEL, created the theory of elliptic functions, and Jacobi also applied them to number theory and developed hyperelliptic functions. He did research on differential equations and determinants, and Jacobian determinants are now used in dynamics and quantum mechanics.

Jacquard, Joseph Marie (1752–1834) French technologist: devised an improved loom incorporating control by punched cards, a basic element of automation.

Jacquard served apprenticeships in bookbinding, cutlery making and typefounding and then, inheriting from his parents a small weaving business, he attempted to make his living as a weaver. However, weaving patterns attractive enough to sell well was slow and difficult: his business failed and he returned to cutlery. His interest in pattern weaving remained, and during the 1790s he devised improvements in looms, as well as taking an active part on the revolutionary side in the defence of his home town, Lyon.

By 1801 he was weaving fishing nets on a loom which combined several improvements, mainly devised by others. Demonstrated in Paris in 1804, it impressed Napoleon and led to a medal, a patent and a pension. Encouraged, he went on to make his major improvement, which was to be of value outside textiles. This was his use of cards with punched holes to control cams that directed pattern weaving. Improved further by others, chains of cards proved reliable enough before 1820 to cause violence by underemployed traditional weavers, and by the 1830s many were in use. Jacquard's method of coding information for manual looms implies that a hole or its absence can correspond to an 'on or off' action - or to 0 and 1 in binary notation. Punched cards were planned by BABBAGE in the 1830s to programme his mechanical calculators, and in 1890 HOLLERITH used 288-hole cards to process the US census returns. The pianola was another example.

Although Jacquard's method became dominant in textile pattern making, the principle of using punched cards or tape had to await the replacement of mechanical sensing by electronic computers and magnetic tape or discs before, in this much altered form, it could take a large part in the powered machine tool industry.

Jansky, Karl Guthe [yanskee] (1905–50) US radio engineer: discovered first astronomical radio source.

Jansky studied physics at Wisconsin and joined the Bell Telephone Co in 1928. While working for them in 1931 Jansky was given the task of investigating sources of interference to shortwave radio communications. (Causes of static include thunderstorms and nearby electrical equipment.) Using a rotatable directional antenna he detected a weak static emission that appeared at approximately the same time every day. He soon demonstrated that it was coming from the direction of the centre of the Galaxy and suggested in 1932 that it was caused by interstellar ionized gas rather than from the stars. His discovery did not receive much attention at the time, despite its potential value (eg radio waves penetrate dust clouds, which obscure the centre of the Galaxy for optical astronomy), but it was to lead to the development of radio astronomy after the Second World War. The unit of radio emission strength is named after him (1 jansky = $1\,\mathrm{W\,m^{-2}\,Hz^{-1}}$). Jansky did not continue in radio astronomy, which was kept alive by REBER until its post-radar expansion after the Second World War.

Jeans, Sir James (Hopwood) (1877–1946) British astrophysicist: pioneer theorist on stellar structure and the origin of the solar system.

As a talented young mathematician, Jeans was appointed to a chair of applied mathematics at Princeton in 1905 and worked there for 5 years before returning to Cambridge. While in the USA he had worked on classical physics, but after 1914 his research was largely in astrophysics. This began with his studies on rotating masses of fluid (gases and liquids, compressible or incompressible) and their calculated shape and stability, which led him to theorize on the structure and stability of stars and on the possible origin of the solar system. His ideas on this have not held up to the passage of time but they were valuable in spurring the ideas of others.

Although he worked at the Mount Wilson Observatory, CA, from the 1920s, by 1930 he had surprised his friends by turning from research to writing popular accounts of astronomy and cosmology, which were highly successful and made him the UK's best-known scientist.

Jeffreys, Sir Alec (John) (1950–) British biochemist: originator of DNA fingerprinting.

Jeffreys studied biochemistry and biochemical genetics at Oxford and in 1977 joined the department of genetics at the University of Leicester. In that year he discovered, with R Flavell (1945–), the existence of introns in mammalian genes (introns are sequences of bases within DNA that do not code for a protein).

Jeffreys later realized that some very variable, and repeated, parts of the human genome (the full DNA sequence) are highly characteristic of individuals and so can be used for identification purposes. The number of these repeats is specific for each individual (except identical twins), half of them being inherited from the father and half from the mother. The method is valuable in criminal cases for linking an individual with a scene-of-crime sample and a variant of it is applicable in some paternity and immigration disputes. First proposed by Jeffreys in 1984, it was much used in the USA from 1987. A DNA profile can be got from a small sample of material (semen, blood or other tissue) which may be on clothing and can be years old. The sample is extracted to give its DNA; this is fragmented by enzyme action, the fragments are separated by gel electrophoresis and these are then radioactively labelled. A barcode-like pattern, with its bars differing from one another in density and spacing and each derived from a DNA fragment, is the result and can be set beside the pattern obtained by similar treatment of a specimen from, for example, a suspected rapist. Bars of the same size and position in the gels point to the samples having the same genotype. The process can be repeated, say, four times, each time using a radioactive label for a specific DNA sequence, to give a 'multilocus genotype' print.

A chance match between the prints of two different people is widely held to have a probability of only about one in a million, although this near-uniqueness has been disputed since 1990 by some population geneticists; the problem is unresolved. Despite this, many lawyers see the technique as 'the prosecution's dream and the defence's nightmare'.

Jeffreys, Sir Harold (1891–1989) British geophysicist and astronomer: theoretical seismologist and co-author of 'tidal' theory of formation of solar system.

After graduating in mathematics in 1913, Jeffreys spent his entire academic career at Cambridge. He is best known for his work on the internal structure of the Earth and in particular on theoretical seismology, his earthquake travel time tables (worked out with K E Bullen (1906–76) in the 1930s) still being standard. Together with JEANS, Jeffreys proposed the 'tidal' theory of the formation of the solar system, in which it was suggested that a passing star might have drawn a filament of material out of the Sun, from which the planets subsequently condensed. The theory has since been abandoned in favour of much revised versions of LAPLACE's nebular theory. Jeffreys also worked on a variety of planetary problems and was a notable opponent of the theory of continental drift.

Jenner, Edward (1749–1823) British physician and naturalist: pioneer of vaccination.

Jenner, a vicar's son, was apprenticed to a surgeon at 13, and at 21 became a pupil of a leading surgeon and anatomist, J Hunter (1728–93) for 3 years. They had similar interests and became life-long friends; Jenner did well, and could have continued in London, but chose to return to his native village (Berkeley, in Gloucestershire) to practise medicine. At that time smallpox was a long-known and feared disease; epidemics were frequent, mortality was high (typically 20% of those infected) and survivors were disfigured. It was known that survivors are immune to reinfection and that inoculation from a patient with a mild attack of the disease protected against it, but this carried the risk of severe or fatal illness.

Jenner was told by a patient that country people who had cowpox (vaccinia; a rather rare disease of the udders of cows, transmittable to humans and not severe) did not afterwards become smallpox victims. Jenner found 10 such cases. In 1796 he took lymph from a cowpox vesicle on a dairymaid's finger and inoculated this into a healthy boy; 6 weeks later he inoculated him with smallpox matter, but the boy did not develop smallpox. He went on to show that matter from the boy (fluid from a vesicle) could be used to inoculate other individuals and passed by inoculation from person to person indefinitely without losing its protective effect. He used a thorn for inoculations. After finding that the dried vaccine retained its potency for a few months, he sent a sample to President Thomas Jefferson, who inoculated his family and neighbours at Monticello.

The new procedure ('vaccination') was widely but not universally welcomed and by 1800 was much used. In Britain it was made compulsory in 1853, and this was enforced by 1872. Smallpox became the first major disease to be fully overcome, at least in civilized communities; by 1980 it was officially extinct, as there were no recorded cases. It is now known that smallpox (variola) is one of a group of related pox diseases, whose causal agent is a rather large virus of high virulence. Inoculation of the cowpox virus causes production of antibodies effective against smallpox for a period of several years.

Jenner was a keen and skilful naturalist, with a special interest in birds. He did major work on the cuckoo and other migratory birds and he classified the plants from Cook's first expedition. The British government awarded him £30 000 for his discovery of vaccination.

Jex-Blake, Sophia (Louisa) (1840–1912) British physician: pioneer for medical education for women in Britain.

Sophia Jex-Blake was born in Hastings, the youngest daughter of Thomas Jex-Blake, a lawyer. Her brother Thomas William became headmaster of Rugby and dean of Wells. She studied at Queen's College for Women in London and became tutor in mathematics there (1859–61). In 1865 she went to the USA to study medicine at Boston with Lucy Sewall and Elizabeth Blackwell, but returned to England on the death of her father in 1869.

She found that medical schools in England would not admit women, and the Society of Apothecaries had closed the loophole that enabled Elizabeth Garrett Anderson to qualify and practise in Britain. The Medical Act of 1858 excluded foreign qualifications from the Medical Register (registration being needed to practise medicine legally in Britain), so preventing followers of Elizabeth Blackwell's course of entry from practice in England.

Sophia Jex-Blake then began a heated campaign for acceptance into Edinburgh University to study medicine. After initial acceptance by the Senate of separate classes for women, five women, including Jex-Blake, matriculated in 1869. Difficulties occurred when tutors were reluctant to teach the women; they were then refused admittance to the Royal Infirmary. Students raised a petition against their admission (a necessary part of their course) and as Sophia Jex-Blake and the other female students attempted to enter Surgeon's Hall a riot broke out among students of divided opinion. A libel action followed in 1871 when Jex-Blake was accused of leading the riot; she was awarded a farthing in damages (and had to pay a legal bill of nearly £1000). After further obstacles she brought a legal action against the university for not honouring its contract to admit the women to degree examinations; judgement in their favour was reversed in 1873 on the grounds that in admitting the women to matriculation the university had acted *ultra vires*.

Sophia Jex-Blake had a tempestuous personality, abrasive and emotional. Her battles were a gift to the newspapers but her struggles and humiliations had the value of changing public opinion. In 1874 she gathered together a group of sympathetic people to form the London School of Medicine for Women, which opened in 1874 with a staff of lecturers including Elizabeth Garrett Anderson. However Jex-Blake's personality made her unsuitable to serve as secretary and she was not qualified to

teach; she became an active spur on its council. It was the quiet, diplomatic Elizabeth Garrett Anderson who was appointed dean in 1883. In 1876 the Russell Gurney Enabling Act went through Parliament allowing medical examining bodies to test women. Clinical work was secured when the (Royal) Free Hospital admitted women students in 1877. Later that year the King and Queen's College of Physicians, Dublin, agreed to act as examiners and in 1877 Sophia Jex-Blake and her four co-students from Edinburgh graduated.

Sophia Jex-Blake settled in Scotland and practised medicine in Edinburgh; in 1886 she organized a medical school for women there, again amid controversy. It was not until 1894 that the University of Edinburgh admitted women to graduate in medicine. (See panel overleaf.)

Johanson, Donald (1943–) US palaeoanthropologist: discovered *Australopithecus afarensis*, the oldest hominid yet found.

After graduating in anthropology from the University of Chicago in 1966, Johanson conducted research in Chicago and Alaska before joining an archaeological expedition to the Omo River, southern Ethiopia, from 1970–2. He then turned his attention to Afar, in north-eastern Ethiopia, where he and Maurice Taieb had found fossil-bearing beds of considerable age. In 1973 he discovered the remains of a knee joint of a previously unknown hominid in deposits over 3 million years old, at that time the earliest conclusive evidence of man's bipedalism. A partial skeleton of *Australopithecus afarensis* was found by Johanson the following year, a female of about 20 years of age and 1.2 m in height which he nicknamed 'Lucy' and which was then the earliest known ancestor of man, about 3.1 million years old and so almost a million years more ancient than any other hominid then known.

Joliot, Frédéric [zholyoh] (1900–58) French nuclear physicist: co-discoverer of artificially-induced radioactivity.

Trained in science in Paris, Joliot became assistant to Marie Curie in 1925 and soon proved his skill as an experimenter. Mme Curie's elder daughter Irène was already her assistant in the Radium Institute, and she and Joliot married in 1926. Their personalities were very different, he an extrovert, she very diffident. Only in 1931 did they begin to collaborate in research, with notable success. They also perpetuated a family tradition; their daughter Hélène (1927–) became a nuclear physicist and married a physicist grandson of Langevin.

Foreseeing the consequences of the nuclear fission of uranium discovered by others in 1939, Joliot secured from Norway the world's major stock (less than 200 kg) of heavy water (D_2O; used as a moderator in early atomic piles) and, when France was invaded in 1940, he arranged for it to be sent to the UK. After the war, as their

THE ENTRY OF WOMEN INTO MEDICINE

The first formal European school of medicine was at Salerno in southern Italy; women both studied and taught there. The most famous among them was Trotula (d.1097?), whose *Passionibus mulierum curandorum* (1547, The Diseases of Women) became a standard medical textbook until the 16th-c. She classified diseases as inherited, contagious and 'other'. Formal education in medicine, especially for women, became increasingly difficult in Europe. In France women had studied and taught medicine at Montpellier, but by 1239 Montpellier had followed the example of the University of Paris, and excluded women. Many universities founded in the 12th- and 13th-c also debarred women; licences to practise medicine were restricted; Pope Sixtus IV (reigned 1471–84) forbade the practice of medicine by those who were not university graduates. However some women midwives, physicians and surgeons did practise in France, Germany, Italy and Britain. Many studied with a relative and followed their profession. In 1390 Dorotea Bocchi succeeded her father as professor of medicine and moral philosophy at the University of Bologna. In England, the dissolution of the monasteries deprived the poor of nursing care and women of an avenue of medical training; while wars, witch-hunts and constitutional crises prevented progress in the medical education of women.

In the liberal University of Bologna, Anna Morandi Manzolini (1716–74) the wife of the professor of anatomy, became his assistant in the construction of anatomical models. During his illness she lectured in his place, with the consent of the university, and on his death was elected, without a degree, as professor and *modellatrice* to the chair of anatomy in 1750. Her work became internationally famous. Maria della Donne (1776–1842) obtained her degree in medicine in 1799; Napoleon appointed her professor of obstetrics in 1802. In 1804 she became director of the School for Midwives and a member of the French Academy in 1807.

In France from the 16th-c, obstetrics began to develop as a science. At the forefront was Louyse Bourgeois (1553–1638) a friend and pupil of Ambroise Paré (*c.*1510–90); her husband was an assistant to Paré. She was midwife to Queen Marie de Medici through seven deliveries and wrote a major treatise on obstetrics in 1609, the most comprehensive since that by Trotula. It was based on personal observations of 2000 cases; 12 presentations and the rules for the delivery of each type were described. She also gave instruction on cleanliness and on avoiding cross-infections. Marie

Louise Lachapelle (1769–1821), succeeded her mother as head midwife at the Hôtel Dieu in Paris; her *Pratique des accouchements* (1821–25, Practice of Obstetrics) contained statistical tables compiled from 50 000 case-studies. Marie Anne Victorine Boivin (*née* Gillain) (1773–1847), who was educated by nuns at a hospital in Etampes, was Lachapelle's student and assistant and her successor. Her work on diseases of the uterus, published in 1833, was used as a textbook for many years. She was awarded an honorary degree by the University of Marbourg, and the Order of Merit by the King of Prussia.

The first woman to be awarded a medical degree in Germany was Dorothea Erxleben, in 1754. Taught by her physician father alongside her brother, she petitioned Frederick II of Prussia for permission to enter the University of Halle. Her success prompted what was to become the familiar expressions of outrage, shock and horror at the idea of women practising medicine. It was 1901 before another woman graduated from the University of Halle's School of Medicine. Charlotte von Siebold Heidenreich (1788–1859) was awarded a degree in obstetrics at the University of Giessen in 1817 and later became professor there; in 1819 she delivered the future Queen Victoria. German universities did not admit women as medical undergraduates until 1908, although they admitted women with 'foreign' diplomas for further study in 1869. In Switzerland the first woman graduated in 1868; Sweden gave women permission to practise medicine in 1870 and its first woman graduated in 1888; the University of Copenhagen's faculty of medicine accepted women in 1877, and its first woman graduated in 1885; Norway, by Act of Parliament, opened its doors to women in 1884 and its first female graduated in 1893. In the Netherlands Aletta Jacobs graduated in 1878. In 1889 the first women graduated in Spain and Portugal.

At least one woman did not wait for the barriers to lift. James Miranda Steuart Barry (1795–1865) was the name used by a woman who, disguised as a man, graduated from Edinburgh University in 1812. She continued the deception as a male army surgeon until her death, attaining the rank of Inspector-General of Army Hospitals. Her identity remains something of a mystery. That she had influential help in her deception is evident in the intervention of the Earl of Buchan on her behalf when the University Senate proposed withholding her degree on the grounds of her apparent youth, and later on her entry into Army service, when the physical examination was waived. During her service at the Cape of Good Hope Barry performed a successful caesarean operation. She was appointed Inspector-General of Hospitals for Upper and

Lower Canada in 1857, and so was the first woman to practise medicine in Canada, albeit as a man.

ELIZABETH BLACKWELL was the first woman to graduate in medicine in America in 1849, as the result of an error by Geneva College, a recognized medical school. The application from a woman to study medicine was thought to be a joke perpetrated by another college and was accepted in like spirit; however when she appeared, the college honoured the contract. Her attempts to practise medicine in the USA were blocked: no dispensary or hospital would allow her to see patients and she turned to Europe for further training in London and Paris. She opened a dispensary for destitute women and children, the New York Infirmary for Women and Children in 1853 and the Women's Medical College (1868–99). In 1858 the British Medical Council was formed, and because she held a medical degree and had practised medicine in England before this date, Blackwell's name appeared on the British Medical Register in 1859. In 1869 she returned to practise medicine in Britain. Nancy Clark (1825–1901) received a medical degree from Cleveland Medical College, the Medical Department of Western Reserve College, in 1852, although when she applied to the Massachusetts Medical Society they refused to license her on the grounds that she was a woman. It was 1876 before the first woman was admitted to the American Medical Association. Across the USA the medical schools began to open their doors to women by the end of the 19th-c. By 1894 about 10% of students at 18 medical schools were women. The first Canadian woman to gain a medical degree from a Canadian school was Augusta Stowe-Gullen, who graduated via the Toronto School of Medicine in 1883.

In England ELIZABETH GARRETT ANDERSON acquired her qualification to practise medicine by finding that the wording of the charter of the Society of Apothecaries did not specifically exclude women. She was awarded a diploma from the Society in 1865, which enabled her to get her name added to the British Medical Register, and so to practise medicine. She became the first woman to gain a medical degree at the University of Paris in 1870, after the university opened its medical school to women in 1868.

By the time SOPHIA JEX-BLAKE and others wished to enter the medical profession the Society of Apothecaries had closed the opening for women and holders of foreign degrees were by then excluded from the British Medical Register. Jex-Blake's battles to enter the Edinburgh University medical course were reported in the newspapers and her treatment gradually changed public opinion. In 1874, with a group of sympathetic people, including Elizabeth Garrett Anderson, she

formed the London School of Medicine for Women (LSMW). It took an Act of Parliament in 1876 to enable medical bodies legally to admit women. No examining body or hospital would accept the women students and the school faced closure in 1877. Then clinical teaching was arranged with the Royal Free Hospital; the King's and Queen's College, Dublin, recognized the teaching course at LSMW and was the examining body. Sophia Jex-Blake and four women co-students gained degrees from the King and Queen's College of Physicians of Dublin in 1877. The University of Edinburgh accepted women as medical students in 1894.

LSMW might not have formed without Jex-Blake's energetic, albeit tempestuous nature, and might not have survived without Garrett Anderson's professionalism and soothing diplomacy among male supporters. Garrett Anderson worked for the students of LSMW to be accepted into the University of London and in 1878 all faculties were opened to women. The first women to qualify in medicine were Mary Scharlieb and Edith Shove in 1883; the LSMW became a college of the University of London. Due to Britain's political position in the world, the LSMW had an influential role in the medical education of the women of India, Burma, South Africa and East and West Africa. From the Far East woman at first trained at LSMW or at the Women's Medical College of Pennsylvania in the USA; they are now accepted in all medical faculties in Singapore, Malaysia and Thailand.

Elizabeth Garrett Anderson also worked against the General Medical Council's proposal for a separate Medical Register for women and she proposed uniform standards in medicine, surgery and obstetrics for all candidates. The first female Fellow of the Royal Colleges of Surgeons of Ireland, England and Edinburgh was admitted in 1910; the Royal College of Physicians of London amended its regulations in 1926 and elected its first female Fellow in 1934.

In New Zealand there were six medical students by 1896 and in Australia the first women graduated in medicine in the 1890s. In India there was a great need for qualified women doctors to treat segregated women. Mary Scharlieb, from Madras, one of the first to graduate from the LSMW, so impressed Queen Victoria with the needs of her Indian subjects that a fund was set up to provide many small hospitals and clinics, all staffed by women. In 1894 the North India School of Medicine for Christian Women was founded and trained hundreds of nurses, midwives, dispensers and doctors. The first Indian girls entered LSMW in 1894 and returned to serve in civil hospitals and the Indian Women's Medical Service. Women are now admitted to all medical faculties in India and

Pakistan, with separate schools where custom demands.

The First World War gave women the opportunity to prove their value; 20% of British women doctors volunteered and served in the medical services, also women from Canada, Australia and New Zealand. They were not, however, accepted by the War Office to work at the front. Louisa Garrett Anderson (daughter of Elizabeth Garrett Anderson) equipped a hospital staffed by women; this was not taken seriously by the establishment but was accepted by the French Red Cross. Dr Elsie Inglis (1864–1917) founded the Scottish Women's Hospitals, staffed entirely by women, including surgeons, and offered its services to the War Office. When this was turned down (with the reply, 'Go sit quietly at home, dear lady'), Dr Inglis took the first Scottish Women's Hospital Unit to Serbia, and her colleagues also served with great efficiency in France, Corsica, Salonika, Romania, Russia and Malta. By 1916 the War Office began to recruit women doctors for service abroad. By the end of the war there were 85 women doctors serving in Malta, 36 in Egypt, 21 in France and 39 in Salonika; however, unlike the men, they were not given commissions. The war made it easier for women to be accepted by the medical hospitals and increased their experience of surgery, epidemic diseases, war wounds and the effects of war gases. Dr Inglis's colleagues received some of the highest decorations from France and Serbia, while she was merely given a magnificent funeral in Edinburgh and a hospital was named after her. The conduct and courage of medical women in the First World War was one of the most important contributions to the granting of women's suffrage in the UK.

After the war, the now increasing number of women qualifying in medicine (78 in 1917, 602 in 1921) were finding, like CICELY WILLIAMS, that getting a medical post was almost impossible, because returning ex-servicemen were given priority. Williams eventually joined the Colonial Service and went to the Gold Coast. After 7 years in Africa she was moved to Malaya, was trapped by the Japanese invasion of Singapore and was imprisoned in Changi jail, where she survived its form of interrogation.

In the Second World War the Royal Army Medical Corps women had uniforms, equivalent ranks and equal pay. The British women civilian doctors had testing medical duties in many theatres of war; some served with the Resistance in the occupied territories. These women demonstrated the worth of their claim to be treated with equality. The present position is that women are strongly represented among medical students and graduates, less strongly in junior posts, and hardly at all at senior levels in the medical profession in the UK.

MM

leading nuclear physicist, he directed work on France's first atomic pile, which operated in 1948; but his successes became confused by his showmanship, his need for adulation and his communist sympathies, and he was removed from his post as high commissioner for atomic energy in 1950.

Joliot-Curie, Irène [zholyoh küree] (1897–1956) French nuclear physicist: co-discoverer of artificially-induced radioactivity.

Irène, daughter of PIERRE and MARIE CURIE, had a unique education; she was taught at home, in physics by her mother, in maths by LANGEVIN and in chemistry by PERRIN. In the First World War she served as a radiographer, then inadequately protected against radiation; later, she became a victim, like her mother, of leukaemia, fairly certainly because of exposure to radiation, which eventually killed them both.

In the 1930s she did notable work on artificial radioactivity with her husband F JOLIOT, for which they shared the Nobel Prize for physics in 1935. In the Second World War she escaped to Switzerland and in 1946 became director of the Radium Institute and a director of the French Atomic Energy Commission. Her work in the 1930s with Joliot led them in late 1933 to make the first artificial radioelement by bombarding aluminium with alpha-particles (helium nuclei, $^4_2$He), which gave a novel radioisotope of phosphorus. Similar methods then led them and others to make a range of novel radioisotopes, some of which have proved of great value in research, in medicine and in industry.

Joly, John (1857–1933) Irish geologist and physicist.

Joly studied a range of sciences and engineering at Trinity College, Dublin, became professor of geology there in 1897 and held this post until his death. Early in his career he devised the steam calorimeter always linked with his name; he used it to find the specific heat of minerals and, for the first time, the specific heat of gases at constant volume.

HALLEY had proposed from 1693 that the age of the Earth (since water condensed on it) could be deduced from the rate at which the salt content of the oceans is increased by leaching of the salt of land masses by water, which rivers carry to the seas; he doubted both the age of a few thousand years deduced from the Bible and the idea that the Earth was eternal. In 1899 Joly used Halley's methods, by measuring the rate of increase of oceanic sodium content, and arrived at an age of $80–90 \times 10^6$ years. Although now seen as far too short a time, Joly's estimate was valuable then in supporting the geologist's need for a much longer period than that given by KELVIN's work from 1862 based on the supposed rate of radiative cooling. In 1903 Joly pointed out that radioactivity (then recently discovered) would provide terrestrial heating and so affect Kelvin's calculation.

Soon afterwards new estimates of the Earth's age were made based on the rate of radioactive decay of uranium to lead and the U:Pb ratio in old rocks. Joly aided this work in 1907 by showing that the dark rings (pleochroic halos) found in some minerals had been formed by radioactive inclusions within them. He went on to use these microscopic halos present in rocks of differing geological age to show that radioactive decay had occurred at a constant rate during geological time. Only after Joly's work could geological dating by radioactivity have a logically secure basis.

Jordan, (Marie-Ennemond) Camille [zhordã] (1838–1922) French mathematician: a major contributor to group theory.

Jordan trained as an engineer at the École Polytechnique, and as an engineer pursued mathematics in his spare time. At 35 he joined the mathematical staff of the École Polytechnique and taught at the Collège de France.

Jordan absorbed the ideas of the ill-fated GALOIS and developed a rigorous theory of finite, and then infinite, groups. He linked permutation groups and Galois's study of permuting the roots (solutions) of equations to the problem of solving polynomial equations. Jordan published a classic on group theory, *Traité de substitutions*, in 1870. He advanced symmetrical groups, and reduced the linear differential equations of order n to a group theoretic problem. Finally, he generalized HERMITE's work on the theory of quadratic forms with integral coefficients. Jordan inspired KLEIN and LIE to pursue novel research on group theory. Topology also interested Jordan and he devised homological or combinatorial topology by investigating symmetries in polyhedra.

Jordan, Ernst Pascual [yordan] (1902–80) German theoretical physicist and one of the founders of quantum mechanics.

Jordan grew up in Hanover, where he took his first degree, moving to Göttingen for his doctorate. He gained a post at the University of Rostock in 1929 and became professor of physics there in 1935. Chairs of physics at Berlin (1944) and Hamburg (1951) followed.

At 23 Jordan collaborated with BORN and then HEISENBERG, setting out the theory of quantum mechanics using matrix methods (1926). Later he contributed to the quantum mechanics of electron–photon interactions, now called quantum electrodynamics, while it was in its early stages of development. Another area in which he published significant research was gravitation.

Josephson, Brian David (1940–) British theoretical physicist: discovered tunnelling between superconductors.

Josephson studied at Cambridge and remained to become a professor of physics (1974). He discovered the Josephson effect while still a research student (1962), by considering two superconducting regions separated by a thin region of insulator (perhaps 1–2 nm thick). He showed theoretically that a current can flow between the two with no applied voltage as a result of electron tunnelling and that when a DC voltage is applied an AC current of frequency proportional to the voltage is produced. Experimental verification of this effect by J M Rowell and P W ANDERSON at Bell Telephone Laboratories supported the BCS theory which Josephson had used. The application of a small magnetic field across the junction sensitively alters the current. Such Josephson junctions have been used to measure accurately h/e, voltage and magnetic fields, and in fast switching devices for computers. Josephson shared a Nobel Prize in 1973. Since then his interests have moved to psychic phenomena and to music.

Joule, James (Prescott) [jowl] (1818–89) British physicist; established the mechanical theory of heat.

Joule grew up in a wealthy Manchester brewing family, a shy and delicate child. He received home tuition from DALTON in elementary science and mathematics. He was attracted to physics and especially to the problems of heat and began experimental work in a laboratory near the brewery. Joule's skill enabled him to measure heat and temperature changes accurately and he was later encouraged to pursue his work by W THOMSON.

When he was 18, Joule began his study of the heat developed by an electric current and by 1840 he had deduced the law connecting the current and resistance of a wire to the heat generated (Joule's Law). Between 1837 and 1847 his work established the principle of conservation of energy and the equivalence of heat and other forms of energy. J R Mayer (1814–78) arrived at the idea of conservation in the 1840s but in an unclear form, and W Thomson and HELMHOLTZ also were major contributors, but Joule made it a precise and explicit concept. The amount of mechanical work required to produce a given amount of heat was determined by Joule in 1843. He measured the small amount of heat produced in water by the rotation of paddles driven by falling weights.

Thomson and Joule collaborated for 7 years from 1852 in a series of experiments, mainly on the Joule–Thomson effect, whereby an expanding gas is cooled as work is done to separate the molecules. Joule also produced a paper on the kinetic theory of gases that included the first estimation of the speed of gas molecules (1848).

He was over-modest and made himself into an assistant to Thomson rather than following his own lines of thought; and he became unwell when he was 55 and did little more afterwards.

Joule remains one of the foremost experimentalists of his century; his main work was done before he was 30, on one problem of great importance, the mechanical equivalent of heat. He attacked this with ingenuity, made precise measurements and tenaciously located sources of error. The SI unit of energy, the joule (J) (pronounced 'jool') is the energy expended when the point of application of a force of 1 newton moves through 1 metre in the direction of the force, so $1\,J = 1\,Nm = 1\,kg\,m^2\,s^{-2}$. Heat and work are measured in the same units.

Kamerlingh-Onnes, Heike [kamerlingk awnes] (1853-1926) Dutch physicist: liquefied helium for the first time and discovered superconductivity.

Kamerlingh-Onnes studied physics at Groningen and then spent 2 years in Heidelberg studying under BUNSEN and KIRCHHOFF. His special interest was then in finding new proofs of the Earth's rotation, but after he became professor at Leiden in 1882 he concentrated on the properties of matter at low temperatures. DEWAR had liquefied nitrogen, and by cooling hydrogen with this and using the Joule–Thomson effect he had obtained liquid hydrogen. Kamerlingh-Onnes used an improved apparatus and similar principles. In 1908, by cooling helium with liquid hydrogen to about 18 K, and then using the Joule–Thomson effect (the cooling of a gas when it expands through a nozzle), he obtained liquid helium, which he found to boil at 4.25 K. If it was boiled rapidly by pressure reduction the temperature fell to just below 1 K, but it did not solidify. In 1911 he found that metals such as mercury, tin and lead at very low temperatures became superconductors, with near-zero electrical resistance. He won the Nobel Prize in 1913 for his work in low temperature physics, which he dominated until he retired in 1923. A theoretical explanation of superconductivity had to wait until the work of BARDEEN and others in 1957.

Kant, Immanuel (1724-1804) German philosopher: had influential ideas on cosmology.

Although primarily known for his work in philosophy, in 1755 Kant proposed the nebular hypothesis for the formation of the solar system that was later to be developed and made famous by LAPLACE. More influential was Kant's work on gravitation, in which he argued that attractive forces could act at a distance without the necessity for a transmitting medium. He also suggested (correctly) that the Milky Way was a lens-shaped collection of stars and that tidal friction slowed Earth's rotation. Less wisely, he was convinced that the planets were populated, with the most superior intellects on planets most distant from the Sun.

Kapitsa, Piotr Leonidovitch [kapitsa] (1894-1984) Russian physicist: experimenter on high magnetic fields and low temperatures.

Son of one general (an engineer) and grandson of another, Kapitsa studied electrical engineering at the Petrograd Polytechnic (St Petersburg), graduated in 1918, and continued there as a lecturer for 3 years. In 1919 his first wife and two children died in the famine following the revolution, and in 1921 the unhappy young man visited England and secured a place in RUTHERFORD's Cambridge laboratory. Both were energetic, outspoken and talented experimentalists, who formed a high regard for each other. Kapitsa became a popular figure, adventurous, ingenious and with wide-ranging interests.

After completing his PhD he began to work independently, on the problem of obtaining very high magnetic fields. For this he designed and built circuits which passed currents of 10 000 A or more through a small coil for 0.01 s or less, a time shorter than it would take the coil to burn out. By 1924 he could in this way obtain field strengths of up to 50 T, which he used in studies on the properties of materials in high fields. The electrical resistance of metals increases in high fields, and this effect increases at low temperatures, so Kapitsa turned his ingenuity to the design of an improved liquefier for helium. His work went so well that a new laboratory building (the Mond) was built for his work and opened in 1933. (He had the sculptor Eric Gill carve a crocodile over its entrance; it was not until 1966, revisiting his laboratory, that he admitted that this represented Rutherford.) By 1934 his new method for making liquid helium allowed him to study its strange properties; he discovered that it conducts heat better than copper, and it shows superfluidity, ie apparent complete loss of viscosity, below 2.2 K. (Later, his friend LANDAU was able to explain these properties of helium II.)

Also in 1934 he returned to the USSR to visit his mother and he was not allowed to leave, despite protests from the West. Soon, he learned (from reading *Pravda*) that he had been made head of a new and luxurious Institute for Physical Problems near Moscow. His equipment was sent to him from Cambridge, and his wife (he had remarried) was allowed to return to the UK for their children. One of his colleagues was Landau, who was arrested in the 1930s and accused of being 'an enemy of the state'. Kapitsa protested forcefully and successfully: Landau, unusually, survived.

From 1939 Kapitsa worked on liquid air and oxygen, which aided Soviet steel production. He did not work on atomic weapons, and wrote to Stalin in 1946 criticizing the competence of the notorious Beria, head of the NKVD (secret police) as director of the programme. For this he was exiled to house arrest in the country, where he worked on high temperature physics and on ball lightning (a type of plasma) aided by his sons; after 8 years Stalin was dead and Beria executed, and Kapitsa was restored to his post and worked on plasmas. Soon he was permitted visitors, and in his 70s was allowed to travel. In 1929 he had been elected a Fellow of the Royal Society (the first foreigner for 200 years) and in 1978 he shared a Nobel Prize for his work in low temperature physics.

Kármán, Theodore von kah(r)man] (1881–1963) Hungarian–US physicist: discovered Kármán vortices.

Von Kármán studied engineering at the Budapest Technical University and at the University of Göttingen. He subsequently became director of the Aachen Institute and of the Guggenheim Aeronautical Laboratory of the California Institute of Technology; he was instrumental in setting up the Jet Propulsion Laboratory at CalTech and was a leading figure in a number of international scientific organizations. After 1930 he was largely in the USA.

An outstanding theoretical aerodynamicist, von Kármán discovered the two rows of vortices in the wake generated by fluid flow around a cylinder, known as Kármán vortices, and together forming a Kármán vortex street. Kármán vortices are important factors in aerodynamics, as they can create destructive vibrations. (The destruction by wind of the Tecoma Narrows Bridge in 1940 also had this cause.) His work, which included long-range ballistic missiles, aerofoil profiles, jet-assisted take-off and supersonic flight, was important in enabling the USA to become a world leader in the aerospace industry.

Karrer, Paul [karer] (1889–1971) Swiss organic chemist: best known for work on carotenoids and vitamins.

Karrer was born in Moscow. His father was a Swiss dentist and he returned with his parents to the Swiss countryside when he was 3; in 1908 he began to study chemistry at Zürich under A WERNER. His DPhil work with him was in inorganic chemistry, and afterwards he worked with EHRLICH in Frankfurt on organo-arsenic compounds for use in chemotherapy. In 1919 he succeeded Werner and remained in Zürich thereafter, working on organic natural products of several kinds.

In 1926 he began work on natural pigments, concentrating on the carotenoids, which are complex molecules containing many C=C bonds and including lycopene ($C_{40}H_{56}$, from tomatoes), the carotenes, the vitamins A_1 and A_2 and retinene, the light-sensitive pigment of the eye. In the 1930s his masterly studies of these closely related compounds revealed their structures, which he confirmed in many cases by synthesis (eg of vitamin A_1, $C_{20}H_{29}OH$, in 1931). He also worked on other vitamins (B_2 and E), on alkaloids and on coenzymes. He shared a Nobel Prize in 1937.

Kekulé, Friedrich August [kaykuhlay] (1829–96) German organic chemist; founder of structural organic chemistry, and proposer of ring structure for benzenoid compounds.

Kekulé began his student career in architecture at Giessen, but he heard LIEBIG give evidence in a murder trial and was attracted to his chemistry lectures. Later he studied in Paris and London. He claimed that the key idea of organic molecular structure came to him in a daydream, on the upper deck of a London bus. His theory (1858) adopted FRANKLAND's idea of valence; that each type of atom can combine

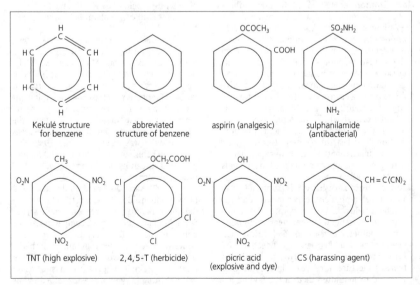

Kekulé structure for benzene

abbreviated structure of benzene

aspirin (analgesic)

sulphanilamide (antibacterial)

TNT (high explosive)

2, 4, 5 - T (herbicide)

picric acid (explosive and dye)

CS (harassing agent)

Structure diagrams for benzene and some derivatives

with some fixed number of other atoms or groups. Kekulé proposed that this 'valence' for carbon atoms is four. He also proposed that carbon atoms could be linked together to form stable chains, which was a new and vital idea.

It only needed the further idea, gradually developed by many chemists, that these structural formulae based on carbon chains represented molecular reality, for organic chemical theory to make the largest step in its history: for these structure diagrams, as the formulae became, could be deduced from the chemical and physical properties of the compound. From the diagram, in turn, new properties could be predicted; and so these molecular structures became the fruitful focus of every organic chemist's thinking. In the hands of BAEYER especially, experimental work on this basis pushed ahead. The ideas of structural theory became well known through Kekulé's lectures in Ghent (from 1858) and then at Bonn (from 1867) and from his textbook (1859).

Kekulé contributed little as an experimentalist, but his second gift to theory did much to form a basis for the new organic chemical industry making dyes and drugs from coal tar products. This was his proposal of 1865, that the six-carbon nucleus of benzene consisted of not a chain but a closed ring of carbon atoms. In benzene each carbon atom carries a hydrogen atom, but one or more of these can be replaced by other atoms or groups to give a vast range of compounds (see diagram). As with his structure theory, Kekulé was anticipated by COUPER in the idea of a cyclic molecule, but the latter's illness ensured Kekulé's superior place in developing the theory and securing his reputation.

Kelvin, William Thomson, Baron see **Thomson**

Kendall, Edward Calvin (1886–1972) US biochemist: pioneer of corticosteroid biochemistry.

Kendall studied chemistry at Columbia University, New York and afterwards worked mainly at the Mayo Foundation in Rochester, MN. During the First World War he isolated from the thyroid gland a new amino acid, thyroxin. This contains iodine, and it is a component of the thyroid hormone (thyroglobulin) that partly controls the rate of the body's metabolism. Kendall went on to study the hormones of the cortex (outer part) of the adrenal glands. In the 1930s he isolated a series of steroids from this source; one of them (Kendall's compound E, later named cortisone) was shown by his co-worker P S Hench (1896–1965) to relieve the symptoms of rheumatoid arthritis. During the Second World War there was a belief that the Germans were buying adrenal glands from Argentinian slaughterhouses and using extracts from them to help their pilots fly at great heights. The rumour was false, but it led to intensified study and by 1943

no less than 23 corticosteroids had been isolated in the USA or in Switzerland, and Kendall and others had devised synthetic routes to make related compounds. Since then, corticosteroids have been much used to treat inflammatory, allergic and rheumatic diseases. In 1950 the Nobel Prize for medicine or physiology was shared by Kendall, Hench and T Reichstein (1897–   ), who had worked on these compounds at Zürich.

Kepler, Johannes (1571–1630) German astronomer and physicist: discovered laws of planetary motion.

Kepler had smallpox at the age of 3, damaging his eyesight and the use of his hands, which makes his achievements, as the son of a mercenary, the more remarkable. Kepler read and approved the work of COPERNICUS while studying theology at Tübingen, intending originally to go into the church. He became a teacher of mathematics in the Protestant Seminary at Graz in Austria. In 1600, having been forced to leave his teaching post due to the religious persecution of Protestants, he was invited to join BRAHE in Prague, who assigned him the task of working out the orbit of Mars. Brahe died 2 years later, leaving Kepler his 20-year archive of astronomical observations. In 1609, having failed to fit Brahe's observations of Mars into the perfectly circular orbits of the Copernican cosmology, he formulated Kepler's first two laws of planetary motion: that planets follow elliptical orbits with the Sun at one focus and that the line joining a planet to the Sun, as it moves, sweeps through equal areas in equal times. At last understanding the principles of planetary motion, Kepler then proceeded with the more onerous but valuable task of completing the 'Rudolphine Tables', the tabulation of Brahe's results for his sponsor Emperor Rudolph, a job which he did not complete until 1627. A fine scientist, he was also a mystic and astrologer, cast horoscopes and believed in the 'music of the spheres' (from the planets).

In 1611 civil war broke out, Rudolph was deposed and Kepler's wife and child died. He moved to Linz, where he propounded his third law of planetary motion: that the squares of the planetary periods are proportional to the cubes of their mean distances from the Sun. These three laws gave a sound basis for all later work on the solar system.

Kepler also wrote a book on the problems of measuring the volumes of liquids in wine casks that was to be influential in the evolution of infinitesimal calculus.

Kerr, John (1824–1907) British physicist: discovered the electro-optical and magneto-optical Kerr effects.

Kerr was educated at the University of Glasgow, becoming a research student with Lord Kelvin (W THOMSON), and working with him in the converted wine cellar known as 'the coal hole'. He later became a lecturer in mathematics at the Free Church Training College for Teachers in Glasgow, continuing his research in his free time.

Kerr is best remembered for the effect that bears his name. In 1875 he showed that birefringence (double refraction) occurs in some materials such as glass when subjected to a high electric field. With great experimental skill he showed that the size of the effect is proportional to the square of the field strength; the electro-optical Kerr effect is used today as the basis for ultra-fast optical shutters ($c.10^{-10}$ s), using a Kerr cell in which a liquid (eg nitrobenzene) undergoes birefringence. He also discovered the magneto-optical Kerr effect, in which plane-polarized light reflected from the polished pole of an electromagnet becomes elliptically polarized; the effect has been used in the study of domain structure and other magnetic properties of ferromagnetic materials.

Kettlewell, Henry Bernard Davis (1907–79) British lepidopterist and geneticist: experimented to confirm Darwin's theory of natural selection.

A medical graduate who practised in England and worked on locust control in South Africa, Kettlewell's best-known work was done as an Oxford geneticist in the 1950s. He noted that many species of peppered moth that in the mid-19th-c were light in colour had became dark by the 1950s. He deduced that the darkening (melanism) was related to the darkening of the tree stems on which the moths remained by day, by industrial smoke. This would cause dark forms to survive predation by birds more successfully. To test his idea, he released light and dark forms of one moth (*Biston betularia*) in large numbers in both a polluted wood near Birmingham and an unpolluted wood. Recapture of many of the moths after an interval confirmed that the light form survived best in the unpolluted wood and the converse in the Birmingham wood. The result provided some experimental confirmation of DARWIN's theory of natural selection.

Khorana, Har Gobind [korahnah] (1922–) Indian–US molecular biologist: co-discoverer of genetic code and first synthesizer of a gene.

Educated at universities in the Punjab, at Liverpool, Zürich and Cambridge, Khorana moved to Vancouver in 1952 and extended his work in the area he had studied in Cambridge; the synthesis of nucleotide coenzymes. From 1960–70 he was at Wisconsin, and while there he carried out valuable syntheses of polynucleotides with known base sequences. These were of great value in establishing the 'genetic code word dictionary'. This refers to the fact that the four bases (A,C,G and T) present in DNA chains are 'read' in linear groups of three (codons), as was known by the late 1950s. It was

also known that the sequence is non-overlapping and 'comma-less'. Since four bases in groups of three allow $4^3 = 64$ combinations, but these code for only 20 amino acids which make up proteins, it would appear that some codons are 'nonsense codons' and/or some amino acids are coded by more than one codon.

Khorana had a major part in the work which established the dictionary, by his synthesis of all the 64 codons. This was an essential step in the further development of molecular biology. It turns out that the first two bases in a codon triplet are the main determinants of its specificity. Khorana's continued work on RNA and DNA has included the synthesis in 1970 of a DNA from *Escherichia coli*, a gene with 126 nucleotide base pairs. He was at the Massachusetts Institute of Technology from 1970 and shared a Nobel Prize in 1968 with M W Nirenberg (1927–) and R W Holley (1922–93), who also made major contributions to this area.

Khwarizmi, Al *see* **Al-Khwarizmi**

Kipping, Frederick Stanley (1863–1949) British chemist: discoverer of silicones.

A student at Manchester and Munich, Kipping became a co-worker and friend of W H Perkin Jr (1860–1929), and a professor at Nottingham from 1897–1936. Early in his research career he attempted to make compounds of silicon analogous to some of the familiar organic compounds based on carbon; specifically he tried to make ketone analogues. His actual product was a polymer mixture, soon named 'silicone', and over a period of years he published extensively on compounds of this type. He foresaw no practical use for them as late as 1937, but wartime needs for new materials linked with his methods led to their production in the 1940s for use eventually as specialist lubricants, elastomers, hydraulic fluids, sealing compounds, insulators and surgical implants; specialist uses include the soles of moon boots for astronauts. This family of polymeric organosilicon compounds contain chains of Si–O–Si links, and are inert and water-repellent. A key step was the discovery in 1940 by E G Rochow (1909–) of an easy route to methyl silicones.

Kirchhoff, Gustav Robert [keerkhhohf] (1824–87) German physicist: pioneer of spectroscopy; devised theory of electrical networks.

Kirchhoff was educated at the University of Königsberg (later Kaliningrad, USSR) and spent his professional life at the universities of Breslau, Berlin and Heidelberg. An early accident made him a wheelchair user but did not alter his cheerfulness.

Kirchhoff was still a student when in 1845 he made his first important contribution to physics, formulating Kirchhoff's Laws, which enable the current and potential at any point in a network of conductors to be determined. The two laws are extensions of OHM's Law, and state

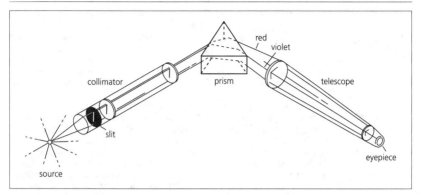

Basic components of a prism spectrometer. Light from the source passes through an adjustable slit, and the collimator lens forms a parallel beam. This is refracted and dispersed by a prism and the resulting spectrum is observed through a telescope fitted with cross-hairs and mounted to rotate horizontally, so that line positions in the emission spectrum can be measured.

that (1) the sums of the currents in a network must be zero at circuit junctions, $\Sigma I = 0$, and (2) $\Sigma IR = \Sigma V$ when applied to a closed loop in the network. Kirchhoff's other contributions to the study of electricity include demonstrating that oscillating current in a conductor of zero resistance propagates at the speed of light, and the unification of static and current electricity.

Kirchhoff was a lifelong friend and collaborator of the chemist BUNSEN and it was with him that much of his work on spectroscopy was done. They established spectroscopy as an analytical technique, using the nearly colourless flame of the Bunsen gas burner and a prism system designed by Kirchhoff. They saw that a continuous spectrum is produced by a glowing solid, liquid, or a gas under high pressure. An emission-line spectrum is given by a glowing gas under low pressure. An absorption-line spectrum is shown when a cooler gas is placed between a continuous source and the observer. In their hands, the spectrometer joined the telescope and microscope as a dominant scientific instrument.

In 1860 they demonstrated that when metal compounds are heated in a flame they emit spectral lines that are characteristic of the metal concerned, a fact which led Bunsen to discover the elements caesium and rubidium shortly afterward. Kirchhoff had discovered in 1859 that the dark FRAUNHOFER spectral lines in the Sun's rays were intensified when sunlight passed through the burner flame containing certain salts, leading him to the realization that they were absorption lines corresponding to elements found in the Sun's atmosphere. He also showed that the ratio of the emission and absorption powers of radiation of a given wavelength from all bodies is the same at the same temperature (Kirchhoff's Law of Emission), from which he later developed the concept of the black body. The study of black body radiation was the key in the development of quantum theory.

Kirkwood, Daniel (1814–95) US astronomer: discovered and explained the gaps in the asteroid belt.

Kirkwood observed that the orbits of the asteroids are not evenly distributed within the asteroid belt, there being a number of bands in which no asteroids are found (the Kirkwood gaps). He demonstrated that these bands correspond to orbital periods that are simple fractions of Jupiter's orbital period and that any asteroids lying within the 'gaps' would eventually be gravitationally perturbed into other orbits. Kirkwood was similarly able to explain the CASSINI division in the rings of Saturn as being due to the effect of its satellite Mimas.

Kitasato, Shibasaburo (1852–1931) Japanese bacteriologist: co-discoverer of antitoxic immunity and of the plague bacillus.

Kitasato grew up in an isolated mountain village, where his father was mayor. He studied medicine at Tokyo and in 1886 was sent by his government to study bacteriology with KOCH in Berlin. He proved an exceptionally good student and became a close friend, and in 1889 he grew the first pure culture of the tetanus bacillus, which A Nicolaier (1862–1934) had described in 1884. In 1890, working with BEHRING, they showed that animals injected with small doses of tetanus toxin developed in their blood the power of neutralizing the toxin, and that their blood serum could protect other animals for a time. This discovery (antitoxic immunity) quickly led to the use of serum (made in horses) for treating tetanus, and a similar antitoxin was developed for treating diphtheria and for protection against the diseases. The theory of these immunological reactions was developed especially by EHRLICH.

In 1892 Kitasato returned to Japan, and in 1894 he was sent to Hong Kong to study the bubonic plague epidemic there. He succeeded in identifying the plague bacillus, at nearly the same time as A Yersin (1863–1943) from Paris.

Klaproth, Martin Heinrich [klaproht] (1743–1817) German chemist: a founder of analytical chemistry and discoverer of new elements.

Klaproth came into chemistry from an apprenticeship as an apothecary, as did a number of chemists of his time. By 1810 his fame was such that he had left his pharmacy and was appointed the first professor of chemistry in the new university of Berlin. Before then he had done much to develop analytical chemistry, and the standard methods of gravimetric analysis (eg heating precipitates to constant weight) owe much to him. His analyses led him to deduce that new elements must be present in various minerals; eg uranium in pitchblende (1789; named in honour of the new planet, Uranus) zirconium in zircon (also in 1789), strontium in strontianite (1793) and titanium in rutile (1795). In each case, the free element was later isolated by others. He began the study of the rare earth metals, and he showed that nickel is present in meteorites. He was an early supporter of LAVOISIER'S ideas and did much to ensure that they were taught in Germany.

Klein, Christian Felix (1849–1925) German mathematician: the founder of modern geometry unifying Euclidean and non-Euclidean geometry.

Klein studied at Bonn, Göttingen, Berlin and Paris, and began research in geometry, although he had at first wished to do physics. Work on transformation groups with LIE followed in 1870, and he became professor of mathematics at Erlangen at 23, having just finished service as a medical orderly in the Franco-Prussian war. In his inaugural lecture of 1872 at Erlangen he put forward the audacious 'Erlanger Programm', a unification of mathematics to be achieved by considering each branch of geometry as the theory of invariants of a particular tranformation group. This was well received and influenced his colleagues to unify geometry. During most of his career he held a professorship at Göttingen, and helped to make it a centre for all the exact sciences, as well as mathematics.

Euclidean geometry comes from the metrical transformation, projective geometry from linear transformations, topology from continuous transformations and non-Euclidean geometries from their particular metrics.

Klein developed projective geometry, taking it from three to n dimensions and applied group theory widely, for example to the symmetries of regular solids (1884). He invented the Klein bottle in topology, which is a one-sided closed surface with no boundaries (it has no 'inside').

Klein also added to number theory and the theory of differential equations, and recast RIEMANNIAN geometry as a part of function theory. In the 1890s Klein and SOMMERFELD worked out the theory of the gyroscope and produced a standard textbook on it; he was against the tendency of mathematics to become highly abstract and liked engineering applications. He was also uninterested in detailed calculations, which he left to his students.

Klitzing, Klaus von (1943–) German physicist: discovered the quantum Hall effect.

Von Klitzing was born in Schroda/Posen and studied at Braunschweig and Würzburg. He became a professor in 1980 at Munich, and in 1985 director of the Max Planck Institute, Stuttgart. In 1977 he presented a paper on two-dimensional electronic behaviour in which the quantum Hall effect can be clearly seen. However few realized the significance of the measurements, and it was only when working one night at the high magnetic field laboratory in Grenoble in 1980 that von Klitzing appreciated what had occurred.

An electronic gas that is confined into a flat layer can be made by depositing a very thin layer of semiconductor upon a base material. Under a magnetic field electrons will perform circular orbits, with only particular energy states allowed (called LANDAU levels). At certain values of the field the Landau levels become filled and the conductivity and resistivity fall to zero. Others had put forward a theory that the (Hall) resistance under such conditions should rise in units of h/e^2; von Klitzing demonstrated that the resistance did rise in steps and accurately obeyed this condition. He had discovered the quantum Hall effect (QHE), and for this won the 1985 Nobel Prize for physics.

The effect has caused new thinking on electrical conduction in high magnetic fields; and it has allowed resistance to be measured with exceptional accuracy.

Klumpke, Dorothea [klumpkuh] (1861–1942) US astronomer: the first woman to make astronomical observations from a balloon.

Dorothea Klumpke was born in San Francisco, where her father had made a fortune from real estate. Her mother believed that their two sons and five daughters should have the same educational opportunity and so, as they could find no suitable schools in post-gold-rush San Francisco, they moved to Europe, where Dorothea was to live for the next 50 years. She was educated in Germany, Switzerland and France. She gained her bachelor of science degree in mathematics in 1886 from the Sorbonne, and joined the staff of the Paris Observatory and worked on photographic star charts. In 1891 she became director of a bureau of the Paris Observatory for the measurement of the plates of the astro-photographic catalogue of stars down to 15th magnitude. She became the first woman to gain a doctorate in mathematics at the Sorbonne –

for a thesis on Saturn's rings (1893). In 1899, France, Germany and Russia made plans to launch balloons to observe the Leonid meteor shower of mid-November and Klumpke was chosen to ascend in *La Centaure*. She married Isaac Roberts in 1901; he was a Welsh businessman and amateur astronomer 30 years her senior; they settled in London. After his death in 1904 she returned to France with her husband's large photographic plate collection, and published in 1928 a *Celestial Atlas* in tribute to him. She was made a chevalier of the Legion d'Honneur in 1934 for her long service to French astronomy. Together with her sister Anna she returned to the USA in the late 1930s.

Koch, Robert [kokh] (1843–1910) German bacteriologist: a founder of medical bacteriology.

Koch was one of the 13 children of a mine official; he studied medicine at Göttingen, served in the Franco-Prussian war of 1870 and became a district medical officer in a small town in east Germany. He became interested in anthrax and worked on the disease in a room in his house, using a microscope given to him by his wife. Anthrax is a deadly disease of cattle, which caused huge losses in France at that time. It is highly contagious, can be passed to man, and can infect animals in fields from which cattle have been excluded for years. It was known to be caused by a bacterium. Koch found by 1876 that the anthrax bacilli can form spores (if the temperature is not too low, and if oxygen is present) and that these are resistant to heat and to drying. These spores can re-form the bacillus. Koch was able to isolate the anthrax bacillus from the blood of infected cows and he produced pure cultures, able to cause the disease; for the first time, a laboratory culture was shown to cause disease. (PASTEUR was working competitively on similar lines and in 1882 made an anthrax vaccine which protected against the infection.)

Koch much improved techniques in bacteriology. He used dyes to stain bacteria and so make them more visible under the microscope; he used a solid medium (agar gel) to grow them conveniently and separately on plates, or in the flat glass dishes designed by his assistant J R Petri (1852–1921); and he aided surgery by showing that steam kills bacteria on dressings and instruments more effectively than dry heat. From 1879 he worked in the Health Office in Berlin, where in 1882 he identified the tubercle bacillus. This was difficult work; the bacillus is small and slow-growing, but human tuberculosis ('TB') was responsible for one in seven of all European deaths at that time. In 1890 Koch was persuaded to announce a vaccine against it, but the claim was premature and his 'cure' survived only as a test method to show whether a patient had experienced tuberculosis.

From the 1880s he travelled widely, much enjoying his position as one of the first of the

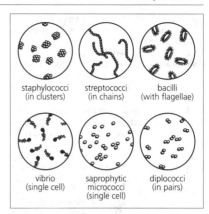

| staphylococci (in clusters) | streptococci (in chains) | bacilli (with flagellae) |
| vibrio (single cell) | saprophytic micrococci (single cell) | diplococci (in pairs) |

Bacteria, magnified a few hundred times

'international experts'. He did major work on cholera in Egypt, in India on bubonic plague, in Java on malaria, in East Africa on sleeping sickness and in South Africa on rinderpest. Although not as wide-ranging a biological genius as Pasteur, he is the greatest figure in medical bacteriology and many of its leaders after him were his pupils. His Nobel Prize in 1905 was for his work on tuberculosis. His criteria for deciding that an organism causes a disease (Koch's postulates) remain as critical tests: they require (1) the presence of the organism in every case of the disease examined; (2) the preparation of a pure culture; (3) the re-production of the disease by a pure culture, removed by several generations from the organisms first isolated. Use of these principles of 1890 established modern medical bacteriology.

Kolbe, (Adolf Wilhelm) Hermann [kolbuh] (1818–84) German organic chemist: developed useful routes in organic synthesis.

Kolbe was the eldest of the 15 children of a Lutheran pastor. He studied chemistry under WÖHLER and BUNSEN, and with L Playfair (1819–98) in London. He succeeded Bunsen at Marburg in 1851 and moved to Leipzig in 1865. He was an inspiring teacher and a talented researcher, despite holding firmly to outdated theories (mainly BERZELIUS'S) and his intemperate criticism of newer ideas (he vigorously abused KEKULÉ'S structure theory, for example).

His many successes in synthesis include the Kolbe reaction, in which a hydrocarbon is made by electrolysis of an alkali metal salt of an organic acid (this was the first use of electrolysis in organic synthesis):

$$2RCO_2K + 2H_2O \rightarrow R{-}R + 2CO_2 + 2KOH + H_2$$

At anode   At cathode

For example: $2Br(CH_2)_{11}CO_2K \rightarrow Br(CH_2)_{22}Br$

Also named after him is the Kolbe or Kolbe–Schmitt reaction, in which an alkali metal phenoxide is heated with carbon dioxide under pressure; a carboxyl group enters the ring and the product is a phenolic acid; eg a synthesis of salicyclic acid:

$$C_6H_5ONa + CO_2 \rightarrow o\text{-}(NaO)C_6H_4CO_2H$$

Kolmogorov, Andrei Nikolaievich [kolmogorof] (1903–87) Russian mathematician: advanced the foundations of probability theory.

Kolmogorov graduated from Moscow in 1925 and in 1933 became director of the Institute of Mathematics. In 1933 his book *Foundations of the Theory of Probability* became the first rigorous treatment of the subject. The 'additivity assumption' basic to probability is set out (due originally to Jakob Bernoulli (1654–1705)): that if an event can occur in an infinite number of ways its probability is the sum of the probabilities of each of these ways. Kolmogorov then explored Markov processes – those where a probability of a variable depends on its previous value but not its values before that. He constructed such processes by analytic means.

Another of his interests was the theory of algorithms (or mathematical operations, such as division), and he showed its relationship to computing and with cybernetics, which is concerned with communication and control, and 'feedback'. Kolmogorov produced a theory of programmed instructions and of how information is conveyed along communication channels.

Kornberg, Arthur (1918–) US biochemist: devised artificial synthesis of DNA using enzyme.

Kornberg graduated in chemistry and biology at City College New York in 1937 and in medicine at Rochester in 1941. For the next 10 years he worked at the National Institutes of Health at Bethesda; he was also a lieutenant in the US Coast Guard. From 1959 he was professor of biochemistry at Stanford. His main concern was always with enzymes, and in 1956 he made an outstanding discovery. This was his isolation of an enzyme from *Escherichia coli*, now called DNA polymerase I, which he showed was able to synthesize DNA from nucleotide molecules (which can themselves be made synthetically) in the absence of living cells, provided that the reaction mixture included some natural DNA to act as a template and primer. This last idea was both novel and of great importance in later work on DNA. Kornberg shared the 1959 Nobel Prize with S Ochoa (1905–) for this work.

Kovalevsky, Sonya (Sofya) Vasilyevna, Sonya Vasilyevna Kovalevskaya (*Russ*), *née* Korvin-Krukovsky (1850–91) Russian mathematician.

Although Sonya's parents were well-educated members of the nobility, they were not sympathetic to the 'liberated' ideas of their young daughters. However, one of the children's rooms was temporarily wallpapered with the student lecture notes on calculus of their father, a general of engineers. (Years later her tutor in mathematics was surprised at the speed with which Sonya grasped the concepts of the calculus.) A family friend, recognizing her ability, advised that she should study mathematics, which she did through tutorials at the naval academy at St Petersburg during visits there with her mother.

When she was 18 years old Sonya followed a path then favoured by Russian women in search of study abroad: she contracted a marriage of convenience. Her husband, the young palaeontologist V Kovalevsky, took her to Germany. Women there were not admitted to university lectures, but she was tutored by leading physicists at Heidelberg, and in Berlin studied with WEIERSTRASS, who recognized her talent and persuaded the university of Göttingen to consider her published research papers for a doctorate *in absentia*: she was granted the degree in 1874. She was unable to find an academic post open to women, and so returned to Russia, where her daughter was born. In debt, her husband became involved in fraud, was disgraced and committed suicide in 1883. Soon afterwards, with the help of Weierstrass, she was appointed lecturer at the university of Stockholm and in 1884 was granted a professorship.

Her work in mathematical analysis ranks her as the leading woman mathematician before the 20th-c, with her major contributions being on partial differential equations and on Abelian functions. In applied mathematics she worked on the structure of Saturn's rings, on propagation of light, and notably on the rotation of a solid body about a fixed point. Her paper on the last of these subjects won prizes from both the French and Swedish Academies, and in 1889 she was elected a corresponding member of the Russian Academy of Science. She died 2 years later, at the height of her career, following an influenzal infection.

Krebs, Sir Hans Adolf (1900–81) German–British biochemist: discovered energy-generating cycle in living cells.

Krebs followed his father in studying medicine and, after the German practice of the time, did so at five universities; he then spent 4 years working in Berlin on biochemical problems with WARBURG. The latter had developed a method for studying metabolic reactions by using thin tissue slices and measuring their gas exchange manometrically, and Krebs used and improved this technique in 1932 to show how, in the liver of most animals, amino acids lose nitrogen to give urea in a process now known as the ornithine cycle.

The next year he escaped from Germany to England and, after a short period in Cambridge, settled in Sheffield for 20 years; here most of his work was done, notably his work on the Krebs cycle (also known as the citric acid or tricarboxylic acid cycle). This cycle is the central energy-generating process in cells of most kinds, occurs in their mitochondria and generates energy for the entire organism. It was already known that foods in general are broken down to glucose and then to pyruvic acid; but those stages yield little energy. Krebs showed how the glucose is broken down in a cycle of changes to give carbon dioxide, water and energy. For this fundamental study of metabolism, Krebs shared a Nobel Prize in 1953 with F Lipmann (1899–1986), who worked out important details of the cycle.

Krogh, (Schack) August [krawg] (1874–1949) Danish physiologist: studied physiology of respiration.

A shipbuilder's son, young Krogh was always an enthusiast for experiment, and soon after beginning medicine at Copenhagen he moved to zoology and to medical physiology. He spent his life in Copenhagen as a zoophysiologist. As a student he worked in his room on the hydrostatic mechanism of *Corethra* larvae, devising methods for analysing gas in their air bladders and showing that they 'function like the diving tanks of a submarine'. He went on to study gas exchange in the animal lung, and the whole problem of how an animal responds to a 'call for oxygen', both in vertebrates and in insects. In this, the behaviour of the capillaries is critical, and Krogh studied their movement and expansion in the frog's tongue, and from this developed a general picture of the behaviour and response of the capillary system and its regulatory mechanism, which involves both nerves and hormones. He won the Nobel Prize in 1920.

Kronecker, Leopold (1823–91) German mathematician: developed algebraic number theory and invented the Kronecker delta.

Kronecker was born into a rich Jewish family and was taught at school by E E Kummer (1810–93), who became a lifelong friend. From this time sprang his interest in number theory and arithmetic, and he went on to take his degree at Berlin (1843), receiving his doctorate in 1845 for research on complex units. Kronecker spent the next 10 years managing the family estate and an uncle's banking business: he prospered and married. When he returned to Berlin to do mathematics he was financially independent, and only lectured at the Berlin Academy from 1861 for his own pleasure. He declined the chair at Göttingen (1868) but accepted Kummer's old chair at Berlin, holding it until his death. Kronecker was a man of wide culture, supporting the arts and interested in philosophy and Christian theology, he became a Christian shortly before his death.

In his early mathematical research on complex units Kronecker nearly anticipated Kummer's famous concept of ideal numbers, and his work on number theory, algebra and elliptic functions unified much previous research. On algebraic numbers he rederived much of existing theory without referring to what he (incorrectly) claimed were the ill-defined complex and irrational numbers. He once made the comment in an after-dinner speech: 'God made the integers, all else is the work of man'. He was often in debate with WEIERSTRASS and G Cantor (1845–1918) and this gave rise to his system of axioms (1870) to support a formalist viewpoint. In linear algebra Kronecker invented the Kronecker delta ($\sigma_{mn} = 1$ if $m = n$ and $\sigma_{mn} = 0$ otherwise), and established its use when evaluating determinants. It is famous as the first example of a tensor quantity being used.

Kühne, Wilhelm [künuh] (1837–1900) German physiologist: discovered reversible photosensitivity of animal eye pigment.

Although Kühne was a medical man, taught by VIRCHOW and BERNARD, and a professor of physiology (mainly at Heidelberg), his selection and approach to problems was rather that of a chemist. He worked on trypsin from pancreatic juice and coined the name enzyme (Greek, 'in yeast') for the class of 'ferments' which activate chemical change in living cells. Other compounds that interested him were proteins; he studied post-mortem change in muscle and found the protein myosin to be the cause of rigor mortis. In the 1860s he separated various types of egg albumen; and when in 1876 F Boll (1849–79) discovered a photosensitive protein pigment in the retina of a frog's eye, Kühne took up the study of this 'visual purple' (now known as rhodopsin) and showed that it is bleached by light and regenerated in the dark. The retina works, he showed, like a renewable photographic plate; he obtained a pattern of cross-bars of a window on the retina of a rabbit that had been kept in the dark, exposed to a window and killed. It needed new techniques, in the 1930s, for G Wald (1906–) and others to advance knowledge of the mode of action of rhodopsin.

Kuiper, Gerard Peter [kiyper] (1905–73) Dutch–US astronomer: discovered Miranda and Nereid.

Educated in Leiden, Kuiper moved to the USA in 1933. He discovered two new satellites: Miranda, the fifth satellite of Uranus, and Nereid, the second satellite of Neptune, in 1948 and 1949. He also studied the planetary atmospheres, detecting carbon dioxide on Mars and methane on Titan, the largest Saturnian satellite. Kuiper was involved with the early American space flights, including the Ranger and Mariner missions.

In the 1950s Kuiper predicted that a belt of debris, left after the formation of the Solar System, could exist beyond the most distant planet, Pluto. In 1988 it was suggested that this

Kuiper belt could explain the origin of short-period comets such as HALLEY'S, which appear and return at intervals of a few years or decades; and in 1993 the first two such objects, a few score miles across and named 1992 QBI and 1993 FW, were discovered. Their existence, with possibly thousands of similar mini-planetary objects, could also explain the small aberrations that appear to exist in the orbits of Neptune and Uranus.

Laënnec, René Théophile Hyacinthe [laynek] (1781–1826) French physician: invented the stethoscope.

Laënnec studied at the Charité hospital in Paris and qualified as a doctor in 1804. Despite ill-health he did much to advance clinical diagnosis and his writing on disease is modern in approach. It was well known that listening to chest sounds (auscultation) was useful in diagnosis, and tapping the chest (percussion) had been shown by J L Auenbrugger (1722–1809), in 1761, to be informative also.

In 1816 Laënnec met a difficulty in hearing the heart action of a plump and shy young woman, and solved it by connecting his ear to her chest with a paper tube. He was surprised to find that the heart sounds were then louder and clearer. He soon replaced his paper tube with a wooden tube 30 cm long – the first stethoscope. It was not until the end of the century that the binaural stethoscope, with rubber tubes to both ears, replaced the simple tube in general use and became the physician's most readily identifiable instrument. Laënnec was able to link chest sounds with a range of diseases, mainly of the heart and lungs, and described his results in his book *On Mediate Auscultation* (1819). He was an outstanding clinician.

Lagrange, Joseph Louis, comte (Count) [lagrãzh] (1736–1813) French mathematician: revolutionized mechanics.

Born of a French father and Italian mother in Turin, Lagrange saw the family wealth frittered away when he was a teenager. He took to mathematics early and became a professor at the Royal Artillery School in Turin at 19. He moved in 1766 to succeed EULER as Director of the Berlin Academy of Sciences. In 1797 he was in Paris as professor of mathematics at the École Polytechnique. His first wife died young; when he was 56 he married a teenage girl, remaining happily married until his death. He was a highly productive mathematician but his health broke down as a result of overwork and he suffered periodically from intense spells of depression; he virtually gave up mathematics by his late 40s. He was modest and widely liked.

The great book for which he is known, *Analytical Mechanics,* was started when he was 19, but despite early progress was only finished and published when he was 52. It developed mechanics, and used a powerful combination of the calculus of variations and the calculus of four-dimensional space to treat mechanical problems generally. The book does not use geometric methods as NEWTON did (there are no diagrams!).

Lagrange made contributions to the gravitational three-body problem and to number theory; he proved some of FERMAT'S unproven theorems and solved the ancient problem of finding an integer x such that $(nx^2 + 1)$ is a square where n is another integer (not a square).

Lagrange worked with LAVOISIER on weights and measures, and in effect was the father of the metric system. Napoleon thought very highly of him.

Lamarck, Jean (Baptiste Pierre Antoine de Monet) (1744–1829) French naturalist: proposed early ideas on variation and on evolution.

Lamarck's fame is peculiar. Part of his work, on classification and on variation, was widely approved in his own time and later; but some of his ideas on evolution were strongly attacked then and since, and partly so through misunderstanding of his emphasis and meaning.

The 11th and youngest child of poor aristocrats, he joined the army at 16, served in the Seven Years' War, and then (for health reasons) gave up the army and, after working in a bank, began to study medicine. His interest first focused on botany, and his writing on this (especially his introduction of an easy key for classification) impressed the famous naturalist BUFFON; he became botanist to the King in 1781 and, after the Revolution, a professor of zoology in Paris in 1793. After that he worked mainly in zoology, especially on the invertebrates (a term he introduced; he was also a very early user of the word 'biology'). After 1800 he put forward general ideas on plant and animal species, which he began to believe are not 'fixed'. One reason for his view is that domesticated animals vary greatly from their wild originals. He proposed that in Nature it is the environment which produces change; his most quoted example is the giraffe's neck which he thought was a result, over generations, of the animal reaching up for food. A facet of his views was that such a change could be inherited; this attracted ridicule and was largely abandoned after the work of DARWIN and MENDEL. Somewhat unfairly, 'Lamarckism' is linked with the idea of inheritance of a characteristic acquired in life. In this form Lamarck has few supporters (the Soviet botanist T D Lysenko (1898–1976) was one), although recent claims have been made that acquired immunological tolerance in mice can be inherited. More broadly, Lamarck and Buffon can now be seen as having views on common descent, a 'chain of being' for living things, which formed a precursor to the theory of evolution offered by Darwin and WALLACE.

Lamb, Hubert Horace (1913–) British climatologist: pioneered study of palaeoclimates.

After graduating in natural science and geography from Cambridge, Lamb spent most of his professional career at the Meteorological Office before establishing the Climate Research Unit at the University of East Anglia in 1973.

Lamb developed the view, controversial at the time, that there have been long-term variations in climate. Using a variety of historical sources (such as ship's logs and the variations in tree-ring width) he produced a number of definitive studies of past climate, including daily weather classifications for Britain back to 1861, monthly temperature charts back to 1750 and a list of major volcanic eruptions and a corresponding dust veil index back to 1500. These indices have since been extended by other palaeoclimatic indicators, such as oxygen isotope measurements on lake and deep sea sediments. Together, such data have proved invaluable in the study of past climates and have shown beyond doubt that definite trends in climate occur on a variety of time scales.

Land, Edwin Herbert (1909–91) US inventor: invented Polaroid® and a fast photographic process.

To obtain polarized light (ie light in which the electromagnetic vibrations are all in one plane) the early method was the use of a NICOL prism. Later it was realized that passing light through some organic crystals gave polarized light (ie the crystals are dichroic) but the crystals could not be grown to large size. While Land was a Harvard student, he realized that very small crystals would serve the purpose if they were all aligned together and not randomly orientated; and he found a way of doing this, with the aligned crystals (of quinine iodosulphate) embedded in a clear plastic sheet of any required size. The result was given the trade name Polaroid®, and it is widely used in scientific instruments requiring polarized light and in sunglasses, where it is useful because reflected sunlight is partly polarized. Land abandoned his degree course to develop his inventions; the best-known is an ingenious camera in which a multi-layered film is used with developing chemicals included, which are released to process the film in seconds and give an acceptable colour print. He never graduated in the usual sense but Harvard awarded him an honorary doctorate in 1957 to add to his exceptionally large collection of honorary degrees.

Landau, Lev Davidovitch [landow] (1908–68) Russian theoretical physicist: explained remarkable properties of liquid helium.

Landau was the son of a petroleum engineer and a doctor, and attended the universities of Baku and Leningrad. In 1929 he met BOHR in Copenhagen, forming a long-lasting and productive working friendship. In 1932 he moved to Kharkov, becoming professor of physics in 1935. In 1937 KAPITSA, setting up the Institute of Physical Problems in Moscow, asked Landau to be director of theoretical physics. A professorship at Moscow State University followed in 1943. As a notable teacher and personality Landau (known as Dau) created a strong school of theoretical physics in Moscow, contributing to statistical physics, thermodynamics, low-temperature physics, atomic and nuclear physics, astrophysics, quantum mechanics, particle physics and quantum electrodynamics. He published from 1938, with E M Lifshitz (1917–69), a famous series of textbooks. Landau explained the superfluidity (vanishing viscosity below $2.19\,\mathrm{K}$) and superconductivity properties of helium II using the concepts of a phonon (quantized vibrational excitation) and a roton (quantized rotational excitation). He was awarded the Nobel prize in 1962 for this work and other contributions to condensed matter physics. Sadly, he was critically injured in a motor accident in 1962 and never recovered, dying 6 years later.

Landsteiner, Karl (1868–1943) Austrian–US immunologist: discoverer of human blood groups.

Born and educated in Vienna, Landsteiner graduated there in medicine in 1891, and spent the next 5 years in university research in chemistry, partly with E FISCHER in Würzburg. He held posts in pathology in Vienna until 1919, then moved to the Netherlands and finally, in 1922, to the Rockefeller Institute in New York. His work in medical science was wide-ranging, but his results in immunology and especially on blood groups outshine the rest.

Before 1900, blood transfusion had an unpredictable outcome. In that year Landsteiner showed that the blood serum from one patient would often cause the red blood cells of another to 'clump' (agglutinate). He went on to show that all human blood can be grouped in terms of the presence or absence of antigens (A and B) in the red cells and the corresponding antibodies in the serum. Either antigen may be present (blood groups A and B) or both (AB) or neither (O), giving four groups of this kind. Using this idea, simple tests for grouping blood samples and R Lewisohn's discovery in 1914 that sodium citrate prevents clotting – and helped by refrigeration – blood banks and blood transfusion were widely used by the time of the Second World War. Other blood antigens were later found, for example the MNP system (in 1927) and the Rhesus factor (1940), both discovered by Landsteiner and his co-workers and both (like the A, B, AB and O types) inheritable. Other blood group systems have since been found. The complexity of blood types now known leads to millions of blood-type combinations.

Landsteiner's work won him a Nobel Prize in 1930 and has been valuable not only for safe transfusion but also in paternity cases, in forensic work and in anthropology for tracing race migration.

Langevin, Paul [lăzhvĭ] (1872–1946) French physicist: established modern theory of magnetism and invented sonar.

Langevin was a student of PERRIN in Paris and later worked there with PIERRE CURIE; in between he spent nearly a year with J J THOMSON in Cambridge. His interests in physics were wide-ranging and he became the leading French physicist of his time.

Work on ionized gases led him to study the magnetic properties of gases; most are feebly diamagnetic (repelled by a magnetic field) but ozone (O_3) is paramagnetic (weakly attracted into the field). Langevin showed in 1905 that magnetic behaviour could be understood in terms of the electrons present in atoms; electrons had recently (1895) been discovered by Thomson.

In the First World War, he worked on a method for detecting U-boats by echo-sounding, using the reflection of ultrasonic waves (ie sound waves of very high frequency, and not audible). Curie had studied the piezoelectric effect – the small change in the size of some crystals produced by an electric field. Langevin used radio circuitry to produce rapid changes in electric potential in a crystal, so that it vibrated and formed an ultrasonic generator. Reflection of the waves for submarine detection was developed too late for the First World War, but was used in the Second World War as 'sonar'. It is used also to survey the seabed, to detect fish shoals, and in medical scanning.

Shortly before the Second World War, Langevin worked out how to slow down fast neutrons, a method essential for the later work by others on atomic reactors. After France fell to the Germans in 1940, he was outspoken in his anti-Fascist views and was soon under house arrest; his daughter was sent to Auschwitz and his son-in-law was executed. Langevin escaped to Switzerland and survived to return to his Paris job, as director of a research group.

Langley, Samuel Pierpont (1834–1906) US astronomer and aviation pioneer: invented the bolometer and pioneered infrared astronomy.

Langley was mainly self-educated as an engineer and astronomer. His most important contribution to astronomy was the invention of the bolometer, a device consisting of a thin, blackened platinum wire which, when used in conjunction with a spectrometer and a galvanometer, allows very precise measurement of the energy of radiation at different wavelengths. This enabled him to study the solar spectrum at wavelengths of up to 5.3 μm in the far infrared, to measure variations in the solar flux and also to quantify the selective absorption of energy by the Earth's atmosphere.

Being interested in the possibility of manned flight, in 1896 Langley constructed a steam-powered model aircraft, which achieved flights of up to 1200 m in length, and which was the first heavier-than-air machine to fly; however a full-sized version failed to leave the ground.

Langmuir, Irving (1881–1957) US chemical physicist: inventor of ideas and devices, often related to surfaces.

Langmuir was the third of four sons and was only 17 when his father died; but the latter worked in insurance and the family was financially secure. The young man attended the School of Mines at Columbia (New York) and then studied at Göttingen with NERNST. His work there, on the dissociation of gases by a hot platinum wire, began an interest in surface chemistry which he never lost. In 1901 he joined the General Electric Company research centre at Schenectady, NY, and worked there for 41 years. An early success for him was the improvement of tungsten filament lamps, by filling them with inert gas (argon) at low pressure to reduce evaporation and by using a coiled-coil filament. Further work on hot filaments led to the discovery of atomic hydrogen and the invention of a welding torch using its recombination to H_2 to achieve 6000°C.

In 1919–21 he worked on ideas of atomic structure, developing the ideas of LEWIS to form the Lewis–Langmuir octet theory of valence, which was simple and useful in explaining a range of chemical phenomena. The words electrovalence and covalence were first used by him. His interest in hot surfaces moved to thermionic emission, where his work advanced both theory and practice. His study of surface films on liquids allowed some deductions on molecular size and shape; and his work on gas films on solids led to the Langmuir adsorption isotherm, the first important theory of the adsorption of gases on solid surfaces.

His ideas on surface adsorption advanced understanding of heterogeneous catalysis. He was awarded the Nobel Prize for chemistry in 1932 largely for this, becoming the first scientist fully employed in industry to receive a Nobel Prize. He went on to work on electric discharges in gases, and made the first full studies of plasmas (a word he coined). He was a keen sailor and flyer, and his studies of atmospheric physics led to trials in weather control (eg rainmaking by seeding clouds with solid CO_2). He had a wide range of interests, including music, conservation and Scouting, and his distinctions include having Mount Langmuir in Alaska named after him.

Laplace, Pierre Simon, marquis de [lahplas] (1749–1827) French mathematician, astronomer and mathematical physicist: developed celestial mechanics; suggested hypothesis for origin of the solar system.

Although from a poor family, Laplace's talent led him to become an assistant to LAVOISIER in thermochemistry. Later he moved to astronomy, and became a minister and senator, skilfully contriving to hold a state office despite violent political changes. Laplace's most important work was on celestial mechanics. In 1773 he showed that gravitational perturbations of one planet by another would not lead to instabilities in their orbits (NEWTON had believed that such small irregularities would, without divine intervention, eventually lead to the end of the world). He later proved two theorems involving the mean distances and eccentricities of the planetary orbits and showed that the solar system has long-term stability. In 1796 Laplace proposed in a note that the Sun and planets were formed from a rotating disk of gas; he did not know that KANT had made a similar suggestion; modified forms of this nebular hypothesis are still accepted. Between 1799 and 1825 he published his five-volume opus *Mécanique céleste* (Celestial Mechanics), which incorporated developments in celestial mechanics since Newton as well as his own important contributions. (The book has its oddity: frequently the phrase 'it is obvious that' occurs, in mathematical equations, when it is far from obvious. And Napoleon is said to have remarked, critically, that it made no mention of God.) Laplace is also remembered for putting probability theory on a firm foundation, and for developing the concept of a 'potential' and its description by the Laplace equation. In the fields in which Laplace worked and where Newton had worked previously, he is seen as second only to Newton in his talent.

Larmor, Sir Joseph (1857–1942) British physicist: worked out the electrodynamics of electrons.

Larmor was educated at Queen's University Belfast and Cambridge and took posts at Queen's College Galway and Cambridge. In 1903 he became Lucasian Professor of Mathematics there. He was also a member of Parliament for 11 years.

Larmor was active in the final phase of classical physics that laid the ground for the breakthroughs in relativity and quantum mechanics. He incorrectly believed in the ether (ie an absolute space-time frame) and that it was involved in all wave propagation. He contributed to electromagnetic theory, optics, mechanics and the dynamics of the Earth. In electrodynamics he showed (1897) that the plane of an orbiting electron in an atom wobbles (or precesses) when in a magnetic field (Larmor precession). The expression for the power radiated by an accelerated electron (proportional to the square of its charge and the square of its acceleration) is also due to him.

Lartet, Edouard Armand Isidore Hippolyte [lah(r)tay] (1801–71) French palaeontologist: demonstrated that man had lived in Europe during the Ice Age, and discovered Cro-Magnon Man.

The son of a wealthy landowner, Lartet studied law at Toulouse before taking over the management of the family estates. He became interested in fossils and discovered two early primates, *Pliopithecus* (an ancestor of the gibbon) in 1836 and *Dryopithecus* (an early ape) in 1856. In 1863 he discovered the first evidence that man had been living in Europe during the Ice Age when he found, in a cave at La Madeleine in southern France, a piece of ivory with the figure of a woolly mammoth carved on it. This was also the first evidence, provided it was not a forgery, that man had lived at the same time as animals now extinct, and 5 years later, at Cro-Magnon in the Dordogne, he found several skeletons of Cro-Magnon Man, the earliest known fossilized man in Europe (although a relatively recent ancestor by hominid standards).

la Tour, Charles Cagniard de *see* **Cagniard de la Tour**

Laue, Max (Theodor Felix) von [low-uh] (1879–1960) German physicist: suggested a classic experiment to show diffraction of X-rays by atoms in crystals.

Von Laue studied physics at four German universities and was also an art student for 2 years. He then taught physics at three universities before settling in Berlin in 1919. He remained professor of theoretical physics there until 1943, when his long-standing antagonism to the racist policy of the National Socialist party led him to resign. From 1946 he worked to rebuild German science. His early work on optics gave support for EINSTEIN's relativity theory, but he is now best known for his work with X-rays.

Early in this century it had been suggested that X-rays were electromagnetic waves like light but of very short wavelength, although some physicists thought otherwise. It was also believed that the atoms in crystals were in regular array, in accord with their external regularity. Von Laue realized that if both these ideas were true, then the spacing between layers of atoms in a crystal should be of the order of size (10^{-10} m) to bring about diffraction of X-rays. In 1912 he tested this idea; an assistant W Friedrich (1883–1968) and a student, P Knipping (1883–1935), passed a narrow beam of X-rays through a crystal of $CuSO_4 5H_2O$ and obtained a diffraction pattern of spots on a photographic film placed behind it; a crystal of ZnS served even better. The experiment proved the wave-nature of X-rays and also gave the basis on which the BRAGGS later created X-ray crystallography. Einstein called the experiment 'one of the most beautiful in physics' and von Laue was awarded the Nobel Prize in 1914.

Laurent, Auguste [lohrã] (1807/8–53) French organic chemist: classifier of organic compounds.

Laurent was an organic chemist of much talent and energy, who studied under DUMAS. Thereafter his life was fraught with misfortune to an operatic extent; employers swindled him, posts he hoped for were unavailable or were found to lack facilities, a business venture failed and his contributions to theory brought him abuse until almost the end of his life. His last post, as underpaid assayer to the Mint, provided a damp cellar as laboratory and he died of lung disease just before his book *Methods of Chemistry* (1854) was published, leaving a near-destitute family. He was a skilful experimenter with a passion for classification; in particular, he developed Dumas's ideas on organic substitution. He recognized that organic compounds could be classed in 'types', and he used this, and his idea of a nucleus of carbon atoms within an organic compound, to organize much of the organic chemistry of his time. This led to vigorous debate, from which a clearer view of organic compounds emerged by 1860. Laurent also did valuable work in benzene and related chemistry. His work in organic chemical theory is interwoven with GERHARDT'S work.

Lavoisier, Antoine Laurent [lavwazyay] (1743–94) French chemist and social reformer; creator of the Chemical Revolution and victim of the French Revolution.

Lavoisier's father was a prosperous lawyer in Paris and the boy studied law after leaving school. However, he had been interested in science at school and later a family friend, GUETTARD, the geologist, took him on field trips. Lavoisier worked on the first geological map of France; this work, and his competition essay on a method of street-lighting, was so good that he was elected to the Royal Academy of Sciences in 1768, when he was only 25. In the same year he bought a part-share as a 'tax-farmer', to give him an income while he followed his new interest, chemistry. The tax-collecting company had leased from the Government the right to collect some indirect taxes for 6 years. The investment proved reasonable; he worked hard on company business; and at 28 he met and married Marie Anne Paulze, the 14-year-old daughter of a fellow tax-farmer (see next entry). She became the expert assistant in his chemical work. His involvement in the tax-farm was to prove unfortunate.

Lavoisier worked on a scheme for improving the water supply to Paris and on methods of purifying water. He showed in 1770 that water cannot be converted into earth, as was then widely believed. In this, as in all his work, he used the law of conservation of matter: that, in chemical operations, matter is not created or destroyed. He went on to show that air is a mixture of two gases: oxygen, which combines with reactive metals on heating and which supports combustion and respiration; and the unreactive nitrogen. He found that metals combine with oxygen to give oxides which are basic ('alkaline'), whereas the non-metals (S,P,C) give acidic oxides. In this work he used the sort of logic he admired in JOSEPH BLACK'S studies on

lime; and he was helped by information from PRIESTLEY; but he used his own work, and that of others, to form a general theory of combustion, oxidation and the composition of the air, in an original way. His new theory soon displaced 'phlogiston' from most chemists' minds and directed chemistry into new and valuable paths. He showed that water was a compound of hydrogen and oxygen; CAVENDISH'S work on this was skilful, but it was Lavoisier who first explained the results. (Similarly, SCHEELE and Priestley had made oxygen before him, but failed to understand its significance.)

From 1776 he lived and worked happily at the Royal Arsenal, in effective charge of gunpowder production and research. It was there, with LAPLACE as Lavoisier's co-worker, that Black's early work on calorimetry was extended; an ingenious ice calorimeter was made for this, and heats of combustion and respiration were measured; this was the beginning of thermochemistry and also showed that animal respiration is essentially a slow combustion process. (A young assistant to Lavoisier at the Arsenal, E I du Pont (1771–1834), emigrated to America and in 1802 began making gunpowder on the banks of the Brandywine River in Delaware. The venture prospered and founded a major US chemical industry.) In 1787, with three other French chemists, Lavoisier introduced the method of naming chemical compounds which has been used ever since. His main contributions to chemistry were elegantly set out in his *Elementary Treatise on Chemistry* (1789), with fine plates by his wife. In it he gave his definition of a chemical element, as 'the last point which analysis can reach'; this was BOYLE'S view, but Lavoisier used it experimentally and gave a working list of elements. The book had enormous influence on chemistry, comparable with NEWTON'S *Principia* in physics a century earlier.

Lavoisier had remarkable energy: from 1778 he ran an experimental farm near Blois to improve the poor level of French agriculture; he developed schemes for improving public education, equitable taxation, savings banks, old age insurance and other welfare schemes. His liberal and generous views found too few imitators, however, and by 1789 revolution had begun. All might have gone well for Lavoisier for, although the tax-collecting firm was a natural target, its affairs were in good order and charges against the tax-farmers could be refuted. But revolution followed its usual pattern of moving to extremism, and Marat, a leading figure in the Terror, had early in his career pursued scientific ambitions – his worthless pamphlet *Physical Researches on Fire* had been condemned by Lavoisier. A new charge of 'counter-revolutionary activity' was speedily contrived, which ensured a guilty verdict, and France's greatest scientist was guillotined the next day.

Lavoisier, Marie Anne Pierrette, *née* Paulze [lavwazyay] (1758–1836) French illustrator, translator and assistant to Lavoisier.

Marie Paulze's mother was a niece of the Abbé Terray, France's controller general of finance in 1771 and one of the most powerful men of the kingdom. Terray proposed a marriage between the 13-year-old Marie and the 50-year-old penniless brother of a valued acquaintance. To save her from this unwelcome alliance Marie's father, a parliamentary lawyer and financier, quickly arranged a marriage for her with a colleague in the Ferme Générale, the 28-year-old ANTOINE LAVOISIER. It was to be a very successful marriage.

Marie Lavoisier assisted her husband's scientific work; she became his laboratory assistant, kept the laboratory records, made sketches of his experiments and illustrated his classic *Traité de chimie* (1789, Elementary Treatise on Chemistry). Marie Lavoisier was a skilled artist, engraver and painter, having studied under Louis David (1748–1825) (who painted the only known portrait of Lavoisier from life). She learned English, and sought tuition in Latin from her brother, in order to translate the new chemical treatises from England, which included the works of PRIESTLEY and CAVENDISH; Lavoisier was not a good linguist.

Because of his involvement with the Ferme Générale (a tax-gathering consortium) Antoine Lavoisier was arrested and imprisoned during the 'Terror' in November 1793. His estate was confiscated, including his library and laboratory instruments. Marie was imprisoned, but later released, and took refuge with a family servant. Despite her efforts to gain his release and the difficulty the National Convention faced in finding a supportable charge, Lavoisier was executed in May 1794, together with Marie's father.

Marie Lavoisier clearly understood the position her husband should hold in science and was determined that his reputation should not be overlooked and his claims should be known and recognized. She petitioned for the return of the estate and, having obtained his books and papers, she edited and privately published Lavoisier's unfinished memoirs. She presented copies to the great scientific societies and eminent scientists around Europe.

Marie Lavoisier blamed friends and scientific associates of Lavoisier, especially members of the Convention, for not protesting against her husband's imprisonment and for not pointing out his past valuable work for France and his future scientific worth. Bitterly, she held them responsible for his death.

She again opened her home as a meeting place to the leaders of science in France, to DELAMBRE, CUVIER, LAGRANGE, LAPLACE, BERTHOLLET, ARAGO, BIOT, HUMBOLDT and others; she did not receive those who had failed to use their political influence to try to save her husband.

In 1805 Marie married Benjamin Thomson, Count RUMFORD, but the marriage was not a success and they separated 4 years later.

Lawes, Sir John Bennet (1814–1900) British agriculturalist: founder of Rothamsted Experimental Station.

Lawes had an amateur interest in chemistry, but after inheriting a farm estate at Rothamsted in 1834 he became an enthusiast for agricultural chemistry. He found that ground bones ('mineral phosphates') were effective in some fields but not in others, and soon discovered that acid treatment of bones made a universally effective fertilizer (it converts the insoluble tricalcium phosphate into soluble monocalcium phosphate, a conversion that acidic soils perform naturally). Despite bitter opposition from his mother (who was against 'trade') he began to make and sell 'superphosphate' prepared from bone or mineral phosphate and sulphuric acid, and used the profits to finance further experiments at Rothamsted. Aided by a chemist, J H Gilbert (1817–1901), much valuable work was done there, including the demonstration by 1851 that, as well as minerals, plant growth generally requires nitrogenous manure (in conflict with LIEBIG's views). In 1889 Lawes put Rothamsted under control of a trust and its scientific studies of agriculture continued. Eight field experiments are still running there after 150 years.

Lawrence, Ernest Orlando (1901–58) US physicist: invented the cyclotron and produced new radioactive elements.

Lawrence's father was head of a teacher's college and his mother had taught mathematics. The boy grew up in South Dakota; he was tall, energetic, fond of tennis and physics, and impatient of 'culture' and of inactivity throughout his life. He studied at South Dakota, Minnesota and Yale and in 1928 moved to a post at the University of California at Berkeley, becoming director of the Radiation Laboratory in 1936.

From 1929 Lawrence worked to produce sufficiently energetic particles for nuclear reactions, having noted EDDINGTON's suggestion that stars may be 'powered' by nuclear reactions. Linear accelerators for making high-energy particles were awkwardly long and used high voltages. Lawrence decided to accelerate particles on a spiral path within a pair of semicylinders ('dees') mounted in a vacuum between the poles of an electromagnet. An AC voltage at high frequency applied to the dees gave the particles their impetus. The first small cyclotron (using a 10 cm magnet) operated in 1931. Later and larger cyclotrons achieved proton beams of 8×10^4 eV, and converted lithium nuclei to helium nuclei to confirm COCKCROFT and WALTON'S first nuclear transformation (1932). Hundreds of new radioactive isotopes were eventually produced, including most of the transuranium elements; Lawrence investi-

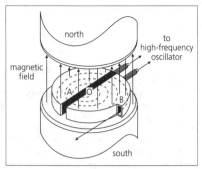

Lawrence's cyclotron. The oscillator reverses the PD between the dees several million times per second. Positive ions (eg protons, H⁺) are released at the centre and are accelerated into nearly circular paths until they emerge with a high energy. The spiral path is actually many kilometres long; ie the spiral is 'tightly wound'.

gated their use in medicine. Mesons and antiparticles were generated and studied, with Lawrence coordinating the efforts of a team. Lawrencium (Lw, atomic number 103) was named for him, and he received the 1939 Nobel Prize for physics. In 1940 his team isolated plutonium and neptunium, and he contributed to the development of the atomic bomb.

Leakey, Louis (Seymour Bazett) (1895–1972) British–Kenyan archaeologist and palaeoanthropologist: discovered several hominids.

The son of a British missionary working in British East Africa (now Kenya), Leakey became interested in Stone Age man while he was young. He studied anthropology at Cambridge, reading French and Kikuyu (the language of the Kenyans among whom he had been brought up) and taking part in a British Museum expedition to Tanganyika in his second year. Between 1926 and 1935 he organized a series of archaeological and palaeontological expeditions to East Africa, the later ones visiting Olduvai Gorge, Tanganyika (now Tanzania), where a German lepidopterist had found fossils in 1911. During expeditions there he found hominid skulls and stone tools, which he believed to represent an early ancestor of man. Further discoveries followed, including in 1960 the remains of *Homo habilis*, a tool-making hominid with a relatively large brain. Leakey's findings established East Africa as the possible birthplace of man and traced his ancestry further back than had been possible previously. His son Richard (1944–) is also a noted East African palaeoanthropologist. (See panel overleaf.)

Leakey, Mary (Douglas), *née* Nicol (1913–) British palaeoanthropologist: discovered several hominids.

Mary Nicol had a somewhat unconventional upbringing, travelling a good deal and lacking a regular formal education. Her father was a

landscape painter. Interested in archaeology and early man, she attended lectures at University College, London, and between 1930 and 1933 worked on several archaeological digs in England. Her ability as an illustrator brought her into contact with LOUIS LEAKEY, whom she joined in 1935 on an archaeological expedition to Olduvai Gorge, East Africa, and married the following year. She spent much of her life searching for hominids in East Africa, and made many of the discoveries for which she and her husband became well-known. In 1959 she found the skull of *Zinjanthropus boisei*, a species of *Australopithecus* and a possible ancestor of man. It is notable as the first hominid to be reliably dated, by the K/Ar method, at 1.75 million years. In 1976 Mary Leakey led an expedition to Laetoli, Tanzania, on which the earliest evidence of man's ancestors yet found was discovered: two sets of hominid footprints in a layer of volcanic ash provided indisputable evidence that man's predecessors walked upright 3.75 million years ago. (See panel overleaf.)

Leavitt, Henrietta (Swan) (1868–1921) US astronomer: discoverer of the period-luminosity relation of Cepheid variable stars.

Henrietta Leavitt did well as a student at Radcliffe (despite her deafness) and joined the Harvard Observatory in 1895, originally as a volunteer assistant. While studying photographic plates made at Harvard's field station in Peru, she deduced in 1912 that Cepheid variable stars have a simple relationship between the period of a given star and its luminosity; based on this, HERTZSPRUNG suggested that their distances could be calculated. The method soon proved invaluable for measuring stellar distances. From 1915 SHAPLEY used this method to obtain the first rough estimates of the size and shape of our Galaxy. It was not until 1924 that EDDINGTON found an acceptable theoretical reason for the relationship that Leavitt had first observed.

Lebedev, Pyotr Nicolayevich (1866–1912) Russian physicist.

Lebedev studied physics at Strasbourg and Berlin and became professor of physics at Moscow in 1902. MAXWELL's theory of electromagnetic radiation implied that radiation should exert a very small pressure on matter. CROOKES's radiometer appeared to confirm this, but it was soon shown that the apparent effect of light pressure in it was due to traces of gas. However, with better vacua Lebedev was able to measure the effect of light pressure, and to show that it is twice as great on reflecting as on absorbing surfaces.

For the small particles of cosmic dust, light pressure can exceed gravitational attraction, and Lebedev concluded that this is why a comet's tail points away from the Sun. However, the effect of the solar wind on a comet's tail is now known to be much greater in this case than the light pressure.

THE QUEST FOR HUMAN ORIGINS

Homo habilis   Homo erectus   Homo sapiens

The search for the earliest human ancestors has been an area of science very much dominated by individuals, often with strongly held views about the interpretation of their finds. The story of the discoverers tells much about the history of the subject.

Not surprisingly, many of the earlier finds were made in heavily populated Europe, with several discoveries in the 19th-c of evidence of early habitation in caves. In 1863 LARTET discovered the first evidence that man had lived in Europe during the last ice age and soon afterwards found several skeletons of Cro-Magnon Man, the earliest human predecessor found in Europe. although a relatively recent ancestor in evolutionary terms.

The publication of DARWIN's ideas on evolution in 1858 led to a conscious search for a 'missing link' between humans and the apes, and in 1891 DUBOIS discovered fragments of *Homo erectus* (Java Man), a hominid (see end of LINNAEUS entry) that walked upright about 0.5–1.5 million years ago. Like many after him, Dubois's interpretation of his finds was widely ridiculed at the time, but accepted later after Otto Zdansky found other examples of *Homo erectus* near Peking in 1926. So strongly was opinion divided about Darwin's theory and the nature of human ancestry that some were tempted to fabricate evidence, such as the 'discovery' in 1912 of Piltdown Man, widely accepted as one of our early ancestors until conclusively shown to be a fake in 1953.

It is commonly accepted today that the human family, the Hominidae, originated in Africa from ape-like ancestors, whence they migrated across the globe. This Out of Africa theory is reflected by the subsequent pattern of hominid discoveries. In 1924 DART had announced the discovery of a hominid twice as old as Java Man; *Australopithecus africanus*, from Taung, Botswana, was 1.2–2.5 million years old. The 'Taung baby' was the first truly primitive ancestor within the human family to be found; soon afterwards BROOM found a second example at Sterkfontein, followed in 1938 by the first specimen of *Australopithecus robustus* (1–2 million years old).

The oldest hominids known to date come from east Africa. From the late 1950s onwards LOUIS and MARY LEAKEY made a series of exciting discoveries of progressively earlier hominids. In 1959 Mary Leakey found *Zinanthropus boisei*, a species of *Australopithecus* dated to 1.75 million years ago and for many years the earliest hominid known. The following year Louis Leakey discovered *Homo habilis*, a large-brained hominid that made and used tools.

A major step back in time was achieved in 1974 when JOHANSON discovered 'Lucy', a female *Australopithecus afarensis* about 3.1 million years old, in the Afar valley of north-eastern Ethiopia. Tantalizingly, 2 years later Mary Leakey discovered the footprints, but no remains, of a hominid walking upright at Laetoli on volcanic ash dated as 3.75 million years old.

The oldest hominid currently known is 4.4 million years old and again hails from the Afar valley. *Australopithecus ramidus* was found in 1994 by a Japanese/American/Ethiopian team; far from being a complete skeleton, the remains consist of 50 fragments of bone and teeth from a group of 17–20 individuals, including jaw and skull fragments and a complete left arm. Pre-dating the use of stone tools by almost 2 million years, *A. ramidus* lived in a woodland area also inhabited by monkeys, antelopes, sabre-toothed cats, rhinos and elephants, and was probably predominantly vegetarian. Although these hominids walked upright like humans, it is thought that they would have slept in trees, and resembled apes in other ways.

The classification of these finds, and the interpretation of their interrelationships and migrations, can be assisted by DNA analysis. In 1995, DNA studies at Yale on part of the Y chromosome of 38 living men from around the world showed the samples to be surprisingly similar. The simplest explanation for this lack of 'genetic scrambling' is that modern humans are genetically young, ie only a few hundred thousand years old, and that they developed from a single homogeneous colony and not from separated and scattered centres.

Current thinking, on the basis of both fossil finds and genetic evidence from modern human and ape populations, suggests that the split between humans and apes occurred some time between 5 and 8 million years ago and that hominids who could merge in a human crowd today without much difficulty first appeared not more than 100 000 years ago.

DM

Lebesque, Henri Leon [luhbeg] (1875–1941) French mathematician: introduced the modern definition of the integral.

Lebesque was a product of the École Normale Supérieure and from 1921 taught at the Collège de France. His main contributions were to set theory, the calculus of variations and function theory. He and E Borel (1871–1956) built the modern theory of functions of a real variable, and Lebesque in particular produced a new general definition of the integral (1902), developed beyond the Riemannian definition. This led to important advances in calculus, curve rectification and trigonometric series, and initiated measure theory.

Le Chatelier, Henri Louis [luh shatlyay] (1850–1936) French physical chemist and metallurgist: devised a much-used but doubtful principle.

As a young man, Le Chatelier was much influenced by his father Louis, an engineer who was Inspector General of Mines for France. Tuition from his father and family friends such as DEVILLE aided him and shaped his interests, and he became a professor at the École des Mines in 1877. His early research was on cement (his grandfather operated lime kilns); he worked also on the structure of alloys, on flames and on thermometry. In the 1880s he developed the idea known as Le Chatelier's principle: this states that if the conditions (temperature, pressure, or volume) of a chemical system initially at equilibrium are changed, then the equilibrium will shift in the direction that will tend to annul the change, if possible. The principle has been much criticized, and it is best replaced by two laws due to VAN 'T HOFF; they are (1) increase in pressure favours the system having the smaller volume, and (2) rise in temperature favours the system formed with absorption of heat. Thus for the equilibrium $N_2 + 3H_2 \Leftrightarrow 2NH_3$ in which the volume diminishes when the reaction proceeds to the right, an increase of pressure will shift the equilibrium in favour of ammonia formation. Also, as ammonia formation is exothermic, rise in temperature favours the reactants.

Leclanché, Georges [luhklāshay] (1839–82) French engineer: devised carbon–zinc electrical cell.

Educated in Paris, Leclanché was employed as a railway engineer from 1860. By 1866 he had devised his carbon–zinc electrical cell, which was soon adopted by the Belgian telegraphic service. Modified to the non-spillable form of the familiar dry cell, it has been greatly used; this has a carbon rod as the positive pole, surrounded by a wet paste of carbon black, manganese dioxide and ammonium chloride, with a thickener such as sawdust, inside a zinc container which is the negative pole.

Lederberg, Joshua [layderberg] (1925–) US geneticist: pioneer of bacterial genetics.

A New Yorker almost from birth, Lederberg graduated in biological science at Columbia University and then enrolled there in 1944 as a medical student during his service in the US Naval Reserve. At that time bacteria were not thought to have genes, or sex. During his course on medical bacteriology, Lederberg began experiments to test this and in 1946 went to Yale to work on it with the experienced microbiologist E L Tatum (1909–75). They were skilful and lucky in the choice of the intestinal bacterium *Escherichia coli* strain K-12 for their work, and within weeks showed that mutants of this strain crossed; in a large colony, a few reproduced by sexual mating ('conjugation'). Lederberg went on to show that this is not uncommon and can be used to map bacterial genes; bacterial genetics had begun and its methods became valuable to geneticists, as had earlier use of the fruit fly *Drosophila* and the fungus *Neurospora*.

His next major discovery, made with ZINDER in 1952, was that bacteriophage (a bacteria-infecting virus) could transfer genetic material between strains of bacteria ('transduction') to produce recombinant types. For the first time genes had been deliberately inserted into cells, a basis for 'genetic engineering'. In 1957, with G Nossal (1931–), Lederberg showed that immune cells produce single types of antibody, a result which was basic to the development of monoclonal antibodies by others. With his first wife, Lederberg obtained the first firm evidence that adaptive mutations in bacteria can occur spontaneously; this had been an unproved assumption in the theory of evolution. After Yale, Lederberg taught genetics at Wisconsin and at Stanford and became president of Rockefeller University in 1978. At age 33, he had shared a Nobel Prize with BEADLE and Tatum in 1958. Aside from his work on bacterial genetics, he researched on artificial intelligence and the specific problem of computerizing some of the work of organic chemists by devising a linear notation for organic molecular structures. In collaboration with E A Feigenbaum, these studies pioneered the development of 'expert systems'.

Lee, Tsung Dao (1926–) Chinese–US theoretical physicist: demonstrated that parity is not conserved in the weak nuclear interaction.

Lee studied for his degree in China. His work was interrupted by the Japanese invasion during the Second World War; he fled to another province. In 1946 he won a grant to Chicago, studying astrophysics under FERMI; work at Princeton (1951–3) and Columbia followed. He held a post at Columbia from 1956.

While the electromagnetic and strong nuclear interactions conserve parity (ie are identical in a mirror-image of the physical system) Lee and YANG showed in 1956 that this is not so for the weak nuclear interaction. They deduced this extraordinary result, with far-reaching implications, by considering nuclear beta-decay (electron emission). They suggested a number of experiments, and in the ensuing months their conclusion was verified by Wu. Lee and Yang also argued (1960) that the very light neutral particle called the neutrino (see FERMI) produced in electron emission was different from the neutrino associated with muon emission. This was verified by experiment in 1961. In the same paper they predicted the existence of the W-boson as the heavy particle conveying the weak nuclear force, and this has since been shown experimentally. They also indicated the existence of neutral weak currents, first observed in 1973. Lee and Yang became the first Chinese to win a Nobel Prize, in 1957.

Leeuwenhoek, Antony van [layvenhook] (1632–1723) Dutch microscopist: observed blood corpuscles, protozoa, bacteria and spermatozoa.

Leeuwenhoek had no formal training in science and rather limited schooling. Apprenticed to a draper, he later had his own shop in Delft and a paid post in local government. He became an enthusiastic user of microscopes. The compound microscope was in use before 1650 but was optically poor, and Leeuwenhoek preferred to use a small single lens, doubly convex and of very short focus (1–3 mm). He ground these himself and mounted them between metal plates; in all he made some hundreds of these magnifiers. He was a passionate microscopist, ingenious, secretive and with the advantage of having very unusual eyesight, so that he could use magnifications of 50× to 200× with his ultra-small lenses (possibly he used a second lens, as an eyepiece). He also used 'a secret method', which may have been dark-ground illumination, or the enclosure of his specimens in a drop of liquid in some cases. His results were mostly sent to the Royal Society in illustrated letters (375 of them); unsystematic, enthusiastic and written in Nether-Dutch, and his fame attracted visits by other microscopists and even royalty.

He was the discoverer or an early observer of blood capillaries, red blood cells, protozoa, bacteria (in 1683), rotifers, *Hydra, Volvox* and spermatozoa (of dog). He was opposed to the idea of spontaneous generation, which was not disproved until PASTEUR's work a century and a half later. He ground 419 lenses and lived, actively researching, to age 90.

Lehmann, Inge [layman] (1888–1993) Danish seismologist: discovered solid inner core within Earth's outer liquid core.

Inge Lehmann studied mathematics in Copenhagen and in Cambridge, and afterwards worked in insurance until she was 30, before returning to study science in Copenhagen. From 1928 she headed seismological work at the Danish Geodetic Institute, dealing with seismic data recorded from all parts of Europe. At that time the Earth was believed to consist simply of a liquid core, shown to exist by GUTENBERG, surrounded by the mantle and then the crust. But in 1936 Lehmann observed that compressional P waves travelling through the Earth's core from an earthquake undergo a marked increase in velocity at a depth of about 5150 km. She argued from this that there is a solid inner core within the liquid core, extending about 1200 km from the Earth's centre. It is believed that this inner core consists of solid iron and nickel, the very high pressure making the metal solid despite its high temperature. (See diagram at GUTENBERG.)

Leibniz, Gottfried Wilhelm [liybnits] (1646–1716) German mathematician: one of the greatest polymaths in history.

The son of a Lutheran professor of moral philosophy, Leibniz developed an interest in a wide range of subjects from his father's library. He attended the universities of Leipzig, Jena and Altdorf, where he received his doctorate in law in 1666. Leibniz was to show talent in law, religion, statecraft, history, literature and philosophy as well as mathematics.

He took up a career as a somewhat shady lawyer and diplomat, working initially for the elector of Mainz. During two trips to London in 1673 and 1676 HUYGENS and BOYLE interested him in current work in mathematics, and in his spare moments Leibniz proceeded to make the immense discoveries of both the calculus (independently from NEWTON) and combinatorial analysis. Leibniz was at the same time much involved with establishing the legal rights of the legitimate and many illegitimate members of the household of the three electors whom he served in succession. Frequently on the move and prolifically noting his thoughts on many subjects, he was involved in diplomacy and in making plans for a French invasion of Egypt. His talents were dissipated in the sordid tasks of his master's power-broking. He also became involved in an unsuccessful attempt to unite the Catholic and Protestant churches in 1683 and in the founding of the Berlin Academy of Sciences (1700). When his last employer, the elector of Hanover, had been steered into becoming George I of England, Leibniz was discarded and left behind to write the Brunswick family history. He died neglected, dogged by illness and in the midst of controversy over his invention of the calculus.

In mathematics Leibniz had tremendous flair. He invented a calculating machine (1672) far beyond PASCAL's, which could only add and subtract; Leibnitz's could also multiply, divide and find square roots. When a young man he conceived of a universal language for logic and began the study of symbolic logic. Later came his construction of the differential and integral calculus, and a fierce priority dispute on this with NEWTON; Leibniz did his work following Newton (after 1665) but independently. The notation now used in calculus is that due to Leibniz. A minor part of his work was on infinite series, where he discovered in 1674 an amusing relation between π and all the odd numbers: $\pi/4 = 1 - 1/3 + 1/5 - 1/7 + 1/9...$ which had earlier been found by GREGORY.

Lemaître, Georges Edouard (Abbé)

[luhmaytruh] (1894–1966) Belgian astronomer and cosmologist: originator of the 'Big Bang' theory for the origin of the universe.

Lemaître studied at the University of Louvain, and afterwards trained and was ordained as a Catholic priest. He then spent some time at the Cambridge and Harvard observatories before becoming professor of astronomy at Louvain in 1927, where he remained for the rest of his career.

Lemaître was an originator of the 'Big Bang' theory for the origin of the universe. In 1927 he found a solution to EINSTEIN's equations of relativity that resulted in an expanding universe (Einstein's own solution was a static one), and 2 years later HUBBLE showed observationally that this was indeed the case. Independently, the Russian FRIEDMANN came to similar conclusions. However, Lemaître further suggested that, by backward extrapolation, the universe must at one time have been small and highly compressed, which he referred to as the 'primal atom'. He conjectured that radioactive decay had resulted in an explosion, the 'Big Bang' (so named, disparagingly, by HOYLE in the 1950s). Although the importance of Lemaître's work was not fully appreciated at the time, the 'Big Bang' theory is now accepted as the best model for the origin of the universe.

Lenard, Phillipp Eduard Anton [laynah(r)t]

(1862–1947) German physicist: investigated the photoelectric effect and cathode rays.

Lenard, the son of a wine-merchant, was educated at Budapest and in Germany, where he became professor at Heidelberg in 1907.

Before 1914 Lenard made a series of fundamental contributions to physics. He took the known fact that ultraviolet light falling on some metals causes electron emission (the photoelectric effect) and showed that this occurred only with light below a critical wavelength; that the electron velocity increases with falling wavelength and is independent of light intensity; and finally that increasing the light intensity produces a larger number of emitted electrons (1902). EINSTEIN explained all these observations in 1905 and, with PLANCK, introduced light quanta (photons) into physics, preparing the way for the development of quantum theory.

Lenard showed that cathode rays are an electron beam and received the 1905 Nobel Prize for physics for this work. The cathode rays would penetrate air and thin metal sheets and he deduced that atoms contained much empty space and both positive and negative charge (1903). RUTHERFORD's work confirmed and extended this picture of the atom (1911).

Lenard had disputes over priority with RÖNTGEN (Lenard having narrowly failed to discover X-rays) and with J J THOMSON, but his case does not appear strong. Lenard's book *Great Men of Science* (1934) is marred by his omission of contemporaries with whom he had quarrelled. He was distressed that Germany lost the First World War, and afterwards by the death of his son and the loss of his savings by massive inflation. He developed an extreme dislike of the increasing mathematical sophistication of physics through the influence of Einstein and others. From 1919 Lenard argued for the establishment of 'German physics' untainted by Jewish theories, attacking Einstein as a socialist, pacifist and a Jew, but above all for being a theoretician. As the only leading scientist who was a Nazi supporter, Lenard acquired increasing power. In the 1930s a generation of disillusioned scientists left Germany, most of Germany's capacity to achieve creative physical science departing with them.

Lenz, Heinrich Friedrich Emil [lents] (1804–65)

Russian physicist: discovered Lenz's Law.

Lenz studied chemistry and physics at the University of Dorpat (later Tartu), served as geophysicist on a voyage around the world when he was 19 and, on his return, was appointed to the staff of St Petersburg Academy of Science, eventually becoming dean of mathematics and physics.

On his voyage around the world Lenz made some important investigations of barometric pressure and of sea temperature and salinity, establishing (and explaining) the difference in salt content between the Atlantic and Pacific Oceans and the Indian Ocean. However, he is best remembered for his work on electromagnetism; Lenz's Law states that the current induced by an electromagnetic force always flows in the direction to oppose the force producing it. This is a special case of the more general law of conservation of energy. He also showed that the resistance of eight metals increases with temperature and discovered (independently from JOULE) the proportionality between the production of heat and the square of the current flowing in a wire.

Lepaute, Nicole-Reine, *née* Etable de Labrière

[luhpoht] (1723–88) French astronomical computer.

Married to Jean-André Lepaute (1720–1789), the royal clockmaker, Nicole-Reine Lepaute investigated oscillations of pendulums of different lengths; the result of this work was included in her husband's *Traité d'horlogerie* (1755). She was employed by J J Lalande (1732–1807), director of the Paris Observatory, to assist

A-C Clairaut (1713–65) to determine the extent of the gravitational attraction of Jupiter and of Saturn on HALLEY'S comet and the exact time of its return in 1759. Lalande gave Lepaute full credit for her work. She calculated the path of the 1764 eclipse of the Sun for all of Europe and the resultant chart was published by the French government. During 1759–74 she helped Lalande with the annual *Connaissance des temps* (an almanac for the use of astronomers and navigators published by the Académie des Sciences), and from 1774–83 she worked on the seventh and eighth volumes of the *Ephemeris*, making the computations for the positions of the Sun, Moon and planets covering the decade to 1784 and the period up to 1792. A crater on the Moon is named for her.

Levene, Phoebus (Aaron) [luhveen] (1869–1940) Russian–US biochemist, who showed that 'nucleic acid' is of two kinds (RNA and DNA) and defined the difference between them.

Levene's father was a prosperous Jewish shirtmaker in St Petersburg, and the boy was able to become a student in the Imperial Medical Academy there. PAVLOV taught physiology, and BORODIN chemistry, in the Academy; the latter apparently influenced Levene the most, because he afterwards inclined more towards chemistry than medicine. In 1891 the family emigrated to the USA and after completing his MD degree in St Petersburg Levene began his medical practice in the Russian-Jewish colony on New York's East Side. He and a brother-in-law shared a small office; the brother-in-law, a socialist lawyer, was very similar in appearance to Levene and, as neither of them could afford to lose a client, it was their custom to 'bluff it out' for one another when alone in the office.

Levene continued to spend much time studying chemistry and in about 1900 decided to abandon medical practice for medicinal chemistry. He worked for a year with EMIL FISCHER in Berlin; by 1905 he had a reputation in biochemical research and was appointed to the new Rockefeller Institute for Medical Research in New York, where he spent the rest of his life. He was small, energetic, artistic and multilingual, prone to toss his heavy shock of hair about and devoted to his own experimentation despite having a team of co-workers. His work ranged over a large area of tissue constituents and was notably productive in sugar chemistry.

His best-known work concerns nucleic acids, which had first been isolated in HOPPE-SEYLER'S laboratory, in 1869. It is now known that nucleic acids are long, chain-like molecules, constructed from repeating units:

```
    base    O    base    O
     |      |     |      |
~ sugar-O-P-O-sugar-O-P-O ~
            |            |
            O            O
```

in which the bases are of four different kinds.

Levene's important contribution was to show that the sugar component came in two kinds, both of them unknown until he isolated them by breaking down nucleic acids. The first sugar, isolated in 1909, was soon shown to be ribose. The second sugar, 2-deoxyribose, was not discovered for another 20 years. This was because 2-deoxyribose is destroyed by acid used to break up the chain, and a non-destructive enzyme for the purpose could not be found. Success was eventually achieved by passing a solution of this nucleic acid through a gastrointestinal segment of a dog, by introducing it through a gastric fistula and withdrawing it through an intestinal fistula, a difficult procedure. With Levene's identification of 2-deoxyribose, it was clear that the nucleic acid found in cells is of two kinds, ribonucleic acid and deoxyribonucleic acid, and their abbreviations (RNA and DNA) became familiar.

The more detailed structure of the nucleic acids (notably the precise mode of linkage of base and phosphate groups to the sugar) was elucidated especially by the work of A R Todd (1907–   ); and then in 1953 CRICK and WATSON virtually created molecular biology by showing that the sequence of the four bases (in groups of three) along the DNA chains form a code of genetic information which directs the synthesis of RNA, which in turn directs the synthesis of proteins. They also showed that DNA exists in cells in the form of a double helix of two entwined strands of DNA, whose uncoiling provides templates for their own replication. With their work genetics and heredity had found its basis at the molecular level, with Levene's work (from a quarter of a century earlier) forming a key part of these later developments.

Leverrier, Urbain Jean Joseph [luhveryay] (1811–77) French astronomer: predicted position of Neptune and discovered advance of perihelion of Mercury.

Leverrier was a student and then a teacher at the École Polytechnique, initially in chemistry and later in astronomy. Realizing that the irregularity of the orbit of Uranus was due to the influence of an undiscovered planet further out, Leverrier succeeded in computing the mass and orbit of the perturbing body. He sent his prediction of the missing planet's position to Johann Galle (1812–1910) in Berlin, who discovered Neptune on his first night of looking, 23 September 1846. Although Leverrier initially received the credit for the discovery, it soon became clear that ADAMS had made the same prediction a year earlier; this led to a celebrated dispute, not made easier by Leverrier's arrogance and violent temper.

Leverrier was the first to appreciate the advance of the perihelion (the point of its orbit nearest the Sun) of Mercury, and predicted the existence of a planet between Mercury and the Sun to explain it, even going so far as to name it

Vulcan. No such planet has ever been found, and the advance of the perihelion was subsequently explained using EINSTEIN's theory of general relativity.

Levi-Montalcini, Rita [layee montalcheenee] (1909–) Italian neurophysiologist: discovered nerve growth factor.

Montalcini's training was difficult; her Italian-Jewish family long opposed her entry to medical school, and when she graduated the Second World War began and as a non-Aryan she had to go into hiding. Her early research, with her medical school instructor G Levi on the neuroembryology of the chick, was done in her bedroom; eggs were easy to secure (by pretending she had young children) and could be eaten after experimentation. In 1947 she went to Washington University in St Louis and in 1949, with V Hamburger, showed that the embryonic nervous system produces many more nerve cells than are needed; the number of survivors depends on the volume of tissue they need to serve. From this clue, she went on to discover the nerve growth factor (NGF), which appears to be critically involved in the growth of nerves of all kinds, including those of the central nervous system.

With the biochemist S Cohen (1917–) she showed that male mouse saliva is a good source of NGF; and he went on to discover the related epidermal growth factor (EGF). In 1979 Levi-Montalcini retired from directing the Laboratory for Cell Biology in Rome, and in 1986 she shared the Nobel Prize with Cohen; she was then 77. Her work may well prove fundamental to the understanding and treatment of senile dementia.

Lewis, Gilbert Newton (1875–1946) US physical chemist: major contributor to theory of chemical bonding.

A Harvard graduate, Lewis studied in Germany for 2 years and then went to the Philippines as a government chemist. From 1905–12 he was at the Massachusetts Institute of Technology and then spent the rest of his career at the University of California at Berkeley.

Lewis developed GIBB's ideas on chemical thermodynamics and made the experimental measurements that allowed the outcome of a range of chemical reactions to be predicted by calculation. He was also a pioneer in taking ideas concerning electrons from physics and applying them in chemistry. From 1902 he shaped his ideas on this subject and published them in 1916; they were then publicized and expanded by LANGMUIR and later by SIDGWICK. In developed form (as in his work of 1923) Lewis's ideas focused on the arrangement of electrons around atomic nuclei. He assumed that elements heavier than the lightest two (H and He) had a pair of electrons surrounding the nucleus, with further electrons (in number to balance the nuclear charge) in groups, with a group of eight as especially stable. Bonding between atoms of the lighter elements occurred in such a way that atoms gained or lost outer electrons to create octets, either by transfer (electrovalence) or by sharing (covalence). Noting that nearly all chemical compounds contain an even number of electrons, he concluded that the electron pair is especially important, and a shared pair can be equated with a covalent bond. The familiar 'dot diagrams' showing the electronic structure of many simple compounds are devised on this simple theory.

Lewis also saw the importance of electron pairs in another context. He defined a base as a substance that has a pair that can be used to complete the stable shell of another atom; and an acid as a substance able to accept a pair from another atom to form a stable group of electrons. This very general concept of Lewis acids and bases has proved valuable.

Probably no man has done more to advance chemical theory in this century, but he was always a diffident as well as an attractive and engaging person, with an unorthodox mind.

Libby, Willard Frank (1908–80) US chemist: developed radiocarbon dating technique.

Libby taught at the University of California at Berkeley until 1941, when he joined the Manhattan Project developing the atom bomb. After the war he moved to the Institute of Nuclear Studies at the University of Chicago, returning to California in 1959.

In 1939 Serge Korff (1906–) discovered carbon-14, a radioactive isotope of carbon with a half-life of 5730 years, and showed that it is produced in the upper atmosphere by the action of cosmic rays on nitrogen atoms. In 1947 Libby and his colleagues used this discovery to develop their radiocarbon dating technique, which has proved to be invaluable in archaeology and Quaternary geology. The technique is based on the fact that living biological material contains carbon-14 and carbon-12 in equilibrium with the atmosphere (which contains a very small but approximately constant proportion of carbon-14 to carbon-12). However, when the organism dies it stops taking up carbon dioxide from the atmosphere and so the proportion of carbon-14 to carbon-12 starts to diminish as the carbon-14 undergoes radioactive decay. By measuring the proportion of carbon-14 to carbon-12, therefore, the time since death may be determined. The technique is applicable with reasonable accuracy in dating organic objects up to about 40 000 years old, but greater accuracy can be achieved by calibrating the technique with objects of known age, and this has been done back to about 5000 years ago. This calibration is desirable because the rate of production of carbon-14 in the atmosphere varies slightly with time. Libby was awarded the 1960 Nobel Prize for chemistry for his work.

A HISTORY OF AGRICULTURE IN THE DEVELOPED WORLD

Until the middle of the 18th-c the authorities on agriculture were Roman writers such as Lucius Columella (1st-c) and Rutilius Taurus Aemilianus Palladius (4th-c); the ability of agriculture to support a non-agricultural community remained strictly limited. Innovation began in the Low Countries at the very end of the 17th-c, but new ideas were taken up and developed in Britain where, because of the lower population density, there was greater opportunity for them to produce dramatic increases in productivity.

Agricultural reforms were of three types. The first were innovations in husbandry, such as the selective stock-breeding methods developed by Robert Bakewell (1725–95) in Leicestershire in the middle of the 18th-c, and the four-year rotation of crops devised slightly earlier by Charles, Viscount Townshend – 'Turnip' Townshend (1674–1738). Samuel Marsden (1764–1838) pioneered the breeding of Australian sheep for wool, and this work was carried on in the middle of the 19th-c by John MacArthur (1767–1834) and especially by his wife Elizabeth (1766–1850), who introduced the merino sheep to New South Wales. William James Farrer (1895–1906) emigrated to Australia in 1870 and, by his scientific breeding of specialist strains of wheat, was almost single-handedly responsible for the success of the Australian wheat industry.

Except in undeveloped areas, the advances that could be made by improvements in methods were limited. In the middle of the 18th-c, however, advances in technology began to make a great difference to agriculture. One of the first areas of improvement was plough design. The plough with an iron coulter was invented by ancient Egyptians, and had not been much altered until several improved types were brought out in the mid-18th-c. In 1771 James Arbuthnot introduced the use of a mould-board, which was much more efficient. James Anderson (1739–1808) invented the Scotch plough for use on heavy ground and James Smith (1789–1850) designed the subsoil plough for use on land with poor drainage. One of the most influential figures was Robert Ransome (1753–1830), who invented a self-sharpening plough and also designed one that could be dismantled and modified, thus obviating the necessity for small farmers to have several expensive pieces of equipment.

Other inventors produced other pieces of machinery. Jethro Tull (1674–1741) invented the seed drill in 1701, and the horse hoe a few years afterwards. The effect of these two inventions, although not felt until two generations later because of the scepticism with which they were received, revolutionized the way in which cereals were cultivated and greatly improved yield. James Meikle (c.1690–1717), an East Lothian miller, produced a winnowing machine around 1720 and his son Andrew Meikle, besides inventing the fantail which allowed windmills to turn into the wind automatically and thus to work more efficiently, also invented the first effective threshing drum. His design, which used a revolving drum and longitudinal beater bars, is essentially the same as that used in modern combine harvesters.

At the same time, steam power was coming into use on the farm, following the inventions of Richard Trevithick (1771–1833) and WATT. Steam threshing was introduced by the Shropshire ironmaster John Wilkinson (1728–1808) in 1798, and the use of steam power rapidly spread. Steam ploughing came into use during the 1850s in places where the fields were long and flat enough to make it economic, and continued until steam power was superseded by the diesel engine and by the ubiquitous tractors produced by such manufacturers as Harry George Ferguson (1884–1960) and Henry Ford (1863–1947).

Where horse power was still necessary, however, other inventions improved productivity. James Smith's (1789–1850) experimental reaper of 1811 did not work because the speed of the horses affected the action of the gathering drum, but an

Lie, Marius Sophus [lee] (1842–99) Norwegian mathematician: discovered the theory of continuous transformation groups.

Lie was inspired to study mathematics by reading PONCELET and J Plücker (1801–68) on geometry, and spent his life fruitfully developing the latter's idea of creating geometries from shapes as elements of space rather than points. Research in Berlin (with KLEIN) and in Paris was somewhat marred by being arrested as a German spy (1870), but this false charge was soon dropped and he left Paris just before the Germans besieged it. A year in a mental hospital interrupted later work at Leipzig, and he then returned to a post created for him at Christiania in Norway (now Oslo University).

Lie, along with his close friend Klein, introduced group theory into geometry, using it to classify geometries. Lie discovered the contact transformation which maps curves into surfaces (1870). Work on transformation groups followed (1873) and he invented Lie groups, which use continuous or infinitesimal transformations. Lie used these groups to classify partial differential equations, making the traditional methods of solution all reduce to a single principle. The Lie group also gave the basis for the growth of modern topology.

Liebig, Justus, Freiherr (Baron) **von** [leebikh] (1803–73) German organic chemist; the greatest chemical educator of his time.

As a druggist's son, Liebig was attracted early

effective reaper was designed in 1827 by Patrick Bell (1799–1869). When sent to America, this machine enabled the production of the first commercially successful reaping machines by Obed Hussey (in 1833) and Cyrus Hall McCormick (in 1834). Mechanical reapers caught on very quickly thereafter, and by 1870 a quarter of all the harvest in Britain was being cut mechanically. In the 20th-c the development of diesel and electric power, and machinery in general, came together with the production of ever more complex and efficient pieces of equipment, such as the combine harvester and the electric milking machine. As in other areas, however, these were the products of research teams and commercial companies rather than individual pioneers.

The above advances were largely technological, or the direct result of farming experience, such as the use of liming to reduce the soil acidity that results from sustained crop removal. Agricultural science began with LAVOISIER's work, was enhanced by DAVY's lectures and his book on the subject (1813), and emerged fully with LIEBIG's books *Chemistry and its Applications to Agriculture and Physiology* (1840) and *The Natural Laws of Husbandry* (1862). By the middle of the 19th-c studies by BOUSSINGAULT, LAWES and others had shown the importance of nitrogen (N), phosphorus (P) and potassium (K) in plant nutrition, chemical fertilizers were supplementing the use of manure and Lawes had founded the agricultural research centre at his Rothamsted estate in 1842. By the early 20th-c the long-neglected fundamental work in genetics by MENDEL was being applied in agricultural botany by BIFFEN and others. S M Babcock (1843–1931) in the USA had developed scientific dairying. ELEANOR ORMEROD had effectively created the study of agricultural entomology and given a scientific basis for insect pest control.

The Second World War intensified studies on food production, and on animal and human nutrition. Some insecticides had been used since the 1870s; but DDT, introduced by P H MÜLLER from 1939, was highly effective and widely used, until its injurious effects in food chains was fully appreciated, notably by RACHEL CARSON in the 1960s, and bans on its use in the UK and USA soon followed. In Germany organophosphorus esters were made and tested for use in chemical warfare, and some of these (eg Parathion) found major uses as insecticides after 1945. Rodenticides to reduce food losses were also needed, and after 1939 work by K P Link in Wisconsin led to warfarin which was highly effective: its anticoagulant effect on animal blood also causes it to find important use in medicine.

Another wartime effort was directed to finding chemicals to destroy the enemy's crops. One result was '2,4-D', which has a valuable selective action in attacking dicotyledons but hardly affects monocotyledons (which include cereals). Although not used in war, it has been much used as an agricultural herbicide as a result. Unselective destruction of all above-ground plant growth without soil toxicity is achieved by paraquat, which was marketed as able to replace the plough and is widely used. Unselective crop destruction, notably by 'Agent Orange', was employed during the Vietnam War.

More recently, emphasis in agricultural science has moved from newer pesticides and fungicides towards 'greener' methods of biological control, by methods such as the use of one insect species to control another, or the use of insect hormones to achieve deception and facilitate trapping. Modern genetics has opened up a range of techniques with applications in agriculture, and the 'green revolution' has led to great interest not only in improved strains of animals and food crops but also in the better understanding and management of ecosystems. The importance of such advances, especially in the less developed countries, was signalled by the award of the Nobel Prize not for science but for peace in 1970 to Norman E Borlaug (1914–   ) of the USA, in recognition of his work on the development of a new short-stemmed wheat.

Sukie Hunter

to chemistry. In 1822 he went to study in Paris (then the centre for chemistry) and became assistant to GAY-LUSSAC. By 1825 he became professor in the very small university at Giessen, near Frankfurt. He stayed there for nearly 30 years, and set up his famous laboratory for students of practical chemistry. It was not the first, as he claimed; but, like his research group of graduate students, it was the model on which systematic training in chemistry was afterwards based elsewhere. His university is now the Justus von Liebig University.

In 1826 his work showed that the fulminates, and the very different cyanates (made by WÖHLER) had the same molecular formulae. This sort of phenomenon (isomerism) could not then be explained, but it showed that a molecule was not merely a collection of atoms; they were arranged in particular ways, with each arrangement corresponding to one compound and one set of properties. The work led also to his friendship with Wöhler and their valuable joint work on the benzoyl group. The friendship survived when Liebig's combative nature had eventually spoiled all his other chemical friendships.

By 1830, Liebig had developed a method for the analysis of organic compounds which was quick and accurate, by burning them in a stream of air and oxidizing the products fully to CO_2 and H_2O; collecting and weighing this CO_2 and H_2O gave a direct way to find the percentages of carbon and hydrogen in the organic compound. Liebig and his students used this method to analyse hundreds of organic compounds, and the results were basic for the great advances to be made (notably by KEKULÉ) in organic chemistry after about 1850.

In his middle age, from 1840, Liebig worked on what we would now call biochemistry. He argued (correctly) that carbohydrates and fats are the fuel of the animal body and (incorrectly) that fermentation did not involve living cells. In agriculture, he argued (rightly) for the use of potassium- and phosphorus-containing fertilizers but underrated the importance of nitrogen, and of soil structure, in fertility. He always played a vigorous – sometimes ferocious – part in debates on chemical theory; his pupils dominated organic chemical teaching; and his views moved agriculture towards chemistry. He made a good deal of money out of his scientific work and attracted criticism for this.

Lind, James (1716–94) British physician; treated scurvy after making first controlled experiment in clinical nutrition.

A surgeon's apprentice at 15, Lind later became a Naval surgeon. He was interested in scurvy ('the plague of the sea'), first seen in sailors at the end of the Middle Ages when, for the first time, sea voyages lasted some months. Vasco de Gama had noted in 1498 that oranges were temporarily curative. In 1747, Lind made his excellent nutritional experiments, dividing a crew of scorbutic sailors into small groups given different dietary supplements for 14 days. He found that citrus fruit with the diet gave much improvement in 6 days; and in 1754 he published *A Treatise on the Scurvy*. Adoption of the treatment and use of fruit to prevent the disease was slow, although COOK used the method in his great southern explorations of the 1770s, losing only one man (out of 118) to scurvy in 3½ years. Only by 1795 was lime juice given regularly to sailors; even so, cases were still reported in the following century, from prisons, the Crimean War and polar expeditions (possibly including Scott's). By 1907, Norwegian workers had induced scurvy experimentally in guinea pigs, HOPKINS's classic work on vitamins had begun and in 1928 SZENT-GYORGY isolated vitamin C (ascorbic acid), present in citrus fruit, deficiency of which leads to scurvy.

Lindblad, Bertil (1895–1965) Swedish astronomer: proposed rotation of our Galaxy.

Lindblad graduated at Uppsala and spent 2 years in research in astronomy in the USA before becoming director of the new Stockholm Observatory in 1927 and spending his career there. Jacobus Kapteyn (1851–1922) had discovered from a survey of stellar motion in 1904 that most stars fell into two groups, or streams, moving in opposite directions in the sky. Kapteyn's interpretation of this was that our Solar System lies near the centre of our Galaxy

(the Milky Way). However, SHAPLEY proposed that the centre of the Galaxy was some 50 000 light years away, in the direction of the constellation Sagittarius, and so a lively debate between astronomers ensued. Lindblad studied Kapteyn's results and concluded that Shapley's idea was correct, provided that the speed of rotation of stars about the centre of the Galaxy depends on their distance from it (the 'differential rotation' theory). Soon afterwards, in 1927, OORT's study of stellar motion provided support for Lindblad's views, which inspired a number of Swedish astronomers (including his son Per Olaf Lindblad) to work on stellar motion.

Lindemann, Frederick Alexander, 1st Viscount Cherwell (1886–1957) British physicist: personal scientific adviser to Churchill in the Second World War.

Lindemann's background, talents and position were all unusual. His father, a prosperous engineer, emigrated to the UK from Alsace rather than become a German citizen after the Franco-Prussian War of 1870. Lindemann studied physics in France and Germany, took a doctorate with NERNST in 1910 and in the First World War worked at the Royal Aircraft Establishment at Farnborough on the problem of how to take an aircraft out of an uncontrolled spin, a situation often fatal for the pilot. He learned to fly, despite poor sight, and personally tested and proved his theory on the spin problem.

In 1919 he became professor of physics at Oxford and head of its run-down Clarendon Laboratory, which he built up to effectiveness and a leading position in low temperature physics. Although he made valuable contributions to the theory of specific heats (he proved that the melting point of a crystal depends on the amplitude of the atomic vibrations), to several laboratory instruments and even to pure mathematics and chemical kinetics, his attitude to science was that of a keen amateur.

As an amateur tennis player he had to leave his first prize behind after the European Tournament in Germany in July 1914 in order to return hastily to the UK. He competed at Wimbledon while he was an Oxford professor. His Rolls-Royce cars formed a travelling office, and his aristocratic friends included Churchill from 1921. The novelist Vita Sackville-West wrote in 1925 that at a large houseparty at Blenheim she sat between Churchill and 'a scientist called Lindemann who is absolutely thrilling'. When Churchill became Prime Minister in 1940, Lindemann became his independent scientific adviser, at times in conflict with his old friend H Tizard (1885–1959), the government's senior scientist. With the advantage of hindsight, it is clear that Tizard was right to give very high priority to radar in air defence, and Lindemann's opposition was wrong; the latter was probably in error also in his belief that heavy bombing of Germany was a direct route to victory (Tizard believed that air defence of Atlantic shipping was more valuable).

Churchill frequently preferred 'the prof's' advice, including for example Lindemann's enthusiasm for heavy area bombing. He was a minister (Paymaster-General) from 1942–5 and 1951–3, and he largely created the UK Atomic Energy Authority in 1954, having been involved in the decision to make the atomic bomb. He became Baron Cherwell in 1941, resumed his professorship in 1953 and was made a viscount in 1956.

Linnaeus, Carl [linayuhs], from 1762 **Carl von Linné** (1707–78) Swedish botanist: the great classifier of plants; popularized binomial nomenclature.

Linnaeus began his training in medicine at the University of Lund in 1727 but his father, a pastor and enthusiastic gardener, was unable to maintain his education. Linnaeus became interested in plants, and moved to the university at Uppsala with the help of a benefactor. Here he investigated the newly proposed theory that plants exhibit sexuality. O Rudbeck (1630–1702) (of *Rudbeckia*) arranged that Linnaeus should take over his unwanted lectures on botany, and attendance rose from 80 to 400. He began to form a taxonomic system based on the plant sex organs, stamens and pistils. In 1732, Linnaeus undertook a visit to Lapland and in 1733–5 to mainland Europe, in order to examine its flora and animal life.

Deciding to earn his living as a physician (out of necessity), he went to Holland to qualify (1735). While there he published *Systema naturae*, in which he divided flowering plants into classes depending on their stamens and subdivided them into orders according to the number of their pistils. This system, though useful for ordering of the many new species being discovered, only partly showed the relationship between plants.

Linnaeus returned to Sweden as a practising physician in 1738, gaining patients in court circles. In 1741 he was appointed professor of medicine and botany at Uppsala and was able to extend his teaching, and his collection and investigation of plants. Linnaeus's passion for classification led him to list the species and gather them into related groups (genera).

Linnaeus's lasting service to taxonomy was his introduction in 1749 of binomial nomenclature; he gave each plant a latinesque generic noun followed by a specific adjective. This became the basis for modern nomenclature. Until that time plants had been given a name and short Latin description of their distinguishing features, unsatisfactory both as a name and description and leading to a tangled overgrowth, strangling further development. The Linnaean system helped pave the way towards notions of evolution, an idea Linnaeus rejected emphatically; he insisted no new species had been formed since Creation and that none had become extinct.

Linnaeus was an excellent teacher and his students travelled widely, imbued with his enthusiasm, in search of new forms of life; it is estimated that one in three died in the search.

Linnaeus had a complex, self-conscious personality. He had a tidy mind and absolute belief in the value of his system. He was skilful in getting others to accept his system, even though that meant setting aside much of their own work. He cleared the way for development in biology without taking part in it. The Linnaean system based on plant sexual organs was completely artificial, but convenient. By his success he stifled some aspects of botanical development for a century. After Linnaeus's death his collection was bought by Sir James Smith (1759–1828). The London-based Linnean Society, founded by Smith in 1788, purchased the books and herbarium specimens in 1828.

Modern classification of living organisms has the species at the lowest level with (in order of increasing generality) the genus, family, order, class, phylum (for animals) division (for plants) and kingdom. Some of these categories may be further subdivided.

Thus modern man (*Homo sapiens sapiens*) belongs to the species *Homo sapiens* and subspecies *sapiens*; his genus is *Homo*; family Hominidae; order Primates; class Mammalia; phylum Chordata; kingdom Animalia.

Each category has its own definition: eg Primates, which includes gorillas, chimpanzees and humans, has defining features that include upright posture, opposable thumbs, large brain and similar blood plasma proteins. The family Hominidae includes modern man and fossil man species from the Pleistocene onwards.

Liouville, Joseph [lyooveel] (1809–82) French mathematician: developed the theory of linear differential equations.

Liouville held a professorship at the École Polytechnique for many years (1838–51), then moved to the Collège de France (1851–79). He was briefly elected to the constituent assembly in 1848, but his political career only lasted a year.

He is famous for developing Sturm–Liouville theory, which is part of the theory of linear differential equations and important in physics; he worked also on boundary-value problems. He made further contributions in differential geometry, conformal transformation theory and complex analysis, influencing developments in measure theory and statistical mechanics. He was the first to prove the existence of transcendental numbers (and an infinite number of them) and he suggested that e is transcendental (1844); this was proved by Hermite. Liouville was also editor of the influential *Journal de mathématiques pures et appliquées* from 1836–74.

Lipscomb, William Nunn (1919–) US inorganic chemist: developed low temperature X-ray crystallography and devised structures for boron hydrides.

Educated at Kentucky and California, Lipscomb became a professor at Harvard in 1959. The boron hydrides had first been made by Stock, but their structures had proved mysterious; in terms of established theory they appeared to be electron-deficient. Lipscomb deduced their structures in the 1950s by X-ray diffraction analysis of their crystals at low temperatures, a technique which he and others were to use later on a variety of chemical problems. He went on to theorize on the bonding in the boron hydrides, to extend his ideas to the related carboranes and to use nuclear magnetic resonance methods to examine molecules. His approaches to these problems have proved highly fruitful in chemistry; he was awarded a Nobel Prize in 1976.

Lister, Joseph, Baron Lister (1827–1912) British surgeon: introduced antiseptic surgery.

Lister was the son of a Quaker wine merchant who was also a skilful microscopist; his achromatic microscope design (1830) marks the beginning of modern microscopy. As an Arts student in London, young Lister attended as a spectator the first surgical operation under a general anaesthetic, in 1846. He then turned to medicine, qualified as a surgeon in 1852 and worked in Edinburgh, Glasgow and later in London. His main work on antisepsis was done in Glasgow. At that time, surgery was usually followed by inflammation and 'putrefaction'; of limb amputations about half were fatal from this sepsis, and abdominal surgery was largely avoided because of it. It was widely (but erroneously) thought that sepsis was due to air reaching moist tissues, and awkward but ineffectual attempts had been made to exclude air from surgical sites. The work of Semmelweis in Vienna, in which he showed in 1846 that sepsis after childbirth in hospital could be avoided by cleaning the hands of surgical operators, had been ignored.

In 1865 Lister read Pasteur's work on fermentation. Lister concluded that sepsis was akin to fermentation and was initiated by infectious agents, some air-borne. By 1867 he had shown that antiseptic procedures are very successful; his methods were quickly adopted in Germany (eg in the Franco-Prussian War of 1870) and more slowly in Britain. Lister used crude phenol solution as his preferred antiseptic for dressings and instruments, and as a spray in the air of the operating theatre. Later (from 1887) he gave up the spray and increasingly used aseptic methods, with steam as a sterilizing agent. His work enormously reduced the incidence of fatal post-surgical infection and encouraged surgeons to develop abdominal and bone surgery. Lister's scientific work was largely related to his 'antiseptic system'; he published on inflammation, bacteriology (he was probably the first to grow a microorganism in pure culture) and on surgical ligatures and their sterilization (his preference was catgut). He revolutionized general surgery by making it safer and was widely honoured for his work.

Lobachevski, Nikolai Ivanovich [lobachefskee] (1793–1856) Russian mathematician: discovered one of the first non-Euclidean geometries.

Lobachevski's father died when he was about 6, and after his mother had moved the family to Kazan he attended the new university there in 1807. He joined its staff in 1814 and in 1827 became its rector. He was honoured by his government, but in 1846 fell into disfavour for reasons which are unclear; he had done much for his university and his country.

From 1827 onwards Lobachevski developed the first non-Euclidean geometry to be published, although J Bolyai (1802–60) was doing similar work at the same time and Gauss had done so decades before, but without publication. Euclid's fifth postulate ('axiom XI') could not be proved. The fifth postulate is that, given a straight line and a point, just one straight line can be drawn in their plane passing through the point and never meeting the other line. Euclidean geometry was widely thought adequate to describe the world and the universe. Lobachevskian geometry accepts Euclidean postulates, except the fifth, and occurs on a curved surface with two lines always meeting in one direction and diverging in the other. The angles of a triangle no longer add up to two right angles but sum to less than that. Lobachevski's work was only widely accepted as important when Einstein's general theory of relativity showed that the geometry of space-time is non-Euclidean; it also prepared the way for the systematic exploitation of non-Euclidean geometry by Riemann and Klein. Euclidean geometry is now seen as a special case, adequate for all everyday purposes, within a more general system.

Lockyer, Sir (Joseph) Norman (1836–1920) British astronomer: discovered helium in the Sun.

As a young civil servant at the War Office, Lockyer developed an interest in astronomy and made it his career. He was particularly interested in the Sun and in the use of the recently-introduced methods of spectral analysis. Following the eclipse of 1868 Lockyer discovered, independently of the French astronomer P Janssen (1824–1907), that solar prominences could be seen with a spectroscope at any time, not merely during eclipses, and that the forms of the prominences slowly changed with time. He identified an unknown element in the Sun's spectrum, which he named helium and which was subsequently isolated in the laboratory by Ramsay in 1895. In

1873 he proposed that some unfamiliar solar spectral lines were caused by the dissociation of atoms into simpler substances with their own spectra (the electron was not to be discovered until 20 years later; we now recognize the dissociation as loss of electrons). Lockyer was also interested in archaeology, pioneering the study of possible astronomical alignments with ancient structures. He founded the Science Museum in London and also the science journal *Nature*, of which he was editor for 50 years.

He was a fearless fellow, in debate and as an expedition leader in pursuit of solar eclipses, and he was a founder of solar astrophysics.

Lodge, Sir Oliver (Joseph) (1851–1940) British physicist: pioneer of radio telegraphy.

Lodge was born in Penkhull in the Potteries, in Staffordshire, where his father was a supplier of pottery materials. The family was prolific; Oliver Lodge had 12 children, he was the eldest of 9, and his thrice-married grandfather had 25 children.

After working for his father for 7 years, he studied physics at the Royal College of Science and at University College, London. In 1881 he was appointed professor of physics at Liverpool, and in 1900 became the first principal of the University of Birmingham.

Lodge performed early experiments in radio, showing in 1888 that radiofrequency waves could be transmitted along electric wires. Simultaneously, however, HERTZ demonstrated that such waves could be transmitted through air, establishing the basis for radio communication and somewhat overshadowing Lodge's work. Lodge went on, however, to make useful technical advances, designing an improved radio detector (based on the drop in resistance of some metallic powders when exposed to electromagnetic radiation) and demonstrating a form of radio telegraphy in 1894. He was the first person to attempt to detect radio waves from celestial objects. In the 1880s the MICHELSON–MORLEY experiments had shown that the 'ether' did not exist unless it moved with the Earth, but in 1893 Lodge devised an ingenious experiment that demonstrated that even a stationary ether did not exist. In later years he devoted much effort to the scientific investigation of extrasensory perception and psychic phenomena.

His interest in spark phenomena led him to greatly improve the early internal combustion engine for motor cars by his invention of the spark plug to ignite the mixture; two of his sons set up a company to manufacture it.

Loeffler, Friedrich August [loefler] (1852–1915) German bacteriologist.

Educated in medicine in Würzburg and Berlin, Loeffler became an assistant to KOCH in the 1880s. Bacteriology was then a young science; pure cultures were difficult to secure and new techniques were needed. In 1884 Loeffler devised a new medium (thickened serum) in which he was able to culture the bacillus of diphtheria, then a major killing disease especially of children. He had previously discovered the organism responsible for glanders (a contagious disease, mainly of horses).

In 1898, with P Frosch (1860–1928), he showed that foot-and-mouth disease could be passed from one cow to another by inoculation with a cell-free extract. This demonstration that a disease of animals is due to a virus was a basic step in the founding of virology; it followed the discovery, by others, of viral diseases of plants. Soon after Loeffler's work, REED showed that yellow fever is a viral disease, transmitted by mosquitoes.

London, Fritz Wolfgang (1900–54) German–US physicist: discovered the London equations in superconductivity.

Fritz and Heinz (1907–70) London were the sons of a mathematician at Bonn and became known for contributions to superconductivity published together. Fritz studied classics, and did research in philosophy leading to a doctorate at Bonn. Later he was attracted by theoretical physics, worked with SCHRÖDINGER at Zürich in 1927 and published on the quantum theory of the chemical bond with W Heitler (1904–). In 1930 he calculated the non-polar component of forces between molecules, now called VAN DER WAALS or London forces. Having fled from Germany in 1933 the brothers did research in F E Simon's (1893–1956) group at Oxford. They soon published major papers on superconductivity, giving the London equations (1935). Fritz moved to Duke University in the USA and continued to work on superconductivity and on the superfluidity of helium (see KAPITSA).

Lonsdale, Dame Kathleen, *née* Yardley (1903–71) British crystallographer: applied X-ray diffraction analysis to organic crystals.

The 10th and last child of an Irish postmaster, Kathleen Yardley came to London when she was 5 and graduated there in physics when she was 19. For 20 years she worked at the Royal Institution, and for the next 20 at University College, London, developing methods pioneered by the BRAGGS for finding molecular structure by X-ray diffraction of crystals. In 1929 she worked out her first structure of great interest to organic chemists: it was that of hexamethylbenzene and her work showed that its benzenoid ring is a flat and regular hexagon of carbon atoms, whose carbon–carbon bond lengths she measured. Some 2 years later she worked out the structure of hexachlorobenzene using (for the first time) FOURIER analysis to solve the structure; the method was to become the major technique used by her and others. When she began her work on organic structures, she 'knew no organic chemistry and very little of any other kind'; but her results were of great value to organic chemists, as was her work on the physics of crystals, which gave reality to the concept of molecular orbitals.

A passionate pacifist and Quaker, Lonsdale refused in 1939 to register for civil defence or any other national service, and in 1943 she was fined £2 for the omission; refusing to pay, she spent a month in prison. In 1945 the Royal Society agreed to elect women Fellows, and with MARJORY STEPHENSON she became one of the first two female Fellows of the Royal Society.

Lorentz, Hendrik Anton [lohrents] (1853–1928) Dutch theoretical physicist: contributed greatly to the theory of the electron and of electromagnetism.

Lorentz completed his studies early at Arnhem and Leiden; a thesis on light reflection and refraction won him the first chair of theoretical physics in Holland at Leiden when he was 24. In 1912 he became director of the Teyler Institute, Haarlem. He did a great deal to found theoretical physics as an academic discipline in Europe.

Lorentz's thesis showed how to solve MAXWELL'S equation when an interface between two materials is present. He was then able to predict the FRESNEL formula for the behaviour of light in a moving medium. In 1892 his 'electron theory' was published; it regarded electrons as embedded in the ether, which transmitted Maxwell's electromagnetic fields and obeyed an additional relation for the force of the field on the electron (1895), now known as the Lorentz force. The Lorentz force was proposed independently by HEAVISIDE (1889). Lorentz showed, by averaging microscopic forces on electrons to give macroscopic forces on materials, how Maxwell's 'displacement current' arises and why an additional term is needed. These results were later confirmed by experiment. Lorentz coined the word 'electron' in 1899, and identified electrons with cathode rays. He showed how vibrating electrons give rise to Maxwell's electromagnetic waves, and with ZEEMAN explained the Zeeman effect whereby atomic spectral lines are split in the presence of magnetic fields (1896). For this work Lorentz and Zeeman were awarded the 1902 Nobel Prize for physics. So successful was the 'electron theory' that its failure to explain the photoelectric effect (see LENARD, and EINSTEIN) was a major clue to the need for quantum theory.

Lorentz studied the result of the Michelson–Morley experiment, which gave no indication that the Earth was moving through the hypothetical ether. He showed that if moving bodies contracted very slightly in the direction of motion, the observed results could occur. Derived independently by FITZGERALD, this is known as the Lorentz–FitzGerald contraction. In 1904 Lorentz developed a firm mathematical description of this, the Lorentz transformation, and this was later shown by Einstein to emerge naturally out of his special relativity theory (1905).

Lorenz, Edward (Norton) [lorens] (1917–) US meteorologist.

After serving as a meteorologist in the US Army Air Corps during the Second World War, Lorenz was one of the first to develop numerical models of the atmosphere and to use computers for weather forecasting. He demonstrated the inherent impossibility of long-range forecasting, and helped found the study of chaos.

Lorenz observed that minute differences in the initial conditions of his numerical models of the atmosphere could, after a relatively short time, lead to radically different outcomes. He realized that the differential equations used to describe atmospheric behaviour, while deterministic, were also highly dependent on initial conditions and that this limited the usefulness of practical weather forecasts to about a week. This phenomenon has become known as the butterfly effect, from the idea that the small air movement caused by a butterfly flapping its wings in one part of the globe could in theory result in a storm weeks later thousands of miles away.

He went on to investigate other examples of chaotic behaviour, establishing in 1963 that even very simple deterministic systems can show chaotic behaviour. One of his examples was the motion of a waterwheel, which, as he demonstrated, becomes unpredictable and prone to random reversals in direction when the rate of water flow exceeds a threshold value. In order to illustrate the chaotic dynamics of such systems, Lorenz devised the Lorenz attractor, a three-dimensional curve in which the location of a point represents the motion of a dynamical system in phase space. The curve shows how the motion of the system oscillates aperiodically between the two directions and never settles into a steady state.

Lorenz, Konrad (Zacharias) [lohrents] (1903–89) Austrian ethologist: a founder of modern ethology.

A surgeon's son, Lorenz studied medicine in New York and Vienna and graduated in 1928. Afterwards he taught anatomy in Vienna but by the mid-1930s his interest had moved to animal psychology; in fact he had collected animals and recorded their habits from childhood. In the late 1930s he made close studies of bird colonies, and in 1935 described 'imprinting'. An example of this is the way a young bird regards the first fair-sized moving object it sees as a representative of its species. This is usually a parent, but Lorenz showed it could be a model, a balloon, a tractor or a human being. In Lorenz's view, much behaviour is genetically fixed or innate; this was in conflict with the ideas of most psychologists of the 1930s, who

CHAOS

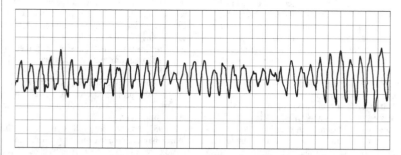

Electrocardiograph indicating chaotic rhythm (fibrillation) in the ventricles of a human heart: untreated this leads rapidly to death

In the traditional world of Newtonian physics, dynamical systems are described by equations which allow the future motion of an object to be predicted with great certainty. For example, the movement of the planets can be reliably computed years ahead to within a fraction of a second. For centuries it was assumed that the dynamics of all systems were inherently calculable, even if some are so complicated as to be beyond our practical computational ability.

Contrary to intuition, however, there are many natural systems whose motion turns out to be inherently chaotic. The first example of such a system to be recognized as such was the weather, or rather equations used to model it. These never settle into a steady state, but constantly vary in an aperiodic, apparently random manner. E Lorenz showed that they also exhibit an extreme dependency on their initial conditions, a factor that makes long-range weather forecasting effectively impossible.

Many other phenomena in all branches of science have since been recognized to be chaotic. Examples are the motion of a simple dynamo,

which can undergo unpredictable reversals and which may model the erratic reversals of the Earth's magnetic field throughout geological history. In biology, cardiac arrhythmias and erratic nerve impulses are chaotic, and in astronomy, once the showpiece of Newtonian physics, the motion of some objects is now known to be chaotic, such as the moon Hyperion, which orbits Saturn in an unpredictable tumbling motion. Turbulence is a classic example, as are wildlife populations, which undergo unpredictable cycles.

Chaos has been defined as the irregular, unpredictable behaviour of deterministic, non-linear dynamical systems. As such, fractals (see Mandelbrot) are highly visual examples of chaotic systems, where apparently simple shapes are seen, upon closer inspection, to reveal an infinity of detail on progressively finer scales.

DM

saw behaviour as entirely flexible or learned. Their emphasis was on laboratory experimentation on animal learning, while Lorentz valued studies of species-specific behaviour in the wild. In 1942 he joined the German army (as a motor cycle instructor and then as a psychiatrist), was captured and spent 4 years as a prisoner in the USSR (he studied the courtship rituals of his fleas). Later, working in Austria, he continued his studies on birds and other animals, and his generalizations did much to found ethology as a particular branch of animal behaviour study. Lorenz has been criticized for his emphasis on innate patterns and for his extrapolations from animals to man. His views on human aggressiveness, population expansion and environmental deterioration are pessimistic. He shared a Nobel Prize in 1973 with Tinbergen and Frisch.

Loschmidt, Johann Joseph [lohshmit] (1821–95) German physical chemist: early worker on valence theory and on molecular size.

Born into a peasant family in what is now the Czech Republic, Loschmidt studied in Prague and Vienna. In the 1840s he tried to establish himself in business but the times were difficult and in 1854 he became bankrupt. He taught science in Vienna, and became a friend of Stefan. His book *Chemical Studies I* (1861; there was never a Part II), included some novel and correct ideas: that sugar is an ether-like compound, that ozone is O_3, that benzene is cyclic and that double and triple bonds can usefully be shown as connecting lines. He assumed variable valences for some atoms (eg 2, 4 or 6 for sulphur) but fixed values for C (4), O (2) and H (1). His book had little influence, and Loschmidt moved to work on the kinetic theory of gases; he calculated the first accurate value for the size of air molecules. From this he calculated in 1867 the number of molecules of gas per cm^3, but his value is about 30 times too small. However, for this pioneer attempt to obtain a value for the AVOGADRO constant N_A, the constant is sometimes named as the Loschmidt number (L); Avogadro never gave any pertinent numerical calculations on this.

Lovell, Sir (Alfred Charles) Bernard (1913–)
British physicist: pioneer radio astronomer.

After studying physics at the University of Bristol, Lovell was appointed to the staff of the physics department at the University of Manchester in 1936, where he was to spend his career. He became professor of radio astronomy and later director of the Nuffield Radio Astronomy Laboratories at Jodrell Bank, near Manchester.

Lovell was an outstanding experimental physicist who played a key part in the wartime development of radar in Britain, and went on to pioneer the study of radio astronomy, constructing the world's largest steerable radio telescope at Jodrell Bank.

After early university work on the detection of cosmic ray showers with BLACKETT, Lovell was conscripted to the war effort in 1939 and worked on the development of airborne radar systems to enable British nightfighters to locate German bombers. This was highly successful and resulted in heavy casualties being inflicted on the bombers, saving many lives. In 1942 he was directed to take charge of the development of a radar system to help bombers locate their targets by ground returns, thus greatly increasing the effectiveness of Allied bombing raids against Germany. In March 1943 a modification of the system enabled aircraft to detect German U-boats at night, again with great success, dramatically reducing Allied shipping losses in the Atlantic.

After the war Lovell returned to Manchester; between 1946 and 1951 he used war surplus radar equipment to detect meteors via their ionized trails in the upper atmosphere. He made two significant discoveries: first, that many hitherto unknown meteor showers occur during daylight hours, often exceeding the known seasonal showers in number and intensity; second, by measuring meteor velocities using the radar he showed that their orbits were 'closed' (ie confined to the solar system) and that the meteors were therefore not of interstellar origin, as had been suggested.

In 1947 Lovell completed the construction of a 218 ft (66 m) aperture fixed parabolic aerial. This aerial was used to detect the radio emission from the Andromeda galaxy in 1951 by R Hanbury-Brown (1916–) and C Hazard. This work stimulated the plans made by Lovell in 1949–50 for what is undoubtedly his greatest and most lasting contribution to science, the

building of a 250 ft (76 m) diameter steerable parabolic radio telescope. The actual building of the telescope became a *cause célèbre* of British science, greatly exceeding original cost and time estimates. Although Lovell was criticized for this, the Jodrell Bank telescope was breaking new ground as the first 'big science' project; it required large funding but it has been used by numerous teams of scientists for studying a wide range of phenomena for many years. Today, such projects are often funded on an international basis and are administered by sizable committees.

In 1957, almost before it was completed, the Jodrell Bank telescope caught the public imagination by tracking the carrier rocket of the world's first artificial satellite, Sputnik. In 1959 the telescope measured the descent of Lunik 2 to the impact on the Moon and in 1966 recorded the first photographs of the lunar surface transmitted by the Russian Luna 9 probe. Although the telescope was used in the radar mode for studies of the Moon and measurement of the distance of Venus, the emphasis has been on radio astronomy. The identification of radio sources with optical objects was an early goal and this led directly to the discovery of the objects that became identified as quasars in 1963. Much of the early work on pulsars followed and the gravitational lens effect for quasars was discovered in 1979. Lovell's own research with the telescope led to the discovery of the radio emission coincident with the optical flares on the red dwarf stars.

While Lovell's personal scientific achievements were noteworthy, his lasting impact on science has been through his administrative and political efforts, which led to the building of the great scientific instrument that today bears his name.

Lovelock, James (Ephraim) (1919–) British
environmental scientist: devised the Gaia hypothesis.

After studying chemistry and medicine in England, Lovelock held medical research positions in the UK as well as posts in chemistry, medicine and space science in the USA. After 1964 he conducted an independent scientific career. Lovelock's most significant contributions have been in the field of environmental science.

In 1957 Lovelock invented the electron capture detector, an extremely sensitive device that revolutionized the study of environmental chemistry by enabling man-made chemicals to be detected in the environment in very low concentrations. The device led to the discovery that pesticide residues had become widely distributed in nature, and also to the discovery of significant concentrations of chlorofluorocarbons and nitrous oxide (N_2O) in the atmosphere, both discoveries leading to increased concern about mankind's pollution of the environment.

In the early 1970s Lovelock proposed the theory for which he is best known, the Gaia hypothesis. Named after the Greek Earth goddess, it proposes that the Earth can be viewed as a single living system in which complex feedback mechanisms act to regulate the environment and maintain optimal conditions for life. According to the Gaia hypothesis, therefore, life on Earth is bountiful not by chance but because the Earth's feedback mechanisms have evolved a complex and interwoven pattern of life that not only makes optimal use of the environment but also transforms the environment into a state that best supports life.

While the Gaia hypothesis is, through its global and all-embracing nature, difficult to 'prove', it is a way of viewing the Earth that has proved fruitful. One result was Lovelock's own prediction of oceanic dimethyl sulphide emissions in order to balance the global sulphur cycle, which has since been observationally confirmed, as have his proposals on aspects of the carbon cycle in nature.

Lovelock's inventions and theories have significantly contributed to the advancement of the environmental sciences and have increased awareness of the global effects of man's polluting activities. Early scepticism about the Gaia hypothesis has diminished as its value has been appreciated.

Lowell, Percival (1855–1916) US astronomer: predicted existence and position of Pluto.

Son of a wealthy Boston family, Lowell travelled extensively after graduating from Harvard. His sister Amy was a major poet and his brother became president of Harvard. Lowell's interest in astronomy was first stimulated by SCHIAPARELLI's report in 1877 of 'canali' on Mars. He became convinced that Mars was inhabited by an intelligent race and wrote books on the subject. At the beginning of this century such views were not so ridiculous as they appear today, and the excellent observatory he built in Arizona in 1894 at a height of 2200 m became an important centre for planetary studies. However, Lowell's most important contribution was the prediction (based on its gravitational influence on Uranus), of a ninth planet beyond Neptune. Although he himself searched for it from 1905 until 1914, Pluto was not detected until 1930 by TOMBAUGH, working at Lowell's own observatory. It was named after the Greek god of outer darkness.

Lower, Richard (1631–91) English physiologist: made first successful direct blood transfusion.

Lower qualified in medicine at Oxford, where he assisted his teacher WILLIS with dissections, and then moved to London to practise; he was an early Fellow of the Royal Society and had belonged to the Oxford group that founded it. In Oxford in 1665 he demonstrated transfusion of blood from the artery of one dog to the vein of another. Pepys in his diary for 1667 reports on a successful human blood transfusion performed as a Royal Society experiment by Lower: 'I was pleased to see the person who had his blood taken out. He speaks well ... and as a new man.' This was probably the first human blood transfusion in the UK. Later attempts by others to transfuse from animals to humans led to some deaths, and only after LANDSTEINER's work from 1900 on blood groups did human transfusion become useful. Lower's *Treatise on the Heart* (1669) gives a good account of the structures of the heart. He recognized that it is not 'inflated by spirits' but acts as a muscular pump, with systole as the active phase and diastole a 'return movement'. He studied the colour change between dark venous blood and red arterial blood, experimented with dogs and deduced that the red colour results from mixing the dark blood with inspired air in the lungs; he realized that the purpose of respiration is to add something to the blood. After 1670 he concentrated on his medical practice.

Lubbock, (Sir) John, Baron Avebury (1834–1913) British biologist: contributor to archaeology, entomology and politics.

Lubbock's father was a successful banker and amateur mathematician, and the boy joined the family bank at 15. He was successful enough, but his main interest was in biology, in which he was self-taught. Fortunately he could concentrate fully on different matters at short intervals, and in his adult life banking, biology, politics and education all engaged him. He was lucky that DARWIN lived near the family home, made a friend of the boy and developed and used his talent for drawing. Lubbock early became an enthusiast for Darwin's ideas and helped to expound them, and he was one of the few men whose opinions mattered to Darwin.

In 1855 he found the first fossil musk-ox in Britain, which gave early evidence of an ice age. In the 1850s and 1860s he began to link ideas on evolution with studies in archaeology and human prehistory and he travelled widely in Europe to study lake village sites and tumuli there. He coined the words 'neolithic' and 'palaeolithic' for the New and Old Stone Ages. His books on prehistory were pioneers in the field, but his work here did not obstruct another of his pursuits, entomology. He worked especially on the social insects, and devised the 'Lubbock nest', in which he could examine colonies in a movable glass-sided container. His methods (including the first use of paint-marking of insects for their identification, and obstacles and mazes to test intelligence) led to much new knowledge of their habits, instincts and intelligence. With a device designed for him by GALTON he showed that ants can distinguish colours and see ultraviolet light; and he discovered bees' colour preferences.

He became well-known as an MP by introducing bills, on Bank Holidays ('St Lubbock's Days', in 1871), also on wild bird protection, open spaces, ancient monuments and dozens more. He belongs to the rare group, including FRANKLIN, distinguished in science, politics and commerce. Few later amateur scientists can compete with that distinction.

Ludwig, Karl Friedrich Wilhelm [ludveekh] (1816–95) German physiologist: pioneer of modern physiology.

Ludwig enrolled as a medical student in Marburg in 1834, but he had a stormy student career. He soon had a heavily scarred lip through duelling, and conflict with the university authorities sent him to study elsewhere; but he returned to Marburg in 1840 and was teaching there by 1846. Later he taught in Zürich, Vienna and Leipzig. His work helped to create modern physiology; he saw no place for 'vital force' and he sought explanations of living processes in terms of physics and chemistry. In this he was much influenced by his friend, the chemist BUNSEN. When Ludwig began, physiology had few experimental instruments. He developed the kymograph (1846) and used it to discover much about respiration and the circulation; he devised the mercurial blood pump (1859), the stream gauge (1867) and a method of maintaining circulation in an isolated organ, perfusion (1865). His blood pump allowed the study of blood gases and respiratory exchange. Later he worked on the action of the kidneys and the heart, on salivary secretion and on the lymphatic system. During 30 years he and his students did much to create modern physiology and when he died almost every leading physiologist had at some time studied with him.

Lummer, Otto [lumer] (1860–1925) German physicist: experimentalist on black body radiation.

After a period as assistant to HELMHOLTZ, in 1904 Lummer became professor at Breslau. His early work was on photometry and later he worked on spectrometry, but his best-known work is on radiant heat. Since a black body approximates to a perfect absorber, it follows that a black body should form an ideal radiator; but this at first seemed to be an abstract concept. However, in the 1890s Lummer and WIEN realized that a small aperture in a hollow sphere heated to the required temperature should be equivalent to a black body at the same temperature. Experimental work on this basis was carried out with Wien and later with E PRINGSHEIM on the distribution of energy in black body radiation, and was important in leading to PLANCK's quantum theory of 1900.

Lyell, Sir Charles [liyl] (1797–1875) British geologist: established the principle of uniformitarianism in geology.

Lyell at first embarked on a legal career, but his interest in geology led in 1823 to his appointment as secretary of the Geological Society of London. During the first part of the 19th-c geology had made great advances in the collection of information, but most geologists still believed in one or more world-wide 'catastrophes' to account for the creation of what they found. Lyell was responsible for the general acceptance of the principle of uniformitarianism, the idea that rocks and geological formations are the result of the ordinary processes that go on every day, but acting over very long periods of time. This principle was first advocated in a general way by HUTTON, but was much more convincingly illustrated and argued by Lyell. In 1830 he published his popular *Principles of Geology*, in which he applied his ideas in explaining many of the geological features that he had discovered on his extensive travels through Europe and America. This classic work greatly influenced DARWIN in developing his theory of evolution, a concept which Lyell, strangely enough, never accepted.

Lyon, Mary Frances (1925–   ) British geneticist.

After graduating from Cambridge in 1946 and taking a doctorate there in 1950, Mary Lyon joined the Medical Research Council's Radiobiology Unit at Harwell, leaving it as Deputy Director in 1990. Her extensive work on mouse genetics led to better knowledge of the mammalian genome, the problems of inherited disease and the genetic risks of low radiation exposure. The 'Lyon hypothesis' of 1961 proposed that one of the two X chromosomes in female mammals can be randomly inactivated, so that the females become mosaics of different genetic cell lines. This idea has been confirmed and is important in chemical genetics and in genetic imprinting.

Lyot, Bernard Ferdinand [lyoh] (1897–1952) French astronomer: invented the coronagraph.

Lyot worked at the Paris Observatory at Meudon from 1920. He invented the coronagraph, a device that allows the Sun's corona to be observed without the necessity for a total solar eclipse, in 1930. This is achieved by creating an artificial eclipse inside a telescope with very precisely aligned optics. Rocket-borne solar observatories such as Skylab have, in the last 20 years, allowed coronagraphs to be used outside Earth's atmosphere, adding much to knowledge of the corona.

Lyot also pioneered the study of the polarization of light reflected from the surface of the Moon and of the planets, allowing him to infer something of their surface conditions.

MacArthur, Robert Helmer (1930–72) Canadian –US ecologist: developed theories of population biology.

Born in Canada, MacArthur moved to the USA when he was 17 and studied mathematics at university, moving to Yale for his doctorate. However, in the second year of his PhD work he changed to zoology. After 2 years military service he returned to Yale and then concentrated on ecology. From 1965 until his early death from cancer he was professor of biology at Princeton. His first research was on five closely similar species of warbler which co-exist in the New England spruce forest and which were thought to violate the competitive exclusion principle, ie that in 'equilibrium communities' no two species of the same animal occupy the same niche. He found that the birds tend to occupy different parts of the trees and that the principle was followed. From then on, he studied population biology and the strategies used to form multi-species communities. He devised ways to quantify ecological factors and to predict mathematically the level of diversity of bird species in a given habitat. His ideas have proved influential, including his division of animals into r and K species. The r species are opportunistic, with high reproductive rates, heavy mortality, short lives and rapid development. The K-strategists are larger, develop more slowly and are more stable; and he was able in 1962 to show that natural selection principles apply to both groups.

McClintock, Barbara (1902–92) US plant geneticist and discoverer of 'jumping genes'.

Barbara McClintock's early success in science at school disappointed her mother, who thought that her daughter was not developing 'appropriate feminine behaviour'. She graduated from Cornell and had early success in her postgraduate work; she found she could identify individual maize chromosomes under the microscope, which led to the integration of plant-breeding experiments with chromosomal analysis. She gained her PhD in 1927 and soon became recognized as a leading scientist in her field. Failing to get professional advancement, she left Cornell and went to the University of Missouri in 1936. Promotion still eluded her and in 1941 she went to work at the Cold Spring Harbor Laboratory, where she discovered and studied a class of mutant genes in maize. Her experiments in maize genetics led her to the very novel idea that the function of some genes is to control other genes; and that some are able to move on the chromosome and control a number of other genes. This concept of 'jumping genes' is now familiar and accepted, even though it is far from fully understood; as she demonstrated, it must involve physical movement of DNA from site to site. When she presented the work at a symposium in 1951 the significance and implications were not understood, and her discoveries were neglected by most geneticists for many years. Disappointed, she stopped publishing the results of her continuing experiments. One aspect of McClintock's work, on promotor and suppressor genes, was to be much extended by MONOD in the 1960s.

In the 1970s a series of experiments by molecular biologists proved that pieces of bacterial DNA do indeed 'jump' on the chromosomes, and McClintock's work was finally recognized. She was awarded the first unshared Nobel Prize for physiology or medicine to be given to a woman, in 1983. WATSON described her as one of the three most important figures in the history of genetics. Asked if she was bitter about the lack of recognition of her work, McClintock answered 'If you know you're right you don't care. You know that sooner or later it will come out in the wash.'

Macewen Sir William [muhkyooan] (1848–1924) British surgeon: pioneer of aseptic surgery, neurosurgery and orthopaedic surgery.

Macewen was very much a Glaswegian, graduating there in 1869 and afterwards working there until his death. He was a student under LISTER at the time antiseptic methods were started in the Glasgow Royal Infirmary, but he soon modified these methods and became a pioneer of aseptic techniques, giving up the carbolic spray by 1879 and using boiling water or steam to sterilize gowns, dressings and surgical instruments. A full surgeon by age 28, his forceful personality allowed him to impose rigorous aseptic routines, and his surgery was bold and effective. In the 1880s he operated on abscesses and tumours of the brain with success; surgery of the skull was ancient, but work on the brain was novel and called for skilful diagnosis and localization and precise surgery. In 1893 he reported on 74 brain operations; 63 succeeded. At the same period he developed successful bone surgery, including bone grafts. His interest in bone growth led to his work on the growth of deer antlers, published when he was over 70; he devised methods and instruments for corrective bone surgery on acute deformities such as those resulting from rickets, then common in Glasgow children.

Mach, Ernst [mahkh] (1838–1916) Austrian theoretical physicist: fundamentally reappraised the philosophy of science; 'the father of logical positivism'.

Mach was mainly educated at home until age 15, but later studied at Vienna. There he became interested in the psychology of perception as well as in physics. An appointment as professor of mathematics at Graz followed (1864), and he later moved to Prague (1867) as professor of experimental physics and to Vienna in 1895 as a professor of philosophy. A slight stroke in 1897 caused partial paralysis and he had to retire from the university in 1901. Thereafter for 12 years he was a member of the upper chamber of the Austrian parliament.

The theme of Mach's work was his belief that science, partly for historical reasons, contained abstract and untestable models and concepts, and that science should discard anything that was not observable. Mach argued that all information about the world comes through sensations, and that the world consists of data; that which may not be sensed is meaningless. A historical view of science also convinced Mach that discoveries are made in many ways, not particularly related to the scientific method, and that accidents and intuition play a role. Mach influenced the authors of quantum mechanics, particularly the 'Copenhagen' school of BOHR, and the theory sharply distinguishes between observable quantities and the abstract mathematical wavefunction from which it is derived and which has a higher information content. Mach and his book *Mechanics* (1863) greatly influenced EINSTEIN. What is now known as Mach's principle states that a body has no inertial mass in a universe in which no other mass or bodies are present, as inertia depends on the relationship of one body to another. Einstein's efforts to put this on a sound footing led to his theory of relativity. This result was not to Mach's liking and he rejected it.

Mach also did some experimental work, and investigated vision, hearing, optics and wave phenomena. In 1887 he published photographs of projectiles in flight showing the accompanying shock waves; he found that at the speed of sound the flow of a gas changes character. In supersonic flow, the Mach angle is that between the direction of motion of a body and the shock wave. In 1929 the Mach number was named as the ratio of the projectile speed to the speed of sound in the same medium. At Mach 1, speed is sonic; below Mach 1, it is subsonic; above Mach 1, it is supersonic.

Mackenzie, Sir James (1853–1925) British cardiologist: developed instrumental methods for study of heart disease.

Mackenzie was an Edinburgh medical graduate who did most of his work in Burnley, Lancashire. Following the unexpected death of a pregnant girl from a heart attack, he began to keep regular detailed records of heart action.

For this he devised improved instruments to record ink-tracings on paper of the pulses in arteries and veins, which he correlated with heart action. He soon found that some irregularities of rate and rhythm were common and appeared unrelated to disease (previously all such disorders were thought to be signs of disease), while other arrhythmias did point to disease. His book *Diseases of the Heart* (1908) described his polygraph and its use, and was a milestone in cardiology. His recognition that advanced mitral valve disease leads to auricular fibrillation was a step towards its later treatment. He did much to reintroduce digitalis as a heart drug; it had been used by WITHERING but had fallen into disfavour because the dose needs careful regulation. The chemically pure digoxin gives much better control and is widely used.

McLaren, Anne Laura (1927–) British geneticist.

As a daughter of Lord Aberconway, horticulturalist and a creator of the famous garden at Bodnant in North Wales, Anne McLaren might have been expected to work in botany. In fact she studied zoology at Oxford and afterwards specialized in developmental biology and genetics at London and Edinburgh. From 1974–92 she directed the Medical Research Council's Mammalian Development Unit in London, working mainly in embryology and using mice. For example, she showed that a cell type from the testes of mice embryos can develop to give a variety of cell types in culture, depending on its environment. Such work has implications in sex determination in germ cells, in IVF, in twinning and in the development of malignancy. She became a Fellow of the Royal Society in 1975, and in 1991 its foreign secretary, a major post in British science; she was the first woman to become an officer of the Society. She has been a leading and successful advocate for continuing research in the controversial field of human embryology.

McMillan, Edwin Mattison (1907–91) US physicist: discoverer of neptunium.

Educated at Caltech and Princeton, McMillan joined the University of California at Berkeley in 1935 and was there for the rest of his career. In the late 1930s he was mainly concerned with nuclear reactions and the design of cyclotrons.

In 1940 with P Abelson (1913–) he showed that, when uranium is bombarded with neutrons, one nuclear reaction that occurs leads to formation of a new element, the first discovered to be heavier than uranium; it was named neptunium. The nuclear reactions are:

$$^{238}_{92}U + n \rightarrow\ ^{239}_{92}U + \gamma \text{ (gamma radiation)}$$
(an instantaneous reaction)

and $$^{239}_{92}U \rightarrow\ ^{239}_{93}Np + e^- \text{ (beta radiation)}$$
(half-life of $^{239}_{92}U$ = 23 min)

SUPERHEAVY CHEMICAL ELEMENTS: A LIMIT TO THE PERIODIC TABLE?

The chemical element with the heaviest atoms occurring in nature is uranium, with atomic number 92.

The atomic number is the number of positively charged protons in the atomic nucleus. The remaining nuclear mass consists of neutrons, each with nearly the same weight as a proton, but with no charge. In uranium most of the atoms have 146 neutrons, but about 0.7% have 143 neutrons. The relative atomic mass of the two kinds of atom will therefore be 238 or 235. Since the heavier sort of uranium atom predominates, the average relative atomic mass of a quantity of uranium is close to 238. Atoms with the same atomic number but different atomic mass have the same chemical properties, but their nuclear properties – such as their radioactivity if the nucleus spontaneously breaks down – are different. They are known as 'isotopes' of that element: uranium exists therefore as two isotopes, uranium-235 and uranium-238.

The first transuranium elements (ie those with an atomic number above uranium, higher than 92) were neptunium (Np, 93) and plutonium (Pu, 94), made by McMILLAN and his co-workers in 1940 at the University of California by bombarding uranium nuclei with nuclear particles from a cyclotron or a linear accelerator. Later in the 1940s SEABORG and his colleagues made americium (Am, 95), curium (Cm, 96), berkelium (Bk, 97) and californium (Cf, 98) by essentially similar methods.

When in 1952 the first hydrogen bombs were exploded, the debris of a thermonuclear explosion was found to contain traces of two even heavier elements, named as einsteinium (Es, 99) and fermium (Fm,100). Like all the transuranics, these elements are highly radioactive; evidently the mutually repulsive forces between the large number of nuclear protons, even though diluted by neutrons, leads to nuclear instability.

Then, between 1955 and 1975 at the University of California, still heavier nuclei were made, by bombarding heavy nuclei (californium in most cases) with nuclei from lighter elements (boron, carbon, nitrogen or oxygen).

In this way mendelevium (Md, 101), nobelium (No, 102), lawrencium (Lr, 103), and the elements with atomic number 104, 105, and 106 were made. The last of these has a radioactive half-life of only 0.9 s, supporting the view held until the early

1970s that, as these elements are formed in only a minute number of internuclear collisions and then survive for such a short time as to render detection difficult, it seemed likely that a practical limit to the periodic table must exist, perhaps near to the then-undiscovered element 108. However, from the 1970s some theories of nuclear structure were developed that proposed that relatively stable shells of protons and neutrons exist in the nucleus; and specifically that in the region with atomic number 110–118 an 'island of stability' should allow superheavy nuclei with a detectable life-span to exist.

Experimental work on this basis was at first directed to the possibility of detecting a superheavy element in nature, but the search proved barren. However, in the early 1980s at the Laboratory for Heavy Ion Research at Darmstadt, Germany, elements 107, 108 and 109 were made by nuclear synthesis, by bombarding the nuclei of bismuth or lead with relatively heavy nuclei of chromium or iron. Only minute amounts of these elements were made and their nuclei quickly disintegrated; in the case of 109 only a few atoms were detected and the radioactive half-life is a mere 3.4 ms. In 1994 the same team made element 110 by bombarding lead with atoms of nickel-64, and element 111 by similar attack on a bismuth-209 target. In each case only a few atoms of the new element were detectable and they quickly decayed.

Similar methods continue to be applied in the search for yet heavier elements; the problem is seen by many physicists as among the most intriguing targets in physics. Its resolution will continue to spur theorists in their efforts to refine the theory of nuclear structure, and experimentalists in their search for more powerful particle accelerators and more sensitive detectors for possibly very short-lived superheavy nuclei.

In late 1994 a ruling was made by IUPAC (the International Union of Pure and Applied Chemistry) on the names of the six heaviest elements. Except for the first of them, the names commemorate major contributors in nuclear chemistry. The names are: 104, dubnium (symbol Db; after Dubna, a research centre near Moscow); 105, joliotium (Jl); 106 rutherfordium (Rf); 107 bohrium (Bh); 108 hahnium (Hn); 109 meitnerium (Mt).

The question remains: does the periodic table of the elements have an end? And if so, where is it?

IM

McMillan obtained evidence that the radioactive neptunium decayed to form a new element, plutonium (number 94), but in 1940 he moved to defence work on radar and sonar and the new transuranic elements were studied by SEABORG, with continuing success. For their

work on this Seaborg and McMillan shared the 1951 Nobel Prize for chemistry.

LAWRENCE'S cyclotron had met a limit to its performance in the early 1940s; particles accelerated in it above a certain speed increased in mass in accord with EINSTEIN'S theory of relativity and this put them out of phase with the electric impulses. McMillan in 1945 devised a solution to this by use of a variable frequency for the impulses, adjusted to keep in phase with the particles. This machine, the synchrocyclotron, could be designed to give results up to 40 times more powerful than the best cyclotrons.

Magendie, François [mazhãdee] (1783–1855) French physiologist: pioneer of experimental pharmacology.

Magendie graduated in medicine in Paris in 1808, and afterwards practised and taught medicine in Paris. In 1809 he described his experiments on plant poisons, using animals to find the precise physiological effect and then testing out the compounds on himself. In this way he introduced into medicine a range of the compounds from plants now known as alkaloids, which contain one or more nitrogen atoms within ring structures; many have striking pharmacological properties and Magendie showed some of the medicinal uses of strychnine (from the Indian vomit-nut), morphine and codeine (from opium) and quinine (from cinchona bark). Magendie's studies were remarkably wide-ranging. He showed in 1816 that protein is essential in the diet, and that not all kinds of protein will suffice. He studied emetic action; the absorption of drugs; olfaction; and the white blood cells. In 1822 he showed that spinal nerves have separate paths controlling movement and sensation, confirming and extending C BELL's work. His enthusiasm for vivisection sacrificed hundreds of animals, mainly dogs, which was much disapproved in England (but not in France); he pursued data, avoided theory and did much to found the French school of experimental physiology.

He also made some major errors: he claimed that cholera and yellow fever were not contagious, and he was against anaesthesia (induced by ether) in surgery.

Maiman, Theodore Harold [miyman] (1927–) US physicist: constructed the first laser.

Maiman was the son of an electrical engineer and, after military service in the US Navy, he studied engineering physics at Colorado University. Later he did his doctorate in electrical engineering at Stanford and joined the Hughes Research Laboratories in Miami in 1955. The maser (producing coherent microwave radiation) had been devised and induced to work by TOWNES in 1953, and Maiman improved the design of the solid-state version. He then constructed the first working laser (Light Amplification by Stimulated Emission of Radiation) in the Hughes Laboratories in 1960, although Townes and SCHAWLOW had published a theoretical description. A ruby crystal with mirror-coated cut ends was used and this resonant cavity was stimulated by flashes of light to produce a coherent, highly monochromatic, pulsed laser beam. The first continuous wave

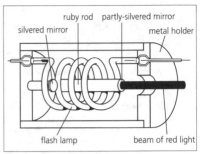

Section through a ruby laser

(CW) laser was constructed by A Javan (1926–) of Bell Telephone Laboratories in 1961.

Since then, lasers have found use in a variety of applications, including spectroscopy, repair of retinal detachment in the eye, compact disc (CD) players and check-out scanners. Maiman left Hughes to found Korad Corporation in 1962 which became a leading developer and manufacturer of lasers; he also founded Maiman Associates in 1968 and Laser Video Corporation in 1972. In 1977 he joined TRW Electronics of California.

Malpighi, Marcello [malpeegee] (1628–94) Italian biologist; discovered capillary blood vessels.

Born in the year in which HARVEY published his *De motu cordis* (1628, On the Motions of the Heart) describing the circulation of blood in mammals, Malpighi graduated at Bologna in philosophy, and then in medicine in 1653. From 1666 he was professor of medicine there, and the Royal Society of London began to publish his work, largely based on his microscopy and carried out in the 1660s and 1670s. In 1660 he began his studies of lung tissue, and the next year used frog lung. This was well suited to the early microscope, which had developed in GALILEO's time after 1600 but was optically poor; much of the best work in the 1650s was done using a single lens rather than a compound system. Frog lung is almost transparent, with a simple and conspicuous capillary system. Malpighi was able to observe the latter for the first time and to see that it was linked to the venous system on one side and to the arterial system on the other, thereby vindicating and completing Harvey's work on the animal circulation. Later he studied the skin, nerves, brain, liver, kidney and spleen, identifying new structures. In 1669 he gave the first full account of an insect (the silkworm moth) and then began his work on the chick embryo. In the 1670s he turned to plant anatomy, discovering stomata in leaves and describing the development of the plant embryo.

Mandelbrot, Benoit (1924–) Polish–French mathematician: initiated the novel geometry of fractional dimensions and fractals.

Born into a Lithuanian-Jewish family, Mandelbrot was educated at the École Polytechnique in Paris, before visiting the USA and obtaining a research position at IBM's Thomas J Watson Research Centre. Mandelbrot's uncle, Szolem, had been a founder member of the innovative French group of mathematicians who worked under the collective name of NICHOLAS BOURBAKI.

Mandelbrot's career as an applied mathematician included teaching economics at Harvard, engineering at Yale and physiology at the Einstein College of Medicine. He worked on mathematical linguistics, game theory and economics, before being asked to investigate the problems of noise on telephone wires used for computer communications. He discovered that, contrary to his intuition that the noise would be random in timing, it occurred in bursts and that, as he studied these bursts on shorter and shorter time scales, the distribution of the noise spikes always remained a scaled-down version of the whole. He was able to model the noise distribution as a Cantor dust, which has the property of containing infinitely many spikes, while being infinitely sparse.

His studies of the scalability of such time series led to a famous paper 'How long is the coast of Britain?', in which he showed that the answer depended upon the scale at which you measured it. The finer the scale, the greater amount of detail is resolved and the longer the coastline appears. Even stranger, as the scale of measurement becomes smaller the answer does not tend to a fixed value, as one might expect, but to infinity.

Mandelbrot went on to show that it is an inherent property of nature to contain roughness at all scales, and to describe this mathematically he devised a geometry with fractional dimensions, rather than the usual integral 1,2,3,4... A well-known example is the Koch snowflake, a curve of infinite length and a fractional dimension of 1.2618. Mandelbrot coined the term fractal to describe such objects, which require fractional dimensions to properly describe them; snowflakes and fern leaves are familiar examples of fractals from nature.

Initially a mathematical curiosity, fractals and fractal geometry have increasingly provided insights into natural phenomena such as the distribution of earthquakes, and have found application in many areas of human activity such as polymers, nuclear reactor safety and economics.

Manson, Sir Patrick (1844–1922) British physician: pioneer of tropical medicine.

Manson qualified in medicine at Aberdeen in 1865 and then worked in China for 23 years. He virtually founded the specialty of tropical medicine, in part by his studies of tropical parasitic infection. He studied the life-cycle of the para-

site causing filariasis and deduced that it is passed to man by a common brown (Culex) mosquito. These experiments gave the first proof of the necessary involvement of an insect vector in the life cycle of a parasite, but his report on this to the Linnean Society in 1878 was received with ridicule.

In London from 1890, he met R Ross and discussed the role of mosquitos in malaria. As a result, Ross went to India and the two collaborated in studying the life-cycle of malaria parasites in the mosquito; Manson modestly gave Ross the main credit for their results. Meanwhile, G B Grassi (1854–1925), working independently in Rome, showed in 1898 that human malaria is transmitted by mosquito bites and in 1901 he described the complete, complex life cycle of the parasites causing the disease. Grassi was able to show that mosquitos of the genus Anopheles are exclusively responsible, but Ross claimed priority in the overall work on human malaria and in 1902 a Nobel Prize was, somewhat unjustly, awarded solely to him.

Manson also studied other parasitic diseases and effectively founded the London School of Tropical Medicine in 1899.

Mantell, Gideon Algernon (1790–1852) British geologist: discovered first fossil dinosaurs.

Son of a shoemaker in Lewes, Mantell studied medicine in London. In 1811 he began work as a surgeon in Lewes, but his interest in geology increased and, after moving to Brighton in 1833, the interest became obsessive: his fossil-filled house became a public museum and his wife and children were displaced. He wrote much on the small fossils of the Downs, but his major discovery is that of the first dinosaur; aquatic saurian remains had been described previously, but great land saurians (dinosaurs) were unsuspected until Mantell's discoveries in 1822 at Tilgate Forest in the Cretaceous rocks of the English Weald. Mary Anne Mantell, his wife, first noticed the teeth, with some bones, and Mantell named the large herbivorous reptile Iguanodon, because of its relation to the much smaller modern lizard, iguana. In 1832 Mantell discovered the armoured dinosaurs. He was essentially an enthusiastic and expert amateur, aided in vertebrate palaeontology by his surgeon's knowledge of anatomy. A full-scale model Iguanodon was shown at the Crystal Palace in 1854. Public interest in the massive Mesozoic creatures (up to 35 m long) has remained high ever since.

Manton, Irene (1904–88) British botanist: set up the first laboratory for the ultrastructural study of plants.

Irene Manton was the younger sister of SIDNIE MANTON; both became Fellows of the Royal Society, the first case of sisters gaining this distinction. She was an undergraduate and a postgraduate at Cambridge, and after a short time in Stockholm she gained a lectureship at the

University of Manchester in the early 1930s. Her research at this period was on the spiralization of chromosomes in *Lilium* and *Tradescantia*, and on the cytotaxonomy of ferns.

In 1946 she was appointed to the chair of botany in Leeds and, while maintaining an interest in her previous internationally known work, her research changed direction. Realizing the possibilities of the electron microscope, she visited the Rockefeller Institute in New York and, on her return to England, borrowed time on electron microscopes in medical schools and published remarkable results. She set up a laboratory for the ultrastructural study of plants and maintained a place at the front of this research. She and her colleagues discovered ultrastructural classics such as the '9 plus 2' structure of cilia, thylakoid organization of choroplasts and scale formation in Golgi bodies. She had begun her pioneering work with the use of the electron microscope in the study of algal flagella, with observations on the spermatozoids of brown algae. She had a fruitful collaboration with Mary Winifred Parke (1908–89), resulting in 14 papers adding to knowledge of the smaller marine flagellates and revealing many novel features of fine structure.

In 1961 she was elected a Fellow of the Royal Society and was president of the Linnean Society of London from 1973–6.

Manton, Sidnie Milana (1902–79) British zoologist; a classical zoologist of the first rank.

Sidnie Manton was the elder of two daughters of a dental surgeon and his Scottish–French wife; both daughters became Fellows of the Royal Society, the first case in its history of two sisters achieving this distinction. At both her London schools there was an emphasis on biology and she was encouraged by her parents to collect, study and draw butterflies, moths and fungi. In 1921 she went to Girton College, Cambridge, where she took the Natural Science Tripos and in Part 2 came top of the final list, but was not awarded the University Prize as women were not then accepted as full members of the university. She was appointed as the first female university demonstrator in comparative anatomy in Cambridge (1927–35) and in the following year obtained her PhD. In 1934 she became the first woman to be awarded a Cambridge ScD, and from 1935–42 was director of studies at Girton College. She married J P Harding in 1937; he became keeper of zoology at the British Museum and she moved to King's College, London, becoming Reader in 1949.

Her research was on the structure, physiology and evolution of the arthropods, covering also the functional morphology and feeding mechanisms of the Crustacea, arthropod embryology, the evolution of arthropodian locomotor mechanisms, the mandibular mechanisms of arthropods and arthropod evolution. She was the author of *The Arthropoda: Habits, Functional Morphology and Evolution* (1977). Her influence on zoology was profound and long-lasting.

She was elected a Fellow of the Royal Society in 1948, being among the first women to be so honoured. The Linnean Society awarded her its Gold Medal in 1963 and the Zoological Society the Frink Medal in 1977.

Marcet, Jane, *née* Haldimand [mah(r)set] (1769–1858) British writer of introductory science books, widely read.

Jane Haldimand was born in London to Swiss parents, and married the Swiss physician Alexander Marcet (1770–1822), whose interest was in chemistry. They made their friends among the London scientific circle; BERZELIUS was a frequent visitor to their home. Jane Marcet inherited a large fortune from her father, which enabled her husband to give up his post at Guy's Hospital and devote his time to experimental chemistry; he was elected a Fellow of the Royal Society. She attended the public lectures at the Royal Institution given by HUMPHRY DAVY, but found she needed further explanation and gained this in conversations with 'a friend'. Her most successful book was *Conversations on Chemistry*, published anonymously in 1805 to assist others to understand scientific discussions; chemistry was advancing rapidly during the early 19th-c. It was presented in the form of a discussion between the teacher, Mrs Bryan, and two attentive pupils, the more serious Emily, and Caroline, who enjoyed spectacular experiments. The work was an immediate success, and went into 16 editions, each corrected and updated with the latest discoveries and their applications. The book was also widely sold in the USA. A copy of *Conversations on Chemistry* was left at the bookbinder where FARADAY was an apprentice, and he credited Jane Marcet 's book with the start of his interest in chemistry.

Marconi, Guglielmo, marquese (Marquis) (1874–1937) Italian physicist and engineer: pioneer of radio telegraphy.

Of mixed Italian and Irish parentage, Marconi was privately educated and later studied at the Technical Institute of Livorno.

At the age of 21, intrigued by HERTZ's 'electric waves', Marconi developed improved radio equipment capable of transmitting for a range of over a mile (the length of the family estate). To achieve this he used ground connections for both transmitter and receiver; larger, elevated antenna; and LODGE's coherer as detector, all improvements on Hertz's methods. In 1896 he succeeded in interesting the British government in his invention, and 3 years later transmitted MORSE code across the English Channel. This attracted considerable attention, particularly from the Admiralty, who began to install his equipment on Royal Navy ships. In 1901 he transmitted across the Atlantic from Cornwall to a kite-borne antenna in Newfoundland and became a household name overnight at the age of 27. In 1909 he shared the Nobel Prize for physics.

Although Marconi did not discover radio waves, and may be thought of as primarily an electrical engineer and businessman, he developed much of the technology necessary for its practical use, such as the directional aerial and the magnetic detector. During the First World War he developed short-wave radio equipment capable of directional transmission over long distances, and by 1927 had established a worldwide radio telegraph network on behalf of the British government. He spent most of his life improving and extending radio as a practical means of communication and building a company to commercialize it; from 1921 he used his steam yacht *Elettra* as his home, laboratory and mobile receiving station.

Marey, Étienne-Jules [maray] (1830–1904) French physiologist: ingenious inventor of physiological instruments.

Marey's work as a physiologist gave him scope for his passion for novel mechanical devices. The arterial system of the animal body is a complex arrangement of muscular and elastic tubes; in 1860 Marey devised a portable sphygmograph, which amplified the pulse movement and drew a trace of the pulse wave on smoked paper, and so gave some basic physiological information. He also began in 1876 to study irregularities of heart action, using his polygraph, which recorded the venous pulse and heartbeat simultaneously; these had not been much noticed previously and he found one type in which at varying intervals there are two heartbeats that follow abnormally rapidly (extrasystoles). In the 1890s MACKENZIE improved the polygraph, studied heart irregularities further and both related them to disease and showed that some are non-pathological.

In 1868 Marey showed that insect wings follow a basic figure-of-eight movement, by observing a fragment of gold leaf fixed to the wing tip of a fly held under a spotlight, and also by having the wingtip brush against the smoked surface of a rotating cylinder. He saw the value of scientific photography and in 1881 devised the first useful cine camera, which used a ribbon of sensitized paper with 'stopped motion' synchronized with a rotating shutter that cut off light as the paper moved forward. By 1890 he was using this to analyse human and animal movements by high-speed photography to slow down rapid movements; and he also invented its converse, time-lapse photography to speed up slow changes such as plant growth. From 1868 he was professor of natural history at the Collège de France, succeeding FLOURENS.

Markov, Andrei Andreivich (1856–1922) Russian mathematician: originator of Markov chains.

A graduate of St Petersburg, Markov taught there for 25 years until his political activism led him in 1917 to a self-imposed exile in the small town of Zaraisk. His early work was on number theory and on probability theory, which he worked on from the 1890s. This led him to discover the sequence of random variables now known as Markov chains. A Markov chain is a chance process which has the unusual feature that its future path can be predicted from its present state as accurately as if its entire earlier history was known. Markov appears to have thought that literary texts were the only firm examples of such chains, and applied the idea to an analysis of vowels and consonants in a text by Pushkin, but his method has since been applied in quantum theory, particle physics and genetics.

Mariotte, Edmé (?1620–84) French experimental physicist.

Mariotte lived in the same period as BOYLE, and in 1676 he announced his discovery of the same law for gases that Boyle had discovered in 1662. (In France Boyle's Law is named after Mariotte). Mariotte noted also the effect of a rise in temperature in expanding a gas; and he attempted to calculate the height of the atmosphere. He also studied elastic collisions, colour and the eye (he discovered the 'blind spot' in 1668). He was a founder member of the French Académie des Sciences.

Marsh, Othniel Charles (1831–99) US palaeontologist.

A student at Yale, followed by 3 years' study in Europe, Marsh became in 1882 the first vertebrate palaeontologist of the US Geological Survey, as well as teaching at Yale. He established his subject in the USA, and his four major expeditions to the western USA with his students (and William 'Buffalo Bill' Cody as scout) in the 1870s produced startling fossil discoveries. They included fossil mammals which showed the evolution of the horse, early primates, dinosaurs, winged reptiles and toothed birds, and Marsh traced the enlargement of the vertebrate brain from the Palaeozoic era.

Martin, Archer (John Porter) (1910–) British biochemist: co-discoverer of paper chromatography.

Martin graduated in Cambridge in 1932 and took his PhD there in biochemistry in 1938, working on vitamins (this included looking after 30 pigs, unaided, in work on pellagra). Then he joined the staff of the Wool Industries Research Association at Leeds. There, working with R L M Synge (1914–94) on the problem of separating complex mixtures of amino acids into their components, he developed the technique of partition chromatography. By 1944 the most familiar form had been devised by Martin; it combined with brilliant simplicity both the partition and adsorption methods. This is paper chromatography, in which a small amount of sample applied as a spot to a piece of paper is caused to move and to separate into its components by allowing a solvent front to move across the paper. The method is simple and has been of great value to chemists in analysing a variety of complex non-volatile mixtures; it is especially useful in biochemistry.

From 1948 Martin was on the staff of the Medical Research Council and from 1953 he worked particularly on gas-liquid chromatography. This separates volatile mixtures by means of a column of absorbent (such as silicone oil) on an inert support. Again, the method has proved a hugely successful analytical technique. Martin and Synge shared the 1952 Nobel Prize.

Matthews, Drummond Hoyle (1931–) British geologist: co-discoverer of magnetic anomalies across mid-ocean ridges.

Together with his student VINE, Matthews showed in 1963 that the oceanic crust on either side of mid-ocean ridges is remanently magnetized in alternately normal and reversed polarity, in bands running parallel to the ridge. This, they argued, was consistent with the sea-floor spreading hypothesis proposed by H H HESS the year before, and was seen as powerful support for Hess's hypothesis. Newly-formed crust would become magnetized in the prevailing direction of the Earth's magnetic field at the time of its emergence, but since this field undergoes periodic reversals, the oceanic crust would be expected to be magnetized alternately in opposite directions. Vine and Matthews showed that this was indeed the case, and also showed that the magnetic patterns were symmetrical about the mid-ocean ridges and that the same patterns were found for ridges in different oceans.

Matthews has since turned his attention to the continental crust, using deep seismic reflection techniques to study the lithosphere to depths of 80 km, an order of magnitude deeper than can be reached by drilling.

Matuyama, Motonori (1884–1958) Japanese geologist: discovered reversals in Earth's magnetic field.

The son of a Zen abbot, Matuyama taught at the Imperial University in Kyoto and studied at the University of Chicago before being appointed professor of theoretical geology at Kyoto. In 1929, while studying the remanent magnetization of basalts, Matuyama discovered that the direction of the Earth's magnetic field appeared to have changed its polarity since early Pleistocene times. Further investigations, notably by A Cox (1926–) and R Doell in the 1960s, have revealed that the Earth's field has abruptly reversed over 20 times during the past 5 million years, in an apparently random fashion. Reversals are now believed to be in some way caused by fluctuations in the convection currents within the Earth's liquid core that is the source of the field. The predominantly reversed period between 0.7 and 2.4 million years ago is known as the Matuyama reversed epoch.

Maunder, Edward Walter (1851–1928) British astronomer: discovered long-term variations in solar activity.

Maunder, who had no university training, was appointed photographic and spectrographic assistant at the Greenwich Observatory. In 1893, while checking historical records of sunspot activity, he realized that between 1645 and 1715 there had been little activity, and that in 32 years not a single sunspot had been seen. This event, which coincided with a pronounced period of cooling in the Earth's climate (the Little Ice Age), is now known as the Maunder minimum. He also discovered that the solar latitude at which sunspots appear varies in a systematic way during the solar cycle.

Maunder married Annie Scott Dill Russell (1868–1947), a 'lady computor' at Greenwich Observatory appointed in 1891. She had been a student at Girton College, Cambridge and had won high honours, Senior Optime in the Mathematical Tripos. They worked in close collaboration and published prolifically on the Sun, and on popular astronomy.

Maury, Antonia (Caetana de Paiva Pereira) [mawree] (1866–1952) US astronomer; made important contributions to sidereal astronomy.

Antonia Maury was the niece of Henry Draper (1837–82) who first photographed stellar spectra, and a grand-daughter of J W DRAPER, a pioneer in the application of photography to astronomy. Her great-grandfather had been British physician to Pedro I, emperor of Brazil; his wife was Portuguese.

Maury graduated from Vassar College in 1887 and, at the request of her father, was employed by Edward Pickering (1846–1919) at Harvard College Observatory classifying the bright northern stars according to their spectra. Pickering had just established Mizar as the first spectroscopic binary (ie a binary detected by the DOPPLER shift in spectral lines, due to the relative motion of the pair of stars); Maury determined its period as 104 days. In 1889 she discovered the second spectroscopic binary star, β Aurigae, with a period of about 4 days. In 1890 Williamina Fleming's (1857–1911) *Draper Catalogue of Stellar Spectra* was published. Maury had been assigned a more detailed study of the brighter stars, made possible by placing three or four prisms in front of the 11 in (28 cm) Draper refractor (nearly the last scientific use to which the famous telescope was put before it was sent to China). Maury discovered deficiencies in the Draper Catalogue system and devised her own, published in 1896, in which she classified the spectra by the width and distinctness of their lines: spectra with (a) normal lines, (b) hazy lines and (c) sharp lines, with subdivisions. HERTZSPRUNG used this to verify his discovery of dwarfs (*a* and *b*) and giants (*c*), and considered Maury's system a major advance. Pickering refused to acknowledge its value and it was largely ignored at Harvard in favour of the system used by ANNIE JUMP CANNON. The merits of Maury's system were eventually recognized in 1943 when she received the Annie Jump Cannon Prize from the American Astronomical Society. Her work is now considered an essential step in the development of theoretical astrophysics.

The conflict between Pickering and Maury impeded her career. Her aunt, Henry Draper's widow and benefactor of the Harvard Observatory, urged Pickering to 'bid her goodbye without regret'. Maury left the Observatory in 1891 for a teaching position, returning to resume her work, under Pickering's successor SHAPLEY, on spectroscopic binaries.

Maury, Matthew Fontaine [mawree] (1806–73) US oceanographer: conducted first systematic survey of ocean winds and currents.

Sometimes referred to as 'the father of physical oceanography', Maury was a US naval officer who was forced to retire due to a leg injury. In 1842 he became director of the US Naval Observatory and Hydrographic Office, and organized the first systematic collection of information on winds and currents from merchant ships, greatly improving knowledge about oceanic and atmospheric circulation. In 1847 he began to publish pilot charts, which enabled sailing voyages to be dramatically shortened, cutting as much as a month from the New York to California voyage. Maury also produced the first bathymetric profile across the Atlantic (from Yucatan to Cape Verde), with a view to the laying of a submarine cable. At the outbreak of the American Civil War in 1861 he became commander of the Confederate Navy, a move which later led to a period of exile in Mexico and England.

Maxwell, James Clerk (1831–79) British physicist: produced the unified theory of electromagnetism and the kinetic theory of gases.

Maxwell went to school at the Edinburgh Academy, a harsh institution where his country accent, home-designed clothes and sense of humour gained him the undeserved nickname of 'Dafty', and possibly caused his shyness; he was happier at Edinburgh University, which he entered at 16. The previous year he had developed the known method of drawing an ellipse using pins and thread, to generate a series of novel curves. The work was published by the Royal Society of Edinburgh in 1846.

In 1850 he entered Trinity College, Cambridge and graduated as Second Wrangler, winning the Smith's Prize (1854). Then 2 years later he secured a professorship at Marischal College, Aberdeen, where he married the principal's daughter, but an administrative reorganization made him redundant and in 1860 he

moved to King's College, London. After the death of his father (1865), who had cared for him since his mother died when he was 8 and of whom he was very fond, he resigned his post at King's and remained at the family home in Scotland as a gentleman-farmer doing research. However, he was persuaded to become the first Cavendish Professor of Experimental Physics in Cambridge, setting up the laboratory in 1874. He contracted cancer 5 years later and died soon afterwards, aged 48. In setting up the Cavendish Laboratory, he formed an institution unique in physics, to be headed by a succession of men of genius and producing graduates who dominated the subject for generations.

Maxwell was the most able theoretician of the 19th-c, perfectly complementing FARADAY, who was its most outstanding experimentalist. He began research on colour vision in 1849, showing how all colours could be derived from the primary colours red, green and blue. This led, in 1861, to his producing the first colour photograph using a three-colour process; the photograph was of a tartan. Other early work (1855–9) showed that Saturn's rings must consist of many small bodies in orbit rather than a solid or fluid ring, which he showed would be unstable. He casually referred to this as 'the flight of the brickbats'.

His monumental research on electromagnetism had small beginnings. Faraday viewed electric and magnetic effects as stemming from fields of lines of force about conductors or magnets, and Maxwell showed that the flow of an incompressible fluid would behave in the same way as the fields (1856). Then, in 1861–2, he developed a model of electromagnetic phenomena using the field concept and analogous vortices in the fluid which represented magnetic intensity, with cells representing electric current. Having explained all known electromagnetic phenomena, Maxwell introduced elasticity into the model and showed that transverse waves would be propagated in terms of known fundamental electromagnetic constants. He calculated that the waves would move at a speed very close to the measured speed of light. He unhesitatingly inferred that light consists of transverse electromagnetic waves in a hypothetical medium (the 'ether').

To study electromagnetic waves further, the fluid analogy was taken over into a purely mathematical description of electromagnetic fields. In 1864 he developed the fundamental equations of electromagnetism (Maxwell's equations) and could then show how electromagnetic waves possess two coupled disturbances, in the electric and magnetic fields, oscillating at right angles to one another and to the direction in which the light is moving. The original mechanical model was now rightly cast off.

Furthermore, Maxwell stated that light represented only a small range of the spectrum of electromagnetic waves available. HERTZ confirmed this in 1888 by discovering another part of the spectrum, radio waves, but by this time Maxwell was dead. Maxwell also suggested the MICHELSON–MORLEY experiment (1881, 1887) to search for an absolute electromagnetic medium (the ether). Its proven absence prompted EINSTEIN's research on relativity (1905) and the era of modern physics.

Maxwell also contributed to the kinetic theory of gases, building on the existing picture of a gas as consisting of molecules in constant motion, colliding with their container and with each other; this picture was due to BERNOULLI and to two little-known men, J Herapath (1790–1868) and J J Waterston (1811–83). As gases diffuse into each other rather slowly, CLAUSIUS deduced that although they travel fast, the molecules must have a very small 'mean free path' between collisions.

From 1860 Maxwell (and independently BOLTZMANN) used statistical methods to allow for the wide variation in the velocities of the various molecules in the gas, deriving the Maxwell–Boltzmann distribution of velocities. Maxwell showed how this depends on temperature, and that heat is stored in a gas in the motion of the gas molecules. The theory was then used to explain the viscosity, diffusion and thermal conductivity of gases.

Maxwell and his wife found experimentally (1865) that gas viscosity is independent of pressure and that it is roughly proportional to the temperature, and rises with it (the reverse of the behaviour of liquids). This did not agree with Maxwell's theory, and he could only gain agreement by assuming that molecules do not collide elastically but repel one another with a force proportional to their separation raised to the fifth power. This and further work by Boltzmann from 1868 allowed the full development of the kinetic theory of gases.

Maxwell was a shy and mildly eccentric person, who was deeply religious, with a strong sense of humour and no trace of pomposity. Like Einstein, and in contrast to NEWTON or Faraday, Maxwell made his enormous advances in physics without excessive mental strain. He excelled in his sure intuition in physics, in applying visual models or mathematical methods without being tied to them, and above all in freeing himself from preconceptions and in exercising his creative imagination. Maxwell's summary of electromagnetism in his field equations is an achievement equalled only by that of Newton and Einstein in mechanics.

Mayow, John (1641–79) English physician: early experimenter on combustion.

Mayow studied law and medicine at Oxford, practised medicine in Bath, and experimented in Oxford where he perhaps worked with HOOKE and BOYLE. In a book published in 1674 he gives a theory of combustion similar to Hooke's but supported by new experiments. He burned candles in air in a closed space over water, and found that the reduced volume of gas which remained would not support combustion; he got similar results using a mouse in place of a burning candle to consume part of the air. He concluded that air consists of a least two parts; one ('the nitro-aerial spirit') supports combustion or respiration, which are in this way related processes; the other part of air is inert. Ignited gunpowder continued to burn under water, so its 'nitre' contained the nitro-aerial spirit; it is surprising he did not try heating nitre (KNO_3) alone and so discover oxygen, and this may be because he visualized his 'spirit' as a philosophical principle rather than as a gaseous substance. In experiments with an air-pump (probably Boyle's) he found that venous blood under the pump effervesced only gently, but arterial blood bubbled freely if fresh. He had sensible, if primitive, views on chemical affinity. In many ways Mayow was ingenious both as an experimenter and in ideas, but it can be said also that few of the ideas were new and his theory of combustion was hopelessly confused in comparison with LAVOISIER's clear-mindedness a century later.

Mead, Margaret (1901–78) US social anthropologist.

The eldest daughter in an academic family, Margaret Mead was educated mainly at home until she entered Barnard College, where RUTH BENEDICT was senior to her and became a close friend. In 1923 she married the first of her three anthropologist husbands, with all of whom she collaborated, as the senior partner, in ethnographic studies of Pacific island cultures. She wrote in all over 40 books, some of which, like *Coming of Age in Samoa* (1928), became best-sellers and helped her become the world's best-known anthropologist. Her work examined adolescence, child rearing, gender roles and the rift between generations, and illuminated these and related cultural patterns through comparisons between primitive and developed societies. As a result, social anthropology became accessible to non-specialists and aided them in understanding their own society.

In the Second World War she studied food habits and also worked to reduce British–US cultural conflicts and misunderstanding, and afterwards she returned to senior posts in the American Museum of Natural History and in Columbia University, New York. Her general beliefs were optimistic; in thinking that cultural factors were more important than biological ones in shaping human behaviour, she also believed that this behaviour was essentially alterable in favourable circumstances.

Mechnikov, Ilya Ilich (Russ), Elie Metchnikof (Fr) [mechnikof] (1845–1916) Russian–French biologist: discoverer of phagocytosis.

Educated in Russia and Germany, Mechnikov taught zoology in Odessa from 1872. Some 10 years later he inherited modest wealth and went to Messina in Italy on a research visit. There he studied the conveniently transparent larvae of starfish and noticed that some of their cells could engulf and digest foreign particles; he called these amoeba-like cells 'phagocytes' (cell-eaters). In 1888 he moved to Paris to the Pasteur Institute and continued his search for phagocytic action. He found that in human blood a large proportion of the white cells (leucocytes) are phagocytic and will attack invading bacteria. Infection leads to an increase in the number of white cells, and phagocytosis at the site of a local infection leads to inflammation and a hot, red, swollen and painful region with dead phagocytes forming pus. From 1898 Mechnikov studied human ageing; he believed that phagocytes eventually began to digest the cells of the host (an early idea of auto-immune disease) aided by the effects of intestinal bacteria. If these effects could be resisted, he argued, the normal human life-span would be 120–130 years. For his work on phagocytosis he shared a Nobel Prize in 1908.

Medawar, Sir Peter Brian [medawah(r)] (1915–87) British immunologist: pioneer in study of immunological tolerance.

Born in Brazil, the son of a Lebanese–British business man, Medawar was educated in England at Marlborough (which he much disliked) and then at Oxford, studying zoology under J Z Young (1907–   ) from 1932. In the 1940s he began to study skin grafts in connection with wartime burns victims, and when he moved in 1947 to Birmingham he continued this interest. He was a keen and skilful experimenter; he was aware that grafts are successful between certain types of twins and he knew of BURNET's work, suggesting that an animal's ability to produce antibodies against foreign cells (and hence rejection of a transplanted tissue) is not inherited but is developed in fetal life, and so he believed that 'immunological tolerance' should be achievable. Medawar's ingenious work with mouse skin grafts supported Burnet's idea. From this stemmed the successful human organ transplants achieved by surgeons from the 1960s, using tissue-typing to secure a partial matching between the donor organ and the patient, and also using immunosuppressive drugs to inhibit the normal immune response that would cause rejection. Medawar moved to London in 1951, and shared a Nobel Prize with Burnet in 1960. He did not allow the strokes he had in his last 18 years to much limit his work, and his seven popular books were written in this period.

Meitner, Lise [miytner] (1878–1968) Austrian-Swedish physicist and radiochemist: co-discoverer of nuclear fission.

Meitner studied physics in Vienna under BOLTZMANN and in Berlin with PLANCK. Soon she was attracted into radiochemistry, and worked with HAHN in Berlin in this field for 30 years.

Despite her talents she was a victim of more than one prejudice, being both female and a Jewish Protestant. In academic Vienna she was regarded as a freak; she was only the second woman to obtain a doctorate in science there. In Berlin, E FISCHER could not allow women in the laboratory, although he welcomed her 2 years later when the State regulations changed. In 1912 she began working with Hahn at Berlin-Dahlem, but the war soon interrupted their work. His leaves from the German Army sometimes coincided with hers from nursing duty in the Austrian Army; however, some radiochemistry involves long gaps between measurements; and so they were able to continue some of their work and announce a new radioelement, protactinium, at the war's end. In 1918 she became head of physics in the Institute, and continued her work on radioactivity.

In the 1930s she and Hahn worked on uranium bombarded with neutrons, initially not realizing that fission was occurring. By the late 1930s her Jewishness was a threat to her safety, and friends (including Hahn and DEBYE) helped her escape through Holland to Denmark and then to Sweden. In Stockholm a cyclotron was being built; although aged 60 she learned Swedish and built up her research group again. Hahn sent her the results of his work on neutron bombardment of uranium, begun with her and F Strassmann (1902–80), but which he had erroneously interpreted. She discussed the work with her nephew FRISCH who was visiting her. They shaped their joint and novel ideas on 'nuclear fission' into a paper, which was actually composed over a telephone line since he had by then returned to Copenhagen.

She declined to work on the atomic bomb, hoping that the project would prove impossible, and did no more work on fission. In 1960 she retired to live in England after 22 years in Sweden.

Mendel, Gregor (Johann) (1822–84) Austrian botanist: discovered basic statistical laws of heredity.

In the long term Mendel was certainly successful; he laid a foundation for the science of genetics. In another sense he was a failure; he did not succeed in examinations and his research was largely ignored until 16 years after his death.

A peasant farmer's son, he entered the Augustinian monastery in Brno (then in Moravia, now the Czech Republic) when he was 21 and was ordained 4 years later. He became a junior teacher and during the 1850s twice tried to pass the teachers' qualifying examination. From 1851 he was sent by his order to study science for 2 years in Vienna, and afterwards he began his plant-breeding experiments in the mona-

THE HISTORY OF GENETICS

Although appreciation of inheritance as a determinant of the characteristics of living species probably began with the introduction of agriculture several millennia ago, the scientific study of genetics is largely a 20th-c phenomenon. It has developed along a number of independent but occasionally overlapping avenues. These include breeding studies on whole organisms (and observations of the effects of breeding in humans), microscopic examination of cells and chromosomes, chemical investigation of the material of genes and the biochemical pursuit of gene action.

The first real scientific studies of hereditary factors controlling the phenotype of a species began with the work of the Austrian monk MENDEL, published in 1866. Choosing the garden pea for his breeding experiments, he showed that (1) his plants inherited two factors for each trait he studied, with one derived from each parent; (2) inherited factors did not blend or mix in the offspring, but were segregated; (3) some factors were dominant over others; and (4) factors controlling different traits in the plants assorted independently. He also suggested that it was the germ cells that were responsible for the transmission of the inherited factors.

Mendel's experiments lay largely unnoticed until 1900. During the intervening years, the cell theory was developed, and it was proposed that the physical basis of inheritance lay in the cell nucleus, and in all probability in microscopically visible particles called chromosomes. WEISMANN formulated the idea of the continuity of the germ plasm, and suggested that there must be a reduction in the number of chromosomes in the germ cells to one half of that in the body cells in sexually reproducing organisms. However, it was not until 1903 that Walter Sutton (1877–1916) and BOVERI drew attention to the parallels between Mendel's segregating factors and the behaviour of chromosomes in meiosis. Inherited factors were named genes by W L Johanssen (1857–1927) in 1909. In the following years, MORGAN showed that genes lying close together on chromosomes could be inherited together in linkage groups, which occasionally broke apart in a process called crossing-over or recombination. In 1931, BARBARA MCCLINTOCK correlated the microscopically visible rearrangements of chromosome segments with the redistribution of specific genetic traits in maize. By the mid-1930s it was accepted that genes had physical reality and were determinants of biological specificity.

The chemical nature of genes emerged more slowly. Although Friedrich Miescher (1844–95) had described nucleic acid in 1869, cell nuclei also contain abundant protein. The phenomenon of bacterial transformation was reported in 1928, but the nature of the transforming agent was only revealed as deoxyribonucleic acid (DNA) by the work of AVERY and colleagues in 1944. In 1952 HERSHEY and Martha Chase confirmed these findings and firmly established DNA as the chemical material of the gene.

A curious feature of the history of genetics is that an early clue to the function of the genes was provided by the work of GARROD within a year of the rediscovery of Mendelian principles in 1900. He studied a rare human disease called alkaptonuria, which behaved as though inherited as a recessive trait. Affected children excrete large amounts of homogentisic acid in their urine, and Garrod speculated that this resulted from the absence of a specific enzyme. But although the relationship between gene and enzyme was implicit in his work, it was an idea before its time. Some 40 years later, BEADLE and Edward Tatum (1909–75) reformulated this concept in a precise form, 'one gene, one enzyme' – subsequently modified to 'one gene, one polypeptide chain'. In 1949, SANGER devised a method for determining the amino-acid sequence of protein molecules. It is now accepted that the primary function of genes is to carry the information that specifies the chemical nature and quantity of unique proteins.

By the early 1950s the conceptual framework was in place to launch a more fundamental attack on the structure and function of the hereditary material. The era of molecular genetics began in 1953, when CRICK and WATSON proposed a two-stranded model for DNA, with complementary chains wound around each other in a double helix. An early clue to this structure came from the work of CHARGAFF, who had separated and measured the four nucleic acid bases – thymine (T), cytosine (C), adenine (A), and guanine (G) – in DNA from several species, and found that the amount of A equalled that of T, while the amount of G equalled that of C. On the basis of X-ray crystallographic studies by Maurice Wilkins (1916–   ) and ROSALIND FRANKLIN, Crick and Watson were able to build an accurate model of DNA, with the As in one chain hydrogen-bonded to the Ts in the other, and Cs in one chain bonded to the Gs in the other. Inspection of this self-complementary model suggested a mechanism for the exact replication of these genes in each cell generation, with the helix unwinding and old strands forming the templates for newly synthesized daughter strands. This semiconservative nature of DNA replication was demonstrated experimentally by MESELSON and Franklin Stahl (1929–   ) in 1958.

Because the Watson–Crick model placed no restriction on the sequence of bases in a single strand of DNA molecule, it provided a satisfactory explanation for its role as carrier of the hereditary information. Working out how base sequences were turned into the amino-acid sequences of specific proteins occupied most of the next decade. It became apparent that DNA was transcribed first into messenger ribonucleic acid (RNA) and only then translated into protein. RNA and DNA sequences were complementary. The genetic code was cracked by the elegant experiments of Marshall Nirenberg (1927–) in the USA, and Severo Ochoa (1905–93) and colleagues in Spain. Nirenberg showed conclusively that it was a triplet code, with a sequence of three bases determining each individual amino acid. By 1969, 100 years after the first description of DNA by Miescher, the details of the genetic code were complete.

It was now possible to approach the study of genetics through the gene itself. However, DNA is a large molecule of relative chemical homogeneity and could not be cleaved in a reproducible manner. This problem was solved with the discovery of restriction endonucleases by Werner Arber (1929–), HAMILTON SMITH, and Daniel Nathans (1928–). These enzymes cut double-stranded DNA at recognition sites determined by particular base sequences, and thus allowed the genes to be digested into consistent segments, a process known as physical mapping. Furthermore, the DNA segments were now small enough to be cloned into plasmid vectors and then amplified to a high copy number by growing in bacterial hosts, thus ushering in the era of recombinant DNA technology. Another contribution to the new science was made by Ed Southern (1933–), who showed that DNA fragments could be bound to nitrocellulose filters and identified by hybridization to complementary sequences. In 1977, Sanger, WALTER GILBERT and their colleagues independently described two different and equally reliable methods for determining the base sequences of DNA molecules.

The ability to clone and manipulate selected genes has had a profound influence not only on the science of genetics, but also on the pharmaceutical, agricultural and food industries. It has also begun to provide a route to the understanding and prevention of human diseases. In the past decade, well over 1000 human genes have been identified, many of whose mutant forms are responsible for the types of Mendelian disorder first described by Garrod, and later catalogued and indexed by V A McKusick (1921–). There are real prospects that some of these will be treated successfully by gene therapy – the introduction of normal genes into affected tissues and organs. The discovery of oncogenes and tumour-suppressor genes, both mutant forms of normal cellular genes, is in the process of revolutionizing the understanding of cancer. Even in the common multifactorial diseases, with mixed environmental and genetic origins, the application of molecular techniques has begun to lead to the identification of major susceptibility genes.

Although methods for determining the order of the bases in DNA fragments were reported only in 1977, so rapid was the progress of molecular genetics that by 1985 it was possible to contemplate sequencing the entire human genome (3×10^9 base pairs). This idea, termed the Human Genome Project, was seen as controversial in both ethical and scientific quarters. Some feared that the megascience involved will siphon funds away from most other branches of biology. Others worry that inevitably there will be misuse and abuse of the detailed knowledge of the genetic structure of humans. But as arguments about the virtues of the project continue to rage, DNA sequencing of the human genome proceeds apace. A recent estimate points to 1999 as the completion date. If this is achieved (and most time projections in molecular genetics have been overestimates), genetics will have moved from Mendelian rediscovery to a complete molecular blueprint in under a century.

Professor David Brock, Human Genetics Unit,
University of Edinburgh

---

stery garden. He was elected abbot in 1868, which left him little time to continue this work; in any event his personality and reputation were unsuited to publicize his scientific ideas and most biological interest was directed elsewhere. Mendel's work had to await rediscovery by DE VRIES and others to become appreciated, in 1900. Even then it needed the vigorous advocacy of BATESON, and many plant and animal breeders, to be accepted.

Mendel's famous work on the inheritance of characters was done on the edible pea (*Pisum* spp.), in which he studied seven characters, such as stem height, seed shape and flower colour. The plants were self-pollinated, individually wrapped (to prevent pollination by insects) and the seeds collected and their offspring studied. The characters were shown not to blend on crossing, but to retain their identity. Mendel had a gardener's skill and his experiments were excellently organized. He found that the characters were inherited in a ratio always close to 3:1; he theorized that hereditary elements or factors (now called genes) exist that determine the characters, and that these segregate from each other in the formation of the germ cells (gametes).

His results of 1856 are summarized in two laws, expressed in modern terms as follows. The characters of a diploid organism are controlled by alleles occurring in pairs. Of a pair of such alleles, only one can be carried in a single gamete. This is Mendel's First Law, or the law of segregation. We now know, although Mendel did not, that this law follows from the process of meiosis and the physical existence of alleles as genes. He also found that each of the two alleles (ie the two forms) of one gene can combine randomly with either of the alleles of another gene (Mendel's Second Law, or the law of independent assortment). Genetics has both confirmed and refined Mendel's laws; MORGAN's work showed how linkage and crossing-over modify the second law.

Mendel was disappointed that his work aroused little interest and he sent his paper to NAEGELI, the leading German botanist, who advised him to experiment with more plants; Mendel had already studied 21 000. Curiously, when FISHER in 1936 studied his results, he found they are statistically too ideal; possibly because a few intermediate plants occur, which Mendel classified to accord with his expectations. Or, perhaps, an assistant tried too hard to be helpful.

Mendelayev, Dmitri Ivanovich [mendelayef] (1834–1907) Russian chemist: devised periodic table of chemical elements.

Mendelayev grew up in Siberia, the last-born of a family of 14 children. His father, a teacher, became blind at this time, but his mother was a forceful woman and she reopened and ran a nearby glass factory to give an income. When Dmitri was 14 his father died and the factory was destroyed by fire; but the boy had done well at school and his mother decided that he deserved more education. They made the long trip to St Petersburg and he began to study chemistry, and was so successful (despite much illness) that he was given an award to study with BUNSEN in Germany. Back in St Petersburg from 1861, he began his career as a teacher and researcher in the university.

His career was not smooth; he was irascible and outspoken, supported the students' political ideas and quarrelled with two successive ministers of education. He and his wife divorced and he remarried, without waiting the 7 years then required by Russian law. Officially a bigamist, he was not directly penalized for this, but the priest was; Mendelayev, forced out of the university in 1890, became director of the Board for Weights and Measures in 1893. Mendelayev was more honoured outside Russia than within it, and he was never admitted to the Imperial Academy of Sciences, despite his work for the Russian chemical industry and his great scheme which brought order and prediction to inorganic chemistry; this was his periodic table (or periodic law or classification).

Mendelayev saw the need for a new textbook of chemistry in the 1860s, and in shaping his ideas for this book he prepared a series of cards, each listing the main properties of one chemical element; he liked playing patience as a relax-

ation. In arranging these, he was struck by the fact that if the 60 cards were placed in rows of suitably varying length, with most of the elements in order of increasing relative atomic mass, then elements with similar chemical features were found to lie in the vertical groups (the periodic law). Mendelayev did not know of NEWLANDS's primitive work on similar lines, and his went much further. In his table of 1868–9, he boldly transposed some pairs of elements on the basis that their claimed atomic masses must be in error if they were to fit the scheme; likewise he left spaces for three yet undiscovered elements. He predicted the properties of the latter from those of their known neighbours. By 1886 the predicted elements were discovered by other chemists, and their real properties were found to be in good accord with prediction. Later still, the noble gases and the transuranium elements were fitted into the table. The whole scheme brought order into chemistry by allowing a great range of known facts to be arranged and classified. It stands like NEWTON's work in physics or DARWIN's in biology as one of the great intellectual advances in science. It was devised on an entirely empirical basis and it was half a century later that MOSELEY's work, and that of BOHR, provided an explanation for it in terms of atomic structure. Mendelayev produced his table when he was 34. It made him famous, and he worked on it for a few years, but then moved to a variety of other matters. It has framed and shaped ideas in inorganic chemistry ever since. In 1955 a new element, atomic number 101, was named mendelevium (Md) in his honour.

Mercator, Gerardus (*Lat*), Gerhard Kremer (*Dutch*) [merkayter] (1512–94) Dutch cartographer and geographer: invented Mercator map projection.

Educated at the University of Louvain under G Frisius (1508–55), Mercator set up a centre for the study of geography at Louvain in 1534, issuing a number of maps and also making surveying instruments and globes. Persecuted as a Protestant, he moved to Duisberg in 1552, from where in 1569 he issued a map of the world in the new projection which now bears his name. The Mercator projection was a great advance because it allowed navigators to plot their course as a straight line of constant heading, corresponding to a great circle on the globe. To achieve this meridians of longitude were made parallel, instead of converging at the Poles. This rectangular orthomorphic projection is clearly convenient for navigators, both in allowing compass bearings to be drawn as straight lines and in making dead reckoning easy. Surprisingly, since Mercator's projection distorts reality so greatly, it has dominated maps made for all purposes, ever since his maps appeared in the late 16th-c.

Mercator or his sons, also mapmakers, are

also credited with coining the term 'atlas' to describe a set of maps.

Meselson, Matthew (Stanley) (1930–) US molecular biologist: showed how the DNA double helix replicates.

Born in Colorado, Meselson first studied liberal arts at Chicago and then physical chemistry at Caltech, where he remained to teach physical chemistry. In 1961 he moved to Harvard.

When CRICK and WATSON in 1953 proposed that genes were constructed of a double helix of DNA, they also suggested that, when this duplicated, each new double helix in the daughter cells would contain just one DNA strand from the original helix ('semiconservative replication'). The alternative would be for one daughter cell to contain both the old strands and the other daughter to receive both new strands ('conservative replication'). In 1957 Meselson and F W Stahl (1929–) showed by ingenious experiments using the bacterium *Escherichia coli* labelled with nitrogen-15 that replication is indeed semiconservative; an important result, verifying Crick and Watson's ideas and using intact dividing cells without the use of injurious agents.

Meselson also worked on ribosomes, the cell organelles that are the site of protein synthesis. The ribosomes are 'instructed' on protein construction by m-RNA and if given abnormal instructions will produce abnormal protein. When a virus invades a bacterial cell, the viral DNA releases its m-RNA, which acts on the bacterial ribosomes, causing them to make viral protein rather than bacterial protein.

Meyer, Viktor [miyer] (1848–97) German chemist: wide-ranging chemical experimenter.

Meyer's father, a dye merchant, wished his sons to become chemists; Viktor wanted to be an actor. The family persuaded him to attend some lectures in Heidelberg, and BUNSEN's lectures on chemistry duly converted him. He became an enthusiastic and successful chemistry student, and later a strikingly effective lecturer, perhaps because of his acting skills. After working as assistant to Bunsen and to BAEYER he became professor at Zürich at the early age of 24. Later he succeeded Bunsen at Heidelberg, but in the 1880s he became ill and depressed and later killed himself with cyanide, a fate too common among famous chemists.

His early work on benzene compounds established the orientation of many substituted acids, but his main fame in the 1870s was due to his work on nitroparaffins; he was also the first to prepare oximes, by the reaction of hydroxylamine H_2NOH with an aldehyde or ketone. His name is much linked with a method for finding relative molecular mass by measuring vapour density; he used this first for organic compounds and then (at temperatures up to 3000°C) for inorganic compounds and elements. In 1883 he discovered (through a lecture demonstration which failed) the novel sulphur ring-compound thiophene, parent of a series of sulphur compounds. He also did valuable work in stereochemistry (he invented this word, usefully shorter than VAN 'T HOFF's 'chemistry in space') and he discovered 'steric hindrance', which he first observed in ortho-substituted benzoic acids. He made a novel range of aromatic iodine compounds and he studied what later came to be seen as electronic effects on acidity in organic molecules.

Meyerhof, Otto Fritz (1884–1951) German-US biochemist: elucidated mechanism of lactic acid formation in muscle tissue.

Meyerhof studied medicine at Heidelberg and began to specialize in psychiatry. However, he became attracted to biochemistry and in 1909 worked with WARBURG and studied his methods; afterwards in Kiel, Berlin and Heidelberg, he used similar techniques to examine the chemical changes linked with muscular action. F G HOPKINS had shown that lactic acid is formed in a working muscle, and Meyerhof showed how this is formed and how it is removed when the muscle rests. He became increasingly unhappy in Nazi Germany, moved to France in 1938 and when France fell in 1940 escaped to the USA and worked in Philadelphia until his death. He shared a Nobel Prize in 1922 with A V Hill (1886–1977), who had worked in Cambridge and Manchester on the heat evolved in muscle action. Hill was able to measure this with delicate thermocouples and to deduce from his results that oxygen is taken up only after (and not during) the action of muscle.

Michaelis, Leonor [meekaylis] (1875–1949) German-US biochemist: made early deductions on enzyme action.

Educated in Germany, Michaelis worked in Berlin until 1922 when he went to Japan; in 1926 he moved to the USA, first to Johns Hopkins and then to the Rockefeller Institute. He showed in 1913 that an expression (the Michaelis–Menten equation) will describe the change in the rate of an enzyme-catalysed reaction when the concentration of substrate is changed; and from this and other studies he deduced that reaction between an enzyme and its substrate was preceded by their combination to form a complex. It was 50 years before this was confirmed by direct experiments; rate studies on enzymes and on the transport of substances through cell membranes are now normally based on these ideas.

Michell, John (*c.*1724–93) British astronomer: discovered double stars, estimated stellar distances and predicted existence of black holes.

A Cambridge graduate in divinity, Michell was professor of geology there for 2 years before becoming a village rector near Leeds, a post he held for life.

His scientific work was mainly in astronomy, where he made several significant contributions. He proposed the existence of double stars as a way of explaining the large number of apparent close pairs that had been observed, which he argued could not be due merely to the chance of the two stars being near the same line of sight. In 1803 F W HERSCHEL found observational proof for his proposal.

In a particularly far-sighted suggestion, Michell proposed in 1783 that if stars were sufficiently massive and compact then light would not be able to escape from their surface; such objects are today known as black holes. He thought that there might be a large number of black holes, and further suggested that they might be detectable through their gravitational effect on nearby objects.

Michell was also the first to make a realistic estimate of a stellar distance, using a neat argument based on apparent brightness. He deduced a distance of 460 000 AU for the star Vega, about a quarter of today's value.

Michelson, Albert Abraham [mikelsn] (1852–1931) US physicist: devised optical measurement methods of great accuracy; and showed that the hypothetical ether probably did not exist.

Born in Strelno (now in Poland), Michelson emigrated with his parents to the USA as a child of 4. At 17 he entered the Annapolis Naval Academy (after an entry appeal in which he saw President Grant) and, following graduation and a tour of duty at sea, was appointed as instructor in physics and chemistry there.

His interest in science was apparently much increased from this time and, when he needed to demonstrate to the midshipmen how the speed of light can be measured, he applied himself to improving the accuracy of the measurement. It is certainly of fundamental importance for physics (and for navigation) and Michelson was to measure it with increasing accuracy throughout his life. The optical devices he used for this, based on his interferometer, were useful for a variety of purposes in physics.

In the early 1880s he visited Europe for 2 years on study leave, and his first interferometer was built in HELMHOLTZ's laboratory and paid for by A G BELL. It allowed the speed of light to be compared in two pencils of light split from a single beam. One result of this work concerned the so-called ether. Since waves such as sound waves or water waves require a substance or 'medium' for their transmission, it had been widely presumed that light and other electromagnetic waves must likewise require a medium, and a hypothetical ether, invisible, universal and weightless, had been invented for the purpose. However, Michelson's refined results would show the effect on light of the Earth's motion through the ether; but there was no effect and

physicists were forced to doubt if the ether really existed.

In 1881 Michelson left the Navy, and next year became professor of physics in Cleveland, OH. There he continued and improved his optical measurements, and with MORLEY (the professor of chemistry) confirmed the null result on the ether in 1887. This classical Michelson–Morley experiment, a major result in physics, won him a national prize in 1888, 'not only for what he has established, but also for what he has unsettled'. In a sense the problem was not 'settled' until 1905 when EINSTEIN's theory of relativity dispensed with the need for ether.

Michelson went on to apply his ingenuity and skill in optics to measure the metre in terms of the wavelength of light and to solve some astronomical problems (he was the first to measure the angular diameter of a star; it was Betelgeuse, and the margin of error was equivalent to a pinhead's width at a distance of 1000 miles) and to refine his value for the velocity of light (close to $3 \times 10^8 \text{ m s}^{-1}$ in air). He also discovered new features of spectra; and he used his interferometer to measure tidal movement due to the Moon's effect not on the seas, but on the solid Earth. From 1890 until his death he worked at Chicago; in 1907 he became the first American to be awarded a Nobel Prize.

Midgley, Thomas (1889–1944) US engineer and inventor: introduced tetra-ethyl lead (TEL) anti-knock and Freon refrigerant.

Midgley was the son of one inventor and the nephew of another. He studied engineering at Cornell, finishing with a PhD in 1911. Working with C Kettering for Delco in the First World War, he led a team working on the problem of 'knocking' in petrol engines. ('Knocking' or 'pinking' is the metallic noise caused by pre-ignition.) After finding some anti-knock additives as a result of random trials, Midgley realized that their effectiveness could be related to the position of the heaviest atom in the compound, within the periodic table. On this basis, working at General Motors in 1921, he tried tetra-ethyl lead, $Pb(C_2H_5)_4$, and found it to be very effective, used with some 1,2-dibromo-ethane to reduce lead oxide deposits in the engine. The mixture has been extensively used, although there has been rising concern since 1980 that the lead in vehicle exhausts is a health hazard. Midgley also devised the octane number method of rating petrol quality.

In 1930 Midgley introduced Freon-12 (CF_2Cl_2) as a non-toxic non-flammable agent for domestic refrigerators; again, he used the periodic table as a guide to select a suitable compound with the required properties. In the 1980s there has been rising concern that chlorofluorocarbons (CFCs) such as Freon cause destruction of the ozone layer of the upper atmosphere, with potentially damaging climatic and other effects as a result of the increased passage of ultraviolet radiation following ozone loss.

For one man to invent two major environmental hazards is curious; and so was his death. A polio victim, he had devised a harness to help him rise in the morning. Becoming in some way entangled in it, he strangled himself.

Milankovich, Milutin (1879–1958) Serbian climatologist: developed astronomical theory of climatic change.

Milankovich was educated in Vienna, but in 1904 moved to the University of Belgrade, where he spent the rest of his academic career. He is remembered for his work on the cause for long-term changes in the Earth's climate. Following earlier proposals by J HERSCHEL and J Croll (1821–90), he recognized that the major influence on the Earth's climate is the amount of heat received from the Sun. Three astronomical factors can affect this: the eccentricity of the Earth's orbit (which varies on a time scale of about 100 000 years), the tilt of the Earth's axis (time scale of 40 000 years), and a precessional change which determines whether the northern or southern hemisphere receives most radiation (time scale of 20 000 years). Milankovich spent 30 years computing the amount of radiation received at different latitudes for the past 650 000 years, and was able to demonstrate that changes in insolation corresponded with the then known ice ages. Since Milankovich's work in the 1920s, the number of known ice ages has increased, and their history is seen as complex. Although the Earth's orbital variations are important, it is now clear that as well as the cooler summers studied by Milankovich and ascribed by him to these orbital variations, other factors probably contributed to ice age formation. One such is the level of CO_2 in the atmosphere; TYNDALL and ARRHENIUS both studied the way in which loss of CO_2 and therefore of its 'greenhouse effect' could cool the Earth, and interest in this rose in the 1970s after study of air bubbles trapped in polar ice cores gave information on CO_2 levels over a long period. Certainly over the last million years the oceans have varied in temperature and volume, with the global ice mass changing by some 10^{19} kg and resulting in changes in sea level of up to 100 m. Over the same period, CO_2 has varied by up to 30%. A complete theory of factors leading to ice ages must be complex, and good models are not yet available.

Miller, Jacques (Francis Albert Pierre) (1931–) French–Australian immunologist: discovered function of the thymus gland.

Educated in Sydney and London, Miller worked from 1966 at the Hall Institute, Melbourne. Until his work in 1961, the function of the thymus gland was not known. The gland is in the chest of mammals, close to the heart, and becomes relatively smaller from infancy to adulthood. To discover the function of such an organ, one general method is to remove it from a mature experimental animal and to examine the resulting changes; but in the case of the thymus, no significant change could be observed. Likewise its removal in human adults in cases where it had become cancerous produced no obvious physiological change. Miller pointed to an answer by removing the thymus from 1-day-old mice (which weigh only about a gram). Then, thymectomy produced much change; normal growth failed and death followed in 8–12 weeks. Suggestively, the lymph nodes shrink, the lymphocyte blood count falls and immune responses fail, so that skin grafts from unrelated mice (or even rats) are not rejected. From this basis, later work showed that T-lymphocytes are formed in the fetal thymus and fulfil a critical role in the complex cell-mediated immune response.

Miller, Stanley Lloyd (1930–) US chemist: experimented to simulate production of prebiotic biochemicals from simple gas mixtures.

A graduate of California and Chicago, Miller worked at the University of California at San Diego from 1960. His most familiar work was done when he was a research student with UREY in Chicago in 1953. Interested in the possible origin of life on Earth, he devised an experiment using an early planetary reducing atmosphere as proposed by Urey in 1952; it contained water vapour, methane, ammonia and hydrogen. This simple gas mixture (H_2O, CH_4, NH_3, H_2) was passed for some days through an electric spark discharge (to simulate a thunderstorm's energy input) and Miller then analysed it. He found traces of hydrogen cyanide, methanal, methanoic, ethanoic and other acids, urea and a mixture of amino acids. The result is certainly suggestive, bearing in mind the short period of the experiment in comparison with 'prebiotic time'. Since Miller's work, others using similar methods and other intense energy sources (eg ultraviolet light and gamma radiation) have produced more complex organic molecules (including the nucleic acid base, adenine) but it remains very unclear how a mixture of organic compounds might have evolved into something like a living system as we now know it.

Millikan, Robert Andrews (1868–1953) US physicist: determined e and h accurately for the first time.

Millikan was the son of a Congregational minister and small farmer and grew up in the still romantic age of the American midwest. His talent at school was mainly in classics, and he did little physics; but in his second year at Oberlin college he was invited to teach elementary physics and was told 'anyone who can do well in Greek can teach physics'. He learned quickly and was soon immersed in the subject, which was not then much developed in the USA. Then he went to Columbia, where he was the sole graduate student in physics there in 1893-5. Later he studied in Germany, and in 1896 was offered a job in Chicago with MICHELSON. He took it, and was there until he went to the California Institute of Technology in 1921.

Between 1909 and 1913 Millikan determined the charge on an electron with considerable accuracy, not surpassed until 1928. Between two horizontal plates a cloud of fine oil droplets was introduced and irradiated with X-rays so as to introduce varying amounts of charge on some of them. By adjusting the voltage on the plates, electric force and buoyancy could be made to just counterbalance gravity for an oil-drop viewed by a microscope. Calculation then revealed its charge; a long series of measurements showed that the measured charge always occurred in multiples of a single value, the charge (e) on a single electron. Millikan was lucky to use a field strength (about $6000\,V\,cm^{-1}$) within the narrow range in which the experiment is possible.

He then studied the photoelectric effect (1912-16) confirming EINSTEIN's deduction of 1905 that the energy E of an electron emitted from a metal by light is given by $E = hv - E_0$ where v is the frequency of the incident radiation, E_0 is the energy required to leave the metal (the work function) and h is PLANCK's constant. For his accurate measurements of e and h Millikan was awarded the 1923 Nobel Prize for physics.

During the 1920s Millikan researched on cosmic rays, showing in 1925 that they come from space. Millikan argued that they are uncharged and consist of electromagnetic radiation, but COMPTON showed them to consist of particles. However, Millikan was responsible for directing C D ANDERSON to view cosmic rays in a Wilson cloud chamber, which led to Anderson's discovery of the positron.

Milne, Edward Arthur (1896-1950) British astrophysicist and mathematician: proposed the cosmological principle.

Milne was educated at Cambridge, where he later became assistant director of the Solar Physics Observatory. In 1929 he was appointed professor of mathematics at Oxford, and stayed there for life, except during the Second World War, when he worked on ballistics, rockets and sound ranging for the Ordnance Board.

Milne's early work was on stellar atmospheres, in particular the relationship between stellar class and temperature. In 1932 he turned to cosmology, proposing the cosmological principle - that the universe appears (on the macroscopic scale) the same from whatever point it is viewed. This remains a basic axiom of much modern cosmological thought. Milne later attempted to deduce a complete model of the universe from somewhat philosophical 'first principles', but was not successful.

Milstein, César (1927-) [milstiyn] Argen-
tinian-British molecular biologist: co-discoverer of monoclonal antibodies.

Born and educated in Argentina, Milstein was a chemistry student, research student and staff member in Buenos Aires before his first stay in Cambridge from 1958. Back in Buenos Aires in 1961, he returned to Cambridge in 1963 to join SANGER on the Medical Research Council staff. There he worked on the structure of an immunoglobulin (an antibody) and then on a corresponding m-RNA, which led him towards monoclonal antibodies (MCAs). An MCA is a single, specific and chemically pure antibody produced in cloned cells, ie cells that are genetically identical, by a method devised in 1975 by Milstein and G Köhler (1946-95). Such MCAs are of great value in diagnosis and testing - for example they are now routinely used in the UK for typing of blood before transfusion - and they are potentially valuable in therapy, eg an MCA against Rhesus-D antigen (anti-D) can now be used for mothers who are Rh-negative and have just given birth to Rh-positive children.

To make an MCA, formation of the required antibody is first induced in an experimental animal by injection of an antigen. After a few weeks, antibody-rich B-lymphocytes are taken from its spleen. These cells are then fused with a malignant cell line (such as a myeloma cell). In some cases the resulting cell, a hybridoma, combines the lymphocyte's ability to produce a pure antibody with the cancerous cell's immortality, so that after selection and cloning it can be grown in culture indefinitely in laboratory conditions; in this way the antibody (which is normally present in serum only in trace amounts and mixed with other antibodies) can be produced in quantity.

The discovery of MCAs promises a revolution in biological research and in clinical diagnosis, and possibly in treatment for a variety of diseases, including some cancers. Milstein, Köhler and N Jerne (1911-94) shared a Nobel Prize in 1984. Jerne, an immunological theorist, had proposed in 1955 the first selection theory of antibody formation, which was then expanded by BURNET, who proposed the clonal selection theory. This states that each antibody is produced by a lymphocyte and that the effect of antigen is just to increase the number of cells producing specific antibodies. Later G Nossal (1931-) showed that, as predicted, individual lymphocytes produce only one kind of antibody, a necessary result for success in producing useful hybridomas.

Minkowski, Rudolph Leo [mingkofskee] (1895-1976) German-US astronomer: made first optical identification of a radio galaxy.

In 1954 Minkowski, with BAADE, made the first definite optical identification of a radio source beyond our Galaxy, Cygnus A. This object was found to be a very distant galaxy emitting an immense amount of radio energy, about 10 million times that generated by a normal galaxy, and it is thought to be undergoing a violent explosion.

Mitchell, Maria (1818–89) US early female astronomer; observed a new comet.

Maria Mitchell was born on Nantucket Island, MA, one of 10 children of a schoolteacher and amateur astronomer. Her education in private elementary schools finished when she was 16. However, Nantucket was an area where it was usual to become familiar with some mathematics, astronomy and navigation. Mitchell assisted her father with his astronomical observations and learned to use his instruments. In 1836 she was appointed librarian at the Nantucket Atheneum and continued with her studies and astronomy. In 1847 she observed a new comet and won a gold medal which had been offered by the king of Denmark for such a discovery, despite competition from observers in Rome and Britain. From this time she was honoured as a leading astronomer. In 1849 she became a computor for the *American Ephemeris and Nautical Almanac* and began to work for the United States Coast Survey. She visited observatories in England and Europe, and met many of the scientists of the time. In 1865 she became professor of astronomy and director of the observatory at Vassar College, New York.

Mitchell, Peter (Dennis) (1920–92) British biochemist: devised new theory of cellular energy transport.

It is rare for a scientist of this generation to create, direct and equip his own laboratory and institute; it is also unusual to develop a theory which first attracts widespread opposition but later achieves full acceptance. Mitchell did both. Educated in Cambridge, he went on to teach biochemistry there and in Edinburgh before taking over the ruined Glynn House near Bodmin in Cornwall in 1963. This he restored, partly with his own skills, and converted it into his family quarters and a research laboratory. There he developed his chemiosmotic theory of energy generation in the mitochondria and chloroplasts of plant cells, using a novel concept of protonic coupling. Working particularly with Jennifer Moyle, his valued technician, his experimental work supported and expanded his theory in detail. After a decade this convinced his peers and led to his Nobel Prize for chemistry in 1978.

Mitscherlich, Eilhardt [mitsherleekh] (1794–1863) German chemist and mineralogist: discovered law of isomorphism.

Mitscherlich's youthful enthusiasm was the Persian language, which he studied at Heidelberg and Paris in the hope of visiting Persia as a diplomat. When this appeared impossible he decided to study medicine, with the intention of travelling as a physician. To begin, he studied science in Göttingen; and was soon so attracted by chemistry that he gave up the idea of visiting Persia. He moved to Berlin in 1818 to study chemistry and soon noticed that potassium phosphate and potassium arsenate form nearly identical crystals. He went to Stockholm to work with BERZELIUS for 2 years, continuing to measure crystal angles and forms; where these were closely similar in different compounds, he described them as 'isomorphous'. His law of isomorphism states that isomorphous crystals have similar chemical formulas. For example, he showed that the manganates, chromates, sulphates and selenates are isomorphous; from this the formula of the newly discovered selenates could be deduced (as the formulas of the first three were known) and from the formula the relative atomic mass of selenium was found by analysis. From 1825 Mitscherlich was professor of chemistry at Berlin.

Berzelius and others found the law useful in deducing formulas, and therefore relative atomic masses, for several elements and for correcting earlier erroneous formulas. Mitscherlich continued his work on crystallography; he found that some substances (eg S) can crystallize in two forms (dimorphism) or even more (polymorphism). From 1825 Mitscherlich was professor in Berlin. As well as crystallography, he worked on organic chemistry, microbiology, geology and catalysis. He first made benzene (and named it *Benzin*) by heating calcium benzoate; and also nitrobenzene, azobenzene and benzenesulphonic acid. He helped his youngest son develop an important industrial process for obtaining cellulose from wood pulp by treatment with hot aqueous calcium hydrogen sulphite solution.

Möbius, August Ferdinand [moebyus] (1790–1868) German mathematician and astronomer: inventor of barycentric calculus and of the Möbius strip.

Möbius studied law at the University of Leipzig, before abandoning it in favour of mathematics and astronomy. In 1816 he was appointed professor of astronomy at Leipzig, and in 1848 became director of its observatory.

Although in mathematics Möbius developed barycentric calculus, which simplifies a number of geometric and mechanical problems, he is better known for his work on topology. In this field he invented, to illustrate the idea of a one-sided surface, the Möbius strip. This can be formed by taking a strip of paper, rotating one end through 180° and connecting the ends together. Möbius's description of it was only discovered in his papers after his death. Möbius is also remembered for posing the 'four-colour problem' in a lecture in 1840: what is the least number of colours needed in a plane map to distinguish political regions, given that each boundary must separate two differently-coloured regions? No such map requiring five colours has been found, despite attempts, and in 1976 a computer-based proof has been offered showing that four colours will always suffice.

Mohorovičić, Andrija [mohhorohvuhchich] (1857–1936) Croatian geophysicist: discovered the boundary between the Earth's crust and the mantle.

A talented young man who spoke eight languages when he was 15 years of age, Mohorovičić was educated at the University of Prague, being appointed professor at the Zagreb Technical School in 1891, and later at Zagreb University. In 1909, while studying a nearby earthquake, Mohorovičić discovered that the P and S waves from it were each of two types, with one of each pair having, he decided, travelled through deeper and denser rocks and arrived before waves travelling through the Earth's crust. He realized that the Earth's crust must therefore overlay a denser mantle, and measured the depth to this transition (the Mohorovičić discontinuity, or 'Moho') to be about 30 km. This corresponds to the thickness of the continental crust. The depth of the Moho has now been extensively mapped using reflection seismic techniques and is known to vary between only 10 km under the oceans and about 50 km at some places beneath the continents.

Mohs, Friedrich [mohz] (1773–1839) German–Austrian mineralogist: devised a scale of mineral hardness.

Born appropriately at Gernrode in the Hartz mountains (which are rich in minerals), Mohs became professor of mineralogy at Graz in 1812 and at Vienna in 1826. He is now remembered for the scale of relative hardness named after him. This runs from talc (1) to diamond (10); the mineral with the higher number scratches anything beneath it or equal to it in hardness. The scale is not linear (the true hardness differences do not coincide with the simple intervals of Mohs's scale); and the hardness of crystals is usually different in different crystal directions. The scale is useful in field mineralogy.

Moissan, (Ferdinand Frederic) Henri [mwasā] (1852–1907) French inorganic chemist: first isolated fluorine; pioneer of high-temperature chemistry.

Coming from a poor family, Moissan's pursuit of education and his enthusiasm for chemistry proved difficult until marriage and a generous father-in-law eased his financial position. After his first successes in chemistry, he held posts in Paris at the university. In the early 1880s he began to experiment on ways to isolate the element fluorine from its compounds. Earlier attempts by DAVY and others had shown only that fluorine must be highly reactive, and some attempts had fatal results. Moissan succeeded in 1886 by electrolysis of a solution of KF in HF, at −50° in an apparatus made of platinum and calcium fluoride. Fluorine was isolated at the anode as a yellow gas, and, as the most chemi-

cally reactive of all elements, afforded Moissan a rich seam of new chemistry.

Later he explored boron chemistry (he was the first to make pure boron) and he attempted the synthesis of diamond by crystallizing carbon from molten iron under pressure. He was the first to make a range of metal hydrides, which proved to be highly reactive. His interest in high-temperature chemistry led him to devise electric furnaces in which a carbon arc gave temperatures up to 3500°C. In this way another new area of chemistry was opened up and Moissan was able to make synthetic gems such as ruby, and silicides, borides and carbides of metals, as well as metals such as Mb, Ta, Nb, V, Ti, W and U which were then little known. He was awarded a Nobel Prize in 1906.

Moivre, Abraham de see **de Moivre**

Mond, Ludwig (1839–1909) German–British industrial chemist.

Son of a prosperous Jewish merchant, Mond studied chemistry under KOLBE and BUNSEN. From 1858 he had jobs in chemical industry, and devised a rather unsatisfactory process for recovering sulphur from the offensive 'alkaliwaste' of the Leblanc soda-making process. He operated this in Widnes in the 1860s. In 1872, with J T Brunner (1842–1919), he began to use the new Solvay process in his own works at Winnington, Cheshire. This made soda (Na_2CO_3) from common salt, ammonia and carbon dioxide, and displaced the Leblanc process.

In 1889 the corrosion of warm nickel by CO was noted and found to be due to the formation of volatile $Ni(CO)_4$. Mond saw this as a novel way of purifying nickel, by forming and purifying the tetracarbonyl and then decomposing it by heat; he set up the Mond Nickel Company to do this. Mond became very wealthy and his benefactions included the re-equipping of the Royal Institution's laboratory and the gift of his valuable art collection to the National Gallery. Brunner, Mond and Company in 1926 became a major component in the merger which formed ICI.

Monge, Gaspard [mõzh] (1746–1818) French mathematician: founder of descriptive geometry.

Monge was educated the Collège de la Trinité in Lyon, and later at the military academy at Mézières, subsequently becoming professor of mathematics there. He was an active supporter of Napoleon, becoming minister of the navy in 1792, official recorder of Louis XVI's trial and execution in 1792–3 and accompanying Napoleon to Egypt in 1798. An inspired teacher, he helped found the École Polytechnique in 1795, becoming its director.

Monge is remembered as the founder of descriptive geometry, the basis of modern engineering drawing, and for his work on the curvature of surfaces. The theory of the class of Monge equations (equations of the type $Ar + Bs +$

$Ct + D = 0$), was developed by him. He was wide-ranging in his interests, tackling problems as diverse as partial differential equations, the composition of nitrous acid and capillary phenomena. His interest in chemistry led him to synthesize water from hydrogen and oxygen in 1783 independently of LAVOISIER, although the two later collaborated on the same problem.

Monod, Jacques (Lucien) [monoh] (1910–76) French molecular biologist: devised theories on the control of gene action.

A graduate of Paris, Monod taught zoology there from 1934, served in the French Resistance in the Second World War, and joined the Pasteur Institute in 1945, becoming its director in 1971. He worked particularly with F Jacob (1920–) on the problem of how gene action is switched 'on' and 'off', especially in the enzyme syntheses they control in mutant bacteria, which in turn control the bacterial metabolism. In this area he introduced the idea of operons, groups of genes with related functions which are clustered together on a chromosome and are controlled by a small end-region of the operon called an operator. This in turn can be made inactive by a repressor, which combines with and switches off the operator. The scheme was developed in 1961 to include the idea of messenger RNA (mRNA), which carries genetic information from the DNA of the chromosomes (the operon) to the surface of the ribosomes, where protein synthesis occurs. These ideas found much support in experiments on microorganisms; their extension to more complex plants and animals is less firmly established. Monod and Jacob, with A Lwoff (1902–94), shared a Nobel Prize in 1965. Monod was talented and active as a sportsman, musician and philosopher. His work on the origin of life led him to argue that it arose by chance and evolved by Darwinian selection through necessity, with no overall plan.

Monro, Alexander [muhnroh] (*primus*, 1697–1767; *secundus*, 1733–1817; *tertius*, 1773–1859) Scottish anatomists: an anatomical dynasty.

This dynasty dominated the teaching of medicine and surgery in Edinburgh for 126 years, and did much to create and then to increase the fame of the 'Edinburgh School'. The first of the Alexanders, himself the son of a surgeon, was the first professor of anatomy (or any medical subject) there, and under him the number of students increased fourfold; his son ('*secundus*') was the most able of the three and wrote on the distinction between the lymphatic and circulatory systems and on the physiology of fishes; when he retired he had taught 13 404 students, including 5831 from outside Scotland. His son, *tertius*, wrote much but made no real anatomical discoveries.

Montagnier, Luc [mõtanyay] (1932–) French virologist: discoverer of HIV, the generally accepted cause of AIDS.

Montagnier studied science at Poitiers and science and medicine at Paris, becoming MD in 1960. From 1960–4 he worked in virology as a research Fellow for the Medical Research Council in the UK, initially at Carshalton and then in Glasgow. Returning to France, he headed research groups at Orsay, at the Pasteur Institute and for the National Centre for Scientific Research (CNRS), specializing in virology and becoming professor at the Pasteur Institute in 1985.

Recognition of Montagnier as the discoverer of the virus causing AIDS (see panel overleaf) followed a period of some confusion in work on the cause of this disease. An early claim for the discovery was made in 1984 by R Gallo (1937–) of the US National Cancer Institute. However, later and rather prolonged investigations on priority made in the USA showed that the retrovirus causing AIDS had been described and photographs of it published by Montagnier in May 1983, but the discovery was inadequately appreciated in the USA at that time. It seems clear that virus samples loaned by the French group to Gallo's laboratory in early 1984 contaminated isolates in that laboratory and that Gallo's discovery was in fact the French virus. By the 1990s it was also widely agreed that this virus (now known as HIV, with several strains recognized) is the central cause of AIDS, although Montagnier has argued that other organisms must be present together with this virus if the death of immune-system cells, characteristic of AIDS, is to occur in the patient.

Montgolfier, Joseph(-Michel) de (1740–1810) and (Jacques-)Étienne de Montgolfier [mõgolfyay] (1745–99) French inventors of the hot air balloon.

These brothers, who both worked in the family paper-making business, became interested in the possibility of balloon flight about 1782. Their earliest paper models were hydrogen-filled, but the gas soon escaped. Their first large model, which reached 25 m, had an envelope of silk taffeta and was lifted by hot air, heated by burning a mix of chopped hay and wool. In 1783 they made a much larger balloon of canvas covered with paper and used it to raise a sheep, a cock and a duck in a wicker cage, and later in the year two friends ascended, with a brazier to maintain the heat. They remained airborne for half an hour, reaching about 100 m and travelling across Paris; this was the first human flight. Étienne never ascended in a balloon, and Joseph only once, under a huge balloon more than 30 m in diameter in which he flew with six friends. Lighter-than-air dirigibles soon began to use hydrogen, and in this century helium, but hot air balloons have again become popular since the Second World War. Joseph also designed a parachute and tested it with a sheep dropped from a tower in 1784, but afterwards did no more in aeronautics. The 'ballooning

AIDS AND HIV

Total of adult HIV infections from the late 1970s to late 1994

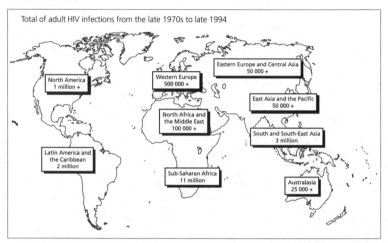

In 1981, the first reports of a new disease came from medical centres on the USA's West Coast. Its key feature was collapse of the body's immune system, leading to low resistance to some cancers and to infections, such as pneumonia, which are usually treatable, but which in these cases were often fatal. By 1982 it was named as Acquired Immune Deficiency Syndrome (AIDS).

The AIDS epidemic spread, with deaths from it in the USA increasing from 2000 in 1984 to over 10 times as many in 1986. Its victims were then almost entirely among the 'five H's': homosexual men, heroin addicts, hookers, Haitians and haemophiliacs. At a fairly early stage it was judged to be viral in origin, with its spread associated with the transfer of body fluids such as blood or semen during sexual contact, or from mother to child before birth, or during blood transfusion or by sharing of contaminated hypodermic needles. The 100% fatality rate and rapidly increasing number of victims (especially in Africa) clearly warranted major investigation.

However, the expansion of AIDS research soon exposed some unusual aspects. These included intense and damaging rivalry between two leading research groups (those led by MONTAGNIER at the

---

craze' he had begun spread rapidly to the USA; and the English Channel was crossed in 1785.

Moore, Stanford (1913–82) US biochemist: co-inventor of method for analysing amino acids.

Moore spent his career at the Rockefeller Institute, having graduated in chemistry from Vanderbilt University in 1935 and followed this with a PhD from Wisconsin. A central problem in protein chemistry is to determine which amino acids are present in a protein chain, and in what amount. Only when this is known can work begin on the sequence of the amino acids in the chain. With W Stein (1911–80), Moore devised in the early 1950s a general method of analysis. First, the protein is hydrolysed completely (eg by warm acid) to give a mixture of amino acids. These are then separated from one another by applying the mixture to the top of a column of ion exchange resin and then eluting the column with a series of buffer solutions of progressively changing acidity. The amino acids emerge separately from the column, can be identified by their rate of emergence, and

the quantity of each is measured by the intensity of the blue colour it gives on reaction with ninhydrin. By 1958 Moore and Stein had devised an ingenious automated analyser to carry out all these steps on a small sample. The problem of finding the sequence of amino acid groups in the chain can then be attacked by methods such as those used by SANGER in his work on insulin (1905). The Moore–Stein analytical method was soon used for cases varying from the simple heptapeptide evolidine (seven amino acid groups) to the enzyme ribonuclease (124 amino acid groups). Moore, Stein and ANFINSEN shared the Nobel Prize for chemistry in 1972.

Morgagni, Giovanni Battista [mawrganyee] (1682–1771) Italian anatomist: pioneer of pathology.

Graduating in Bologna, Morgagni taught anatomy there and later in Padua. Although active in anatomical research throughout his life, his great work was not published until he was 80. This was a survey of about 700 cases,

Pasteur Institute in Paris and by Robert Gallo (1937–) of the National Cancer Institute in Washington, DC); the moralistic overtones associated with studies on a mainly sexually-transmitted disease; and the large sums of money linked with research on AIDS, on patentable test methods for identifying its victims and on drugs for its treatment.

By the early 1990s it was widely agreed that AIDS follows infection by one of the strains of a retrovirus, human immunodeficiency virus (HIV), which attacks the T-lymphocytes (a type of white blood cell). Then, after a period of normally some years, AIDS develops. Treatment was attempted from 1986 with the costly drug AZT (now called zidovudine), but a large-scale trial begun in 1988 showed by 1993 that this alone gave no clear benefits, whether the drug treatment began early on patients shown to be HIV-positive or was delayed until the full symptoms of AIDS were present.

Despite intensive efforts by epidemiologists to predict the pattern of spread of the disease, their projected figures have recurrently needed revision; the number of cases in the early 1990s was much below earlier projections. Predictions that the disease would spread among heterosexuals have largely been unfulfilled in the USA and northern Europe, but in parts of Africa have been realized, with up to 50% of the general population infected in countries such as Kenya, while significant infection in some less-developed European countries such as Romania has occurred through poor medical hygiene. Some geographical areas and population groups have remained unaffected by the epidemic (eg parts of the Middle East), for

reasons that are unclear. The scale and intensity of research on AIDS is remarkable; by 1992, over 36 000 research papers had been published on it, over $6000 million had been spent on the study of the disease and an estimated 500 000 victims of it had died.

As a result of this massive programme, HIV is one of the most fully studied of all disease-causing microorganisms. It is known to be highly variable, with its genetic make-up changing frequently to give new strains; this adds to the difficulty of devising a protective vaccine. But many questions are unanswered, such as: where and when did the virus originate? Why are so few T-lymphocytes in the blood of most HIV-positive people infected? Why is there a delay of typically 7 years between infection by HIV and the onset of AIDS? And why do a very small number of HIV carriers fail to develop AIDS?

Problems also arise with test animals. Chimpanzees are used because they can be infected with HIV. However, they are costly and not easy to work with, and their immunology does not model human patients closely – they are not made ill by HIV and do not proceed to develop AIDS. Trials of protective vaccines in human subjects will offer obvious problems. Despite the difficulties, the potential problem of AIDS is so important that great efforts will continue to be made both to develop a vaccine against HIV, and to secure an effective drug for treatment: AIDS is seen by many workers as the greatest threat to public health of our time.

IM

written in the form of 70 letters to an unknown medical friend. For each case, he describes first the clinical features of the illness in life and then the post-mortem findings. His object was always to relate the illness to the lesions found at autopsy. He did not use a microscope. After Morgagni's book, physicians increasingly related symptoms to 'a suffering organ' rather than to 'an imbalance of the four humours' and developed methods such as percussion in 1761 (L Auenbrugger, 1722–1809) auscultation (LAËN-NEC, 1819) and radiography (RÖNTGEN, 1896) to locate lesions causing disease. Morgagni was the first to describe syphilitic tumours of the brain and tuberculosis of the kidney, and to recognize that, where paralysis affects one side of the body only, the lesion is on the other side of the brain. The modern science of morbid anatomy, central to pathology, begins with him.

Morgan, Thomas Hunt (1866–1945) US geneticist: established chromosome theory of heredity.

Morgan was a product of two prominent

American family lines (including his great-grandfather F S Key, who composed the national anthem) and he grew up in rural Kentucky with an interest in natural history. He studied zoology there at the State College and then at Johns Hopkins University. His later career was at Columbia, and then at the California Institute of Technology from 1928.

A quick, humorous and generous man, he is linked especially with the use in genetics of the fruit fly, *Drosophila melanogaster*, and with establishing the chromosome theory of heredity, and the idea that genes are located in a linear array on chromosomes. When he began his work on genetics he was doubtful of the truth of MENDEL's views, but his studies with *Drosophila* soon convinced him and he became a vigorous supporter. W Sutton (1877–1916), in 1902, had suggested that Mendel's 'factors' might be the chromosomes; Morgan proved him right, and showed that the units of heredity (the genes) are carried on the chromosomes. With his co-workers he established sex-linkage, initially through the observation that the mutant variety 'white-eye' occurs almost exclusively in fruit flies that are male; he also discovered crossover (the exchange of genes between chromosomes) and he and his team devised the first chromosome map, in 1911 (it showed the relative position of five sex-linked genes; by 1922 they had a map showing the relative positions of over 2000 genes on the four chromosomes of *Drosophila*). He won a Nobel Prize in 1933.

Morley, Edward Williams (1838–1923) US chemist and physicist.

Morley's work in science is marked by his passion for precise and accurate measurements. Like his father he was a Congregational minister, but from 1882 he taught science in the college that became Western Reserve University in Ohio. His early research was on the oxygen content of air; his purpose was to test a meteorologist's theory of the atmosphere. From this he moved to a study of the relative atomic mass of oxygen, which he measured to within 1 in 10 000. (Only after his retirement was it known that such measurements represent the weighted average of the stable isotopes of the element concerned.) Also, he worked with MICHELSON in their famous experiments to detect the 'ether-drift'.

Morse, Samuel (Finley Breese) (1791–1872) US artist and inventor.

After education at Yale, Morse concentrated on painting portraits, which he studied in Europe before his appointment as professor of literature and design in New York's City University. Always interested in novelties, he devised an electric telegraph system in 1832 and improved it with technical help from JOSEPH HENRY and financial help from Alfred Vail. In 1843 he set up a 40-mile line from Washington to Baltimore, fortunately completed one day before the Democratic Convention in the latter city. He had in 1838 devised a code of dots and dashes, widely used thereafter, but only of historical interest by the 1990s. Highly religious, his first message on the new line, to Vail on 24 May 1844, was 'what God hath wrought'. He later made a fortune through his patents, despite costly litigation. He earned little in Europe: WHEATSTONE had set up a telegraph system in England in 1838, which was quickly extended, and linked with France by cable by 1851.

Another of Morse's interests was photography. He learned early from DAGUERRE of his work, probably made the first photographs in America and, with his fellow-professor J W DRAPER, set up a portrait studio on the university roof; on sunless days he instructed students in the new process. His Morse code remains his main claim to fame.

Moseley, Henry Gwyn Jeffreys [mohzlee] (1887–1915) British experimental physicist: showed identity of atomic number and nuclear charge of a chemical element.

Moseley came from a family of scientists and graduated from Oxford in physics in 1910. At once he joined RUTHERFORD in Manchester, but in 1913 he returned to Oxford to work. In 1914 he visited Australia, and on the outbreak of the First World War he joined the Royal Engineers and later fought at Gallipoli. He was shot through the head by a Turkish sniper there during the battle of Suvla Bay.

Moseley's major work was on the characteristic X-rays which W H BRAGG and others had shown to be produced from metals used as targets in an X-ray tube. VON LAUE'S work had shown that X-ray frequencies could be measured by crystal diffraction, and Moseley was instructed in this by W L BRAGG. In 1913, using a crystal of potassium hexacyanoferrate(II) to measure the X-rays, he used over 30 metals (from Al to Au) as targets, and found that the X-ray lines changed regularly in position from element to element, in the order of their position in the periodic table. He suggested that this regular change must mean that the nuclear charge can be equated with what he called 'the atomic number'. His work allowed prediction that six elements were missing from the table and, from their position, their properties and likely association could be predicted. As a result, these new elements were soon sought and found. The relation between an element's X-ray frequency and its atomic number is known as Moseley's Law. He was also able to resolve confusion over the identity of the rare earth metals, but his major achievement was to link chemical behaviour (as shown by an element's place in the periodic table) with the physics of atomic constitution. Rutherford called him 'a born experimenter' who, as SODDY described it, 'called the roll of the elements'.

Mössbauer, Rudolf Ludwig [moesbower] (1929–) German physicist: discovered the Mössbauer effect, which he used to verify Einstein's general relativity theory.

Mössbauer was taking his doctorate at the Max Planck Institute for Medical Research in Heidelberg (1955–7) when he discovered what is now called the Mössbauer effect. When an atom absorbs a gamma ray it recoils and, by energy conservation, the wavelength of the re-emitted gamma ray is altered. However, Mössbauer found that at low temperatures one can avoid (for a certain fraction of gamma ray processes) the excitation of vibrational motions in a solid. In such processes, the lattice recoils as a whole and recoil shifts of the energy are absent in the associated gamma-lines; in effect recoil-less nuclear resonance provides very high precision (1 in 10^{12}) in the absorbed and re-emitted gamma ray wavelength. This precision allows detection of different electronic environments surrounding particular nuclei (Mössbauer spectroscopy). Thus Fe^{2+} and Fe^{3+} can be separately detected in Fe_3O_4. Also, EINSTEIN'S general relativity theory was verified by measuring the change in wavelength of a gamma ray due to its moving from one point of gravitational potential to another (1960).

Mössbauer shared the 1961 Nobel Prize for physics, and afterwards held professorships at the California Institute of Technology (until 1965) and at Munich.

Mott, Sir Nevill Francis (1905–) British physicist: discovered aspects of the electronic structure of disordered materials.

Mott's parents both worked at the Cavendish Laboratory and he studied mathematics at Cambridge. He became a lecturer and Fellow there, working with RUTHERFORD, and later with BOHR in Copenhagen. With H Massey (1908–83) he applied the new quantum mechanics to the scattering of particles in atomic physics, and established this field. At 28 Mott moved to a professorship at Bristol and, influenced by H Jones, became interested in solid-state physics. Close collaboration between theoreticians and experimentalists led to rapid progress. Metal and alloy behaviour (with Jones) and ionic crystals (with R W Gurney) formed the subject of books by Mott. Work during the Second World War led to research on dislocation, defects and material strengths. In 1954 he moved to the Cavendish professorship and started research on the transition between metallic and insulating behaviour (the Mott transition). He decisively shaped the Cavendish Laboratory's research activities, and 'retired' in 1965.

Then at 60 he returned to full-time research, choosing to work on the new area of non-crystalline semiconductors and immediately recognizing the significance of P W ANDERSON's papers on electronic localization. Once again he published a classic text on his interest (with E A Davis), which established a complex but rapidly growing area of research. Mott was knighted in 1962 and shared the 1977 Nobel Prize for physics for his work on the electronic properties of disordered materials.

Mott is one of the major theoretical physicists of this century, opening new and difficult areas of solid-state physics and materials science. He influenced a generation in showing how to model the complexity of physical problems such as fracture of metals or electronic processes in disordered semiconductors.

Mueller, Erwin Wilhelm [müler] (1911–77) German-US physicist: invented the field-ion microscope.

A graduate in engineering from Berlin, Mueller worked for industrial laboratories in Berlin and for the Fritz Haber Institute until 1952 when he joined Pennsylvania State University. In 1936 he invented the field-emission microscope, in which a high negative voltage is applied to a fine metal tip held in a vacuum near a phosphorescent screen. Electrons emitted from the tip travel to the

screen and form a highly magnified image of the tip's surface, allowing study of conditions at that point of atoms and molecules provided they are stable enough to survive the conditions at the tip. In 1951 he devised a field-ion microscope, in which the metal tip is held positive in gas at a low pressure. Gas adsorbed on the tip becomes ionized, and the resulting positive ions are repelled from the tip and form the image. The resolution is improved by cooling the tip with liquid helium, and in this way in 1956 Mueller was able to obtain well-resolved images from atoms for the first time.

Müller, Hermann Joseph [muler] (1890–1967) US geneticist: discoverer of use of X-rays to induce genetic mutation.

Müller became an enthusiast for genetics at 16; he had already founded the first science club at his Harlem school, and won a scholarship to Columbia University. By 1915 he began his experiments on spontaneous gene mutation, a key effect in the study of genetics, under the guidance of T H MORGAN, the leading American geneticist. Müller already saw natural mutations as not only rare but usually both detrimental (often lethal) and recessive; and he saw the gene itself as the true basis of both evolution and of life itself, as its ability to reproduce itself was the central property of living matter. In 1926 he achieved the abundant and easy production of mutations by use of X-rays, which hugely increased the scope of genetic studies. He concluded, correctly, that mutation is in essence no more than a chemical reaction. He used the fruit fly *Drosophila* for much of his work, but the method can be applied to reproductive cells of any kind. Since then, chemicals (such as colchicine, and mustard gas) have been used in place of X-rays or other high-energy radiation to induce mutations.

Müller moved to Germany in 1932 but in 1934 he moved on to the USSR, only to find its political climate even more unhappy for him; Edinburgh was his home from 1937–40 and thereafter the USA. His influence in genetics was very great and his X-ray work won him the Nobel Prize in 1946. He was much concerned both that environmental radiation from many sources can injure human genes and that modern medicine tends to preserve mutants in the population, and he advocated sperm banks to maintain and improve the human gene pool.

Müller, Johannes Peter [müler] (1801–58) German physiologist: made wide-ranging discoveries, contributing also to anatomy, zoology and neurology.

The Müller family were Moselle wine-growers, but Johannes's father was a prosperous shoemaker in Coblenz. The boy entered the newly founded university at Bonn in 1819, and his combination of talent and great ambition soon attracted attention. In 1826 he became a professor there. When a post became vacant in

Berlin, he took the remarkable step of proposing himself for the job, and got it. He was a frequent victim of depression and his death was probably due to suicide, but this is uncertain because he had forbidden an autopsy.

When well, he was intensely productive as a physiologist. His first work covered problems of locomotion in animals. Then, in 1820, he attacked the Bonn prize question: does the fetus breathe in the womb? Experiments on a ewe showed that the blood-colour entering and leaving the fetus indicated that it did respire. (Afterwards Müller was antagonistic to vivisection on warm-blooded animals, although he was a great user of frogs.) His later work was wide-ranging; he studied electrophysiology, the sensory system of the eye, the glandular system, the human embryo and the nervous system. He showed the value of microscopy in pathology, developing procedures now used in daily clinical work, especially on tumours. He worked on classification in zoology, especially on marine animals. He proposed the law of specific nerve energies in 1840. This states that each sensory system will respond to a stimulus (whether this is mechanical, chemical, thermal, or electrical) in the same way, specific to itself. Thus the eye always responds with a sensation of light, however it is stimulated; the ear with a sensation of sound, and so on. Man does not perceive the external world directly, but only the effects on his sensory systems. 'In intercourse with the external world we continually sense ourselves' – an important statement for philosophy.

His many pupils include SCHWANN, DU BOIS-REYMOND, HELMHOLTZ, J Henle (1809–85) and VIRCHOW. He is widely regarded as the greatest of all physiologists.

Müller, Paul Hermann [müler] (1899–1965) Swiss chemist: developed DDT as an insecticide.

Educated in Basle, Müller spent his career from 1925 with the Swiss chemical company of J R Geigy. From 1935 he attempted to find an insecticide that would be rapid and persistent but harmless to plants and to warm-blooded animals. By 1940 he had patented as an insecticide a chemical first made in 1873; it was dichlorodiphenyltrichloromethane ('DDT'), which is cheaply and easily produced. This was highly effective, eg in killing lice (the carriers of typhus fever) and so preventing epidemics at the end of the Second World War. For 20 years it was much used, but fell into partial disfavour when the emergence of resistant insect species limited its effectiveness, while it was also found to have a damaging effect on some other animals; its persistence in food chains was seen as a disadvantage and by the 1970s its use was banned or limited in some advanced countries. Müller was awarded a Nobel Prize in 1948.

Mulliken, Robert (Sanderson) (1896–1986) US chemical physicist: developed molecular orbital theory and investigated molecular spectroscopy.

An organic chemist's son, Mulliken graduated in chemistry at MIT in 1917, went on to study poison gases and then in 1919 began work in Chicago on a problem in chemical physics (isotope separation). Except for some research visits he was to spend the rest of his long career at Chicago, working on a variety of topics in chemical physics involving molecular spectra and quantum theory. By 1932 BOHR, F Hund (1896–   ) and others had done much to show how energy levels in atoms could be understood in theory and related experimentally to atomic spectra. Mulliken extended these ideas to molecules. His central idea was that, in a molecule, the electrons that bind the nuclei together move in the field produced by two or more nuclei; the atomic orbitals (a word he devised) become molecular orbitals extending over these nuclei. He showed how the energies of these orbitals could be found from the spectra of the molecules. These ideas formed the basis of molecular orbital (MO) theory and were developed by Mulliken, and by COULSON, HÜCKEL and others in Europe, to become the major approach to understanding the bonds between atoms in molecules. He was awarded the Nobel Prize for chemistry in 1966.

Mullis, Kary (Banks) (1944–   ) US biochemist: devised the polymerase chain reaction for amplification of traces of DNA.

Mullis studied chemistry at the Georgia Institute of Technology and then biochemistry at Caltech, and afterwards joined a biotechnology company, Cetus.

His fame and the Nobel Prize for chemistry which he shared in 1993 stem from an idea that came to him during a 3-hour night drive in 1983. This concerned the problem of identifying genes (or other fragments of DNA), especially when only a small sample is available. His scheme was to first break up the DNA by a known technique using restriction endonucleases, which split the DNA at specific base-pairs to give a mixture of oligonucleotides: these are of modest molecular size (a few dozen base-pairs) and contain the genes, in the form of specific base sequences.

The Mullis idea was to take these oligonucleotides and use a chemical method whereby they replicated themselves, employing relatively uncomplicated reagents and allowing the synthesis to proceed repetitively on the products. His method for doing this was subtle but experimentally simple (a 'one-pot' cycle of reactions whose result would be to continually amplify the starting oligonucleotide). Each cycle, doubling the DNA fragments, took only 1–2 minutes, so in a few hours one molecule gave 100 billion replicas of itself. A key material, used as a catalyst, was DNA polymerase, discovered by KORNBERG in 1955. Within a year

Mullis had tested and improved the method; he named it the polymerase chain reaction (PCR); it has been widely used ever since. These uses include very sensitive tests for genes, based on PCR's ability to amplify a trace of DNA to a handlable quantity, valuable in forensic work to identify a small hair or semen sample and in studies on DNA traces from extinct fossil animals, such as insects trapped in ancient amber, and from dinosaur bones.

His recorded interests include artificial intelligence, computing, photography and cosmology. Mullis has acquired a notable reputation as an eccentric. Never a conformist, he doubts if the HIV virus alone causes AIDS. Now an independent consultant, he set up a company (Stargene) to make and market amplified samples of the DNA of entertainment stars, in lockets and bracelets. He believes Nobel Prizes should be home-delivered by royal messenger.

Munk, Walter Heinrich [munk] (1917–   ) US oceanographer and geophysicist: improved understanding of the Earth's rotation.

Born in Vienna, Munk emigrated to America when he was 16. He was educated at the California Institute of Technology and the University of California, and held positions at Scripps Institution of Oceanography of the University of California.

Munk's special interest was the rotation of the Earth and its variability. In 1961, together with G J F MacDonald (1929–   ), he showed how a variety of geophysical factors – including tides, air circulation, and glaciation – all have measurable effects on the length of the day, which varies by roughly 0.002 s between summer and winter.

Murchison, Sir Roderick Impey (1792–1871) British geologist: first identified the Silurian, Devonian and Permian periods.

Murchison entered the army at 15, served briefly in the Peninsular War and then married and settled near Durham to follow his interest in foxhunting. At 32 he became friendly with DAVY and his enthusiasm moved to science and particularly to geology, where he had a flair for stratigraphy. He made a series of arduous geological field explorations, at times with SEDGWICK or LYELL, and in 1839 he produced his major book *The Silurian System*, based on his study of the greywackes of South Wales. With Sedgwick he established the Devonian system in southwest England, and an expedition to Russia in 1841 led him to define another worldwide system, the Permian, based on rock stratification in the Perm area.

After about 1840 he became arrogant and intolerant, and in disputes with Sedgwick and others he treated the Silurian as personal property. He was always totally opposed to DARWIN's theory of evolution. In 1855 he succeeded DE LA BECHE as director-general of the Geological Survey.

Muybridge, Eadweard James, originally Edward James Muggeridge [moybrij] (1830–1904) British–US photographer: pioneered use of photography to study animal locomotion.

Born Edward James Muggeridge, Muybridge believed the adopted spelling was the Anglo-Saxon form of his name. Muybridge emigrated to California when he was 22 and became a professional photographer. His 'composite' landscapes were impressive, and by about 1870 he was the official photographer to the US Government.

In 1872 ex-governor Leland Stanford of California commissioned Muybridge to photograph his horses in motion, to resolve an argument about the horse's gait, but his first efforts failed to give decisive results. (Stanford was rich; his will bequests included 2½ million dollars to 'the Leland Stanford Junior University', named in memory of his son). Muybridge was interrupted in this work by his trial for the murder of his wife's lover. He was acquitted, and after a prudent absence he returned to the problem in the late 1870s and soon proved that a galloping horse has all its feet off the ground at times (as Leland had claimed). He used a battery of up to 24 small cameras with shutters speeded to 1/500 s and released by clockwork or by threads successively broken by the horse.

His books and lectures on animal locomotion broke new ground and attracted both scientific and popular audiences. By the 1880s, sponsored by the University of Pennsylvania, he was using up to 36 cameras and the new faster dry plates to study running and jumping men, as well as animals; his book *Animal Locomotion* (1887) contains over 20 000 figures. Despite its high price, the book sold well, probably because the largest category of photographs showed nude women engaged in such actions as falling, jumping and throwing water at one another. Muybridge devised projection equipment for sequential pictures (the Zoopraxiscope, 1879) and his Zoopraxographical Hall in Chicago in 1893 has been claimed as 'the world's first motion picture theatre'. However, a cinema as we now know it was first opened by the Lumière brothers in Paris in late 1895.

Naegeli, Carl Wilhelm von [nayglee] (1817–91) German botanist: made early studies of cell division and plant growth.

Educated in Zürich, Naegeli gave up medicine to study botany under CANDOLLE and then SCHLEIDEN. From 1857 he was professor at Munich. In 1842 he had studied pollen formation with great care, accurately describing cell division, including division of the nucleus. He saw the chromosomes but regarded them as unimportant; and when MENDEL sent him a copy of his classic paper on peas, Naegeli disregarded it. Naegeli's views on evolution were broadly Darwinian, but he supported LAMARCK in believing that evolution occurs in jumps rather than by gradual variation. Oddly, he also believed in spontaneous generation of a rather special sort. In plant taxonomy he did good work, but his best work was on plant growth; he recognized the distinction between meristematic tissue and structural tissue, and he worked on cell ultrastructure.

Nansen, Fridtjof (1861–1930) Norwegian explorer and biologist: pioneer of arctic exploration.

Nansen graduated in zoology from the University of Christiania (now Oslo), later being appointed professor of zoology and then of oceanography at Christiania. Nansen is known as a pioneer of Arctic exploration. In 1888 he crossed the Greenland ice sheet for the first time, demonstrating that it covered the entire island. Between 1893 and 1896 he made an epic voyage in the *Fram*, a specially strengthened ship which he allowed to freeze into the ice pack and to be carried by the currents around the Arctic Ocean, reaching 87°57'N, further north than anyone had been before. He later did further oceanographic work in the Barents and Kara Seas and in the north-east Atlantic. An ardent nationalist, Nansen played an important part in the separation of Norway from Sweden in 1905 and became the first ambassador to Britain of the newly independent state. In 1922 he was awarded the Nobel Peace Prize in recognition of his humanitarian work for famine relief and refugee aid after the First World War.

Napier, John (1550–1617) Scottish mathematician: inventor of logarithms.

Napier was educated in France and at the University of St Andrews. He came from a landed family and pursued mathematics as a hobby, his other interests being religious controversy and the invention of machines of war.

His studies of imaginary roots led him to develop the principle of the logarithm, and he then spent 20 years computing tables of them (in the course of which he also developed modern decimal notation), publishing his results in 1614. His work was enthusiastically received, but the base that he had chosen was not always convenient, leading BRIGGS to calculate, in 1617, a table of logarithms to base 10. In the same year Napier also described a system of rods ('Napier's bones') designed for practical multiplication and division. KEPLER, then involved in the tedious process of calculating planetary orbits, was largely responsible for the introduction of logarithms to the continent. The Swiss J Bürgi (1552–1632) had the idea of logarithms about the same time as Napier but did not publish until 1620. Electronic calculators have now displaced 'log tables' and slide rules for calculation.

Natta, Giulio (1903–79) Italian polymer chemist: developed theory and technology of stereospecific polymerization.

Natta graduated in chemical engineering at Milan Polytechnic and after short periods at three Italian universities returned to Milan in 1938 as professor of industrial chemistry. From 1938 his work was directed to new polymers, initially synthetic rubbers. In 1953 he began work with the catalysts shown by K Ziegler (1898–1973) to polymerize alkenes under mild conditions. By 1954 he had shown that these catalysts can give polymers that are stereoregular, ie the repeating unit in the polymer chains has a recurring and regular space-arrangement (stereochemistry). This feature (named as 'tacticity' by Natta's wife) is important because a suitable stereoregular form has commercially useful properties of high strength and melting-point. He found that propene could be made to give an isotactic polypropylene well-suited for moulded products. Both the ideas and the products devised by Natta have been much used; with Ziegler, he shared a Nobel Prize in 1963.

Néel, Louis (Eugène Félix) [nayel] (1904–) French physicist: discovered antiferromagnetism.

Néel graduated from the École Normale Supérieure and worked under P Weiss (1865–1940) at the University of Strasbourg. In 1940 he moved to Grenoble and became the driving force in making it one of the most important scientific centres in France, becoming director of the Centre for Nuclear Studies there in 1956.

Néel's research was concerned with magnetism in solids. He predicted in 1936 that a special type of magnetic ordering called 'antiferromagnetism' should exist. Whereas unpaired electron spins align in a ferromagnet (eg Fe), they are arranged up-down-up-down from site to site in an antiferromagnetic lattice (eg in FeO). Above a critical temperature, the Néel temperature, the antiferromagnetic substance then becomes paramagnetic. This was experimentally confirmed in 1938, with full neutron diffraction confirmation in 1949. Néel also first suggested (1947) that antiferromagnetism could occur with unequal up-and-down moments (ferrimagnetism) as in some ferrites (ceramics important in electronics; lodestone is an ancient example). Néel was awarded the Nobel Prize for physics in 1970. He also studied the past history of the Earth's magnetic field.

Ne'eman, Yuval (1925–) Israeli particle physicist: developer of particle theory.

Ne'eman is unusual in having careers in engineering, physics, politics and military intelligence, usually being active in at least two of these apparently disparate areas concurrently. He graduated in engineering at the Technion (the Israel Institute of Technology) in Haifa in 1945 and then began a military career, becoming deputy head of the intelligence branch of the Israeli forces in 1955. After the Israeli–Arab war, wishing to work in theoretical physics, he arranged with General Dayan to become a military attaché at the Israeli Embassy in London, which enabled him to work also at Imperial College with Salam (his first intention had been to work at King's College with Bondi on relativity, but the time required to travel across London to King's was too great, whereas Imperial College is close to the Embassy).

His interest in particle classification and in group theory developed under Salam's guidance, and in 1961 he published a valuable general classification of nuclear particles; Gell-Mann arrived at a similar scheme at about the same time. The quark model of nuclear structure made the new concepts widely known.

After founding and heading the physics department at Tel-Aviv University Ne'eman became president of the university from 1972–5 and concurrently director of the Centre for Particle Theory at the University of Texas at Austin (1968–). Then in 1981 he founded the Tehiya political party, and as its chairman was elected to the Knesset (the Israeli parliament), becoming minister of science and development in the cabinet 1982–4, and head of the Israeli Space Agency from 1983.

Nernst, (Hermann) Walther (1864–1941) German physical chemist: pioneer of chemical thermodynamics and discoverer of the third law of thermodynamics.

Nernst studied physics in four German universities and worked in two more, becoming increasingly concerned with the application of physics to chemical problems. He was appointed to a professorship in Berlin in 1905. His early researches on electrochemistry, and on thermodynamics, established his fame. In 1904 he devised an electric lamp which he sold for a million marks. The lamp was soon superseded for lighting by Edison's, but it had made Nernst rich. He acquired a country estate and indulged his passion for the new enthusiasm, the motor car.

Of his many contributions to chemical thermodynamics, the best-known is his 'heat theorem' which became the third law of thermodynamics: all perfect crystals have the same entropy at absolute zero. He argued that it was the last law of thermodynamics; because the first law had three discoverers, the second two, and the third, one (Nernst). In fact the zeroth law was yet to be formally enunciated and has no single discoverer.

Nernst's widespread research in physical chemistry included much electrochemistry; the concept of solution pressure is due to him, and much of the thermodynamic treatment of electrochemistry, as well as contributions to the theory of indicators and buffer action. In photochemistry he proposed the now familiar path for the fast reaction between hydrogen and chlorine, involving a chain reaction based on atomic chlorine.

Nernst was kindly but immodest, and he and his family were known as the most hospitable academic family in Berlin. In the First World War he saw early that Germany must lose, and tried to persuade the Kaiser and others to seek peace, without success. Both his sons were killed in the war. Afterwards he declined an offer to become ambassador to the USA. He was awarded a Nobel Prize in 1920. From the beginning he opposed Hitler's policies; unsuccessful, he retired to his country estate. Anecdotes about him, from his many distinguished pupils, are legion. No-one in his time held a wider or deeper grasp of physical chemistry or did more to advance it.

Neumann, John (János) Von see **Von Neumann**

Newlands, John Alexander Reina (1837–98) British chemist: devised primitive form of periodic classification of chemical elements.

Newlands studied chemistry in London under Hofmann and later became an analytical chemist, specializing in sugar chemistry. In 1860 he spent a period in Italy as a volunteer in Garibaldi's army; his mother, Mary Reina, was of Italian descent.

In 1864, and during the next 2 years, Newlands showed that if the chemical elements are numbered in the order of their atomic weight and tabulated, then 'the eighth element starting from a given one is a kind of repetition of the first, like the eighth note in an octave of music'. Thus his law of octaves (as he called it) showed the halogens grouped together and the alkali metals in another group. He did not leave gaps for undiscovered elements and his rigid scheme had some unacceptable features and was much criticized; one critic asked him derisively if he had tried an alphabetical arrangement. In 1869 MENDELAYEV published a table which is essentially modern; Newlands then tried to claim priority and was so persistent that the Royal Society awarded him its Davy Medal in 1887. They did not elect him to their Fellowship. Newlands certainly had a part of the periodic classification in mind, but he did not develop the idea as effectively as did Mendelayev or J L Meyer (1830–95).

Newton, Sir Isaac (1642–1727) English physicist and mathematician: discovered the binomial theorem, invented calculus and produced theories of mechanics, optics and gravitation.

Newton was born prematurely in the year GALILEO died, 3 months after the death of his father, the owner of Woolsthorpe Manor in Lincolnshire. He was left in the care of his grandmother at Woolsthorpe when his mother remarried, and came under the influence of his uncle, who recognized his talents. Newton went to the grammar school in Grantham and after farming at Woolsthorpe for 2 years was sent to Trinity College, Cambridge, in 1661. He remained there for nearly 40 years. It is clear that as a young man Newton was what we would now describe as rather a hippy. A contemporary recalled that he often 'dined in college ... stockings untied, head scarcely combed'. His portrait by Kneller, made when he was 46, still shows him in casual dress and with very long hair.

As a student Newton attended Isaac Barrow's (1630–77) lectures on mathematics. In 1665 the Great Plague caused him to return to his isolated home at Woolsthorpe. Here he worked on many of the ideas for which he is so famous, during what became known as his 'miraculous year'. Later (c.1716) Newton wrote in his notebooks:

'In the beginning of the year 1665 I found the method for approximating series and the rule for reducing any dignity [power] of any binomial to such a series [ie the binomial theorem]. The same year in May I found the method of tangents of Gregory and Sulzius, and in November had the direct method of Fluxions [ie the elements of the differential calculus], and in the next year in January had the Theory of Colours, and in May following I had entrance into the inverse method of Fluxions [ie integral calculus], and in the same year I began to think of gravity extending to the orb of the Moon ... and ... compared the force requisite to keep the Moon in her orb with the force of gravity at the surface of the Earth.... All this was in the two years of 1665 and 1666, for in those years I was in the prime of my age for invention, and minded Mathematics and Philosophy more than at any time since.'

On returning to Trinity College, he was elected a Fellow (1667) and succeeded Isaac Barrow as Lucasian Professor in 1669 at the age of 26. He was made a Fellow of the Royal Society in 1672. During 1669–76 Newton presented many of his results in optics and became engaged in controversies concerning them. In 1679 he began to correspond with HOOKE, renewing his interest in dynamics and solving the problem of elliptical planetary motion discovered by KEPLER. HALLEY visited Newton in 1684 and persuaded him to write a work on dynamics, which was written within 18 months; his *Philosophiae naturalis principia mathematica* (1687, The Mathematical Principles of Natural Philosophy) – the 'Principia'. It is the most important and influential scientific book ever written.

From that point Newton's mathematical interests waned, giving place to theology (ironically, bearing in mind his college, he seems to have been anti-trinitarian) and involvement in political life. He also spent much time and effort on alchemy, without result. In 1687 he courageously accompanied the vice-chancellor to London to defend the university against illegal encroachments by James II. In 1692 Newton 'lost his reason', as he phrased it; probably he suffered a period of severe depression.

Then his interests turned to London and, via his friendship with Charles Montague, Fellow of the Royal Society and first earl of Halifax, Newton became Warden and then Master of the Mint in 1696 and 1698 respectively, skilfully reforming the currency. He was knighted for this in 1705. In London, a young niece became his housekeeper. She was Catherine Barton, described as beautiful, charming and witty, and the mistress of Charles Montague. Newton's duties at the Mint included supervising the recoinage and, rather oddly, the capture, interrogation and prosecution of counterfeiters.

In 1701 he resigned the Lucasian professorship and his fellowship at Trinity, although he remained president of the Royal Society from 1703 until his death. He was elected a Whig member of Parliament for the university, but was not very active politically.

Much of Newton's last 20 years were spent in acrimonious debate over priority in scientific discoveries with FLAMSTEED and LEIBNIZ, and in this Newton showed both ruthlessness and obsessiveness. Following a painful illness (due to a gallstone), he died in 1727, and is buried in Westminster Abbey. He had been remarkably fit, lacking only one tooth and not needing spectacles, even in old age, and reasonably wealthy. He seems to have had no romantic or imaginative life outside science. Asked his opinion on poetry, he replied 'I'll tell you that of Barrow; he said that poetry was a kind of ingenious nonsense.'

Newton's researches on mechanics display great mastery and established a uniform system based on the three laws of motion: (1) a body at rest or in uniform motion will continue in that state unless a force is applied; (2) the applied force equals the rate of change of momentum of the body; (3) if a body exerts a force on another body there is an equal but opposite force on the first body. From these Newton explained the collision of particles, Galileo's results on falling bodies, Kepler's three laws of planetary motion and the motion of the Moon, Earth and tides. The deductions were made using calculus, but were proved geometrically in the *Principia* to clarify it for contemporary readers. The general theory of gravitation, that any two bodies of mass m_1, m_2 at a distance d apart, attract each other with a force F:

$$F = G m_1 m_2 / d^2$$

where G is a universal constant, was developed by Newton from his original work on the Moon's motion of 1665. It may well be correct (as Newton's niece maintained) that the idea stemmed from seeing an apple fall from a tree beside his Woolsthorpe home. Three centuries ahead of the technology he showed that the escape speed s of an artificial satellite from a planet of mass m and radius r is given by $s = (2Gm/r)^{\frac{1}{2}}$. At speeds less than this, the projected satellite will return to the planetary surface.

Newton published another celebrated treatise, the *Opticks* of 1704, which was an organized and coherent account of the behaviour of light. Based on his own ingenious experimental work, it proposed a corpuscular theory of light but added ideas of periodicity (which were missing even in Hooke's and HUYGEN's wave theory). Such phenomena as refraction of light by a prism (with the production of colours by dispersion) and Newton's rings of coloured light, about the point of contact between a lens and mirror, were considered. Also named after him is the Newtonian telescope, which used mirrors rather than lenses to gather light and achieve magnification (see panel on Telescopes, p. 86).

Newton's name is linked with a variety of matters in physics, in addition to those already noted (eg the laws of motion). Thus the SI unit of force, the newton (N), is based on the second law, in the form which defines the force F which produces a constant acceleration a in a body of mass m, by the relation $F = ma$. The newton is the force which produces an acceleration of 1 m s^{-2} when it acts on a mass of 1 kg.

In fluid mechanics, Newtonian fluids are those whose viscosity is independent of the rate of shear or the velocity gradient. Colloids and some other solutions form non-Newtonian fluids. Newton's law of cooling states that the rate at which a body loses heat to its surroundings is proportional to the temperature difference between the body and its surroundings. It is empirical, and applies only to small differences of temperature, and to forced convection.

Newton exerted a unique and profound influence on science and thought. As a mathematician he discovered the binomial theorem (1676) and the calculus; the latter, together with his law of universal gravitation are the peaks of his achievement and the basis of his colossal stature. His work established the scientific method and placed physics on a new course, giving mathematical expression to physical phenomena and permanently altering modern thought.

Einstein wrote of him: 'Nature was to him an open book, whose letters he could read without effort.... In one person he combined the experimenter, the theorist, the mechanic and, not least the artist in exposition. He stands before us strong, certain and alone: his joy in creation and his minute precision are evident in every word and every figure.'

Newton's view of himself at the end of his life has a different emphasis: 'I do not know what I may appear to the world; but to myself I seem to have been only like a boy playing on the seashore, and diverting myself in now and then finding a smoother pebble or a prettier shell than ordinary, whilst the great ocean of truth lay all undiscovered before me.' Some have suggested this was conventional false modesty.

His birthplace, Woolsthorpe Manor, is maintained by the National Trust.

Nicholson, William (1753–1815) British chemist: discovered electrolysis.

Nicholson had a wide-ranging career, being variously an agent for the East India Company and for Josiah Wedgwood (1730–95) the pottery manufacturer, a schoolmaster, a patent agent and a waterworks engineer.

Within months of VOLTA's invention of the first electric battery, Nicholson had built the first one in Britain. Soon afterwards he discovered that, if the leads from it were immersed in water, bubbles of hydrogen and oxygen were produced. His discovery of the phenomenon of electrolysis was followed up by DAVY, who was to become a pioneer of the new field of electrochemistry. Nicholson also invented a hydrometer, and both wrote and translated a number of of well-respected chemical textbooks.

Nicol, William (1768–1851) British geologist and physicist: inventor of the Nicol prism.

Nicol lectured in natural philosophy at the University of Edinburgh. In 1828 he invented the Nicol prism, which utilizes the doubly refracting property of Iceland spar and proved invaluable in the investigation of polarized light.

However, this usage was much delayed; his invention did not appear in print until 1831 and then only in a little-read book on fossil woods. SORBY knew of Nicol's invention and by 1861 applied it to the study of mineral structure, and so made the polarizing microscope an essential instrument in petrology and later in metallurgy. Organic chemists also applied it in polarimeters, to measure the polarization of solutions. More recently, LAND's invention of Polaroid® has largely replaced the Nicol prism, which was so valuable for a century.

Nicol also developed the technique of grinding rock specimens, cemented to a glass slide, to extreme thinness so that they could be examined by transmitted light; this major advance in petrology also did not appear until 1831.

Nicolle, Charles (Jules Henri) [neekol] (1866–1936) French microbiologist: identified the louse as the vector of typhus.

Nicolle became director of the Pasteur Institute in Tunis in 1902 and soon began to study the epidemic typhus fever there. The typhus fevers are a group of related diseases, long-known and world-wide, with a mortality of 10–70%; their causal pathogens are the rickettsias, which lie between bacteria and viruses in size and type. Nicolle's success in combating typhus began when he noted that the victims infected others before they entered hospital but did not infect others when in hospital. He deduced that the path of infection was broken when they were separated from their clothing and cleaned, and guessed that the body louse was the vector. Experiments with monkeys proved his guess to be correct, and showed that the louse is only infective after taking blood from a victim and that it spread infection through its faeces. After this work (1909) vigorous attack on the lice has led to effective control. Nicolle's later work on typhus showed that antibodies exist in recovered patients; and also after influenza and measles. He also discovered the 'carrier' state, important in immunology. Other rickettsial diseases (eg Rocky Mountain fever) are carried by ticks and by mites. Nicolle was awarded a Nobel Prize in 1928.

Niépce, Joseph Nicéphore [nyeps] (1765–1833) French inventor of photography.

Although several individuals made major contributions to black-and-white photography as we now know it, Niépce has the best claim to be regarded as the first photographer. From 1792 he attempted with his brother Claude to record the image formed in a camera obscura by chemical means, assisted by Joseph's son Isidore. By 1816 he had some success, using a paper sensitized by impregnation with silver chloride and an exposure time of about an hour. The resulting image showed light and shade reversed and had to be viewed by candlelight to avoid further darkening. Even so, this result was an advance on WEDGWOOD's and was some years ahead of comparable results by DAGUERRE and TALBOT. By the 1820s Niépce turned from silver salts to the oddly named bitumen of Judaea as his light-sensitive material. A coating of this on a glass or metal plate is hardened and made insoluble by light, and he used this to obtain photocopies by superposing an engraving, made transparent by oiling, on a glass plate coated with the bitumen.

The first photograph in the modern sense was probably made by Niépce in 1826 or 1827, using a bitumen-coated pewter plate in a camera obscura, and showed the view from his workroom window. The image was fixed by dissolving away the unhardened bitumen in lavender oil. Niépce met Daguerre in 1826 and in 1829 a partnership with him to exploit 'heliography' was agreed and a contract signed. Niépce's cameras are the earliest known and include a focusing tube and iris diaphragm for the lens, and bellows to adjust the length. Hoping to convert his result into a printing-plate, Niépce began in 1829 to use a silvered copper plate as base and blackened this with iodine vapour to improve its contrast. In this way, rather indirectly, he was led to a satisfactory photosensitive surface of silver iodide, which his partner Daguerre later found by chance in 1835 could be developed with mercury vapour and in 1837 fixed with a solution of common salt: this gave the 'daguerrotypes' which soon became famous. But Niépce had died in 1833 before this success, and his efforts in 1827 and later to interest King George IV and the Royal Society in his ideas had failed, largely because he kept his methods secret. It was left to Daguerre, combining his improvements in method with his skill as a publicist, to first achieve real renown, and to Talbot and his friend JOHN HERSCHEL to devise a widely useable system.

Nobel, Alfred Bernhard [nohbel] (1833–96) Swedish chemist: inventor of dynamite.

Nobel's father was an inventive engineer who travelled widely. The family moved to Russia in 1842 (where Nobel's father was supervising the manufacture of a submarine mine he had devised) and Alfred was educated there by tutors; his studies included chemistry and five modern languages. He went in 1850 to study chemistry in Paris, and then travelled in Europe before visiting the USA to work with J Ericsson (1803–99), the Swedish–American inventor of the marine screw propeller. He returned to Russia and then to Sweden in 1859. He was much interested, like his father, in the use of explosives in civil engineering, especially in the developing US market. In 1865 he began to manufacture glyceryl trinitrate ('nitroglycerin', discovered in 1847 by A Sobrero (1812–88)), but the dangerously explosive liquid caused accidents in handling; his factory blew up the same year with five deaths (including that of his brother Emil). In 1866 he found that it was a safe high explosive if absorbed in kieselguhr; the mixture was sold in waxed card tubes as 'dynamite'. In 1875 he invented blasting gelatine or gelignite (nitroglycerin in nitrocellulose), an even better blasting agent; and he profited from oil wells he owned in Russia. His inventions were wide-ranging, and covered by 355 patents. His fortune was large and much of it was left to endow the Nobel Prizes. Element 102 is named nobelium after him.

Noddack, Ida (Eva), *née* Tacke (1896–) German inorganic chemist: co-discoverer of rhenium.

Ida Tacke was educated in Berlin and then worked in the Physico-Chemical Testing Laboratory there with W Noddack (1893–1960), whom she married. With O Berg they searched for the missing element 75, predicted by MOSELEY'S results. They found it in 1925 in traces in the mineral columbite and named it rhenium after the Rhine. They also claimed to have found element 43, but were mistaken in this claim.

In 1934 FERMI had obtained unclear results by bombarding uranium with slow neutrons. Ida Noddack suggested that nuclear fission had occurred, but Fermi and others were unconvinced and her idea was rather passed over. The same idea offered by FRISCH 5 years later was speedily examined and accepted, and dramatic results soon followed.

Noether, Emmy (Amalie) [noeter] (1882–1935) German mathematician of distinction.

The daughter of Max Noether, professor of mathematics at Erlangen, Emmy Noether was a member of a talented family; her three brothers were scientists and her mother a musician. She was allowed to attend lectures in mathematics and foreign languages (she planned to become a teacher) as a non-matriculated auditor at the University of Erlangen (1900–02). In 1903 she moved to Göttingen to specialize in mathematics, but almost immediately the University of Erlangen changed its policy and allowed women to matriculate, so she returned there in 1904. She studied at Erlangen under Paul Gordan (1837–1912), a family friend, and gained a PhD *summa cum laude* in 1907 for a dissertation on algebraic invariants. Paid employment was impossible and from 1908–15 she occasionally lectured at the university for her father. In 1911 Ernst Fischer visited the university and introduced her to the ideas of the 'new' algebra.

After the retirement of her father and Gordan and the death of her mother in 1915, HILBERT invited her to Göttingen. After 7 years she was given the title of 'unofficial associate professor' and later a small salary. She applied her knowledge of invariants to problems Hilbert and KLEIN were considering and was able to provide an elegant pure mathematical formulation to

aspects of EINSTEIN'S general theory of relativity. She taught at Göttingen 1922–33, with visiting professorships at Moscow (1928–9) and Frankfurt (1930). In April 1933 she and other Jewish professors were dismissed. Through the efforts of WEYL she was offered a visiting professorship at Bryn Mawr College in the USA, and she worked at Bryn Mawr and at the Institute for Advanced Study, Princeton. In 1935 she had surgery for an ovarian cyst and died 4 days later.

Emmy Noether's most important contributions to mathematics were in the area of abstract algebra. Weyl divided her career into three periods: relative dependency 1908–19; investigations around the theory of ideals 1920–6, when she was influenced by the work of DEDEKIND and profoundly changed the appearance of algebra; and non-commutative algebras 1927–35. Her work was original and creative, and inspired her successors in abstract algebra to create a 'Noether school'.

Northrop, John Howard (1891–1987) US biochemist: obtained a range of crystalline enzymes.

Educated at Columbia University, New York, Northrop worked throughout his career at the Rockefeller Institute in the same city, starting in 1916. In 1926 J B Sumner (1877–1955) had for the first time crystallized an enzyme (a biochemical catalyst) and showed it to be a protein. However, the value of this work, on the enzyme urease, was insufficiently grasped by most researchers. Northrop saw its importance and used similar methods to obtain pure crystalline samples of other enzymes; he did this in the early 1930s. His pure enzymes included the digestive enzymes trypsin and pepsin, and also ribonuclease and deoxyribonuclease. They were found to be proteins (as, in fact, all enzymes appear to be). This work changed both attitudes and techniques; enzymes were no longer regarded as mysterious, and the availability of pure enzymes was of great value in laboratory work. Later, in 1938, Northrop isolated a bacterial virus and showed this also to be a type of protein (a nucleoprotein). He shared the 1946 Nobel Prize for chemistry with Sumner and STANLEY, who had first crystallized a plant virus.

O

Occhialini, Guiseppe Paolo Stanislao [oh-kahleenee] (1907–   ) Italian physicist: discovered the pi-meson.

Occhialini graduated from the University of Florence and taught there until 1937. He belonged to the school of physics around FERMI and, like others, he left Italy with the growth of fascism. After Sao Paulo, Brazil (1937–44), Bristol University (1944–7) and the University of Brussels (1948–50), Occhialini held professorships at Genoa and Milan (1952).

Working with BLACKETT, he obtained cloud-chamber photographs that showed the positron for the first time (1933), thus confirming DIRAC's theory, but they did not publish their results until after those by C D ANDERSON. In 1947 Occhialini and POWELL discovered the pi-meson (or pion) by observing a cosmic ray particle of mass 300 times that of an electron.

Oersted, Hans Christian (1777–1851) [oersted] Danish physicist: discovered that an electric current produces a magnetic field.

Oersted studied physical science and pharmacy at the University of Copenhagen and, after a period of travel, journalism and public lecturing he was appointed professor of physics there in 1806, later becoming Director of the Polytechnic Institute in Copenhagen.

Oersted is remembered for his discovery that an electric current flowing through a wire induces a magnetic field around it, something that he felt on intuitive grounds ought to be true. In a famous experiment first performed in front of his students in 1820, he placed a magnetic compass needle directly below a wire; when the current was switched on the needle moved slightly. His discovery led to a surge of activity by other physicists interested in electricity and magnetism. He also obtained the first accurate value for the compressibility of water, in 1822.

Ogilvie Gordon, Dame Maria Matilda, *née* Ogilvie (1864–1939) British geologist: one of the first professional women geologists.

Maria Ogilvie, the eldest daughter of the Rev Alexander Ogilvie, a prominent educationalist, attended the Ladies College, Edinburgh until she was 18. Her first interest was music and she became a student at the Royal Academy of Music in London, with success. She was then a student at University College, London, studying geology, botany and zoology, and gained her BSc in 1890. She produced a thesis on the geology of the Wengen and St Cassian strata in southern Tyrol and for this, in 1893, she obtained the London DSc, becoming the first woman to do so.

From 1891–5 she studied at the University of Munich and was the first woman to obtain a PhD there in 1900, passing her examinations in German with highest honours in zoology, geology and palaeontology and with a thesis on recent and fossil corals. She published on the microscopic skeletal structure of recent Medreporarian corals in *Transactions of the Royal Society* and on the Upper Jurassic corals of the Stramberg fauna through the Bavarian Royal Academy of Sciences.

She married Dr John Gordon, a physician in Aberdeen, who sometimes accompanied his wife on her excursions. She continued her researches in the Dolomites; her later work appeared in the Austrian Geological Survey. In 1932 she was awarded the Lyell Medal of the Geological Society of London. She was one of the first women to be appointed a Justice of the Peace.

Ohm, Georg Simon (1789–1854) German physicist: discovered relationship between current and voltage in a conductor.

Ohm was educated at the University of Erlangen. He held rather minor academic posts in Cologne, Berlin and Nuremberg before being appointed professor of physics at Munich in 1849.

Ohm formulated the law for which he is now best known early in his career, in 1827, but received little recognition for 20 years. Ohm's Law states that the current flowing in a conductor is directly proportional to the potential difference across it, provided there is no change in the physical conditions (eg temperature) of the conductor; the constant of proportionality is known as the conductance of the conductor, the reciprocal of the resistance. Ohm came to this conclusion by an analogy with FOURIER's work on heat flow along a metal rod. He also discovered that the human ear is capable of breaking down complex musical sounds into their component frequencies, an important conclusion but again one that was ignored at the time. The SI unit of electrical resistivity, the ohm (Ω), is named in his honour. It is defined as being the resistance of a conductor through which a current of one ampere is flowing when the PD across it is one volt, ie $1\,\Omega = 1\,V\,A^{-1}$. (The unit of conductivity, the inverse of resistivity, was formerly known as the mho and now the siemens.)

Olbers, Heinrich (1758–1840) German astronomer: discovered Pallas and Vesta, and presented the Olbers paradox.

Olbers, a physician and amateur astronomer, discovered two asteroids, Pallas and Vesta, and rediscovered the asteroid Ceres (discovered by

PIAZZI, but lost again). He suggested that the asteroids originated in a small planet in the same orbit which had exploded. He also found five comets, one of which is named after him, and devised an accurate method of calculating their orbits.

He is now best known, however, for his phrasing in 1823 of a deceptively simple, but important question: 'why is the sky dark at night?' This quandary, noticed by KEPLER in 1610 and discussed by HALLEY, became known as the Olbers paradox. He assumed that the stars are evenly distributed and infinite in number (as NEWTON proposed) and presented the thought that, in whatever direction we look in the night sky, it would be expected that the line of sight will end on the surface of a star. In which case, he argued, the entire night sky ought to have a brightness comparable to that of the Sun.

In modern cosmology the problem has been re-examined and the present answer seems to be that the expansion of the universe has the effect that, at a certain distance, objects are receding from Earth at the speed of light. This limits the seeable size of the universe, and within this limited radius there are not sufficient stars in all directions to yield a bright night sky. Alternatively, we can conclude that the universe is of finite size and contains a finite number of stars.

Oldham, Richard Dixon (1858–1936) British seismologist and geologist: first observed P and S waves and discovered the Earth's core.

Oldham was educated at the Royal School of Mines, in 1879 joining the Geological Survey of India (of which his father was director). Upon retirement in 1903 he became director of the Indian Museum in Calcutta. Following the violent Assam earthquake of 1897, Oldham was able clearly to distinguish in the seismograph record for the first time the P (primary, or compressional) and S (secondary, or shear) waves (predicted theoretically by POISSON) and the tertiary (surface) waves. In 1906 he discovered that at points on the Earth's surface opposite to the epicentre of an earthquake the P waves arrive later than expected, when compared with their arrival times at other places on the globe. He correctly recognized this as clear evidence for the existence of a relatively dense Earth's core, through which P waves travel more slowly than in the mantle.

In 1913 GUTENBERG showed that the core is at least partly liquid; and in 1936 INGE LEHMANN deduced that within it there is a solid inner core. Earlier than this, MOHOROVIČIĆ had shown that the Earth's outer crust must be less dense than the mantle, and had calculated the thickness of the crust. All of these geophysical studies depended on seismological results.

Oort, Jan Hendrik [oh(r)t] (1900–92) Dutch astronomer: detected the rotation of the Galaxy and proposed theory for origin of comets.

Oort worked mainly in Leiden, as director of the Observatory from 1945–70. In 1927 he proved, by extensive observation of the proper motion of stars, that our Galaxy is rotating, with the nearer stars to the centre having higher angular velocities than more distant ones (recalling that the inner planets of the Solar System move more rapidly than the outer ones). He established that the Sun is about 30 000 light years from the centre of the Galaxy and that it completes an orbit in about 225 million years, moving at 220 km s^{-1}. The Galaxy has a mass about 10^{10} times that of the Sun. He was influential in the discovery in 1951 of the 21 cm radio emission from interstellar hydrogen, which has allowed the distribution of interstellar gas clouds to be mapped. This technique has also revealed a large 'hidden' mass of stars at the centre of the Galaxy. In 1956 Oort observed that light from the Crab supernova remnant was strongly polarized, implying that it was synchrotron radiation produced by electrons moving at relativistic velocities through a magnetic field.

In 1950 Oort suggested the existence of a sphere of incipient cometary material surrounding the solar system, at a distance of about 50 000 AU and with a total mass of perhaps 10–100 times that of the Earth. He proposed that comets occasionally detached themselves from this Oort cloud and went into orbits about the Sun. Because the cloud is spherical comets can approach the Sun at any angle, and not just in the plane of the ecliptic.

Oparin, Alexandr Ivanovich (1894–1980) Russian biochemist: pioneered chemical approach to the origin of life.

Although his training and his work was in plant physiology in Moscow, Oparin's name is most familiar through his initiation from the 1920s of modern ideas on 'the origin of life', a phrase now much linked with him. He emphasized that early in the Earth's history its atmosphere did not contain oxygen (which was generated later, by plant photosynthesis); that simple organic substances could have been present in a 'primeval soup' before life began; and that the first organisms were probably heterotrophic, ie used organic substances as food and were not capable, as present-day autotrophs are, of feeding on simple inorganic substances. Oparin believed that the key characteristics of life are its organization and integration, and that the processes which led to it should be susceptible to reasonable speculation and experiment; he did much to make these attitudes respectable.

Oppenheimer, (Julius) Robert [openhiymer] (1904–67) US theoretical physicist: contributed to quantum mechanics and the development of the atomic bomb.

Oppenheimer was born into a wealthy New York family and was educated at Harvard and

SCIENCE AND THE SECOND WORLD WAR (1939–45)

The Second World War initiated projects of great importance both to the outcome of the war and to improvement of everyday life afterwards. Innovations directly attributable to the pressures of the war effort include radar, which was begun before 1939 in England and Germany with the British team led by WATSON-WATT; air and sea transport has benefited since and the microwave oven is a side product. Radio astronomy was shaped by radar equipment and methods. Operations research (a mathematical modelling technique) grew out of the radar programme and protection against submarine warfare. Penicillin, the first antibiotic, discovered by A FLEMING in 1928, was developed for clinical use by 1944 by FLOREY and CHAIN. Penicillin and related antibiotics have dominated the treatment of many infections since. Computers were initially developed in the USA to calculate artillery trajectories, and in the UK by TURING and others for decrypting the German 'Enigma' code messages. This decrypting programme ('Ultra') was notably successful and, by diminishing the effects of the Battle of the Atlantic and the (air) Battle of Britain, was critical in deciding the outcome of the entire war. Thereafter, computerized control has proved central to calculation in business, and developments include home entertainment and the control of space flight.

The German V-2 missile (the first ballistic missile) was developed by VON BRAUN for warfare; he later headed the NASA space probe programme in the USA. Pesticides (notably DDT, due to P H MÜLLER), curbed typhus during and after the war but their later use to control agricultural insect pests caused environmental problems; the work of RACHEL CARSON and others led to limitations in their use. During the war a new class of poison gases was developed in Germany and made on a large scale. These are the intensely toxic organophosphorus esters ('nerve gases') such as Sarin (the lethal dose for humans is below 1 mg). Although not used in the war (the reasons are unclear) related compounds have been much used since, as insecticides in agriculture.

Turbojet aircraft engines were developed by WHITTLE in England and in Germany by P von Ohain (1911–), and were first used in British military aircraft in 1941. After the war, jet propulsion largely replaced propellers in powering aircraft.

The Manhattan Project, which led to the atomic bomb and nuclear power, was work of massive scale and significance. At the time, its financial cost and the scientific and technical effort were vast, as were the military, political, energy-generating and environmental consequences. As those working on the project foresaw, the world was grossly changed after a controlled fission reactor, and later weapons based on both fission (the A-bomb) and fusion (the H-bomb), became practical realities from 1945. If the First World War was a chemist's war, the Second was a physicist's war.

Some novel schemes were only partial successes: for the Allied invasion of France in 1944, PLUTO, a fuel pipeline under the Channel, was partly successful; the huge transportable harbours (eg Mulberry) were valuable; the idea of a floating airfield of ice (refrigerated and reinforced with wood pulp), code-named Habakkuk, proved a false trail.

IM

Göttingen. In 1929 he took up posts at both the University of California at Berkeley and California Institute of Technology, having studied under RUTHERFORD, HEISENBERG and DIRAC whilst travelling in Europe. When the Manhattan Project (to develop an atomic bomb) was set up in 1942, Oppenheimer was asked to become director of the Los Alamos laboratories where much of the work was done. Having carried out this role with great skill, leading to the rapid development of the bomb, he attempted to remain a Government adviser on nuclear weapons, but was forced to resign in 1953. He became director of the Institute for Advanced Study at Princeton in 1947 and remained there after his retirement in 1966.

Oppenheimer's early success in research began in 1930 when he analysed DIRAC's relativistic quantum mechanics and theory of the electron (1928). He showed that a positively charged anti-particle with the same mass as the electron should exist, and this positron was first seen by C D ANDERSON in 1932. During the 1930s Oppenheimer built up a formidable team of young theoretical physicists around him, the first time that the subject had been studied intensely outside Europe. In 1939, working on stellar structure, he showed that any massive star, when its thermonuclear energy is exhausted, will collapse to form a black hole, which has mass but from which light cannot escape.

After 1942 as director of Los Alamos he concentrated on gathering scientists and generating an atmosphere of urgency, skilfully handling the interface between his military superior, General Groves, and the unorthodox research scientists under him.

Oppenheimer's wife and brother were left-wing sympathizers and possibly communists, and he ran into difficulties in 1943 when Groves demanded the name of a communist agent who had approached Oppenheimer; after much delay he finally gave it. The first atomic test explosion took place in July 1945, and two atomic bombs ended the war with Japan a month later.

After the war Oppenheimer initially continued his important role in atomic energy, but he opposed the development of the hydrogen (fusion) bomb. In 1953 his political background and his support for the Super program (the hydrogen bomb project) were questioned, and President Eisenhower removed his security clearance, ending his Government service. However, the Fermi Award was conferred on him by President Johnson in 1963, implying that doubts about his integrity had been resolved.

Ormerod, Eleanor (Anne) (1828–1901) British economic entomologist.

Eleanor Ormerod was born at Sedbury Park, her father's large estate in Gloucestershire, which provided the insects that prompted her interest in entomology and in the infestation of crops. Apart from an elementary education from her mother, a botanical artist, she was self-taught. One brother became an anatomist and surgeon and from him she gained experience of using a microscope. She became an expert on insect infestations.

She got in touch with the Royal Horticultural Society in 1868 and offered to compile a collection of insects injurious or helpful to farmers. The resulting collection was awarded the Flora Medal in 1870. She corresponded with entomologists throughout the world and published on insect pests and their control, often subsidizing and distributing the pamphlets free. In 1881 she published her *Manual of Injurious Insects, with Methods of Prevention and Remedy*, and in 1898 the *Handbook of Insects Injurious to Orchard and Bush Fruits, with Means of Prevention and Remedy*. Between 1877 and 1900 she undertook an Annual Report on economic entomology and during the 1880s she became a successful public lecturer. From 1882–92 she was consulting entomologist to the Royal Agricultural Society and in 1898 was recommended for a newly created lectureship in agricultural entomology at Edinburgh University, but women were not yet acceptable. In 1900 Edinburgh University awarded her their first honorary LLD offered to a woman. Eleanor Ormerod was a prime mover in making economic entomology an important specialty within biology and agricultural science.

Ostwald, Friedrich Wilhelm [ostvahlt] (1853–1932) German physical chemist: pioneer of modern physical chemistry.

Modern physical chemistry was largely created by three men; VAN 'T HOFF, ARRHENIUS and Ostwald. His parents were German but they had settled in Latvia, then under Russian domination. He had a happy childhood, with some hobbies of a fairly chemical kind; painting (he ground his own colours), photography (he made his own wet plates) and firework-making. He had to repeat one school year, and he had problems with the compulsory Russian language. He studied chemistry at Dorpat (now Tartu) University, did well and became professor at the Riga Polytechnic in 1881. His fame spread, and in 1887 he was called to Leipzig University, where he remained. His work in physical chemistry was wide-ranging; he studied the rates of hydrolysis of salts and esters, the conductivity of solutions, viscosities, the ionization of water and catalysis; he was awarded the Nobel Prize in 1909. He took up new ideas in physical chemistry with enthusiasm, did much to unify and expand the subject, and saw the study of the energetics of chemical reaction as central to the subject.

For a long time Ostwald believed that atoms were only a convenient hypothesis and had no real existence; but in the 20th-c direct evidence for them had arrived and by 1908 he was a late convert to 'atomism'.

Otto, Nikolaus August (1832–91) German engineer: effectively devised the four-stroke internal combustion engine.

Although he lacked conventional engineering experience, Otto became fascinated by the gas engine devised by J Lenoir (1822–1900); this was a double-acting low-compression engine and the first internal combustion engine to be made on any scale. Otto and two friends began to make similar engines; and then in 1876 he described the system usually called the Otto cycle, in which an explodable mixture of air and gas is drawn into the cylinder by the piston (the induction stroke), compressed on a second (compression) stroke, ignited near the top dead centre piston position (the combustion stroke, in which the hot expanding gases provide the power to drive the piston) and the burned gases are then driven out of the cylinder on a fourth (exhaust) stroke (see diagram, p. 252). The new engine was quiet and fairly efficient and sold well. However, in 1886 his competitors showed that A B de Rochas (1815–93) had suggested the principle in an obscure pamphlet, although he had not developed the idea, and this invalidated Otto's patent. Soon Otto's gas engine, mainly used in small factories, was developed for use with gasoline (petrol) vapour and air, using a carburettor to control the mixture and improved ignition systems; the resulting engine was well suited for the motor car. In these developments K Benz (1844–1929), G Daimler (1834–1900), W Maybach (1846–1929) and F W Lanchester (1868–1946) all played important parts, and the result is still dominant for this purpose. (See panel overleaf.)

THE HISTORY OF
THE HEAT ENGINE

An engine is a mechanical contrivance by means of which some form of energy is converted to useful work. The first mechanical utilization of an energy source to do work dates back to the 1st-c BC, when simple water wheels were used to lift water from rivers and to mill grain. Although it might be considered that these were examples of simple hydraulic 'engines', it is generally accepted that the term engine is associated with a much later period of technological development; that of the Industrial Revolution. Engines are normally associated with the conversion of fossil fuels (thermal energy sources) into useful work, and so we may fully refer to them as heat engines.

The most significant heat-engine development was that of the steam engine. The use of steam to produce a mechanical effect has its origin in the 1st-c AD, when the mathematician and inventor HERO of Alexandria described a steam-operated 'wheel' which utilized the thrust effect of escaping steam jets. This device was really only a toy, and no attempt was made to find a use for it. Hero demonstrated other interesting arrangements and, although none of them led to the development of a practical heat engine, he did go some way to demonstrate that when a fluid such as water or air is heated, it is possible to use it to bring about some form of mechanical effect.

In 1661, GUERICKE, in Magdeburg, invented a device consisting of a close-fitting piston within a pipe with a closed end. Using a vacuum pump (which he had also developed), he created a partial vacuum inside the pipe, the pressure there falling below that of the atmosphere. Atmospheric pressure acting on the other side of the piston then forced it into the pipe. The piston was connected by a system of ropes and pulleys to a weight which was then raised. This was the forerunner of the various heat engines which were generally categorized as atmospheric engines, since they all made use of the fact that atmospheric pressure, in conjunction with a partial vacuum created within a piston–cylinder arrangement, could do useful work.

It was the Frenchman Denis Papin (1647–1712) who first had the idea of constructing an atmospheric engine which made use of the evaporation and condensation properties of water. In 1689 he demonstrated what was in effect an embryo steam engine. It consisted of a vertical hollow cylinder with a base, above which was a piston. A small amount of water within the piston-cylinder arrangement was heated; steam was generated and it was allowed to expand, thereby pushing the piston up the cylinder. It was then held

by a catch while the steam cooled and condensed, with a consequent reduction in its pressure. When the catch was released, atmospheric pressure acting on the top of the piston forced it back to the bottom of the cylinder. Although Papin's arrangement had limited application, it was a significant step towards the goal of obtaining a practical steam engine.

In 1698 Thomas Savery (c.1650–1715), an English engineer, patented a steam pump which could be used to pump water out of mines. His machine was a simple arrangement with no piston and requiring the hand operation of valves. It used steam above atmospheric pressure. Since it did not have automatic safety valves, and as reliable boilers had not been perfected, the Savery pump proved to be dangerous and it was eventually abandoned. About the same period Thomas Newcomen (1663–1729) was also engaged in the design of a steam-operated pumping engine. Newcomen's design was different from Savery's, but as the latter's patent was very general and protected the rights for the raising of water 'by the impellent force of fire', it was necessary for Newcomen to liaise with Savery in 1705 to come to some arrangement regarding the manufacture of his engine. In 1712 the first practical Newcomen steam engine was constructed. It was an atmospheric engine and was thus dependent for its operation on obtaining a pressure within the cylinder below that of atmospheric pressure; this was achieved by injecting cold water into steam in the cylinder. This engine proved to be a significant breakthrough in atmospheric engine development, and Newcomen-type engines were manufactured and sold in numbers for mine pumping.

While preparing a model of a Newcomen engine for Glasgow University in the winter of 1763–64, WATT, a Scottish instrument-maker, realized that a considerable amount of heat was wasted by successively heating the cylinder to produce steam and subsequently cooling it to condense the steam. He proposed a major improvement to Newcomen's design, by the use of a cooling chamber (a condenser) separate from the steam cylinder. His design also incorporated an air pump which, by sucking air out of the condenser, created a partial vacuum which assisted condensation of the steam. This allowed a large increase in the overall efficiency of the engine. Watt obtained a patent for his engine in 1769. In order to exploit his invention commercially, he joined forces with the leading Birmingham manufacturer, Matthew Boulton (1728–1809). This was a very significant industrial partnership, and the widespread use of Watt's engine began in 1776–77. Watt's engine was much lower in coal consumption than Newcomen's, which was an attractive selling point. In 1782, Watt patented a double-acting engine in which steam was used alternately below and above the piston to produce a power stroke in both directions, thereby making it more suitable for the eventual use of a rotative motion, required to operate factory machinery. Both the Newcomen and Watt engines transmitted work through massive beams.

Further developments in steam engine design did not take place until the early 1800s, when steam above atmospheric pressure was investigated. The Cornish engineer Richard Trevithick (1771–1833) realized that the need to create a vacuum within the cylinder, requiring a cumbersome and heavy condenser, could be replaced by making use of high-pressure steam. This was a departure from the atmospheric type of engine. He perfected the first high-pressure steam engine in about 1803. In the same year he also built the first steam locomotive at the Coalbrookdale ironworks, thereby giving a further boost to the Industrial Revolution. From the 1830s steam locomotives dominated rail transport for over a century.

In parallel with the early development of the steam engine, several inventors pursued the idea of an engine in which the combustion of a fuel would take place within a piston-cylinder arrangement. In 1673 the Dutch scientist Huygens demonstrated a piston engine to the French Académie des Sciences. It was an atmospheric-type engine which made use of a small quantity of explosive. It worked by displacing cold air from the cylinder by means of the hot gases from the explosion. When the hot gases remaining in the cylinder cooled and contracted, the cylinder gas pressure was lower than the atmospheric pressure outside the arrangement and, like other atmospheric-type engines, movement of the piston was effected. This invention was unfortunately thwarted by the inability to find a suitable fuel. It was much later that it was found that when coal is heated in a closed vessel a combustible gas (coal gas) is given off, which forms a suitable fuel.

In 1816, Robert Stirling (1790–1878), a Scottish clergyman, developed another concept; that of an engine with air as the operating medium. His engine consisted of two cylinders. In one of them air was heated (by an external source) and cooled alternately. When the air expanded, it effected a power, or working, stroke in the other cylinder. This engine was a closed-circuit, hot-air type, and its operating principle was later seen to be based on excellent thermodynamic considerations. Stirling obtained a patent for his design in 1827. Some engines were manufactured industrially in 1844, but they never attained mass production. Later, Stirling engines used helium and hydrogen. (Even up to the present time, these engines, owing

to their stringent manufacturing requirements, are confined to special applications, such as submarines and spacecraft.)

In 1824 Carnot published his work which, in discussing steam engine efficiency, created the new science of thermodynamics. The next 30 years saw several designs and patents for gas engines, but most of them were never constructed; those that were had limited success. The first internal combustion engine able to operate reliably was built by the Belgian-born inventor Jean Joseph Lenoir (1822–1900). In 1860, he patented a well-thought-out design for a gas engine. The fuel was lighting gas (derived from coal) mixed with air. This engine was a two-stroke, double-acting design, with the fuel-air mixture fed into the cylinder alternately at either end of the piston. The Lenoir engine was slow-running (200 rpm), and the gas-air mixture was not compressed before ignition. It lacked power, and it had a very high fuel consumption. Despite this, it sold in reasonable numbers, and so became the first internal combustion engine to compete with the long-established dominance of steam. In 1862, the French engineer Alphonse Beau de Rochas (1815–93) described the much more efficient four-stroke cycle. However, his proposed engine was never constructed, and he allowed his patent to lapse. It was the German engineer Otto who constructed the first four-stroke internal combustion engine in 1876. This design was a significant one from the viewpoint of the development of the motor car. These early internal combustion engines all operated on gas. The use of liquid fuels was not introduced until near the end of the 19th-c. In 1883, the German engineers Gottlieb Daimler (1834–1900) and Wilhelm Maybach (1846–1929) designed an engine that could operate on petrol. It ran faster than Otto's engine and was capable of obtaining more power for a given weight of engine. In 1889 the engine was installed in a car designed by Maybach, which is considered to be the first modern motor car. The heavy-oil engine was pioneered in Britain in 1890 by Herbert Akroyd-Stuart and improved by the German engineer Diesel. In 1893, Diesel patented a prototype four-stroke engine. This engine differed from the petrol engine in that ignition of the fuel occurred spontaneously without a spark. The high compression attained within the cylinder resulted in sufficiently high temperatures to bring about ignition of the fuel and effect a pressure stroke. The Diesel engine has of course found wide use for both marine and land transport.

Another, more recent development in internal combustion engine design was the rotary engine, invented in 1956 by the German engineer Wankel. In this engine the conventional reciprocating action of pistons in cylinders is replaced by the

Four-stroke engine cycle. (1) Fuel and air drawn into cylinder. (2) Fuel/air mixture compressed and ignited by spark. (3) Power stroke. (4) Exhaust gases expelled.

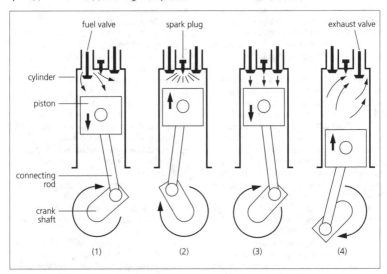

rotary motion of an ingeniously designed rotor within a specially shaped chamber. This design has not found widespread application within the automobile industry, in which conventional petrol engines and Diesel engines seem destined to continue their dominance.

Dr W K Kennedy, Open University

Palade, George Emil (1912–   ) Romanian–US cell biologist: discoverer of ribosomes.

Palade qualified in medicine at Bucharest and became professor of anatomy there. When Soviet forces entered Romania in 1945 he moved to the USA, working first at the Rockefeller Institute and from 1972 at Yale. His work was particularly on the fine structure of cells as revealed by electron microscopy. He showed beyond doubt that one type of organelle, the mitochondrion (typically 1000 of these small sausage-shaped structures are present in each animal cell) form the sites where energy (in the form of adenosine triphosphate, ATP) is generated by enzymic oxidation, to meet the energy needs of the cell. In 1956 he discovered smaller organelles (now called ribosomes) found to be rich in ribonucleic acid (RNA) and showed that they are the sites of protein synthesis. Subsequently he worked out in detail the pathway followed by secretory proteins in glandular cells. He shared a Nobel Prize in 1974.

Pappus (of Alexandria) (lived *c*.300) Greek mathematician: made major contributions to geometry.

The name of a son is the only detail known of Pappus's personal life; but his account is the main source of knowledge of parts of Greek mathematics before his time, and his own contributions to geometry are substantial. Theorems named after him deal with the volume and surface generated by a plane figure rotating about an axis in its own plane. His work was the high point in the field of Greek geometry.

Paracelsus (*Lat*), Theophrastus Bombastus von Hohenheim (*Ger*) (1493–1541) Swiss alchemist and physician: pioneer of medical chemistry.

Paracelsus's father was a physician working near Zürich, who gave his son his early medical training. The young man travelled widely before settling in Strasbourg. There he achieved cures for some influential people and as a result was appointed City Physician in Basle. In lectures and books he pressed the view that alchemy should be directed not only to transmuting base metal into gold, but principally to preparing effective medicines. His ideas found support, but he used such offensive language in abusing opponents that, following a legal case which he lost, he had to leave Basle, and he died in Salzburg. He was certainly a loud-mouthed and often drunk and boastful mystic; but he probably did much to deflect alchemy towards improving medical chemistry.

His theoretical ideas were too clouded in mysticism to be useful, but in practical medicine he was more effective. He was one of the first to study occupational diseases and he recognized silicosis as a hazard for miners. He realized that goitre and cretinism are related; and he used morphine, sulphur and lead in medicine, and mercury, with which he treated the then new disease syphilis. He gave good descriptions of several types of mental disease, which he saw as an illness and not as due simply to demons. However, he firmly claimed that it is possible to create human life in the laboratory and gave full experimental detail on how to achieve this, starting with the fermentation of a sample of semen.

Parsons, Sir Charles Algernon (1854–1931) British engineer and inventor: designed the first effective steam turbine.

Some of Parsons's talents can be seen in his parents; his mother was a talented modeller and photographer and his father (the Earl of Rosse) was an astronomer who made and used some outstanding telescopes (his 65 in (1.65 m) reflector at Parsonstown in Ireland was without rival, and he used it to discover much about nebulae) and was president of the Royal Society.

Charles studied in Dublin and Cambridge, did well and then became an engineering apprentice. His firm had interests in electric lighting and he saw the need for a high-speed engine to drive dynamos. For this he devised the multistage steam turbine, which he patented in 1884; it used high pressure steam and ran at up to 20 000 rpm. Parsons set up his own company to make turbo-generators, and from the early 1890s developed his turbines also for marine use. For this he devised reduction gearing, and he also studied the cavitation due to propeller blades and improved their design. At the 1897 Naval Review on the Solent to celebrate the Queen's Jubilee, his 48 m *Turbinia*, with Parsons at the wheel, created a sensation; its 2000 hp moved it at an unheard-of 34½ knots. By 1906 his turbines were fitted in the warship HMS *Dreadnought*, and soon the great Cunarders followed suit; some of his turbines generated 70 000 hp.

He went on to design searchlights for naval use, and large telescopes. The company of Grubb Parsons have retained their special position in this field. Parsons used his scientific and mathematical skill to stride ahead of existing engineering practice and to become a leading engineer of his time, the first to join the Order of Merit and the most original British engineer since WATT. Power generation and marine propulsion were never to be the same after his work.

Pascal, Blaise [paskahl] (1623–62) French mathematician, physicist and philosopher: pioneer of theory of probability.

Educated by his father, Pascal showed early intellectual ability, proving one of the most important theorems of projective geometry by the age of 16. He was fervently religious, belonging to the rigorous Jansenist sect of the Roman Catholic church. He was also neurotic, dyspeptic and humourless.

Much of Pascal's early work was on projective geometry, developed from his paper on conic sections of 1640 from which he deduced 400 propositions, deriving most of those put forward by APOLLONIUS. The most notable was Pascal's theorem (for any hexagon inscribed in a conic, the intersections of opposite pairs of sides are collinear), also known as the problem of PAPPUS, which had been used by DESCARTES as a test case for the power of his own analytical geometry. At the age of 19 Pascal invented a calculating machine that could add and subtract, in order to help his father with his business; he built and sold about 50 and several survive. Later, his interest moved towards physics, demonstrating with the help of his brother-in-law that air pressure decreased with altitude as TORRICELLI had predicted, by taking a mercury barometer to the summit of Puy de Dôme (a height of 1200 m, near Clermont Ferrand) in 1648. His interest in hydrostatics also led him to demonstrate that pressure exerted on a confined fluid is constant in all directions (Pascal's Law). Together with FERMAT, he also developed the mathematics of probability and combinatorial analysis, using the familiar Pascal triangle to obtain the coefficients of the successive integral powers of the binomial $(p+q)^n$. In 1655, after a profound religious experience, Pascal entered the Jansenist retreat at Port Royal, where his sister was already a nun, and he did little further mathematical work. His philosophical work *Pensées* was published in 1670. The SI unit of pressure (or stress), the pascal (Pa), defined as a force of one newton per square metre, commemorates his work on hydrostatics, and the modern programming language Pascal marks his contribution to computing.

Pasteur, Louis [paster] (1822–95) French chemist and microbiologist: founder of stereochemistry and developer of microbiology and immunology; exponent of germ theory of disease.

Pasteur is one of the greatest figures in science, who made major changes in all the fields in which he worked. He was enormously talented, with great powers of scientific intuition; he was also ambitious, arrogant, combative and nationalistic.

His father served in the Peninsular War and then returned to the family tanning business in Dôle, near the Swiss border. Louis was the only son; there were three daughters. His school record was only moderate, but just good enough for him to go to Paris and to hope for entry at the teacher training college, the École Normale. In preparing for this for a second time (the first time his physics was classed as 'passable' and chemistry 'mediocre'), he went to lectures on chemistry by DUMAS, along with 700 other students. The subject captured him, he became a 'late developer', and all his future work showed a chemical approach even to biological problems.

His first major research, done at the École Normale, concerned tartaric acid (a by-product in wine making). BIOT had shown that one form of the acid is optically active (ie it rotates polarized light when in solution). Pasteur examined a salt of the optically inactive form of tartaric acid and showed that the crystals were of two kinds, which were non-superposable mirror-images of each other (ie were dissymmetric). He separated these and showed they were both active, with equal and opposite rotation. He deduced, correctly, that the molecules themselves must therefore be dissymmetric, a fundamental idea and one that was more fully explored by VAN 'T HOFF. It was the beginning of stereochemistry.

This work had interested Pasteur in fermentation, and when he became professor of chemistry at Lille in 1854 he found this useful, because alcohol-making was Lille's main industry. Back at the École Normale from 1857 he continued this interest, which was to carry him from chemistry to biology and from there to medicine. In becoming a microbiologist and improving wine- and beer-making technology in his early middle-age, Pasteur became convinced that spontaneous generation did not occur. Work by SPALLANZANI and by SCHWANN should have established the view expressed by VIRCHOW: 'all cells come from cells'. But the experimentation is not easy and the debate continued. However, Pasteur's now-classic studies showed in the early 1860s that putrefaction of broth and fermentation of sugar did not occur spontaneously in sterile conditions, but could be readily initiated by airborne microorganisms. He put his view with typical force, and it has been generally accepted since. He introduced pasteurization (brief, moderate, heating) to kill pathogens in wine, milk and other foods. Since fermentation, putrefaction and suppuration of wounds were fairly widely regarded as kindred processes, it was reasonable for LISTER to use Pasteur's principles to revolutionize surgery, but Pasteur himself was not involved in this. However, in 1860 he said he planned to work in medicine, and he eventually achieved his own revolution there.

His first experience with animal diseases was with silkworms, then a major French industry but much threatened by infections. Pasteur, helped by a microscope and his fermentation experience and with wife, daughter and several assistants acting with him as novice silk-growers, fairly soon established procedures to deal with the two infections then rife. Then, in 1868 when he was 46, he had a stroke; he was fully paralysed for 2 months and partly paralysed thereafter. His work habits were unchanged, but his irritability increased. Most experiments now had to be performed under his direction but not with his own hands. In one way this suited him; he disliked vivisection, but saw it as essential for some of his research, and preferred others to perform the work.

Only in the late 1870s did Pasteur achieve success against a disease in a larger animal. In 1879 by a fortunate chance he noted that if a chicken-cholera bacillus culture was 'aged' or 'attenuated' by storage, it failed to produce the disease in chickens on injection; but the injected chickens (after an interval) were resistant even to infection by a fresh culture. He deduced that the change in virulence of the culture on attenuation, so that it protected but did not infect, could be compared with the use of the mild cowpox vaccine against the virulent smallpox, studied by JENNER before 1800. Pasteur used the idea to make a vaccine against anthrax (a major disease of cattle, and sometimes found in man). The scheme worked well, and Pasteur staged a demonstration for agriculturists in May 1881 on a farm. A herd of 50 sheep, cows and goats was divided into two equal groups, and one group of 25 was inoculated with 'attenuated' vaccine. After 2 weeks all 50 were given an injection of a strong anthrax culture. Two days later, a crowd formed to see the dramatic result: the protected 25 were all healthy; of the others, 22 were dead, two dying, one sickening.

In 1880 he began to study rabies. The work was dangerous, and difficult because the first step he wished to take – isolation of the pathogen – was not then possible; it is now known to be a virus. However, he could inject dogs, guinea pigs and rabbits with rabid saliva and thereby infect them; and he found that spinal cord from a rabid rabbit, if kept in dry air for a few days, formed an attenuated vaccine which could be used to protect and to treat other animals. However, he was understandably fearful of human trials. Then in 1885 he was brought 9-year-old Joseph Meister, who had been bitten 14 times by a rabid dog. The child was treated with the vaccine, survived and later became a caretaker at Pasteur's institute. In 1886 2671 patients were treated, and only 25 died. This success made Pasteur world-famous, and an Institute was built for his research, by public subscription; it was opened in 1888. But by then he was old and ill, and rabies was his last success. Pasteur's notebooks were not opened to the public until 1971. It then emerged that some of his trials were very inade-

quate and some claims improved on the record; for example, the vaccine used on Meister appears to have been tested on only a few dogs and not on 50 as Pasteur later claimed.

Medicine was never the same after his work; infectious disease could now be combated by established techniques and research guided by a general theory. Vaccines were sought against most major diseases, but only in some cases could a vaccine be made. Pasteur had all the marks of genius, including an intuition on when to continue with a study of details until success followed and when to leave a field for other workers to explore. He had a number of distinguished co-workers; the best was his wife. He was buried in the chapel of the Pasteur Institute in Paris. In 1940 the invading Nazis ordered Meister to open the ornate crypt for inspection, but the gatekeeper chose to kill himself rather than do so.

Pauli, Wolfgang [powlee] (1900–58) Austrian–Swiss–US physicist: discovered the Pauli exclusion principle in quantum mechanics.

The son of a professor of physical chemistry at the University of Vienna, Pauli obtained his PhD at Munich in 1921. Encouraged by SOMMERFELD, Pauli had written (when he was only 19) an article, subsequently published as a small book, on relativity that was admired by EINSTEIN for its 'deep physical insight'. Pauli studied further with BOHR in Copenhagen and BORN in Göttingen. He then taught at Hamburg and gained a professorship in 1928 at the Federal Institute of Technology, Zürich, remaining there until his death, except for 5 war years spent at Princeton (when he became a US citizen).

Working on quantum mechanics, he contributed the Pauli exclusion principle (1924), which explained much about atomic structure. The principle requires that no two electrons in an atom can be in the same quantum state. The original Bohr–Sommerfeld model of the atom (1915) specified for each electron in an atom three quantum numbers (n,l,m) and Pauli additionally required the electron to have another, called the spin quantum number $s = \pm\frac{1}{2}$. Pauli's principle that no more than one electron is able to occupy a state described by n, l, m and s then gave the correct formation of electronic shells in atoms, gave a theoretical basis for the periodic classification, and explained the ZEEMAN effect of atomic spectra. This concept of spin, able to have one of two values, was verified experimentally by GOUDSMIT and UHLENBECK in 1926. For his idea of the exclusion principle Pauli was awarded the Nobel Prize for physics in 1945. It was much overdue.

Pauli also studied the relation between the spin of a particle and the statistics of energy level occupancy (quantum statistics); the paramagnetic properties of gases and metals (including electrons in metals); the extension of quantum mechanics from one to a large number of particles; the explanation of the meson and the nuclear binding force.

Furthermore, Pauli solved a major problem concerning beta decay, in which atomic nuclei eject electrons and apparently contravened the conservation of energy principle. The energies of emitted electrons cover a continuous range up to a maximum value, and it was unclear what happened to the 'missing' energy if an electron had less than the maximum. Pauli realized that this energy could be carried off by an undetected, very light neutral particle (named the neutrino, Italian for 'little neutral one', by FERMI) emitted at the same time as the electron. This was correct and the neutrino was first observed more directly by F Reines (1918–) in 1956.

Pauli had a caustic wit; he was not a good lecturer and he was notoriously bad as an experimentalist; but he is one of the giants of 20th-c theoretical physics.

Pauling, Linus (Carl) [pawling] (1901–94) US chemist: the outstanding chemist of the 20th-c.

Pauling's chemical beginnings were very ordinary. He grew up in a country area in Oregon, and his father (a pharmacist) died when he was 9; the boy began experimenting with chemicals when he was 11 and continued at school. By 15 he had decided to become a chemical engineer. He attended the small Oregon Agricultural College and did well enough (especially in chemical analysis) to be paid to teach first-year students. He went on to the California Institute of Technology, working for his PhD on X-ray studies of inorganic crystals. He read intensively and his memory was remarkable. He began to develop a scheme to assign sizes to atoms in crystals, and used these dimensions to work out the structures of a wide range of minerals, including eventually the silicates and some other major groups which had previously been seen as a structural mystery. After his PhD in 1925, he spent 2 years studying in Europe, mainly in Germany with SOMMERFELD. This led him to an extensive study of the use of quantum theory in understanding chemical bonds. By the early 1930s he had largely developed the valence-bond (VB) approach to bonds, using concepts such as 'hybridization' of bonds and 'resonance' for calculating bond energies, lengths and shapes and magnetic properties. He also devised an electronegativity scale, valuable in predicting bond strength. From his return to CIT in 1927 he held a position there for 35 years, together with others in California.

In the 1930s he began to work on biochemical problems, beginning with X-ray studies on the precise shape of amino acids and peptides. From this he went on to deduce two model structures for proteins: these are the 'pleated sheet' and 'α-helix' types, both found in important biological structures. Other ventures in biochemistry included theories of the chemical basis of anaesthesia and of memory; and with DELBRÜCK he studied the structure and action of antibodies. Here he used the new idea of 'complementary structures in juxtaposition'. This last idea, together with his ideas on helical structures in biomolecules and on hydrogen-bonding as an important determiner of their shape, form the key aspects of CRICK and WATSON's model of DNA as a self-replicating double helix. Also in the 1940s he proposed that sickle-cell disease (a genetic anaemia) resulted from a change in the normal amino acid content of haemoglobin; proof of this gave the earliest example of a disease being traced to its precise origin at the molecular level.

Pauling's work has generated some controversy: his views on the value of a high vitamin C level in the diet in combating a range of ills from the common cold to old age are not universally accepted and his political views in pursuit of world peace led to problems (his passport was withheld for a time). He won two Nobel Prizes; for chemistry in 1954, and for peace in 1962. His elementary texts remain among the best available.

His work in science is exceptional in its range, covering inorganic and organic chemistry, theoretical chemistry and practical devices, work on minerals and in biology. His work in chemistry is without peer in the 20th-c in its vitality, vision and significance. His contributions to novel chemical theory continued in his 80s.

Pavlov, Ivan Petrovich (1849–1936) Russian physiologist: discoverer of the conditioned reflex.

When Pavlov graduated in medicine at St Petersburg in 1875 he had already done useful research in physiology, and his interest was mainly in this field rather than in practical medicine. In the 1880s he studied in Germany and from 1890 he held research posts in St Petersburg, finally building up a very large research centre.

In the 1890s Pavlov studied digestion, using great surgical skill to modify dogs so that, for example, part of the stomach (a 'Pavlov pouch') could be separated from the rest and its gastric juice collected. He discovered the secretory nerves to the pancreas, and he studied the nerves and action of the salivary glands. During this work he noted the way in which dogs salivate when stimulated by the routine of feeding, even before the arrival of food; this led him to the study of reflexes which became his life-long and best-known work, although his Nobel Prize of 1904 was for his work on digestion. He knew food in a dog's mouth causes gastric juice to flow; this is an example of an unconditioned reflex. If a bell is always rung before food appears, the dog will soon salivate when the bell is rung even without the food.

Such conditioned or trained reflexes are easily induced in dogs, but can be established in other animals; they can be linked with stimuli other than sound and depend on a response in the cerebral cortex. Pavlov did extensive and ingenious work on such reflexes, and since then others have studied conditioning both in the laboratory and in the wild, and in vertebrates and invertebrates. Pavlov was a critic of Soviet communism and tried to move abroad in 1922 but failed. Despite this his work was well funded, and it remains more highly regarded in Russia than outside it, where it has certainly given a psychological dimension to physiology but even more has contributed to behaviourist approaches to psychology.

Payne-Gaposchkin, Cecilia (Helena), *née* **Payne** (1900–79) British–US astronomer: the most eminent woman astronomer and a founder of modern astrophysics.

As a very English middle-class child, Cecilia Payne was sent to a church primary school. Precocious as a pupil in mathematics, languages and music, she also tested the efficacy of prayer by dividing her exams into two groups, and prayed for success in one group only; she got better marks in the group without divine help, and was agnostic thereafter. She went on to St Paul's Girls School, Hammersmith, and to Newnham College Cambridge in 1919. In that year, EDDINGTON lectured on the 'eclipse expedition' he had led whose results had verified EINSTEIN's predictions on the deflection of stellar images close to the Sun, and thereby provided experimental support for relativity theory. Cecilia Payne by chance attended the lecture; the experience changed her life and created the devotion to physics and astronomy which dominated it thereafter.

After Cambridge she took a fellowship to the Harvard College Observatory at Cambridge MA and spent her career there. The vast collection of stellar spectra gathered and classified by Williamina Fleming (1857–1911), ANTONIA MAURY and ANNIE CANNON was available to her and provided rich material for her PhD thesis (the first in astronomy approved by Harvard) which was published in 1925 as *Stellar Atmospheres*. In it she used their data to deduce temperature, pressure and composition for a variety of stars, concluding that their composition was surprisingly constant, with helium and hydrogen as dominant constituents (although she hardly believed this herself for a time). Her work gave a new basis for astrophysics, although it was written too early for nuclear reactions to have a place in her thinking. Before she was 30 she had written another major book, on *The Stars of High Luminosity* (1930), the stars much used to find stellar distances. She never lost her passion for stellar spectra, but by 1942 she turned to the study of variable stars.

Valued by her peers but always underpaid, she became the first woman professor at Harvard only in 1956. By then she had married (a fellow astrophysicist, Sergei Gaposchkin, in 1934), had a family, travelled widely and been widely honoured. She was an active researcher until shortly before her death.

Peano, Giuseppe [payahnoh] (1858–1932) Italian mathematician: introduced the Peano axioms into mathematical logic.

Peano grew up on a farm and from the age of 12 was taught privately in Turin. On winning a scholarship to Turin University his talent was fully revealed, and by 32 he held a professorship there. The lack of rigour in mathematics provoked him (like DEDEKIND) to try to unravel the areas where intuition had concealed the logic of analysis. During the 1880s he looked at the integrability of functions; he proved that first order differential equations $y' = f(x,y)$ are always solvable if the function f is continuous. In 1890 this was generalized to the first statement of the axiom of choice. In the same year he demonstrated a curve that is continuous but filled space, indicating that graphical methods are limited in the analysis of continuous functions.

Peano worked on the application of logic to mathematics from 1888, producing a new notation. He also wrote down a set of axioms that covered the logical concept of natural numbers (Peano axioms), later acknowledging that DEDEKIND had anticipated him in this. Peano's work in this area was probably more important than that of the other great figures – BOOLE, F L G Frege (1848–1925) and B Russell (1872–1970).

Peano also initiated geometrical calculus by applying the axiomatic method to geometry. However after 1900 his interests shifted and he did no more creative mathematics, working on a general European language (Interlingua) and on the history of mathematics.

Pearson, Karl (1857–1936) British statistician: pioneer of statistics applied to biology.

Pearson's father was a barrister and Karl also qualified in law but never practised; but he did well in Cambridge in mathematics and afterwards studied physics and biology in Germany. At his Cambridge college he successfully rebelled against compulsory chapel attendance and then infuriated the authorities by occasional appearances there. In 1884 he became a professor of mathematics at University College, London and was soon influenced by two colleagues there, GALTON and W Weldon (1860–1906) – both enthusiasts for the application of arithmetic to the study of evolution and heredity. In the 1890s Pearson developed statistical methods for a range of biological problems, largely published in a journal he did much to found, *Biometrika*. His forceful and effective work led him to define standard deviation (an idea already well established) and to break new ground on graphical methods, probability theory, theories of correlation and the theory of random walk. In 1900 he devised the chi-square test, a measure of how well a theoretical probability distribution fits a set of data, that is valuable in showing, for example, whether two hereditary features (eg height and eye colour) are inherited independently; or whether one drug is more effective than another. His productivity was enormous, right up to his death, and his work largely founded 20th-c statistics.

Peierls, Sir Rudolf Ernest [payerlz] (1907–95) German–British theoretical physicist: contributed to solid-state physics, quantum mechanics and nuclear physics.

Peierls was educated at Berlin, and then studied under SOMMERFELD in Munich, HEISENBERG in Leipzig and as PAULI's assistant in Zürich. Research in Rome, Cambridge and Manchester followed and in 1937 he was appointed professor at Birmingham. In 1963 he moved to Oxford and from 1974–77 to the University of Washington, Seattle.

Peierls began research in physics during the dawn of quantum mechanics in 1928; the basic theory was complete but its applications to almost every physical system hardly begun. Peierls studied the theory of solids and analysed how electrons move in them, concentrating on the effect of magnetic fields (notably the Hall effect). In 1929 he explained heat conduction in non-metals, predicting an exponential growth of their thermal conductivity at low temperatures, as was verified in 1951. He also developed the theory of diamagnetism in metals.

Turning in 1933 to nuclear physics, he began to work out how protons and neutrons interact, and in 1938 showed how resonances (or dramatic increases in interaction) occur at particular beam energies in nuclear collisions. After the Second World War had begun, Peierls and FRISCH studied uranium fission and the neutron emission that accompanies it with a release of energy. In an influential report (1940) they showed that a chain reaction could be generated in quite a small mass of enriched uranium, giving an atomic bomb of extraordinary ferocity. The British Government took this up, and Peierls led a theoretical group developing ways of separating uranium isotopes and also calculating the efficiency of the chain reaction. The work was moved to the USA as part of the combined Manhattan Project (1943), and by 1945 yielded the first atomic weapons and quickly brought the war with Japan to an end.

Pelletier, Pierre-Joseph [peltyay] (1788–1842) French chemist: founder of alkaloid chemistry.

Pelletier followed the family profession of pharmacy and studied at the École de Pharmacie in Paris, and afterwards taught there. From about 1809 he began to examine natural products, using mild methods of separation (mainly solvent extraction) rather than the older methods such as destructive distillation. An early success was his isolation of chlorophyll from green leaves; and in 1817 with MAGENDIE he isolated the emetic substance from ipecacuanha root and named it emetine. In the next few years he isolated a series of 'vegetable bases' from plants; they are the cyclic nitrogenous bases now known as alkaloids, which often show potent physiological properties. With his friend J Caventou (1795–1877) he isolated strychnine, brucine, colchicine, veratrine and cinchonine, and (later) piperine and caffeine. Their most valuable discovery was quinine from cinchona bark (1820), which for a century was the only effective treatment for malaria and was almost the earliest chemotherapeutic agent in the modern sense.

Penrose, Sir Roger (1931–) British theoretical physicist: major contributor to theories on black holes.

Son of a distinguished geneticist and expert on mental defects, Penrose studied at University College, London, and Cambridge. After posts in London, Cambridge and the USA he became professor of applied mathematics at Birkbeck College, London (1966), and Rouse Ball Professor of Mathematics in Oxford (1973).

Penrose revealed many of the properties of black holes by his research. Black holes occur when large stars collapse and reach a density such that even light (photons) cannot escape from the intense gravitational attraction. The 'event horizon' marks the region within which light cannot escape. HAWKING and Penrose proved that a space-time singularity (a point, having mass but no dimensions) arises at the centre of a black hole, and Penrose established that event horizons always prevent us from observing these singularities from the outside. They also argued, in 1970, that the universe must have begun as a singularity, from which the 'Big Bang' developed.

However, if a black hole is rotating but uncharged, (a Kerr black hole) it possesses a region around it in which matter will always be broken into one mass that falls inside the hole and the remainder, which is ejected. Curiously, the ejected mass-energy must exceed that of the original matter, so that the Kerr black hole has lost mass-energy on accreting matter. Overall, Penrose's research adds much to our knowledge of gravitation, and he added to the efforts to formulate a satisfactory quantum theory of gravity, which are as yet unsuccessful.

Penzias, Arno Allan (1933–) US astrophysicist: discovered the 3 K microwave background radiation.

A refugee from Nazi Germany, Penzias was educated in New York and joined Bell Telephones in 1961.

In 1948 GAMOW, ALPHER and R Herman (1914–) hypothesized that the radiation released during the 'Big Bang' at the creation of the universe ought to have permeated the universe and progressively cooled, to a present-day temperature of about 5 K. In 1964 DICKE and P J Peebles (1935–) at Princeton repeated and extended this theoretical work. At the same time, but unknown to them, Penzias and his colleague R W WILSON were exploring the Milky Way with a radio telescope having a 6 m horn reflector at the Bell Telephone Laboratories in New Jersey, a mile or two away. Working at a wavelength of 7 cm they found more radio noise than they had expected, or could account for from any known terrestrial source (they even excluded the effect of pigeon droppings on the radio telescope's surface). The signal was equally strong from all directions (including apparently empty sky), and corresponded to that emitted by a black body at about 3.5 K. Their discovery provided some of the strongest evidence for the 'Big Bang' theory for the origin of the universe, and is arguably the most important discovery, bearing on cosmology, made in this century. Penzias and Wilson were awarded the Nobel Prize for physics in 1978.

Perey, Marguerite [pairay] (1909–75) French nuclear chemist: discoverer of actinium K (francium); the first woman to be elected to the Académie des Sciences.

Marguerite Perey's first wish was to study medicine, but the death of her father made that impossible. She joined MARIE CURIE'S staff as a junior laboratory assistant in 1929, and almost left when she found the conditions of work so austere, but she remained at the Institute for 20 years. In 1939 she discovered a new chemical element, which she named francium (Fr) after her country. It had been intensively looked for by many experienced researchers, and was found by the modest 29-year-old technician. She found this by a careful study of the radioactive disintegration of the rare natural radioelement actinium, showing that, by a series of changes, this is converted into francium. The element is one of the isotopes with atomic number 87, and is the heaviest of the alkali metal group and the most electropositive element known. It is exceptionally rare in nature, is itself radioactive and is now made artificially by atomic bombardment. It has been estimated that the entire Earth's crust (to a depth of 1 km) contains only 15 g of francium, and its rarity and intense radioactivity both point to the elegant work and skill of its discoverer. Encouraged to take a degree course during the Second World War, she obtained her degree in 1946.

Perey became head of research and, subsequently, administrator and director of the nuclear research centre at Strasbourg, and held a chair there from 1949. She received the Legion d'Honneur, Grand Prix Scientifique de la Ville de Paris (1960), Lauréate de l'Académie des Sciences (1950 and 1960) and the Silver Medal of the Chemistry Society of France (1964). She was elected to the Institut de France as the first woman member of the Académie des Sciences in 1962.

Perkin, Sir William Henry (1838–1907) British chemist: made first synthetic dye and founded organic chemical industry.

Young Perkin's interest in chemistry began in a familiar way. He records that when he was about 12, 'a young friend showed me some chemical experiments and the wonderful power of substances to crystallize in definite forms especially struck me... and the possibility also of making new discoveries impressed me very much.... I immediately commenced to accumulate bottles of chemicals and make experiments.' Despite his father's opposition Perkin entered the Royal College of Science to study chemistry at 15. At 17 he was assisting HOFMANN there, and also doing some research at home. Hofmann had mentioned the desirability of synthesizing quinine. Perkin, at home for Easter in 1856, tried to make quinine by oxidizing aniline. The idea was quite unsound, but Perkin noticed that the dark product contained a purple substance, later named mauve, that dyed silk. At age 18, helped by his father, he set up a small factory to make his 'mauve' and, later, other synthetic dyes based on coal tar products. Remarkably, they dealt successfully with the novel problems of chemical manufacture and marketing, although little commercial equipment or material was available (they even had to make nitric acid, and re-purify coal tar benzene) and their skills were those of the 18-year-old boy and his retired builder father. Young Perkin even maintained his academic research, solving by 1860 some important problems on organic acids and synthesizing the amino acid glycine. Mauve manufacture went on for 10 years; it was used for textiles and the Victorian 1d lilac postage stamp. Later Perkin manufactured magenta and alizarin dyes. By age 36, he was able to retire as a dyemaker and pursue his research exclusively. He developed a general synthesis of aromatic acids (the Perkin reaction) and studied magnetic rotatory power.

Chemical interests seem to run in the family; a grandfather had a laboratory in his Yorkshire farmhouse and Perkin's three sons were all distinguished organic chemists. His own venture began the synthetic organic chemical industry, in which leadership soon passed to Germany. Academic organic chemistry was maintained in Britain especially by his son Professor W H Perkin Jr (1860–1929) and his pupils at Edinburgh, Manchester and Oxford.

Perrin, Jean Baptiste [perï] (1870–1942) French physical chemist: gave first definitive demonstration of the existence of atoms.

Perrin studied in Lyon and Paris, and in 1910 became professor of physical chemistry at the Sorbonne, but fled to America in 1941.

While studying for his doctorate, Perrin investigated cathode rays, showing them to be negatively charged and obtaining a rough value of their charge/mass ratio by measuring the negative charge required to stop them illuminating a fluorescent screen. His work showed that the rays are particles and not waves, and J J THOMSON was soon to improve upon his results and show them to be electrons. Perrin is better known, however, for his classic studies of Brownian motion in 1908, in which he measured the distribution of particles of gamboge (a yellow gum resin from a Cambodian tree) suspended in water. His results confirmed a mathematical analysis of the problem by EINSTEIN, and enabled Perrin to give accurate values for AVOGADRO's number and for the size of the water molecule. His work was widely accepted as final proof of the existence of atoms, and he was awarded the Nobel Prize for physics in 1926.

Perutz, Max (Ferdinand) (1914–) Austrian–British molecular biologist: showed structure of haemoglobin.

Both sides of Perutz's family were textile manufacturers. After studying chemistry in Vienna he came to Cambridge in 1936 to work for a PhD in crystallography with J D Bernal (1901–71). The latter had shown in 1934, with DOROTHY CROWFOOT (HODGKIN), that a wet crystal of a protein (pepsin) would give an X-ray diffraction pattern, thereby implying that it might be possible to use the X-ray method that the BRAGGS had used for inorganic compounds to deduce the structure of proteins. However, Bernal gave Perutz some dull work on minerals and it was 1937 before Perutz secured some crystals of haemoglobin and found them to give excellent X-ray patterns. Haemoglobin is the protein of red blood cells, which carries oxygen to the tissues and CO_2 to the lungs.

The invasion of Austria in 1938, and the Second World War, diverted Perutz from protein work for a time; but from 1947 he directed a Medical Research Council Unit in Cambridge, consisting at first only of himself and his student J C Kendrew (1917–). In 1953 Perutz showed that the haemoglobin structure could be solved by comparison of two or more X-ray diffraction patterns, one from the pure protein and the others from the same protein with heavy atoms such as mercury attached to it at specific positions. This method led to the solution of the first two protein structures; Perutz and his colleagues solved haemoglobin (relative molecular mass 64 500) and Kendrew and his colleagues solved the related (but simpler) myoglobin from sperm whale muscle. Their methods have been adopted and extended to several hundred other proteins, including enzymes, antibodies and viruses.

The Unit became the Medical Research Council Laboratory of Molecular Biology, chaired for many years by Perutz and a focus of world talent in its field, attracting BRENNER, CRICK, H E HUXLEY, MILSTEIN and SANGER among others. Perutz and Kendrew shared a Nobel

Prize in 1962. Perutz went on to study the structural changes in haemoglobin that occur when it takes up oxygen, and its mutant forms, characteristic of some inherited diseases. Before 1950 Perutz also did research on the crystallography and mechanism of flow of glaciers. He joined the Order of Merit in 1988 (a UK decoration for those providing valued service to the country, restricted to 24 members).

Peters, Sir Rudolf Albert (1889–1982) British biochemist: wide-ranging discoverer of metabolic paths in cellular metabolism.

Peters studied science in Cambridge and medicine in London, qualifying in 1915, and serving with distinction in France in the First World War before his recall to the UK in 1917 to join the chemical warfare defence laboratory at Porton Down. After the war he began his studies on the vitamin B complex in Oxford, and was the first to isolate thiamine (vitamin B_1), and he began to elucidate its function as a precursor of a metabolic enzyme (co-carboxylase). He held the Oxford professorship of biochemistry from 1923–54.

In the Second World War he again worked on chemical warfare defence. Having noted that arsenical poisons inhibit enzymes, he led the work which showed that 2,3-dimercaptopropanol condenses with the arsenical poison gas Lewisite to form a relatively harmless product. As a result of this, 2,3-dimercaptopropanol ('British anti-Lewisite', BAL) became available for use both in war and peace for the treatment of poisoning by arsenic and some other metals. As the value of BAL depends on the presence of mercapto (-SH) groups in BAL and in enzymes, this stimulated valuable world-wide studies on the importance of -SH groups in biochemistry generally.

During the Second World War some compounds of fluorine were a major potential chemical warfare threat, and then and afterwards Peters showed how these compounds act, by inhibiting an essential part of the KREBS metabolic cycle. Again, the work had valuable spin-off results in pure biochemistry. He was also able to show in 1969 how fluorine is incorporated, in organic form, as an important component of bone.

Retiring from Oxford in 1954, he had two further careers, firstly with the Agricultural Research Council's Animal Physiology Unit and then from 1959 in Cambridge, where he was an active researcher until 1981. Uniquely, and through the accidents of war, he had used chemical warfare problems as a source for unwarlike and valuable contributions to biochemistry.

Phillips, Peregrine (c.1800– ?) British vinegar manufacturer: devised contact process for sulphuric acid.

Phillips's position in science is curious; almost nothing is known about him, and the process he patented in 1831 was not used by him on any scale. The old method of making sulphuric acid was by burning sulphur in a lead chamber, and then oxidizing the resulting SO_2 with nitrogen oxides and absorbing the resulting SO_3 in water. The Phillips patent proposed to pass SO_2 and oxygen over a platinum catalyst to make SO_3 and offered great advantages, but initial difficulties delayed its use until 1876 when R Messel (1847–1920) used it. Now, the process makes over 90% of the world's sulphuric acid.

Piazzi, Guiseppe [pyatsee] (1746–1826) Italian astronomer: discovered the first asteroid, Ceres.

Piazzi, a monk of the Theatine Order who taught mathematics, became first director of the Palermo Observatory in 1890. In 1814 he published a catalogue of 7646 stars visible from Sicily and established that proper motion was a common property of stars, and not only of a few nearby ones. He discovered the first asteroid, Ceres, in 1801, but after only three fixes of its position lost it, being temporarily prevented from observing due to illness. However it was soon recovered (by OLBERS), following a remarkable calculation of its orbit by GAUSS, based on the three observations. The 1000th asteroid was named Piazzia in his honour.

The asteroids have planetary orbits between Mars and Jupiter, in the region assigned for a planet by BODE's mysterious 'law', although it was not found through this by Piazzi, but by chance.

Picard, Charles Emile [peekah(r)] (1856–1941) French mathematician: advanced analysis and analytical geometry.

Soon after entering the École Normale Supérieure in 1874 Picard made some useful discoveries in algebra and earned his doctorate. At 23 he became a professor at Toulouse, and from 25 taught at the Sorbonne and at the École Normale Supérieure.

Picard proved two theorems (known as his 'little theorem' and 'big theorem') which show that an integral function of a complex variable takes every finite value, except for possibly one exception. Picard also developed a theory of linear differential equations which paralleled GALOIS's theory of algebraic equations using group theory. By studying integrals associated with algebraic surfaces he created areas of algebraic geometry with applications in topology and function theory.

Pincus, Gregory Goodwin (1903–67) US biologist: introduced the oral contraceptive pill.

Pincus followed in his father's footsteps by graduating in agriculture at Cornell; his father was a lecturer in the subject. Then he studied genetics and physiology at Cambridge, Berlin and Harvard, and later founded his own consultancy in experimental biology. In 1951 he was influenced by the birth control campaigner Margaret Sanger (1883–1966) to concentrate on

reproductive physiology. With M C Chang (1908–91) he studied the antifertility effect of steroid hormones (notably progesterone) in mammals, which act by inhibiting ovulation. In this way, refertilization is prevented during pregnancy. Synthetic hormones similar in their effects to progesterone became available in the 1950s, and Pincus saw that they could be used to control fertility. He organized field trials of suitable compounds in Haiti and Puerto Rico in 1954 which were very successful, and oral contraceptives ('the pill') have been widely used ever since. His success is a pharmaceutical rarity – a synthetic chemical agent that is nearly 100% effective, and one that has had remarkable social results.

Planck, Max (Karl Ernst Ludwig) (1858–1947) German physicist: originated quantum theory, making 1900 the transition between classical and modern physics.

The son of a professor of civil law, Planck attended university at Berlin and Munich, finishing his doctorate in 1880. He then went to Kiel, becoming a professor there in 1885. A move to Berlin in 1888 followed. In 1930 he became president of the Kaiser Wilhelm Institute; he resigned in 1937 in protest at the behaviour of the Nazis towards Jewish scientists. At the end of the Second World War the Institute was moved to Göttingen and renamed the Max Planck Institute and Planck was reappointed president.

In 1900 Planck published a paper which, together with EINSTEIN's paper of 1905, initiated quantum theory. KIRCHHOFF, STEFAN, WIEN and RAYLEIGH had studied the distribution of radiation emitted by a black body as a function of frequency and temperature. Wien found a formula that would agree with experiments at high frequencies, while Rayleigh and JEANS found one for low frequencies. Planck discovered one that worked at all frequencies v, but this needed the assumption that radiation is emitted or received in energy packets (called quanta); these have an energy $E = hv$ where h is the Planck constant (6.626×10^{-34} J Hz^{-1}). This assumption is counter to classical physics and its adoption began the modern age of using quantum theory in physics. The conservative Planck immediately recognized how revolutionary the result was, and on a New Year's day walk in 1900 with his small son told him how the age of classical physics had just passed away. Rapid acceptance of the idea came with its use in Einstein's prediction of the photoelectric effect (1905) and in BOHR's successful theory of the electronic structure of atoms (1913). A full quantum theory arrived in the 1920s, when Planck and others had shown how to express all the new concepts consistently. Planck was awarded the Nobel Prize in 1919 for his discovery of the energy quanta.

Planck bore the tragedies of his second son

dying in the First World War, his twin daughters both dying in childbirth and finally his first son Erwin being executed for his part in the plot against Hitler of July 1944. He was always anti-Nazi, but the other founder of 20th-c physics, his friend Einstein, never forgave Planck for not showing firmer opposition; it is not clear that he could have achieved more. Planck is one of the very few scientists to be immortalized on a coin (the German DM 2 piece of 1958).

Pliny (the Elder), in full Gaius Plinius Secundus (*c*.23–79) Roman writer on natural history: the first encyclopedist.

Pliny was a child of a wealthy Roman family and so was well educated; at 23 he began an official career as a member of the second great Roman order, the equestrian order. His early duties, as was usual, were in the army: he served in the cavalry on the Rhine frontier and his first writing was on the use of javelins by cavalry.

In about AD 57 he left the army and wrote on grammar and on Roman history, and travelled in the Roman empire as a financial controller. He must then have begun his most famous work, the 37 books of his *Natural History*. His last post was as commander of the fleet at Misenum near Naples. He probably never married, but adopted a nephew (Pliny the Younger) as his heir. His energy as a writer was remarkable: he needed little sleep, and his motto was 'to live is to be awake'. He saw the eruption of Vesuvius in AD 79, and was killed near Pompeii by its fumes.

He was a passionate gatherer of the scientific and technical knowledge of his time, ever-anxious to record the facts for posterity. He wrote that 'it is god-like for man to help man'; and his curiosity was boundless. His *Natural History* covers astronomy, geology, geography, zoology, botany, agriculture and pharmacology, and the extraction of metals and stone and their uses, especially in art. His emphasis was on facts, and his theorizing spasmodic. His fact-gathering was uncritical and myth and legend are mingled with observation. Pliny has a unique place as a source of information on the science and technology of his time.

Pogson, Norman (Robert) (1829–91) British astronomer.

A keen youthful astronomer, Pogson by age 18 had calculated orbits for two comets. In 1851 he became assistant at the Radcliffe Observatory, Oxford and by 1860 was government astronomer at Madras, where he stayed. Although he discovered nine asteroids and 21 new variable stars and prepared a massive star catalogue, his best-known achievement was devising (in 1854) Pogson's ratio to define stellar magnitudes. It was known that first magnitude stars are about 100 times brighter than sixth magnitude stars, so Pogson proposed that each magnitude interval should correspond to a brightness ratio of exactly $100^{\frac{1}{5}}$, ie 2.512. The system was used by E C Pickering (1846–1919)

and his 'Harvard ladies' in their work on stellar photometry, a massive project which during 25 years recorded over 1.5 million photometric readings. The magnitude scale is still used.

Poincaré, Jules Henri [pwĩkaray] (1854–1912) French mathematician: discovered automorphic functions and contributed independently to relativity theory.

Poincaré was the son of a physician and was educated at the École Polytechnique and École des Mines. After teaching at the University of Caen he spent his life from 1881 as a professor at the University of Paris. He became a member of the Académie des Sciences (1887) and also of the Académie Française (1909). Poincaré earned a reputation as 'the last universalist', producing about 500 papers and 30 books, which contributed to a wide variety of branches of mathematics and allied subjects. In contrast, he was clumsy, absent-minded and inept with simple arithmetic.

In pure mathematics Poincaré discovered automorphic functions, which are a generalization of periodic functions in being invariant under an infinite group of linear fractional transformations. This led to work on parameterization of curves, the solution of linear differential equations with rational algebraic coefficients and topology. He also did significant work on the theory of numbers, on probability theory and ergodicity.

In mathematical physics Poincaré published a paper on the dynamics of the electron (1906) which independently obtained within electromagnetic theory several of the results of EINSTEIN's theory of special relativity (1905). In celestial mechanics he made important contributions to the theory of orbits, the shape of rotating fluids, the gravitational three- and *n*-body problems and the origination of topological dynamics. In the course of this work Poincaré developed powerful new techniques such as asymptotic expansions and integral invariants.

Poisson, Siméon-Denis [pwasõ] (1781–1840) French mathematician: contributed to electrostatics, magnetostatics, probability theory and complex analysis.

Poisson's talent in creative mathematics was recognized by LAGRANGE while he was at the École Polytechnique. He had begun training as a surgeon, but found he had neither taste nor talent for the work. In 1800 he was appointed to a post at the École. Poisson's mathematical contributions were to mathematical physics, and he added to this by conducting experiments in sound and heat. He developed the theory of heat and elasticity (Poisson's ratio is the ratio between the lateral and longitudinal strain in a wire).

In 1812 Poisson adopted an early 'two-fluid' theory of electricity which was later superseded. He used LAGRANGE's potential function, originally applied in gravitation, and showed that it could be used for electrostatic problems. Using a suggestion by LAPLACE he used the technique to prove the formula for the force at the surface of a charged conductor, and to solve for the first time the charge distribution on two spherical conductors a given distance apart. COULOMB'S experimental results were in close accord with Poisson's formula.

His paper of 1824 constructed a 'two-fluid' theory of magnetism, and expressed the magnetic potential at any point as a sum of volume and surface integrals of magnetic contributions (magnetostatics).

Poisson built upon Laplace's work in probability theory. Poisson's formula gives the probability of a given number of events if its probability is low, and it has wide applicability. The Poisson distribution, which is related to this, was later shown to be a special case of the general binomial distribution. Poisson also wrote an important memoir in 1833 on the Moon's motion.

In pure mathematics he advanced complex analysis, being the first person to integrate complex functions along paths or contours in the complex plane (now called contour integration).

Poncelet, Jean-Victor [põslay] (1788–1867) French mathematician: substantially advanced projective geometry.

In November 1812 the bedraggled rearguard of the French Grande Armée under Marshal Ney was overwhelmed at Krasnoi on the frozen plains of Russia; Poncelet, a young engineer, was left for dead on the battlefield. A search party who found him took him for questioning, as he was an officer, and so he survived to be marched through a Russian winter for 5 months before entering a prison at Saratov in March 1813.

He passed what turned out to be 2 years of captivity recalling all the mathematics he could from his 3 years at the École Polytechnique (1807–10); and he went on to contribute new mathematics to projective geometry. After he was released his sense of duty led him to put his creative urge on one side and to do routine military engineering tasks, and later to work on water-power. However, in the notes that he produced under such difficult conditions are the principle of duality (the equivalence of various geometric theorems) and the first use of imaginary points in projective geometry.

Popov, Aleksandr Stepanovich [popof] (1859–1906) Russian physicist: radio pioneer.

Planning to enter the clergy, Popov studied at a seminary, but after graduation his interests in physics and engineering led him to study at the university at St Petersburg and concurrently to work at the Elektrotekhnik artel which generated electric power. In 1883 he joined the naval Torpedo School at Kronstadt, as an instructor in electrical engineering.

Following the discovery of electromagnetic radiation in the radio band by HERTZ in 1888, and the development of a crude detector for this by E Branly (1844–1940), Popov became interested in radio and like LODGE he worked to improve Branly's 'coherer' of 1894. By 1895 Popov was able to demonstrate a detector that was effective up to 80 m from a spark generator, and in the same year he used it in modified form to record atmospheric discharges (his 'storm indicator'). MARCONI in 1896 staked his claims for wireless telegraphy, using equipment patented in 1897 which accorded with Popov's description of 1896. During 1897–1900 Popov's apparatus was used by the Russian army and navy, and he became professor at the St Petersburg Institute of Electrical Engineering and director there in 1905. In that role he refused to follow instructions to repress firmly the students' political activism, and died soon after this conflict with authority.

The Russian Physico-chemical Society early concluded that Popov should 'be recognized as the inventor of wireless telegraphy' and this has remained the view of his countrymen. However, this recognition has not been generally conceded elsewhere in Europe, where priority has been accorded to Marconi; Popov, like Lodge, is seen as active in the field but not the foremost innovator. It is widely accepted that Popov was the first to use antenna aerials for radio transmission and reception.

Porter, George, Baron Porter (1920–) British physical chemist: developed flash photolysis for detection of short-lived photochemical entities.

Porter took his first degree in Leeds, and in his final year took a course in radiophysics; in the Second World War, as a naval officer, he worked with radar. In Cambridge from 1945, he worked with R G W Norrish (1897–1978) on the detection and study of the short-lived radical intermediates involved in photochemical gas reactions. Porter developed the idea of using a flash technique to produce the radicals, by discharging a large bank of capacitors to produce a short, high-energy flash (the principle of the photographic flash gun) and using this flash (lasting 10^{-3} s or less) to break up the gas to form radicals and excited molecules. A second flash, after a brief delay, served to give a spectrum of the contents of the reaction tube, so that the radicals could be detected and their lifetimes calculated. Porter developed these ideas in ingenious ways; by 1975 he could detect molecules with a life of only a picosecond (10^{-12} s); he extended the method to liquids; he showed that radicals can be trapped in a supercooled liquid (a glass); and he applied laser beams to photochemical studies. His work did much to develop photochemistry, including its application to biochemical problems. In 1966, after 10 years in Sheffield, he became director of the Royal Institution; he shared a Nobel Prize in 1967 with Norrish and M Eigen (1927–   ).

Porter, Rodney (Robert) (1917–85) British biochemist and immunologist: deduced general structure of antibodies.

Born and educated in Liverpool, Porter had just graduated there when his career was diverted by military service from 1940–6; afterwards he worked on proteins with SANGER in Cambridge. Then, in London from 1949, he developed his interest in antibodies. He showed in 1950 that some could be partly broken down without total loss of their antigen-binding ability; and by the early 1960s he was able to show that antibodies contain both 'heavy' and 'light' protein chains; and that they have three distinct regions, of which two are alike and serve to bind antigens, leading to 'clumping' (agglutination). Aware also of EDELMAN's results, and of data from electron microscopy, Porter made a brilliant guess at the overall molecular architecture of antibodies (see diagram). His scheme could incorporate the facts then known, and it inspired further work by MILSTEIN and others that has refined it further. Ideas on antibodies, which had begun with EHRLICH and had developed with LANDSTEINER and PAULING, at last took on a firm outline that fruitfully linked their biochemistry with their immunology. Porter became professor of biochemistry at Oxford in 1967 and shared a Nobel Prize with Edelman in 1972.

Powell, Cecil Frank (1903–69) British physicist: used photographic emulsion to detect new elementary particles.

The son of a gunsmith, Powell studied at Cambridge and obtained his PhD there in 1927 for work with C T R WILSON on condensation in cloud chambers. In that year he went to Bristol, and spent his career there. Miss Blau and Miss Wainbacher of Vienna had shown that photographic emulsion is affected not only by light but also by fast particles, which leave a track. In the late 1930s Powell began to use photographic plates (later, films) to record the tracks of fast nuclear particles. These tracks (due to ionization, which leaves blackened silver grains) can be studied under a microscope and Powell showed that the mass, charge and energy of a particle can be estimated from the tracks it produces. In this way he discovered a new particle in 1947; this was the pi-meson or pion, with mass 273 times that of an electron, which had been predicted by YUKAWA in 1935. Since 1947 other unstable particles have been discovered by the same method, and Powell's work marks the start of modern high-energy particle physics. He used the photographic method with 'stacks' of plates, both at mountain height with cosmic rays as the source of particles (as in his pion work) and carried by free balloons above the atmosphere. He won the Nobel Prize in 1950.

Poynting, John Henry (1852–1914) British physicist: demonstrated existence of radiation pressure.

Poynting gained his qualifications from Manchester and Trinity College, Cambridge, becoming professor of physics at Birmingham in 1880. He held this post until his death. He researched on electromagnetic waves, and Poynting's vector (1884, based on MAXWELL's theory) gives the direction and magnitude of energy flow from the electric and magnetic fields at a point. He showed that radiation has momentum and exerts a pressure (1904), which can be large under astronomical conditions. He also (from 1878) used a balance to measure NEWTON's gravitational constant, essentially by CAVENDISH's method, and by 1891 had a result close to the modern value.

Pratt, John Henry (1809–71) British cleric and geophysicist: proposed isostatic principle to account for gravity anomalies.

Pratt went to India in 1833 as a chaplain with the East India Company, and in 1850 became archdeacon of Calcutta. As an amateur scientist, he recognized that the cause of some surveying errors found by George Everest (1790–1866) near the Himalayas was that the mountains were failing to exert as great a gravitational attraction as Everest had allowed for. In 1854 he suggested his isostatic principle, in which the higher a mountain range the lower its density, so that the effective pressure in the lithosphere beneath the crust remains constant. AIRY suggested a similar idea soon afterwards, and their principle of isostasy has since been found to account for gravity anomalies in a wide range of situations.

Priestley, Joseph (1733–1804) British chemist: discoverer of gases, including oxygen.

After a difficult Yorkshire childhood during which he was orphaned and often ill, Priestley began training as a non-conformist minister

antigen molecule
recognised by
variable region
of light and heavy
chains

light chain

heavy
chain

'hypervariable'
regions

variable
region

disulphide
bonds holding
heavy and
light chains
together

constant regions
which interact
with complement
to break down
viruses and cells

The structure of an antibody molecule

NO – MOLECULE OF
THE YEAR 1992

The prestigious US journal *Science* chose nitric oxide (NO) as its Molecule of the Year in 1992, citing it as 'a molecule of versatility and importance that has burst on to the scene in many guises. In the atmosphere it is a noxious chemical, but in the body in small controlled doses it is extraordinarily beneficial'.

The first point about NO is not to confuse it with nitrous oxide (N_2O, 'laughing gas'), one of the other six nitrogen oxides and a long-used and popular anaesthetic for brief surgical procedures.

Nitric oxide, NO, was discovered by PRIESTLEY in 1774. Pure NO is a colourless gas, acrid and very toxic, used in the bulk production of nitric acid for industry and nitrate fertilizers for agriculture. It is the simplest stable molecule known to have an odd number of electrons, which contributes to its intense chemical reactivity. It reacts readily with air or oxygen to form the brown gas nitrogen dioxide NO_2, and it also reacts with metals, one of its aspects which have been much studied by inorganic chemists. Until recently it had no interest for biochemists or physiologists, except perhaps as a component of the nitrogen oxide mixture ('NO_x') forming part of the atmospheric pollution from petrol engines and adding to the troubles of asthmatics.

Until the later 1980s it was never expected that a molecule that is small, light, gaseous and reactive would have a previously undiscovered and subtle role in physiology: but since then it has been found to be essential in digestion, blood-pressure regulation and antimicrobial defence.

In the body, it is made from an amino acid (arginine) by an enzyme, NO synthase. When released by cells in the wall of blood vessels, it relaxes nearby muscle cells, the vessel dilates and blood pressure falls. The effect has been used (without understanding its origin) since the discovery in the 1860s that nitroglycerin has a dramatic effect in providing relief to victims of coronary artery narrowing. However, too much NO, in response to a bacterial infection, causes septic shock, a major cause of death in intensive care wards; from 1992 NO inhibitors have saved such cases. In the body's defence system NO acts as an antitumour agent. It also combats bacteria, a reminder that nitrates and nitrites that release NO have been used for centuries in curing meat. In the brain, NO acts as a neurotransmitter, usually desirably, but in stroke cases its release in excess is toxic and can be fatal. Also in the brain, there is some evidence that it has a key place in learning and memory. In the digestive system, the relaxation component in peristalsis (the wave-like movement of the gut that propels the food) depends upon NO. Lack of it is the cause of infantile pyloric stenosis, which can be fatal.

The intensive studies on NO by neuroscientists in the 1990s have shown conclusively that in male mammals this is the molecule that converts sexual excitement into potency. The brain causes NO to be released in penile blood vessels, and erection is the result; with some 10% of human males suffering from impotence, the possible clinical use of this knowledge makes the discovery of this sexual neurotransmitter the subject of intensive research. The 'discovery phase' for NO in physiology clearly will run for some years to come.

IM

and 3 years later (in 1755) was appointed Unitarian minister in a Suffolk village. Later, he taught in schools and was librarian-companion to Lord Shelburne, afterwards Prime Minister. His stutter and his radical views on theology, and in particular on politics, made him unpopular as a preacher. As a vociferous supporter of the French Revolutionary idea he became very unpopular and, after his house had been burned in 1791 by a Birmingham mob, he felt forced to seek refuge in the more liberal USA in 1794.

He had a simple character, much personal charm and exceptional intelligence, and wrote on theology, education, history, philosophy, politics, physics, chemistry and physiology and knew at least nine languages. He was an amateur in science, with little use for theory, but he was the greatest English-speaking experimental chemist of the 18th-c, even though his interpretation of his results was usually unsound.

He met FRANKLIN in London in 1766 and suggested that he would write a *History of Electricity* if Franklin would lend him some books; he had earlier done some reading in science and had shown his school pupils in Nantwich, in Cheshire, some experiments in electricity and optics. It was a good book, published in 1767 and included new experiments; Priestley became a Fellow of the Royal Society in 1766. After this his interest in science turned to chemistry; but theology was always more important to him than science.

His work as a minister in Leeds in 1767 was near a brewery, and he became interested in the 'fixed air' (CO_2) generated by fermentation, and then in other gases, although only three were then known: air, CO_2 (studied by BLACK) and H_2 (studied by CAVENDISH). All these formulae and modern names came much later, of course. Priestley used the pneumatic trough invented by HALES to collect gases, filling it with mercury if the gas was water-soluble. He used a large lens and the Sun to provide a clean heat. Within a few years he had discovered and examined the gases we now know as HCl, NO, N_2O, NO_2, NH_3, N_2, CO, SO_2, SiF_4, and O_2; as DAVY said, 'no single person ever discovered so many new and curious substances'. His results were described in papers and books, notably *Experiments and Observations on Different Kinds of Air and other Branches of Natural Philosophy* (1790).

His most famous experiment was carried out on 1 August 1774 at Bowood (Shelburne's house near Calne, Wiltshire). He had been given a large (12 in/30 cm) lens, and used it to try to make gases by heating various chemicals given by his friend J Warltire (1739–1810). When he heated mercury oxide (HgO) he was surprised to find that it gave a colourless gas that was not very soluble in water and in which a candle burned with a dazzling light. A few months later he wrote that 'two mice and myself have had the privilege of breathing it' and recommended its use in medicine. In October 1774 in Paris he talked with LAVOISIER about this; later Lavoisier repeated and improved the experiment, and saw (as Priestley had not) the full significance of the discovery. SCHEELE had in fact made oxygen (O_2) earlier, in this and other ways, but did not publish until 1777, after Priestley. However, Priestley showed that O_2 is given off by plants, and that it is essential for animals. He also studied hydrogen and used it to reduce metals, noticing that water is formed in this reaction; and that water is also formed by exploding hydrogen with oxygen.

Priestley had exceptional energy and skill but he remained always a firm believer in the erroneous phlogiston theory in chemistry, although his own results did much to refute that theory.

He also studied the densities of gases, their thermal conductivity and electrical discharges in gases. After he joined his sons in the USA he continued to work in chemistry until 1803, although he declined the professorship of chemistry at Philadelphia.

Prigogine, Ilya [prigogeenay] (1917–) Russian-Belgian theoretical chemist: developed irreversible thermodynamics.

Living in Belgium from age 12, Prigogine was educated in Brussels and was a professor there from 1951; he also had posts in the USA.

Classical thermodynamics is concerned with reversible processes and, in chemistry, with equilibrium states. In fact such situations are rare in the real world; eg the Earth's atmosphere receives energy continuously from the Sun and living cells are also not in equilibrium with their surroundings. Inanimate systems tend in general to a state of increasing disorder (ie their entropy increases) whereas living systems achieve an organized and ordered state, from relatively disorganized materials. Prigogine developed mathematical models of these

non-equilibrium systems and was able to show in general terms how such dissipative structures (as he named them) are created and sustained. His ideas have application in studies on the origin of life and its evolution, and on ecosystems in general. He was awarded the Nobel Prize in 1977.

Pringsheim, Ernst [pringshiym] (1859–1917) German physicist: measured the wavelength of thermal radiation as a function of temperature.

Pringsheim studied at the universities of Heidelberg, Breslau and Berlin, holding the position of professor of physics at Berlin, before eventually returning to Breslau in 1905 as professor of experimental physics.

Pringsheim developed the first accurate infrared spectrometer in 1881. His subsequent work with LUMMER, on black body radiation in the infrared region, enabled him to confirm experimentally the Stefan–Boltzmann law relating radiated energy of a body to its temperature, and WIEN's displacement laws, which describe the wavelength at which maximum energy is emitted at a given temperature. His observations encouraged PLANCK to formulate his quantum theory in order to account for his results.

Pringsheim, Nathanael (1823–94) German botanist: made important studies of algae and cell reproduction.

Pringsheim's father wished him to become an industrialist like himself, but the boy was attracted to science. A course in medicine was a compromise, but on graduation he escaped to research in botany. He did little teaching and inherited enough money to follow his interests in his home laboratory in Berlin and his Silesian estate.

Pringsheim contributed to the revival of scientific botany in the later 18th-c, mainly by his work on lower plants. He was an early observer of sexual reproduction in algae, and of the alternation of generations between the two sexual forms of zoospores and the asexual spore resulting from their fusion. His work on marine algae led him to the view that natural selection is unimportant in evolution; he thought (like NAEGELI) that variations are spontaneous, without survival value, and tend always to greater complexity of form. His studies supported the view that cells result from the division of preexisting cells, and not from a process of free-cell formation as claimed by SCHLEIDEN. With J VON SACHS he first described the plastids, granules found only in plant cells and containing either starch or chlorophyll.

Proust, Louis Joseph [proost] (1754–1826) French analytical chemist: defender of the law of constant proportions.

Proust followed his father in becoming an apothecary in Paris, but in his 30s he moved to Spain. From 1789 he taught in Madrid, but his well-equipped laboratory was pillaged by Napoleon's troops during the siege of Madrid in 1808 and he returned to France.

Proust was a skilled and prolific analyst. He opposed BERTHOLLET's view that chemical compounds could vary in composition over a wide range. Proust's extensive work led him by 1797 to the law of constant proportions: that different samples of a pure substance contain its elementary constituents (elements) in the same proportions. Thus malachite $Cu_2CO_3(OH)_2$, whether from nature or synthesized in various ways, had the same composition. In a courteous conflict of views, Proust showed that Berthollet's samples were in fact mixtures. By 1805 Proust's view prevailed, and soon after, the law was seen to relate directly to DALTON's atomic theory. However, 130 years later it was found that some compounds (eg some intermetallic compounds, and some sulphides) can have slightly variable compositions, and are sometimes called berthollides.

Prout, William [prowt] (1785–1850) British physician and chemist: proposed that the hydrogen atom is 'primary matter'.

Like many chemists of his time, Prout was trained in medicine and, like most English physicians of his century, he studied in Scotland, qualifying at Edinburgh in 1811. He began his medical practice in London, and from 1813 he also researched and gave lectures on 'animal chemistry'. He is best known for Prout's hypothesis, which appeared anonymously in 1815. This suggested that (1) the relative atomic masses of all elements are exact multiples of that of hydrogen and (2) that hydrogen is a primary substance or 'first matter'.

The idea stimulated analytical work, which showed that Prout was wrong; for example chlorine has an atomic mass close to 35.5 times that of hydrogen. Nevertheless, over a century later, ASTON's work on isotopes revealed a real basis for Prout's idea; and in modern terms the hydrogen nucleus (the proton) is a kind of primary substance, as indicated by its name, which also recalls Prout's.

Ptolemy (of Alexandria), Clausius Ptolemaeus (*Lat*) [toluhmee] *c*.90–170) Egyptian–Greek astronomer: wrote classic summary of Greek astronomy, geography and optics.

Little is known of Ptolemy's life or original work. He was probably born in Egypt, and became Hellenized; he should not be confused with the kings of Egypt of the same name. His fame lies in his four books, which summarize 500 years of Greek astronomical ideas and which dominated Western thought on astronomy until the time of COPERNICUS, 14 centuries later. His *Almagest* (Arabic for 'the Greatest'; he called it 'the mathematical collection') described the motions of the heavens on a geocentric basis, making use of various devices such as epicycles (80 in all) to obtain a plausible match with observations (the Ptolemaic system; it

owed much to work by HIPPARCHUS, now lost). He gave distances and sizes for the Sun and the Moon, a catalogue of 1028 stars, descriptions of astronomical instruments, and computed π to be 377/120 (3.1417). In the *Geography* he described a system of determining latitude and longitude, and a map of the world (based largely on the travels of merchants and Roman officials). Ptolemy's world did not include the Americas, or extend below the Equator (which he placed too far north); but his view that the Earth was spherical, and his exaggeration of Asia eastward, encouraged Columbus in his famous attempt to reach Asia by sailing to the west. Ptolemy's *Optics* dealt with the basics of reflection and refraction, and the *Tetrabiblios* is the origin of much modern astrology. Many of his accounts are due to Hipparchus, including the trigonometry, whose basis remains today.

Purcell, Edward Mills [persel] (1912–   ) US physicist: developed nuclear magnetic resonance; and first detected the interstellar 21 cm microwave emission.

Purcell graduated from Purdue in electrical engineering and then studied physics at Karlsruhe and at Harvard, where he taught from 1938. During 1941–5 he worked on the development of microwave radar at Massachusetts Institute of Technology, returning to Harvard as professor of physics.

Purcell was instrumental in the late 1940s in developing nuclear magnetic resonance (NMR) methods for measuring the magnetic moments of atomic nuclei, in solids and liquids. Any atomic nucleus with spin (such as hydrogen or fluorine) will, if held in a powerful magnetic field, absorb radiation in the radiofrequency range by a resonance effect, and measurement of this has given valuable information on features of the absorbing nuclei and their molecular environment. This NMR method has since become a dominant method in chemistry for a variety of analytical purposes. For his work on NMR, Purcell shared the 1952 Nobel Prize for physics with BLOCH.

In radio astronomy Purcell was in 1951 the first to report observation of the 21 cm wavelength microwave radiation emitted by interstellar neutral hydrogen. This was predicted theoretically by VAN DE HULST and has been used in the mapping of much of our Galaxy, and in deducing the temperature and motion of the interstellar gas.

Purkinje, Johannes (Evangelista) [poorkinyay] (1787–1869) Czech histologist and physiologist: advanced cell theory and observed cellular division.

Educated by monks, Purkinje first trained for the priesthood; then he studied philosophy, and lastly medicine. He graduated in 1818, with a famous thesis on vision which gained him the friendship of Goethe, the poet-philosopher. Helped by this, he became professor at Breslau and later in Prague. In 1825 he first observed the nucleus in bird's eggs; in 1832 he began to use a compound microscope, and soon made many new observations. In 1835 he described ciliary motion; in 1837 he outlined the key features of the cell theory, which was to be fully propounded by SCHWANN in 1839. Also in 1837 he described nerve cells with their nuclei and dendrites and the flask-like cells (Purkinje cells) in the cerebellar cortex. In 1838 he observed cell division, and the following year he was the first to use the word 'protoplasm' in the modern sense. His improvements in histology included early use of a mechanical microtome in place of a razor, to obtain thin tissue slices.

Pythagoras (of Samos) [piythagoruhs] (c.560–480 BC) Greek mathematician, astronomer and mystic: founder of a cult united by the belief that 'the essence of all things is number'.

Although his name is so familiar, rather little is known of Pythagoras's personal life. Born on the Greek island of Samos in the eastern Mediterranean, he travelled widely before settling about 530 BC at Croton, then a Greek colony, in south-east Italy. There he founded the sect that survived for a century after his death. With Pythagoras as their cult leader, the sect was devoted to a life of political and religious mysticism, in which astronomy, and especially geometry and the theory of numbers, was central. Number was seen as pure, magical and the key to religion and philosophy. This secret society became powerful and aroused hostility, which eventually destroyed it.

Pythagoras himself is said to have discovered the theorem on right-angled triangles named after him; and to have begun the science of acoustics with his work on the tones produced from a stretched string, which are perceived as harmonious to the ear provided that the lengths of string for the two tones have a simple number relation (for example a length ratio of 2:1 corresponds to a musical octave). This was probably the first mathematical expression of a physical law, and the beginning of mathematical physics. A variety of arithmetical and geometrical relations were discovered by members of the sect, which inspired their belief in number as a basis for astronomy and even for morality. In their view the Earth was a sphere, and it and the stars moved in circles in a spherical universe, because these were 'perfect' forms in a mystical sense. They were dismayed by the discovery that the square root of two is irrational (ie is not expressible as a perfect fraction) and are reputed to have put to death a member of the sect who revealed this secret to others. The political ambitions of the sect led to its persecution and Pythagoras was exiled to Metapontum about 500 BC. At an unknown date later in the century the Pythagoreans were involved in a democratic rising in which many were killed, and the rest dispersed.

Although Pythagoras's ideas on the significance of numbers were erroneous, his contributions were important in mathematics: few ideas are more fundamental than that of irrational numbers.

Rabi, Isidor Isaac (1898-1988) Austrian-US physicist: developed molecular beam experiments.

Rabi grew up in the Yiddish community of New York, studying at Cornell and Columbia Universities. He obtained a professorship at Columbia in 1937 and remained there until retirement.

During a brief period (1927-9) working with STERN, Rabi was greatly impressed by the recent Stern-Gerlach experiment. Starting a research programme at Columbia, Rabi invented the atomic- and molecular-beam magnetic resonance methods of observing spectra. In this a constant magnetic field excites the molecules into a set of states and a radio wave signal of the right frequency can resonantly flip the molecule from one magnetic state to another. The magnetic properties of the molecule or atomic nucleus may then be found accurately. The magnetic moment of the electron was measured in this way to nine significant figures, thereby testing the theory of quantum electrodynamics (QED). The technique was also a precursor of the NMR (nuclear magnetic resonance) method developed by PURCELL and by BLOCH, and has also been applied to an atomic clock, nuclear magnetic resonance, the maser and the laser. Rabi won the 1944 Nobel Prize for physics for this work.

During the war Rabi worked on microwave radar and afterwards was concerned with administration and scientific policy-making, serving as chairman of the General Advisory Committee of the Atomic Energy Commission and as a member of the delegation to UNESCO which founded CERN (laboratory in Geneva for high-energy physics).

Rainwater, Leo James (1917-86) US physicist: unified two theoretical models of the atomic nucleus.

After studying at the California Institute of Technology, Rainwater went to Columbia University, and remained there to become professor of physics in 1952. During the war he contributed to the Manhattan (atomic bomb) Project.

In 1950, two theories were available to describe the atomic nucleus and each was in accord with some experimental results. In one model the nuclear particles were arranged in concentric shells; the other model described the nucleus as analogous to a liquid drop. Neither model accounted for experiments that appeared to show that some nuclei are not electrically spherically symmetrical. In explaining this, Rainwater in 1950 produced a collective model, in which the two ideas were combined.

In association with Aage Bohr (1922- ; son of NIELS BOHR) and B R Mottelson (1926-), Rainwater developed this theory and they secured experimental evidence in its support. The three shared a Nobel Prize in 1975.

Raisin, Catherine Alice (1855-1945) British geologist; one of the first professional women geologists.

Catherine Raisin went to the North London Collegiate School, one of the earliest schools to provide serious education for girls. In 1878 the University of London opened its degree course to women and she took her BSc with honours in geology and zoology in 1884. She remained there as honorary research assistant to Professor Bonney, working in microscopic petrology and mineralogy. Her 24 published papers appeared in the *Quarterly Journal of the Geological Society*. Her best known work was a detailed investigation of the serpentines. The Geological Society of London awarded her the Lyell Fund in 1893, the first woman to receive the honour, but as women were not then allowed to attend meetings the award had to be accepted by Bonney on her behalf. She was awarded a DSc in 1898, was a demonstrator in botany at Bedford College for Women (1889-90) and became head of the geology department in 1890, the first woman to do so in a British university, holding the post until her retirement in 1920. She became a Fellow of University College in 1902 and a Fellow of the Geological Society of London in 1919, when women were admitted.

Raman, Sir Chandrasekhara (Venkata) [rahman] (1888-1970) Indian physicist: showed that light scattered by molecules will show lower and higher frequency components (the Raman effect).

Raman gained a distinguished first class honours degree from Madras, but the lack of scientific opportunities in India prevented him then starting a career as a physicist. Instead he worked as an auditor in the Indian Civil Service for 10 years, continuing his research in his leisure time. The work he produced in sound and on diffraction secured him the professorship of physics at Calcutta. During his time there (1917-33) he discovered the Raman effect (1928), established the *Indian Journal of Physics* (1926), became president of the Indian Science Congress, was knighted (1929) and received the 1930 Nobel Prize for physics (the first awarded to an Asian). He had an important influence in building up the study of physics in India.

Viewing the blue colour of the Mediterranean in 1921 he concluded that RAYLEIGH's explanation of the colour in terms of light scattering from suspended particles was inadequate. Raman realized that the scattering and frequency shift of the light is due to the water molecules themselves. Light is generally scattered by molecules in solids, liquids or gases, lower and higher frequency components being added as the molecular bonds absorb or impart energy to the deflected photons (the Raman effect).

Raman's discovery led to one of the earliest confirmations of quantum theory, and also gave a powerful method of analysing molecular structure (Raman spectroscopy).

Ramón y Cajal, Santiago [ramon ee kakhal] (1852–1934) Spanish neurohistologist: a founder of modern neurology.

After doing well at school, Ramón y Cajal was apprenticed to a barber, then to a shoemaker, and lastly followed his father (at the latter's insistence) in studying medicine. He began to practise in 1873, and then served in the army in Cuba. From 1884 he was a professor in Spain, at Madrid from 1892–1922. He was often unwell, having contracted malaria in Cuba and then tuberculosis in Spain. About 1885 he was shown a microscope section of brain tissue, stained with silver by GOLGI's method. He was fascinated by it, and proceeded to improve the method. Within a few years he had added greatly to knowledge of the nervous system, and his work on it filled the rest of his life. Ramón y Cajal worked on the connections of the cells in the brain and spinal cord and showed the great complexity of the system. (The human brain contains about 10^{10} nerve cells, each connected with about 50 others, giving a total of 5×10^{11} synapses.) In opposition to Golgi, he argued that the nervous system consisted only of discrete nerve cells and their processes, with the axons ending in the grey matter of the brain, and not joining other axons or the cell bodies of other nerve cells (the neuron theory). He worked also on the difficult problem of the degeneration and regeneration of nerve cells, on the neuroglia, and on the retina. He shared a Nobel Prize with Golgi in 1906.

Ramsay, Sir William (1852–1916) British chemist: discovered the noble gases, a new group of elements.

Ramsay was proud of the fact that his ancestors included several scientists, and he moved easily into science at Glasgow and then into chemistry with BUNSEN at Heidelberg. In 1880, at Bristol, he began to make exact measurements of gas densities, and became an expert glassblower. In 1887 he moved to London and in 1894 began the work that made him famous. RAYLEIGH had shown that nitrogen from the air is 0.5% denser than nitrogen made chemically. Ramsay thought that this might be due to a previously unknown heavier gas in the air, and he set to work to find it. He sparked dry air to remove oxygen and then passed the nitrogen repeatedly over hot magnesium, when most of

the gas was slowly absorbed: $3Mg + N_2 \rightarrow Mg_3N_2$. He was left with a denser, monatomic gas: an inert new element, which was named argon and which makes up less than 1% of the air. Looking for it elsewhere, Ramsay found another new gas in the mineral cleveite, whose spectrum showed CROOKES that it was helium, previously observed only in the Sun. Ramsay went on to examine air for traces of other new inert gases, and by distilling liquid air he discovered three more: krypton, xenon and neon. Ramsay proposed that the five new gases formed a new group of 'zero-valent' elements in the periodic table. In 1910 he found the sixth of these gases, radon, which is formed with helium by the radioactive decay of the metal radium. The group are now known as the noble gases.

All are very rare (except argon, A) and chemically unreactive, although in 1962 BARTLETT found some reactivity for Xe and Kr. This inertness was important in early theories of chemical bonding. They all show striking spectra: except for the ultra-rare radon (Rn), they are used in discharge and fluorescent tubes and for work requiring an inert atmosphere. Liquid He is used as a cryogen (an extreme refrigerant); it has the lowest boiling point known (−269°C). Although uncommon on Earth, helium is the second most common element (23%) in the universe as a whole. Ramsay was awarded the Nobel Prize for chemistry in 1904. He is the only man to have discovered an entire periodic group of elements. He was much liked, and all his best work was done with a co-worker.

Raoult, Francois Marie [rah-oo] (1830–1901) French physical chemist: pioneer of solution chemistry.

Little is known of Raoult's early life, but his family was poor and although he began to study in Paris he could not afford to complete his course. He worked as a teacher, and began his research in physical chemistry in difficult circumstances, but this allowed him to gain a degree from Paris, in 1863. From 1867 he taught in the university at Grenoble. In the 1870s he tried to devise a new method for finding the alcohol content of wine, and this led him to study the freezing points of solutions of organic substances. He found that the depression of freezing point of a solution (compared with that of a pure solvent) was simply related to the quantity of dissolved solute and to its relative molecular mass (Raoult's Law, 1882). A few years later he showed a similar relation for the effect of a dissolved solute on the vapour pressure of a solution, and therefore on the elevation of its boiling point. In 1889 BECKMANN showed that this elevation of boiling point, for a measured amount of substance in a suitable solvent, is a very convenient method for measuring the relative molecular mass of the substance; the method has been in routine use ever since.

Ray, John (1627–1705) English naturalist: pioneer of plant taxonomy.

Ray's father was the village blacksmith and his mother a herbalist at Black Notley in Essex; the boy went to Cambridge at 16 and taught classics there after his graduation. His university career was ended after the Civil War when he refused to conform to new laws on religious observance. He was already a keen naturalist, and from 1662 he was supported by his wealthy ex-pupil and fellow-naturalist, F Willughby (1635–72). They toured Europe as well as England to study both flora and fauna. Ray used a taxonomic system which emphasizes the division of plants into cryptogams (flowerless plants), monocotyledons and dicotyledons, the basic scheme used today. His major work on botany covers some 18 600 species, with much information on each. He saw the species as the fundamental unit of taxonomy, although he eventually realized that species are not immutable. His taxonomy was not surpassed until the work of LINNAEUS. Ray was ahead of his time in his view that fossils are petrified remains of plants and animals, an idea not accepted until a century later. Willughby died in 1672, and Ray lived on in his house, but eventually quarrelled with his widow and returned to Black Notley to write on a variety of matters, including travel, proverbs and natural history.

Rayleigh, John William Strutt, Baron [raylee] (1842–1919) British physicist: did classic work on sound, light and electricity.

Rayleigh had high ability as a mathematician, which he found useful in the unusually wide range of problems in physics which attracted him, and he was a skilful experimenter. When his father died in 1873 Rayleigh inherited the title, and continued to work in his laboratory in the family mansion, Terling Place, in Essex. He agreed to succeed MAXWELL in Cambridge, but only for 5 years; in that time his physics students increased in number from six to 70. His first researches were on waves, both in optics and in acoustics; his book *The Theory of Sound*, written in part in a houseboat on the Nile, is a masterpiece of classical physics. His enthusiasm for precise measurement led him, in Cambridge, to work on the standardization of the ohm and ampere. Interest in PROUT's hypothesis caused his work on gas densities and led RAMSAY to the discovery of argon. His interest in radiation and spectra led him to study black body (isothermal) radiation, and the Rayleigh–Jeans formula for this represented the best that classical theory could achieve in this area. However, although the formula agrees well with experiment for long wavelength radiation, it fails entirely for shorter wavelengths. The problem was solved by PLANCK's novel idea that energy is emitted in small packets or quanta; this concept was to revolutionize physics, but it was never fully accepted by Rayleigh. He won the Nobel Prize for physics in 1904, for his work on gas densities and on argon. He was married to the sister of Britain's most intellectual prime minister, A J Balfour.

Réaumur, René-Antoine Ferchault de [rayohmür] (1683–1757) French technologist and naturalist: pioneer entomologist.

A member of the lesser nobility, Réaumur was probably educated by the Jesuits before studying law; but soon he was attracted to mathematics and then to metallurgy and biology. From 1713, he had the huge task of compiling an encyclopedia of technology, commissioned by Colbert (Louis XIV's finance minister) and intended to aid French industry. Soon he was engaged in studying iron and steel making, porcelain and thermometry. His researches in these areas were of value to his successors, rather than to his contemporaries.

His lasting fame rests on his work as a naturalist, where he worked on molluscs and especially on insects. His work on bees was particularly detailed. He also wrote on regeneration in hydra and in marine animals, and he showed that digestion is a chemical process. In all these matters, his work did much to spur research by others. He inherited a castle in 1755, but 2 years later had a fatal fall from his horse.

Reber, Grote [rayber] (1911–) US radio astronomer: discovered first discrete radio sources in Milky Way.

Reber was a radio ham before he studied the subject at the Illinois Institute of Technology and so became a formally qualified radio engineer. Stimulated by the discovery by JANSKY of radio emission from the Milky Way, Reber built a steerable 30 ft (9.3 m) parabolic radio antenna in 1937 in his back yard and began to investigate in more detail. He and HEY were for many years the world's only radio astronomers, since others had not followed up Jansky's reports. His discovery of strong, discrete sources in Cygnus, Taurus and Cassiopeia persuaded them of the value of observations at radiofrequencies and led to the development of radio astronomy after the Second World War.

In 1954 he moved to Tasmania, set up a large radio telescope and worked to find evidence in support of his view that the 'Big Bang' theory is a fallacy: he was still doing so in the 1990s.

Reed, Walter (1851–1902) US epidemiologist; established the cause of yellow fever.

Reed trained in medicine in the University of Virginia and in New York, joined the US Army Medical Corps in 1875 and served in a series of frontier posts before specializing in bacteriology in the 1890s. In 1900 he was appointed to lead a small commission to study yellow fever, based in Cuba.

Yellow fever has a dramatic history; it has frequently proved a devastating epidemic disease, especially when non-immune groups (usually

Europeans or North Americans) entered new areas (as in central Africa or the Caribbean) and became exposed. It is now known to occur in two forms; the long-known type is urban yellow fever; largely by Reed's work, this is known to be due to a virus carried only by the female *Aëdes aegypti* mosquito. Reed's group tested theories on the transmission of yellow fever using army volunteers, and fairly soon were able to prove a theory (due to G W M Findlay (1893–1952)) that the mosquito was the transmitting vector.

By 1901 Reed had shown that the pathogen was a non-filterable microorganism, again using army volunteers; for the first time a virus was deduced as the cause of a specific human disease (LOEFFLER had already shown that foot-and-mouth disease in cattle is a viral disease). Reed's work was quickly followed by vigorous attacks, by drainage or addition of kerosene, on the mosquito's breeding places; one notable success, that of W C Gorgas (1854–1920) in Panama, proved to be a major factor in the completion of the Canal there. Since 1937 a vaccine has been available, due to THEILER, and a large measure of control has been achieved; but in Africa especially, jungle yellow fever (which is transmitted by a variety of mosquito vectors) remains a major problem, with mass vaccination as the preferred public health strategy. Although urban yellow fever and jungle yellow fever are often referred to as two forms, the virus is the same in both; but the cycle of transmission is different and the jungle form is sporadic in man, while the urban form occurs as an epidemic.

Regnault, Henri Victor [renyoh] (1810–78) French physical chemist: made experimental contributions to study of thermal properties of gases.

Regnault's father was an officer in Napoleon's army and died in 1812 in the Russian campaign; the boy was soon orphaned, but a friend of his father found him a job in a Paris shop. He worked hard for entrance to the École Polytechnique, graduated and became first GAY-LUSSAC's assistant and then his successor there in 1810. From 1854 he was director of the famous Sèvres porcelain factory, until his laboratory was destroyed by the Prussians in the war of 1870. Regnault had no clear plan of research, but he did valuable work in several areas. He discovered a series of organo-chlorine compounds, including the chlorinated ethenes and CCl_4. He worked on specific heat capacities and measured deviations from DULONG and Petit's Law; he measured the thermal expansion coefficient of gases accurately and found it varied slightly with the nature of the gas; and he studied the deviations from BOYLE's Law. He was cautious and no theorist, but his careful measurements made over 30 years were used by physical chemists and engineers for a generation.

Reid, Harry Fielding (1859–1944) US geophysicist: proposed elastic rebound theory of earthquakes.

Reid has been claimed to be the first American geophysicist. His mother was the greatniece of George Washington, and his prosperous parents took Reid as a child to Switzerland. There began his love for mountains and glaciers. He graduated from Johns Hopkins University in 1876 in physics and mathematics, and returned there in 1894 to teach until his retirement. His major work was his 'elastic rebound' theory of the source of the earthquake waves. The theory proposed that strain developed in the Earth's crust due to forces acting from below, of unknown origin. The strain leads eventually to breaks (faults) in the crust. The sudden release of strain energy by faulting, when the fracture strength is exceeded, can lead to an offset across the fault of up to 15 m. The rupture travels as a wave at about 3.5 km s^{-1}, for a distance up to 1000 km. The theory was soon accepted (about 1911) in the USA, but more slowly (up to 50 years later) elsewhere.

Remak, Robert (1815–65) Polish–German physician: advanced understanding of the structure of nerves.

A student of J P MÜLLER at Berlin, Remak remained there to work in general practice and in the university, although he was denied a senior teaching post because he was Jewish. In his early 20s he did notable work on the microscopy of nerve; he discovered the myelin sheath of the main nerves in 1838 and also showed that the axis-cylinder (axon) of a peripheral nerve arises from a nerve cell in the spinal cord and runs continuously to the terminal branch of the nerve. In this and his further work he saw that nerves have a flattened solid structure and are not merely structureless hollow tubes as they had been viewed for centuries. He was also a pioneer embryologist and one of the first to fully describe cell division and to argue that all animal cells came from pre-existing cells.

Reynolds, Osborne [renuhldz] (1842–1912) British engineer and physicist: gave definitive analysis of turbulent flow.

Reynolds studied mathematics at Cambridge, before being appointed as the first professor of engineering at Owens College (now Manchester University), where one of his students was J J THOMSON.

Reynolds was one of the outstanding theoretical engineers of the 19th-c. Most of his work concerned fluid dynamics, problems such as the flow around ships' propellers, vortex production by moving bodies and the scaling up of test results from models in which he used dyes to reveal flow patterns and vortices. He is remembered particularly for his work on the turbulent and laminar flow of liquids, and for defining (in 1883) a dimensionless quantity, the

Reynolds number, to determine the type of flow regime. The Reynolds number depends upon the viscosity, velocity, density and linear dimensions of the flow. He also carried out definitive work on lubrication, explained why radiometers rotate and performed a classic determination of the mechanical equivalent of heat.

Richards, Theodore William (1868–1928) US analytical chemist: famed for his accurate determination of relative atomic mass by quantitative chemical analysis.

From age 14 Richards was keenly interested in astronomy, but poor eyesight caused him to change his college studies to chemistry. After doing well at Harvard he visited Europe to learn the latest chemical methods; despite an offer at Göttingen he returned to a professorship at Harvard and stayed there. His particular interest became the exact determination of 'atomic weights' (ie relative atomic masses) and he carried classical gravimetric analysis to a level of high refinement in this work. He obtained accurate values for 25 elements, and his coworkers secured atomic weights for another 40, so giving a firm basis for quantitative analytical chemistry. He showed in 1913 that the atomic weight of ordinary lead differs from that of lead derived from uranium by radioactive decay; this work confirmed ideas on radioactive decay series and SODDY's prediction of the existence of isotopes. Richards's values for relative atomic mass were the best available until the widespread use of physical methods based on mass spectrometry gave even greater accuracy and precision, after the Second World War. He won the Nobel Prize for chemistry in 1914.

Richter, Burton (1931–) US particle physicist: experimentally demonstrated existence of charm quarks.

Richter studied at the Massachusetts Institute of Technology, then joined the high-energy physics laboratory at Stanford University, becoming a professor in 1967.

Richter was largely responsible for the Stanford Positron–Electron Accelerating Ring (SPEAR), a machine designed to collide positrons and electrons at high energies and to study the resulting elementary particles. In 1974 a team led by him discovered the J/psi hadron, a new heavy elementary particle whose unusual properties supported GLASHOW's hypothesis of charm quarks. Many related particles were subsequently discovered, and stimulated a new look at the theoretical basis of particle physics. Richter shared the 1976 Nobel Prize for physics with TING, who had discovered the J/psi almost simultaneously. Richter has been a strong proponent of the trend in particle physics towards building larger and larger particle accelerator rings.

Richter, Charles Francis (1900–85) US seis-

mologist: devised Richter scale of earthquake strength.

Richter worked at the Carnegie Institute before moving to the California Institute of Technology in 1936, becoming professor of seismology there in 1952. In 1935 he devised the scale of earthquake strength which bears his name. Unlike earlier, qualitative, scales, the Richter scale is an absolute scale based on the logarithm of the maximum amplitude of the earthquake waves observed on a seismograph, adjusted for the distance from the epicentre of the earthquake. Generally, earthquakes of magnitude 5.5 or greater cause significant damage. The largest earthquakes observed this century registered magnitude 8.9 on the Richter scale, and the earthquake that destroyed San Francisco in 1906 would have registered magnitude 8.25.

Riemann, Georg Friedrich Bernhard [reeman] (1826–66) German mathematician: originated Riemannian geometry.

Riemann, the son of a Lutheran pastor, studied theology to please his father, and then studied mathematics under GAUSS at Göttingen to please himself. In 1859 he became professor of mathematics there. At the age of 39 he died of tuberculosis. His friend DEDEKIND said of Riemann 'The gentle mind which had been implanted in him in his father's house remained with him all his life, and he served his God faithfully, as his father had, but in a different way.'

Riemann's papers were few but perfect, even in Gauss's eyes, producing profound consequences and new areas of mathematics and physics. Riemann's earliest publication was a new approach to the theory of complex functions using potential theory (from theoretical physics) and geometry to develop Riemann surfaces, which represent the branching behaviour of a complex algebraic function. These ideas were extended by introducing topological concepts into the theory of functions; this work was developed by POINCARÉ to advance algebraic geometry. In another paper Riemann defined a function $f(s)$, the Riemann zeta function, where

$$f(s) = 1 + \frac{1}{2^s} + \frac{1}{3^s} + \frac{1}{4^s} + \dots$$

where $s = u + iv$ is complex, and conjectured that $f(s) = 0$ only if $u = \frac{1}{2}$ for $0 < u < 1$. No-one has proved Riemann's hypothesis and it remains one of the important unsolved problems in number theory and analysis. Another of Riemann's contributions to analysis was the introduction of the Riemann integral, defined in terms of the limit of a summation of an infinity of ever smaller elements.

In 1854 Riemann gave his inaugural lecture, 'Concerning the hypotheses which underlie geometry': a mathematical classic. The content was so fruitful that it altered mathematics and physics for a century afterwards. Riemann considered how concepts like distance and curvature could be defined generally in n-dimensional space, extending Gauss's work (1827) on non-Euclidian geometries. He foresaw how important this was for physics and provided some of the mathematical tools for EINSTEIN to construct his general theory of relativity (1915).

Robin, Gordon de Quetteville (1921–) British glaciologist: made investigations into the thickness and flow of polar ice sheets.

Educated in Melbourne, where he graduated in physics, Robin then had 4 years of naval service, first on convoy escort and then in the submarine service. On demobilization in Britain he joined Oliphant's nuclear physics department in Birmingham University, where the vice-chancellor was Raymond Priestley of Shackleton's and Scott's Antarctic expeditions. Both helped to start Robin's polar career. In 1958 he was appointed director of the Scott Polar Research Institute, which he built into one of the world's leading glaciological research centres.

Robin advanced our understanding of polar ice sheets in a number of significant ways. In 1949–52 he led the first long seismic traverse to the Antarctic plateau, which crossed a mountainous terrain buried by ice up to 2.4 km thick. This was one of the last epic journeys across the continent. Later he and his group in Cambridge pioneered the use of airborne radio-echo-sounding to survey large areas of the Antarctic ice sheet. This revealed unexpected features such as subglacial lakes under ice depths exceeding 4 km in situations where his early theory of temperature distribution in polar ice sheets had predicted basal melting. Further theoretical contributions based on interpretation of field data included modifications to flow theory, and the dating and interpretation of palaeoclimatic evidence such as layering caused by volcanic dust within the ice that produced radio echoes. The layers extend over hundreds of kilometres within otherwise pristine Antarctic snows.

Robinson, Sir Robert (1886–1975) British organic chemist: master of organic synthesis and pioneer of the electronic theory of organic chemistry.

His family had a prosperous business making surgical goods, but young Robinson hoped to become a mathematician. However, his father wished to construct a bleach works (on the information supplied by Chambers *Cyclopaedia*) and so pressed him to study chemistry. He was sent to Manchester, did well and was afterwards successively professor at Sydney, Liverpool, St Andrews, Manchester, London and Oxford. He also had links with ICI and Shell. He was highly productive; his name is on more than 700 papers (20 after his 80th birthday) and 32 patents. One area of his talent was the chemistry of natural products; he worked on natural dyes such as brazilin, on the anthocyanins (plant petal pigments) and on alkaloids (he established the structure of the complex plant alkaloids strychnine and morphine) and steroids and antibiotics. His work carried organic chemistry to its highest points of achievement in the period before complex equipment came in, from 1960. Typically he would both show the structure of a natural product and devise elegant methods for its laboratory synthesis. In some cases he devised a method which neatly imitates a natural biosynthetic route.

In Manchester, Robinson took up the ideas of his teacher Arthur Lapworth (1872–1941) on the electronic mechanism of organic reactions and, at first with him and later alone, he offered an electronic theory that helps to explain and predict the course of reactions of organic molecules. However, his interests moved on and he left others (such as INGOLD) to expand the subject. Similarly, his seminal work on the biogenesis of organic compounds in plants was largely developed by others.

Robinson had a keen intuition in chemistry, as well as a highly analytical mind displayed both in his synthetic schemes and in chess (he was a powerful player). He was also a mountaineer, a keen traveller and an alarming motorist. In personality he was forceful and abrasive, and an irascible defender of his priorities.

He was awarded a Nobel Prize in 1947 and received most of the other honours open to him, including the Order of Merit (a UK decoration awarded for particular service to the country and limited to 24 persons).

Roche, Edouard Albert [rosh] (1820–83) French mathematician: proposed limits on the stability of planetary satellites.

Roche studied at Montpelier and Paris, subsequently being appointed professor of pure mathematics at Montpelier, a post he held for his entire working life.

In 1850 Roche calculated that a satellite orbiting a planet of equal density would break up under the influence of gravity if it were to approach closer than 2.44 times the radius of the planet. This limit is now known as the Roche limit and is thought to be the reason why the particles in the rings of Saturn, which extend out to 2.3 times Saturn's radius, do not aggregate into a moon.

Römer, Ole Christensen [roemer] (1644–1710) Danish astronomer: discovered finite velocity of light.

In 1675, while working with CASSINI on tables of the eclipses of Jupiter's moons, Römer noticed that the moons reached their predicted eclipse positions later than expected when Earth was distant from Jupiter, and earlier when it was closer to it. He realized that this discrepancy (about 10 minutes) must be due to the finite time that light took to reach Earth. Using Cassini's recent determination of Jupiter's distance he was thus able to calculate the speed of light to be 140 000 miles per second (2.25×10^8 m s^{-1}), about 75% of the correct value. Although this was the first proof of the finite speed of light, it was not until BRADLEY confirmed Römer's result in 1729 by the measurement of stellar aberration that it became widely accepted. Römer became Astronomer Royal in Copenhagen, and its mayor in 1705.

Röntgen, Wilhelm Konrad [roentgen] (1845–1923) German experimental physicist: discoverer of X-rays.

Originally a student of engineering at Zürich Polytechnic, Röntgen was attracted to physics, which he studied and later taught in several German universities. He was professor of physics in Würzburg when he made his famous discovery, in 1895. While using a discharge tube (in which an electric discharge is passed through a gas at low pressure) in a darkened room, he noticed that a card coated with BaPt(CN)$_4$ glowed when the tube was switched on. Röntgen soon found that the radiation causing this was emitted from the discharge tube at the region where the cathode rays (now known to be a stream of electrons) struck the glass end of the discharge tube. The new rays were found to have a much greater range in air than cathode rays; they travelled in straight lines and were not deflected by electric or magnetic fields; they passed through card, and even through thin metal sheet, and could be detected by a fluorescent screen or photographic plate. He named them X-rays. If passed through a human hand on to a photographic plate, the bones were seen as shadowed areas against the lighter flesh, and metal objects (for example a ring) gave opaque shadows. Röntgen suggested that the new rays were an electromagnetic radiation akin to light but of shorter wavelength, and this was proved by VON LAUE in 1912.

Röntgen was awarded the first Nobel Prize in physics, in 1901, 'for the discovery of the remarkable rays subsequently named after him'; in fact they are still known as X-rays. Their study added much to physics, gave a new technique for use in medicine and, after the work of the BRAGGS in 1915, led to X-ray crystallography as a new and immensely valuable method for the study of crystal and molecular structure. A modern X-ray tube uses a hot wire to generate electrons, which are accelerated by a high voltage and then strike a metal target, emitting X-rays (see diagram). Röntgen did excellent work on other areas of experimental physics. He took out no patents on his work, and died in some poverty in the period of high inflation in Germany.

Rohrer, Heinrich (1933–) Swiss physicist: invented the scanning tunnelling microscope (STM).

Rohrer joined IBM at Zürich in 1963 and later began to collaborate with Gerd Binnig (1947–), who joined in 1978. Together they took up work that Russell Young at the National Bureau of Standards in Washington had initiated: to build a scanning tunnelling microscope. A tungsten electron field emitter tip was moved across a surface by precision piezoelectric transducers and the tip was raised or lowered to keep it the same distance above the surface. The result could be plotted as a contour map of the surface. Binnig and Rohrer achieved a working microscope by reducing the tip to a single atom and bringing it within a couple of atomic diameters of the surface; as a result by 1981 they could even produce images of single atoms. By reducing vibration, a horizontal resolution of ≈ 2 Å and a vertical resolution of ≈ 0.1 Å (about one-30th the size of an average atom) was possible. Application of the technique to the study of semiconductor surfaces, microelectronics, chemical reactions on surfaces and biochemistry has occurred rapidly. As a result Rohrer and Binnig shared the 1986 Nobel Prize for physics with RUSKA (a key figure in the invention of the electron microscope).

Roscoe, Sir Henry Enfield (1833–1915) British chemist: pioneer in photochemistry, vanadium chemistry and chemical education.

Son of a Liverpool lawyer, Roscoe's enthusiasm for chemistry began at school. As a Dissenter, he went to University College, London and studied under GRAHAM and WILLIAMSON and then in Heidelberg with BUNSEN. Back in London in 1855, he juggled several modest chemical jobs in teaching and consultancy to make a living, but in 1857 he became professor in Owens College in Manchester, which gave him great scope. The college was unpopular in 1857, but Roscoe soon attracted students and convinced manufacturers of their value. His Manchester school of chemistry, then the best in Britain, did much to convert Owens College into the Victoria University. In 1885 he became an MP for Man-

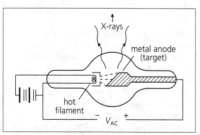

Diagram of an X-ray tube. A current heats the tungsten filament to a temperature high enough for it to emit a stream of electrons. The potential difference VAC is large (> 20 000 volts) so the electrons are strongly attracted to it and strike the metal anode target at high speed, emitting X-rays. Ancillary equipment to evacuate the tube, water-cool the anode and protect the operator from X-rays is not shown.

ANTARCTICA: THE CONTINENT FOR SCIENCE

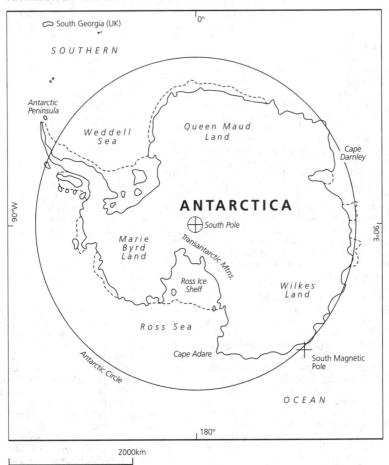

The Antarctic is a continent of extremes, one of which is that its inhabitants are almost exclusively scientists. The unique properties of this remote and ice-bound continent have made it a natural laboratory for, in particular, the earth sciences, life sciences, space science and environmental studies.

The place names of Antarctica are a record of its discoverers – men like Bellingshausen, Weddell, Dumont d'Urville and Amundsen. Most were adventurers and explorers; others such as Cook and Ross could be considered scientists through their additional scientific observations. The exploration of Antarctica continued through the efforts of men such as Shackleton, Byrd and Mawson in the early 20th-c, the so-called 'heroic era' characterized by the outstanding physical efforts of individuals.

Routine scientific studies largely began in 1957 when the International Geophysical Year provided the impetus for many nations to establish permanent bases and year-round scientific programmes, as well as convenient means of adding credibility to territorial claims. The logistics of simply maintaining bases has meant that most Antarctic science since has been a team effort. Most such work has concentrated on mapping the geology and geophysics of the Antarctic, and studies of the unique and abundant marine life. While richly rewarding, such studies have been of limited wider relevance.

Ironically it has been the most obviously unique feature of the Antarctic that has been the most informative about the rest of the world - the ice sheet which covers 98% of the continent. Ice cores drilled into the 4 km thick ice sheet preserve a wealth of climatic information covering the past 150 000 years, revealing information about air temperature, precipitation levels, atmospheric composition and CO_2 levels, meteorite falls, volcanic activity through the deposited acids and dust, and past solar flares through the resulting deposits of beryllium-10. (This is formed when high-energy particles associated with solar flares act on nitrogen in our upper atmosphere to give beryllium, later deposited on Earth.) The continuity and purity of Antarctic ice cores has thus provided one of the most comprehensive records of past climate available. Analysis of more recent snowfalls, over a scale of a few years, has provided unique baseline data on man-made pollutants such as pesticides, lead from car exhausts, fallout from bomb tests, etc, stretching back to before the industrial revolution.

One of the most recent discoveries made in the Antarctic has been the most dramatic, possibly so important that the whole future of human existence may depend upon it. In 1984 FARMAN observed a 40% depletion in ozone levels in the stratosphere above the Antarctic. Constant studies since have shown that ozone levels around both the South and North Pole have continued to drop alarmingly. Since the ozone layer protects life from the more harmful effects of the Sun's ultraviolet radiation, its disappearance may well have dramatic and severe effects for all life forms over the coming years. The cause has been identified as the manufacture and use of chlorofluorocarbons (CFCs) during the past 20–30 years, and Farman's discovery directly resulted in worldwide government action to limit the use of CFCs.

In 1995 a dramatically large iceberg (2888 km²/ 1115 sq mi) broke away from N W Antarctica into the Weddell Sea: a major event that may link with a rise of 2.5°C in average temperatures during the last 50 years.

As concern grows about global environmental issues, the Antarctic is likely to become an increasingly important laboratory for studying marine populations such as whales, and for taking a more careful look at other environmental issues such as climatic warming. A deep drilling project from 1996 should give information on past global climate, and so aid prediction of future change.

DM

---

chester for 10 years, and then vice-chancellor of London University.

His research with Bunsen was on the chemical action of light, using particularly the reaction: $H_2 + Cl_2 \rightarrow 2HCl$. This was the first research in quantitative photochemistry. In 1865 he heard that vanadium ores had been found in a Cheshire copper mine and this led him to explore vanadium chemistry. He was the first to make the metal, by reduction of VCl_2. He always had an interest in industrial chemistry, and in technical education.

Ross, Sir James Clark (1800–62) British polar explorer: located the north magnetic pole and explored the Antarctic Ocean.

Entering the Royal Navy when he was 12, James Ross served under his uncle John Ross in surveys of the White Sea and the Arctic, and later with W E Parry (1790–1855) in four attempts in the 1820s to reach the North Pole over the ice. From 1829–33 he was with his uncle on a private expedition (financed by the distiller F Booth) to explore the Arctic, and in 1831 he located the north magnetic pole. Back in the Navy as a captain from 1834, his expertise in magnetic measurement led to him being employed by the Admiralty in 1838 to make a magnetic survey (declination and dip) of the UK, and the next year to command an expedition to the Antarctic. This voyage lasted four years; he discovered Victoria Land, the 4000 m/13 000 ft volcano he named Mount Erebus, and 'the marvellous range of ice cliffs barring the approach to the Pole'. When he returned he had made the greatest survey of its kind, covering magnetic, geological and meteorological observations and studies of marine life at great depths; and only one man had been lost through illness, largely because Ross ensured good supplies of a mixed diet. In 1848 he commanded his last expedition, searching for the Arctic explorer Sir John Franklin (1786–1847), who had disappeared looking for the North-west Passage; he found no trace of Franklin, but new observations were made. He left the Navy with the rank of rear admiral.

Ross, Sir Ronald (1857–1932) British physician: discovered major steps in life-cycle of malarial parasite.

Ross was born in India, where his father was a British army officer; he returned to the UK to school when he was 8, studied medicine in London, and joined the Indian Medical Service in 1881. His interest in medicine increased from that time, but his interest in poetry, fiction and mathematics was life-long and he published in all these fields. From 1890 he studied malaria, and when on study-leave in London in 1894 he was shown the malarial parasite by MANSON, who suggested that it was transmitted by mosquitoes. Malaria was a long-known disease, and in the 1880s had been shown by C L A Laveran (1845–1922) and others to be due to a protozoon (a single-celled animal parasite), which invaded the red blood cells. The life-cycle of the protozoon (the genus *Plasmodium*) is complex, with several stages in human blood and liver and other stages in the stomach and salivary gland of a species of mosquito. Ross, back in London in 1895, dissected over 100 infected mosquitos before he saw, in 1897, the same stages that Laveran had seen in human blood; Ross found them in the *Anopheles* mosquito. By 1900 he and others had shown that human malaria is passed by the bite of *Anopheles*. Although still a major problem, control of malaria in many areas followed this knowledge of its transmission.

Ross worked in England after retiring from the Indian Medical Service in 1899, first in Liverpool and then in London; he was awarded a Nobel Prize in 1902.

Rossby, Carl-Gustaf (Arvid) (1898–1957) Swedish–US meteorologist: discovered large-scale waves in the upper atmosphere, and the jet stream.

Rossby was educated at the University of Stockholm and at the Bergen Geophysical Institute. In 1926 he emigrated to America, where he subsequently held professorships at the Massachusetts Institute of Technology and the University of Chicago. In 1940 Rossby demonstrated that large-scale undulatory disturbances exist in the uniform flow of the westerly winds in the upper atmosphere, developing in the zone of contact between cold polar air and warm tropical air; such waves are inherent features of a rotating fluid with a thermal gradient. There are usually three to five such Rossby waves in each hemisphere, with wavelengths of up to 2000 km. Rossby also showed that the strength of the westerly winds has an important influence on global weather, either allowing the normal sequence of cyclones and anticyclones to develop when the westerlies are strong or allowing cold polar air to sweep south when they are weak. He is further credited with the discovery of the jet stream, the broad ribbon of upper westerly winds travelling at about 45 m s^{-1} in the mid-latitudes.

He set up a weather service for airways in 1927 in California; and his work has been important in numerical weather prediction, especially since computers became available.

Rossi, Bruno Benedetti (1905–94) Italian–US physicist: discovered cosmic rays to be positively charged particles and found first astronomical X-ray source.

In 1934 Rossi demonstrated that many cosmic rays are positively charged particles by an experiment in the Eritrean mountains, using two sets of Geiger counters pointing east and west. A 26% excess of particles travelling in an eastward direction was found, indicating that these cosmic rays were positively charged particles and were deflected eastwards by the Earth's magnetic field. In fact, later work has shown that cosmic rays consist mainly of protons and a small proportion of heavy positive nuclei, together with electrons, all of high energy and coming from the Sun, probably with a contribution from supernovae.

Rossi also contributed to the birth of X-ray astronomy in 1962: he led a team which discovered the isotropic flux of X-rays incident on the Earth, and the first astronomical discrete X-ray source, Scorpio X-1, by use of a rocket-borne probe.

Rous, Francis Peyton [rows] (1879–1970) US pathologist and oncologist: showed that some cancers are caused by a virus.

During his second year as a medical student at Johns Hopkins University in Baltimore, MD, Rous scraped his finger on a tuberculous bone while doing an autopsy and became infected. After surgery he spent a year working as a cowboy before returning to medicine and graduating in 1905. He had a long career of over 60 years at the Rockefeller Institute in New York, working mainly on cancer. In 1911 he showed that a spontaneous cancerous tumour in a fowl could be transplanted by cell grafts and (remarkably) that even cell-free extracts from it would convey the tumour. This pointed to the cause being a virus, and by the 1930s several types of animal cancer were shown to be due to a virus. The Rous chicken sarcoma remains the best-known example. Initially the idea of a virus causing cancer was hard to believe, as the pattern of the disease is so different from that of typical viral infections. Rous developed methods for culturing viruses and cells; and he proposed that cancer-formation (carcinogenesis) typically involves one or both of two processes, initiation and promotion, which can require two different agents that may be chemical, viral, radiological or even mechanical. He shared a Nobel Prize in 1966. He was an active researcher until he was 90.

Roux, Pierre Paul Emile [roo] (1853–1933) French bacteriologist: co-discoverer of first bacterial toxin.

Even before he graduated in medicine in Paris, Roux assisted PASTEUR, working with him on anthrax and rabies, and in 1904 he suc-

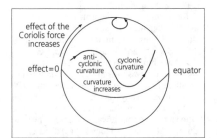

Rossby waves – the mechanisms of development of high-altitude westerly winds

effect of the Coriolis force increases

effect = 0

anti-cyclonic curvature

cyclonic curvature

curvature increases

equator

ceeded him as director of the Pasteur Institute. With A Yersin (1863–1943) he cultured the diphtheria bacillus in broth, and in 1888 showed that, if this was filtered through unglazed porcelain, the cell-free extract still produced the symptoms of the disease when injected into test animals. For the first time it was shown that, at least in this case, the effects of a bacterial infection are largely due to a potent toxin produced by the bacteria; he wrote 'snake venoms themselves are not as deadly'. BEHRING and KITASATO went on to show that blood serum from infected guinea pigs contained a counterpoison, an antitoxin; Roux used horses in place of guinea pigs, which gave enough serum to use on human patients, from 1894.

Rubner, Max [rubner] (1854–1932) German physiologist: made important investigations of animal metabolism and energy balance.

Rubner was professor of physiology at Marburg and later at Berlin. Before 1800 LAVOISIER had shown by experiments using a guinea-pig in a calorimeter that its heat production was the same as that given by burning a quantity of carbon equal to that in the CO_2 it expired, and so he concluded that metabolism is equivalent to burning at body temperature. Rubner developed such experiments on mammalian heat production, and proposed his surface law: that the rate of metabolism is proportional to the superficial area of the mammal and not to its weight. He also found that recently fed animals lost heat more quickly than fasting animals, pointing to a cellular regulatory system. He confirmed the results obtained by Lavoisier and others that metabolic energy production is equal to ordinary combustion despite the temperature difference; this implies that the law of conservation of energy applies to animate as well as inanimate objects. Rubner compared the energy available from various foods and showed that carbohydrates, fats and proteins were broken down equally readily, and that a mammal's energy usage for growth purposes is a constant fraction of its total energy output.

Rumford, Benjamin Thompson, Count (1753–1814) Anglo-American adventurer, social reformer, inventor and physicist: measured relation between work and heat; founded Royal Institution.

It would be hard to name a scientist who had a more extraordinary life than Rumford. He was a store apprentice, then a part-time teacher, gymnast and medical student with an interest in electrical machines. At 18 he married a rich young widow of 30 and decided to become a gentleman-soldier and farmer; he so impressed his seniors that at 19 he had become a major in the militia and squire of Concord.

However, excitement was on the way: New England was the centre of the American Revolution. His family had been there since 1630 and

he had good prospects if he cast in his lot with the revolutionaries, but Thompson supported the 'loyalist' view and did so by acting as a secret agent for the British army. There is no good evidence that he was a double agent, but by 1776 he was prudent enough to leave America for England, where he took up his scientific interests again (on projectiles, appropriately), was elected Fellow of the Royal Society in 1779, and next year became undersecretary of state in the Colonial Office at 27. In 1782, seeking active service, he went back to America, did well as a soldier and was shocked when peace was declared the following year.

This left him with no clear future, and he had no wish to 'vegetate in England', but soon, through carefully nurtured contacts, he was appointed adviser to the elector of Bavaria, being knighted, rather surprisingly, by George III as a preface to his new career. He was highly effective in Bavaria, reforming the conditions of the army, setting up welfare schemes for the poor which were well ahead of their time and creating a large park in Munich still treasured as the English Garden. He became a wealthy and respected public figure, with the title of Count, and minister for war. He was there for 14 years before moving to London in 1798, where he was welcomed as a great philanthropist and an expert on new methods for heating and feeding the poor.

Scientific work was always a part of Rumford's life and in Munich he made his greatest contribution to physics. He visited the arsenal and 'was struck by the very considerable degree of heat which a brass cannon acquires in a short time in being bored'. At the time, heat was thought to consist of a subtle fluid, 'caloric', which was squeezed out of the metal on boring, but Rumford showed, by using a blunt borer, that an apparently limitless amount of heat could be got from one piece of metal and that the supposed 'caloric' seemed to be weightless. He concluded that caloric was non-existent and that heat was 'the motion of the particles of a body'. He went on to measure the relation between work and heat, getting a result within 30% of the modern value. This concept was fundamental to modern physics, and the quantitative relation between heat and work was soon studied with great care by JOULE.

Rumford was always an enthusiast for the application of science, himself designing improved stoves, lamps and carriages; and in London in 1800 he planned and largely created the Royal Institution, which has been so valuable in British science ever since. Rumford's appointment in 1801 of DAVY, aged 22, to work at the Royal Institution was a happy and fruitful choice. Soon Rumford was travelling again, partly in exasperation as a result of disputes with the Royal Institution's managers. He settled in Paris, and in 1805 married MARIE LAVOI-

SIER, widow of the great chemist and reformer. The marriage quickly proved unfortunate: their quarrels were dramatic, and Rumford spent much time in his laboratory, studying the heat generated in combustion and using for this an improved Lavoisier calorimeter. He separated from Marie, but he had friends, money, a resident mistress, visits from his American daughter Sally and a substantial reputation. F D Roosevelt rated him with FRANKLIN and Thomas Jefferson, as 'the greatest mind America has produced'. He may well be the most colourful character in 18th-c science.

Ruska, Ernst August Friedrich (1906–88) German physicist: pioneer in development of the transmission electron microscope.

The electron microscope has had such a revolutionary effect in science (especially biology) that it must rank with the telescope, optical microscope and spectroscope as an outstanding device. Like these devices it has no universally agreed single discoverer; claims and counterclaims have been made, but the award of a Nobel Prize to Ruska in 1986 for the discovery makes him a central figure.

He studied high voltage and vacuum methods in Munich and Berlin, the appropriate background for his pioneer work on electron optics. By the mid-1920s it was known that electrons could behave not only as particles but, in appropriate experiments, as waves; and H Busch found that a magnetic coil could focus a beam of electrons, rather as a convex lens could focus a light beam. In 1928 M Kroll and Ruska (then a research student in Berlin) made a microscope giving 17× magnification using these methods of electron optics, and by 1933 Ruska made an instrument giving 12 000×, and commercial models were in use by 1938. However, G R Rüdenberg secured the first patent, which was upheld in a law suit in the USA (but not in Germany). Ruska's work, supported by the Siemens and Halske company, was continued in a converted bakery in Berlin during the Second World War, until Soviet troops looted the laboratory. His transmission electron microscope eventually achieved up to 10^6×, compared with 2000× for an optical microscope. Ruska shared his Nobel Prize with G Binnig (1947–) and ROHRER of IBM, who from 1978 worked in Zürich on a complementary device, the scanning tunnelling microscope, using an ultrasharp tip at high voltage to explore conducting surfaces and valuable for the study of metal surfaces, giving resolution down to atomic size.

Russell, Henry Norris (1877–1957) US astronomer: inferred stellar evolution from spectral type/luminosity relationship.

Russell's interest in astronomy probably began when, as a 5-year-old, his parents showed him a transit of Venus. After graduating from Princeton and a period in Cambridge, Russell spent his working life in Princeton. In 1913 he

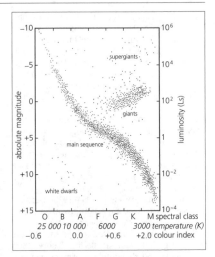

Hertzsprung-Russell diagram – schematic colour-magnitude diagram for many stars. The Sun, spectral type G, lies in the main sequence.

discovered that the absolute magnitude (the magnitude a star would have at 10 parsec distance from the Sun) of stars correlate well with their spectral class, which is indicative of surface temperature and related to their colour. He displayed his findings on a diagram of absolute magnitude *vs* spectral class, now known as the Hertzsprung–Russell (or H–R) diagram – HERTZSPRUNG had obtained similar results a few years earlier but his publication in an obscure journal had received little attention. Russell went on to suggest that the diagram represented an evolutionary path, with stars evolving into hot, bright blue-white giants and ending as cooler red dwarfs. Although this simple theory was soon abandoned, the diagram has remained a key tool in astrophysics, leading to research which has provided the modern and more complex views on stellar evolution.

Russell also studied chemical abundances in the Sun (a yellow dwarf) from the solar spectrum, concluding in 1929 that hydrogen makes up 60% of the Sun's volume, at the time a surprisingly large figure, but now known to be an underestimate by over 20%.

Rutherford, Ernest, 1st Baron Rutherford (of Nelson) (1871–1937) New Zealand–British physicist: founded nuclear physics.

Born near Nelson in New Zealand into a wheelwright's large family, Rutherford showed wide-ranging ability at school, won a scholarship to Canterbury College, Christchurch, and in his final years there concentrated on mathematics and physics. In his last year there he invented a sensitive radio-wave detector, just 6 years after HERTZ had discovered radio waves and in the same year that MARCONI began to use radio for practical purposes.

In 1895 he won a scholarship to Cambridge to work under J J THOMSON; he borrowed money for his passage to England and soon began research on the conductivity produced in air by X-rays, recently discovered by RÖNTGEN. He was J J Thomson's first research student. Rutherford became a professor at McGill University, Montreal 3 years later. Within a short period he greatly extended the foundations of nuclear physics, taking advantage of collaboration with the chemist SODDY and a good supply of the costly radium bromide. They produced nine papers in 18 months.

Radioactivity had been discovered in uranium in 1896 by BECQUEREL, and in thorium by G C Schmidt (1865–1949); PIERRE and MARIE CURIE had discovered two more radioactive elements, radium and polonium. Rutherford's studies revealed (1898) that the radioactive emission consisted of at least two kinds of rays; those which were less penetrating he called alpha rays (helium nuclei), and the others beta rays (electrons); 2 years later he discovered a third, and even more penetrating kind, gamma rays (electromagnetic waves). Together with Soddy he proposed in 1903 that radioactive decay occurs by successive transformations, with different and random amounts of time spent between ejection of each of the successive rays, sometimes years and sometimes fractions of a second. While the process is random, it is governed by an average time in which half the atoms of a sample would be expected to decay. This idea that atoms of some elements are not permanent, but can disintegrate, was then revolutionary.

A skilful set of experiments was then designed with T Royds (1884–1955) to examine alpha rays; having found that the mass and charge were correct for helium nuclei, this was

finally proved by sending the rays into an evacuated thin glass vessel and observing the build-up of helium gas inside.

In 1907 Rutherford returned to Britain, and in Manchester he extended his study of alpha particles, working with GEIGER on the detector named after him, and inventing the scintillation screen for observing them. Geiger and E Marsden (1889–1970) made the surprising discovery that about one in 8000 particles striking a gold foil was deflected back from it. As Rutherford put it '... quite the most incredible event that has ever happened to me in my life ... It was almost as if you fired a 15-inch shell at a piece of tissue paper and it came back and hit you.' Knowing that collision with a comparatively light electron could not produce such a large deflection, he deduced (1911) that atoms possess a very small but massive nucleus at their centre, holding all the positive charge to balance that of all the electrons about them. This was the first correct model of the atom, and BOHR developed it during a 3-month visit by showing which electronic orbits would be allowed by the 'old' quantum theory (which PLANCK had introduced in 1900).

The First World War caused Rutherford to work on sonic methods for detecting submarines, but in 1919 he returned to his research on succeeding J J Thomson as Cavendish Professor of Physics at Cambridge. Another major discovery occurred within months: he observed that nuclei could be made to disintegrate by artificial means, rather than waiting for their natural disintegration. This work he did himself, as young men were not yet back from the war. Alpha particles striking atomic nuclei, such as nitrogen, would knock out a proton to leave a different and lighter nucleus. Between 1920 and 1924 Rutherford and CHADWICK showed that most light atoms could be broken up by using alpha particles. Chad-

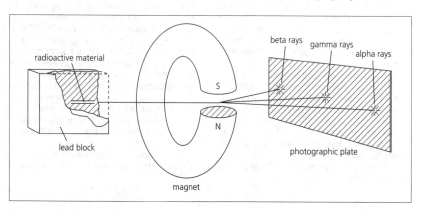

When radiation from a radioactive source such as radium passes through a magnetic field, it is split into three types. The gamma radiation (short X-rays) is undeflected and the alpha rays (helium nuclei) and beta rays (electrons) are deflected in opposing directions.

wick's later discovery of the neutron, and the nuclear disintegrations of heavier atoms achieved by COCKCROFT and E T S Walton (1903–95) using a linear accelerator, owed much to discussions with Rutherford. These and other major discoveries in his laboratory made 1932 a 'marvellous year' for nuclear physics. He also was involved, together with M Oliphant (1901–) and P Harteck, in the first nuclear fusion reaction (1934) by bombarding deuterium with deuterium nuclei to produce tritium. He had predicted the existence of the neutron, and the deuteron, as early as 1920.

Rutherford initiated and directed the beginnings of nuclear physics. He received the Nobel Prize for chemistry in 1908 and the simplicity and power of his work gives him a place as one of the greatest experimental physicists of all time. In personality he was forceful, exuberant and enormously likable, with physicist friends worldwide, many of them former students of his. He was not, of course, always right; his response to a suggestion that nuclear energy might one day be useful was that the idea was 'all moonshine'. But that was in the early 1930s; by 1936 he saw some prospect of useful atomic energy.

Ružička, Leopold [rutsichka] (1887–1976) Croatian–Swiss organic chemist: devised the isoprene rule.

Ružička grew up in Croatia when it was part of Austria-Hungary and, after an education in Germany and Switzerland, he finally became professor in Zürich. In 1916 he began work on perfumes; one famous study concerned the costly perfume fixatives in musk (from a Himalayan deer) and civet (from an African wildcat). He found that the key compounds, muscone and civetone, are cyclic ketones. What was remarkable is that both contain large rings of carbon atoms (16- and 17-membered respectively). It had been thought that such large rings could not exist. In fact, Ružička's work showed them to be stable, and he devised methods of synthesis. He also worked on steroids, specially the male sex hormones; his synthesis of testosterone made him rich and he became a major collector of early Dutch paintings. His work on terpenes led to the isoprene rule, which proposes that terpenes have structures based on units of the five-carbon isoprene molecule, joined head-to-tail (1920). In this form, or in its later version (which proposes the biogenesis of these compounds from isopentenyl pyrophosphate) the rule has proved a valuable guide to structure. Ružička shared a Nobel Prize in 1939.

Rydberg, Johannes (Robert) [rüdberg] (1854–1919) Swedish spectroscopist: early theorist on atomic number.

Educated at Lund, Rydberg stayed there for his entire career, as professor of physics from 1901. He was fascinated by MENDELAYEV's periodic table of the elements and he had the brilliant and valuable intuition that the periodicity was related to atomic spectra and atomic structure. Working on atomic emission spectra in 1890, he found a simple general formula for the frequency of some of the spectral lines. He introduced the useful idea of wave numbers ($1/\lambda$, where λ is the wavelength); and showed that BALMER's equation giving the frequencies of many lines in the hydrogen spectrum could be generalized in the form $1/\lambda = R(1/m^2 - 1/n^2)$. Within 30 years successive spectral series were found fitting this formula, with simple integral values for n and m; the Balmer series is that having $m = 2$ and $n = 3,4,5$ etc. The constant R is known as the Rydberg constant. Rydberg himself never reached his goal of relating spectra to atomic structure, but his view that a relation between structure and spectra must exist was valuable and reached fruition with BOHR's work on atomic structure in 1913, from which Rydberg's formula emerges with a value of R calculated in excellent agreement with experiment.

Rydberg's study of the periodic table led him in 1897 to see the importance of atomic number (rather than atomic weight), a view confirmed by MOSELEY in 1913; and in 1906 he stated for the first time that 2 ,8 and 18 (ie $2n^2$ where $n = 1$, 2, 3) are the numbers of elements in the early periods. He also corrected the number of lanthanides to 32.

Ryle, (Sir) Martin (1918–84) British astronomer: produced first detailed map of the radio sky.

Ryle was a key figure in the development of radio astronomy after the Second World War, following the pioneering discoveries of JANSKY and REBER. He began a series of surveys at Cambridge in the 1950s, culminating in 1959 in a definitive catalogue of the strengths and positions of 500 radio sources (increased to 5000 in 1965). By using radio dishes separated by up to 5 km, Ryle was able to obtain a large effective aperture and to produce detailed radio maps of areas of sky. In this way pulsars, quasars and radio galaxies could be studied. In showing that the distant parts of the universe appear different from the nearer parts, Ryle's results supported the 'Big Bang' view of its origin (rather than a 'steady-state' theory) and led to major public disputes with HOYLE.

He was honoured as Astronomer Royal (1972–4) and shared the Nobel Prize for physics in 1974.

Sabin, Albert (Bruce) [saybin] (1906–93) US virologist and developer of an oral poliomyelitis vaccine.

Polish by birth, the Sabin family emigrated to the USA in 1921, and Albert Sabin studied at New York Medical School. During the Second World War he devised vaccines for the US army, notably one against dengue fever. After the war, working at the University of Cincinnati, he attacked the problem of finding a safe and effective live attenuated vaccine against polio. This disease was then a major problem, with 57 000 mainly young cases in the USA in 1952. J E Salk (1914–95) used culture methods due to ENDERS to develop in the 1950s a dead vaccine, killed by formaldehyde, which was widely used despite the difficulty that it needed several injections to give protection and was only 80–90% effective.

Sabin's live vaccine was attenuated by culture in monkey kidney tissue and could be given by mouth as a single dose on a sugar lump. After trials with volunteers in an Ohio reformatory, Sabin persuaded the USSR to use it on a large scale in the late 1950s. It was quickly seen as better than Salk's vaccine: the US public health service approved it in 1960 and the UK changed to Sabin's vaccine in 1962. The two men were highly competitive. Sabin continued to work on viruses, and was awarded the US National Medal of Science in 1970.

Sabin, Florence (Rena) [saybin] (1871–1953) US anatomist and histologist: ensured the passage of vital public health legislation.

Florence Sabin's mother died when she was 4 and she was brought up largely by her uncle and grandparents. She went to Smith College and gained a BS degree in 1893, taught for 3 years to save money, then entered Johns Hopkins Medical School and received the MD degree in 1900. After a year's internship she became the school's first woman faculty member (1902) as an assistant in anatomy. She was appointed professor of histology in 1917, and had a long association with Johns Hopkins. She worked on the lymphatic system, an area at the time little understood. Her findings were that the lymphatic channels represented a one-way system, closed at their collecting ends, where the fluids entered by seepage, and that they arose from pre-existing veins; a view at first highly controversial but proved correct. She also studied the origin of blood vessels and plasma and the development of blood cells in embryos. In 1925 Sabin moved to the Rockefeller Institute in New York where she remained until her retirement in 1938. Her research there

was a study of tuberculosis and the development of immunity to it.

Retiring from research in 1938, she returned to her home in Denver and in 1944 became chairman of a public health subcommittee, and an energetic supporter of public health reforms.

She was the first woman elected to the National Academy of Sciences of the USA in 1925.

Sabine, Wallace Clement [sabin] (1868–1919) US physicist: the founder of architectural acoustics.

A midwesterner from a farming family, Sabine did well as a physics student at Harvard and became an instructor there in 1890. Except for war work in the First World War (he was in effect the first chief scientist for the US Air Force, in 1917–18) he never left Harvard. Some 5 years later Harvard president C W Eliot asked for his help: the acoustics of a major new lecture theatre were so bad as to make it useless. Little was then known of architectural acoustics, but Sabine saw that such problems must arise from the size, shape, and materials of a room, which affect the reverberation time. For speech this should be 1–2 s and for music about 25% longer. In the disastrous new theatre this time was 5.5 s, so that a speaker had the first and last words of a sentence mingled. Since size and shape are not easily altered, Sabine worked on the materials; he and two helpers spent many nights (after midnight, when the street was quiet) moving cushions from another large theatre to the disaster area, making tests, and returning them before dawn. Authority was dismayed by the delay and after 2 years demanded action; Sabine prescribed 22 hair-felt blankets, and the place was rendered usable for the next 75 years.

He had devised an electrically-blown organ pipe and drum recorder to measure reverberation time at 512 Hz, and he now worked hard to develop a formula that would allow calculation of the acoustics of an unbuilt hall. With the aid of more massive cushion-moving experiments, he derived, in 1898, the Sabine formula: $T = KV/Sa$, where T is the time in seconds for a sound to decay 60 decibels in the hall; K is a near-constant, inversely proportional to the speed of sound; V the volume of the room; S is the total area of all the room surfaces; and a is the average sound absorption coefficient of these surfaces (it varies from about 0.01 for plaster to 1.0 for an open window). With its aid, he was able to advise on a projected new Boston Symphony Hall, opened in 1900. However, musicians were critical of the result; this was probably because an orchestra of 90 sounded thin in such a large hall; today with 104 players the sound is fuller. From 1904 he was much in demand to advise on architectural acoustics, and his methods have been in use ever since.

Sachs, Julius von (1832–97) German botanist: pioneer of plant physiology.

Sachs was an assistant to PURKINJE at Prague, and after various posts in Germany became professor at Würzburg from 1868. During nearly half a century he made massive contributions to plant physiology, which before him was largely neglected. Much apparatus and technique now familiar is due to him, and his many pupils continued to develop the subject. His early work was on stored nutrients in seeds and on the culture of plants in nutrient solutions (hydroponics). He studied the uptake of minerals by plants, and the influence of temperature on plant growth, and discovered the 'law of cardinal points'. In 1861 he showed that photosynthesis actually occurs in chloroplasts and that the first 'visible' product of carbon dioxide uptake is starch, deposited in the chloroplasts. He studied etiolation and the formation of flowers and roots; and geotropism, phototropism and hydrotropism. When, after about 1880, he moved more towards theory, he was less successful; and his authority held back new views and delayed discoveries. In his last years he strongly attacked DARWIN's views on evolution.

Saha, Meghnad (1894–1956) Indian astrophysicist: demonstrated that elements in stars are ionized in proportion to their temperature.

The son of a small shopkeeper in Dacca, Saha was educated at Presidency College, Calcutta and afterwards visited Europe. He taught at the University of Allahabad and in 1938 he was appointed professor of physics at Calcutta. The absorption lines in the spectra of stars vary widely, with some showing only hydrogen and helium lines and others showing numerous metal lines. In 1920 Saha demonstrated that this did not necessarily represent a true variation in elemental composition, but a different degree of ionization of the metal atoms, which was related to temperature by Saha's equation for a monatomic gas. At higher temperatures the metal atoms exist only in ionized form and the absorption lines of neutral metal atoms become very weak. The proportions of the various ions of the same metal can be used to estimate stellar temperature. RUSSELL used Saha's results to estimate the amount of hydrogen in the Sun.

Sakharov, Andrei Dmitriyevich [sakarof] (1921–89) Russian nuclear physicist and political activist.

The son of a physics teacher, Sakharov was born and educated in Moscow, where he graduated in 1942. From that time he worked in the Lebedev Institute of Physics, and from 1948 he was the key figure in developing the Soviet hydrogen bomb, which was exploded in 1962, when he received the highest honours from the state for this work. But he had long been concerned about the pollution effects of nuclear weapon testing, and after he had published his underground essay *Thoughts on Progress, Peaceful Co-existence and Intellectual Freedom* (1968) he was seen by the state as a possibly subversive critic and was excluded from secret work.

From 1965 Sakharov became interested in cosmology, and especially in the problem of 'what was before the "Big Bang"?' In a major paper of 1967 he attempted to explain why the universe is built of protons, neutrons and electrons while the corresponding anti-particles are rare. To explain this 'baryon asymmetry of the universe' he proposed that the original universe was neutral and had no such asymmetry; the asymmetry built up following the 'Big Bang', as a result of non-stationary processes in the expansion of the early universe, together with his idea that protons are inherently unstable (with a lifetime perhaps of 10^{50} years) and the fact that particle–antiparticle symmetry can be violated. This last fact had been shown experimentally by others in 1964, but confirmation of the ultimate instability of protons (whose decay would involve disposing of three quarks) has yet to be observed, perhaps because it is too rare an event: the proposed proton lifetime is 10^{20} times the present age of the universe.

During the 1970s the Soviet state became more repressive, while Sakharov became increasingly active as a human rights supporter, especially after his marriage to his Jewish second wife in 1971. He was awarded the Nobel Peace Prize in 1975 but not permitted to collect it, and his outright condemnation of Soviet aggression in Afghanistan in 1979 led to his arrest in 1980 and exile in the 'closed city' of Gorky. In 1986 the new leader of the USSR, Gorbachev, recalled him to Moscow and to relative freedom, but by then he was exhausted and ill, and in his remaining years his role as an advocate for change in his country to a democratic multi-party state with a free-market economy was mainly symbolic. There is no doubt that for many years he had been a moral leader of the opposition to the communist regime in Russia, but he was never a direct and fully effective political figure and he was elected to the Congress of People's Deputies only shortly before his death.

Salam, Abdus (1926–   ) Pakistani theoretical physicist: developed unified theory of the weak nuclear force and electromagnetism.

Salam's early career shifted between Punjab University, Cambridge University and Lahore, where he became a professor at the Government College and Punjab University. He then lectured at Cambridge (1954–6) and in 1957 became professor of theoretical physics at Imperial College of Science and Technology, London. Salam's concern for his subject in developing countries led to his setting up the International Centre of Theoretical Physics in Trieste in 1964.

Physicists recognize four basic forces in nature: gravity, electromagnetism and the 'strong' and 'weak' nuclear forces, which are active only within nuclear range. In 1979 Salam won the Nobel Prize for physics together with WEINBERG and GLASHOW. Independently each had produced a theory explaining both the 'weak' nuclear force and 'electromagnetic' interactions. This led to the prediction of neutral currents, later found by experiments at CERN (European Organization for Nuclear Research) in 1973, and 'intermediate vector bosons', first seen in 1983.

Sandage, Allan Rex (1926–) US astronomer: made first identification of an optical object with a quasar.

Sandage studied at the University of Illinois and the California Institute of Technology before joining the Hale Observatories, initially as an assistant to HUBBLE. Although radio galaxies had been discovered in the 1940s, it was some years before compact radio emitters ('quasi-stellar radio sources' or quasars) were found by use of improved radio telescopes, and only in 1960 that the first strong compact radio source (3C 48) was matched with the position of an optical bright star, by Sandage. Others were located in the 1960s, and M SCHMIDT showed they have massive red shifts, implying high speeds of recession; this is now regarded as a defining feature, rather than intense radio emission, which Sandage showed is absent for 90% of quasars (so the word is better now regarded as defining 'quasi-stellar objects'). Typically quasars are bluish objects, intensely luminous, double-lobed in shape, emitting synchrotron radiation which is variable over weeks or months.

Their origin and huge energy generation is mysterious: consensus opinion regards them as the most distant visible objects in the universe, so that they are now seen as they were billions of years ago. They are best explained as the cores of infant galaxies, having a massive black hole at their centre. The SEYFERT galaxies may represent a more mature stage. Intensive study of quasars continues; over 1500 are now known.

Sanger, Frederick (1918–) British biochemist: pioneer of chemical studies on the structure of proteins and nucleic acids; the only double Nobel laureate in chemistry.

Sanger, a physician's son, graduated in Cambridge in 1939 and researched there through his career, on the staff of the Medical Research Council laboratories from 1951. In the early 1940s he devised a method using 2,4-dinitrofluorobenzene (Sanger's reagent) to label the amino acid at the 'free amino' end of a protein chain. By combining this method with the acid or enzymic break-up of the longer protein chains to give shorter, identifiable fragments, Sanger was able to deduce the sequence of amino acids in the chains of the protein hormone insulin; by the early 1950s he had worked out the sequence of the 51 amino acids in its two-chain molecule and found the small differences in the sequence in insulins from pig, sheep, horse and whale.

After he was awarded a Nobel Prize for this, in 1958, he moved on to the bigger problem of the structure of nucleic acids. These biological macromolecules have double helical chains of nucleotides whose base sequence determines the information carried by the genes. He worked first on RNA, whose chains are of modest length, and then moved to DNA, which has very long chains with up to 10^8 units in a chain. Sanger used a highly ingenious combination of radioactive labelling, gel electrophoresis and selective enzymes which can split or grow DNA chains at specific points.

By 1977 he and his group were able to deduce the full sequence of bases in the DNA of the virus Phi X 174, with over 5400 bases. Mitochondrial DNA, with 17 000 bases, soon followed. Such methods, by 1984, led to the full base sequence in Epstein–Barr virus (EBV), whose genome (the complete set of genes of an organism) is over 150 000 bases long. For his nucleic acid work Sanger shared the 1980 Nobel Prize and became the first to win two Nobel Prizes for chemistry. His work has given new, surprising and detailed knowledge of both proteins and genes and has stimulated others in this field.

Santorio, Santorio (*Ital*), Sanctorius (*Lat*) (1561–1636) Italian physician: applied physics to medicine.

A graduate of Padua, Santorio was physician to the king of Poland for 14 years before returning to Padua as professor of theoretical medicine in 1611. He was a colleague of GALILEO and Santorio's medical research was doubtless influenced by him: Galileo worked on pendulums and thermometers, but it was Santorio who used a pendulum to compare pulse rates (no clocks were then available), and he invented the clinical thermometer in 1612 and later explained its use. He was the first to apply quantitative methods in medicine and is best known for his work on metabolism. He examined the change in body weight with diet, sleep, activity and disease, and for 30 years spent much time suspended from a steelyard, weighing himself and his solid and liquid input and output and deducing the amount of 'insensible perspiration' lost through the skin and lungs.

Sargant, Ethel (1863–1918) British botanist who suggested a new interpretation of the relationship between mono- and dicotyledons.

Ethel Sargant studied natural science at Girton College, Cambridge (1881–5), and spent a year at the Jodrell laboratory at Kew Gardens training in research methods; in 1897 she visited several laboratories in Europe. For many years she cared for her elderly mother and an invalid sister, and worked first from a laboratory built in the grounds of her mother's house and later from her own home in Cambridge. She acted as research adviser to the Cambridge students of botany.

She worked in the cytology and the anatomical morphology of plants, and her earliest work concerned the presence of centrosomes in higher plants. She moved to a general study of oogenesis and spermatogenesis in *Lilium martagon*. Her work demonstrating the existence of the synaptic phase in living cells was published in the *Annals of Botany* in 1896 and 1897.

Sargant's study of monocotyledonous seedlings resulted in her suggestion that both mono- and dicotyledons evolved from a common ancestral stock and that the single seedleaf in the monocotyledon was homologous to the pair in the dicotyledon. These findings were discussed in *A Theory of the Origin of Monocotyledons Founded on the Structure of Their Seedlings, The Evolution of Monocotyledons* and *The Reconstruction of a Race of Primitive Angiosperms*, published between 1903 and 1908.

She was elected a Fellow of the Linnean Society in 1904, and was the first woman to serve on its council. At the 1913 Birmingham meeting of the British Association for the Advancement of Science she was elected president of the botanical section, becoming the first woman to preside over a section. She was elected to an honorary fellowship of Girton College in 1913.

Schaudinn, Fritz Richard [showdin] (1871–1906) German zoologist and microbiologist: identified the organism responsible for syphilis.

After starting university work as a student of philology at Berlin, Schaudinn turned to science and specialized in zoology. During his short career he worked in Berlin, especially on those protozoa (notably trypanosomes) that cause human disease. He demonstrated the alternation of generations in Foraminifera and worked out the life-cycle of the Coccidiae (scale insect). He distinguished between the amoeba causing tropical dysentery (*Entamoeba histolytica*) and its harmless relative *Entamoeba coli* which lives in the human intestinal lining; his work included experimental self-infection with both organisms. His best-known discovery, made in 1905 with the dermatologist P E Hoffmann (1868–1959), was of the pale threadlike undulating spirochaete, now named *Treponema pallidum*, that causes syphilis. Proof that this causes venereal syphilis was not easy and its acceptance was delayed. He also worked on malaria, proved an earlier guess that the parasite of human hookworm enters through the skin of the feet, and described many new animals first observed by him on expeditions to the Arctic.

Schawlow, Arthur (Leonard) [showloh] (1921–) US physicist: co-inventor of the laser.

Schawlow's early work was at Toronto and his postdoctoral research was with TOWNES at Columbia University. The two remained in contact; Schawlow married Townes's sister. After 10 years at Bell Telephone Laboratories, a professorship for Schawlow at Stanford followed in 1961.

Townes and Schawlow collaborated to extend the maser principle to light by devising the laser, although the first working laser was constructed by MAIMAN in 1960. (For an account of masers and lasers see Townes's entry.) From the early 1970s Schawlow used laser methods to simplify atomic spectra, and to give improved values for basic physical quantities, such as the RYDBERG constant, and extraordinarily precise values for the electronic energy levels in the hydrogen atom. Schawlow shared a Nobel Prize in 1981 for his work on laser spectroscopy.

Scheele, Carl Wilhelm [sheeluh] (1742–86) Swedish chemist: a discoverer of chemical elements (chlorine, and oxygen) and of many chemical compounds.

Scheele was trained as an apothecary, at a time when they made most of their own drugs and had available a range of minerals, plants and simple equipment for chemical operations. He had a passion for chemical experimentation which is probably unsurpassed, but he was unlucky in that some of his major discoveries were also made by others at nearly the same time and they published sooner. Nevertheless his renown led to prestigious job offers, which he refused, preferring to take a series of posts as an assistant apothecary. This left him able to experiment freely in his limited leisure time but the overwork, the poor conditions and the absorption of hazardous chemicals may have led to his early death.

Scheele first made the reactive green gas chlorine in 1774 from hydrochloric acid and MnO_2, but the fact that it is an element became known by DAVY's work of 1810. Earlier, in 1773, Scheele had shown that air is a mixture (of 'fire air' and 'foul air' as he named its components) and he made oxygen in several different ways (e.g. by heating HgO, or KNO_3, or $Hg(NO_3)_2$; or MnO_2 with H_2SO_4), but his book on this did not appear until 1777. Before then, in 1774, PRIESTLEY had published his discovery of oxygen. The two men were similar in their skill as experimenters, their limited interest in theory and their adherence to the phlogiston theory.

Scheele made a variety of new acids in the 1770s (phosphoric, molybdic, tungstic and arsenic acids) as well as HF, SiF_4, AsH_3 and other reactive and very toxic compounds. In the 1780s he made a number of new organic acids, and fairly pure hydrocyanic acid (recording its taste!). His death at 43 is unsurprising; he was a fanatical, prolific and probably unwise chemical discoverer.

Schiaparelli, Giovanni Virginio [skyaparelee] (1835–1910) Italian astronomer: demonstrated that meteors follow cometary orbits; observed Martian surface features.

Educated in Italy, Germany and Russia, Schiaparelli was director of the Milan Observatory for 40 years. In 1866 he showed that meteors follow cometary orbits and he identified the comets associated with the annual Leonid and Perseid meteor showers. In 1877 Mars made one of its closest approaches to Earth, enabling Schiaparelli to discover its southern polar ice cap and to identify the direction of the axis of rotation. He also saw what he thought to be many dark lines crisscrossing the planet and termed them 'canali' (channels). Schiaparelli himself believed these to be natural features, but the American astronomer LOWELL suggested that they were irrigation features constructed by a Martian civilization (mistranslating 'canali' as 'canals'), and thus started a long controversy that was only finally laid to rest by the Mariner space probes.

Schleiden, Jakob Mathias [shliyden] (1804–81) German botanist: a founder of cell theory.

Schleiden studied law in Heidelberg and practised it in his birthplace, Hamburg; but his interest in botany grew and he studied it at three universities, graduating at Jena in 1831. He lectured on botany there from 1839, and became a private teacher of botany after 1864. He was a skilful microscopist, and from about 1840 good compound microscopes became available that were largely achromatic. By 1880, thanks to improved designs by ABBE, oil-immersion objectives and better sectioning and staining of specimens, the instrument reached a high point, with good resolution at magnifications up to 2000×. Plant cells (ie delimited spaces within walls) had been observed two centuries earlier by HOOKE and others, but Schleiden's studies convinced him of their importance, and by 1838 he argued that all the various plant structures are composed of cells or their derivatives. He accurately observed many features and activities in plant cells (eg cytoplasmic streaming); he recognized the importance of the nucleus in cell division, but believed (wrongly) that new cells were formed by budding from its surface.

Despite some uncritical attitudes and a quick temper his ability was great; he has been named as 'the reformer of scientific botany' and he initiated SCHWANN's work, which led to their joint creation of cell theory. He was a popular writer and lecturer on a wide range of matters and frequently engaged in harsh combative debates on scientific theories.

Schmidt, Bernhard Voldemar [shmit] (1879–1935) Estonian–German optical engineer: designer of a novel telescope system.

As a young man, Schmidt had several dull jobs before he took a course in engineering and set up his own workshop to build telescopes. He ground and polished the mirrors unaided, despite having lost his right arm in an accident with explosives. From 1926 he worked with the Hamburg Observatory staff; he was an alcoholic and died in a mental hospital.

Before his work, large telescopes gave good images only in the centre of the field; further from the optical axis the images of stars showed a tail (coma), which made making star maps difficult. Schmidt designed in 1930 a reflecting telescope with a spherical mirror, and with a smaller glass corrector plate in front, at the centre of curvature. The specially shaped aspherical plate provides the telescope with good definition over a wide field at low power. The telescope is usually used as a camera, and it revolutionized optical astronomy. From this design other catadioptric systems were developed, such as the Maksutov, which has a spherical meniscus corrector plate; it is compact and, like the Schmidt camera, combines useful features of reflecting and refracting telescopes.

Schmidt, Maarten [shmit] (1929–) Dutch–US astronomer: explained optical spectra of quasars as due to their relativistic velocities.

Schmidt studied at the universities of Groningen and Leiden, moving to the California Institute of Technology in 1959. He became director of the Hale Observatories in 1978.

Following the identification of a quasar with a faint optical object with a highly unusual spectrum in 1960 by SANDAGE, Schmidt studied the spectrum of another optically identified quasar, 3C 273, and discovered that the peculiarities of its spectrum were caused by a massive red shift. The quasar appeared to be receding at nearly 16% of the speed of light, leading to what is normally the ultraviolet part of the spectrum being observed in the visible region. Such high velocities are now interpreted as implying that quasars are very distant objects. Schmidt also found that the number of quasars increases with distance from Earth, a finding that provided evidence for the 'Big Bang' theory for the origin of the universe rather than the rival steady-state theory.

Schrieffer, John Robert [shreefer] (1931–) US physicist: contributor to the BCS theory of superconductivity.

After studying electrical engineering and physics at Massachusetts Institute of Technology and the University of Illinois, Schrieffer started working for his PhD under BARDEEN. Initially he worked on electrical conduction on semiconductor surfaces, but moved to superconductivity as his thesis topic. A close and fruitful collaboration with Bardeen and COOPER gave rise to the BCS (Bardeen, Cooper,

Schrieffer) theory in 1957, for which all three shared the 1972 Nobel prize for physics (see account under Bardeen and Cooper). Schrieffer became professor of physics at the University of Pennsylvania in 1964. He has also done research on dilute alloys, ferromagnetism and surface physics.

In the formulation of BCS theory Schrieffer contributed particularly to the generalization from the properties of a single Cooper pair to that of a solid containing many pairs. Using a statistical approach he found a suitable quantum mechanical wave function that possesses the correct properties.

Schrödinger, Erwin [shroedinger] (1887–1961) Austrian physicist: the founder of wave mechanics.

Schrödinger was the son of a prosperous oilcloth manufacturer; he was educated by a private tutor and by his father before going to the University of Vienna. After his doctorate in physics in 1910 he joined the staff there. During the First World War he served as an artillery officer in an isolated fort, which gave him time to read physics; from 1920 he spent short periods of time as a student at Jena, Stuttgart, Breslau and Zürich. He began to produce inspired work, and early in 1926 published a series of papers founding wave mechanics. As a result he succeeded PLANCK as professor of theoretical physics at Berlin (1927), but chose to leave once Hitler had assumed power in 1933. He moved to Oxford, but became homesick and returned to Graz in Austria in 1936. When the Germans moved into Austria in 1938 Schrödinger fled for his life, settling in Dublin and working at the Institute for Advanced Studies created for him there, as a result of the Irish Taoiseach (prime minister) de Valera's mathematical interests. After 17 happy years in Eire, he became a professor at the University of Vienna, having refused to return until Soviet occupation ceased. Shortly after arriving he became ill and never fully recovered.

Schrödinger began to think about the consequences of Broglie's ideas when they were published in 1924. The latter had postulated that any particle has a wave associated with it and the properties of the particle result from a combination of its particle-like and wave-like nature. Schrödinger and Broglie both realized that a partial differential equation called a wave equation would describe the motion of a particle, and deduced an equation of this type. This approach of considering the wave function alone avoided the difficulties which BOHR's old quantum theory of particles had involved, but was difficult to apply in practice. Schrödinger then used HAMILTON's method for describing particle motion, and wrote this in wave form to give 'Schrödinger's equation'. Unlike the previous equation this ignores relativistic effects, but is much easier to visualize (as standing waves around the nucleus) and to apply to real situations. When applied to the hydrogen atom the equation gives the correct energy levels of an electron in the atom, without the *ad hoc* assumptions of Bohr's model of the atom. These energy levels had been measured experimentally by using the lines observed in the hydrogen spectrum. For this considerable achievement he shared the 1933 Nobel Prize for physics with DIRAC.

Schrödinger's theory was known as wave mechanics (1926), and was shown by Dirac to be mathematically equivalent to matrix mechanics, devised in 1925 by BORN, JORDAN and HEISENBERG. The combined theory, together with PAULI's exclusion principle, was used by Dirac to set out quantum mechanics in virtually complete form by the year's end.

While quantum mechanics had great predictive power and correctly described a wealth of previously unexplained phenomena, Schrödinger saw in it an awkward problem. Relating the wave function to the particle (for example an electron) is difficult. Born put forward the now-accepted explanation that the wave amplitude describes the probability of finding the particle at that point. Schrödinger, like Broglie and EINSTEIN, opposed this, and together they argued against a probabilistic quantum mechanics. Born's view condemns physics to describing only the likelihood of one event following another and is not able to definitely predict cause and effect, as classical theories sought to do.

Schrödinger had an informal manner much liked by his colleagues and students; throughout his life he travelled with walking-boots and rucksack, which caused him some problems in gaining entrance to the Solvay conferences for Nobel laureates.

Schwabe, Samuel Heinrich [schvahbuh] (1789–1875) German astronomer: discovered the sunspot cycle.

Schwabe studied pharmacy in Berlin but, although he took over his mother's pharmacy business, his real enthusiasm was in astronomy. When he was 40 he sold the business and expanded to full time his interest in astronomy. From 1826 he had observed the Sun, projecting its image from a small (2 inch/5 cm) telescope, in an effort to find a hypothetical planet within the orbit of Mercury, hoping to see it in transit across the Sun's disc. In this way he became interested in sunspots, and observed them daily for the rest of his long life.

By 1843 he could announce that they appeared to increase and decrease in number over a 10-year period. Later study of records by the Swiss astronomer J R Wolf (1816–93) confirmed this, and refined the periodicity to 11.1 years. At about the same time J Lamont (1805–79) showed that the Earth's magnetic field also varied over about a 10-year cycle, which alternated between weak and strong, and that the trend of the cycle could also be observed in Earth's weather conditions and in plant growth.

All attempts to observe a new 'innermost planet' have failed. LEVERRIER's prediction of it in 1845 (he even named it Vulcan) was based on perturbations in Mercury's orbit, which by 1916 were deduced to result from relativity effects.

Schwabe was the first to record Jupiter's Great Red Spot, in a sketch of 1831.

Schwann, Theodor [shvahn] (1810–82) German physiologist: the major figure in the creation of cell theory in biology.

Schwann went to school in Cologne and then studied medicine, graduating in Berlin in 1834. He stayed there as assistant to J MÜLLER for 4 years, when most of his best work was done. Early in this period he studied digestion, and isolated from the stomach lining the proteolytic enzyme pepsin; it was the first enzyme to be isolated from an animal source. He went on to study fermentation and showed in 1836 (independently of CAGNIARD DE LA TOUR) that it was a result of the life processes of the yeast cells; this led him to doubt the idea of spontaneous generation, and so he repeated and improved the experiments on this done by SPALLANZANI. He confirmed that no microorganisms appeared and no putrefaction occurred in a sterile broth to which only sterile air was admitted. His results did not prevent all belief in spontaneous generation, which persisted until PASTEUR's work a few years later.

Schwann's work on fermentation led to quite vicious (if comical) attacks in papers by the chemists LIEBIG and WÖHLER, to the extent that Schwann saw no prospect of a career in Germany, and in 1838 he emigrated to Belgium. There he became a mystic, solitary and depressed, and did little more in science.

He had in 1838 discovered the Schwann cells composing the myelin sheath around peripheral nerve axons; and he showed that an egg (whether large or small) is a single cell which when fertilized develops into a complex organism, a central idea in embryology. Schwann's most famous work, on the cell theory, began through discussions with his friend, the botanist SCHLEIDEN, who had argued that all plant structures are cells. Animal tissues are more difficult for the microscopist, being soft, of low contrast and subject to rapid decay; and even more than plant cells, those of animals show great diversity. However, Schwann became convinced that animal tissues, like plants, are based on cells and he became, with Schleiden, a principal advocate for the cell theory, whose main points are that (1) the entire plant or animal is made up of cells or of substances thrown off by cells; (2) the cells have a life that is to some extent their own; and (3) this individual life of the cells is subordinated to

that of the organism as a whole. This theory, well defined in Schwann's book in 1839, soon became dominant in biology; the cell has been seen ever since as a natural unit of form, of function and of reproduction, at the microscopic level. This last dominance is summarized in VIRCHOW's phrase of 1855, 'all cells arise from pre-existing cells'. Virchow saw the study of affected cells as central to pathology and physiology.

In shaping much biological research, the theory was beneficial; but it was unfortunate that Schwann accepted the older Schleiden's erroneous idea that new cells are formed by 'budding' from a nucleus, which was to prove a false trail for many biologists for half a century. Neither of them had any notion of the formation of cells by division; and neither regarded the cytoplasm as important.

Schwarzschild, Karl [shvah(r)ts-shilt] (1873–1916) German astronomer: predicted existence of black holes.

Schwarzschild became interested in astronomy as a schoolboy and published papers on binary orbits at 16. He became director of the Potsdam observatory in 1909. Although an excellent observational astronomer who made great advances in photographic methods, Schwarzschild's lasting contributions are theoretical and were largely made during the last year of his life. In 1916, while serving on the Russian front, Schwarzschild wrote two papers on EINSTEIN's recently published general theory of relativity, giving the first solution to the complex partial differential equations of the theory.

He also developed the idea that when a star contracts under gravity, there will come a point at which the gravitational field is so intense that nothing, not even light, can escape. The radius to which a star of given mass must contract to reach this stage is known as the Schwarzschild radius. Stars that have contracted below this limit are now known as black holes. As an example, the Sun would become a black hole if it shrank until its radius was only 2.5 km. The idea that a star with its mass sufficiently concentrated would emit no light and so become invisible was foreseen in an unsophisticated way by PRIESTLEY and by MICHELL in the 18th-c.

Schwarzschild's son Martin (1912–   ), also an astronomer, is distinguished for his work on the evolution of stars and galaxies. He was an early user of balloons carrying telescopes to study the Sun, at heights, in 1959, of over 24 000 m/80 000 ft. He has also worked on the range of mass shown by stars: the range is now thought to be from one-100th that of the Sun, up to about 65 solar masses.

Schwinger, Julian Seymour [shwinger] (1918–94) US physicist: one of the founders of quantum electrodynamics (QED).

Schwinger was a graduate of Columbia

University, New York, gaining his doctorate at age 20. Research with OPPENHEIMER at the University of California at Berkeley followed and he joined the work on the atomic bomb during 1943–5. In 1946 he became one of Harvard's youngest-ever professors. As a talented 18-year-old, he had published elegant work on neutron scattering by magnetic materials. Even when he became world-famous among physicists, he remained rather unworldly.

Schwinger studied the papers by DIRAC, HEISENBERG and PAULI on the quantum mechanics of the electron and attempted to produce a fully quantum mechanical electrodynamics that was consistent with EINSTEIN's theory of relativity. FEYNMAN, DYSON and TOMONAGA, as well as Schwinger, all arrived independently at a correct theory within a short time. The details of how electrons interact with electromagnetic fields were then understood for the first time. Feynman, Tomonaga and Schwinger shared the 1963 Nobel Prize for physics.

Thereafter Schwinger studied synchrotron radiation, which is the particular sort of electromagnetic radiation emitted when a charged particle changes speed or direction in a magnetic field.

Seaborg, Glenn Theodore (1912–) US nuclear chemist: discoverer of transuranic elements of the actinide series.

After he obtained his PhD in chemistry at the University of California at Berkeley, in 1937, Seaborg worked with LEWIS, joined the staff and became professor of chemistry in 1945. He was to be linked with the University of California for the rest of his career, with absences on government work during the Second World War and from 1961–71, when he was chairman of the US Atomic Energy Commission.

The heaviest element that occurs in nature in fair quantity is uranium, with atomic number 92. In the 1930s more than one research group bombarded uranium with neutrons and examined the results, and believed that elements heavier than uranium ('transuranic elements') had been formed. Then came the proposal in 1939 by MEITNER and FRISCH that fission had occurred; the uranium nucleus accepts the neutron and then breaks up to give two or more nuclei of middle-range mass. But in 1940 McMILLAN and P Abelson (1913–) showed that transuranic elements can in fact be made; some of the uranium nuclei struck by neutrons do not undergo fission but form a new element, the first to be discovered beyond uranium, which they named neptunium. From 1940 Seaborg became involved in the work, found a new way of making an isotope of neptunium (atomic number 93) and went on to extend the research by making heavier transuranic elements. Seaborg was a key figure in the work

which resulted in making and identifying nine of them, from plutonium (atomic number 94) through to nobelium (102). In much of this a cyclotron was used to generate the bombarding particles, and the work was directed in part to study the basic chemistry of the new elements, and in part to produce an atomic bomb (of the first two of these, exploded in 1945, one was fuelled by uranium and one by plutonium). Seaborg realized in 1944 that the series of elements from actinium (89) onwards, could be classed within the periodic table as a new transition series, akin to the lanthanides; he named the new series (now seen as numbers 89–103) the actinides. They are all radioactive and the transuranic members (numbers 93 onwards) occur only in minute traces in nature. Seaborg and McMillan shared the Nobel Prize for chemistry in 1951.

Sedgwick, Adam (1785–1873) British geologist: identified the Cambrian period.

After graduating in mathematics in 1808, Sedgwick remained at Cambridge for the rest of his life, being appointed professor of geology in 1818. In 1835 he worked out the stratigraphic succession of fossil-bearing rocks in North Wales, naming the oldest of them the Cambrian period (now dated at 500–570 million years ago). In South Wales his friend MURCHISON had simultaneously worked out the Silurian system, some strata of which overlapped with Sedgwick's Cambrian system. In a celebrated dispute the two were to argue about which system these common strata should be assigned to for almost 40 years; the matter was only resolved after their deaths when C Lapworth (1842–1920) proposed in 1879 that the Upper Cambrian and Lower Silurian be renamed the Ordovician. Sedgwick and Murchison also identified the Devonian system in south-west England.

Sedgwick's skill was in palaeontology and stratigraphy, and his expert fieldwork greatly illuminated the geology of the British Isles, despite his rejection of uniformitarianism, theories of evolution and ideas on ice ages. DARWIN was a pupil of his and became a friend, but this did not affect Sedgwick's antagonism to his ideas, in large part on religious grounds.

Seebeck, Thomas Johann [zaybek] (1770–1831) Estonian–German physicist: discovered the thermoelectric effect.

A member of a wealthy merchant family, Seebeck went to Germany to study medicine. He qualified in 1802 but thereafter spent his time in research in physics. His best-known work was done in Berlin in 1822, when he showed that, if a circuit is made of a loop of two metals with two junctions, then when the junctions are at different temperatures a current flows (eg if copper and iron are used and the junctions are at 0°C and 100°C, the circuit has an EMF of about a millivolt). Seebeck himself did not grasp that a current was generated (he saw only that a nearby compass needle was affected) and called the effect 'thermomagnetism'. Later it was realized that whenever two different metals are in contact, an EMF is set up. whose magnitude depends on the temperature (the thermoelectric or Seebeck effect). The effect is used in the thermocouple for temperature measurement.

In 1834 the watchmaker J C A Peltier (1785–1845) found the converse effect: when a current is passed through a junction of two different conductors, a thermal effect occurs (heating or cooling, depending on the direction of the current, ie whether the current adds to or opposes the EMF of the junction). The related THOMSON effect is the development of an EMF between the ends of a single metal rod when these ends are at different temperatures. These two effects are mainly of theoretical interest.

Segrè, Emilio Gino [segray] (1905–89) Italian–US physicist: discovered the antiproton.

Segrè attended school and became a university student of engineering in his home city of Rome. Then in 1927 he changed over to physics and became the first research student to work with FERMI, obtaining his doctorate at Rome in 1928. He rose to hold a laboratory directorship in Palermo, but was dismissed for racial reasons by the Fascist government in 1938. He took a post at the University of California at Berkeley, and had an active part during the Second World War in the Manhattan Project to develop the atomic bomb at Los Alamos.

Segrè has the distinction of being involved in the discovery of three elements: technetium (1937), astatine (1940) and plutonium (1940). Technetium was made by using LAWRENCE'S cyclotron to irradiate molybdenum with deuterium nuclei, and was the world's first purely artificial element to be made. It is radioactive, and one isotope is much used in medical diagnosis.

After the war Segrè joined in the hunt for a novel particle predicted by DIRAC, the antiproton. In 1955 the Berkeley bevatron proton accelerator reached the threshold energy for producing antiprotons by proton–proton collision, 6 GeV. A beam of particles obtained by proton collisions on a copper target contained a few antiprotons, among many other secondary particles. Segrè and his group devised an apparatus and performed an experiment for detecting these antiprotons. Segrè and CHAMBERLAIN received the 1959 Nobel Prize for physics 'for the discovery of the antiproton'.

Semmelweis, Ignaz Phillip [zemelviys] (1818–65) Hungarian physician: pioneer in treatment of sepsis.

Semmelweis graduated in medicine in Vienna in 1844 and stayed there to specialize in obstetrics, working under J Klein in the General Hospital's obstetric wards. There 10–30% of the

women died from puerperal (childbirth) fever. Semmelweis studied the records and found that, in the clinic staffed by medical students, the mortality due to the fever was three times that in the clinic staffed by midwives. He also noted that the mortality had risen since Klein's appointment as head of the unit. Then, when a colleague died in 1847 from a scalpel wound made in a post-mortem dissection, Semmelweis noted that his illness was closely similar to puerperal fever. He deduced that something was conveyed on the hands of the medical staff from the dissecting rooms to the patients. By insisting that they washed their hands in disinfectant, Semmelweis reduced the mortality to 1%. However, his success produced opposition rather than imitation; discouraged and persecuted he returned to his native Buda in 1850 and a few years later became insane. He died from a septic finger infection of the same kind as his colleague in Vienna in 1847. Only after LISTER'S success in antiseptic surgery, in the 1870s, did antiseptic procedures in midwifery become widespread.

Puerperal fever was due to a streptococcal infection, but this was not fully understood until the 1880s. Strangely, the advance in obstetrics could have been made much earlier; the procedures to avoid the fever had been proposed by A Gordon (1752–99) of Aberdeen in 1795 and by the literary anatomist Oliver Wendell Holmes (1809–94) of Harvard in 1843, but the former was largely ignored and the latter abused.

Serre, Jean-Pierre [sair] (1926–) French mathematician: revolutionized homotopy theory.

A graduate of the École Normale Supérieure, and a teacher at the universities of Nancy and Princeton, Serre held a professorship at the Collège de France from 1956. He was a contributor to the BOURBAKI group.

H Cartan (1904–) and Serre collaborated in recasting the theory of the complex variable in a new way, in terms of the branch of mathematics called cohomology theory. Serre also advanced homotropy theory, particularly with reference to the homotropy of spheres. As a result Serre was able to link homotropy and homology theory for the first time. The principal advantage gained was that algebraic methods could then be used on the problems of homotropy theory, producing powerful new results.

Seyfert, Carl Keenan [sayfert] (1911–60) US astronomer: discovered Seyfert galaxies.

Seyfert studied at Harvard and subsequently held appointments at a number of American observatories. In 1951 he became director of the Dyer Observatory at Vanderbilt University, Tennessee.

In 1943 Seyfert discovered a class of spiral galaxies that have small bright blue nuclei in relation to their spiral arms and that show broad emission lines in their spectra, indicating the presence of hot ionized gas. These Seyfert galaxies also emit large amounts of energy throughout the electromagnetic spectrum, from X-rays through to radio wavelengths, and are believed to be related to quasars. About 2–5% of all galaxies are Seyfert galaxies. In several ways Seyferts resemble quasars, but they are less luminous; it is clearly possible that they are mature descendants of quasars.

Shannon, Claude Elwood (1916–) US mathematician: pioneer of communication theory.

Shannon graduated from the University of Michigan in 1936, going on to conduct research at the Massachusetts Institute of Technology before joining Bell Telephone Laboratories. In 1948, by quantifying the information content of a message and analysing its flow, he established the foundations of communication theory. Communication theory is concerned with the best way to transmit messages, and the ways in which the signal may be degraded or misunderstood. It is of central importance to the design of electronic circuits and computers, as well as to communications systems. Shannon also worked on the binary logic of digital circuits. He coined the term 'bit' for a unit of information (short for binary digit, 0 or 1).

Shapley, Harlow (1885–1972) US astronomer: discovered structure of our Galaxy.

The son of a farmer, Shapley was a teenage crime reporter on two newspapers before entering the University of Missouri, intending to study journalism; he soon changed to astronomy. In 1915, using LEAVITT's 'Cepheid variable' method of estimating stellar distances, Shapley was able to provide the first reasonable picture of the structure and size of our own Galaxy. He studied the distribution of globular star clusters, by means of the Cepheids within them, and showed that they are concentrated disproportionately in the direction of Sagittarius. This, he argued, must be the centre of our disc-shaped Galaxy, and in 1920 he estimated the Sun to be about 50 000 light years from the galactic centre. The overall diameter of the Galaxy he believed to be about 300 000 light years (both figures have since been shown to be overestimates by a factor of three or more). He directed the Hale Observatory at Harvard from 1921–52.

Sherrington, Sir Charles Scott (1857–1952) British neurophysiologist: made important studies of the nervous system.

A Cambridge graduate in medicine, Sherrington studied also in Germany under VIRCHOW and KOCH, researched in bacteriology and afterwards taught physiology at London and Liverpool, and at Oxford from 1913–35. He was a sports enthusiast, including Sunday-morning parachute jumps (from the tower of a London hospital) among his many activities. His main work was on reflex motor activity in vertebrates, detailing the nature of muscle operation at the spinal level. He began with a close study of the knee-jerk reflex and its control, and the hind-limb scratch reflex in the dog. A whole range of concepts and words in neurology are due to him (including synapse, proprioceptor, motor unit, neuron pool and others) and his book *The Integrative Action of the Nervous System* (1906) is a classic of neurology. He continued to be an active experimenter, especially on the reflex system, until 1935, publishing over 300 papers on this. He shared a Nobel Prize with ADRIAN in 1932. He has been called 'the William Harvey of the nervous system'.

Shockley, William Bradford (1910–89) US physicist: invented the junction transistor.

The son of two American mining engineers, and born in London, Shockley was educated at the California Institute of Technology and Massachusetts Institute of Technology. He began work at the Bell Telephone Laboratories in 1936, directed US anti-submarine warfare research (1942–4) and served as consultant to the secretary for war in 1945.

After returning to Bell Laboratories at the end of the war Shockley collaborated with BARDEEN and BRATTAIN in trying to produce semiconductor devices to replace vacuum tubes. It had long been known that some crystals (eg PbS) would act as rectifiers (ie would pass current in only one direction). However, J A FLEMING's diodes (then known as valves in the UK) had replaced these. The new work showed that germanium crystals were better rectifiers, their effect depending on traces of impurity. Using a germanium rectifier with metal contacts including a needle touching the crystal, they invented the point-contact transistor (1947). A month later Shockley developed the junction transistor (**transfer** of current across a re**sistor**) which uses the junction between two differently treated parts of a silicon crystal; such solid-state semiconductors can both rectify and amplify current. These small, reliable devices led to the miniaturization of circuits in radio, TV and computer equipment. Shockley, Bardeen and Brattain shared a Nobel Prize in 1956. From 1963 Shockley was a professor of engineering at Stanford.

After 1965 Shockley became a controversial figure through his support of the view that intelligence is largely hereditary and that the rapid reproduction of some racial groups can damage the intelligence of the overall population.

Sidgwick, Nevil Vincent (1873–1952) British theoretical chemist: systematizer of valence theory.

Sidgwick maintained a family tradition for intellectual virtuosity by getting a First in science at Oxford and then another First in classics 2 years later, a feat which has probably proved unrepeatable. After further study in Germany he took a post in Oxford in 1901 and kept it for his lifetime. He was an odd person in several ways; he looked middle-aged when young and changed only slowly afterwards; his own experimental work was unimportant; his best ideas came after he was 50; and his great influence in chemistry is largely due to three books. In each of them, he brought together a mass of work by others, added his own ideas and produced a coherent and unified account that clarified chemical ideas and pointed the way to new work. The first was his *Organic Chemistry of Nitrogen* (1910); the second was the *Electronic Theory of Valency* (1927), in which his own major contribution to chemical theory was to develop the idea of bonds in which one atom donated both electrons forming the covalent bond, a concept which gave new life to the whole chemistry of metal complexes. His last book was *Chemical Elements and their Compounds* (1950), a vast review that he was able to produce because of the cessation of normal published research in the Second World War. In British science he was distinguished by his pungent wit and his wealth, and in the USA he was for years the best-known British scientist.

Siemens, Sir Charles William [zeemuhns] (1823–83) German–British engineer: co-inventor of the Siemens–Martin open-hearth steel furnace.

Born Karl Wilhelm in Hanover, Siemens was one of a large family of brothers (see below for E W SIEMENS), several of whom were outstanding engineers. Charles came to England at the age of 20 and worked on a method of using waste heat from blast furnaces to improve efficiency, by pre-heating the air blast. In 1861, together with his brother Frederick (1826–1904), he designed an open-hearth steel furnace, which both utilized waste heat and incorporated a gas-producer, which allowed the use of low-grade coal as fuel. By the end of the 19th-c more steel was produced by this method (the Siemens–Martin method) than any other. Frederick Siemens used a similar furnace in glass manufacture.

Siemens, (Ernst) Werner von [zeemuhns] (1816–1892) German electrical engineer: developed electricity generation through application of the 'dynamo principle'.

Siemens was the eldest of the four Siemens brothers and was educated at Lübeck, and later at the army engineering school in Berlin, where he was imprisoned for duelling. His interests were wide-ranging – electroplating, discharge tubes to generate ozone, a standard of electrical resistance using mercury, and electrolytic refining, to name a few. He made several innovations to existing telegraphs, including seamless insulation for the wire, which enabled the company he founded (Siemens & Halske, 1847) to become a leading supplier of such systems,

including the London–Calcutta line in 1870. In 1867 he revolutionized electricity generation by using self-generated electricity to power electromagnets (the 'dynamo principle'), doing away with the expensive permanent magnets previously used in such generators. This enabled his company to become a pioneer in the fields of electric traction and electricity generating equipment.

The SI unit of electrical conductance, the siemens (S), is named after him.

Simpson, Thomas (1710–61) British mathematician: contributed to calculus.

Simpson's life involved a series of strange episodes, but his total contribution to 18th-c mathematics is substantial. Growing up in the weaving trade in the Midlands, he married at 20 a widow of 40 (who outlived him, and drew a Crown pension for his work until her death at 102). After seeing a solar eclipse Simpson became obsessed by astrology and soon acquired some local good reputation in it, but this changed when he 'raised a devil' from a girl who then had fits, and the Simpsons left the area hurriedly. In London he combined weaving and mathematics, and his reputation in the latter secured him the professorship at Woolwich in 1743 and fellowship of the Royal Society in 1745. He was also editor of *The Ladies Diary*. He wrote on calculus, probability, statistics, geometry and algebra, but his most enduring result is the method for finding the area under a curve known as Simpson's rule. The result is exact if the curve is parabolic, and can be used as a close approximation in other cases. He also devised a method for finding the volume of any solid bounded by planes, if two of them are parallel.

Sitter, Willem de (1872–1934) Dutch cosmologist and mathematician: proposed expanding universe solution to equations of general relativity.

De Sitter studied at the University of Groningen and, after a time at the Cape Town Observatory, was appointed professor of astronomy at Leiden in 1908 and director of the Leiden Observatory in 1919.

In 1916 EINSTEIN published his theory of general relativity and found a solution to the relativity equations that yielded a static universe. De Sitter showed soon afterwards that there was another solution, an expanding universe that contained no matter, the de Sitter universe. In 1927 LEMAÎTRE and FRIEDMANN both found a further possible solution, an expanding universe containing matter. Soon afterwards it was shown that a transformation of the de Sitter universe yielded a similar, but mathematically much simpler, solution now known as the Einstein–de Sitter universe.

Slipher, Vesto Melvin (1875–1969) US astronomer: discovered the general recession of other galaxies from Earth.

Slipher studied at the University of Indiana,

and then spent over 50 years working at the Lowell Observatory, Arizona, becoming its director in 1926.

Slipher used spectroscopic techniques to measure the very small DOPPLER shift in light reflected from the edges of planetary discs, thereby determining the periods of rotation of Uranus, Jupiter, Saturn, Venus and Mars in 1912. He extended his methods to spiral galaxies, discovering that the Andromeda galaxy is approaching our own Galaxy at a speed of about $300 \, km \, s^{-1}$, and went on to measure similar radial velocities for a further 14 spiral galaxies, almost all of which are receding from the Earth, and at at even greater speeds. His results later led HUBBLE to propose that all galaxies (outside the Local Group, which includes the Andromeda galaxy) are moving away from one another at velocities proportional to their separation.

Smith, Hamilton (Othanel) (1931–) US molecular biologist: isolated and studied restriction enzymes.

Smith graduated in mathematics in 1952, and in medicine in 1956 at Johns Hopkins University, where in 1973 he became professor of microbiology. In the early 1970s he obtained an enzyme from the bacterium *Haemophilus influenzae* (the strain later known as Hind II) that cleaved DNA at specific sites in relation to the sequence of bases. An early example of such a restriction enzyme had been obtained in the 1960s from *Escherichia coli* by W Arber (1929–), and Smith confirmed and amplified this work before extending it to Hind II. By the later 1970s Smith and other workers, especially D Nathans (1928–) (who collaborated with Smith in some of the research) had isolated many such enzymes. By allowing the controlled splitting of genes to give genetically active fragments, the restriction enzymes allowed the possibility of genetic engineering and of DNA sequencing to be developed. Smith, Arber and Nathans shared a Nobel Prize in 1978.

Smith, Theobald (1859–1934) US microbiologist: studied modes of transmission of cattle diseases.

Smith graduated in medicine in 1883; he chose not to enter medical practice but to move into veterinary work in the new US Bureau of Animal Industry founded to combat infectious diseases in farm animals. At his parents' home German was spoken, and young Smith's fluency in it gave him the advantage of being able to read the reports of KOCH and EHRLICH. Smith became the leading American bacteriologist of his generation and the first of distinction not to be trained in Europe. His successful studies on the nature and control of animal diseases began with his work on hog cholera in 1889; in 1896 he distinguished between bovine and human tubercle bacilli; and in 1893 he published his work on Texas cattle fever, showing that it is

transmitted by a tick. The complex cycle of transmission he had carefully worked out was doubted by many, but was never refuted; it led both to control of the fever and to easier acceptance, within 10 years, of ideas on the place of the mosquito in human malaria and yellow fever. In 1895 he moved to Harvard and in 1914 to the Rockefeller Institute, developing his work in animal pathology and in parasitology. He was an austere, hardworking and self-effacing person; many of his peers thought him comparable as a scientist with Koch, but he carried little fame in the minds of the mass of his countrymen.

Smith, William (1769–1839) British geologist and surveyor: pioneer of geological mapping; proposed principle of superposition.

The son of a blacksmith, Smith became a canal surveyor, an occupation which gave him ample opportunity to study the varying geology of much of England and Wales. He discovered that geological strata could be reliably identified at different places on the basis of the fossils they contained, and he also proposed the principle of superposition – that if a stratum overlies another then it was laid down at a later time. In 1815 he published the first stratigraphic map of England and Wales, at a scale of 5 miles per inch, following this with more detailed geological maps of over 20 counties. Although the value of his work was only recognized late in his life, he was awarded the Geological Society's highest honour in 1831 and is today considered to be 'the father of English geology'.

Snel or **Snell, Willebrord** (1580–1626) Dutch physicist: discovered law concerning refraction of light passing between media.

Snel studied law and mathematics in a number of European universities and in 1613 succeeded his father in the new university of Leiden, as professor of mathematics. He continued to publish translations of mathematical work, but also became involved in geodesy, and has been described as 'the father of triangulation'. Starting with his own house, and with the spires of town churches as reference points, he soon mapped a substantial local area. His best-known discovery, Snel's Law, or the second law of refraction of light waves, was probably made in about 1621 after much experimentation. In modern form, it states that for light passing from one isotropic medium to another, the ratio of the sine of the angle of incidence to the sine of the angle of refraction is constant for light of a particular wavelength; ie $\sin \theta_1 / \sin \theta_2$ = constant. The law can be deduced from FERMAT's principle. It is found that the value of the constant is equal to the ratio of the velocity of light in material (1) to that in material (2), and this is known as the relative refractive index.

Snow, John (1813–58) British physician: pioneer of anaesthesiology and epidemiology.

As a very young medical apprentice Snow was sent to Sunderland to work on victims of England's first cholera epidemic, which entered through that seaport in 1831. The disease was not curable, and was often fatal, and the experience gave Snow an interest in cholera that led him to study its epidemiology in the London outbreak of 1854. He was then practising in Soho; at that time PASTEUR's work on microorganisms had not appeared and the cholera vibrio was not to be described by KOCH until 1884. Snow believed, however, that the disease was due to a living, water-borne organism, he had sensible views on disinfection and he surveyed the incidence of cases and their relation to water supply, concluding that faecal contamination of Thames water was a major culprit. A plot of cases in his own parish pointed to the Broad Street pump as a focus; a sewer pipe passed close to its well. Snow persuaded the council to remove the pump handle, and dramatic improvement followed. From then on, contamination of water by faeces was seen to be a key factor in the spread of cholera. A tavern, the 'John Snow', now marks the place of the pump.

From 1840 Snow had been interested in the physiology of respiration and so, when anaesthetic inhalation methods came into the UK in 1846, in the form of knowledge of the use of diethyl ether $((C_2H_5)_2O)$ by dental and general surgeons in the USA, Snow was well placed to experiment on the new technique. He devised an apparatus for its use which gave proper control, and divided the stages of anaesthesia into five degrees. In 1847 J Y Simpson (1811–70) introduced trichloromethane (chloroform, $CHCl_3$) as an anaesthetic in obstetrics; again, Snow applied physiological principles and devices to its use, and became an expert operator with it and the first specialist anaesthetist. As such, he was called to give it to Queen Victoria in 1853 for the birth of her seventh child, Prince Leopold, and her use of it gave the procedure respectability and did much to overcome religious and medical prejudice.

Snyder, Solomon Halbert (1938–) US pharmacologist: suggested existence of endorphins.

Educated at Georgetown University, Snyder worked at Johns Hopkins University from 1965. His work which led to the discovery of the body's natural pain relievers, the endorphins, began with the knowledge that some drugs are effective in very small concentration; thus the synthetic drug etorphine (an analogue of morphine) relieves pain in doses of only 0.0001 g. To be so effective, the drug must act on some highly selective receptor sites, and Snyder and Candace Pert *née* Beebe (1946–) used morphine-like drugs with radioactive labels to locate these sites. Success in a difficult search came in 1973 when they reported that receptor sites are located in the mammalian limbic system, which is in a region in the centre of the brain associated with the perception of pain. Clearly these receptors have not evolved in order to accept synthetic drugs, and their existence implies that natural morphine-like substances must also exist. Within a few years such substances, endorphins, were found by other workers; they are highly potent analgesics (pain-relievers) and are now known to be peptides formed in the pituitary gland. Their existence may be relevant to the analgesia obtained in acupuncture.

Soddy, Frederick (1877–1956) British radiochemist; proposed theory of radioactive decay (with Rutherford); pioneer theorist and experimenter in radiochemistry.

The youngest of seven children, Soddy grew up to become a forceful, talented and eccentric individual. After graduating in chemistry from Oxford, he found a job as demonstrator at McGill University in Montreal. RUTHERFORD was there as professor of physics, and his work needed a chemist. Together during 1900–03 they offered a brilliant and simple answer to the question: what is radioactivity? Their disintegration theory proposed that heavy atoms are unstable; that such an element could undergo spontaneous atomic disintegration, losing some mass and charge from its atoms and forming a new element. The process could recur, so that a series of such changes occurred. They went on to predict that helium gas should be a decay product of radium. In 1903, working with RAMSAY in London, Soddy used 52 mg of radium bromide, collected gas from it and showed this to contain helium.

In 1913, Soddy gave the clearest of the statements of the radioactive displacement law that emerged about that time: that emission of an alpha-particle (helium nucleus) from an atom reduces its atomic number by two; whereas the emission of a beta-particle (an electron) increases the atomic number by one.

It was Soddy who gave the name 'isotopes' to atoms with the same atomic number (and therefore the same chemical properties) but differing in mass. In 1920 he foresaw their value in finding the age of rocks; back in 1906 he had foreseen the use of atomic energy from uranium, which he also lived to see (1945). In 1919 he was appointed professor in Oxford. He was frustrated there in his efforts to change chemical teaching and research arrangements, and his interest in chemistry faded after his Nobel Prize award (1921). His new concern was for political and economic schemes which would ensure that the benefits of science became widely available, but he was unsuccessful as an advocate for his ideas.

Somerville, Mary, *née* Fairfax, formerly **Greig** (1780–1872) Popular writer on scientific subjects.

THE ENTRY OF WOMEN INTO CHEMISTRY IN BRITAIN

Any amateur's participation in science before 1800 was gained through public lectures, membership of an appropriate society, with its lectures and publications, and the availability of basic instruments and a home laboratory. HUMPHRY DAVY'S lectures at the Royal Institution created a large new audience of potential scientists. Among them was JANE MARCET, who found that she needed help in understanding the lectures and in following the discussions. With her husband, a physician and Fellow of the Royal Society whose hobby was chemistry, she lived among the scientific circle in London. She decided to share her new knowledge with others struggling to understand the public lectures, and *Conversations on Chemistry* (1805) was the result. This was an immense success both in Britain and in Europe and the USA, and went into 16 updated editions. Later, MARY SOMERVILLE was to continue the explanation of science to amateurs with her very successful book *The Connexion of the Physical Sciences* (1835).

The great interest in science generated in the early 19th-c did not result in active participation by amateur female chemical scientists. Unlike those women interested in astronomy, the biological sciences or geology, who followed their own research and published in the appropriate journals of provincial and national scientific societies, there appear to be no papers submitted by women to the British scientific journals on chemical subjects before the entry of women into universities.

In Britain Elizabeth Fulhame, about whom little is known, appears to be the first female independent researcher in chemistry. Her interest began with a search for a method of depositing thin layers of metal on cloth and paper. She enthused over LAVOISIER'S new theory of combustion and devised her own variation on it. Her ideas on combustion, although erroneous, show strange pre-echoes of modern Gaia theory: she wrote that her ideas 'may serve to show how nature is always the same, and maintains her equilibrium by preserving the same quantities of air and water on the surface of our globe'. Her *Essay on Combustion* (1794) was reprinted in the USA in 1810 and she was elected an honorary member of the Philadelphia Chemical Society.

Access to laboratory materials may have been as difficult for the potential female chemists as was access to instruments for the female astronomers. MARIE LAVOISIER was introduced to work in the laboratory as an assistant to her husband in much the same way as CAROLINE HERSCHEL was introduced to the telescope as an assistant to her brother. Apart from her help to her husband in notetaking and translations, Marie Lavoisier rescued her husband's papers from his killers, preventing their suppression or incorporation in others' work, printed them at her own expense and presented them to eminent scientists around Europe.

MARIE CURIE had her own well recorded struggles. Together with PIERRE CURIE and HENRI BECQUEREL she was awarded a Nobel Prize for physics, and later the unshared Nobel Prize for chemistry. The possible applications of radium as a treatment for cancer as well as the value of X-rays and the novelty of a woman scientist made her a public figure and a source of encouragement for those women who wished for a scientific career.

The opening of universities to women in the later part of the 19th-c gave them the opportunity to work in a laboratory, but to gain an opening in the male-dominated world of research in physical science often required the assistance of men in the field. The biochemist F G HOPKINS provided research places for women in his biochemical laboratory in Cambridge, despite much criticism. In the 1920s and 1930s almost half the research places in his laboratory went to women. Not surprisingly, it was to his laboratory that the women trained in chemical science gravitated, as well as the biologists. Similarly, they went to the Cambridge laboratories of Sir Michael Foster (1836–1907), the physiologist, who provided research places and to the Cavendish (Cambridge physics) laboratories of W H AND W L BRAGG, who also enabled women to find places. Another enabler was ERNEST RUTHERFORD. The early women in chemical sciences tended to be in biological, physical and inorganic chemistry rather than organic chemistry. These early enablers may have had an influence in the flow of women into biological sciences. In the USA women at first researched mostly in applied analytical chemistry.

Many of the early British women biochemists went into teaching and the administration of the women's colleges; the pressure to do so must have been great. One such was Marion Greenwood (1862–1932), who obtained Class 1 in both parts of the Cambridge Natural Sciences Tripos Examinations of 1882 and 1883 and became demonstrator in physiology at Newnham College, Cambridge. She researched in Foster's physiology laboratory on the role played by acid in protozoan digestion, and became head of the women's laboratory (the Balfour Laboratory) in 1890.

Eleanor Balfour Sidgwick (1845–1936) had an interest in mathematics, encouraged by her mother, and studied with her brother-in-law John Strutt (later LORD RAYLEIGH). She was also interested in education for women and in the new women's colleges. She was first secretary to the principal of Newnham and assisted in the teaching there and became treasurer of the college 1876–1919, vice-principal in 1880 and principal 1892–1911. She never completed her studies at Cambridge, but in 1880 began research work with Rayleigh, who was then professor of experimental physics. She worked in electrochemistry and the result was published jointly in two papers in *Philosophical Transactions.*

Ida Freund (1863–1914) came to Britain from Vienna in 1881, studied at Girton College, Cambridge and gained a Class 1 in her examinations. She became lecturer in chemistry at Newnham College and ran the chemistry laboratory there despite being confined to a wheelchair, having lost a leg in a cycling accident in her youth. Her impact, however, was, like Jane Marcet's, in science writing, but unlike Marcet not at a popular level. Her book *The Study of Chemical Composition* (1905) became a definitive classic in its area of chemical history.

Frances Mary Gore Micklethwait (1868–1950) was able to pursue a career in chemistry. At the age of 30 she became a student at the Royal College of Science (now Imperial College, London) and graduated in 1901. She carried out research there for the next 13 years at a time when it was difficult for a woman to do so. She worked mainly with Sir Gilbert Morgan (1870–1940), doing valuable work on diazo reactions and organo-arsenic compounds. Her work in the First World War was undisclosed, but was probably on explosives, and led to her MBE in 1919.

Rutherford said of HARRIET BROOKS: 'next to Mme Curie she is the most pre-eminent woman physicist in the department of radioactivity'. She gained a first class honours degree in 1898 at McGill University, Montreal and joined Rutherford's research group there. Her work led Rutherford and SODDY to the realization that a transmutation of one element to another had occurred; this entirely novel idea was central to the whole development of nuclear physics and chemistry.

One of the 'Bragg pupils', KATHLEEN LONSDALE, is strictly a physicist but her contributions to chemistry (by X-ray diffraction methods) led to her becoming the first woman Fellow of the Royal Society, and professor of chemistry at University College London .

May Sybil Leslie (1887–1937) moved into industrial chemistry. She graduated with first class honours from the University of Leeds in 1908 and worked with Marie Curie in Paris and later with Rutherford in Manchester, where she continued research on thorium and actinium. During the First World War she worked in a war factory in Litherland (and later in North Wales), initially as a research chemist, but later as chemist in charge of the laboratory. This unusual position for a woman no doubt came about through the departure of the men to the war. Her work was on the chemical reactions involved in the formation of nitric acid and the best conditions for making it in large quantities for munitions.

Ida Smedley MacLean (1877–1944) attended the King Edward VI High School for Girls in Birmingham, which had a high reputation for science and a number of Cambridge-trained staff. She was a student at Newnham and took her examinations in 1898. She worked on the chemistry and metabolism of fats at the Lister Institute of Preventative Medicine in London and was one of the founders of the British Federation of University Women. She worked for the admission of women to the Chemical Society of London, which, like the Geological Society of London, had been slow in admitting them to its fellowship. She was the first woman to be elected in 1920.

Resistance to employing women in the chemical industry was forceful and prolonged. In the interwar years the number of posts in science-based industry was limited, and was further reduced by a long period of recession. In these circumstances, it is unsurprising that in chemistry (as in other professions and trades) a male workforce protected its job opportunities as best it could. The phrase 'the already overcrowded state of the profession' is recurrent in arguments rejecting the admission of women.

MM

Mary Fairfax was born in Jedburgh, Scotland, the daughter of an officer in the Royal Navy. She had little education, but taught herself French, Latin and Greek late at night to avoid parental criticism. Introduced to algebra by a ladies' magazine article, she began her mathematical education by studying EUCLID's work, smuggled to her by her younger brother's tutor. Her parents objected to her mathematical studies and removed the candles so that she could not read at night, so she memorized the problems and worked them in her head.

This self-education was halted in 1804 with her first marriage but, widowed after 3 years, she returned to Scotland and renewed her studies. William Wallace, professor of mathematics at Edinburgh University, taught her further mathematics and she studied NEWTON's *Principia*. Her second husband, William Somerville, a former army doctor, approved and encouraged her scientific interests. They moved to London and she began to study current work in mathematics and astronomy. Through her husband, a Fellow of the Royal Society, she met the major international scientists, who appreciated her ability. The Somervilles were popular hosts and she combined caring for her six children with her studies, entertaining, socializing and writing.

Mary Somerville, able to understand and discuss recent scientific developments with researchers, and sometimes to suggest useful directions for future investigation, was ideally placed to write in the field of popular science. MAXWELL said of her that she 'put into definite, intelligible and communicable form, the guiding ideas that are already working in the minds of men of science... but which they cannot yet shape into a definite statement'. She published on astronomy, physics, mathematics, chemistry and geography.

Her work began when Henry Brougham asked her to translate LAPLACE'S *Mécanique céleste* for his Society for the Diffusion of Useful Knowledge. It was said that perhaps 20 men in France and 10 in England could read and understand Laplace's work. Somerville translated, condensed and simplified it, and her *Mechanism of the Heavens* was published in 1831. Its high level of mathematics reduced its general sale. However, it became the recommended reading for mathematical students at Cambridge and was a standard text for the rest of the century. She had popular success with *On the Connexion of the Physical Sciences* (1834), in which she stressed the increasing interdependence of the various branches of science, covering physical astronomy, mechanics, magnetism, electricity, heat and sound. In a later edition she repeated a suggestion current among astronomers that analysis of the perturbations of Uranus might yield the orbit of an unseen planet. ADAMS, while an undergraduate, read her comment and computed the orbit of the hypothetical planet, later to be found and named as Neptune. Her third book *Physical Geography* (1848) surveyed geology, topography, hydrography, meteorology, oceanography, geographical botany and zoology and became her most popular book; it was used widely as a textbook. In 1869 Somerville received the Victoria Medal of the Royal Geographical Society. After her death one of the first women's colleges in Oxford was named in her memory.

Sommerfeld, Arnold [zomerfelt] (1868–1951) German physicist: developed Bohr's theory of atomic spectra.

Educated at Königsberg, Sommerfeld spent most of his career at Munich. He worked on a variety of problems, including gyroscopes, X-ray and electron diffraction and radio waves, but his well-known work is especially on the theory of atomic spectra. Here he developed BOHR's theory of atomic structure, replacing the idea of circular electron orbits by elliptic orbits (with the nucleus at a focus) and introducing a new azimuthal quantum number; the ellipticity should result in relativistic effects being shown in the fine structure of atomic

spectra, as was confirmed in some detail by F Paschen (1865–1947) the next year (1916). Sommerfeld was influential in physics not only for his application of relativity and quantum theory to the understanding of a variety of spectra (X-ray, atomic and molecular spectra) but through his pupils, who included BETHE, DEBYE, HEISENBERG, HEITLER and PAULI.

Sorby, Henry Clifton (1826–1908) British geologist and metallurgist: created petrological microscopy and metallography.

Sorby was born and lived in Sheffield, Yorkshire, then a town and dominated by steel-, tool- and cutlery-making. His family had been cutlers since the 16th-c, but the business was sold in 1844; as a result, Sorby was wealthy, allowing him to become an archetypal Victorian gentleman-amateur scientist, with wide-ranging interests largely in geology but including forensic work, archaeology, optics and marine biology. Most notably, he fathered metallography, sedimentology and microscopical petrography. His first published research, done in his early 20s, was on the deposition of water-borne sediments, which form sedimentary rocks, on which he experimented, in part using the River Rother on his estate. At this time he also studied limestones (notably present in his home area) by microscopy, concluding that they are formed from minute organisms (mainly coccoliths). Like his other geological ideas, these are now fully accepted; then they were controversial.

However, it is his work in making and studying thin rock sections and polished metal surfaces that is best known and that created petrography and metallography respectively. By the 1850s he had developed cutting methods used for fossil woods and gemstones to make for himself rock sections down to one-1000th of an inch thick. Examined by transmission microscopy (especially using a microscope fitted with the new NICOL polarizing prisms, which he had by 1861), minerals gave Sorby much new information, especially on the origin of slates and granites. Petrologists have used his methods ever since, despite a slow early acceptance of them.

In the 1860s he became interested in meteorites, as igneous rocks from outside the Earth. Many contain an iron–nickel alloy, and at about the same time Sorby was examining locally made iron and steel samples, whose polished or acid etched surfaces revealed their structure. He was later the first to show, in the 1880s, that carbon steel is a two-phase alloy, with its phases identified by him as iron and iron carbide: metallography had really begun.

In the 1860s he devised spectroscopic methods for detecting traces of blood. However, spectroscopy also led him into error. Studying zircons, and especially jargons from Ceylon, he found absorption lines, which he at first thought arose from a new element, jargonium. Strongly pressed by CROOKES to publish this idea, he did so in 1869. Within the year, he realized that 'jargonium' was in fact uranium. The 'jargonium fallacy' was his only major error in research, and it injured the credibility of his blood tests for forensic use.

Sorby bought a yacht in 1878 and afterwards spent 6 months each year sailing, with marine zoology and work on tidal flow and estuary sedimentation included. On shore, he was the central figure in creating Firth College, which became Sheffield University.

When he died he had published over 150 papers on a great range of topics, over 62 years: and ridicule was no longer directed to 'the strange idea of studying mountains and railway lines with a microscope'.

Sörensen, Sören [soerensen] (1868-1939) Danish biochemist: invented pH scale for measuring acidity.

Sörensen studied chemistry at Copenhagen. From 1901 he was director of the Carlsberg Laboratory and worked on amino acids, proteins and enzymes. His fame rests on his invention in 1909 of the pH scale. The pH of a solution is defined as $\log(1/[H^+])$ where $[H^+]$ is the concentration of hydrogen ions in moles per litre. The scale is rigorous enough for physical chemists but is also useable by non-specialists, and it is universally employed. For a solution, a pH of 1-7 indicates diminishing acidity; 7-14 shows increasing alkalinity and pH 7 neutrality, at 25°C. The value is measured practically by a meter with a glass electrode or by suitable indicator papers.

Spallanzani, Lazzaro [spalantsahnee] (1729-99) Italian biologist: pioneer of experimental physiology.

Spallanzani first studied law, but his cousin LAURA BASSI, who was professor of physics at Bologna, encouraged his interest in science. He became a priest and eventually professor of natural history at Pavia, and was an enthusiastic traveller in pursuit of specimens for the natural history museum there. His other enthusiasm was experimental physiology and especially reproduction. The older biologists had largely believed (with ARISTOTLE) in spontaneous generation (eg from mud or an animal corpse). A fellow Italian, F Redi (1626-97), had shown in the 17th-c that insects developed on meat only from deposited eggs; Spallanzani showed in 1765 that well-boiled broth, hermetically sealed, remained sterile. (Despite this, it was not until PASTEUR's work, a century later, that the idea of spontaneous generation was largely abandoned.) He also studied digestion, which GALEN had thought to be a kind of cooking by stomach heat, while RÉAUMUR had experimented with buzzards and concluded it was solvent action. Spallanzani experimented with many animals and one man (himself) and showed that gastric juice is the active digestive agent. He was the first to observe blood passing from arteries to veins in a warm-blooded animal (the chick). He achieved artificial insemination of amphibians, silkworms and a spaniel bitch, although he did not grasp the importance of spermatozoa (which had been discovered much earlier) and believed that the ovum contained all the parts that appeared later in the embryo. He worked on the senses of bats, and found that blinded bats could still catch insects and navigate well enough to avoid even thin silk threads. (L Jurine showed they lost their skill if their ears were covered, which was not explained until 1941 – bats use sonar.) His interest in zoology was mainly in marine biology, including sponges and *Torpedo*, and in rotifers and tardigrads.

Spencer Jones, Sir Harold (1890-1960) British astronomer: discovered slowing down of Earth's rotation.

Spencer Jones was educated at Cambridge University, becoming chief assistant to the Astronomer Royal. He later spent 10 years as astronomer at the Royal Observatory at the Cape of Good Hope before being appointed Astronomer Royal. Spencer Jones organized an international project to accurately determine the Earth–Sun distance (the astronomical unit) in 1931, using a close approach of the asteroid Eros, the result being a great improvement over previous values. More importantly, he discovered in 1939 that the Earth's rotation was slowing down by about a second per year, thus explaining some anomalies that had been observed in the orbits of the Moon and planets.

Sperry, Roger Wolcott (1913-94) US neurobiologist: made important studies of brain function.

Sperry was a student of psychology at Oberlin College and of zoology at Chicago, and worked in several centres before joining the California Institute of Technology in 1954 as professor of psychobiology. His ingenious experimentation challenged previous theories of brain function and led to new ones.

Although a mammal cannot repair a severed optic nerve, an amphibian can. Sperry studied this regeneration and found that, even with obstacles in its path, the new nerve would find its way to its original synaptic connection in the brain. Again, if the optic nerve of a salamander was severed, and the eye removed and replaced after a rotation of 180°, the animal when offered food on its right side would aim to the left, showing that the fibres had remade their old functional connection.

Sperry did much work on the brain in higher animals. The brain consists of two similar halves (containing roughly 10^9 interconnected cells) with many nerve fibres (commissures) linking the two sides. It was well known that the two halves controlled muscles on the opposite side of the body. Sperry examined animals in which all commissures were severed to give a 'split brain', and found (surprisingly) that in many ways monkeys and cats so modified behave as if they had two brains. Human patients are sometimes commissurotomized to prevent severe epilepsy spreading, and Sperry found that such split-brain patients, although normal in many ways, show that usually the right hemisphere specializes in non-verbal processes (eg emotions and spatial relationships) while the left (as has long been known) is dominant in language processing. If a split-brain person picks up an unseen object (for example, a pencil) in the left hand, the 'feel' goes to the right hemisphere; but the person could not say what was held, as 'putting it into words' calls for links with the left hemisphere. A woman shown a picture of a nude woman in her left visual field only said she saw nothing, but she blushed and giggled. Sperry's results point to discrete pathways in the brain carrying specific types of information, and have implications for theories of consciousness. He shared a Nobel Prize in 1981.

Stahl, Georg Ernst [shtahl] (1660–1734) German chemist and physician; developed phlogiston theory of combustion.

A clergyman's son, Stahl trained in medicine at Jena and later taught medicine and chemistry. He took a theory of combustion due to J J Becher (1635–82) and developed it well enough to dominate chemical theory for a century. The theory was that when a substance burned it was losing 'phlogiston'. This was a principle of fire, perhaps akin to heat; and the idea was used to explain what we now call the oxidation of metals and the reduction of ores to metal. To account for the observed weight changes, phlogiston had to have negative weight. (Later, Cavendish and Priestley thought hydrogen might be pure phlogiston). The theory was erroneous but not ludicrous. It could be made to account for a range of chemical reactions in a consistent way, and it pointed to further experiments. It was displaced, but not easily, from chemistry by Lavoisier's work on oxygen, oxidation and reduction.

In medicine, Stahl's best work was on mental illness. He was one of the first to see that some mental states are of physical origin, while others are functional; and he recognized the influence of the body on the mind and vice versa.

Stanley, Wendell Meredith (1904–71) US biochemist and virologist: isolated the first crystalline virus.

In his early years as a student Stanley's main interest was football and he planned to become a coach. However, his interest in chemistry increased and after graduating at Illinois in 1929 he worked in Germany with Wieland, and then joined the Rockefeller Institute at Princeton in 1931. At about that time Northrop

proved that several enzymes are crystallizable and are proteins. Stanley set out to find if the viruses could be purified by similar methods. Viruses are infective agents, too small to be filterable like bacteria and able to reproduce themselves in living cells. Stanley worked on the virus causing mosaic disease in tobacco plants ('TMV') and in 1935 he obtained it in fine needle-like crystals. By 1938 others found that it is a nucleoprotein. His work showed that a crystalline 'chemical' could also be 'living' (a new concept) and, since its infective part is the nucleic acid portion of the molecule, the work also suggested that reproduction in living systems might be understandable in chemical terms, involving nucleic acid – as was later shown by others to be the case.

Stanley's later work in virology included the isolation of an influenza virus, and the preparation of a vaccine against it, during the Second World War. He shared a Nobel Prize in 1946.

Stark, Johannes [shtah(r)k] (1874–1957) German physicist.

Stark had an extraordinary career, marked by changes of his views in physics, conflict with colleagues and his involvement in a racist political movement and ultimate imprisonment.

His training followed conventional lines and he became a lecturer at Göttingen in 1900, afterwards holding chairs in four German universities. In 1902 he predicted that rapidly moving positive ions in a discharge tube ('canal rays') should show a Doppler effect (a change of frequency of the light emitted from them, due to their rapid movement). In 1905 he demonstrated this for hydrogen ions, showing high skill as an experimenter: it was the first detection of the effect from a terrestrial light-source. Then in 1913 he looked for an electrical analogy to the Zeeman effect, in which spectral lines are split by a magnetic field. He found it, again using hydrogen, and a very high-voltage field. This discovery of the Stark effect, along with his earlier work, led to the award to him of the Nobel Prize for physics in 1919.

Meanwhile he had embarked on the strange reversals of his opinion on theory, and the ferocious attacks on other physicists, that are characteristic of his career. Soon after his own work had aided proof of relativity theory, quantum theory and Bohr's atomic theory, Stark attacked them. Then in 1920 he resigned his chair at Würzburg and, using his Nobel Prize money, attempted a career in the porcelain industry. Failing in this, he tried to return to academic life, despite having antagonized almost all his fellow physicists. By 1934 he was in opposition to most of modern theoretical physics, with his condemnation of 'all Jews and their theories in science' and his admiration for Nazi politics, shared by Lenard. Eventually, with his attempt to control all German physics, his position even within the Nazi party deteriorated, and he retired to his Bavarian estate in 1939. He had done notable work in physics when young; his later influence was malign. In 1947 a German Denazification Court sentenced him to 4 years in a labour camp.

Starling, Ernest Henry (1866–1927) British physiologist: a pioneer of endocrinology and of modern cardiovascular physiology.

Starling was very much a Londoner and, except for short periods in Germany and during his work for the Royal Army Medical Corps on poison gases in the First World War, his career was spent at University College, London. Much of his best-known work was done with his friend, brother-in-law and co-worker W M Bayliss (1860–1924), who was a professor of physiology in the same college. Their joint work would predictably have gained them a Nobel Prize but for the war; and Starling's acid public comments on Britain's leaders largely excluded him from the honours awarded to Bayliss, including a knighthood. In their early work together, they discovered the peristaltic waves of the intestine. Then in 1902 they showed that the pancreas still produces pancreatic digestive juice when food enters the duodenum, even when all the nerves to the pancreas are cut. Pavlov's work had indicated this to be a nerve-controlled process. They concluded that a chemical messenger (they named it secretin) must be carried by the blood from the duodenal wall to the pancreas, stimulating its activity. They found that an extract from the duodenum has this effect; and in 1905 Starling used the word hormone (from the Greek *hormeo*, to excite) to describe such potent biochemical messengers. They had created the subject of endocrinology, which was later to prove so fruitful. Soon it was realized that one hormone had already been found (adrenalin, by Takamine in 1901); another was discovered by Kendall in 1914 (thyroxin), and the subject expanded strongly after 1930.

Starling's other work was largely on the cardiovascular system. Starling's law of the heart (1918) states that for cardiac muscle (as for voluntary muscle) the energy of contraction is a function of the length of the muscle fibres. So the more the heart is filled during diastole (relaxation) the greater is the following systole (contraction); this allows change in output without change in rate. An impaired heart enlarges to maintain its output, in accord with this law, and the enlargement (detectable by X-radiography) is an indication of heart damage.

Staudinger, Hermann [shtowdinger] (1881–1965) German organic chemist: the founder of polymer chemistry.

Staudinger's career in chemistry began with work of a classical organic kind and included the discovery of a new group, the ketenes, and work on the aroma agents in coffee. But in the 1920s he began to study rubber. At that time, rubber and other apparently non-crystalline high-molecular mass materials were supposed to be merely disorderly aggregates of small molecules; linked with this concept, their chemistry was held in low regard. From 1920, working in Zürich and Freiburg, Staudinger took the view that these polymers are giant molecules held together by ordinary chemical bonds and frequently forming long-chain molecular strands. His view was at first strongly opposed, but he devised methods for measuring their relative molecular mass by viscometry, and chemical methods for modifying polymers, and soon X-ray studies also supported his views. When accepted, these ideas formed a philosophy for the new macromolecular chemistry – the chemistry of 'high polymers' (ie having high molecular mass). This has proved fundamental for an industry using synthetic polymers as rubbers, mouldable plastics, fibres, adhesives and so on. He also foresaw the importance of natural biopolymers in biochemistry, and from 1936 had some prophetic insights in that area, (eg 'every gene macromolecule possesses a definite structure which determines its function in life' – correct, but not provable for another two decades). Belatedly, he received a Nobel Prize in 1953. He had then worked in Freiburg since 1926.

Stefan, Josef [shtefan] (1835–93) Austrian physicist: discovered the Stefan–Boltzmann black body radiation law.

After 4 years at the University of Vienna, Stefan became a school-teacher for 7 years, researching in physics in his spare time, but in 1863 he secured the professorship of physics at Vienna and remained there throughout his life.

A skilful experimentalist, Stefan measured the thermal conductivity of gases accurately and thereby gave early confirmation of Maxwell's kinetic theory. In 1879 he considered the heat losses of very hot bodies, which were reputed to cool faster than Newton's law of cooling predicted. Using Tyndall's results obtained with a platinum wire made incandescent by passing a current, Stefan showed that the rate of heat loss per unit area is $E = \sigma T^4$, a relation known as Stefan's Law. Here σ is now known as Stefan's constant and T is the absolute temperature. In 1884 his ex-student Boltzmann used the kinetic theory and thermodynamics to derive this law, and showed that it only held for bodies radiating perfectly at all wavelengths, called black bodies. It became known as the Stefan–Boltzmann Law. Stefan used the law to make the first satisfactory estimate of the temperature of the Sun's surface (the photosphere), arriving at 6000°C for this.

Steinberger, Jack [stiynberger] (1921–) US nuclear physicist: a major contributor to the 'standard model' of particle physics.

Steinberger went to the USA in 1934 as a

THE ENTRY OF WOMEN INTO THE BIOLOGICAL SCIENCES

The biological sciences were relatively more welcoming to women than the other sciences, and fewer obstacles were placed in their path. Unlike medicine, botany was regarded as an acceptable occupation for women, combining fresh air and exercise with the accomplishments of drawing and painting. Painting led a number of women towards botany and zoology. Maria Sibylle Merian (1647–1717), one of the earliest female entomologists, had a business in Amsterdam selling silk, hand-painted with flower designs. She initially studied caterpillars in order to find other varieties than silkworms which could be used to provide fine thread. She spent two years in Surinam collecting and painting insects and plants; her *Metamorphosis Insectorum Surinamensium* was published in 1705. Many women took up botanical illustration, as accurate illustrations were needed to distinguish the different species and varieties. Madeleine Frances Basseport (1701–80) was an illustrator for the French Royal Gardens 1735–80, and there are over 700 drawings of fungi by Anna Worsley Russell (1807–76) in the British Museum.

Botany began to be popular among women as a hobby with the publication of a number of books written by women for women. Priscilla Bell Wakefield (1751–1832) wrote an *Introduction to Botany* as letters from one sister to another, explaining the Linnaean system of classification; it ran in 11 editions by 1840. When the Botanical Society of London was formed in 1836, about 10% of the members were women. One member, ANNA ATKINS, led the use of photography in place of drawing to illustrate scientific books. She used J F W HERSCHEL's 'cyanotype' process (1842) to make contact prints of her collection of algae and produced *Photographs of British Algae: Cyanotype Impressions* (3 vols, 1843–53). Elsewhere in Europe women were taking an interest in botany. Amalie Konkordie Dietrich's (1821–91) special interest was in the alpine flora of Europe and she

was later to be appointed curator of the Hamburg Botanical Museum.

Unlike astronomy, women in the biological sciences could provide for themselves, if self-financing, the laboratory equipment needed. ELEANOR ORMEROD gained experience in the use of the microscope through her brother's interests and became an expert on insect infestations. Although largely self-taught, she became recognized as an authority on agricultural entomology by Edinburgh University with an honorary LlD in 1900.

During the later part of the 19th-c university education became available to women. In the USA women such as the twin sisters Agnes and Edith Claypole (1870–1954/15) gained science degrees and moved into teaching in higher education. Edith died of typhoid fever contracted during her research on the typhoid bacillus. Mary Brandegee (1844–1920) gained an MD degree in 1878 at the University of California and became interested in medicinal plants; with her husband she founded a journal of botanical observations. Clara Cummings (1853–1906) studied at Wellesley College and joined the staff; she specialized in cryptogamic (spore-producing) flora. NETTIE STEVENS was a student at Stanford University and gained her PhD in 1903 under T H MORGAN at Bryn Mawr. She found the chromosomal basis of sex-determination, but died before the importance of her work and her reputation were fully established. In Britain, ETHEL SARGANT went to Girton College, Cambridge and took the natural science tripos examination in 1885. She carried out research in cytology and in anatomical morphology, working from a laboratory at home. Her best-known work concerned the anatomy of seedlings; she suggested a new interpretation of the relationship between mono- and dicotyledons.

Cambridge occupied a key position in the education and training of the first generation of British women scientists. Those at Girton and Newnham were only accepted by their colleges if they were candidates for the tripos (honours) examination, and so formed a body of especially able students.

teenage Jewish refugee, and later studied chemistry at Chicago. In the Second World War he worked in the radiation laboratory at Massachusetts Institute of Technology and his interest moved to physics; as a result, he worked for his PhD in Chicago on the muons present in cosmic rays. He showed that a muon decays to give an electron and two neutrinos; and he continued his work in this field with M Schwartz (1932–   ) and L Lederman (1922–   ) at Columbia University, New York, from 1959. They theorized that neutrinos should be of two kinds, the electron neutrino and the muon neutrino. In the early 1960s a new high-energy proton accel-

erator became available at the Brookhaven National Laboratory, which could provide enough neutrinos to test their theory. To exclude other particles, a filter was used consisting of a stack of steel plates from a scrapped battleship, 13.5 m thick. Behind this a detector located a few nuclear reactions that confirmed that two kinds of neutrino exist.

Working in Europe at CERN near Geneva from 1968, Steinberger continued to use neutrinos to study nuclei and nuclear forces. The current 'standard model' for nuclei proposes two types of component as fundamental units of matter: the quarks and the leptons. Of the six kinds of

Although at this stage the women were not granted degrees, only certificates of proficiency, they were attracted to the high quality of teaching as the women were permitted to attend the university lectures if they had the consent of the lecturers. They were not officially part of the university and so were barred from university prizes and scholarships and restricted in their use of the library. Such opposition seemed to create a challenge to which the women rose, often gaining higher marks than the male prizewinners. The presence at Cambridge of Emily Davies (1830–1921), Girton's founder and a prominent figure in the reform of female education, put Cambridge in a prime position to provide first-class women science teachers for the best of the newly opened girls' schools, which in turn brought a steady supply of well-taught, able women students to the Cambridge tripos courses. Opportunities for research in biochemistry and physiology at Cambridge were the result of such good training. F G HOPKINS, regarded as 'the father of biochemistry', at a time when the opportunities for research biochemists in British universities was minimal and when there were few women research workers in other university departments, provided research places for women in his department despite the criticism this caused.

The 20th-c has seen the emergence of women among the highest achievers in the biological sciences. In 1945 MARJORY STEPHENSON, who researched under Hopkins at Cambridge, became the first woman to be elected a Fellow of the Royal Society in the biological sciences; she worked on bacterial metabolism. The following year AGNES ARBER was honoured; her researches in Cambridge were concerned with the anatomy and morphology of monocotyledonous plants. In 1947 Muriel Robertson (1883–1973) was elected an FRS; she worked on protozoa, especially the trypanosomes that cause sleeping sickness, and on bacteriology and immunology. The only sisters to both achieve an FRS were SIDNIE MILANA MANTON, elected in 1948, who was the world's expert on the higher level classification of arthropods, and IRÈNE

MANTON in 1961, who set up the first laboratory in the world for the ultrastructural study of plants using the electron microscope.

By now the pathway to honours was well trodden by women in the biological sciences. Honor Bridget Fell (1900–86) became an FRS in 1952, and studied the cellular interactions of cartilage and bone. Helen Kemp Porter (1899–1987), elected in 1956, was interested in starch metabolism and became one of the first investigators to apply chromatography and radioactive tracers to prepare radioactive biochemicals and use them to study the intermediate metabolism of plants. Sheina Macalister Marshall (1896–1977), a marine biologist, was elected in 1963. JEAN HANSON, the co-deviser of the sliding-filament theory of muscle contraction, was elected an FRS in 1967. Mary Winifred Parke (1908–89), elected in 1972, was an all-round expert on algae. In 1991 ANNE McLAREN, an embryologist elected an FRS in 1975, became the first woman to hold office in the Royal Society's 330-year history, by taking the post of Foreign Secretary.

During the later half of the 20th-c women began to win Nobel Prizes in physiology or medicine, despite their difficulty in obtaining senior academic posts. GERTY THERESA CORI, née RADNITZ, shared a Nobel Prize with her husband in 1947 for their discovery of the course of the catalytic conversion of glycogen in animal cells. In 1977 ROSALYN YALOW received hers for the development of radioimmunoassays of peptide hormones, now routinely used in clinical diagnosis. In 1983 BARBARA McCLINTOCK received the prize, unshared, for her discovery of mobile genetic elements ('jumping genes'). RITA LEVI-MONTALCINI was awarded hers jointly with Stanley Cohen for their discoveries of nerve growth factors in 1986. GERTRUDE BELLE ELION was awarded the 1988 Nobel Prize for physiology or medicine jointly with GEORGE HITCHINGS for 'introducing a more rational approach based on the understanding of basic biochemical and physiological processes' to the synthesis of novel drugs.

MM

quark, three have electric charge 2/3 and three have −1/3 of a proton's charge. All are subject to the strong nuclear force. They also have a new form of charge ('colour') and 'spin'. None have been found free; they are permanently confined in nuclei. All matter is made of quarks (eg the proton and the neutron are made of three quarks, of different kinds) together with leptons, which are also of six kinds. Leptons are light particles (electrons, muons, tau and their neutrino partners). The neutrinos interact only through the weak nuclear force: muons and tau particles are short-lived.

The present picture of the fundamental particles of matter will certainly change in the future, probably with the development of higher energy particle accelerators, as in the past. Steinberger, Lederman and Schwartz shared the Nobel Prize for physics in 1988.

Steno, Nicolaus (*Lat*), Niels Steensen (*Dan*) [steenoh](1638–86) Danish anatomist and geologist: made early studies of crystals and fossils.

An anatomist by training, Steno was also interested in crystals and fossils. He showed that a pineal gland like that of man is found in other animals, and used this and other arguments to refute DESCARTES's claim that it is the seat of the soul and uniquely human. His study of quartz crystals revealed that, although the shapes varied, the angle between corresponding faces is fixed for a particular mineral. This constancy, sometimes called Steno's Law, is a consequence of the internal ordering of the constituent molecules in the crystal. Steno also accepted the organic nature of fossils and recognized that sedimentary strata were laid down in former seas, having found fossil teeth far inland that closely resembled those of a shark that he had dissected. His geological sections were probably the first to be drawn.

He became a priest in 1675, was ordained a bishop in 1677 and gave up science thereafter.

Stephenson, Marjory (1885–1948) British biochemist and microbiologist: the first female Fellow of the Royal Society in biological sciences.

Marjory Stephenson's father was a Cambridgeshire farmer with an interest in applied science; he read DARWIN and MENDEL and began fruit growing in a region previously without orchards. It was he who explained nitrogen fixation to Marjory as a child when they walked together in a clover field, but it was her mother who oversaw her education and pressed for her to enter university at a time when this was a novel path for girls.

She studied at Newnham College, Cambridge (1903–06) and became a teacher of domestic science for a short period, but then began biochemical research in London in 1911. When the First World War broke out she joined the Red Cross, serving with distinction in France and in Salonika. Afterwards she joined F G HOPKINS in research back in Cambridge, and spent the rest of her life in the Biochemical Laboratory there, working at first with him on vitamins but from 1922 making her own mark in the study of bacterial metabolism and becoming the leading expert on the enzymes that control this. She was able to show that these enzymes were essentially similar in nature and activity to those of higher organisms.

She did much to establish bacterial chemistry as a valuable branch of biochemistry. Many researchers were trained by her in microbiology, and it was no surprise that she was one of the first two women to be elected Fellows of the Royal Society in 1945, together with the crystallographer KATHLEEN LONSDALE. (See panel on p. 302.)

Stern, Otto (1888–1969) German–US physicist: showed that magnetic fields of atoms are quantized.

Otto Stern, the son of a grain-merchant, completed his doctorate in Breslau in 1912. He travelled and attended lectures by SOMMERFELD, LUMMER and E PRINGSHEIM and became a postdoctoral associate and friend of EINSTEIN in Zürich. Following military service during the First World War he worked with BORN in Frankfurt on statistical mechanics.

In 1920 Stern and W Gerlach (1889–1979) collaborated in a historic experiment. A molecular beam of silver atoms (produced by heating the metal in a vacuum) was used to investigate whether space quantization (proposed by Sommerfeld) occurs or not. A silver atom should possess a magnetic moment (spin) and when placed in a non-uniform magnetic field should be found in two (spin-up or spin-down) configurations, so that the beam would be split into two distinct beams by such a field. Such quantum mechanical space-quantization was proved by the Stern–Gerlach experiment, and Stern was awarded the 1943 Nobel Prize for physics.

Stern took a professorship at Hamburg and set up a large molecular-beam laboratory, collaborating with PAULI, BOHR and P Ehrenfest (1880–1933). Stern determined the magnetic moment of the proton and found it to have two to three times the value predicted by DIRAC. In 1933 Stern moved to Carnegie Institute of Technology, Pittsburgh, but the momentum of the Hamburg laboratory was not regained and he retired early. He enjoyed luxury, good food and the cinema, and it was in a cinema that he died of a heart attack at 81.

Stevens, Nettie (Maria) (1861–1912) US cytologist: elucidated the chromosomal basis of sex determination.

For over 2000 years speculation and some experiments had been directed to an obvious biological question: what determines whether a living organism (including the human) is male or female? Only in the early 20th-c was it shown that this is fixed at the point of fertilization and depends on the chromosomes of ovum and sperm, and not, for example, on the external conditions of early growth. Of the several biologists concerned, Nettie Stevens was almost certainly the first to carry out the crucial experimental work and to clearly appreciate its result.

She was 35 when she became a student at Stanford University, having saved money from working as a teacher, and after graduation she went to Bryn Mawr in 1900 to research for a doctorate. She was fortunate that the small college for women then had the eminent geneticist T H MORGAN to teach biology: later he was to claim her as the most talented of his many graduate students.

Although LEEUWENHOEK had observed spermatozoa with a primitive microscope in the 1670s, it was not until the late 19th-c that the use of new staining and fixing methods, together with improved compound microscopes allowed cytologists to observe the entry of the sperm nucleus into the ovum (egg) cell and its fusion with the egg nucleus. Sexual reproduction could now be seen as involving the fusion of two sets of chromosomes, one from each parent (meiosis).

Nettie Stevens obtained her PhD in 1903,

studied in Europe for a year, partly with BOVERI, and returned to Bryn Mawr to work on 'the question how sex is determined in the egg' as she put it, at first by studying the chromosomes of several insects and comparison with the sex of the progeny. Her success came in 1904 with the common mealworm *Tenebrio molitor*. She found that the sperms were of two kinds; their nuclei had either 10 large chromosomes, or nine large and one small. The egg nuclei all had 10 large chromosomes. The somatic cells of the female offspring have 20 large chromosomes, those of the male have 19 large and one small; Stevens concluded that the former result from X,X fusion and the latter from X,Y fusion. Studies of some other species gave similar results, since X,X fusion to give a female embryo and X,Y to give a male is a widespread pattern. E B Wilson (1856–1939) found essentially similar results at about the same time, but was later than Stevens in seeing their generality and significance.

This understanding of sex determination, and its linkage with MENDEL's work, which she recognized, is a basic result in biological science of Nobel-prizeworthy importance. But Stevens died from breast cancer in 1912, before her reputation or the importance of sex chromosomes in genetics was fully established. By the time science in this area was clarified, credit had perhaps inevitably focused on T H Morgan, who brought together ideas on genes, chromosomes and genetics, and it was he who received a well-deserved Nobel prize in 1933. (See panel on p. 302.)

Stevin, Simon (*Dutch*), **Stevinus** (*Lat*) (1548–1620) Flemish engineer and mathematician: introduced decimal notation to Europe.

Stevin entered the Dutch government service, rising to the rank of Quartermaster-General to the Army, where he developed a system of sluices to defend parts of Holland from invasion by flooding them with water.

He is noted for his demonstration that hydrostatic pressure in a liquid depends only on the depth of liquid and not on the shape of the containing vessel. His booklet *De Thiende* (1585) publicized the decimal system for representing fractions and for weights, measures and coinage; NAPIER inventing the decimal point soon afterwards. Stevin wrote an excellent book on statics, giving the law of the inclined plane. He was also an advocate of writing scientific works in the vernacular, rather than Latin, as was the custom of the day.

Stibitz, George Robert (1904–95) US computer scientist.

Stibitz attended colleges in New York, emerging with a PhD in physics from Cornell and then joining Bell Telephone Laboratories in 1930. He was there until 1941 when he moved to defence work, and from 1945 worked on the computer modelling of biomedical systems. He designed the first genuinely binary calculator in 1937, followed by a series of machines for Bell including the first multi-user machine, which was demonstrated as a remotely controlled device in 1940 using telephone lines between Hanover, NH and New York City. As well as this introduction of remote job-entry, he built the first machine capable of floating point arithmetic in 1942.

Stock, Alfred [shtok] (1876–1946) German inorganic chemist: a pioneer of silicon hydride and boron hydride chemistry and of vacuum handling methods.

After graduating in Germany, Stock went to Paris in 1899 to join MOISSAN's research group. They were a happy international team, although 'one was constantly in danger of losing one's life'. Later, as professor in Breslau, he began work in 1909 on the dangerously explosive boron hydrides; in this work he developed the vacuum-line methods so much used for volatile materials by later inorganic chemists. His work on the boron hydrides led to later work on their strange, electron-deficient structures and to their use as rocket propellants.

Stock became a victim of mercury vapour poisoning; he was not the first chemist to suffer this, but he was unusual in being aware of the cause of his illness, and from 1923 he worked on mercury poisoning and methods of avoiding it. He also devised a method for making beryllium which is used commercially; and the use of P_4S_3 in place of phosphorus in match heads is also due to him.

Stokes, Sir George Gabriel (1819–1903) Irish physicist: a contributor to fluid dynamics.

Educated in his native Ireland and at Cambridge, Stokes became Lucasian professor at Cambridge in 1849 and in the next half-century did much to rescue physics teaching there. He worked in most areas of theoretical and experimental physics except electricity. One of his enthusiasms was hydrodynamics and another was fluorescence, and both have laws named after him. Stokes's Law in hydrodynamics is that the frictional force (ie drag) on a spherical body of radius r moving at its terminal speed v through a viscous fluid of coefficient of viscosity n is $6\pi n r v$; this holds only for a restricted range of conditions. Stokes's law of fluorescence states that the wavelength of fluorescence radiation is greater than that of the exciting radiation; again, the law does not always hold.

He pioneered in 1849 studies of gravity variations over the Earth's surface: such geophysical methods are now used in stratigraphic studies to assist oil prospecting. An ultrasensitive spring balance (gravimeter) is used, allowing the acceleration due to gravity (g) to be measured; a low value indicates a low density material (oil or water) below.

Stoney, George Johnstone (1826–1911) Irish

physicist: suggested the name electron for the smallest unit of electricity.

For most of his working life Stoney was secretary to Queen's University, Dublin. Rightly believing that science would be simplified by a wise choice of fundamental units, he argued this in 1874 and proposed the charge on a hydrogen ion as a unit, calculating its value from the mass of hydrogen liberated on electrolysis. This idea that negative electricity has a 'smallest unit' was also advanced by HELMHOLTZ in 1881, and 10 years later Stoney introduced the word 'electron' for the unit. Later the word came to be used for the 'corpuscles' discovered by J J THOMSON.

Stopes, Marie (Charlotte) (1880–1958) British palaeobotanist and early advocate of birth control.

Marie Stopes studied botany, geography and geology and gained her degree at University College, London. She then went to Munich and took a doctorate in 1904, and was awarded a London DSc in 1905. In 1904 she became the first woman member of the science staff of Manchester University. Stopes soon had a leading position in research on fossil plants, and was responsible for a useful classification of coals on this basis.

Her first marriage was annulled for non-consummation in 1916, and led to her book *Married Love* (1916). In it she argued that women were as entitled as men to physical pleasure; it also discussed birth control methods. The book was much attacked, and was banned in the USA, but it led to many enquiries on contraception which she answered in *Wise Parenthood* (1918). She married the aviator and industrialist H V Roe (1878–1949) who gave her support to open the first British birth control clinic in 1921, in north London. Her robust attitude to opposition and her forceful and eccentric personality helped to change attitudes to sex and contraception in the UK.

Strachey, Christopher [straychee] (1916–75) British computer scientist: pioneering worker on programming languages.

Strachey came from a literary family (Lytton Strachey was his uncle) and was educated at Cambridge, after which he joined Standard Telephones and Cables Ltd. The Second World War hastened the development of electronic computers, and Strachey wrote some of the largest programs for them. In 1951 he joined the National Research and Development Corporation, designing the Ferranti Pegasus computer. In 1962 he became a research fellow at Cambridge and worked on the design of the high-level language CPL, which led to the more common BCPL (Basic Computer Programming Language). He later moved to Oxford and established the Programming Research Group, working largely on a comprehensive theory of programming language semantics.

Strasburger, Eduard Adolf [shtrasburger] (1844–1912) German botanist: demonstrated capillary action as cause for sap rising in trees.

A friend and student of N PRINGSHEIM and an early enthusiast for DARWIN's ideas, Strasburger taught at Jena and later at Bonn, and made the latter the major centre for research in plant cytology. He was the first to fully describe the embryo sac in gymnosperms (conifers) and in angiosperms (flowering plants) and to recognize the process of double fertilization in the latter. In 1875 he described the principles of mitosis and deduced that the nucleus was responsible for heredity, and a little later he proposed a basic principle of cytology: that new nuclei arise only from the division of existing nuclei. In 1891 he demonstrated that physical forces (eg capillarity) are largely responsible for the rise of sap in a tree stem, rather than physiological forces.

Strutt, John William, Baron Rayleigh *see* **Rayleigh**

Sturgeon, William (1783–1850) British inventor: much improved electromagnet design.

After a few years as an apprentice shoemaker, Sturgeon joined the army and began to study science at night, until he became an expert on electrical instruments. After he left the army in 1820 he became a bootmaker and itinerant teacher of science for the army, and for some schools and societies, and he published popular accounts of science. In 1821 he much improved the electromagnet by using a bar of soft iron coated with shellac varnish to insulate it from the bare wires carrying the current (J HENRY and FARADAY later insulated the wires, so allowing many more turns and greater improvement in performance). For his work on electrical apparatus Sturgeon received a prize in 1825 from the Society of Arts; it consisted of a silver medal and 30 guineas. In 1836 he invented a moving-coil galvanometer and the first commutator for a workable electric motor. He published the first journals on electricity in English: his monthly *Annals of Electricity* ran from 1836–43.

Sturtevant, Alfred Henry [stertevant] (1891–1970) US geneticist: pioneer of chromosome mapping.

As a boy Sturtevant drew up pedigrees for his father's farm horses, and as a student at Columbia University his older brother encouraged this interest through books on heredity. A book on Mendelism spurred his enthusiasm, since he felt that some horse coat colours could be explained on a Mendelian basis. He wrote on this to the leading American geneticist T H MORGAN and in 1910 joined the group of enthusiasts in the crowded 'fly room' at Columbia, working with Morgan on the genetics of the fruit fly, *Drosophila*. In 1928 Sturtevant became professor of genetics at the California Institute of Technology, where he remained, except for research visits, until his death.

In the 'fly room' he had the germ of the idea of chromosome mapping and 'went home, and spent most of the night (to the neglect of my undergraduate homework) in producing the first chromosome map'. This was based on his idea that the frequency of crossing-over between two genes gives an index of their relative distance on a linear map of the genes on the chromosome. His paper of 1913 located six sex-linked genes, as deduced from the way they associated with each other; it forms a classic paper on genetics. He later developed a range of related ideas, discovering the 'position effect', ie the way in which the expression of a gene depends on its position in relation to other genes; and he showed that crossing-over between chromosomes is prevented in regions where a part of the chromosome material is inserted the wrong way round. The position effect was to prove of great importance in F Jacob (1920–   ) and MONOD's work on gene clusters (operons) in bacteria. Although Sturtevant's main work was in genetics (where he worked with a range of animals on an assortment of problems, including the curious effect of direction of shell-coiling in snails) he was also a knowledgeable naturalist, with a special interest in social insects.

Suess, Eduard [züs] (1831–1914) Austrian geologist: proposed former existence of Gondwanaland supercontinent.

Suess was educated at the University of Prague, moved to Vienna in 1856 and became professor of geology there in 1861. In addition to being an academic he served as a member of the Reichstat (parliament) for 25 years. On the basis of geological similarities between parts of the southern continents, including the widespread occurrence in Africa, South America, Australia and India of the fossil fern *Glossopteris* during the Carboniferous period, Suess proposed that there had once been a great 'supercontinent' made up of the present southern continents, and he named it Gondwanaland, after a region of India. Subsequent work has established the former existence of Gondwanaland beyond doubt, and Suess's ideas, as extended by WEGENER, led to modern theories of continental drift.

Svedberg, Theodor [svayberg] (1884–1971) Swedish physical chemist: devised the ultracentrifuge.

Svedberg entered Uppsala in 1904, hoping to apply chemical methods to biological problems; he stayed in the university for life and had fair success in his objective. His main work was in developing the ultracentrifuge, in which a centrifugal force much above gravitational force is produced, which is powerful enough to 'pull down' large molecules such as proteins. Svedberg's ultracentrifuges ran at up to 140 000 rpm, giving fields up to 900 000 g, and could be used to purify proteins (and other colloids) and

to confirm STAUDINGER's view that these were giant molecules, of high relative molecular mass (eg Svedberg found relative molecular mass = 68 000 for haemoglobin). Since his work ultracentrifuges have been in routine use for separating large biological molecules. He was awarded a Nobel Prize in 1926. The unit of sedimentation velocity, the svedberg (S), is named after him.

Swallow, John (Crossley) (1923–94) British physical oceanographer: discoverer of strong eddy fields in oceanic interiors.

Swallow was a student of physics at Cambridge when the Second World War interrupted his studies and naval service introduced him to the sea. Back at Cambridge after the war, lectures by BULLARD inspired him to do his PhD in the Department of Geodesy and Geophysics, spending the early 1950s on the survey vessel HMS *Challenger*, studying deep ocean floors by seismic methods. From 1954 onwards he was with the Institute of Oceanographic Sciences and frequently at sea. To study undersea currents he devised the 'Swallow float', a neutrally buoyant float which would transmit information on current and temperature at varying depths and whose location could be tracked from a ship. (The first models were made of discarded builders' aluminium scaffold poles.) The results confirmed theories due to H Stommel (1920–92) on deep currents in the Atlantic and revealed a quite unexpected and important feature of ocean interiors. This was the presence of strong eddy fields, rather akin to atmospheric weather systems. This discovery has much changed ideas about deep seas generally.

Swammerdam, Jan (1637–80) Dutch naturalist and microscopist; a pioneer of modern entomology and discoverer of red blood cells.

Swammerdam's father, an apothecary, had a 'museum of curiosities' and the boy helped with this and became a keen insect collector. He studied medicine at Leiden (with STENO as a fellow student) and graduated in 1667 but he never practised medicine, despite his father's protests and financial pressure. When only 21, he discovered the red blood cells of the frog. Also from the frog, he introduced the nerve-muscle preparation into physiology. This consists of a leg muscle with its nerve, dissected from a recently killed frog; when the nerve is stimulated, the muscle contracts. By immersing the preparation in water in a container with a narrow outlet, Swammerdam was able to show that when the muscle contracts, there is no change in its volume, contrary to earlier belief.

For the second half of his fairly short life he was a victim of mental illness, but this did not stop his skilful pioneer work on insects and their microanatomy. His minute dissections of the mayfly, bee, tadpole and snail were not surpassed until, in the 18th-c, the compound microscope was much improved. His work showed the complexity of small animals (eg the compound eye, sting and mouth of the bee). Much of this work was not found until 50 years after his death, when it was published as *The Bible of Nature* (1737).

Sydenham, Thomas [sidenam] (1624–89) English physician: made early studies in epidemiology.

After 2 months as a student at Oxford, Sydenham left to join the Parliamentary army, and to serve in the Civil War under his older brother (who was commander-in-chief in Dorset). After 3 years he returned to his studies and graduated in 1648, but in 1651 was again in the war as a captain of horse. He was wounded at the battle of Worcester, and in 1655 he married and began his career as a London physician. He began there his researches on smallpox and other fevers, then prevalent in London. He pioneered the use of quinine to treat malaria, opium for pain relief and iron compounds in anaemia. His scientific approach to the natural history of disease was new and valuable; he saw infections as specific entities, best treated conservatively, as described in his influential book *The Method of Treating Fevers* (1666), which is dedicated to his friend BOYLE. He is the major 17th-c clinician; 'the English Hippocrates'.

Sylvester, James Joseph (1814–97) British mathematician: with Cayley founded the theory of invariants.

Sylvester was born into a Jewish family of nine children and in 1828 went to the new University of London, founded for Dissenters. Vociferous and hot-headed, Sylvester fought back against the strong anti-Semitism which he encountered and was sent down for threatening a fellow student with a table-knife. He was admitted to St John's College, Cambridge (1831) and became Second Wrangler; however, as someone unable to accept the 39 articles of the Church of England he could not obtain a degree and only received his Cambridge MA in 1871, when this restriction was lifted. He moved to Trinity College, Dublin and gained his BA in 1841.

After a short period teaching science in London (1837), he decided he preferred mathematics and started in 1841 a disastrous few months as professor at the University of Virginia. He resigned because of the authority's failure to discipline a student who insulted him. For some years he abandoned university life and worked in London as an actuary and then as a barrister, qualifying in 1850. He also took private pupils; one of the best was Florence Nightingale. Fortunately, in 1850 he met A CAYLEY, who rekindled his interest in mathematics, and the two became life-long close friends. He became professor of mathematics at the Royal Military Academy at Woolwich (1855–70) and at the newly founded Johns Hopkins University at Baltimore (1877–83), producing a flood of new ideas in mathematical research and teaching. When he was over 70, he became Savilian Professor at Oxford (1883–94); failure of his eyesight forced his retirement to London.

He was enthusiastic and inventive to the end of his life; at 82 he worked out the theory of compound partitions. Sylvester's mathematical style was brilliant but not methodical, and his creativity was unfettered by rigour. With Cayley, he inspired many of the basic ideas of algebraic invariance. He also published on the roots of quintic equations and on number theory. In 1850 he coined the term matrix for an array of numbers from which determinants can be obtained. Invariance assumed great importance after his lifetime, as much use was made of it in quantum mechanics and relativity theory.

Szent-Györgi, Albert von [sent dyoordyee] (1893–1986) Hungarian-US biochemist: worked on vitamin C and on the biochemistry of muscle.

Szent-Györgi had four generations of scientists in his mother's family, and in 1911 began to study medicine at Budapest. By 1914 he had already published some research on the eye before he was called into the Austro-Hungarian army. He was soon decorated for bravery but, in order to return to his studies, he shot himself in the arm. Later he was redrafted, but again proved an awkward soldier by protesting against the treatment of prisoners. As a result he was sent to a base in Northern Italy where a malaria epidemic was raging, but within weeks the war was over and he returned to complete his medical course. Afterwards he researched in five countries and received his PhD in Cambridge for work on vitamins with F G HOPKINS. Back in Hungary in the 1930s, he showed that vitamin C (the anti-scorbutic vitamin, ascorbic acid) was in fact a compound he had first isolated in Cambridge in 1928. He also showed that paprika (Hungarian red pepper) is a rich source of it; in 1937 he won the Nobel Prize for medicine or physiology for his work on vitamin C. By 1935 he was working on the biochemistry of muscle; he began the work which was later developed by KREBS on the metabolism of muscle. He also isolated two proteins from muscle (myosin and actin) and showed that they combined to form actomyosin. When ATP (adenosine triphosphate) is added to fibres of this, it contracts. 'Seeing this artificial bundle contract was the most exciting moment of my scientific career', he wrote. This work was extended by H E HUXLEY.

He also had an exciting Second World War, working for the Allies and the underground resistance. Afterwards, he was offered the presidency of Hungary, but he emigrated to the USA in 1947 and directed muscle research at the Marine Biological Laboratory at Woods Hole. In the 1960s he worked on the thymus gland and on cancer, which had killed his wife and daughter. He was a man who had novel and daring research ideas; he 'thought big' and, a keen fisherman, claimed he liked to use an 'extra-large hook'.

Szilard, Leo [zilah(r)d] (1898–1964) Hungarian–US physicist: recognized the significance of nuclear fission.

Szilard has been described as 'a difficult child who grew up to become an impossible adult'. He was imaginative, volatile and immodest. A versatile and creative physicist, Szilard had an extraordinarily wide-ranging and original mind. He first studied electrical engineering, trained in the Austro-Hungarian army during the First World War and later took a doctorate in physics at Berlin (1922). Work with VON LAUE on thermodynamics followed, and led to a paper foreshadowing modern information theory (1929). Moving to Oxford and London in 1933, and to the USA in 1938, he began to work on nuclear physics at Columbia University.

In 1934 he had taken the earliest patent on nuclear reactions, covering in general terms the use of neutrons in a chain reaction to generate energy, or an explosion. On hearing of HAHN and MEITNER's fission of uranium (1938), he immediately approached EINSTEIN, in order to write together to President Roosevelt warning him of the possibility of Germany making atomic bombs. Together with FERMI, Szilard organized work on the first fission reactor, which operated in Chicago in 1942. He was a central figure in the Manhattan Project leading to the successful Allied atomic bomb, despite the fact that General Groves, in overall charge of the project, attempted at one point to have him interned for the duration of the war, judging him to be a security risk. Szilard opposed the direct use of the bomb against the Japanese, wishing to use it in a demonstration only; and he forecast the nuclear stalemate after the war in a 1945 report to the secretary for war. His enthusiasm for physics, politics and food continued.

After the war Szilard moved to research in molecular biology, doing experimental work on bacterial mutations and biochemical mechanisms and theoretical work on ageing and memory.

Takamine, Jokichi [takameenay] (1854–1922) Japanese–US biochemist: isolated first hormone (adrenalin; adrenaline, epinephrine).

Born in the year that Japan opened its ports to the West, Takamine became a product of two cultures: brought up in the strict Samurai code, he graduated in chemical engineering at Tokyo and then at Glasgow. Back in Japan, he worked as a government scientist for 4 years and then opened his own factory, the first to make superphosphate fertilizer in Japan. He married an American and in 1890 went to live in the USA and set up a laboratory to make biochemicals. In 1901 he isolated crystalline adrenalin from adrenal glands. After 1905 when STARLING first used the word 'hormone' to describe the animal body's 'chemical messengers' it was realized that adrenalin was the first hormone to be isolated in pure form from a natural source.

Talbot, William Henry Fox (1800–77) British inventor of negative/positive photographic process.

Talbot, the son of a relatively impoverished officer of dragoons, stepson of a rear admiral and grandson of an earl, was educated at Harrow and Cambridge, where he studied classics and mathematics and did well in both. He was 12th wrangler in 1821, and his later researches in mathematics secured him the Royal Society's Royal Medal in 1838. He also had success as a decipherer of Assyrian cuneiform inscriptions. He began a political career, becom-

THE DEVELOPMENT OF MODERN PHOTOGRAPHY

Photography uses two basic principles, both long-known: firstly, a lens to form an image, as in the 'camera obscura' (a box with a lens in one end that forms an image on a screen at the other end for an artist to trace or copy); secondly, the sensitivity to light of some chemical compounds, mostly silver halides. Around 1800 THOMAS WEDGWOOD, aided by his friend DAVY, tried to use these two principles to make stable images, but had little success. He used paper soaked in silver nitrate, but it was not sensitive enough to use in a camera, and his best results were shadow pictures of leaves and similar objects, obtained by placing them in contact with his paper and exposing to sunlight, when the uncovered areas blackened. He failed to find a way of 'fixing' the result, which blackened entirely on viewing in daylight.

In France, NIÉPCE had more success using a pewter sheet covered in bitumen, although it was his partner DAGUERRE who gained fame. Daguerre used a silvered copper sheet treated with iodine, which was sensitive enough to use in a camera. It was slow to respond and gave an image rich in detail but fragile and laterally reversed. It was also unique, in the sense that copies could not be made from it. Although used for some 15 years, the method was a dead end and is now extinct.

The scheme which led to modern photography and photomechanical printing was devised by TALBOT and his friend JOHN HERSCHEL in the late 1830s. Talbot used paper sensitized with silver salts, and found that the slight or invisible 'latent' image formed on it in the camera could be 'developed' by chemicals such as gallic acid. The result was fixed by soaking in 'hypo' solution, a method due to Herschel. The dried paper negative could then be oiled or waxed to made it transparent and used to make any number of positive prints (which would be the right way round) by contact printing on sensitive paper, which would be developed, fixed and washed as before. Herschel saw that a glass plate would be better in many ways than a paper base for the negative, but the film of photosensitive silver salts on glass was fragile and difficult to prepare, even when improved (as proposed by a cousin of Niépce) by the addition of egg albumen. The albumen was replaced by collodion by ARCHER in 1851. Even with this improvement the business of making a photographic negative was burdensome, because the film was only sensitive if freshly prepared and still moist. It then had to be developed immediately, so a portable laboratory was needed. The photography of mountains, wars and tropical scenes on wet collodion plates required heroes for its achievement.

The next advance was due to R L Maddox (1816–1902), a doctor and an enthusiastic photographer whose asthma was irritated by the ether fumes from collodion (which is guncotton dissolved in ether and alcohol). After many trials he found that a suspension ('emulsion') of silver salts in gelatin would serve, and a film of this on a glass plate could be used dry. This made photography much easier for the amateur. The realization in the 1870s that storing and warming the emulsion, 'ripening' it, made it more photosensitive was important. It allowed exposure times of 1/25 s or less in place of seconds or minutes, and so moving subjects could be photographed. By the 1880s MUYBRIDGE could use cameras to study the rapid animal movement of horses, men and women.

ing MP for Chippenham in 1833, but apparently found it unsatisfying – he spent most of that year honeymooning on the Continent, sketching with his wife. Like many others, he used a camera obscura as an artist's aid, but found this frustrating and laborious and, as he said later, 'the idea occurred to me – how charming it would be if it were possible to cause these natural images to imprint themselves durably and remain fixed upon the paper!' He experimented on these lines, using writing paper impregnated with silver salts, whose sensitivity to light was well known. In this way he made, in 1835, a very small picture of the lattice window of the small library at his family home, Lacock Abbey. It is the second oldest surviving photograph, the oldest being by NIÉPCE and also showing a window of a family home, photographed from the inside.

Talbot turned to other interests, but was spurred back to action by news of DAGUERRE'S work in 1839. Talbot exhibited his own 'photo-genic drawings' within weeks at the Royal Institution, but most of these were merely silhouettes made by superposition without a camera, and all his results were much inferior to the daguerrotypes, especially in their brilliance and detail. However, Talbot persisted in seeing his efforts as competitive with Daguerre's. In 1840 he discovered, by chance, the latent image; exposure of his sensitized paper in a camera for only a few minutes gave a result which could be made visible by 'development' in warm gallic acid, and then the picture was fixed with 'hypo' which dissolved out unchanged silver. These last two steps were due to others; both had been publicly suggested by his friend J F W HERSCHEL in 1839 and used by him, but these facts did not deter Talbot from later patenting both uses. Talbot made some of his photographs translucent by waxing them and then used this negative to make positive contact prints, and in this way he obtained lateral reversal and reversal of light and shade as

Despite these advances, the sensitive film was still supported on glass, heavy and breakable, and cameras were usually wooden, large and tripod-mounted. Easy popular photography is due to George Eastman (1854–1932) of the USA, who made film-handling simple by mounting the sensitive gelatin emulsion on a strip of flexible plastic (originally celluloid), which could be protected from light with a length of black paper and then loaded into a lightweight camera as a rolled cartridge. From about 1900 this 'Kodak' system dominated popular photography. With EDISON, Eastman also devised the perforated plastic film that made cine-photography a practical success. From 1927 a track on one side of the perforated film met the need for sound to be linked with the moving image, ensuring a major place for the cinema in entertainment.

The key element in photography has always been the photosensitive silver component in the film, leading to images in which the dark areas are formed by small particles of metallic silver. Black-and-white photographs were the main outcome of both still and cine photography in its first century, but images in colour were always a target. Herschel and later MAXWELL had some early success in colour photography. Only after the Second World War did colour become popular, using ingenious thin sandwiches of coloured organic dyes linked with the sensitive silver layer and with development methods complex enough to be carried out normally in a commercial laboratory. The Polaroid® camera due to LAND uses rapid in-camera colour processes that are rather costly but provide 'instant' results.

Cameras have also increased greatly in sophistication. Modern lenses with several glass elements give good correction of optical defects while still admitting enough light for short exposures, and often allow a change of focal length (zoom designs) to vary the apparent perspective. Autofocus devices use infrared sensors to allow automatic adjustment of the lens-to-film distance, suitable for subjects at different distances. Photosensitive electronic devices measure light intensity and adjust lens aperture and/or shutter speed to ensure that an optimum level of light reaches the film.

The early photographers were as much concerned with contact photography as with camera work, and the main process used became Herschel's cyanotype method, which gave the 'blueprints' used by ANNA ATKINS in the 1840s and which dominated the production of engineer's drawings in the present century, until in the 1950s a very different method, 'xerography', based on electrostatic principles, was developed by CARLSON.

The next major advance in conventional photography will involve a reusable sensitive surface and an electronic system, avoiding 'wet chemistry' entirely. But present methods based on silver compounds are highly perfected and will not readily be displaced. The situation recalls the use of the internal combustion engine in vehicles, another device which, although inherently rich in disadvantages, has been highly developed and now links a major industry with many users whose conversion to a new technology, when it arrives, will not be easy.

However, in 1995 Kodak announced a scheme of digital image processing, with a computer disk replacing film in the camera and electronic image transmission as one of its assets.

IM

in the original scene; as well as the ability to produce prints in any number. Talbot was almost alone in quickly seeing the advantages of this. His results were published in his *The Pencil of Nature* (1844), the world's first book illustrated by camera photographs. ANNA ATKINS had used Herschel's 'cyanotype' process to illustrate her book on algae published in 1843, and was the first to use a photographic method to illustrate a book.

Talbot patented his calotype process in 1841 and thereafter collected licence fees from other users of the method, vigilantly guarding his rights by frequent lawsuits. His aggressive attitude retarded the work of others, generated much ill-will, and by the 1850s was stifling both amateur and commercial photography in the UK, but by 1852 the pressure of protest led him to relax his grip somewhat. Even so, he controlled for some years in the UK all ways of making pictures by light except the daguerrotype, and it was said to be surprising that his claims did not include the Sun as a light source. Only when in 1854 he failed in his claim to cover ARCHER's collodion process did photography really advance.

In 1851 Talbot founded electric flash photography, when at the Royal Institution he made a sharp photograph of a rapidly rotating page of *The Times* by using a battery of Leyden jars, which gave an intense spark lasting less than 10^{-5} s. By 1887 MACH was using this method to photograph bullets in flight. Talbot was always interested in photomechanical printing, and from 1858 a method he had devised using a piece of gauze as a screen to break up the picture into small dots gave good results for half-tone printing. As usual he was a belligerent litigant in this field, in part because of his erroneous belief that abstract ideas are patentable and his view that a successor had no right to use a novel method to achieve a result similar to that he had already obtained.

In his lifetime as a gentleman scholar, he had also prospered by the sale of photographic materials and by the sale of licences to use his methods.

Tartaglia, Niccolo [tah(r)talya] (c.1501–57) Italian mathematician: found a method for solving cubic equations.

Tartaglia's real name was probably Fontana, but in the French attack on Brescia in 1512 he suffered sword wounds in the face which left him with a speech defect and led to the adopted nickname *Tartaglia* ('stammerer') thereafter. He taught mathematics in Verona and in Venice, and wrote on the mathematical theory of gunnery and on statics, arithmetic, algebra and geometry. The array of binomial coefficients now known as Pascal's triangle was first published by him, as was the first translation of EUCLID into Italian and of ARCHIMEDES into Latin. By 1535 he knew a method for solving cubic equations, which he confided to CARDANO under a pledge of secrecy. Cardano much improved and extended the method and published it (crediting Tartaglia) in his book *Ars magna* (The Great Skill, 1545). Cardano found all three roots of a cubic and suspected (correctly) that three roots always exist. Their friendship ended, and controversy over priority followed, with Tartaglia emerging as the loser.

Keen on most branches of mathematics, Tartaglia's greatest enthusiasm was in military applications. 'Tartaglia's theorem' of 1537 states that a firing elevation of 45° gives the maximum range for a projectile regardless of the speed of projection, and that its trajectory is everywhere a curved line. A century passed before GALILEO showed that the trajectory is a paraboloid.

Taussig, Helen (Brooke) [towsig] (1899–1986) US physician and a pioneer in the treatment of 'blue babies'.

Helen Taussig was born in Cambridge MA, the daughter of a Harvard professor. She attended Radcliffe College and the University of California, and obtained her MD at Johns Hopkins University medical school in 1927. She joined the faculty in 1930 and took charge of the cardiac clinic of the Harriet Lane Home for Invalid Children, remaining there until her retirement in 1963. She specialized in congenital malformations of the heart. In 1959 she became the first woman to be appointed a full professor at Johns Hopkins medical school.

Infants whose skin had a bluish hue caused by a lack of oxygen in the blood were termed 'blue babies' and had a restricted and short life. At the time there were few medical remedies for children with congenital heart defects. Deducing that the reason for the lack of oxygen was a blockage or constriction of the artery connecting the heart to the lung, Helen Taussig and Alfred Blalock (1899–1964) devised a procedure, in 1944, to take a branch of the aorta that normally went to the arm and connect it to the lungs. Soon 80% of the operations performed were a success, and a modification of the procedure is still in use today to gain time so that the 'blue baby' can reach an age where it is strong enough to withstand the very major surgery needed to provide a long-term solution.

Hearing of the large numbers of babies being born with heart defects in West Germany, Helen Taussig investigated and traced a link with the drug thalidomide taken during pregnancy, widely in use there in the late 1950s and early 1960s. Returning to the USA she reported her findings to medical associations and to the Food and Drug Administration, which had not yet approved thalidomide, and thereby saved American babies from the same fate. For this work she was awarded the President's Medal of

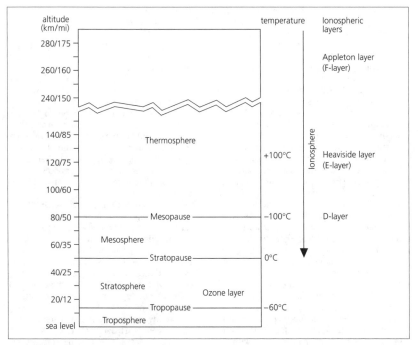

Layers of the atmosphere

Freedom, the highest civilian honour.

Taylor, Sir Geoffrey Ingram (1886–1975) British physicist: discovered how dislocations allow solids to deform under shear.

Taylor qualified at Cambridge, and was only absent from there during the two World Wars throughout his career, when he worked on aircraft design in the first, and on explosive shock waves in the second. From 1923 until his retirement in 1952 he held a professorship of physics. He conducted a great range of research in classical physics, always with originality, and chiefly on the mechanics of fluids and solids.

Having done major work on fluid turbulence he applied it in meteorology, aerodynamics and the planetary physics of Jupiter's Great Red Spot. In 1934 it occurred to him that metals and other crystalline solids might deform under shear because faults due to planes of atoms being 'misarranged' are propagated through the crystal. These faults he named dislocations, and their presence and movement does indeed determine how easily most solids are deformed, by comparison with the resilience of a perfect crystal. Work on the strength and the deformability of metals has since been much shaped by these ideas.

Taylor was renowned for his experimental skill. Very much liked, he was a keen sailor and his many inventions included a highly original anchor which became very popular for small boats. His proudest award was the Royal Cruising Club Cup for 1927, won for his trip to the Lofoten Isles, with his wife and a friend, in his 14.6 m cutter *Frolic*. When he carried out an experiment to see whether electrons are diffracted like waves the apparatus was sealed into a light-excluding box and left untouched for 2 weeks, so that he could go on a sailing holiday and return to the results.

Teisserenc de Bort, Léon Philippe [taysuhrŏk] (1855–1913) French meteorologist: discovered the stratosphere.

Teisserenc de Bort worked for several years as chief meteorologist at the Central Meteorological Bureau in Paris, before setting up his own observatory near Versailles in 1896. In 1902, using unmanned recoverable instrumented balloons, he discovered that above an altitude of about 11 km the temperature of the atmosphere ceases to decrease with height, and remains relatively constant in this region. He named this part of the atmosphere the stratosphere (believing that its steady temperature was due to undisturbed layers, or strata, of air), and the region below it the troposphere (since here the temperature varies and mixing of the air takes place). He went on to show that the tropopause (where troposphere and stratosphere meet) is much higher in the tropics.

He had earlier shown that European weather is very dependent on the atmospheric pressure at certain centres, notably the 'Azores high' and the 'Iceland low' (See diagram on p.313).

Teller, Edward (1908–) Hungarian–US physicist: major figure in development of thermonuclear energy.

Educated in Budapest and in Germany, Teller was one of the many scientists who left Germany in 1933. In 1935 he settled in the USA, becoming professor of physics at George Washington University. In 1940 he, SZILARD and WIGNER met with a Government committee to discuss the possibility of an atomic bomb; afterwards he worked with FERMI and Szilard in Chicago on the first atomic reactor, and then in 1943 he went to Los Alamos to work on the first atomic (fission) bombs. By then, he and others had considered the possibility of a fusion bomb (H-bomb), but he discontinued this work when the war ended in 1945. However President Truman approved H-bomb production in 1950 and the work was largely guided by Teller; a successful device was exploded in 1952.

The H-bomb depends on the energy released when light nuclei are fused to give heavier nuclei, following 'priming' by a fission implosion, using a design due to S Ulam (1909–85), to generate intense X-rays. Much of Teller's work in the 1950s and later was concerned with the theory of nuclear fusion and its use for peaceful purposes; he also supported nuclear arms for the protective defence of the west. He served in a series of senior posts at the University of California from 1952 onwards, and in the 1980s did much to convince President Reagan the need for the 'Star Wars' initiative intended to destroy incoming nuclear missiles from the USSR. After massive expenditure the programme contracted in the 1990s in the face of technical snags and the political collapse of the USSR. In so far as this collapse owed much to the cost of competing with the USA in military matters, the programme was arguably a success.

Tesla, Nikola [tesla] (1856–1943) Croatian–US physicist and electrical engineer: a pioneer of alternating current and inventor of the AC induction motor.

Tesla studied engineering at Graz and Prague before commencing work as an electrical engineer. In 1884 he emigrated to the USA, where he worked for EDISON before quarrelling and setting up on his own. Soon afterwards he developed the alternating current induction motor, eliminating the commutator and sparking brushes required by DC motors. He also made substantial improvements in the field of AC power transmission and generation, realizing that it could be transmitted and generated far more efficiently than the then commonly used direct current. He patented his inventions and set up a partnership with George Westinghouse (1846–1914) to commercialize them. His interest then turned to high-frequency alternating current, developing the Tesla coil, an air-core transformer with the primary and secondary windings in resonance to produce high-frequency, high-voltage output. In 1899 he used this device to produce an electric spark 41 m/135 ft long, and claimed to illuminate 200 lights over a distance of 40 km/25 mi without intervening wires by using the Earth as a transmitting oscillator. He was increasingly interested in transmitting power over large distances without wires; he became an eccentric recluse after 1892. The SI unit of magnetic flux density, the tesla (T), is named after him.

Today, as Tesla foresaw, alternating current and induction motors remain dominant in the power industry.

Thales (of Miletos) [thayleez] (c.625–c.550 BC) Ionian (Greek) merchant and philosopher; an early geometer; pioneer seeker of general physical principles underlying nature.

Thales was born in Miletos (in modern Turkey) 'the most go-ahead town in the Greek world'. Thales was probably a successful merchant who visited Egypt and there learned something of Egyptian geometry; but little is firmly known of his life and achievements, despite his high reputation, and some of the many accounts of his skill lack credibility. But it is certainly possible that he developed some general theorems in geometry, understood similar triangles and was able to find the distance of a ship from shore and the height of a building from its shadow. Such deductive mathematical arguments were systematized 250 years later by EUCLID. Thales was said also to be aware of the lodestone's magnetic attraction for iron.

According to ARISTOTLE, Thales believed that the Earth and all things on it had once been water and had changed by some natural process (akin to the silting-up of the Nile delta), and that the Earth as a whole was a flat disc floating in water. He offered natural (and not supernatural) explanations for phenomena such as earthquakes, and he attempted to derive theories from observed facts. He was thus a pioneer of later Greek science, and his attitudes are still with us. Later Greek thinkers gave him the highest place in their lists of wise men; and it can be argued that he is the earliest 'scientific' thinker we can name.

Theiler, Max [tiyler] (1899–1972) South African–US virologist: developed 17D vaccine against yellow fever.

Theiler studied medicine in South Africa and in London, and joined the staff of the Rockefeller Foundation in New York in 1930. Before that he had shown that yellow fever was certainly due to a virus, which could be cultured in mice, and he went on to show that by transmitting it from mouse to mouse many times, a safe and standardized vaccine could be mass-produced. On injection into human patients this induces yellow fever infection in an extremely mild form, with the benefit that it is followed by immunity from the full disease because antibodies have been generated in the blood. This vaccine (17D) effectively eliminated yellow fever as a major human disease and completed that work begun by REED, who had shown in 1901 that the disease was probably due to a virus and was transmitted through mosquito bites. Theiler won the Nobel Prize for physiology or medicine in 1951 'for his discoveries concerning yellow fever and how to combat it'.

He was fortunate to survive yellow fever himself in 1929, afterwards having the advantage of immunity.

Theophrastus [theeohfrastuhs] (*c*.372–*c*.287 BC) Greek philosopher and botanist: often described as 'the father of botany'.

Born in Lesbos, Theophrastus studied with Plato in Athens and then became ARISTOTLE's assistant, friend and successor as head of the Lyceum in 335 BC. The school did well under him, and may have had 2000 students; he secured botanical information from their home areas to add to his own observations. He wrote on many subjects, but his most important surviving books are on botany. He described over 500 plant species and understood the relation between fruit, flower and seed, and the differences between monocotyledons and dicotyledons and between angiosperms (flowering plants) and gymnosperms (cone-bearers). Plant propagation methods, and the effect on growth of soil and climate, are also accurately described by him. His ideas were usually sound, although he believed in spontaneous generation, in accord with the views of his time.

His classificatory interests extended to people; his best known book, *Characters*, is a series of humorous essays dealing with 'city types'. Each essay first defines an attribute and then illustrates from life the conduct of the Coward, the Mean Man, the Spreader of False Rumours, and so on. The result provides much information on city life in Athens in Theophrastus's time.

Thompson, Benjamin, Count Rumford *see* **Rumford**

Thomson, Sir Charles Wyville (1830–82) British marine biologist and oceanographer: postulated the existence of a mid-Atlantic ridge and of oceanic circulation.

Educated at the University of Edinburgh, Thomson held a number of academic posts in Scotland and Ireland before being appointed professor of natural history at Edinburgh in 1870. He took part in several deep-sea expeditions, leading the 5-year circumnavigational *Challenger* expedition of 1872–6, a landmark in marine exploration which resulted in the discovery of 4717 new marine species.

Thomson disproved two accepted beliefs: that

the oceans were lifeless below 300 fathoms (about 550 m); and that a fairly constant temperature of about 4°C existed at these depths. He found many animals by dredging at depth, and diverse temperatures at similar submarine depths in different regions. The expedition also showed that a clay bottom is usual at great depths and that nodules of nearly pure manganese dioxide (MnO_2) are found on the sea-floor.

All this work revitalized oceanography. On the basis of temperature measurements Thomson postulated the presence of oceanic circulation and the existence of a mid-Atlantic ridge, but the latter was not confirmed until 1925.

Thomson, Sir George Paget (1892–1975) British physicist: discovered experimentally the interference (diffraction) of electrons by atoms in crystals.

G P Thomson, the only son of J J THOMSON, had an outstanding college career. He survived the first year of the First World War in the infantry and in 1915 was attached to the Royal Flying Corps to work on problems of aircraft stability. In 1919 he returned to Cambridge, working first in his father's field of positive rays. He then went to the University of Aberdeen, being appointed to a professorship there at the age of 30. With his student Alex Reid he observed (in 1927) electron diffraction of electrons passing through thin celluloid film or metal foil, recorded on photographic plates as concentric rings of varying intensity about the incident beam. While not undertaken for that purpose they recognized the experiment as confirming BROGLIE's postulate of wave-particle duality. Thomson received the 1937 Nobel Prize for physics jointly with DAVISSON, who had also achieved diffraction of electrons but by use of a nickel crystal rather than a metal foil.

In 1930 Thomson moved to Imperial College, London. By 1939 he was aware of the possibility of a uranium fission bomb being developed, very possibly by Germany. During the Second World War he chaired the Maud Committee, advising the British Government on the atomic bomb, and in July 1941 reported that such a bomb could be made using separated uranium-235. The co-ordinating role on the bomb project then passed to CHADWICK, while Thomson became scientific advisor to Canada, and then in 1943 to the Air Ministry in Britain. After the war, in 1952, he returned to Cambridge as Master of Corpus Christi College. Always known as 'GP', he had considerable intuition in physics. Widely liked and good company, his life-long enthusiasms were sailing and model boats, which he made himself, including their armament; he was particularly pleased with his working model submarines.

Thomson, Sir Joseph John (1856–1940) British physicist: discovered the electron.

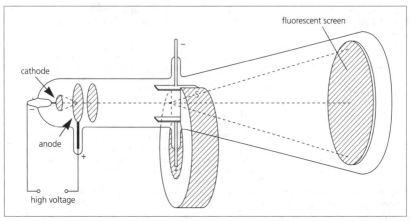

J J Thomson's apparatus for finding the ratio of charge to mass (e/m) for the electron. In the evacuated tube, a high voltage applied to the cathode and anode causes cathode rays (a stream of electrons) to be emitted from the cathode. A narrow beam is selected by two small holes, and forms a bright spot on the screen. The beam is deflected by a magnetic field, but can be restored to its normal position by an electric field applied to the two horizontal plates. From the field strengths, e/m can be calculated.

J J Thomson was a bookseller's son who first studied at Owens College (later Manchester University), hoping to become an engineer. Poverty caused by his father's death in 1872 led him to study mathematics, physics and chemistry instead, as he could not afford the charge then made to become an apprentice engineer. He did well and won a scholarship to Trinity College, Cambridge (1876). As a mathematician he graduated Second Wrangler (1880), and subsequently became a Fellow of Trinity College, Cavendish Professor (1884–1919, succeeding RAYLEIGH) and Master of Trinity in 1918. In modern terms 'JJ' was an experimentalist; but his hands were clumsy and his best work was actually performed by assistants. He was exceptionally well-liked.

Thomson had carried out an excellent mathematical analysis of vortex rings in 1883 and speculation that atoms might be vortex rings in the imagined electromagnetic 'ether' led him to investigate cathode rays (the electrical discharge emitted from an electrode under high fields in a gas at low pressure). Several German physicists believed that cathode rays were waves, and HERTZ had tried to show that they could not be particles, because in his experiments the cathode rays were not deflected by an electric field. However, Thomson repeated the experiment in a better vacuum, in which there was no polarizable air to mask the electric field, and demonstrated that electric fields would deflect cathode rays (1897). Having shown that the rays were made up of negatively charged particles, he proceeded to use their deflection under combined electric and magnetic fields to find the charge to mass ratio (e/m) of the particles, which did not vary from one cathode material to another. In April 1897 he revealed this discovery of a new particle. Developing this classic series of experiments, Thomson then measured the charge e by allowing the particles to strike water droplets and observing the droplet's rate of fall in an electric field (see MILLIKAN). He obtained the same value as the charge on a hydrogen atom, but using both results he found a mass m for the new particle about 1000 times lighter than hydrogen. Shortly afterwards Thomson's particle was named the 'electron' by STONEY. Its discovery opened the way for the study of atomic structure by RUTHERFORD, who succeeded him as Cavendish Professor. His device for measuring e/m is essentially the cathode ray oscillograph, so much used afterwards in both research and in television receivers.

Thomson also examined E Goldstein's (1850–1930) positive rays obtained when a perforated anode was used in the discharge tube, whose nature depended on the gas in the discharge tube; the cathode rays were the same whatever gas was present, and in 1912 Thomson showed how to use positive rays to separate atoms of different mass. This was done by deflecting the positive rays in electric and magnetic fields (a method now called mass spectrometry). The method allowed him to discover that neon had two isotopes, neon-20 and neon-22, and ASTON then developed the technique.

Thomson also showed that electrons are emitted from a hot, negatively charged metal wire, and from a negatively charged zinc plate if it is exposed to ultraviolet radiation.

Thomson received the 1906 Nobel Prize for physics for research on conduction through gases. One of his achievements was to have built up the Cavendish Laboratory as the foremost in experimental physics, with seven of his research assistants subsequently winning Nobel Prizes. To his great pleasure his son (G P THOMSON) was also a Nobel Prize winner, for demonstrating that the electron possessed both particle-like and wave-like behaviour.

Thomson, William, 1st Baron Kelvin (of Largs) (1824–1907) British physicist and electrical engineer: a pioneer of thermodynamics and electromagnetic theory; he directed the first successful project for a transatlantic cable telegraph.

Thomson's father had been a farm labourer who became professor of mathematics at Belfast and from 1832 at Glasgow. Two of his children became distinguished physicists and another did well in medicine. Young William studied science in Glasgow from the age of 10, and later was sent to Cambridge. He graduated when he was 21 and went to Paris to work on heat with REGNAULT, returning the next year to become professor of natural philosophy (ie physics) at Glasgow. He held the job for 53 years.

While still an undergraduate he gave a mathematical demonstration of the analogy between the transmission of electrostatic force and heat flow in a uniform solid (MAXWELL later brought this into his full theory of electromagnetism). Thomson reorganized the theory of magnetism, developing FARADAY's ideas, and introduced the ideas of magnetic susceptibility and permeability, and of the total energy of a magnetic system. Thomson was not only a theoretician but put his knowledge to practical effect, showing that low voltages were better than high ones for the transmission of signals along submarine cables and inventing the mirror galvanometer for the detection of the resulting small currents. He directed work on the first successful transatlantic cable (there had been two previous attempts), which became operational in 1866, bringing him considerable personal wealth. In 1892 he was made a baron, and chose his title from a small river, the Kelvin, passing through the university. He was a major figure in the creation of the Institute of Electrical Engineers. He liked sailing and bought 'a schooner of 126×10^6 g' (ie 126 t), which prompted him to develop navigational instruments; and his large Glasgow house was among the first to be lit by electricity (in 1881).

As a young man Thomson discovered G GREEN's work, then hardly known, and publicized it; he found that Green's and his own theorems gave valuable mathematical methods for attacking problems in electricity and in heat.

Thomson did much to develop heat theory. He heard of CARNOT's work when he was in Paris, but it was three years before he secured his paper; he then made its ideas widely known, and used and developed them further. Thomson proposed in 1848 an 'absolute' scale of temperature now known as the Kelvin or thermodynamic scale; it is independent of particular substances, but corresponds practically to the CELSIUS scale with 273.16 K as the triple point of water, 0°C. The SI unit of temperature is the kelvin (K). Independently of CLAUSIUS he formulated the second law of thermodynamics, which states that heat cannot flow spontaneously from a colder to a hotter body. He worked with JOULE on the relation of heat and work (the first law of thermodynamics), and also with him found the Joule–Thomson effect. This is the drop in temperature shown by most gases when they emerge from a fine nozzle, as a result of the work done to pull the mutually attracting gas molecules apart, and it is the basis of modern methods for cooling gases for liquefaction. Thomson worked on the theory of the cooling of a hot solid sphere and applied his theory to calculate ages for the Earth and the Sun. He recognized that his method assumed that no continuing heat supply was present: his results were about 10 times lower than present values, which now take into account heat due to radioactivity, which was not discovered until many years after Thomson's early work.

Thomson was an unusual scientist; his energy, enthusiasm and talent made him dominate British physics in the later 19th-c and he did much to move the focus of physics from Europe to Britain. He was always generous with ideas and in giving credit to others. His productivity was vast; 661 papers, many books and patents, covering the whole of physics (no-one since has ranged so widely) with sundry excursions into other sciences. He had some oddities, of course. He detested vector methods and gave himself much mathematical toil in avoiding them. He did not normally work on one problem for more than a month, and his results were worked out in green pocket books from which he tore sheets for publication. After his first wife died in 1870 he continued to have a great flow of ideas in physics, but he lost the ability to select good ideas from bad, and he pursued some strange notions (like the theory of the ether; and his opposition to the admission of women to Cambridge). He was probably the first scientist to become wealthy through science.

Tinbergen, Nikolaas [tinbergen] (1907–88) Dutch ethologist: a pioneer in study of animal behaviour.

Tinbergen graduated in zoology at Leiden, and afterwards taught there, except for 3 years in the Second World War spent in a hostage camp in occupied Holland. In 1947 he moved to Oxford and developed there his work on animal behaviour. The emphasis of this was to examine the patterns of behaviour shown by animals in natural conditions as well as in the laboratory; it included work on digger wasps, arctic foxes, seals, sea birds and snails. His now-classic studies of the social habits of herring gulls and the mating of sticklebacks showed that key elements of behaviour follow a stereotyped pattern and can depend largely on particular features. For example, the gull chick pecks for food at the parent's beak largely in response to a red spot on the latter. Also in these gulls, Tinbergen found that aggression between males is shown not only by calls but also by gestures, which in part seem designed to avoid actual fighting and injury. His wide-ranging work included study of learning behaviour, animal camouflage, instinct, and autism and aggression in human beings. He shared a Nobel Prize in 1973. His elder brother Jan also shared a Nobel-style Prize, the first awarded by the Swedish National Bank for economics, in 1969.

Ting, Samuel Chao Chung (1936–) US physicist: discovered the J/psi particle.

Ting had an unusual upbringing, he was born in the USA but educated in China and Taiwan, and finally at the University of Michigan (1956–62). His work in elementary-particle physics began at the European Organization for Nuclear Research in Geneva (CERN) and Columbia University, where he became an associate professor at 29. He also led a research group at DESY, the German synchrotron project in Hamburg, and from 1967 worked at Massachusetts Institute of Technology.

Ting conducted an experiment at the Brookhaven National Laboratory synchrotron in which protons were directed onto a beryllium target, and long-lived product particles were observed (1974). This newly discovered particle was named the J particle; it was observed at the same time and independently by B RICHTER at Stanford, who called it the psi particle. It is now known as the J/psi particle. Soon, other related particles of the J/psi family were detected, and Ting and Richter shared the Nobel Prize for physics in 1976.

Tiselius, Arne (Wilhelm Kaurin) [tisayleeus] (1902–71) Swedish physical biochemist: developed technique of electrophoresis for separating proteins.

Tiselius was a pupil and then assistant to SVEDBERG, and like him had his career in Uppsala, working mainly on proteins. Since these carry electric charges, they can be made to migrate in solution by applying an electric field. Tiselius developed this method (electrophoresis) to separate proteins, and the method has since been widely used. He showed that blood serum proteins can be separated into four groups of related proteins; they are the albumens and the α-, β- and γ-globulins. He was awarded a Nobel Prize in 1948.

Tombaugh, Clyde William [tombow] (1906–) US astronomer: discovered Pluto.

Too poor to attend college, Tombaugh was a keen amateur astronomer and built his own 9 in (23 cm) telescope. He was appointed as an assistant at the Lowell Observatory, Arizona, in 1929, and took over LOWELL's search for a trans-Neptunian planet; Lowell had died in 1916. After repeatedly photographing the sky along the plane of the ecliptic, and checking for differences between successive photographs with a blink comparator, Tombaugh finally announced the finding of the new planet (Pluto) on 13 March 1930. Afterwards he continued the search for possible further planets, but although he discovered over 3000 asteroids, he found no other planets.

Pluto, the outermost of the planets, is in several ways unusual: its orbit is highly eccentric and is tilted, it is much smaller than the other outer planets and it has a relatively large satellite, Charon, with which it is locked in synchronous rotation, each keeping the same side facing the other. Pluto is named after the Greek god of the underworld, and the name also honours (through his initials) PERCIVAL LOWELL, who predicted its existence.

Tomonaga, Sin-Itiro [tomonahga] (1906–79) Japanese theoretical physicist: a founder of quantum electrodynamics (QED).

Tomonaga graduated in physics at Kyoto University and studied with HEISENBERG at Leipzig before becoming professor of physics at Tokyo (1941) and then president of the university (1956).

Like others, Tomonaga started to develop a relativistic theory of the quantum mechanics of an electron interacting with a photon (1941–3). During the Second World War, he, FEYNMAN and SCHWINGER were unaware of each other's work, and it was not until 1947 that it was realized that all three had arrived independently at solutions to the problem of linking special relativity with quantum physics, solutions which were shown to be identical by DYSON. Tomonaga realized the value of a theory that could describe high-energy subatomic particles and he was responsible for the idea that two particles interact by exchanging a third virtual particle between them. The interaction is then like the momentum exchange when one rugby player passes the ball to another. The resulting theory is called quantum electrodynamics (QED) and for its discovery Feynman, Schwinger and Tomonaga shared the 1965 Nobel Prize for physics.

Torricelli, Evangelista [torichelee] (1608–47) Italian physicist: inventor of the mercury barometer and discoverer of atmospheric pressure.

An orphan, Torricelli was educated by the Jesuits and by B Castelli (1578–1643), for whom he worked on the dynamics of falling bodies. This led to his being appointed assistant to GALILEO, whom he subsequently succeeded as mathematician to the court of Tuscany.

Torricelli's interests covered pure mathematics and experimental physics; he worked on conic sections and other curves, deriving the area of the cycloid. He is best remembered, however, for his discovery of atmospheric pressure and for his invention, in 1643, of the mercury barometer. This appears to have come about from an attempt to solve the problem of why water could not be pumped out of a well more than 33 feet (≈ 10 m) deep. Deducing that the reason was that the atmosphere possessed weight and therefore exerted a pressure, he set about verifying the idea by sealing a glass tube at one end, filling it with mercury and inverting it with the open end in a dish of mercury. He found that the height of the mercury column fell to about 760 mm, and reasoned that a vacuum was formed above it; the weight of the column was being balanced by the weight of the atmosphere. He later noticed the small daily variations in the height of the column, which he related to changes in the weather.

Tour, Charles Cagniard de la see **Cagniard de la Tour**.

Townes, Charles Hard (1915–) US physicist: discovered the theory of the maser and produced the first working examples.

Townes was the son of a lawyer and he completed his education in 1939, having attended Furman University, Duke University and California Institute of Technology. He began work at the Bell Telephone Laboratories and spent the war years developing radar-assisted bomb sights. Radar uses microwave radiation, with a frequency between that of radio and infrared light. In 1947 he joined the physics department at Columbia University. He spent 1961–7 at Massachusetts Institute of Technology and then became a professor at the University of California at Berkeley.

Beginning as early as 1945, Townes had studied the absorption and emission of photons when a molecule goes from one configuration to another. This involves very precise photon frequencies because molecular energy states are discrete and precisely defined. Thus the ammonia molecule (NH_3) may flip between two configurations rather like an umbrella blowing inside-out, but with the absorption of a specific microwave frequency, of 1.25 cm wavelength.

In 1951, Townes realized that a wave of photons could be amplified in a practical way by the spontaneous emission process first suggested by EINSTEIN. That is, if a molecule in the higher energy state is stimulated by photons of just the correct frequency, it will fall back to its lower state and emit another photon of precisely the same frequency or energy. If there are fewer molecules in the lower energy state to absorb photons than in the upper state, a net amplification results. He also recognized that if the wave is reflected back and forth in a resonant cavity, it interacts with the molecules for some time, steadily gaining more energy or amplification and resulting in a coherent output signal

consisting of a wavetrain of extremely well-defined frequency.

To obtain ammonia with many molecules in the higher energy state, Townes took advantage of molecular beam techniques, separating out a beam of molecules in a high energy state with a non-uniform electric field. This 'population inversion' method, giving a majority of the high- rather than low-energy molecules, then provided a working amplifier and oscillator (1954). It was called a maser (microwave amplification by stimulated emission of radiation). A somewhat similar idea was suggested in the Soviet Union by A Prokhorov (1916–) and BASOV.

Masers were soon used in atomic clocks and in sensitive receivers, for example for radio telescopes and space communications. In 1958 Townes and SCHAWLOW showed that an optical version of the maser (the laser, for 'light amplification by stimulated emission of radiation') was possible, and discussed its properties and oscillation theoretically. The first operating system was constructed, however, by MAIMAN (1960), and now many versions are made.

Townes, Prokhorov and Basov were awarded the 1964 Nobel Prize for physics 'for fundamental work in the field of quantum electronics, which had led to the construction of oscillators and amplifiers based on the maser–laser principle'.

Trumpler, Robert Julius (1886–1956) Swiss–US astronomer: discovered interstellar light absorption.

Trumpler was educated at Zürich and Göttingen. He moved to the USA in 1915, spending most of his career at the Lick Observatory in California. In 1930, while measuring the distance and size of over 300 open star clusters, Trumpler discovered that the more distant ones generally appeared to be larger than the nearer ones. Since there was no apparent reason for this, he assumed that he was observing the result of the interstellar absorption of light by dust grains, and measured the effect to be a decrease in brightness of approximately 20% for every 1000 light years travelled by the starlight. His conclusions had important effects on ideas about the scale of the universe.

More recent work has shown that starlight is polarized by the interstellar dust, which implies that its grains are elongated and partly aligned, probably as a result of magnetic effects.

Tswett, Michel, also Mikhail Tsvet [tsvet] (1872–1919) Russian botanist: developed the technique of chromatography.

The son of a Russian father and an Italian mother, Tswett was educated in Switzerland and held posts in Poland and Russia. His research was mainly on plant pigments and it was to separate these that he effectively introduced chromatography, although as with most inventions, precursors of success can be traced. In 1903 Tswett separated his plant leaf colours (chlorophyll a and b, carotenes and xanthophylls) by passing the mixture, dissolved in light petroleum, down a column of powdered chalk. Distinct colour bands developed in the column and could be easily separated with a knife. By the 1930s and especially after the Second World War this method of column chromatography (usually using alumina) and later its variants – notably ion-exchange, thin layers, paper chromatography and gas-liquid chromatography (both the last due to MARTIN) – became essential chemical methods. Chromatography did for chemical analysis what the computer did for calculation.

Turing, Alan (Mathison) [tooring] (1912–54) British mathematician and computer scientist: mathematically formalized the concept of the theoretical computer.

After graduating from Cambridge, Turing was elected a Fellow of King's College there in 1935, and then spent 2 years in Princeton. In 1937 he described a theoretical computer in precise mathematical terms (the Turing machine), an important step that formalized the hitherto vague concept of computability and computable numbers. During the Second World War he worked on code-breaking, playing a dominant part in breaking the German naval code, which enabled the Allies ultimately to win the crucial Battle of the Atlantic against German submarines.

After the war, he put his ideas on computing into practice when he supervised the construction of the ACE (Automatic Computing Engine) at the National Physical Laboratory, and at Manchester where he was assistant director of the work on MADAM (Manchester Automatic Digital Machine). His work on the design of such machines and the way in which they could be programmed was of great significance in the development of the computer, but his concept of an automatic electronic digital computer with internal program storage could not be realized until after his death, when advances in electronics made it possible.

In 1952 Turing attacked the problem of the formation of shapes and patterns in biology; the range includes flower patterns, bone symmetry and the tiger's stripes. He argued that chemicals diffusing through tissue and reacting can explain such pattern formation, and he devised equations that describe a distribution of reactants that can lead from homogeneity to pattern formation. These ideas have been much developed since then, and the problem is far from solved, but Turing's work is a valuable starting point.

Turing was also interested in artificial intelligence and developed a useful criterion for an intelligent machine: that it would be able to answer enquiries over a data-link in a manner indistinguishable from a human being.

Turing was a near-Olympic-standard long-distance runner. He was also an active homosexual at a time when this was a criminal offence. In 1952 he reported to the police that a young male Manchester printworker, with whom he was having an affair, was involved in the theft of some goods from his house. This was unwise of Turing, because it led foreseeably and inevitably to him being charged with gross indecency, and convicted. He was placed on probation and required to accept hormone drug treatment. He seemed remarkably unaffected by all of this, although exasperated by some of the effects of the drug, but a year after the probation and the treatment ended, he was found dead in his bed, with potassium cyanide and a partly eaten apple nearby. A rather casual inquest concluded that his death was due to suicide.

Tuve, Merle Antony (1901–82) US geophysicist: a pioneer of radio techniques for ionospheric studies.

Educated at the University of Minnesota and at Johns Hopkins University, Tuve was appointed to the department of terrestrial magnetism at the Carnegie Institution of Washington in 1926. He is remembered for pioneering radio techniques for studying the upper atmosphere. In 1925 he developed, together with BREIT, an early form of radar in order to determine the height of the ionosphere.

Tuve also investigated long-range seismic refraction by the Earth's upper mantle, his results subsequently providing evidence for the theory of isostasy, the process whereby areas of crust tend to float in conditions of near-equilibrium on the plastic mantle.

Twort, Frederick William (1877–1950) British microbiologist: discovered first virus infection of bacteria (bacteriophage).

Qualifying in medicine in London in 1900, Twort became a professor of bacteriology there in 1919; he was eccentric and reclusive. In 1915 when studying staphylococci he noticed that some cultures became transparent, and he traced the effect to an agent which was infecting the cocci. He offered several possible explanations for the effect, including virus action, and planned to continue the work, but army service in the First World War interrupted him and he did not take it up again, but expended much effort in an attempt to show that pathogenic bacteria were descendants of non-pathogenic types. In 1917 F H d'Herelle (1873–1949) found a similar result with some mixed cultures of dysentery bacilli, and named the infective agent bacteriophage ('phage'); vigorous dispute on priority followed, which was not resolved until adjudication (in Twort's favour) by one Professor Flu of Leiden in the 1930s. Twort's laboratory and records were entirely destroyed by German bombs in 1944. Since the 1950s, these viruses which infect bacteria have been much studied; many strains exist, and some have proved of great value in genetic engineering.

Tyndall, John [tindl] (1820–93) British physicist: made pioneering studies of heat and the scattering of light.

Tyndall lacked a university education, leaving school to work as a surveyor and civil engineer in Ireland. He subsequently studied physical sciences at Marburg in Germany and became first a professor and later director of the Royal Institution. He was a talented lecturer and popularizer of science.

Tyndall's early research was on diamagnetism, but he is chiefly remembered for his studies of heat. He measured the thermal conductivity of crystals along their different axes, investigated the effect of radiant heat on gases and made pioneering studies of glaciers (he was also one of the first men to climb the Matterhorn). His studies of the scattering of light by fine particles in the air and in liquids resulted in his discovery in 1859 of the Tyndall effect, whereby a beam of light is made visible by such scattering. Following RAYLEIGH's work on the frequency-dependence of the scattering of light, Tyndall was the first to realize why the sky is blue: atmospheric dust particles scatter the shorter wavelength (blue) components of sunlight to a greater degree than the longer wavelength (red) components. His interest in airborne particles led him to study airborne microorganisms and to support PASTEUR's arguments against spontaneous generation in the 1870s.

He died tragically. His devoted wife confused two medicines he took routinely and gave him an excessive dose of his sleeping draught (chloral); antidotes failed and he died within hours. She survived him by 47 years.

Uhlenbeck, George Eugene [ulenbek] (1900–88) Dutch–US physicist: discovered that electrons possess spin.

Uhlenbeck emigrated to the USA once he had completed his PhD at Leiden (1927), and worked at the University of Michigan (1927–60), where he became professor of theoretical physics in 1939. From 1960–74 he held a post at the Rockefeller Medical Research Centre in New York.

In 1925 Uhlenbeck and Goudsmit collaborated on an experiment whereby a horizontal beam of silver atoms was split by a vertical magnetic field into two components. This occurred because the electrons in the silver atoms possess 'spin', the name arbitrarily given to their property of having half a quantum unit of momentum directed either up or down in the applied magnetic field. The result was that the silver atoms were deflected according to their spin. This was the first observation of this purely quantum mechanical effect, and was an early piece of evidence that the new quantum mechanics was both necessary and correct.

Urey, Harold Clayton [yooree] (1893–1981) US physical chemist: pioneered isotope separation methods and their application.

Although originally a graduate from Montana in zoology, Urey soon turned to chemistry, first in industry and then at university, and following a year with Bohr he afterwards spent his career in chemical physics at four US universities. In 1932 he isolated deuterium, the heavy isotope of hydrogen, and went on to devise a large-scale process for obtaining heavy water (D_2O) by electrolysis, which slightly concentrates it, and to examine a range of deuterium compounds. His expertise on isotope separation gave him a critical role in the Second World War in the atomic bomb project (which required the separation of uranium isotopes) and afterwards in the work on securing tritium for the H-bomb. The same expertise led him to an ingenious way of measuring the past temperature of the oceans, and to ideas on the origin of the Earth and life upon it. He believed that the Earth was formed by the cold accretion

THE ORIGIN OF LIFE ON EARTH: AN UNSOLVED PROBLEM

There is now a consensus of informed opinion on some important dates in the Earth's history. Several lines of astronomical evidence point to the 'Big Bang', the high-temperature, high-density event when the universe began, as occurring roughly 15 billion years ago. Then about 4.5 billion years ago, the Sun and then the Earth formed, essentially by the accretion of dust and small particles, and without living forms of any kind for its first billion years. About 3.5 billion years ago, life appeared: the oldest known fossilized microorganisms, found in datable rock formations in Western Australia, are of that age. So began life on Earth, with its defining characteristic, the ability to replicate itself through the storage and passing on of genetic information from generation to generation. Rather slowly the first primitive, single-cell microorganism led, through evolutionary changes that are now understood in outline, to the whole range of multicell plants and animals known today. Modern man and woman (*Homo sapiens*) is about 50 000–100 000 years old.

Pasteur's work in the 1860s established that 'spontaneous generation' of life does not occur in laboratory conditions, although his experiments did not exclude all possibility. He 'did not think it impossible' in 1878 and Darwin in 1871 had speculated that protein, a characteristic chemical type found in all living things, might have originated in 'some warm little pond'. In the early 1900s Arrhenius suggested that life on Earth might have begun through the arrival of organisms from elsewhere in the universe (the theory of 'panspermia') and in the 1970s this proposal was revived by Hoyle, but has found little support.

A valuable proposal was made by the Soviet biochemist Oparin and independently by Haldane, who argued that the early atmosphere of the Earth contained little free oxygen (O_2), and that this was generated much later as a result of photosynthesis by green plants, a view now generally accepted.

By the 1950s a good deal of information more or less relevant to the problem of the origin of life had been accumulated. For example, reasonably well-based estimates of the origin, age and composition of the universe and of the solar system and of the Earth were available, and a range of simple organic compounds (but none characteristic of living forms) had been identified in the dust and gas of space beyond Earth's atmosphere and in meteorites arriving on the Earth's surface from the outer parts of the system.

In 1952 Urey argued that the important energy

of mainly metallic particles and that it had a primitive reducing atmosphere; and that the Moon was formed separately. Later work has given broad support for his views, developed by 1952; and the next year MILLER in Urey's laboratory carried out successful experiments on the synthesis of organic compounds from an atmosphere on the Urey model. Urey won a Nobel Prize in 1934. His introduction of isotopically labelled compounds has been of immense value in chemistry, physics, biology and medicine.

source for early geochemical and prebiotic chemical reactions was very probably the stream of solar ultraviolet radiation, together possibly with lightning discharges and meteor impacts; and that the early ('primordial') Earth's atmosphere was composed mainly of CH_4, NH_3, H_2 and H_2O. The next year his co-worker S L MILLER passed electric discharges through this mixture and showed that within a week an impressive range of organic molecules was formed, including no less than 25 amino acids. Amino acids react rather easily together to form protein, whose presence is so ubiquitous in living systems.

However, the discovery by CRICK and WATSON (also in 1953) of the double helix of the nucleic acid DNA as the key material forming the genes, a discovery which effectively created the new science of molecular biology, generated a new difficulty for theorists of the origin of life. By 1958 Crick laid down the 'central dogma' of the new science: that the direction of flow of chemical synthesis in living systems is one-way and irreversible, in the sense: 'DNA makes RNA, and RNA makes protein'. (Exceptions to this have proved to be rare, but cases are known: the retroviruses contain RNA as the genetic molecule, and this makes the more complex DNA, a reversal of the general rule.)

So the idea that a thin 'primordial soup' of amino acids, leading perhaps to proteins and on to the simplest living cells, could form a satisfactory model for the origin of life, is clearly unsatisfactory. In any case, as A G Cairns-Smith pointed out, it is hard to see how a thin soup could organize its amino acids and proteins in a way describable as 'life' rather than merely undergoing unfruitful chemical interactions.

Despite much effort by many investigators (theorists, analysts and experimentalists), the question of how life originated remains very open indeed. There is fairly widespread belief that the simpler class of nucleic acid, RNA, probably formed at an early stage in the path to the first, and doubtless very simple, living cells. From RNA, it is not difficult to see in general terms how protein and then simple cells may have developed; and thereafter, evolution may have followed a broadly understood network of increasing complexity. But how did the RNA, a rather complex biopolymer, originate? A credible route to it from the fairly simple organic molecules that might have been available has yet to be devised.

The mystery remains. It may prove to be unsolvable, permanently. Likewise the question of whether life only exists on Earth and if not, whether it began here or came here, may be unanswerable.

IM

Van Allen, James (Alfred) (1914–) US physicist: discovered the magnetosphere (the Van Allen radiation belts).

Van Allen was educated at Iowa Wesleyan College and the University of Iowa. During the Second World War he served in the US Navy, helping to develop the radio proximity fuse for missiles and anti-aircraft shells. Afterwards he worked at Johns Hopkins University, and was appointed professor of physics at the University of Iowa in 1951.

Van Allen's contributions, to a large extent, reflected his war-time experiences with rocketry and miniaturized electronics. After the war he used left-over German V-2 rockets to carry instruments to measure cosmic radiation into the upper atmosphere, and in 1958 put a Geiger radiation counter on the first American satellite, Explorer 1. This and later Explorer satellites revealed a region of high levels of radiation at a height of several hundred kilometres above the Earth. More detailed investigation has since shown that there are in fact two toroidal (doughnut-shaped) belts, which are created by charged particles (electrons and protons) from the Sun being trapped by the Earth's magnetic field. These Van Allen radiation belts constitute the Earth's magnetosphere.

van Beneden, Edouard *see* **Beneden**

Van de Graaff, Robert (Jemison) (1901–67) US physicist: invented the Van de Graaff generator.

Van de Graaff had a varied education, studying engineering at the University of Alabama and physics at the Sorbonne, where he was attracted to particle physics by MARIE CURIE's lectures, and Oxford. On returning to the USA in 1929 he worked at Princeton and the Massachusetts Institute of Technology, becoming associate professor of physics at the latter in 1934.

While a research student at Oxford, Van de Graaff realized that the conventional means of generating static electricity, the Wimshurst machine, could be greatly improved by storing the charge on a hollow metal sphere. In 1929 his first model of the Van de Graaff generator achieved potentials of up to 80 kV, and he subsequently built versions capable of generating millions of volts. The Van de Graaff generator has been an important tool in atomic and nuclear physics for accelerating charged particles, and in medical and industrial X-ray equipment where high voltages are required. In 1960 Van de Graaff resigned his post at MIT to become chief scientist at the High Voltage Engineering Corporation, a company he had formed in 1946 to develop and market such devices.

van de Hulst, Hendrik Christofell *see* **Hulst**
van der Waals, Johannes Diderik *see* **Waals**
van Helmont, Jan Baptista *see* **Helmont**
van Leeuwenhoek, Antony *see* **Leeuwenhoek**
van 't Hoff, Jacobus Henrikus *see* **Hoff**
Van Vleck, John (Hasbrouck) (1899–1980) US physicist: a major contributor to modern theories of magnetic systems.

Van Vleck's father and grandfather were both eminent mathematicians. Van Vleck emerged from study at Wisconsin and Harvard to take up a post at Minnesota in 1923. He later returned to chairs at Wisconsin and Harvard.

Van Vleck largely founded the modern theory of magnetism, taking DIRAC's quantum mechanics and working out the implications for the magnetic properties of atoms. In 1932 he published *The Theory of Electric and Magnetic Susceptibilities*, which laid out the theory and which remains in use today as a classic text. In it the paramagnetic properties of atoms are discussed; the temperature-independent susceptibility is now called Van Vleck paramagnetism.

He also studied chemical bonding in crystals and developed the crystal field and ligand field theories. These allow one to predict and explain some features of magnetic, electrical and spectroscopic properties of metal compounds. Furthermore, Van Vleck explained how local magnetic moment formation is assisted by electron correlation (the interaction between the motion of electrons). His wartime work resulted in showing how water and oxygen in the atmosphere give rise to absorption of radar signals, and was important in devising useful radar systems.

Van Vleck has been described as 'one of the few true gentlemen and scholars'; he was a quiet man with charm. In 1977 his pioneering research was recognized with a joint award of the Nobel Prize for physics.

Vauquelin, Louis Nicolas [vohklĩ](1763–1829) French analytical chemist: discoverer of chromium and beryllium.

As a boy Vauquelin worked in the fields with his peasant father; he did well at school, and at 14 was sent to work in an apothecary's shop, at first in Rouen and then in Paris. Soon the chemist A F de Fourcroy (1755–1809) heard of his enthusiasm for chemistry and took him on as an assistant, and later as a friend and co-worker. Vauquelin rescued a Swiss soldier from a mob during the French Revolution, and as a result had to leave Paris in 1793; but he soon returned and in 1809 he succeeded Fourcroy as professor there.

In 1797 he examined the rare, brilliant orange mineral crocoite and discovered in it a new metal, which he named chromium. Crocoite is actually lead chromate, $PbCrO_4$; Vauquelin obtained Cr_2O_3 from it and, by strongly heating this with charcoal, secured the metal as a powder. The next year he studied specimens of the minerals beryl and emerald sent to him by the mineralogist HAÜY, who suspected from their crystal forms that they were chemically identical. Vauquelin proved that this was correct and that emerald owes its green colour to traces of chromium; both minerals are beryllium aluminosilicate. He realized that a new metal (beryllium) was present, which he was not able to isolate; this was achieved in 1828 by WÖHLER. Vauquelin was also the first to isolate an amino acid; this was asparagine, which he got from asparagus.

Vavilov, Nikolai Ivanovitch [vavilof] (1887–1943) Russian botanist and plant geneticist: a pioneer of cross-breeding to improve crops.

Trained in Moscow and with BATESON at the John Innes Horticultural Institute at Merton in Surrey, Vavilov returned to Russia in 1914 and quickly rose to become, by 1920, director of the All Union Institute of Plant Industry, controlling over 400 research institutes in the USSR with 20 000 staff by 1934. Between 1916 and 1933 he led plant-collecting expeditions all over the world, the intention being to conserve and use the valuable genetic resources in wild and cultivated plants on which crop improvement depends. He devised useful theories on where centres of genetic diversity are to be found by plant hunters. His programme was very successful, his collection of new plants reaching 250 000 by 1940; it was the largest-scale enterprise of its kind and the model for later work of this sort. Vavilov supported the ideas in genetics due to MENDEL and to MORGAN, and this was to prove fatal. The politically active Marxist botanist T D Lysenko (1898–1976), who had reverted to a Lamarckian view, resented his success; Vavilov was arrested in 1940 while plant collecting, charged with 'right-wing activities' and sentenced to death after a 5-minute trial; his name was erased from all records. He died about 2 years later of starvation in a labour camp, an ironic fate for the man who did most to feed Russia during the war by improved agricultural methods. His seed collections were largely eaten during the siege of Leningrad. At present he is re-recognized in Russia and the Vavilov Institute is named in his honour.

Vening Meinesz, Felix Andries [vayning miynes] (1887-1966) Dutch geophysicist: pioneer of submarine gravity measurements.

After graduating in engineering from the Technical University of Delft in 1910, Vening Meinesz worked on a Government gravity survey of the Netherlands. In 1927 he was appointed professor extraordinary of geodesy, cartography and geophysics at Utrecht, and also professor of geophysics at Delft. His lifelong interest was in gravity and the deductions he could make from its accurate measurement.

Gravity determinations can yield useful information about underlying geological structure, but very accurate measurements are necessary since the variation in gravity is small. However, for the majority of the Earth's surface, that covered by the oceans, the lack of a stable platform makes measurements by the conventional pendulum technique impossible. Vening Meinesz realized that a submarine might provide a sufficiently stable base and, with the assistance of the Dutch navy, he made the first marine gravity determinations in the Pacific in 1923, finding that he could obtain results consistent within 1 mgal, comparable with land-based measurements, using three pendulums in an ingenious cradle. From the measurements made during a total of 10 such voyages Vening Meinesz discovered a belt of negative gravity anomalies beneath the deep submarine trenches associated with island arcs. He correctly interpreted this as being due a subduction zone, ie a compressive down-buckling of the oceanic crust below the continental crust. He did not support ideas on continental drift, but when this became established his results fitted in with modern tectonic plate theory.

Vernier, Pierre [vairnyay] (1584-1638) French mathematician and engineer: devised a precision measuring scale.

Vernier worked as a military engineer in Spain. Requiring an accurate method of measuring small distances for map-making, he devised in 1631 a precision scale, consisting of a movable part with nine divisions which slid past a fixed part with 10 divisions. Observing where the two marks on the two scales most closely coincide effectively adds another decimal place to the accuracy of the measurement. The method is a refinement of a multiple-scale device invented by the 16th-c Portuguese mathematician Pedro Nuñez (1492-1577).

Vesalius, Andreas (*Lat*), Andries van Wesel (*Flemish*) [vuhzayliuhs] (1514-64) Flemish anatomist: the founder of modern anatomy.

A pharmacist's son, Vesalius studied medicine at Louvain, Paris and Padua. He did well, and was made professor of anatomy and surgery at Padua when he was 24. His first lectures were novel; he carried out dissections himself, instead of leaving this to an assistant while reading from a text book as was usual; and he used drawings to help his students. During the next four years, he was busy with his research on anatomy based on human dissection. His results were published in *De humani corporis fabrica libri septem* (1543, The Seven Books on the Structure of the Human Body), which included descriptions and fine woodcuts, some by himself and the rest made under his direction. The book set a completely new level of clarity and accuracy in anatomy and made all earlier work outdated. Many structures are described and drawn in it for the first time (eg the thalamus) and the book also broke with tradition by its critical view of earlier work (eg Vesalius notes that he was unable to find a passage for blood between the ventricles of the heart, as GALEN had assumed). At 29, with his master-work published, Vesalius became a court physician, at first to Charles V and then to Philip II of Spain. His research largely ceased. He found that Spanish doctors were Galenists who were hostile and jealous, and he tried to recover his job in Padua. To leave Spain he needed Philip's permission, which he got by proposing a pilgrimage to Jerusalem. He probably got his job on the way to Jerusalem, but died on the return journey, in Greece.

Viète, François [vyet] (1540–1603) French mathematician: made many early contributions to algebra.

Viète grew up in the Poitou region of France, and in 1556 entered the University of Poitiers to study law. While practising law between 1560 and 1564 he took up cryptography and mathematics as hobbies; the former was useful when he moved to Paris in 1570 and became a court official to Charles IX. The persecution of the Huguenots forced him to go into hiding from 1584, and during this time he absorbed himself in mathematics and did work of historic importance.

After 5 years Henri IV succeeded Charles and Viète returned to the royal court. In the war against Spain, Viète broke the Spanish secret cipher, allowing intercepted dispatches between Philip II of Spain and his embassy to be deciphered. He was dismissed from the court in 1602 and died shortly afterwards.

Viète's mathematical research was in algebra, which he applied to solve geometrical problems. He used letters to denote constants as well as variables, and he introduced the terms 'coefficient' and 'negative'. Using algebraic methods he solved a problem that dated back to the Greek APOLLONIUS, that of constructing a circle touching three given circles.

Viète published a systematic account of how to solve problems in plane and spherical trigonometry, making use of all six trigonometric functions for the first time. The cosine law for plane triangles and the law of tangents were included. He also discovered a new and elegant solution to the general cubic equation using trigonometric multiple-angle formulae. The familiar relations between the positive roots of an algebraic equation, its coefficients and the powers of the unknowns are also due to Viète. Always he preferred to establish his identities and his proofs algebraically rather than geometrically, thereby setting a trend.

Vigneaud, Vincent Du see **Du Vigneaud**

Vine, Frederick John (1939–) British geologist: co-discoverer of magnetic anomalies across mid-ocean ridges.

After graduating from the University of Cambridge in 1965, Vine spent 5 years at Princeton before returning to England in 1970, becoming reader and later professor of environmental science at the University of East Anglia. In 1963, whilst a research student under the supervision of MATTHEWS, Vine showed that the oceanic crust on either side of a mid-ocean ridge was remanently magnetized in alternately normal and reversed polarity in bands running parallel to the ridge. This, they argued, was consistent with the sea-floor-spreading hypothesis proposed by H H HESS the year before, and was seen as powerful support for Hess's hypothesis (see MATTHEWS). Vine became professor of environmental sciences at the University of East Anglia in 1970.

Vine and Matthews's work, combined with the idea of continental drift, effectively created the major concept of plate tectonics.

Virchow, Rudolf (Ludwig Carl) [feerkhoh] (1821–1902) German pathologist and anthropologist: founder of cellular pathology.

Virchow graduated in medicine at Berlin, and then secured a junior post in Berlin's great hospital, the Charité. He was a skilful pathologist who recognized leukaemia in 1845 and went on to study thrombosis, embolism, inflammation and animal parasites. He was always politically active, and his liberal sympathies in the unrest of 1848 helped to lose him his Berlin post; but Würzburg gave him another and 7 years later he returned to Berlin as professor of pathological anatomy. He remained in politics and, as a Reichstag member, opposed Bismarck so forcefully that the latter challenged him to a duel in 1865; Virchow managed to avoid this. In the 1850s Virchow took up SCHWANN and SCHLEIDEN's cell theory with enthusiasm, and applied it to pathology; he saw disease as originating in cells, or as the response of cells to abnormal conditions. His ideas led to much fruitful work, aided by the improvements in microscopes after 1850 and the introduction of the microtome for making thin sections and dyes for selective staining. Modern pathology begins with him, and he became Germany's leading medical scientist.

PASTEUR was his near-contemporary; however, Virchow did not enthuse over the germ theory of disease. He saw disease as a continuous change in the cells, rather than as a result only of an invasive agent (we now recognize diseases of both types, of course). Similarly he saw the theory of evolution as a hypothesis only and voted against its inclusion in school biology.

He was an enthusiast for anthropology and archaeology and worked on the 1879 dig to discover the site of Troy. In practical politics, his efforts in public health in Berlin led to improved water and sewage purification.

Vleck, John (Hasbrouck) Van see **Van Vleck**

Volta, Alessandro (Giuseppe Anastasio), Count (1745–1827) Italian physicist: the inventor of the electric battery.

Born in Como of an aristocratic family devoted to the Church, Volta was professor of natural philosophy at Pavia (1778–1818) and, after some political turbulence, became rector of Pavia. Highly religious, but not prudish, a friend records that he 'understood a lot about the electricity of women'.

Following GALVANI's discovery in the 1780s that an electric spark, or contact with copper and iron, caused a disembodied frog's leg to twitch, Volta (who had long studied electricity) was interested in finding the cause of the phenomenon. Experiment showed him that an electric current could be generated by bringing different metals into contact with one another. In 1799 he succeeded in constructing a battery consisting of metal discs, alternately silver and zinc, with brine-soaked card between them. This voltaic pile produced a steady electric current and was the first reliable source of electricity. Volta did little further work on the device, but it was to transform the study of the subject and was invaluable to men such as NICHOLSON, DAVY and FARADAY. It also laid to rest the contemporary theory that animal tissue was somehow necessary for the generation of electricity. Volta was given the title of Count by Napoleon, who invaded Italy in the 1790s and who had become very interested in electricity and correctly foresaw its importance to science. The SI unit of electric potential, the volt (V), is named after him. If the work done in causing one coulomb of electric charge to flow between two points is one joule, then the potential difference between the points is one volt.

von Baer, Karl Ernst see **Baer**
von Baeyer, Adolf see **Baeyer**
von Behring, Emil (Adolf) see **Behring**
von Braun, Wernher Magnus Maximilian see **Braun**
von Euler, Ulf Svante see **Euler**
von Fehling, Hermann Christian see **Fehling**
von Fraunhofer, Josef see **Fraunhofer**
von Guericke, Otto see **Guericke**
von Haller, Albrecht see **Haller**
von Helmholtz, Hermann (Ludwig Ferdinand) see **Helmholtz**
von Hofmann, August Wilhelm see **Hofmann**
von Humboldt, (Friedrich Wilhelm Heinrich) Alexander, Freiherr (Baron) see **Humboldt**
von Kármán, Theodore see **Kármán**
von Klitzing, Klaus see **Klitzing**
von Laue, Max (Theodor Felix) see **Laue**
von Liebig, Justus, Freiherr (Baron) see **Liebig**
von Naegeli, Carl Wilhelm see **Naegeli**
Von Neumann, John (János) [noyman] (1903–57) Hungarian–US mathematician: suggested the concept of the stored-program computer.

Born in Budapest the son of a Jewish banker,

Von Neumann was a mathematical prodigy as a child and became one of the most eminent mathematicians of his day. Educated at the Universities of Budapest, Berlin and Zürich, he moved to the USA in 1930.

Aside from his work in mathematics (he was professor of mathematics at Princeton 1933–55) Von Neumann is principally remembered for his contributions, during and after the Second World War, to the development of electronic computers. He is widely credited with the concept of the 'stored-program computer', whose two essential components are a memory in which to store information and a control unit capable of organizing the transfers between the different 'registers' in memory in accordance with a program also stored in memory. All modern computers work on this principle, and are sometimes called 'Von Neumann machines'. The credit for the work (which was carried out in wartime secrecy) is now recognized not to be entirely his, however, and ought more properly to be shared with others in the development team.

He also participated in the American atomic bomb project (the Manhattan Project) and, together with others, developed the 'high explosive lens' that was essential to its success. In later years he became a leading proponent of nuclear power.

He founded the theory of games in 1926: its focus is to model strategies leading to success in situations involving chance or free choices, which by 1944 he applied in economics and which has since been used widely in this and other social sciences, and in military applications. (See panel overleaf.)

von Sachs, Julius see **Sachs**
von Siemens, (Ernst) Werner see **Siemens**
von Szent-Györgi, Albert see **Szent-Györgi**
Vries, Hugo de [duh vrees] (1848–1935) Dutch plant physiologist and geneticist: early investigator of plant genetics.

De Vries, son of a Dutch prime minister of 1872, studied medicine in Holland and Germany and taught botany in both countries, mainly in Amsterdam. As a pupil of SACHS at Würzburg he worked on turgor in plant cells, and used the term plasmolysis to describe shrinkage of protoplast from the plant cell wall, with loss of turgidity. He used these studies on water relations in plants both to advance knowledge of plant physiology and to confirm VAN 'T HOFF's views on osmotic pressure. In the 1880s he became interested in heredity, although he did not then know of MENDEL's work of the 1860s, and he began breeding plants in 1892. He got clear examples of the '3:1 ratio' and then came across Mendel's work in 1900, and did much to make the work widely known. His breeding experiments included some on the evening primrose, and the striking results led him to propose a bold general 'theory of mutation'; but we now use the word in a

THE DEVELOPMENT OF THE COMPUTER

Computers today are used to perform a dazzlingly wide range of functions and have become indispensable to modern life. Although most of their development in their current electronic form has happened over the past 20 years, they have their origins in the mechanical calculating machines of the 17th-c. Calculating machines are a very primitive form of computer in that they can only perform one arithmetic operation at a time, whereas computers can be programmed to perform a whole sequence of operations, using the answers from the first calculation as the input to the second and so on. This makes them infinitely more powerful than the humble calculator.

Among the first calculating machines were the 1624 'calculating clock' of Wilhelm Schickard (1592–1635), which could perform addition and subtraction, PASCAL's calculator of 1642 and that of LEIBNIZ in the 1670s. Although Leibniz's invention used a stepped gear principle which became common in future designs, all of these were essentially curiosities rather than practical machines. In 1820 Thomas de Colmar (1785–1870) made a practical calculator which partially mechanized all four basic arithmetic operations, and in 1875 another major advance was made with the invention by the American Frank Baldwin (1838–1925) of the pinwheel, a gearwheel with a variable number of teeth. These developments led in turn to perhaps the zenith of mechanical calculator technology, the 'comptometer' of Dorr Felt (1862–1930) in 1885, which was a reliable desktop calculator with the convenience of entering numbers by striking keys as on a typewriter. The comptometer became a standard office calculating machine until it was superseded by electronic devices in the 1970s.

While these were the forerunners of today's calculators, they still lacked the essential ability of the computer to perform a *sequence* of operations automatically. The first attempt at that was made by BABBAGE in 1834, who conceived, but never built, an 'analytical engine' capable of executing any series of arithmetic operations input via punched cards and to print the answer. Sadly, and despite substantial financial backing and ingenious design, Babbage never saw any of his machines completed, and many of his ideas were subsequently reinvented by the pioneers of electronic computers in the 1940s. However, Babbage's machine was to store its instructions on punched cards, and this concept was turned into reality in the 1890s by HOLLERITH, who developed the idea into a practical means of storing data that could be read by mechanical calculating machines

(for the American census, in his case). Hollerith went on to found a company to market his inventions, which subsequently grew to became IBM.

Even with data storage, mechanical calculating machines were far too slow to be of much practical value, and DE FOREST's invention of the thermionic triode in 1907 sowed the seeds for a potentially much faster type of electronic calculator. A number of transitional machines marked the passage from mechanical devices to purely electronic machines, such as those of Konrad Zuse (1910–95), who between 1938 and 1945 used mechanical parts and electromechanical relays to make several automatic programmable calculators. In 1943 Howard Aiken (1900–73) devised a giant, electrically driven mechanical calculator, the Harvard Mark 1, which helped demonstrate that large-scale automatic calculation was possible.

It took the stimulus provided by the Second World War, however, together with the development at that time of the thermionic valve as a reliable and mass-produced device (for radio and radar), to open up a new range of possibilities for electronic machines. Many scientists and engineers made simultaneous developments in the history of the computer around this time. Colossus, a British computer designed in 1943 specifically for code-breaking work, first established the practical large-scale use of thermionic valves in computers, and the American ENIAC (Electronic Numerical Integrator And Computer) built in 1945 by John Mauchly (1907–80) and John Presper Eckert (1919–95) was designed to compute ballistics tables for the US army. Also involved in the ENIAC project was the mathematician VON NEUMANN, who went on to formalize the two essential components of the modern stored-program computer – a central processing unit (CPU) and the ability to hold the results of calculations in memory and use them in subsequent operations.

After the war many of these experimental machines began to be developed into commercial computers. In Manchester the first electronic stored-program machine was run in 1948, and a collaboration with the Ferranti Company resulted in a number of computers such as Pegasus (1956), Mercury (1957) and Atlas (1962). In Cambridge, WILKES built the EDSAC computer in 1949, which was developed in 1951 via a collaboration with the J Lyons Company into the first machine designed exclusively for business use, LEO (Lyons Electronic Office). In 1946 at the National Physical Laboratory, London, TURING, a mathematician who had been involved in the wartime code-breaking work at Bletchley Park, designed ACE (Automatic Computing Engine). First run in 1950, ACE was commercialized as DEUCE by the General Electric Company in 1955. In the USA, Eckert and Mauchly founded the first electronic computer business, and in 1951 produced their first UNIVAC computer. This was used to correctly predict the results of the US presidential election the following year, a widely televised feat which did much to popularize the computer.

The next step forward came in the early 1960s with the transistor, invented by SHOCKLEY, BARDEEN and BRATTAIN in 1947, which began to be utilized to make a new generation of compact and relatively power-efficient machines. Even so, computer circuit boards were so large that their size and complexity limited overall speed and performance. In 1958 Jack Kilby (1923–) of Texas Instruments established that a number of transistors could be manufactured on the same block of semiconductor material, and the following year Robert Noyce (1927–90) of rival Fairchild Semiconductors devised a way of interconnecting and integrating such components to form an integrated circuit, or 'microchip'. The next stage was to put most of the essential components for a complete computer on a single chip, and the resulting 'microprocessor' was announced by Intel Corporation in 1971. This led to the pocket-sized calculators of the early 1970s and to the development of the desktop personal computer in 1977.

Subsequent development in computer hardware has largely been one of continued refinement and miniaturization of the microprocessor components, with doubling of speed and decreasing price becoming routine. Recent developments in computing have increasingly focused on the software that runs on the computer, rather than the hardware itself. Developments such as the graphical user interface (GUI), pioneered by Apple Computer, Inc., have made sophisticated computer systems accessible and useful to many people. In areas such as in engineering, advanced visualization techniques that use 3D colour graphics to interactively display and analyse problems have become commonplace. The development of high-capacity data-storage devices such as CD-ROM has opened up another role for the computer in publishing and education, and the current development of fast public information networks and multimedia promises yet more uses, which will combine the traditional roles of computer, television and telephone. Today the 'computer' effectively embraces a host of devices and applications based on microprocessor technology, and few are used just for computing.

DM

different sense and de Vries's mutations resulted from changes in chromosome number (to which the evening primrose is prone; its genetic make-up is complex) and not to changes in genes. However, despite being wrongly based, de Vries's ideas on a rapid alternative to Darwinian evolution led to valuable debate and experimentation, which ultimately did much to establish the Darwinian view.

von Wassermann, August *see* **Wassermann**

Waals, Johannes Diderik van der [van der vahls] (1837-1923) Dutch physicist: devised a new equation of state for gases.

Van der Waals, a carpenter's son, became a primary school teacher and a headmaster in The Hague. He trained for secondary school work in 1866, and then studied physics at Leiden. His doctoral dissertation on the physics of gases appeared in 1873. His interest was directed to the observations of T ANDREWS and others, who had shown that real gases deviate from the simple gas law $pV = RT$, deduced from kinetic theory for an 'ideal' gas whose particles have no volume and no attraction for one another. Real gases follow the law only approximately, and not at all at high pressures or low temperatures. Andrews also showed that a critical temperature exists below which a real gas can be condensed to liquid only by pressure. Van der Waals devised a modified gas equation by introducing two new constants; a is related to intermolecular attraction and b to the volume of the molecules themselves. The new equation of state has the form $(p + a/V^2)(V - b) = RT$, (again for one mole of gas) and with suitable values of a and b gives results in fairly good accord with observation for real gases over a range of temperatures above the critical point. Van der Waals was professor of physics at Amsterdam from 1877 and was awarded the Nobel Prize for physics in 1910.

Waksman, Selman (Abraham) (1888-1973) Russian-US biochemist: isolated the antibiotic streptomycin and demonstrated its effectiveness against tuberculosis.

Waksman had a difficult time as a young Jewish boy in the Ukraine, and was glad to emigrate to the USA in 1910; he worked his way through his agriculture course at Rutgers College, gained his PhD in California in biochemistry and returned to Rutgers, becoming professor of soil biology in 1930. From 1939 Waksman began a systematic search for antibiotics from soil organisms. He had rich experience of such organisms, and in 1943 he isolated the new antibiotic streptomycin from the soil organism *Streptomyces griseus* (which he had discovered in 1915). This is active against the human tubercle bacillus and, mixed with two other compounds, it became widely used in treatment. Previously there had been no effective drug for this major killing disease, but by its use tuberculosis became a problem that had largely been solved in developed countries by the 1970s. Waksman won the Nobel Prize in 1952. He and his co-workers found a number of other antibiotics in soil organisms, including neomycin, valuable in intestinal surgery.

Walcott, Charles (Doolittle) (1850-1927) US palaeontologist: discovered in the Burgess Shale of British Columbia a vast range of fossilized animals.

Walcott has a strange place in science. His great discovery, made in 1909, was largely misinterpreted by him, and it was many years before others recognized the exceptional role in evolution represented by the creatures he had collected from the Burgess Shale deposits.

Despite his lack of formal education he became his country's leading scientific administrator. From his first interest in trilobites, when he was a young farm worker, he developed his studies in geology and in 1876 became assistant to the New York State geologist. By 1894 he had risen to become director of the US Geological Survey and, in 1907, he became secretary (ie head) of the Smithsonian Institution and the most powerful figure in science in the USA. Thereafter he added other senior committee posts while retaining his expertise in geology and a special interest in aeronautics; field geology and the problems of the Cambrian rocks were his passionate relaxation.

From the time of his discovery, in the Burgess Shale, of strata rich in the fossils of novel soft-bodied animals, typically a few centimetres long, Walcott and his family spent his vacations there collecting specimens. In total he secured nearly 70 000 and stored them in the Smithsonian in Washington. Nearly 90% are animals and most of the rest algae. Most of the animals are soft-bodied and the remainder have shell-like skeletons; they contain 119 genera in 140 species, with nearly 40% of the genera being arthropods – a phylum of animals with jointed appendages, some specialized for mastication, a well-developed head and usually a hard exoskeleton. They include insects, crustacea, spiders etc.

Walcott studied his specimens with care, but he was highly conservative by nature and this led him to classify his finds within an existing taxonomy and to place these weird animals within an evolutionary sequence continuing from their Cambrian origins to the present. Revision of these ideas began in the late 1960s, with the work of H Whittington, D Briggs and S Conway Morris of the UK, whose laborious dissections gradually allowed three-dimensional reconstructions to be devised from the shale-flattened specimens. The conclusion from their work is that 'the Burgess Shale includes a disparity in anatomical design never again equalled, and not matched today by all the creatures in all the world's oceans' (S J GOULD). In terms of evolution most of these designs are 'losers' and only a few are 'winners' with descendants still existing after 530 million years. Some major problems remain: notably, how did such disparity arise and over such a geologically short time and what factors decided who should win and who would lose? The answers, or even attempts to find them, will inevitably much expand current ideas on the process of evolution.

Waldeyer-Hartz, Wilhelm [vahldiyer harts] (1836–1921) German medical scientist: gave the first modern description of cancer.

After studying science and mathematics, Waldeyer (as he was usually known) graduated in medicine at Berlin and later taught physiology and anatomy at three universities; he moved to Berlin in 1883 and soon made his institute famous. He first used haematoxylin as a histological stain; he introduced the name 'chromosome' for the rods seen in cell nuclei, which are readily stained; and he coined the name 'neuron' in neurology. His anatomical work included a description of the lymphoid tissue of the throat (the faucial and pharyngeal tonsils), known as Waldeyer's ring.

In 1863 he gave an account of the genesis and spread of cancer in essentially modern terms. He classified the types of cancer and concluded that cancer begins in a single cell and may spread to other parts of the body by cells migrating from the original site through the blood or lymphatic system (metastasis). This implied that removal of the initial cancerous cells at an early stage could effect a cure, in contrast with the view that cancer was a generalized attack on the body and that treatment was useless. This approach to oncology (the study of tumours in the animal body) became of great value when radiotherapy and later chemotherapy were available as well as surgery in the treatment of cancerous growths.

Wallace, Alfred Russel (1823–1913) British naturalist: developed the theory of evolution independently of Darwin.

Wallace left school at 14 and after a period as a surveyor became a teacher at a school in Leicester, where he met the amateur naturalist H W Bates (1825–92). The two developed a passion for collecting, especially insects and butterflies, and inspired by DARWIN's account of his travels, they set out on a collecting expedition in tropical South America. After many adventures Wallace started to return to the UK, intending to sell specimens to finance their travels, but his ship was destroyed by fire at sea, with most of his specimens and records. Undeterred, he went to Malaya in 1854 on a similar expedition, and while there wrote up his ideas on species and evolution. Like Darwin, he was convinced that plant and animal species were not fixed, but show variation over time. He concluded that 'we have progression and con-

tinued divergence' of organisms and, with no knowledge that Darwin had closely similar ideas, he decided that competition and differential survival determined the path of evolution. He sent his ideas to Darwin, whose friends arranged concurrent publication in 1858; no conflict over priority occurred, the two were on the best of terms, and Wallace became by his own wish the secondary figure and a leading advocate for 'Darwinism'.

Wallace's career continued on rather mixed lines. He became an enthusiast for spiritualism, socialism and women's rights, and he was also a founder of zoogeography: he recognized that there are some half-dozen regions, each with characteristic fauna, whose separation could be linked with the geology and geography of the regions. Wallace's line is an imaginary line dividing the oriental fauna from the Australian fauna and passing among the Malayan islands.

Wallis, Sir Barnes (Neville) (1887–1979) British engineer: innovative designer and inventor of the 'bouncing' bomb and the geodetic lattice.

Wallis was trained as a marine engineer but he spent most of his professional life at Vickers in aeronautical design, joining them in 1913. After the Second World War he led their aeronautical research and development department.

Wallis's reputation is based on diverse and brilliant inventions of great practical application. He designed a very successful airship, the R100, and the geodetic lattice (a triangular lattice of great strength, which he applied to buildings and aircraft wings), and which led to the Wellington bomber. This was the dominant British bomber of the Second World War and over 11 000 were built; later in the war, the Avro Lancaster with four engines became the main RAF bomber.

Wallis's most famous invention, however, was the 'bouncing' bomb, a spinning cylindrical device developed to enable the RAF to destroy the Möhne and Eder dams in 1943. Wallis was convinced that 'big bombs' were best for deep protected targets such as those housing the flying bombs that attacked London in 1944, and 12 000 lb (26 400 kg) Wallis bombs were used effectively against them, and to destroy the battleship *Turpitz*.

After the war he continued to work on aircraft design, developing the principles of the swing-wing aircraft, employed in the Tornado fighter. He also worked on bridge design, commercial submarines and large radio telescopes, notably the Parkes Radio telescope in Australia, completed in 1961. Characteristically, Wallis wanted it to have a 1000 ft (305 m) diameter dish; cost limits cut this to 210 ft (64 m).

Wallis, John (1616–1703) English mathematician: devised an expression for π as infinite series.

HUMAN INHERITED DISEASE AND THE HUMAN GENOME PROJECT

Some 4000 inherited human disorders are known. None are curable (although many are treatable), and they result from gene or chromosome defects. Human cells, other than gametes (the sex cells), have in their nuclei 46 chromosomes, in the form of 22 pairs of autosomes and a sex-chromosome pair (XX in females, XY in males). The chromosomes can be identified and distinguished by their microscopic appearance, after staining which produces a characteristic banding.

Carried on or within the chromosomes are the genes, the units whereby the development of a new organism by cell division is directed. Since 1950 it has been recognized that genes are composed essentially of large molecules of nucleic acid (DNA), which control the form and function of the new cells made in reproduction, and in 1953 CRICK and WATSON were able to show how the double-helical molecules of DNA are able to do this. An essential feature is that DNA molecules are partly composed of four different kinds of cyclic nitrogen-containing base (designated A, C, G and T) which recur along the helical chains. It is the sequence of the A, C, G, T units which carry the genetic information, by controlling (in 'codons') of three letters) the type of protein made by that gene, and hence the form and function of the protein.

The DNA in the 23 human chromosomes of a gamete contains roughly 100 000 genes, comprising about 3 billion base-pairs in the double-helical DNA chains. The ordinary cells of a female have two copies of every gene; males differ in having only one copy of the genes on the X chromosome.

Some inherited diseases (eg Huntington's disease) are due to a mutation defect in only one gene of the pair in each cell; these are called autosomal-dominant diseases. Autosomal-recessive diseases are less common, and only show when the patient has a double dose of affected genes, one from each parent. Examples include cystic fibrosis and sickle cell anaemia. The frequency of each inherited disorder is low, but the total is 1–2% of all live births. In addition, a genetic predisposition appears to be linked with some diseases which typically appear in later life, such as rheumatoid arthritis, some forms of schizophrenia, and Alzheimer's and Parkinson's diseases.

Inherited diseases will be much better understood when more results are available from the Human Genome Project (HGP) which effectively began in 1990. The genome is the totality of the DNA sequences in the cell nucleus: the genes comprise only about 2% of the genome. The function of the remaining 98%, which used to be called 'junk DNA', is unknown.

When in 1990 Watson wrote that 'the United States has now set as a national objective the mapping and sequencing of the human genome' he went on to note that this was similar to the 1961 decision to send a man to the Moon, although he expected the financial cost to be less. The task is clearly substantial. In 1990 the largest

Wallis had a curious career. A member of a fairly wealthy family, he studied medicine and philosophy at Cambridge, was ordained in 1640 and became a private chaplain. Then in 1649 Cromwell made him professor of geometry at Oxford; his appointment was a surprise, but his work in deciphering intercepted letters for the Parliamentarians in the Civil War was probably influential. From about this time he began to meet with BOYLE and others to discuss science, and these meetings led to the formation of the Royal Society in 1660, with Wallis as a founder-member. He also became a highly creative mathematician.

His book *Arithmetica infinitorum* (1655) made him famous; it is mainly concerned with series, theory of numbers, and conics, discusses infinities (he invented the symbol ∞) and includes the curious formula for $2/\pi = 3\times3\times5\times5\times7\times7.../2\times2\times4\times4\times6\times6\times8\times8....$ He went on to write impressive books on mechanics and on algebra; his *Treatise of Algebra* (1685) includes the first graphical representation of complex numbers $a+bi$. His job continued after the Restoration and Charles II even made him a royal chaplain

(he had always been a royalist and had joined a protest against the execution of Charles I) and he continued to decipher letters for the new government. As he wrote, he was 'willing, whatever side was upmost, to promote... the public good'.

As well as being one of the century's leading mathematicians he wrote on a variety of subjects, and had some pioneering success in teaching deaf-mutes to speak. He was remarkably quarrelsome (he maintained a public dispute with the philosopher Hobbes for over 25 years) and the contemporary biographer Aubrey claims he was a plagiarist and that he was 'extremely greedy of glorie'.

Wankel, Felix [vangkl] (1902–88) German engineer: the inventor of the Wankel rotary engine.

Two types of internal combustion engine have dominated road transport: they use either the OTTO cycle or the compression ignition system devised by DIESEL. Both have the inherent defect of requiring the linear reciprocating motion of a piston to be converted into circular motion, with resultant stress and limitations.

Wankel was born in the Black Forest, the son of a ranger. He never attended university, but fully sequenced DNA was that of a herpes virus, consisting of under 250 000 base-pairs. For larger organisms the most fully known was the bacterium *E. coli*, which has more than 800 000 base-pairs of its 4.8×10^6 base-pair genome established: the human genome is almost 1000 times larger. The HGP will require better sequencing facilities, international collaboration and methods akin to automated industrial production lines if it is to maintain momentum and meet its target within, say, 15 years. The three stages of the job will clearly be gene mapping, sequencing and data analysis. The first stage, genetic mapping, uses methods first employed by MORGAN in his classic work on fruit flies: genes which are close together on a chromosome are usually inherited together and are said to be 'linked'. By identifying linked genes and studying family histories, a map of the relative positions of genes on chromosomes can be constructed. Leading organizations in mapping are Généthon in Paris (which announced a near-complete human genome map in late 1993) and the National Institutes of Health (NIH) based in Bethesda, MD. Sequencing techniques for DNA were first due to WALTER GILBERT and to SANGER and have been increasingly automated and exploited, notably by NIH and, in the UK, at Cambridge.

Early work on HGP was directed by Watson: remarkably, his scientific career had embraced both the co-discovery of the double helix and the early years of work on the 3 billion steps of the human genome, a tribute both to his youth in the former work and his continuing high activity in molecular biology. However, from an early stage in the HGP he took the view that 'the human genome belongs to the world's people, as opposed to its nations'. This led to conflict with the US Government view, which required patent protection to be applied to the HGP's results, and Watson resigned from its direction in 1993.

Some of the implications of the HGP outcome are clear. When, with massive computer aid, the genetic messages within DNA are fully interpreted, a near-ultimate understanding of the chemical basis of human life will be available. The mysterious function of the 98% of the genome that contains no genes is likely to be resolved (present suggestions, among others, are that it exerts some sort of control over the genes; or that it provides a 'clean sheet' on which new genes can evolve; or that it is a dumping ground for abandoned genes). The HGP will certainly provide a basis for detecting the carrier state for inherited diseases in parents, and will allow the diagnosis of such disorders during pregnancy. It may allow improved treatment methods to be devised for some genetic diseases, even though there is a substantial gap between understanding the molecular basis of a disease and developing an effective strategy for its elimination or for its cure. And, intriguingly, it is surely likely that the HGP, even before its target date of completion in 2015, will yield some important results of a kind not yet foreseeable.

IM

---

he showed skill in engineering mathematics and an obsessive interest in vehicle propulsion; however his work on a novel engine was delayed by employment in aircraft development before the Second World War and later by being a prisoner of the French.

From 1929 he had in mind a novel engine using hydrocarbon fuel, and eventually made a prototype in the 1960s. The Wankel rotary engine has an approximately triangular central rotor, geared to a driving shaft and turning in a close-fitting oval-shaped chamber so that the power stroke is applied to the three faces of the rotor in turn as they pass a single spark plug. The engine is valveless. The German car maker NSU used the engine in its RO 80 luxury saloon in the 1960s, but it showed problems of high fuel consumption and exhaust pollution; Mazda used it in sports cars in the 1980s, as have high performance motor-cycle makers. Wider use of the Wankel engine is clearly possible if the above problems are fully solved; it remains the most radical innovation in its field since the familiar reciprocating internal combustion engine was developed in the 19th-c.

Warburg, Otto (Heinrich) [vah(r)boork] (1883–1970) German biochemist: had an important influence on biochemistry through applying chemical techniques.

Warburg was an enormously influential biochemist; his use of chemical methods to attack biological problems led him to ideas and techniques that were widely imitated, and his pupils dominated biochemistry for a generation. He first studied chemistry, at Berlin under E FISCHER, and then medicine at Heidelberg, qualifying in 1911. Except for the years of the First World War, when he served in the Prussian Horse Guards, his life was spent in Berlin, where he headed the Max Planck Institute for Cell Physiology until he retired at 86.

Much of his work was on intracellular respiration, and from 1923 he used the Warburg manometer (or respirometer), in which very thin tissue slices are incubated with a buffered nutrient and their uptake of oxygen is measured by the fall in pressure. With this he studied both normal cellular respiration and model systems, the action of enzyme poisons (such as cyanide) and catalytic metals such as iron, and the activity of cancerous cells. From his work and that of his students (who included MEYERHOF and KREBS) much information emerged on cell chemistry, enzyme action, co-enzymes and the function of nicotinamide adenine dinucleotide (NAD), cancerous cells, and photosynthesis in plant cells. He was an early user of spectroscopy as an invaluable aid to biochemical analysis. Awarded a Nobel Prize in 1931, his later career was marred by his increasingly intolerant attitude to ideas other than his own, which eventually isolated him.

Wassermann, August von [vaserman] (1866–1925) German immunologist: devised the Wassermann test for syphilis.

Wassermann studied medicine in Germany, graduated in 1888, was an assistant to KOCH and in 1910 became head of a new Institute for Experimental Therapy at Berlin-Dahlem. In 1906 he and his group devised a test for the presence of syphilitic infection at any past time in an individual's life; this Wassermann reaction was formerly widely used.

Watson, James (Dewey) (1928–) US molecular biologist; a co-discoverer with CRICK of the double helical structure of nucleic acids and their place in molecular genetics.

Watson's boyhood enthusiasm for bird-watching led him to entry, aged 15, to Chicago University where he graduated in zoology when only 19. He worked for his PhD at Indiana University at Bloomington, studying phages (bacterial viruses), learning much about bacterial viruses and biochemistry and becoming convinced that the chemistry of genes, then little understood, was of fundamental importance for biology. A fellowship took him to Copenhagen in 1950 to study bacterial metabolism, but soon his enthusiasm for DNA led him to Cambridge and to collaboration with Crick in the Cavendish Laboratory. Their talents and personalities were highly complementary; their joint ideas, assisted by X-ray diffraction studies by ROSALIND FRANKLIN and by M H F Wilkins (1916–), achieved a revolution in biology with publication in 1953 of the proposed double helix structure for DNA, together with a suggestion of a path for the replication of genes (the basis of heredity) and the effective beginning of the whole new science of 'molecular biology' (see Crick's entry for a brief account).

Watson, Crick and Wilkins shared the Nobel Prize for physiology or medicine in 1962. From 1955 Watson was at Harvard, from 1976 he directed the Cold Spring Harbor Laboratory of Quantitative Biology and from 1988 directed the Human Genome Research project of the National Institutes of Health, which aims to elucidate the chemistry of the 100 000 genes making up the human genome: he resigned in 1993 in opposition to the principle of patenting genetic information from the project. (See panel on p. 332.)

Watson-Watt, Sir Robert Alexander (1892–1973) British physicist and pioneer of radar.

Watson-Watt was educated at University College, Dundee, concentrating on physics. He remained there as assistant to the professor of natural philosophy, before joining the Meteorological Office in 1915. He subsequently became head of the radio department of the National Physical Laboratory at Teddington.

During the First World War, Watson-Watt worked on the radio location of thunderstorms (detecting the radio pulses produced by lightning discharges) and developed a system capable of detecting storms several hundred miles away. In 1921 he became superintendent of the radio research station at Ditton Park, near Slough, and in 1935 proposed the development of a radio detection and ranging (RADAR) system for aircraft location. Powerful pulses of radio energy at a frequency of about 30 GHz and a duration of 10^{-5} s were transmitted, and reflections from aircraft were detected and displayed with an oscilloscope. The time delay between transmission and receipt of the echo gave the distance to the aircraft and, with the direction from which the signal was received, yielded its position. Under his direction, E G Bowen and A F Wilkins quickly developed equipment capable of detecting aircraft at a range of 130 km. By the beginning of the Second World War a network of radar stations was in place along Britain's channel coasts and proved to be crucial in the country's defence. Portable radar sets were soon fitted to fighter aircraft to help them locate their targets in cloud or at night.

It has to be said that Watson-Watt did not invent radar: the basic principle of the reflection of radio waves had been known for some years and developed in at least five countries, but it was his foresight and direction, coupled with the demand created for such a system in wartime, that produced a working system. He led the successful team, and he led the team of seven who successfully claimed the money for the invention of radar after the Second World War. He was elected a Fellow of the Royal Society in 1941 and knighted in 1942. Today, radar systems are used for navigation, the safe routing of air traffic and shipping, rainfall detection, and many other non-military applications.

Watt, James (1736–1819) British instrument maker and engineer: invented the modern steam engine.

The son of a Clydeside shipbuilder, Watt had little formal education because of his poor health, but his skills enabled him to set up in business as an instrument maker in the University of Glasgow. While repairing a working model of a Newcomen steam engine, Watt realized that its efficiency could be greatly improved by adding a separate condenser, preventing the loss of energy through steam condensing to water in the cylinder. He formed a business partnership with M Boulton (1728-1809) in Birmingham to develop the idea, improved the engine in several other ways, and in 1790 produced the Watt engine, which became crucial to the success of the industrial revolution. Soon it was being used to pump water out of mines and to power machinery in flour, cotton and paper mills. Watt retired, a very rich man, in 1800. The SI unit of power, the watt, is named after him: it is the power producing energy at the rate of $1 \, J \, s^{-1}$.

Wedgwood, Thomas (1771-1805) British inventor: made first attempt to link photosensitivity of silver salts with image formation in the camera obscura, and so create photography.

Of the several people who have places in the prehistory of photography, Tom Wedgwood most clearly perceived its possibility. As the youngest son of the first Josiah Wedgwood (1730-95), the famous Staffordshire potter and pioneer industrialist, Tom grew up with a family interest in science. At some date in the 1790s, it occurred to him that two concepts, already well-known, might with advantage be brought together. These were, firstly, the sensitivity to light of silver salts and, secondly, the camera obscura, a device consisting of a box with a convex lens at one end and a screen at the other. The lens formed an inverted image on the screen and this image could be traced (or simply copied) by an artist desiring to reproduce the scene facing the lens. Thomas's father Josiah used the method often; when he secured an order from the Empress of Russia for a dinner service of over 900 pieces, each to show an English country scene, he used the camera obscura to sketch hundreds of scenes in the course of his travels.

Encouraged by PRIESTLEY and assisted by the youthful DAVY, young Wedgwood attempted to capture the camera image on a sheet of paper impregnated with silver nitrate. However, his papers were insufficiently photosensitive and/or his exposures in the camera were not long enough, and so he was unsuccessful. His 'Account of a method of copying...', published with Davy in 1802, could only describe successful contact prints obtained by pressing leaves, insect wings etc, on sensitive paper and exposing to strong light, and he found no way even of 'fixing' these 'heliotypes' or 'sun pictures', which darkened further in light. Wedgwood's health was always frail, and he died aged only 34, leaving DAGUERRE, NIÉPCE and TALBOT to succeed in converting the basic idea of photography into a practical success.

Wegener, Alfred (Lothar) [vayguhner] (1880-1930) German meteorologist and geophysicist: proposed theory of continental drift.

Educated at the universities of Heidelberg, Innsbruck and Berlin, Wegener obtained his doctorate in astronomy in 1905. Although primarily a meteorologist, Wegener is remembered for his theory of continental drift, which he proposed in 1912. Unable to reconcile palaeoclimatic evidence with the present position of the continents, he suggested that originally there had been a single 'supercontinent', which he termed Pangaea. He then provided a number of arguments to support his hypothesis that Pangaea had broken up in Mesozoic times (about 200 million years ago), and that continental drift had subsequently led to the present continental arrangement. Initially Wegener's ideas met with great hostility, largely due to the lack of any obvious driving mechanism for the movement of the continents, but the suggestion of a viable mechanism by HOLMES in 1929, together with geomagnetic and oceanographic evidence obtained during the late 1950s and early 1960s, has since established plate tectonics as one of the major tenets of modern geophysics. Wegener went on several expeditions to Greenland, and it was while crossing the ice sheet on his fourth visit that he died.

Weierstrass, Karl Wilhelm Theodor [viyershtrahs] (1815-97) German mathematician: introduced rigour into mathematical analysis.

Pressed by his overbearing father, a customs officer, to study law, Weierstrass spent 4 unsuccessful years at Bonn, learning little law but becoming a skilful fencer and reading mathematics. Emerging in disgrace, he was sent to Münster to prepare for the state teacher's examination and had the good fortune to be able to pursue mathematics under the guidance of C Gudermann (1788-1852), whose enthusiasm at that time was that power series could be used as a rigorous basis for mathematical analysis.

Weierstrass developed this approach during his stint of nearly 15 years as a teacher in the small Prussian villages of Deutsch-Krone and Braunsberg, completely isolated from contemporary mathematical research. In 1854 he published in Crelle's *Journal* a paper on Abelian integrals that he had written 14 years earlier. The quality and importance of this work, which completed areas that ABEL and JACOBI had begun, was immediately recognized, and he was appointed a professor at the Royal Polytechnic School and lecturer at the University of Berlin in 1856.

The significance of Weierstrass's work was that he gave the first rigorous definitions of the fundamental concepts of analysis; for example a function, derivative, limit, differentiability and convergence. He investigated under what conditions a power series would converge, and how to test for this. Above all, he made great contributions to function theory and Abelian functions.

Weinberg, Steven [wiynberg] (1933-) US physicist: produced a unified theory of electromagnetism and the weak nuclear interaction.

The son of a New York court stenographer, Weinberg was educated at Cornell and Princeton universities. He held appointments at Columbia, Berkeley, the Massachusetts Institute of Technology and Harvard before becoming professor of physics at the University of Texas at Austin in 1986.

In 1967 Weinberg produced a gauge theory (ie one involving changes of reference frame) that correctly predicted both electromagnetic and weak nuclear forces (such as are involved in nuclear decay), despite the two differing in strength by a factor of about 10^{10}. The theory also predicted a new interaction due to 'neutral currents', whereby a heavy chargeless particle (the Z^0 boson) is exchanged, giving rise to an attractive force between particles. This particle (short-lived when free) was duly observed at CERN in 1983 (generated by proton–antiproton collision), so giving strong support to the theory now called the electroweak or Weinberg–Salam theory. As the work was independently developed by Weinberg and SALAM, and subsequently extended by GLASHOW, all three shared the 1979 Nobel Prize for physics.

Weinberg's book on the early universe, *The First Three Minutes* (1977), has become a classic.

Weismann, August [viysman] (1834–1914) German biologist: devised a theory of heredity.

Weismann qualified in medicine and practised for a few years before the attractions of biological research drew him to university teaching in Freiburg, a town which he greatly liked. He was a skilled microscopist, but failing sight from 1864 eventually pushed him to become a theorist, with a special interest in heredity. Basing his ideas in part on his earlier work on the sex cells of hydrozoa, he proposed that all organisms contain a 'germ-plasm', associated especially with the ovum and sperm cells, which he later located in what are now called the chromosomes. In his view, it was germ-plasm that gave the continuity from parent to offspring. All other cells are merely a vehicle to convey the germ-plasm, and it alone is in a sense immortal; other cells are destined to die. As Samuel Butler the satirist phrased it, 'a hen is only an egg's way of producing another egg'. Weissman saw the major events in reproduction as the halving of the chromosome number in germ-cell (ova and sperm) formation, and in the later union of chromosomes from two individuals; he suggested that variability resulted from the combination of different chromosomes. His ideas are of course broadly correct, and it is surprising that he was able in the 1880s to get so near the modern view. He was wrong in his belief that the germ plasm is unalterable and immune to environmental effects, as others were later to demonstrate.

Early in his work Weissman believed (rather ineptly) he had demonstrated that acquired characters are not inherited: he cut off the tails of a family of mice for 22 generations, mutilating 1592 mice, but they still failed to produce tail-less offspring.

Weizsäcker, Carl Friedrich, Freiherr (Baron) von [viytseker] (1912–) German physicist: proposed theories for stellar energy generation, and for the origin of the solar system.

Weizsäcker studied and later taught physics at both Berlin and Leipzig; from 1957 he was professor of philosophy at Hamburg. Independently of BETHE he suggested in 1938 that the energy of stars is generated by a catalytic cycle of nuclear fusion reactions, whereby hydrogen atoms are converted into helium with much evolution of energy. More specifically, this reaction (the proton fusion reaction) has as its net result the conversion of four hydrogen nuclei (ie protons) into a helium nucleus. The reaction requires a high temperature ($\approx 10^9$ K) and yields also a massive amount of thermal energy, along with gamma radiation. The energy is sufficient to maintain a star's energy output (such as the Sun's) for billions of years. It is widely accepted as the key process in stellar energy generation.

Weizsäcker proposed a scheme in 1944 for the origin of the solar system; this scheme developed the older ideas of LAPLACE that the Sun had been surrounded by a disc of gas, which rotated, became turbulent and aggregated to form the planets. Weizsäcker's theory (like Laplace's) failed to account for the angular momentum of the solar system, but it was developed by ALFVÉN and then by HOYLE, who proposed that the Sun's magnetic field could generate the required momentum.

During the Second World War Weizsäcker worked with HEISENBERG to develop nuclear energy from uranium for power or weaponry, but with trivial success in comparison with the Allies.

Werner, Alfred [verner] (1866–1919) German–Swiss inorganic chemist: founded the modern theory of co-ordination compounds.

Werner was born in Alsace; it was French when he was born, became German when he was 4, and French again in 1919. Werner had allegiances to both French and German culture; he usually wrote in German. He lived in Switzerland from the age of 20, graduating at Zürich, and held a professorship there from 1895 until his death.

From 1892 he worked on the inorganic complexes of metals. This large class of chemical compounds had seemed confused; the sort of structure theory that had served well in organic chemistry did not appear to apply, and neither did ordinary valence rules. Werner brought a new view to them. He proposed that the central atom (usually a transition metal atom) had its normal valence, and also secondary valences that bonded it to other atoms, groups or molecules (collectively, 'ligands') arranged in space around it. This theory of co-ordination complexes allowed two to nine ligands to be co-ordinated to the central atom; the commonest co-ordination number is six, with the ligands arranged octahedrally. During 20 years, Werner worked out the consequences of this theory extensively, and rejuvenated inorganic chemistry as a result. Metal complexes are of great importance also in plant and animal biochemistry. He was awarded the Nobel Prize for chemistry in 1913.

Werner, Abraham Gottlob [verner] (1749–1817) German mineralogist and geologist: proposed Neptunist theory of geology.

Werner came from a well-off family operating ironworks in a traditional mining area, and he was educated at the Freiberg Mining Academy and the University of Leipzig, studying law and languages and returning to Freiberg in 1775 as a lecturer in mining. He became the foremost geologist of his time, now remembered for his Neptunist theory of the origin of the Earth, which was widely accepted for much of the 18th-c. He proposed that all rocks were precipitated as sediments or chemical precipitates in a universal ocean created by the biblical Flood, and that all geological strata thus followed a universal and specific sequence. The lowest layer contained 'primitive' rocks such as granites and slates, the next higher layer included shales and fossilized fish, then followed limestones, sandstones and chalks, and finally alluvial clays and gravels. Although such a scheme fitted moderately well with the geology around Freiberg, increased knowledge of the geology of other parts of Europe revealed the flaws in his ideas, and it was largely modified by HUTTON's uniformitarian theory. However, much of his work was of lasting value, his ideas and his many students having a great influence in shaping modern geology.

Weyl, Hermann [viyl] (1885–1955) German mathematician: contributed to symmetry theory, topological spaces and Riemannian geometry.

Weyl was a student under HILBERT at Göttingen and, on becoming a *Privatdozent* there, also worked with him. In 1913 he declined a professorship at Göttingen and moved to Zürich, where he worked with EINSTEIN. He returned to take up the professorship when Hilbert retired in 1930, but increasing Nazi power led him to move to Princeton with GÖDEL and Einstein, retiring in 1951. As well as his outstanding mathematical work, Weyl published on philosophy, logic and the history of mathematics.

Weyl acquired from Hilbert research interests in group theory and Hilbert space and operators. Once developed, these techniques proved central to the rapidly evolving theory of quantum mechanics and the unification of matrix mechanics and wave mechanics. Weyl showed how symmetry relates to group theory and continuous groups, and how this can be a powerful tool in solving quantum mechanical problems.

When Weyl moved to Zürich, Einstein interested him in the mathematics of relativity and Riemannian geometry. In seeking to generalize this, Weyl developed the geometry of affinely connected spaces and differential geometry. Weyl anticipated the non-conservation of parity in particles, a feature that has since been observed by particle physicists working with leptons.

Weyl produced a small number of highly influential papers on number theory, proving results on the equidistribution of sequences of real numbers modulo 1. This was taken up in later work by HARDY and J E Littlewood (1885–1977).

Wheatstone, Sir Charles (1802–75) British physicist: a contributor to cable telegraphy.

Wheatstone was privately educated and started work in the family tradition as a maker of woodwind and other musical instruments. In 1834 he was appointed professor of experimental physics at King's College, London. His science was self-taught.

Much of Wheatstone's early work was (understandably) concerned with acoustics and the theory of resonance of columns of air, and led to his London appointment. This gave him a wider interest in physics, particularly optics and electricity. He was a prolific inventor, patenting the concertina and other musical devices, and in 1838 invented a stereoscope in which two pictures of slightly differing angles of perspective could be combined to give an impression of three-dimensional solidity. In 1837 he collaborated with W F Cooke (1806–79) on a commercial electric telegraph project, which was a great success, with thousands of miles of telegraph lines being constructed. The telegraph used a DANIELL cell to provide current and a STURGEON electromagnet in the recorder. Wheatstone was responsible for several related inventions, such as the printing telegraph and the single-needle telegraph. He popularized (but did not invent) the Wheatstone bridge, a device invented by S Christie (1784–1865) and utilizing OHM's law for comparing resistance.

Whewell, William [hyooel] (1794–1866) British polymath, now best known for his survey of the scientific method and for creating scientific words.

Whewell was the son of a Lancastrian carpenter; he gained a scholarship to Trinity College, Cambridge and showed his breadth of talent by winning prizes for poetry and for mathematics. He remained there and from 1820–40 taught and wrote on mechanics, geology, astronomy, theology, ethics and architecture. He was also active in the work of the Royal Society, the British Association for the Advancement of

Science and the Geological Society. He was successively professor of mineralogy and of moral philosophy, Master of his college from 1841 and vice-chancellor of the university in 1842 and 1855.

The liking many felt for him was not universal; he was both self-conscious and forceful. As Master, he did not allow Fellows to have keys to their college, or dogs or cigars within it, or to marry (he was twice married himself). Undergraduates could not sit in his presence and he required nude paintings to be removed from view in the Fitzwilliam Museum. A Royal Commission's proposal that all Fellows be allowed to vote at college meetings infuriated him.

More positively, he inspired many able young men and his texts and his teaching in applied mathematics gave the necessary basis for Cambridge's later successes in physics; and he pressed his view that every well-educated man should know something of engineering theory. In mineralogy, he founded mathematical crystallography (on the basis of HAÜY's theory of crystal structure) and developed MOHS'S classification of minerals. He became the authority on names for new scientific concepts, creating the now-familiar 'scientist' and 'physicist' by analogy with 'artist'. They soon replaced the older term 'natural philosopher'. Other useful words were coined to help his friends: biometry for LUBBOCK; Eocine, Miocene and Pliocene for LYELL; and for FARADAY anode, cathode, dia- and para-magnetic, and ion (whence the sundry other particle names ending in -on).

In meteorology Whewell devised a self-recording anemometer. He was second only to NEWTON in his work on tides and tidal theory, including organizing and collating tidal observations worldwide and winning a Royal Medal.

His *History and Philosophy of the Inductive Sciences* (1837–60) examined the nature of scientific discovery, which he saw as requiring imaginative guesses which were capable of disproof or verification. Now a classic, it still has authority in its survey of scientific ideas from the Greeks to the 19th-c.

Rather unusually for a scientist he died as a result of being thrown from his horse.

Whipple, Fred Lawrence (1906–   ) US astronomer: proposed 'dirty snowball' model for comets.

Whipple had a distinguished career in astronomy in California and Harvard. In 1950 he proposed that cometary nuclei consist of a mixture of water ice and dust, frozen carbon dioxide, methane and ammonia. This model, known as the 'dirty snowball' theory, accounts for the fact that comets only develop their characteristic tails as they approach the Sun, when the solar wind vaporizes the volatile components in the nucleus. Radiation pressure is then responsible for the fact that the tail always points away from the Sun. Another feature of comets, their

slight variability of orbital period, was also explained by the formation of an evaporated surface crust, through which jets of volatile material are sometimes ejected. Whipple's ideas were largely confirmed by observations made from space probes during the visit of HALLEY's comet in 1986.

Whipple, George Hoyt (1878–1976) US medical scientist.

From his medical student days at Johns Hopkins, Whipple was particularly interested in the oxygen-carrying pigment of red blood cells (haemoglobin) and in the bile pigments that are formed in the body from it. Working in the University of California from 1914–22, he examined the effect of diet on haemoglobin formation. To do this he bled dogs until their haemoglobin level was reduced to a third of normal, and then studied the rate of red cell regeneration when the dogs were fed various diets; he found that meat, kidney and especially liver were effective in stimulating recovery. Since the fatal human disease of pernicious anaemia is associated with red cell deficiency, it was reasonable to attempt to treat it similarly, and G Minot (1885–1950) and W Murphy (1892–1987) found in 1926 that large additions of near-raw liver in the patient's diet were effective: Minot, Murphy and Whipple shared a Nobel Prize for physiology or medicine in 1934. It was another 20 years before other workers isolated the active curative compound, vitamin B_{12}, and made it available for treatment and study.

Whipple spent the rest of his career at the University of Rochester, continuing to work on blood and especially on thalassaemia, a genetic anaemia due to a defect in the haemoglobin molecule found especially in Mediterranean races.

White, Gilbert (1720–93) British naturalist: author of the first English classic on natural history.

White's enthusiasm for all kinds of natural history is remarkable. He followed a family tradition by becoming a curate and living in the family home, 'The Wakes' at Selborne in Hampshire. He declined more senior posts in order to stay there so that he could study nature in his large garden and the nearby countryside. His accounts of this were shared with friends in his letters to them; shortly before his death the diffident White was at last persuaded to edit 110 of his letters to form *The Natural History and Antiquities of Selborne*. The book so pleased its many readers that it has been in print ever since. White's keenest interest was in birds, whose song and habits he studied; other ornithologists at that time interested themselves only in plumage and anatomy. He studied mammals, bats, reptiles (especially his pet tortoise, Timothy), insects, plants and the weather. His observations gave some evidence for

DARWIN'S theory of evolution, but its main value has been to provide pleasure and inspiration to generations of naturalists. As one zoologist wrote in 1901: 'White is interesting because nature is interesting; his descriptions are founded upon natural fact, exactly observed and sagaciously interpreted'.

Whittle, Sir Frank (1907–) British aeronautical engineer: invented the jet engine.

After entering the RAF as a boy apprentice, Whittle qualified as a pilot at Cranwell College and studied engineering at Cambridge. He served as a test pilot with the Royal Air Force, later working as a consultant for a number of companies. In 1977 he became research professor at the US Naval Academy, Annapolis. He joined the Order of Merit in 1986.

Whittle's principal claim to fame was the invention of the turbojet aircraft engine, on which he took out his first patent in 1930 while still a student. In 1936 he formed his own company to develop the concept, and in 1941 a Gloster aircraft with his engine made its first test flight. Due to the war development was rapid, and the Gloster was in service with the RAF by 1944.

In Germany, Hans Joachim Pabst von Ohain (1911–) completed his designs for a jet engine in 1933 and, better supported by his employers and by industry (notably the aircraft makers Heinkel) than was Whittle, von Ohain had jet-driven aircraft in service with the Luftwaffe in mid-1944, slightly ahead of Whittle. Jet engines made supersonic flight practicable, initially in fighters and then for civil air travel.

Wieland, Heinrich Otto [veelant] (1877–1957) German organic chemist: carried out important work on the structure of cholesterol and other steroids.

The son of a gold refinery chemist, Wieland studied and taught in several German universities before succeeding WILLSTÄTTER at Munich in 1925. His early work was on organic compounds of nitrogen, including the fulminates; and in 1911 he made the first nitrogen free radicals. He also worked on natural products; plant alkaloids, butterfly-wing pigments (pterins) and especially the steroids. In steroid chemistry, he showed that three bile acids can all be converted into cholanic acid, which he also made from cholesterol. It therefore followed that the bile acids and cholesterol had the same carbon skeleton, and Wieland proposed a structure for this parent steroid skeleton. His first structure was shown to be incorrect, but a revised version which he and others produced in 1932 is correct. For his steroid work he was awarded the 1927 Nobel Prize. His other work included studies on toad venom, on curare and on biological oxidation (which he showed is often, in fact, dehydrogenation).

Wien, Wilhelm [veen] (1864–1928) German physicist: discovered the energy distribution formula for black body radiation.

Wien grew up in a farming family and originally planned to spend his life farming. He studied briefly at Göttingen and continued his degree work at Berlin from 1884. In 1886 Wien received his doctorate for research on light diffraction and associated absorption effects, and returned to manage his parents' farm. A severe drought 4 years later forced the sale of the farm, and he became assistant to HELMHOLTZ in Berlin. In 1900 he took up the professorship at Würzburg and after 20 years he was appointed as RÖNTGEN'S successor at Munich.

In 1892 Wien began research on thermal (or black body) radiation (see BOLTZMANN), a study which initiated the transition from classical physics to PLANCK'S quantum theory. Wien showed that the wavelength λ, at which a black body radiation source at absolute temperature T emits maximum energy, obeys a law: λT = constant = 0.29 cm K (the constant was measured by LUMMER and E PRINGSHEIM). This is known as Wien's displacement law; in accord with it a red-hot black body on further heating emits shorter-wavelength radiation and becomes white hot as the wavelength of maximum radiation shifts from the long wavelength (red) end to the centre of the visible spectrum.

Developing this, in 1896 he produced Wien's formula describing the distribution of energy in a radiation spectrum as a function of wavelength and temperature. It was based on an assumption that a hot body consists of a large number of oscillators emitting radiation of all possible frequencies and all in thermal equilibrium. Interestingly, Wien's formula is well obeyed at short wavelengths but is clearly wrong for longer values while RAYLEIGH produced a formula accurate at longer wavelengths but not at lower ones.

Planck gave much thought to these discrepancies and showed that, if one assumed that radiation could be emitted only in 'packets' of a minimum energy (which he called quanta), then a radiation law could be calculated that was obeyed accurately at all wavelengths. Planck published his quantum theory in 1900, aware that its assumptions had no justification in classical physics and yet, as they appeared correct, a major revolution in physical science was inevitable. Wien was awarded the Nobel Prize for physics in 1911.

Wiener, Norbert [weener] (1894–1964) US mathematician: established the subject of cybernetics.

As a child Wiener showed his mathematical talent early, but his career then became erratic. At 15 he entered Harvard to study zoology; he changed to philosophy at Cornell and got a PhD from Harvard in mathematics at 19. He then studied logic briefly under B Russell (1872–1970) and HILBERT. Suffering perhaps from too rapid an education, Wiener drifted through various activities, including journalism and writing encyclopedia entries, before obtaining a post in mathematics at Massachusetts Institute of Technology in 1919. He held this until retirement.

Wiener began research on stochastic, or random processes such as Brownian motion, including work on statistical mechanics and ergodic theory (which is concerned with the onset of chaos in a system). Other areas that he advanced were integral equations, a kind now known as Wiener integrals, quantum theory and potential theory. As part of his war work, Wiener applied statistical methods to control (eg of anti-aircraft guns) and communication engineering. Extending this broadly, for example into neurophysiology, computer design and biochemical regulation, led to his founding cybernetics as a subject. Cybernetics is the study of control and communication in complex electronic systems and in animals, especially humans.

His standing as a mathematician is hardly disputed, but his writings are hard to read and uneven in quality. As a person he was extraordinary; small, plump, myopic, playful and self-praising, he spoke many languages and was hard to understand in any of them. He was a famously bad lecturer, perhaps because his mind worked in a very unusual way.

Wigner, Eugene Paul [wigner] (1902–95) Hungarian–US physicist: applied group theory to quantum mechanics and discovered parity conservation in nuclear reactions.

Wigner was the son of a businessman and took his doctorate in engineering at the Berlin Institute of Technology in 1925. He moved to Princeton in 1930, became professor of theoretical physics there in 1938 and held this post until his retirement in 1971. He was a brother-in-law of DIRAC.

He made major contributions in quantum theory and nuclear physics, in particular by showing the value of symmetry concepts and the methods of group theory applied to physics. In 1927 he concluded that parity is conserved in a nuclear reaction: the laws of physics should not distinguish between right and left; or between positive and negative time. As a consequence a nuclear reaction between particles and the mirror image of those particles will be identical, and this was accepted to apply to all types of reaction. However, very surprisingly, LEE and TING identified a class of exceptions to this law of parity conservation in 1958. Reactions involving the weak nuclear force, such as beta decay when an electron is emitted from a nucleus, do not conserve parity.

Wigner's research during the 1930s mainly concerned neutrons and he investigated the strong nuclear interaction which binds neutrons and protons in the nucleus. He showed that the force has a very short range and does not involve electrical charge. The formula describing how moving neutrons interact with a stationary nucleus was given by BREIT and Wigner in 1936. Using this and other discoveries, Wigner assisted FERMI in constructing the first nuclear reactor to produce a sustained nuclear chain reaction in Chicago in 1942. For his contributions to quantum theory and applying it to nuclear physics Wigner shared the 1963 Nobel Prize for physics.

Wilkes, Maurice Vincent (1913–) British mathematician and computer scientist: designed the first delay storage computer.

Wilkes was educated at Cambridge, subsequently taking positions there as lecturer and director of the Mathematical Laboratory, and head of the Computer Laboratory.

After wartime work on radar and operational research, Wilkes worked on the early development of computers, leading the team which built EDSAC (Electronic Delay Storage Automatic Calculator), the first machine to use delay lines to store information. The delay lines were mercury-filled tubes (1.5 m long) with piezoelectric crystals at either end; incoming signals generated a pressure pulse which was transmitted through the mercury to the second crystal, where it was converted back into an electrical impulse. Several such devices and suitable amplification allowed an electrical signal to be stored indefinitely, an essential requirement of a computer 'memory'. EDSAC ran its first program in 1949 and was a milestone in the development of computers. EDSAC II, in service from 1957, included among its improvements the use of magnetic storage in place of delay lines. Wilkes continued to play a leading role in computer development.

Wilkinson, Sir Geoffrey (1921–) British inorganic chemist: carried out important work on the structure of metallocene compounds and transition complexes.

Wilkinson, born in Todmorden, Yorkshire, studied at Imperial College London. After 13 years in Canada and the USA he returned to London in 1956 as professor of inorganic chemistry. While at Harvard University in 1952 he published with WOODWARD and others a paper on the remarkable compound $(C_5H_5)_2Fe$, ferrocene. They showed that this has a structure with an iron atom sandwiched between two flat five-carbon rings. Thousands of 'sandwich type' molecules have now been made, containing other metals and other-sized rings (the 'metallocenes'); even three-decker sandwiches are known. For his work on metallocenes Wilkinson shared the 1973 Nobel Prize with E O Fischer (1918–) of Munich, who had worked independently on similar lines. Wilkinson also worked on transition metal complexes, and discovered the first homogeneous system for catalytic hydrogenation of C=C bonds by use of the rhodium complex $RhCl[P(C_6H_5)_3]_3$.

Williams, Cicely (1893–1992) British paediatrician: identified the condition known as kwashiorkor.

Cicely Williams was born in Jamaica, of a land-owning family that had been there since the 17th-c. She was educated in England at the Bath High School for Girls and at Somerville College, Oxford, where she read medicine. She was one of the first 50 female undergraduates to have their degrees conferred in the Sheldonian Theatre in 1920. She trained at Queen's College Hospital for Children, Hackney and at King's College, Camberwell, but found it difficult to get a medical post, partially because of the priority given to returning ex-servicemen. After a year as a medical officer in Greece she joined the Colonial Health Service and, after a wait of 2 years, was appointed to the Gold Coast (now Ghana). Here she found a condition that results from dietary protein deficiency due to a high intake of carbohydrate of low nutritional value. This nutritional deficiency disease causes the abdomen to swell, the hair to turn red, the liver to enlarge and life-long ill-effects in children under 2. This disease has ravaged children in drought and war-torn areas of maize-eating Third World countries.

Cicely Williams first described the condition in the 1931–2 volume of the annual medical report of the Gold Coast. The reaction of medical editors in London was to reject her paper; as she said, 'They could not concede that a woman in the Gold Coast of all places had anything to say which concerned them'. However the paper was then published in *The Lancet* and here the condition was named and described for the first time in medical terms. She used a word, *kwashiorkor* (from the local language, Ga), which means 'neglect of the deposed', to describe the condition.

After 7 years in Africa Williams was transferred to Malaya and was in a remote province when in 1941 Pearl Harbour was attacked. After weeks of hardship and danger she reached Singapore just as the Japanese invaded. Imprisoned in the notorious Changi jail she became its chief doctor and was proud of the fact that the 20 babies born there all survived. In 1943 she was taken to the headquarters of the Kempe Tai, the equivalent of the Gestapo, and interrogated as a spy, spending 4 months in a cage in which she could only crouch, along with the dead and the dying.

After the war she was sent to America for recuperation and postgraduate study at Johns Hopkins University. In 1948 she was appointed the first head of the Mother and Child Care unit at the World Health Organization in Geneva. From 1959 she became visiting professor of maternal and child health at the American University of Beirut and from 1964 overseas training adviser to the Family Planning Association. Cicely Williams was the first woman to be given an honorary fellowship of the Royal Society of Medicine (1977). In 1985, aged 92, she became a Fellow of Green College, Oxford.

She was a pioneer of women's place in the medical profession and her ideas and methods of treatment in paediatrics are now followed internationally.

Williamson, Alexander William (1824–1904) British chemist: demonstrated the chemical relationship between alcohols and ethers.

Born in London, Williamson lived with his parents on the Continent, studied chemistry in France and Germany and became head of chemistry at University College, London. He is remembered for the Williamson synthesis of ethers, in which a sodium alkoxide reacts with an alkyl halide, ie $RONa + R^1I \rightarrow ROR^1$ where R, R^1 are alkyl groups.

By use of this reaction (1851) the relation between alcohols and ethers became clear. As he lacked one eye and had only one usable arm, Williamson's prowess as a practical chemist is surprising.

Willis, Thomas (1621–75) English anatomist: made important studies of anatomy of the brain.

Willis studied classics and then medicine at Oxford; for a time he served in the Royalist army in the Civil War. He was one of the small group of 'natural philosophers' (including Boyle) who met in Oxford in 1648–9 and who were founder members of the Royal Society of London. His main work was on the anatomy of the brain; the softness of brain tissue makes study of its circulation difficult, but Willis improved on earlier work by injecting the vessels with wax; he thus saw the ring of vessels now known as 'the circle of Willis'. He also worked on fevers and he described a type of diabetes in which the excessive urine has a sweet taste. In 1776 this was found to be due to sugar, and in the 1920s this disease (diabetes mellitus) was brilliantly explored and effectively treated by Banting and C H Best (1899–1978). Willis was also the first to propose that the essential feature of asthma is spasm of the bronchial muscles.

Willstätter, Richard [vilshteter] (1872–1942) German organic chemist: discovered the structure of chlorophyll.

Willstätter was 11 when his father left Germany for New York to establish a clothing factory, following the successful example of his brothers-in-law. An expected short separation lengthened to 17 years as success came to him slowly; and it was his wife and her family who brought up his two sons. Richard's interest in chemistry was prompted by visits to his uncle's factory for the production of carbon for batteries.

He graduated at the University of Munich, studied under Baeyer and gained his PhD in 1894 for work on alkaloids. He obtained a professorship at the University of Zürich (1905–12) and worked on plant pigments, quinones and the chemistry of chlorophyll. Using the chromatographic technique developed by Tswett he worked out the structure of both the a and b form of chlorophyll. He showed that chlorophyll contains a single atom of magnesium in its molecule, rather as haemoglobin contains a single iron atom. His work on cocaine derivatives, begun in Munich, led to the synthesis of new medicinals and the chemical curiosity cyclo-octatetraene. He was awarded the Nobel Prize for chemistry in 1915 for his work on plant pigments.

He returned in 1912 to the Kaiser Wilhelm Institute at Berlin-Dahlem where he worked on the carotenes and anthocyanins. During the First World War Willstätter worked on gas masks and devised a filling of hexamethylenetetramine to absorb phosgene; layered with active carbon it was effective against the gases of the time. In 1916 he succeeded Baeyer to the chair in Munich and worked on photosynthesis, with A Stoll (1887–?), and on enzymes, notably catalase and peroxidase.

In 1925 he resigned his professorship in Munich in protest against increasing anti-semitism, in particular his faculty's rejection of the appointment of V M Goldschmidt, the geochemist, because he was Jewish. Willstätter had resigned his post at 53 without a pension, and losing the house, the status and the protection that went with it.

He was now alone – his wife had died many years before after only 5 years of marriage, shortly to be followed by the death of a small son. His daughter had married and was living in the USA. Despite many offers of posts in other countries he wished to stay in his own country. The next few years were spent in travelling, lecturing and continuing what research he could, with the help of Margarete Rohdewald who was allowed space in the laboratory and who reported her findings by telephone.

It was she who in November 1938 heard that members of the National Socialist party had been requested to volunteer for the arrest of Jews and warned Willstätter. He was able to avoid an immediate journey to Dachau, but he then knew he had to leave Germany to survive. He was determine to emigrate to Switzerland in a proper manner, but his patience and dignity were to be tested in the following months while he was gradually stripped of his possessions in return for his passport. With help from his former student Stoll and influential friends in Switzerland he crossed the border in March 1939.

Wilson, Charles Thomson Rees (1869–1959) British physicist: the inventor of the Wilson cloud chamber.

Wilson left a Scottish sheep farm as a child of 4, and was educated in Manchester, eventually as a biology student there. Then he went to Cambridge, did well in physics and became a teacher in Bradford for 4 years before returning to Cambridge in 1896 and staying there for a long career.

In 1894 he had been attracted by the brilliant cloud effects he observed from the summit of Ben Nevis, and in Cambridge he examined methods of producing artificial clouds in the laboratory by the sudden expansion of moist air. The expansion drops the temperature of the gas and the water vapour partly condenses as droplets on the walls and on any available nuclei. Wilson showed that if filterable dust is absent, then charged ions (produced eg by X-rays) will serve as nuclei. By 1911 he had devised his cloud chamber, in which the path of an ion is made visible as a track of water droplets. It was soon used to detect and examine the alpha- and beta-particles from radioelements, and it quickly became a favourite device for particle physicists, especially in the 1920s and 1930s. It was also the ancestor of the bubble chamber devised by Glaser in the 1950s.

Another of Wilson's researches also began on Ben Nevis, as a result of his own electrification (his hair stood on end) during a storm in 1895. He studied electrical effects in dry and moist air and he noted that a sensitive, well-insulated electrometer shows slow leakage, by day or night, even underground; he concluded that radiation from sources outside the atmosphere might be the cause. In 1911 V F Hess studied this further, and the discovery of cosmic rays proved that Wilson was right. His interest in atmospheric electricity remained; in his long retirement in Scotland, he flew at age 86 over the Outer Isles to observe thunderstorms, and presented his last paper on this subject, aged 87. He shared the Nobel Prize in 1927 for his cloud chamber.

Wilson, Edward Osborne (1929–) US biologist: creator of sociobiology.

Educated at Alabama and Harvard, Wilson taught at Harvard from 1956. He is best known for his remarkable work on social insects and its wider implications in animal behaviour and evolution. In developing his theory on the interaction and equilibrium of isolated animal populations, he and D S Simberloff (1942–) experimented on some small islands in the Florida Keys. They first surveyed the insect species present (75 of them) and then eliminated all insect life by fumigation. Study of the recolonization of the islands by insects over some months showed that the same number of species became re-established, confirming their prediction that 'a dynamic equilibrium number of species exists for any island'. Wilson went on to consider biological and genetic controls over social behaviour and organization in a variety of species in his book *Sociobiology: the New Synthesis* (1975), which virtually created a new

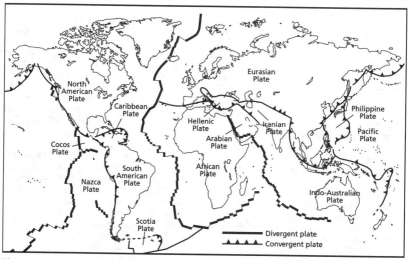

Plate tectonics – major lithospheric plates

subject, integrating ideas on the behaviour of a range of species from termites to man. The work has both stimulated valuable research and provoked vigorous discussion through its extension of ideas on animal behaviour to include human cultural and ethical conduct.

Wilson, John Tuzo (1908–93) Canadian geophysicist: proposed the concept of the transform fault in plate tectonics, and the 'hot spot' theory for the creation of mid-ocean islands.

Wilson worked for the Canadian Geological Survey before being appointed professor of geophysics at Toronto in 1946, a post he held until his retirement in 1974. Although initially a staunch opponent of continental drift, Wilson is now known for his notable contributions to plate tectonics. In 1963 he provided some of the earliest support for the sea-floor spreading hypothesis of H H HESS by pointing out that the age of islands on either side of mid-ocean ridges increases with their distance from the ridge. He subsequently suggested that there exist 'hot spots' in the mantle where plumes of mantle material rise due to convection currents, and that as the lithospheric plates pass over them volcanic islands are formed. His other important contribution has been the concept of the transform fault, introduced in 1965, which occurs where continental plates slide past one another rather than one sinking beneath the other in a subduction zone. Mid-ocean ridges often consist of a series of offsets connected by transform faults.

Wilson, Kenneth Geddes (1936–) US theoretical physicist: discovered the renormalization group technique for treating phase transitions.

While one phase or another of a physical sys-

tem may be easily analysed theoretically, similar analysis of the transition between phases had proved virtually impossible. This is because the length scale on which physical interactions are taking place changes rapidly through many orders of magnitude. In 1974 Wilson developed the first technique able to cope with such transitions, which are called critical phenomena. An example is the onset of ferromagnetism in a magnet cooled below the CURIE point, when the atoms interact with each other and become aligned over large volumes of the magnet. The distance over which ordering of atomic spins occur goes from an atomic diameter to many thousands of diameters under a very small change in temperature.

L Kadanoff (1937–) had suggested that the effective spin of a block of atoms should be found and then a renormalization (or scaling) transformation made to calculate that of a larger block made up of the small blocks. Wilson developed this method and showed how to calculate the properties of large numbers of atoms strongly interacting with each other, as in magnetic systems, metal alloys or liquid-to-gas transitions. He was awarded the Nobel Prize for physics in 1982.

Wilson, Robert Woodrow (1936–) US physicist: co-discoverer of the cosmic microwave background radiation.

Wilson graduated at Rice University, Houston and California Institute of Technology. Thereafter he took up a post at the Bell Laboratories, Holmdel, NJ and became head of the radiophysics research department in 1976.

At Bell, Wilson collaborated with PENZIAS in experiments using a large radio telescope designed for communication with satellites. In

1964 they detected a radio noise background coming from all directions; it had an energy distribution corresponding to a black body at a thermal temperature of 3.5 K. The explanation given by DICKE and P J E Peebles (1935–) was that the radiation is the residual radiation from the 'Big Bang' at the universe's creation, which has been cooled to 3.5 K by the expansion of the universe. This cosmic background radiation had been predicted to exist by GAMOW, ALPHER and R C Herman (1914–) in 1948. Wilson and Penzias together won half the 1978 Nobel Prize for physics for their work.

Windaus, Adolf (1876–1959) German organic chemist: a major contributor to steroid chemistry.

Windaus first studied medicine, but was attracted into organic chemistry by attending lectures by EMIL FISCHER. He became professor at Göttingen in 1915, and remained there. From 1901, when he was 25, he worked (like Fischer) with natural products and became the dominant figure in unravelling the intricate chemistry of the steroid group. His successes here included, especially, his work on the vitamin D group. He also worked on vitamin B_1 and discovered the biogenic amine histamine, a key compound in allergy. He won the Nobel Prize for chemistry in 1928.

Withering, William (1741–99) British physician: made classic study of the medicinal use of digitalis.

A graduate of Edinburgh, Withering practised in Stafford and then moved to Birmingham at the suggestion of Erasmus Darwin of Lichfield (grandfather of CHARLES DARWIN). He was a member of the Lunar Society, a group of Midland scientists including PRIESTLEY, J Wedgwood (1730–95) and M Boulton (1728–1809), who met monthly at the full moon (to assist their homegoing) and he was a keen botanist. Finding that an extract of herbs had long been used to treat 'dropsy' (oedema), Withering made a careful study of the matter and found that the active herb was the foxglove and that some cases of oedema could indeed be treated effectively with foxglove leaf extract. He gave an excellent report of this work in his classic *An Account of the Foxglove...* (1785); it is modern in style, with good case histories and includes failures as well as successes. It was later found that the extract contains digitalis, which steadies and strengthens heart action and which is still used for this.

He was also a mineralogist, and witherite (barium carbonate, $BaCO_3$) is named after him. He suffered greatly from chest disease (probably tuberculosis) and lived for years in a controlled atmosphere. His decline inspired the epigram 'the flower of Physick is Withering'.

Wöhler, Friedrich [voeler] (1800–82) German chemist: achieved a synthesis of urea and first made many novel inorganic and organic compounds.

Young Wöhler was not very successful as a schoolboy; his passion for chemistry distracted him from all else. He graduated in medicine and at once moved to chemistry by joining BERZELIUS for a year. On his return to Germany he began teaching chemistry, which was to fill his life; he was professor at Göttingen from 1836 until his death. Wöhler discovered the cyanates, and in 1828 he showed that ammonium cyanate when heated gave urea: $NH_4CNO \rightarrow CO(NH_2)_2$. Urea is a typical animal product, so that this reaction could be interpreted as marking the end of the idea of a 'vital force' essential for the chemistry of life. In fact several odd features confuse this. Wöhler's cyanate was made by a process which was not wholly inorganic. Also, J Davy (1790–1868) had made urea in 1812 from NH_3 and $COCl_2$ but had not realized what he had made from these truly inorganic reactants. Syntheses by KOLBE and by BERTHELOT in the 1840s and 1850s marked the real logical end of vitalism; but Wöhler's work in 1828 ended it in the minds of many chemists.

In 1832 Wöhler's young wife died and, to distract him, LIEBIG invited him to Giessen for some joint work. This was a study of 'oil of bitter almonds', probably suggested by Wöhler; and from the oil (benzaldehyde) they made the related acid, chloride, cyanide and amide. Structure theory was yet to come; but the two recognized that a group of atoms (the benzoyl group, C_6H_5CO) was present in all these compounds. This was the first substantial 'compound radical' to be recognized, and this recognition was the beginning of the end of a period of confusion in organic chemistry.

Wöhler first made aluminium and beryllium, crystalline boron and silicon, and calcium carbide, and he saw (in 1863) the analogy between compounds of carbon and those of silicon.

In his long and valuable friendship with Liebig, Wöhler displayed none of the enthusiasm Liebig had for controversy. Wöhler had a lighter view; he was the writer of a skit on DUMAS's substitution theory which was published in the *Annalen* under the name S Windler. Later he remarked that he should have given a French name such as Ch Arlaton. He enjoyed writing and teaching even more than research, and probably taught about 8000 students in his life.

Wollaston, William Hyde (1766–1828) British chemist: discoverer of palladium and rhodium and pioneer of powder metallurgy.

Wollaston's father's family included several scientists and physicians and he followed both interests, at Cambridge and in London. However, in 1800 he gave up his medical practice and in partnership with S Tennant (1761–1815) made his income from the sale of platinum and devoted his time to work in chemistry, optics and physiology. He discovered palladium in 1802 and announced this weirdly by anonymous notices offering it for sale; his discovery of rhodium (also from crude platinum ore) he announced in the usual way, in 1804. Malleable platinum had not been made previously, but Wollaston produced it by methods now basic to powder metallurgy. Not unreasonably, he did not give details of his methods until shortly before his death.

Woolaston was very inventive and wide-ranging in his scientific work, and his lasting contributions include a reflecting goniometer (for measuring crystal angles), a modified sextant, an improved microscope and the discovery of the vibratory nature of muscular action.

Woodward, Robert Burns (1917-79) US organic chemist: probably the greatest deviser of organic syntheses.

Woodward's career was marked throughout by brilliance. He went to the Massachusetts Institute of Technology when he was only 16, was 'sent down' for a year for 'inattention to formal studies' but nevertheless emerged with his PhD at 20. Soon he moved to Harvard, and remained there. He did major work in most areas of organic chemistry, but his most striking work was in the synthesis of complex natural products. His successes in synthesis included quinine (1944), cholesterol and cortisone (1951), lysergic acid (the parent of the hallucinogen LSD) and strychnine (1954); the first major tranquillizer, reserpine (1956), chlorophyll (1960) and the tetracycline antibiotics (1962). The high point was the synthesis of vitamin B_{12} (cyanocobalamin) in 1971, after 10 years' work in collaboration with a team of Swiss chemists.

In each case the work was marked by the elegance and ingenuity of the synthesis, in making a valuable and highly complicated product from simple starting materials, using a large number of chemical steps. His methods frequently provided novel general syntheses of other compounds. In 1965 he developed the Woodward-Hoffmann rules concerning the path of a large class of addition reactions.

He had a remarkable memory, an unsurpassed knowledge of organic chemistry and a cool wit. In many ways modest, at conferences he sported a blue silk tie embroidered with the full formula of strychnine, which he had synthesized in 50 stages, each well planned. He was awarded the Nobel Prize in 1965.

Wright, Sewall (1889-1988) US geneticist: discoverer of genetic drift.

A graduate of Illinois and Harvard, Wright joined the US Department of Agriculture in 1916, where he worked particularly on stock improvement. Later he taught at Chicago, Edinburgh and Wisconsin. In his stock improvement studies his aim was to find the best combination of inbreeding and crossbreeding to achieve this. He used guinea pigs as a convenient test animal and developed a mathematical scheme to describe evolutionary development. In his work on small isolated animal populations he found that some genes can be lost randomly, because the few individuals having them may not reproduce successfully. This loss can lead to new species without the normal processes of natural selection being involved, and is known as the Sewall Wright effect, a process of random 'genetic drift' which can be important in small populations.

Wright, Wilbur (1867-1912) and Orville Wright (1871-1948) US aviators: made and flew the first successful aeroplane.

The Wright brothers were a very remarkable pair indeed. Through the second half of the 19th-c a number of individuals in Europe and the USA had attempted flight with heavier-than-air devices but without real success, at best operating unmanned models. The Wrights, sons of a non-conformist bishop in Dayton, OH and owners of a small bicycle-making firm, began to experiment in 1896. They recognized that control was as important as stability and that systematic experimentation was needed, and settled down to a program of wind tunnel experiments on wing sections (aerofoils) and made over 1000 flights with unmanned biplane gliders near Kitty Hawk on the North Carolina coast. One important step was their study of buzzards in 1899, which made it clear to them that three-axis control was needed (to bank, turn, and elevate or descend) and that the bird achieved control over roll by twisting its wings.

By early 1903 they had a biplane with control achieved by warping (twisting) the wings in unison along with a rudder, and with an elevator at the front. They devised an efficient propeller and made a small (12 hp, ie 9 kW) petrol engine, which was fitted to a new biplane, driving two 'pusher' propellers; the craft had skids and not wheels and the pilot lay on the lower plane. On a cold windy December morning in 1903 the first controlled, powered and manned flights were made at Kitty Hawk, two by each brother. They made two more aircraft; the last, *Flyer III* of 1905, could make figures of eight and remain airborne for half an hour. Despite French and British enthusiasm, European results were poor until Wilbur visited Europe and demonstrated their success in the year they ceased to fly, 1908. They deserve great credit; in comparison, the attempts of their predecessors seem inept and those of their successors to be natural developments from the Wright brothers' work. (See panel overleaf.)

Wu, Chien-Shiung (1912-) Chinese-US physicist: confirmed experimentally that parity is not conserved by the weak nuclear force.

Born in Shanghai, Wu moved to the USA in 1936, having completed her degree in China. Under LAWRENCE she obtained her doctorate from the University of California at Berkeley in

THE HISTORY OF AERONAUTICS

As early as the 16th-c, Leonardo da Vinci was designing flying machines based on emulating the flapping motion of birds' wings, but the vehicle which first carried people clear of the Earth's surface was a hot-air balloon constructed by the MONTGOLFIER brothers and flown on an ascent over Paris in 1783 by their volunteer friends de Rosier and d'Arlandes. In France CHARLES was developing a hydrogen balloon (or aerostat) at about the same time, but it was 1852 before the practical problems of controlling direction of travel and coping with adverse weather were overcome by the development of the balloon into an airship by Henri Giffard (1825–82). Early in the 20th-c Zeppelin in Germany had military and commercial success with airships, and small helium-filled versions still find use.

In the intervening period, research was being conducted using models which would be recognized today as a conventional aircraft configuration and, by 1853, SIR G CAYLEY, the father of aeronautics, had established the theory and constructed a man-carrying glider. Also during this period (1847) John Stringfellow (1799–1883) and William Henson built a steam-powered aircraft which, although incapable of sustained flight, influenced subsequent aircraft design. LANGLEY also had limited success.

The development of man-carrying gliders was pursued vigorously by Otto Lilienthal (1849–96) who, from his study of bird flight, recognized the importance of curved aerofoil sections to increase the efficiency of wing surfaces. With the continuing refinement of the internal combustion engine, invented in 1876 by OTTO and successfully harnessed to car propulsion by Karl Benz (1844–1929) in 1885, the way was cleared for the most significant landmark in manned, powered, controlled flight. In 1903, ORVILLE WRIGHT took off and flew for 12 seconds in a 12-hp engined biplane constructed with his brother Wilbur, in their cycle workshop, after a full programme of experimentation. As the reliability of the structures and engines improved, the latter particularly through Lawrence Hargrave's (1850–1915) radical rotary engine design, so pioneers exploited the new medium. Amongst these were Henri Farman (1874–1958), who made the first cross-country flight in 1908, and Louis Blériot (1872–1936), who flew across the English Channel the following year.

The First World War provided a powerful stimulus to all aspects of aeronautical research and development, and in 1919 Alcock and Brown crossed the Atlantic in a twin-engined Vickers Vimy bomber. This feat was followed by flights of ever-increasing range, including the overflight of the North Pole by Richard Byrd (1888–1957) in 1926 in an aeroplane and by Roald Amundsen (1872–1928) in an airship. Later pioneering aircraft flights included Lindbergh's solo crossing of the Atlantic; Amelia Earhart's over the Atlantic and the Pacific oceans; Amy Johnson's from England to Australia, and Mollison's to South Africa and India.

The Second World War created a similar impetus. By the end of that conflict the later Marks of the Spitfire, designed by Reginald Mitchell (1895–1937); the Hurricane, designed by Sidney Camm (1893–1966) and the Messerschmitt, designed by Wilhelm Messerschmitt (1898–1978) and Walter Rethel, were representative of the highest states of evolution of piston-engined aircraft. This war also heralded the arrival of the jet fighter; in 1939 the Heinkel HE 178, built by Ernst Heinkel (1888–1958) became the world's first combat jet aircraft. It was followed less than two years later by the Gloster E 28/39 powered by a jet engine designed by WHITTLE.

Rocket-powered aircraft had been introduced into limited service by the Luftwaffe during the later stages of the war, and research was continued in the USA. In 1947, Charles (Chuck) Yeager (1923–   ) flew a Bell X-1 rocket powered research aircraft to became the first man to fly faster than the speed of sound (1200 km h⁻¹/760 mph at sea level) breaking the so-called sound barrier. This marked a significant advance in aeronautical engineering.

The considerable use of long-range bomber aircraft during the war led directly to the improvement of passenger-carrying aircraft, and to the rapid growth of international air travel. By 1968 the world's first supersonic airliner, the Tupolev TU-144, had flown, followed a little later by the Anglo-French Concorde. The latter continued in scheduled service, with a normal cruising speed twice that of sound.

Among other significant developments were Sidney Camm's Harrier, the first Vertical-Take-Off-and-Landing (VTOL) aircraft successfully to enter service; and the incorporation on some aircraft, notably the General Aerodynamics F-111 fighter bomber and the BAC Tornado, of swing wings. This concept, first conceived by B N WALLIS, enables the angle of sweep-back of the wings to be varied in flight, in order to optimize aerodynamic efficiency during the various stages of flight.

With some military aircraft capable of reaching and operating on the fringes of space, the distinction between aeronautics and astronautics is becoming increasingly blurred. Indeed, future developments will probably draw on both sciences. The concept of aircraft taking off from normal airfields and climbing to achieve Earth orbit, before re-entering the atmosphere prior to a conventional aircraft style approach and landing, has been explored by British Aerospace in the Horizontal-Take-Off-and-Landing (HOTOL) project.

Geoff Leeming

1940 and took up a post at Princeton. From 1946 she taught at Columbia University, becoming professor of physics in 1957.

Wu had intended to return to China after her postgraduate work in physics, but war and afterwards communism stopped that. Columbia became her home, and she and her husband became US citizens in 1954; but she retained her Chinese dress style, her Chinese name and her preference for Chinese food.

From 1946 she became expert on beta-decay in radioactive atoms, the process whereby an electron and a neutrino are ejected from a neutron in the nucleus, leaving behind a proton.

In 1957 she developed her research on nuclear decay by emission of beta particles by observing that the direction of emission is closely tied to the direction of the spin of the emitting nucleus. The critical experiments used radioactive cobalt-60, (half-life 10.5 min) cooled close to absolute zero. In a difficult experiment, she showed that beta particles are not emitted in equal numbers when the cobalt nuclei are aligned in a magnetic field: more are emitted in the direction opposite the field (and therefore opposite to the direction of nuclear spin). The emission process, therefore, is not identical for a mirror image system; and the physical laws do not remain unchanged under a parity change. This extraordinary result had however been predicted by YANG and LEE, who had deduced that the weak nuclear interaction would not be identical under a parity change. The results of this work were far-reaching and many basic assumptions in physics were called into question. Wu won many prizes for her work on parity; her friends were disappointed that she was not included by the Nobel Prize Committee in their award to Lee and Yang in 1957.

Wu then set out to confirm FEYNMAN and GELL-MANN's theory of beta decay (1958), which predicted conservation of a vector current. She confirmed this in 1963. She also observed that electromagnetic radiation that is polarized is released on electron–positron annihilation, as predicted by DIRAC's theory of the electron.

Wurtz, Charles Adolphe (1817–84) French organic chemist: a pioneer of organic synthesis.

Wurtz's father gave him the choice of studying theology or medicine. As he wished to be a chemist Wurtz chose medicine, graduated and diverted to chemistry. He became assistant to DUMAS and succeeded him as professor in the École de Médicine. He was an exuberant lecturer; and his research laboratory in Paris was unique in Europe in attracting as many able young men as the laboratories in German universities. His early research was on the oxoacids of phosphorus, and he also discovered $POCl_3$.

Soon he moved to organic synthesis, where his many successes included the discovery of the first amines, from the reaction of base with an alkyl isocyanate; in this way he made CH_3NH_2 and $CH_3CH_2NH_2$. In 1855 he showed that reactive alkyl halides react with sodium metal to give hydrocarbons (the Wurtz reaction): for example,

$$2n-C_4H_{11}Br + 2Na \rightarrow n-C_8H_{22} + 2NaBr.$$

The reaction has been used to make very long-chain hydrocarbons from, for example, 1-iodo-n-pentacontane:

$$2C_{50}H_{101}I + 2Na \rightarrow C_{100}H_{202}(n\text{-hectane}) + 2NaI.$$

The longest-chain non-polymeric compound known by 1985 was tetraoctacontatrictane $CH_3(CH_2)_{382}CH_3$, ie $C_{384}H_{770}$.

Wurtz was a major supporter of both DUMAS's and GERHARDT's early theories on the nature of organic compounds.

Wynne-Edwards, Vero Copner (1906–) British biologist: proposed animal altruism as basis for population homeostasis control.

An Oxford graduate, Wynne-Edwards taught at McGill University, Montreal from 1930–46 and thereafter at Aberdeen. In Montreal he worked on the distribution of sea birds, making four round trips by Cunarder over the Atlantic in 1933 to see the changes in species with the seasons, and began to gather the results that were to be fully developed in his book *Animal Dispersion in Relation to Social Behaviour* (1962). He proposed that animal populations use hormonal devices and social mechanisms including territoriality, dominance hierarchies and grouping in flocks as methods of controlling population size; and that they will sacrifice their own survival and their fertility for the good of the group, whose survival depends on avoiding overuse of the available resources. This view of animal altruism provoked vigorous discussion and research in ethology and ecology; the book has been highly influential in its proposal for 'population homeostasis' and its ideas have been both criticized and developed by others, including D L Lack (1910–73), J Maynard Smith (1920–) and E O WILSON.

Yalow, Rosalyn, *née* Sussman [yalow] (1921–)
US nuclear physicist: developed the radioimmunoassay method.

A physicist with a special interest in radioisotopes, Rosalyn Yalow turned to nuclear medicine and from 1972 was Senior Medical
Investigator for the Veterans Administration.
Working with S A Berson (1918–72) in a New
York hospital, she developed from the 1950s the
method of radioimmunoassay to detect and
measure peptide hormones (such as insulin) in
the blood. The method has proved of great value
both in locating the origin of hormones in the
body, and in clinical diagnosis and treatment of
a variety of diseases and male and female infertility. Extension of the method in the UK has led
to better control of digoxin therapy in heart disease and diagnosis of neural crest disease (eg
spina bifida) in the fetus. The method can be
used to measure very small amounts (10^{-12} g) of
any substance for which an antibody can be
made. Its value in diagnosis and in the control
of medication is immense.

Rosalyn Sussman had been a forthright child
and her attempts, as a female and Jewish New
Yorker, to enter her chosen career before the
Second World War did not soften her. Experience as a physicist in the Veterans Hospital,
dominated by medical men and service officers,
made her abrasive. After Berson's death she saw
that he had been assumed to be the creative
member of their 22-year, highly effective professional partnership. Stung by this, she increased
her research output in the 20 years before her
retirement and became even more con-
frontational, unsoftened by her Nobel Prize in
1977.

Yang, Chen Ning (1922–) Chinese–US physicist: showed that parity is not conserved by the
weak nuclear force.

Yang, the son of a professor of mathematics,
received his college education in Kunming in
China. Taking up a fellowship for travel and
research in America he completed a PhD under
TELLER at Chicago. He joined the Institute for
Advanced Study at Princeton in 1949. He
became director of the Institute for Theoretical
Physics at the State University of New York,
Stony Brook in 1966.

It is Yang's work in collaboration with LEE
which is justly celebrated as a turning point in
the development of theoretical physics: they
showed that the law of conservation of parity (ie
that physical laws are unaltered in mirror-
image systems) does not hold for the weak
nuclear interactions (1956). The prediction was
confirmed by WU's thorough experimental

study and quickly led to the award of a Nobel
Prize for physics to Yang and Lee (1957).

Yang is also noted for his development of
a non-Abelian gauge theory with R L Mills
(1924–), the Yang–Mills theory. This proved to
be an important new departure in theories of
elementary particles and quantum fields.

Yanofsky, Charles [yanofskee] (1925–) US
geneticist: experimentally verified the hypothesis that the DNA base sequence codes for protein synthesis.

A graduate in chemistry from New York who
went on to work in microbiology at Yale, Yanofsky afterwards worked at Yale and Stanford on
gene mutations. His best-known work was his
demonstration of the validity of CRICK and
WATSON's suggestion of 1953 that the sequence
of bases in the genetic material DNA determines the order of the amino acids that make
up proteins, including of course the enzymes
critical to living systems. Yanofsky secured his
evidence on this by ingenious experimentation
using mutant strains of a bacteriophage; these
were isolated and the positions of the mutations in the gene were mapped. Likewise, the
amino acid sequences were determined in the
various mutant forms of the enzyme produced
by these strains. It could then be shown that the
changes in amino acid sequences correspond
with the mutant sites on the genetic map, in
accord with the theory.

Young, James (1811–83) British chemist: a pioneer petroleum technologist.

Young was a part-time student in Glasgow,
attending 'night-school' classes in chemistry
given by GRAHAM; he became Graham's assistant. From 1843 he was employed in the chemical industry and in 1848 he set up a small works
in Derbyshire to purify oil from a seepage, and
marketed it for lighting and as a lubricant.
When after 3 years the seepage was exhausted
he moved to West Lothian in Scotland and
began to extract oil from oil shale deposits by
distillation. He founded the Scottish oil shale
industry and was one of the first to apply chemical methods to oil handling.

His other ventures included a measurement
of the speed of light by FIZEAU's method; and he
gave financial support for the explorations of
David Livingstone (1813–73), who had been a
fellow student and friend in Glasgow.

Young, Thomas (1773–1829) British physiologist, physicist and Egyptologist: established the
wave theory of light.

Young surely had one of the most acute minds
of his century, but his diversity of interests and
his tendency to move to new ones rather than

consolidate his ideas, caused credit for some of them to go to others. His father was a banker, and for unknown reasons the boy lived largely with his grandfather. He was a precocious child who could read at the age of 2; he had a good knowledge of five languages at 13 and of eight more oriental languages at 14. At this time a young schoolmaster also introduced him to telescope-making.

In 1792 he began to study medicine, intending to follow a friendly and prosperous uncle into his London practice, and in his first year as a medical student in London he published on the physics of the eye. By neat experiments he showed that accommodation (change of focus) is a result of change in the curvature of the lens; at the same time he described and measured astigmatism and in 1801 he devised his three-colour theory of human colour vision. He continued as a medical student, with a full social life, in Edinburgh, Göttingen and Cambridge, and in 1799 set up a practice in London. He was not very successful as a physician, perhaps because (as a friend said) his mind was usually on other matters, and he was not a success as a lecturer at the Royal Institution. During these lectures, when discussing HOOKE's law in 1802, he gave physical meaning to the constant in that law, which has come to be named as Young's modulus, E, defined as the ratio stress/strain; here stress is the force per unit area of cross section of a material, which produces a strain measured as (change in length)/(original length). E is a measure of a material's resistance to change in length, and an average value for natural rubber would be $1 \times 10^6 \, \text{Nm}^{-2}$ while for a mild steel $E = 2 \times 10^{11} \, \text{Nm}^{-2}$.

But his major work was on the wave theory of light, which NEWTON had thought to be corpuscular and HUYGENS wave-like. Young argued in 1800–04 in favour of the wave theory and supported this by clear and detailed accounts of elegant experiments on interference due to superposition of the waves. The current view entirely supports Young's interpretation of these effects, while also using a 'corpuscular' explanation in terms of photons and quantum theory for such results as photoelectric emission.

From 1814 Young busied himself with his medical practice and with Egyptology, where his major contributions to the interpretation of the Rosetta Stone ultimately revealed the ancient Egyptian system of writing, although at the time others were given more credit for this: in part because his work on it was published anonymously, as the entry on Egypt in a supplement to the *Encyclopaedia Britannica* for 1819.

Yukawa, Hideki (1907–81) Japanese physicist: first described the strong nuclear force and predicted the pi-meson (pion).

Yukawa studied at Kyoto University and took his degree there in 1929. He moved to Osaka University to take his doctorate but returned to Kyoto for the remainder of his career, becoming professor of theoretical physics in 1939.

When he was 27, Yukawa developed his theory of nuclear forces. In 1932 CHADWICK had discovered the neutron; Yukawa proposed a strong short-range force between protons or neutrons, which overcame electrical repulsion between the protons in the nucleus without influencing the electrons in the atom. This nuclear 'exchange' force involves the exchange of a particle between the nucleons (nuclear constituents) and from the short range of the force (less than 10^{-8} m) Yukawa inferred that its mass was about 200 times that of an electron.

In 1936 C D ANDERSON discovered a particle of the correct mass and called it the mu-meson (now, muon); but it did not interact with nucleons sufficiently strongly to correspond with Yukawa's prediction. In 1947 POWELL and his co-workers discovered another meson (the pi-meson, now pion) in cosmic rays which did correspond with Yukawa's proposed particle and so established his theory of the strong nuclear force. During the next 10 years many types of pion and muon were discovered, most with short lifetimes. They are now considered (as are protons and neutrons) to be composed of quarks; and the forces in nuclei to be due to interactions between quarks and gluons.

Yukawa successfully predicted (in 1936) that nuclei may absorb one of the innermost electrons (in the 1 K shell) and such K capture by a nucleus was soon observed. He was the first Japanese to be awarded a Nobel Prize, in 1949.

Zeeman, Pieter [zayman] (1865–1943) Dutch physicist: discovered the splitting of spectral lines by magnetic fields.

Zeeman's experiment of 1896 proved to be an early crucial link between light and magnetism, which also gave further identification of the electron and a basis on which to test the quantum mechanical theories of atomic structure. It was performed soon after Zeeman had graduated at Leiden (under LORENTZ) and had become a *Privatdozent* at Amsterdam. He first observed that, when a magnetic field was applied to sodium or lithium flames, the lines in the emission spectrum of the flame were apparently broadened; and this on inspection was due to splitting of the lines into two or three lines. Zeeman's observation agreed with results from Lorentz's classical theory of light as being due to vibrating electrons in atoms.

The normal Zeeman effect is shown when a spectral line splits into two with a strong magnetic field applied parallel to the light path or into three if the field is perpendicular. The old quantum theory and BOHR's model of the atom could explain this. However, in general, atoms show the anomalous Zeeman effect, which involves splitting into several closely spaced lines. The explanation of this required the full quantum mechanics (of 1925) and the concept of electron spin, due to UHLENBECK and GOUDSMIT (1926). Zeeman and Lorentz shared the 1902 Nobel Prize for physics for their work on the magneto-optical properties of atoms.

FARADAY had experimented in 1862 on the application of a magnetic field to emission spectra, but failed to find an effect; Zeeman succeeded and the explanation of his results was a major step for theoretical physics.

The Zeeman effect has been used (first by HALE in 1908) to examine the Sun's magnetic field and (with much greater difficulty) that of other stars.

Zhang Heng (78–139) Chinese astronomer and geophysicist: invented the earthquake seismograph.

Zhang was born in Nanyang, Henan Province, during the Han Dynasty. He was Imperial Historian and official astronomer.

Zhang recognized that the source of the Moon's illumination was sunlight and that lunar eclipses were caused by the Earth's shadow falling upon it. He devised a water-driven celestial globe, which revolved in correspondence with the diurnal motion of the celestial sphere. In mathematics, he calculated π as 365/116 (about 3.1466), a substantial improvement on the hitherto accepted Chinese

value of 3. Perhaps his best remembered contribution, however, was to geophysics: in 132 he invented an early seismograph. This was to help him locate and record earthquakes, one of his official duties as Imperial Historian. It was a bronze device almost 2 m in diameter, containing a mechanism of pendulums and levers, with eight dragon figures arranged around its circumference. Strong seismic tremors caused a metal ball to be released from the mouth of the dragon facing the direction of the shock wave. It is known that the device registered an earthquake in Gansu Province in 138.

Zinder, Norton David (1928–) US geneticist: discovered bacterial transduction.

A graduate of Columbia University, New York, Zinder did graduate work with LEDERBERG and was professor of genetics at Rockefeller University from 1964. Lederberg was the first to observe sexual union (conjugation) in a bacterium (*Escherichia coli*); Zinder looked for it in *Salmonella*. He soon devised a valuable new technique for isolating mutants of this bacterium, but his attempt to observe conjugation led him instead to the discovery of bacterial transduction. This is the transfer, by a phage particle, of genetic material from one bacterium to another; the discovery led to new knowledge of the location and behaviour of bacterial genes.

Zsigmondy, Richard Adolf [zhigmondee] (1865–1929) German colloid chemist; the inventor of the ultramicroscope.

After studying chemistry and physics Zsigmondy joined the Schott glassworks at Jena; it was through his interest in coloured glass that his work on colloids began, and this was continued through his career as a professor at Göttingen. In 1903 he made an 'ultramicroscope' in which the sample is strongly lit from one side against a dark background. This allowed colloid particles to be seen as points of light, even if they were smaller than the resolving power of the microscope. Such studies were of great value before the introduction of the ultracentrifuge and the electron microscope. Zsigmondy examined many colloidal solutions (especially gold sols), deduced particle sizes and concluded that the particles are kept apart by electrostatic charge. Colloidal solutions are of great importance in biochemistry. His pioneer studies did much to advance understanding of sols, gels, smokes, fogs and foams, and he was awarded a Nobel Prize in 1925 for his work.

Zu Chongzhi (429–500) Chinese mathematician and astronomer: improved the accuracy of π and measured the length of the year.

Zu computed π to be 355/113 (about

3.1415929), a value not bettered until 1000 years later by al-Kashi (d. *c*.1430) and Viète, and also gave 22/7 as a simpler value for calculations where less accuracy was sufficient. In astronomy, he measured the length of the year to be 365.2429 days, by extensive observations of the lengths of shadows around the winter solstice, an improvement over contemporary values.

Zworykin, Vladimir (Kosma) [tsvorikin] (1889–1982) Russian–US physicist: invented the electronic-scanning television camera.

Soon after graduating in engineering from St Petersburg Institute of Technology, Zworykin spent the First World War serving as a radio officer in the Russian army. One of his teachers in St Petersburg was B Rosing, who took out the first patent for a television system in 1907; it used a cathode-ray tube as its receiver. After Russia's collapse into revolution in 1917 Zworykin emigrated to America in 1919 and joined the Westinghouse Electric Corporation.

His career developed as he gained a doctorate (1926) and moved to a post with the Radio Corporation of America (1929).

While at Westinghouse in 1923 he reproduced an image from a screen by dividing it into many insulated photoelectric cells, which held a charge proportional to the light falling on them. An electron beam scanning the screen discharged the cells in turn, giving an electrical signal.

Zworykin then took a cathode-ray tube (invented by Braun in 1897), which could produce focused spots on its fluorescent screen using electric and magnetic fields. As the beam scanned the screen the intensity of the spot varied according to the electrical signal and so could reproduce the images from his first device, so that he had a television transmitting and receiving system. By 1929 the camera had been developed to the point of practical use, and it displaced the mechanical system developed by Baird.

Chronology

This lists a number of major events in science, and notes also (in bold type) some other events of historical interest. Titles of important scientific books are given in italics; in general these surveyed earlier work in their field and also initiated fruitful advances.

c.550 BC Anaximander proposes Earth is poised in space.

c.450 BC Empedocles proposes four-element theory of matter.

c.330 BC Theophrastus founds scientific botany.

c.300 BC Euclid systematizes geometry.

c.250 BC Archimedes founds mechanics and hydrostatics.

c.200 BC *On Conic Sections* (Apollonius)

30 *or* 33 **Christ is crucified.**

43 **Romans begin conquest of Britain.**

132 Zhang Heng invents seismograph.

c.825 Al-Khwarizmi gives general solution for quadratic equation.

c.1000 Alhazen's work in optics.

c.1350 **Black Death devastates Europe.**

1440 **Gutenberg and Koster independently introduce printing by movable type.**

1498 **Columbus lands in West Indies and South America.**

1517 **Luther's 95 theses begin the Protestant Reformation in Germany.**

1519–22 **Magellan's ship circumnavigates the world; Cortez conquers Mexico.**

1543 Copernicus publishes heliocentric system in his *De revolutionibus...* Vesalius establishes anatomy by his *De fabrica corporis humani.*

1556 *De re metallica* (Agricola) surveys mining and metallurgy.

1569 Mercator publishes a map of the world based on his new projection.

1572 Brahe observes supernova (Tycho's star).

1576 T Digges proposes that stars are at varying distances, and that universe is infinite.

1579 **Drake, circumnavigating the world, lands in California.**

1588 **Spanish Armada is defeated.**

1600 **Bruno burned for heresy in Rome.** *De magnete* (Gilbert) surveys magnetism, and proposes that Earth is a giant magnet.

1603 Fabrizio describes valves in human veins, but is unclear on their function.

1607 **English settlements in Virginia.**

1608 Lippershey and Jansen make first useful telescope (3 ×) and offer it for military use.

1609 Kepler publishes his first two laws of planetary motion. Lippershey and Jansen make early compound microscope.

1610 Galileo publishes his astronomical observations made with a telescope (30 ×).

1614 Santorio describes first scientific study of human metabolism.

1621 Snel discovers law of refraction.

1624 *Logarithmical Arithmetic* (Briggs) eases computation.

1628 *De motu cordis* (Harvey) describes circulation of the blood.

1630 **Large-scale emigration from England to Massachusetts begins.**

1632 Galileo publishes his *Dialogo...* supporting Copernicus's heliocentric system.

1633 Galileo is charged with heresy, and recants.

1637 Descartes introduces analytic geometry.

1642 **Civil War begins in England.** Galileo dies, Newton is born.

1644 Torricelli constructs mercury barometer.

1646 Pascal experiments with barometers, and later shows that air pressure drops with rising altitude.

1650 Guericke makes an effective air-pump after 15 years work.

1651	Harvey publishes on embryology: ahead of his time, he claims that 'all creatures come from an egg'.
1654	Probability theory initiated by Pascal and Fermat in response to a request from gaming friends. Guericke's hemispheres demonstrate pressure of atmosphere
1658	Swammerdam sees red blood cells in frogs.
1660	Boyle finds law relating pressure to volume for a gas. Royal Society is founded.
1661	*The Sceptical Chymist* (Boyle) defines chemical elements. Evelyn publishes on air pollution: his proposed remedies are not acted upon. Malpighi describes animal capillaries.
1665	*Micrographia* (Hooke) advances microscopy, and also discusses meteorological instruments. **Great Plague** sends Newton home from Cambridge. In 1665–6 he devises the binomial theorem, the calculus, and the theory of gravitation.
1668	Newton makes the first reflecting telescope.
1669	Bartholin describes double refraction by calcite. Steno publishes ideas on mountain-building, strata and fossils.
1673	Leeuwenhoek studies red blood cells.
1675	Roemer shows light has finite speed, and calculates that it takes 11 minutes to cross the Earth's orbit.
1677	Brand discovers phosphorus.
1678	Huygens expounds wave theory of light.
1687	*Principia mathematica* (Newton) creates celestial mechanics.
1694	Camerarius experiments on sexuality in plants.
1699	Amontons observes relation of gas pressure to temperature.
c.1701	Halley produces magnetic map of world.
1702	**First English daily newspaper (*Daily Courant*).**
1704	*Opticks* (Newton) surveys nature and behaviour of light. An appendix pro-

poses that all matter is composed of small, hard particles (atoms).
First American newspaper (*Weekly Review*) published in Boston.

1705	Halley publishes book on comets, and correctly predicts a bright comet for 1758 with a period of about 76 years (Halley's comet).
1707	**Union between England and Scotland creates the kingdom of Great Britain.**
1727	*Vegetable Staticks* (Hales) surveys plant physiology.
1729	Bradley measures speed of light Gray publishes his work on electricity.
1733	Hales's *Haemastaticks* surveys his work on blood circulation and pressure in animals.
1735	*Systema naturae...* (Linnaeus) classifies plants effectively. Hadley's work on trade winds.
1738	D Bernoulli's *Hydrodynamica* initiates the science of fluid flow.
1748	Maria Agnesi publishes her widely-used and comprehensive textbook on mathematics.
1752	Franklin uses kite to show electrical nature of lightning. Desmarest argues that a land bridge between England and France must have existed at one time.
1753	*A Treatise on the Scurvy* (Lind) shows value of citrus fruits in its prevention and treatment.
1756	Joseph Black studies chemistry of carbon dioxide and lime.
1757	Monro distinguishes lymphatic and circulatory systems. Dolland patents his achromatic lens.
c.1763	Black recognizes 'latent heat'.
1767	Spallanzani disproves theory of spontaneous generation.
1768	Lambert proves π and e to be irrational. Cook circumnavigates New Zealand.
1772	Titius proposes Bode's law.
1773	**Boston Tea Party.** Harrison is awarded the balance of the prize for his chronometer, having worked on it since 1728.

1774 Priestley discovers oxygen, but misinterprets his results.

1776 **American Declaration of Independence.**

1779 Ingenhousz's work on photosynthesis.
Iron bridge built across the R Severn at Coalbrookdale, England.

1781 Herschel discovers Uranus.
American victory over British troops at Yorktown.

1783 First manned hot air balloon flight (at Versailles).
F W Herschel shows that the entire solar system is moving in space relative to the stars.

1784 Cavendish publishes his work on the synthesis of water from hydrogen and oxygen.
Michell deduces possibility of black holes in universe.

1785 F W Herschel makes a systematic study of the likely shape of our Galaxy (the Milky Way).

1787 Charles formulates law on gas temperature–pressure relation.

1788 **Penal settlements established in Australia.**

1789 *Elementary Treatise on Chemistry* (Lavoisier) initiates chemical revolution.
Fall of the Bastille; French Revolution.

1791 Galvani publishes on 'animal electricity', which he has studied for 11 years.

1795 *Theory of the Earth* (Hutton) begins much of modern geology.
The metric system officially adopted in France.

1796 Jenner experiments on vaccination against smallpox.

1797 Venturi observes that when a fluid (water) passes through a constriction, its velocity increases and the pressure falls.

1798 Cavendish weighs the Earth.
Rumford publishes his work on the conversion of work into heat.

1800 Volta makes cell for supply of electric current.
Nicolson and Carlisle show that water is decomposed by electricity to give hydrogen and oxygen.
Davy discovers N_2O and suggests its use as an anaesthetic (not used until 1844).
Leslie devises wet and dry bulb hygrometer.
F W Herschel discovers infrared radiation.

1801 Dalton publishes law of partial pressures.
Piazzi discovers first asteroid (Ceres).
Lamarck argues that species change with time by inheritance of alterations made by their adjustment to their environment.
Haüy publishes his studies on crystallography and mineralogy.
Arithmetical Researches (Gauss).
Ritter discovers ultraviolet radiation.
Political Union between Great Britain and Ireland.

1803 Young demonstrates interference of light, supporting wave theory of light.

1804 Dalton publishes law of multiple proportions.

1806 Beaufort proposes his wind scale.
Argand devises diagram for representing complex numbers.

1807 C Bell shows that nerves convey either sensory or motor stimuli but not both.
Young proposes that heat is probably a wave vibration, and not a material substance.
Davy makes sodium and potassium by electrolysis.

1808 W Henry publishes law discovered by him in 1803 on solubility of gases in liquids.
Dalton publishes his atomic theory, first offered by him in 1803.
Gay Lussac's law of combining volumes of gases.
Davy uses electrolysis to isolate Ba, Sr, Ca, Mg.

1809 George Cayley begins to publish his experiments, continued over the next 30 years, which will found the theory of aerodynamics.

1810	Davy shows that chlorine is a chemical element and not a compound.
1811	Avogadro proposes law on atomicity of gases.
	C Bell's *New Idea of the Anatomy of the Brain* proposes that different parts of the brain have different functions.
1812	Cuvier develops his studies on fossil bones, leading to vertebrate palaeontology.
1813	Dulong discovers NCl_3 and loses an eye and two fingers in the process.
1814	Fraunhofer observes dark lines in Sun's spectrum.
	Guy Lussac suggests (correctly) that the *arrangement* of atoms in a compound may be important.
1815	Biot shows that optical activity is a molecular property.
	Napoleon defeated at Waterloo.
	Gay Lussac discovers cyanogen and studies cyano compounds.
	W Smith publishes his stratigraphic map of Britain.
1816	Laënnec invents stethoscope.
	Sophie Germain creates mathematical theory of elastic surfaces.
1817	W Smith shows value of fossils in stratigraphy.
1818	Berzelius publishes his table of relative atomic weights.
	Bichat founds histopathology.
1819	Dulong and Petit find relation between specific heat capacity and relative atomic mass.
	Fresnel and Arago deduce that light vibrates transversely to its direction of forward movement.
1820	Oersted shows that a current in a wire induces a magnetic field around it.
	Ampère begins work on electrodynamics.
	Mitscherlich publishes law of isomorphism.
1822	Seebeck discovers thermoelectric effect.
	Fourier suggests using mass, length and time as fundamental dimensions.
	Olbers's paradox postulated.
	Chevreul deduces nature of fats, and

	begins use of melting point to check purity of a solid substance.
	Macintosh exploits J Syme's rubberizing of cotton fabric.
1824	*Reflections on the Motive Power of Fire* (Carnot) initiates thermodynamics.
	Flourens's work on central nervous system.
	Prévost and Dumas argue that sperm is necessary for fertilization.
	Daguerre produces 'daguerrotype' photographic plates, and announces improved version in 1839.
1825	**First steam locomotive railway, 28 km long, opened in County Durham, England for freight and passengers.**
	Faraday discovers benzene.
	Ohm begins work which leads to Ohm's Law in 1827.
	Balard discovers bromine.
1826	Von Baer begins study of mammalian ovum and embryo.
	Lobachevsky introduces non-Euclidean geometry.
1827	Ampère's Law in electromagnetism.
	Ohm establishes law of electrical resistance.
	Brown observes movement of pollen grains under microscope.
	Friction matches introduced.
1828	Wöhler synthesizes urea.
	Caroline Herschel publishes catalogue of star clusters and nebulae.
	Berzelius lists 'atomic weights' of 28 elements.
1829	Nicol describes his polarizing prism.
	Quetelet analyses Belgian census statistically, and correlates deaths with age, sex, occupation and economic status.
	Babbage writes on scientific frauds; 'cooking', 'trimming' and forging results.
1830	*Principles of Geology* (Lyell) surveys geological changes, seen as due to ordinary effects acting over a long time, rather than 'catastrophes'.
	Faraday begins work on electricity and independently of J Henry discovers electromagnetic induction.

1831	Darwin begins 5-year voyage on HMS *Beagle*.
1833	Babbage makes his 'difference engine' for computing.
	Ada King, Countess of Lovelace writes first computer program for Babbage's more advanced 'analytical engine'.
	Beaumont concludes that digestion is purely chemical.
1834	Wheatstone measures speed of electricity in lengths of wire.
1835	Geological Survey of UK established.
	Morse makes model electric telegraph.
1836	Marc Dax concludes that the brain is asymmetric, eg the left hemisphere controls speech.
1837	Magnus analyses blood gases and finds more oxygen in arterial blood than in venous blood, implying that respiration occurs in tissues.
1838	Remak shows that nerves are not hollow tubes.
	Bessel measures first stellar distance.
1839	Schwann, and Schleiden, expound cell theory in biology.
1840	**Postage stamps introduced in UK.**
	Henle offers essentially correct views on the nature and causes of infectious diseases.
	Agassiz postulates ice ages.
	New Zealand colonized.
1842	Doppler discovers effect named after him.
	W and H Rogers describe the structure and probable origin of the Appalachian Mountains.
1843	Schwabe finds 11-year sunspot cycle.
	Electric telegraph introduced.
	A Cayley invents *n*-dimensional geometry.
	Joule measures mechanical equivalent of heat.
	British Algae: Cyanotype Impressions (Anna Atkins): first scientific book illustrated with photographs.
	Screw steamer *Great Britain* crosses Atlantic.
1845	Faraday classifies magnetic materials and discovers rotation of polarized light by a magnetic field.

1846	Neptune discovered by Galle.
	Mallet theorizes on earthquakes and Dana on volcanoes.
1848	**Marx and Engels publish the *Communist Manifesto*.**
	Joule estimates speed of gas molecules from kinetic theory.
1849	Addison associates a disease with failure of the adrenal gland.
	Snow asserts (correctly) that cholera is transmitted by polluted water.
	Fizeau makes accurate measurement of speed of light.
1850	Clausius develops thermodynamics, using 'first and second laws' as key concepts, ideas further developed by W Thomson in 1851.
1851	**Great Exhibition in London.**
	Hofmeister unifies ideas on plant reproduction.
	Helmholtz invents ophthalmoscope.
	Kelvin proposes an 'absolute' temperature scale, later known by his name.
1852	Frankland introduces idea of chemical valence.
	Foucault demonstrates rotation of Earth by use of pendulum.
1855	Pringsheim confirms sexuality of algae.
	Logan and Hunt describe geology of Canada.
1856	Bessemer patents steel-making by 'converter'.
	Ferrel explains atmospheric circulation using Coriolis force.
1857	Buys Ballot announces law on rotation of cyclones and anti-cyclones.
1858	Kekulé's theory of organic molecular structure.
	Virchow, in his book *Cellularpathologie*, argues that all disease occurs in the cells.
	Wallace sends Darwin his ideas on the origin of species, which spurs Darwin to publish his own similar ideas.
	Sorby shows how microscopy allows deductions on rock formation.
1859	*The Origin of Species* (Darwin) argues that much observation of plant and animal species is consistent with a

belief in their evolution, and with natural selection as the cause of it.

Bunsen and Kirchhoff introduce spectrum analysis.

Drake drills oil-well in Pennsylvania.

British Medical Register created; it includes Elizabeth Blackwell.

1860 **Lincoln becomes president of the USA.**

Selective staining, and use of microtomes, begin to improve microscopy in biology.

Cannizzaro clarifies relation of atoms to molecules.

1861 Gegenbaur shows that all vertebrates' eggs are single cells.

Crookes discovers thallium.

American Civil War begins.

Pasteur ends debate on spontaneous generation.

1862 Alvan Clark and his son discover Sirius B.

Sachs shows that plant starch is a product of photosynthesis, derived from atmospheric CO_2.

Angström discovers hydrogen in Sun.

1863 Waldeyer describes cancer in modern terms.

Tyndall discusses greenhouse effect of Earth's atmosphere.

1864 Maxwell derives equations on electromagnetism, linking magnetism, electricity and optics.

1865 Clausius introduces concept of entropy.

Mendel publishes on heredity, and is largely ignored.

Kekulé proposes ring structure for benzene.

End of American Civil War.

Loschmidt calculates Avogadro constant as $\approx 6 \times 10^{23}$.

Elizabeth Garrett completes a legally qualifying medical course in UK.

1866 **Transatlantic telegraph cable begins operation.**

1867 Lister shows value of antisepsis in surgery, used by him from 1865.

1868 Mendelayev describes periodic table.

Marsh founds American palaeontology.

University of Paris (Sorbonne) admits women.

1869 T Andrews discovers critical state for gases.

Suez Canal opened.

First transcontinental railway opened, in USA.

1870 **Franco-Prussian War.**

1871 *The Descent of Man* (Darwin) surveys possible paths of human evolution, regarding man as an evolved animal.

1874 Perrault deduces that rain and snow provide supply for river water, following close study of Upper Seine.

Van 't Hoff advances stereochemistry.

Braun uses semiconductors as rectifiers.

1875 Hertwig first observes union of sperm and ovum (in sea-urchin).

1876 Draper photographs solar spectrum.

Hertwig shows that genetic material reaches offspring from both parents.

A G Bell patents telephone.

1877 Manson shows insect vectors involved in some diseases.

First British university accepts women medical students (King's and Queen's College of Physicians, Dublin).

1878 Cailletet liquifies common gases.

University of London opens its degree courses to women.

1879 **First electric railway exhibited, in Berlin.**

1881 Cambridge University admits women to examinations, but not to degrees.

1882 Mechnikov describes phagocytosis.

Ants, Bees and Wasps (Lubbock).

Flemming describes mitosis.

First generating station supplying electricity to private consumers opens in New York.

1883 University of London awards first medical degrees to women.

1884 Balmer finds sequence in hydrogen spectra (Balmer series).

Oxford University admits women to examinations, but not to degrees.

1885	Galton demonstrates individuality of human fingerprints.	1901	Bordet experiments on complement fixation.

1885   Galton demonstrates individuality of human fingerprints.

1886   Moissan isolates fluorine.

1887   Tesla makes first AC motor.
Michelson and Morley show that ether is probably non-existent.

1888   Hertz discovers radio waves.
Nansen explores Greenland icecap.

1890   Behring and Ehrlich develop diphtheria antitoxin.
Hollerith punch card system used to process US census.
Dewar improves insulating flask, thereafter widely used in laboratories and on picnics.
London Underground railway opens.

1892   Weismann observes meiosis, and proposes germ-plasm theory of heredity.

1893   Nansen begins *Fram* expedition to Arctic.

1894   Rayleigh and Ramsay discover first noble gas, argon.

1895   Roentgen discovers X-rays.
Diesel invents compression-ignition engine.

1896   Becquerel discovers radioactivity (of uranium).
Arrhenius calculates result of additional CO_2 on greenhouse effect.
Birkeland theorizes on origin of aurora.
Zeeman effect links light with magnetism experimentally.
Boltzmann's equation relates entropy to probability.

1897   Buchner shows that intact yeast cells not needed for fermentation.
J J Thomson studies electrons.

1898   Curies discover activity of radium, and isolate a sample in 1902.

1899   First widely used synthetic drug (Aspirin) marketed by Bayer AG.

1900   J A Fleming invents thermionic diode.
Planck initiates quantum theory.
Marconi transmits by radio across Atlantic.
Mendel's work on heredity rediscovered.
Pearson introduces chi-square test in statistics.

1901   Bordet experiments on complement fixation.

1902   Landsteiner describes ABO blood system.
Richet discovers anaphylaxis.
Heaviside and Kennelly find radio-reflecting layer in atmosphere.
Bateson applies Mendel's laws to animals as well as plants.

1903   Einthoven describes use of ECG.
Wright brothers make first manned, powered and controlled flight.
Rutherford and Soddy propose that radioactivity is due to atomic disintegration, an idea widely ridiculed.
Boveri, Sutton and others argue that 'hereditary factors' (later to be called genes) are located on the chromosomes.

1905   Blackman demonstrates limiting factors in plant growth.
Einstein's theory of special relativity; and $E = mc^2$ relation.

1906   Oldham deduces existance of Earth's core.
Tswett uses chromatography.
Hopkins deduces existence of vitamins.
Nernst states third law of thermodynamics.
Integrative Action of the Nervous System (Sherrington).
Brunhes discovers past geomagnetic reversal in rocks.

1907   De Forest patents triode valve (radio tube).
Ross Harrison devises new tissue culture method using frog nerve cells.
Pavlov publishes his work on conditioned reflexes.

1908   *Inborn Errors of Metabolism* (Garrod).
Kamerlingh-Onnes liquifies helium.
First Model T automobiles made by Ford Motor Co.

1909   Ehrlich begins modern chemotherapy with Salvarsan.
Morgan begins using *Drosophila* in genetics.
C D Walcott discovers fossils of strange soft-bodied animals in the Burgess

Shale of British Columbia; re-studied after the 1960s these small and weird animals, 530 million years old, are seen to form lines of evolution which mostly became extinct.

Mohorovičić describes discontinuity between Earth's mantle and crust.

Blériot crosses English Channel by aeroplane.

Sörensen invents pH scale of acidity.

1910   Millikan measures charge on the electron.

1911   Amundsen reaches South Pole.

Morgan and Sturtevant plot first chromosome map.

Rutherford proposes nuclear structure of atoms.

Millikan measures electron charge.

C T R Wilson devises cloud chamber.

Kamerlingh-Onnes discovers superconductivity.

V F Hess discovers cosmic rays.

1912   Wegener proposes theory of continental drift.

Von Laue shows that X-rays can be diffracted and so behave as waves, and the Braggs develop the concept and create X-ray crystal structure analysis.

Henrietta Leavitt devises method for measuring stellar distances.

Slipher measures speed of rotation of planets and spiral galaxies.

1913   Bohr calculates spectrum of atomic hydrogen, based on his 'Bohr atom'.

Moseley shows meaning of atomic number.

1914   J J Abel isolates amino acids from blood.

Russell publishes H–R diagram.

World War First begins.

1915   Adams identifies first white dwarf star.

Twort discovers bacteriophage.

1916   Schwarzchilde proposes possible existence of 'singularities' ('black holes').

1917   **Bolshevic Revolution in Russia.**

1919   Eddington describes bending of light by Sun, as predicted by Einstein's general relativity theory of 1916.

Adrian uses cathode ray tube to study nerve impulses.

Rutherford discovers the proton, and artificial transmutation of elements.

1920   Michelson measures first stellar diameter (other than the Sun's).

Goddard develops theory of rocket propulsion, and launches liquid-fuelled rocket in 1926.

Oxford University admits women to membership and degrees.

1922   Banting and Best treat diabetic patients with insulin.

Carrel studies white blood cells.

First public radio service begins (BBC).

1924   Pauli states his exclusion principle.

Appleton discovers radio reflecting layer in ionosphere.

1926   First demonstration of TV by Baird.

H J Muller discovers biological mutations induced by X-rays.

Dirac unifies new quantum theory.

The Theory of the Gene (Morgan) begins modern period in genetics.

1927   Heisenberg proposes uncertainty principle.

Lemaitre initiates 'Big Bang' theory of origin of universe.

Electronic Theory of Valency (Sidgwick) surveys chemical bonding, seen as due to electron-transfer (electrovalence) or electron-sharing (covalence).

1928   Alexander Fleming discovers penicillin.

Raman discovers scattering effect of molecules on light.

1929   Berger introduces EEG.

Hubble announces his law on recession of galaxies.

Matuyama discovers remanent magnetization reversals of rocks.

1930   Tombaugh discovers Pluto.

1931   Gödel proves incompleteness of arithmetic.

Van de Graaf builds high voltage electrostatic generator.

1932   C D Anderson discovers positron.

Chadwick identifies neutron.

Cockcroft and Walton induce nuclear reaction artificially.

Deuterium isolated by Urey.

Jansky announces radio emission from stars.

*c.*1933 Ruska develops transmission electron microscope, achieving 12 000 ×.

1934 Beebe reaches 1000 m below sea level
Joliot-Curies discover artificial radioactive isotopes.

1935 Stanley obtains a crystalline virus (TMV).
Richter devises scale of earthquake strength.
Bergeron proposes theory of rain precipitation.
K Z Lorenz describes imprinting in animal development.

1936 First public high-definition (405 line) television service (BBC, London).

1937 Dobzhansky's *Genetics and the Origin of Species* links Darwinian evolutionism with modern Mendelian genetics.

1938 Carlson makes first Xerox copies.
Meitner and Frisch recognize nuclear fission as explanation for Hahn's experiments.

1939 Alvarez measures magnetic moment of neutron.
A L Hodgkin and A F Huxley show that a nerve impulse involves a depolarization wave.
World War Two begins.
The Nature of the Chemical Bond (Pauling).
Energy Production in Stars (Bethe).

1940 Beadle and Tatum begin work in bacterial genetics.
Rossby discovers the atmospheric waves named after him.

1942 Alfvén predicts magnetohydrodynamic waves in plasma.
First nuclear reactor begins operation, in Chicago.

1943 Baade classifies stars as Population I or II.

1944 Martin introduces paper chromatography.
What is Life? (Schrödinger).
Principles of Physical Geology (Holmes).
Avery shows importance of DNA.

1945 **First atomic bombs (using nuclear fission) exploded.**
Royal Society of London accepts women as Fellows.
Beadle and Tatum propose one-gene-one-enzyme hypothesis.

1946 Bloch and Purcell independently develop nuclear magnetic resonance (NMR).
Lederberg and Tatum discover sexuality in bacteria.

1947 Pi-meson discovered by Powell.
Bardeen, Brattain and Shockley develop point-contact transistor.
Libby develops radiocarbon dating.
Cambridge University admits women to membership and degrees.

1948 H W and H D Babcock detect Sun's magnetic field.
Alpher, Bethe and Gamov propose scheme to explain evolution of chemical elements in early universe (alpha-beta-gamma theory).
Alpher and Herman suggest that 'Big Bang' should have left residue of weak radiation.
Shockley develops junction transistor.
Gabor invents holography.
The Steady-State Theory of the Expanding Universe (Bondi and Gold).

1949 Burnet discovers acquired immunological tolerance.

1950 **North Korea invades South Korea.**
Barton introduces conformational analysis in chemistry.

1951 Purcell detects 21 cm radiation from interstellar hydrogen.
Burnet proposes clonal selection theory in immunology.
The Study of Instinct (Tinbergen).
Pauling proposes some protein molecules are helical.

1952 Hershey and Martha Chase prove DNA is genetic information carrier.
Lederberg and Zinder discover bacterial transduction.
Glaser makes first bubble chamber.

1953 Watson and Crick deduce structure of DNA and its implications.

Gell-Mann introduces concept of 'strangeness' in particle physics.

Kettlewell shows industrial melanism in moths.

1954   Backus publishes first high-level programming language (FORTRAN).

Pincus introduces oral contraception.

Townes devises maser.

Seismicity of the Earth (Richter and Gutenberg).

1955   De Duve identifies lysosomes.

Sanger determines amino acid sequence of first protein (insulin).

Thermodynamics of Irreversible Processes (Prigogine).

Segrè and Chamberlain discover antiproton.

1956   Berg discovers first transfer RNA.

Palade discovers ribosomes.

Tijo and Lavan report that human cells have 46 chromosomes (not 48, as believed since 1920s).

Ewing plots mid-Atlantic ridge.

F Reines detects neutrino.

1957   **First artificial satellites (Sputnik I, and II containing dog Laika) in orbit around Earth.**

Meselson and Stahl verify Watson and Crick's ideas on replication.

1958   Explorer I discovers Van Allen belts: other space probes follow.

Dausset discovers human histocompatibility system.

Esaki discovers tunnelling effect in semiconductor junctions.

1960   Sandage and Matthews discover quasars.

Maiman constructs first laser.

Moore and Stein determine sequence of all 124 amino acids in ribonuclease.

1961   **First men in space.**

Brenner and Crick show that genetic code of DNA consists of a string of non-overlapping base triplets.

Good shows importance of thymus gland in the development of immunity.

Gell-Mann and Ne'eman develop scheme for classifying elementary particles.

1962   Josephson discovers effect named after him.

Rossi detects cosmic X-ray source in Scorpio.

Venus examined by passing US space probe Mariner 2.

Bartlett makes first noble gas compounds.

H H Hess proposes sea-floor spreading hypothesis.

Marguerite Perey elected as first woman member of Académie des Sciences.

1963   Cormack and Hounsfield independently develop X-ray tomography (CAT scanning).

Matthews and Vine find evidence for sea-floor spreading.

Gajdusek describes first human slow virus infection.

1964   Penzias and Wilson detect cosmic background radiation, providing important evidence for 'Big Bang' origin of universe.

1965   North Sea gas discovered.

Mars photographed by Mariner 4.

1967   Hewish and Jocelyn Bell discover first pulsar.

First human heart transplant.

1969   **First manned lunar landing.**

Edelman finds amino acid sequence of immunoglobulin G.

1970   Baltimore discovers reverse transcriptase which transcribes RNA into DNA.

Khorana completes synthesis of first artificial gene.

North Sea oil discovered.

1973   Boyer uses recombinant RNA to produce chimera.

1974   J/psi particle first observed.

1975   *Sociobiology* (E O Wilson).

Milstein produces first monoclonal antibodies.

1977   Sanger describes full sequence of bases in a viral DNA.

Quantum Hall effect discovered by von Klitzing.

First personal computer (the Apple II) introduced.

1978   Pluto discovered to have a satellite (Charon) with half Pluto's diameter.

1979   Saturn and Jupiter examined by passing space probes.

Antarctic meteorite (uncontaminated by Earth) found to contain traces of amino acids.

Evidence from deep sea bed cores shows that all fossils are absent for a period of nearly 100 000 years about 65 000 000 years ago; evidence favours asteroid impact resulting in dust clouds cutting off sunlight as cause of destruction.

Microbial fossils from W Australia found to be 3500 million years old.

1980   S B Prusiner proposes that some fatal neurodegenerative diseases of animals are due to infective proteins (prions) free from DNA or RNA. Only after 15 years is this idea accepted: scrapie in sheep, BSE ('mad cow disease') and Creutzfeldt–Jakob disease in humans have this cause.

1981   First reusable spacecraft (space shuttle) launched and recovered.

First flare (lasting ≈1 min) observed on a distant star.

Chain of 11 carbon atoms detected in molecules in star 600 light years distant: biggest space molecule.

Rohrer develops scanning tunnelling microscope, giving $10^8 \times$ and able to show individual atoms.

1982   Probes land on Venus.

1983   Discovery of W vector boson.

Compact disc (CD) system for sound recording introduced

Navigational satellites give speed of tectonic plate movement (eg Europe moving away from USA at ≈ 10 cm year^{-1}).

1984   Alec Jeffreys devises genetic fingerprinting.

Details given of the deepest hole, drilled at Kola Peninsula in Soviet Arctic, and now at 12 000 m and sampling very old rocks.

1985   Laser clock developed with accuracy around 1 s in 50 million years, of value to cosmologists.

Increasing evidence for a black hole at the centre of our Galaxy.

1986   Halley's Comet investigated by space probes and shown to have a rotating nucleus of 'dirty ice', ie dust with water and CO_2 ice; the probe Giotto passes within 605 km of the nucleus, found to be potato-shaped and about 15 km long and 8 km wide.

Fire at a nuclear reactor at Chernobyl, Ukraine, produces massive radioactive contamination.

Voyager 2 examines planet Uranus, and finds strange features such as its large, oddly-shaped magnetic field and satellites with 'bizarre geology'.

1987   Much work done on superconductors, mainly ceramics based on copper, oxygen, barium and a rare earth metal, with critical temperatures up to 240 K (classical superconductivity had been seen only near 0 K): and with possible uses of great interest in electronics.

Canadian astronomers find new and better evidence for planets round stars other than the Sun, and by 1988 believed the case proved.

1988   Increasingly, international scientific efforts directed to problems of worldwide concern, notably: ozone loss in the upper atmosphere over both poles, which allows more UV light to reach Earth with effects on the weather and health; acid rain, now shown to be due as much to lightning (see p. 58) as to fuel burning, and having effects especially on plant life; and the nature and spread of the viral disease AIDS (Acquired Immune Deficiency Syndrome).

New knowledge of dinosaurs result from the discovery at one site near Egg Mountain, MT, USA, of fossil eggs, embryos, juveniles and adult dinosaurs.

1989   The probe Pioneer 10, the first spacecraft to leave the solar system, transmits data that allows the position of the Sun above the plane of our Galaxy.

to be recalculated, as about 40 light years above this plane.

Voyager 2 passes Neptune and photographs rings around the planet and six new moons.

Discovery of a new particle, Z^0, announced by physicists at CERN and at Stanford University, CA; related work indicates that the universe contains only three families of neutrinos.

The wooden track over the Somerset Levels in the UK, 1000 m long and the oldest man-made track known in Europe, shown by dendrochronologists to be made of timber felled in 3807–3806 BC.

Volcanic activity, familiar on Earth and more recently observed on Venus, is seen by Voyager 2 on Triton, the largest of Neptune's moons; an earlier Voyager had detected volcanoes on Io, the innermost of Jupiter's four large moons.

1990   Ball lightning, long known as 'balls of fire' seen during thunderstorms and behaving strangely, is removed from fantasy and made in the laboratory, confirming that the phenomenon results from a plasma discharge caused by interference between radio waves of microwave frequency; Y H Ohtsuki and H Ofuruton in Tokio make typical balls of fire, about 25 cm in diameter and brightly coloured, using a microwave generator and suitable waveguides.

Final collapse of USSR with Russian parliament's declaration of sovereignty.

Fullerene, a carbon molecule C_{60} of soccer-ball shape and composed of six- and some five-membered rings of carbon atoms surrounding an empty centre, was first identified in interstellar space by spectroscopy; H Kroto and others at Sussex University synthesize it.

The death of dinosaurs about 65 million years ago is thought to be due to a major meteorite impact with Earth;

new support for this now comes from study of a buried meteor impact crater some 135 km across, in Mexico.

The oldest known land dwellers are found in the Ludlow bone bed in Shropshire, UK, aged 414 million years and consist of two centipedes and an arachnid. These animals may have left the sea for the land much earlier.

Discovery of a well-preserved 4 t mastodon in a peat bog in Ohio, USA allows study of its last meal some 11 000 years ago.

1991   A man's body found in the alpine glacier on the Italian–Austrian border is dated about 3300 BC (Stone Age). He was arthritic (and so over 30), with bevelled teeth (and so had eaten processed grain). His dagger and arrowheads were flint, with a 6 ft/1.8 m yew bow and a copper-headed axe. His clothes were leather and fur, and he carried flint and tinder.

A cheap and simple method is devised for focussing X-rays by bundles of fine glass tubes.

IBM report moving a single xenon atom from one position on a surface to another position: the ultimate miniaturization, with implications for novel computer switches.

A new application of computer technology – virtual reality or cyberspace – is devised, giving a three-dimensional simulation of an object with its visual, aural and behavioural attributes; the applications of this are unclear.

The space-craft Galileo obtains the first close photograph of an asteroid, 951 Gaspra, at 1600 km; heavily cratered, its size is about 12 × 11 × 20 km.

A new largest prime number is found by a team using computers at Amdahl Corp., CA. It is $(391\,581 \times 2^{216193}) - 1$.

1992   The claim by M Fleischman and S Pons to have achieved a 'cold fusion reaction in a test tube' using D_2O and Pd electrodes to generate neutrons is widely, but not universally, held to be disproved.

One form of Alzheimer's disease is shown by D Selkoe and I Lieberburg of Harvard to result from a defect in a gene on chromosome 21, which leads to over-production of a protein (amyloid) known to congest the brain in victims of the disease.

Measurement of the mass of Pluto's satellite, Charon, show it to have a density of only $1.4 \, \text{g cm}^{-3}$ so it must be mainly water-ice, whereas Pluto itself (density 2.1) is mainly rock.

A fourth black hole is found in our Galaxy, the Milky Way: named Nova Muscae 1991, it is about 18 000 light years from Earth.

Human babies are shown to grow in 'spurts' (up to 2.5 cm in 24 h) rather than steadily.

1993 Human genes are identified which are associated with some human diseases, including cystic fibrosis.

'Gravitational lensing', an effect predicted by Einstein, is detected and confirms that quasars are very distant galaxies.

The Cosmic Background Explorer satellite (COBE) went into orbit round the Earth in 1989, carrying instruments to detect if the cosmic background radiation is at a uniform temperature, or not. This radiation, in the microwave region, represents the residue of radiation from the 'Big Bang' when the Universe began; its discovery in 1964 confirmed this theory of the universe's origin. However, the universe is not uniform; its matter is 'clumped' in galaxies, dust and gas clouds, and so the radiation should not be entirely 'smooth'. Careful analysis of COBE's results shows that the microwave radiation does have variations, corresponding to ripples of about 30×10^{-6} K. However, these small fluctuations give important support for the broad correctness of the 'Big Bang' theory, and are held

by cosmologists to be a major discovery.

The Hubble space shuttle telescope is ingeniously repaired in space to improve its image quality, and highly successful photographs of deep space are received from it in early 1994.

1994 A rail tunnel is completed to give a 31-mile land route between England and France for the first time since the Ice Ages.

M Irwin at the Royal Greenwich Observatory shows that a small galaxy in Sagittarius is being absorbed by our own Galaxy, the Milky Way, as a result of gravitational pull.

A new study of past climatic evidence casts doubt on the reality of the 'Little Ice Age' of 1550–1850: in Europe a sustained cold period then appears absent and a general warming (of less than 1°C) has been concentrated in the 20th-c.

The first nearly complete skeleton of a pliosaur, with a 6 ft/1.8 m skull and 150 million years old, is found in a quarry in Wiltshire, UK.

A hominid shinbone found at Boxgrove, UK, shows that archaic *Homo sapiens* occupied the site about 500 000 years ago; Boxgrove Man was 6 ft/1.8 m tall, robust and used flint hand axes; it is not clear whether he hunted or scavenged for food.

A fragmented comet, in pieces over 1.5 km/1 mi wide, collides with the distant side of the planet Jupiter, causing major impact explosions.

Glycine, the simplest amino acid and a building block for proteins, is detected in Sagittarius B2, a cloud of interstellar gas near the centre of our Galaxy, the Milky Way.

The oldest DNA from a vertebrate is isolated from 80 million year-old bone fragments by S R Woodward of Utah; their genetic sequence matches no known bird, reptile or mammal and may belong to a Cretaceous-period dinosaur.

The spacecraft Ulysses, launched in the USA in 1990 and the fastest (100 000 mph) yet, travels out of the plane of the ecliptic for the first time and passes below the Sun's south pole. For the first time the Royal Institution Christmas Lectures are given by a woman, Susan Greenfield.

Fossilized bones found at Afar, Ethiopia, appear to come from the oldest (4.4 million years) and most ape-like hominid yet found; this human relative is named *Australopithecus ramidus*.

A 'living fossil', the Wollemi pine tree, is found in an Australian rainforest; this newly-found species and genus of conifer is unchanged from the Jurassic period, before Australia became a separate continent.

1995   D H White and co-workers at the Los Alamos National Laboratory, NM, USA, deduce that neutrinos, previously thought to be massless, have about 10^{-5} of an electron mass, but this is later contested. Neutrino abundance is high, so that their contribution to the mass of the universe can be substantial.

E Cornell and C Wieman at Boulder, CO, cool a rubidium-87 sample to 170 nanokelvins ($1 \, nK = 10^{-9} \, K$), a new low. They confirm the Bose–Einstein prediction of 1924: at this temperature the near-stationary atoms are at the same energy level and the condensate behaves as a new state of matter (a 'superatom') without atomic individuality.

Stone tools are found at Orce, Spain, implying that our own genus (*Homo*) reached Europe from Africa over 1 million years ago, about twice as long ago as previously thought.

A US group reports the first full genetic blueprint of a free-living organism, the bacterium *Haemophilus influenzae*, which has 1743 genes (about 1.8 million base-pairs: the human genome has 3 billion base pairs).

UN Panel on Climate Change (IPCC) concludes that global warming is certainly occurring and is partly man-made.

Galileo orbiter and probe reach Jupiter after 6-year trip in December 1995; the first Earthly object to enter a giant planet's atmosphere and report its composition, temperature and pressure, windspeed and lightning bursts.

The spacecraft Pioneer 10, launched from Cape Canaveral, FL, USA in 1972, now 5.8×10^9 miles away and heading into deep space, maintains radio contact. It carries a plaque with atomic, mathematical and solar system diagrams and male and female human images. The most remote object made by man, it should survive after Earth is destroyed.

Nobel Prizewinners in Science

Physics

1901	Wilhelm Konrad von Röntgen	1932	Werner Karl Heisenberg		Maria Goeppert-Meyer
		1933	Paul Adrien Maurice Dirac		Eugene Paul Wigner
1902	Hendrik Antoon Lorentz		Erwin Schrödinger	1964	Charles Hard Townes
	Pieter Zeeman	1934	*No award*		Nikolai Gennadiyevich Basov
1903	Antoine Henri Becquerel	1935	James Chadwick		
	Pierre Curie	1936	Victor Francis Hess		Alexander Mikhailovich Prokhorov
	Marie Curie		Carl David Anderson		
1904	John William Strutt, 3rd Baron Rayleigh	1937	Clinton Joseph Davisson	1965	Julian S Schwinger
			George Paget Thomson		Richard P Feynman
1905	Philipp Eduard Anton Lenard	1938	Enrico Fermi		Sin-Itiro Tomonaga
		1939	Ernest Orlando Lawrence	1966	Alfred Kastler
1906	Joseph John Thomson	1943	Otto Stern	1967	Hans Albrecht Bethe
1907	Albert Abraham Michelson	1944	Isidor Isaac Rabi	1968	Luis Walter Alvarez
1908	Gabriel Lippmann	1945	Wolfgang Pauli	1969	Murray Gell-Mann
1909	Guglielmo, Marchese Marconi	1946	Percy Williams Bridgman	1970	Louis Eugène Félix Néel
		1947	Edward Victor Appleton		Hannes Olof Alvén
	Karl Braun	1948	Patrick Maynard Stuart, Baron Blackett	1971	Dennis Gabor
1910	Johannes Diderik van der Waals			1972	John Bardeen
		1949	Yukawa Hidecki		Leon Neil Cooper
1911	Wilhelm Wien	1950	Cecil Frank Powell		John Robert Schrieffer
1912	Nils Gustav Dalén	1951	John Douglas Cockcroft	1973	Leo Esaki
1913	Heike Kamerlingh Onnes		Ernest Thomas Sinton Walton		Ivar Giaever
1914	Max von Laue				Brian David Josephson
1915	William Henry Bragg	1952	Felix Bloch	1974	Martin Ryle
	(William) Lawrence Bragg		Edward Mills Purcell		Antony Hewish
1916	*No award*	1953	Frits Zernike	1975	Aage Niels Bohr
1917	Charles Glover Barkla	1954	Max Born		Benjamin Roy Mottelson
1918	Max Karl Ernst Planck		Walther Bothe		(Leo) James Rainwater
1919	Johannes Stark	1955	Willis Eugene Lamb, Jr	1976	Burton Richter
1920	Charles Édouard Guillaume		Polykarp Kusch		Samuel Chao Chung Ting
1921	Albert Einstein	1956	William Bradford Shockley	1977	Philip Warren Anderson
1922	Aage Niels Bohr		John Bardeen		Nevill Francis Mott
1923	Robert Andrews Millikan		Walter Hauser Brattain		John Hasbrouck van Vleck
1924	Karl Manne Georg Siegbahn	1957	Tsung-Dao Lee	1978	Pjotr Leonidovich (Peter) Kapitza
			Chen Ning Yang		
1925	James Franck	1958	Pavel Alekseevich Cherenkov		Arno Allan Penzias
	Gustav Ludwig Hertz				Robert Woodrow Wilson
1926	Jean Baptiste Perrin		Ilya Mikhailovich Frank	1979	Steven Weinberg
1927	Arthur Holly Compton		Igor Yevgenyevich Tamm		Sheldon Lee Glashow
	Charles Thomson Rees Wilson	1959	Emilio Segrè		Abdus Salam
			Owen Chamberlain	1980	James Watson Cronin
1928	Owen Williams Richardson	1960	Donald Arthur Glaser		Val Logsdon Fitch
1929	Louis Victor, 7th Duc de Broglie	1961	Robert Hofstadter	1981	Nicolas Bloembergen
			Rudolf Mössbauer		Arthur Leonard Schawlow
1930	Chandrasekhara Venkata Raman	1962	Lev Davidovich Landau		Kai M Siegbahn
		1963	(Johannes) Hans (Daniel) Jensen	1982	Kenneth Geddes Wilson
1931	*No award*			1983	Subrahmanyan

	Chandrasekhar		Alex Müller		Richard Taylor
	William Alfred Fowler	1988	Leon Lederman	1991	Pierre-Gilles de Gennes
1984	Carlo Rubbia		Melvin Schwartz	1992	Georges Charpak
	Simon van der Meer		Jack Steinberger	1993	Joseph Taylor
1985	Klaus von Klitzing	1989	Hans Dehmelt		Russell Hulse
1986	Gerd Binnig		Wolfgang Pauli	1994	Bertram N Brockhouse
	Heinrich Rohrer		Norman Ramsay		Clifford G Shull
	Ernst Ruska	1990	Jerome Friedman	1995	Martin Perl
1987	George Bednorz		Henry Kendall		Frederick Reines

Chemistry

1901	Jacobus Henricus van	1932	Irving Langmuir	1960	Willard Frank Libby
	t'Hoff	1933	*No award*	1961	Melvin Calvin
1902	Emil Hermann Fischer	1934	Harold Clayton Urey	1962	John Cowdery Kendrew
1903	Svante Arrhenius	1935	Jean Frédéric Joliot-Curie		Max Ferdinand Perutz
1904	William Ramsay		Irène Joliot-Curie	1963	Giulio Natta
1905	Johann Friedrich Wilhelm	1936	Peter Joseph Wilhelm		Karl Ziegler
	Adolf von Baeyer		Debye	1964	Dorothy Mary Hodgkin
1906	Henri Moissan	1937	Walter Norman Haworth	1965	Robert Burns Woodward
1907	Eduard Buchner		Paul Karrer	1966	Robert Sanderson
1908	Ernest Rutherford, 1st	1938	Richard Kuhn, declined		Mulliken
	Baron Rutherford	1939	Adolf Friedrich Johann	1967	Manfred Eigen
1909	Friedrich Wilhelm		Butenandt, declined		Ronald George Wreyford
	Ostwald		Leopold Ruzicka		Norrish
1910	Otto Wallach	1940	George de Hevesy		George Porter, Baron Porter
1911	Marie Curie	1944	Otto Hahn	1968	Lars Onsager
1912	(François Auguste) Victor	1945	Artturi Ilmari Virtanen	1969	Derek H R Barton
	Grignard	1946	James Batcheller Sumner		Odd Hassel
	Paul Sabatier		John Knudsen Northrop	1970	Luis Federico Leloir
1913	Alfred Werner		Wendell Meredith Stanley	1971	Gerhard Herzberg
1914	Theodore William	1947	Robert Robinson	1972	Stanford Moore
	Richards	1948	Arne Wilhelm Kaurin		William Howard Stein
1915	Richard Willstätter		Tiselius		Christian Boehmer
1916	*No award*	1949	William Francis Giauque		Anfinsen
1917	*No award*	1950	Otto Diels	1973	Ernst Otto Fischer
1918	Fritz Haber		Kurt Alder		Geoffrey Wilkinson
1919	*No award*	1951	Edwin Mattison McMillan	1974	Paul John Flory
1920	Walther Hermann Nernst		Glenn Theodore Seaborg	1975	John Warcup Cornforth
1921	Frederick Soddy	1952	Archer (John Porter)		Vladimir Prelog
1922	Francis William Aston		Martin	1976	William Nunn Lipscomb
1923	Fritz Pregl		Richard Laurence	1977	Ilya Prigogine
1924	*No award*		Millington Synge	1978	Peter Dennis Mitchell
1925	Richard Adolf Zsigmondy	1953	Hermann Staudinger	1979	Herbert Charles Brown
1926	Theodor Svedberg	1954	Linus Carl Pauling		Georg Wittig
1927	Heinrich Otto Wieland	1955	Vincent du Vigneaud	1980	Paul Berg
1928	Adolf Otto Reinhold	1956	Nikolai Nikilaevich		Walter Gilbert
	Windaus		Semenov		Frederick Sanger
1929	Arthur Harden		Cyril Norman	1981	Kenichi Fukui
	Hans Karl August Simon		Hinshelwood		Roald Hoffmann
	von Euler-Chelpin	1957	Alexander Robertus Todd,	1982	Aaron Klug
1930	Hans Fischer		Baron Todd	1983	Henry Taube
1931	Carl Bosch	1958	Frederick Sanger	1984	(Robert) Bruce Merrifield
	Friedrich Bergius	1959	Jaroslav Heyrovsky	1985	Herbert Aaron Hauptman

	Jerome Karle	1988	Johann Deisenhofer	1992	Rudolph Marcus
1986	Dudley R Herschbach		Robert Huber	1993	Kary Mullis
	Yuan Tseh Lee		Hartmut Michel		Michael Smith
	John C Polanyi	1989	Sydney Altman	1994	George A Olah
1987	Charles Pedersen		Thomas Cech	1995	Paul Crutzen
	Donald Cram	1990	Elias James Corey		Mario Molina
	Jean-Marie Lehn	1991	Richard Ernst		Sherwood Roland

Psychology or Medicine

1901	Emil von Behring		Baron Adrian	1956	Werner Forssmann
1902	Ronald Ross		Charles Scott Sherrington		Dickinson Woodruff
1903	Niels Ryberg Finsen	1933	Thomas Hunt Morgan		Richards
1904	Ivan Petrovich Pavlov	1934	George Hoyt Whipple		André Frédéric Cournand
1905	Robert Koch	1935	Hans Spemann	1957	Daniel Bovet
1906	Camillo Golgi	1936	Henry Hallett Dale	1958	George Wells Beadle
	Santiago Ramón y Cajal		Otto Loewi		Edward Lawrie Tatum
1907	Charles Louis Alphonse	1937	Albert von Nagyrapolt		Joshua Lederberg
	Laveran		Szent-Györgyi	1959	Severo Ochoa
1908	Paul Ehrlich	1938	Corneille Jean François		Arthur Kornberg
	Ilya Ilich Mechnikov		Heymans	1960	Frank Macfarlane Burnet
1909	Emil Theodor Kocher	1939	Gerhard (Johannes Paul)		Peter Brian Medawar
1910	Albrecht Kossel		Domagk, *declined*	1961	Georg von Békésy
1911	Allvar Gullstrand	1940	Carl Peter Henrik Dam	1962	Francis Harry Compton
1912	Alexis Carrel		Edward Adelbert Doisy		Crick
1913	Charles Robert Richet	1944	Joseph Erlanger		James Dewey Watson
1914	Robert Bárány		Herbert Spencer Gasser		Maurice Hugh Frederick
1915	*No award*	1945	Alexander Fleming		Wilkins
1916	*No award*		Ernst Boris Chain	1963	John Carew Eccles
1917	*No award*		Howard Walter Florey,		Alan Lloyd Hodgkin
1918	*No award*		Baron Florey		Andrew Fielding Huxley
1919	Jules Jean Baptiste Vincent	1946	Hermann Joseph Müller	1964	Konrad Emil Bloch
	Bordet	1947	Carl Ferdinand Cori		Feodor Felix Konrad Lynen
1920	Schack August Steenberg		Gerty Theresa Cori	1965	François Jacob
	Krogh		Bernardo Alberto Houssay		Jacques Monod
1921	*No award*	1948	Paul Hermann Müller		André Lwoff
1922	Archibald Vivian Hill	1949	Walter Rudolf Hess	1966	Charles Brenton Huggins
	Otto Fritz Meyerhof		António Caetano de Abreu		Francis Peyton Rous
1923	Frederick Grant Banting		Freire	1967	Haldan Keffer Hartline
	John James Rickard		Egas Moniz		George Wald
	Macleod	1950	Philip Showalter Hench		Ragnar Arthur Granit
1924	Willem Einthoven		Edward Calvin Kendall	1968	Robert William Holley
1925	*No award*		Tadeusz Reichstein		Har Gobind Khorana
1926	Johannes Andreas Grib	1951	Max Theiler		Marshall Warren Nirenberg
	Fibiger	1952	Selman Abraham	1969	Max Delbrück
1927	Julius Wagner-Jauregg		Waksman		Alfred Day Hershey
1928	Charles Jules Henri	1953	Fritz Albert Lipmann		Salvador Edward Luria
	Nicolle		Hans Krebs	1970	Julius Axelrod
1929	Christiaan Eijkman	1954	John Franklin Enders		Bernard Katz
	Frederick Gowland		Thomas Huckle Weller		Ulf von Euler
	Hopkins		Frederick Chapman	1970	Earl W Sutherland
1930	Karl Landsteiner		Robbins	1972	Gerald Maurice Edelman
1931	Otto Heinrich Warburg	1955	(Axel) Hugo Theodor		Rodney Robert Porter
1932	Edgar Douglas Adrian, 1st		Theorell	1973	Konrad Zacharias Lorenz

	Nikolaas Tinbergen		Hounsfield	1988	James Black
	Karl von Frisch	1980	Baruj Benacerraf		Gertrude Elion
1974	Albert Claude		George Davis Snell		George Hitchings
	George Emil Palade		Jean Dausset	1989	(John) Michael Bishop
	Christian René de Duve	1981	Roger Wolcott Sperry		Harold Elliot Varmus
1975	David Baltimore		David Hunter Hubel	1990	Joseph Edward Murray
	Renato Dulbecco		Torsten Nils Wiesel		(Edward) Donnall Thomas
	Howard Martin Temin	1982	Sune Karl Bergström	1991	Erwin Neher
1976	Baruch Samuel Blumberg		Bengt I Samuelsson		Bert Sakmann
	Daniel Carleton Gajdusek		John Robert Vane	1992	Edmond H Fisher
1977	Rosalyn Sussman Yalow	1983	Barbara McClintock		Edwin K Krebs
	Roger (Charles Louis)	1984	Niels Kai Jerne	1993	Richard R Roberts
	Guillemin		Georges J F Köhler		Phillip A Sharp
	Andrew Victor Schally		César Milstein	1994	Alfred G Gilman
1978	Werner Arber	1985	Joseph Leonard Goldstein		Martin Rodbell
	Daniel Nathans		Michael Stuart Brown	1995	Edward B Lewis
	Hamilton Othanel Smith	1986	Seymour Stanley Cohen		Christiane Nüsslein-
1979	Allan MacLeod Cormack		Rita Levi-Montalcini		Volhard
	Godfrey Newbold	1987	Susumu Tonegawa		Eric F Wieschaus

Index

Index

Index